MTI RADAR

MTI RADAR

edited by

D. Curtis Schleher, Ph.d.

Manager of Advanced Development
Raytheon Company
Wayland, MA.

Artech House

610 Washington Street
Dedham, Massachusetts 02026
Printed and bound in the United States of America

Library of Congress Card Catalog Number: 78-10858
Standard Book Number: 0-89006-060-6

Foreword

MTI Radar is usually covered in a single chapter, or in several chapters at most, of books devoted to complete radar systems. Recent advancement in MTI signal processing brought about primarily by the availability of advanced digital components, has made a more complete treatment desirable. This volume is directed at this objective.

The book is aimed at two types of readers: the expert who is interested in having a readily accessible compilation of the significant results in MTI Radars, and the generalist who is interested in an overview of the subject with keys into significant detailed results.

The book is composed primarily of selected reprints, but is liberally interspersed with comments and supporting material that aid the reader in either interpreting the material or appreciating its significance. Each paper, or group of papers, is introduced with an overview that summarizes the results contained in the article, provides background information and points out relationships to the central topic.

The introduction gives a brief history of MTI, describes the different types of MTI, traces the evolution of MTI hardware, and introduces definitions and terminology. In addition, keys are provided into the significant papers included in the volume.

The next section provides a complete Glossary of MTI terms. This section is included to provide an easy reference to terms that are used throughout the reprinted papers and other parts of the book. Each term is fully defined and its significance related to the overall MTI system.

The Outline of MTI Theory provides a compilation of the significant formulas used in the design and analysis of MTI Radars. Many of the formulas were previously available only in research reports. Several of the formulas on Transversal filters and FFT digital filters are new.

The section on Clutter Models and Data is intended to supplement and permit interpretation of the extensive existing radar clutter data base. The discussions on the log-normal and Weibull clutter models have been previously available only in research reports. The presentation of the IIT clutter model illustrates a complex land clutter model suitable for computer simulation of MTI systems.

The discussion of clutter models may seem tangential to the central topic of this volume, but the converse is true in that the posulated clutter model determines the optimum MTI processor. The Gaussian clutter assumption used throughout the reprinted papers in the volume leads to an optimum MTI processor that is linear. It is important to understand that the "optimum" MTI processor designed on the basis of a Gaussian clutter assumption is no longer optimum in non-Gaussian clutter.

The MTI Bibliography tabulates selected MTI literature available in books, journals and reports. This section features the inclusion of selected unclassified reports on MTI available through the Defense Documentation Center (DDC).

The volume brings together much of the significant material available on MTI Radars. However, the scope of the subject is so extensive that many excellent papers have not been reprinted. The reader is encouraged to explore the Bibliography for related material when a more extensive treatment than that available in the book is desired.

Contents

Introduction

Most surface and airborne radar systems operate in an environment where the clutter return obscures targets of interest. If the target is moving relative to the clutter it is possible to filter out the undesired clutter return by exploiting the differential doppler frequency shift produced by the relative target to clutter radial motion. Systems that operate on this principle are called Moving Target Indicator (MTI) radars.

Evolution of Analog MTI Radar

Early MTI radars were only partially effective in suppressing clutter due to hardware limitations. In the mid-1950's, hardware components, such as the quartz ultrasonic delay line, had progressed to a point where the potential of MTI systems could be realized. This led to the incorporation of an MTI mode of operation in most radars operational in this period.

The MTI function was initially implemented in the form of an accessory kit added to the radar after it was designed. Performance suffered from this approach and the MTI mode of operational radars was frequently bypassed.

MTIs of this era were primarily of the analog single canceler type that used an ultrasonic delay line to delay the radar return by a time corresponding to the radar's pulse repetition frequency (PRF). Subtraction of the delayed return from the present return produced the desired MTI effect.

More advanced MTI systems employed a double canceler that was formed by two single cancelers in cascade. This configuration not only produced a more desirable transfer function but also compensated for unbalances in each of the individual cancellation circuits.

The sheer bulk of hardware associated with analog delay line type MTIs of the 1950's and early 1960's limited most applications to heavy surface based radars. Airborne applications were restricted to large aircraft and generally used the noncoherent MTI principle. In this type of MTI the amplitude fluctuation resulting from the beat between the target and clutter is processed (See Paper 1.6).

The state-of-the-art in MTI systems of this era is described by Emerson in the first paper (Paper 1.1). The military implications associated with this paper are reflected in its original secret classification. However the real importance of this paper is the treatment of the MTI radar from a system viewpoint. Many of the system definitions and optimization techniques used in this benchmark paper are in current use.

Application of the system approach to MTI radars resulted in a number of new applications. Military air-defense radars were outfitted with an interference suppressing signal processing unit (See Paper 1.5) that included several forms of MTI processing. Air-Traffic Control radars were designed for MTI operation (e.g., Paper 6.1). Combat surveillance MTI radars were designed for ground vehicle and personnel detection (See Paper 4.4). Airborne early warning (AEW) radars such as the E-2A radar were designed using advanced motion compensation techniques that allowed detection of airborne targets over water from an airborne platform (See Paper 7.2). Simple AMTI s were designed for airborne intercept (AI) and close air support operation (See Paper 7.1).

During this period attention was focused on methods to implement better MTI operation within the existing hardware constraints. It was found that applying feedback around a double canceler type MTI allowed excellent control of the shape of both the passband and rejection band of the MTI (See Papers 1.2, 1.3, 1.4 and 1.6). This velocity shaping approach became standard for the MTI designer and could be used to adapt the MTI to varying clutter conditions such as those caused by changing the radar's scan rate. However, when this technique was incorporated into operational radars a significant problem was observed: any asynchronous interference present caused a transient that circulated around the velocity shaped canceler for many PRF periods resulting in unacceptable operation. A pulse interference separator and blanker (PISAB) was developed (See Paper 1.5) to correct this problem; but current MTI design practice is to use nonrecursive type MTI filters wherever possible (See Papers 2.1, 2.3, 2.4 and 2.5) to take advantage of their superior transient response.

The introduction of semiconductor electronics changed the whole philosophy of MTI radar design. The MTI designer was no longer hardware limited and could implement designs that approached theoretical limits.

At first many existing analog designs were converted to transistor designs resulting in smaller lightweight systems. However the performance of these units were no better than their tube counterparts and the use of the bulky quartz ultrasonic delay line was a definite encumbrance. This led to microelectronic range-gated filter (See Papers 1.1, 1.6 and 4.4) and digital MTIs.

The range-gated filter MTI, as the name implies, range-gates the coherent MTI signal into a filter that separates the target from the clutter. The optimum MTI processor is formed when a clutter filter is cascaded with a bank of contiguous doppler filters matched to the target spectrum (See Paper 5.3). This implementation simultaneously performs doppler filtering and coherent integration, but has the disadvantage that a large number of filters is required. Usually a balance is effected between coherent and noncoherent integration thereby significantly reducing the total number of filters. The simplest implementation uses a range-gated clutter rejection filter followed by a noncoherent integrator (See Paper 4.4) and is used in a number of Combat Surveillance Radars. However, the relative high cost of range-gated filter MTIs and the development of low cost digital components results in a general preference for Digital MTIs (DMTI).

Digital MTI (DMTI)

Digital MTIs can be made to implement any of the analog or range-gated filter configurations. An early digital MTI whose design resembles that of an analog delay line MTI is described in Paper 5.2. The critical element of this design is the analog-to-digital (A-D) converter (See Paper 5.1). This is contrasted to an analog design where the ultrasonic delay line is the critical element. The result of this shift of emphasis is that DMTIs are designed using a relatively larger number of delay sections and employ in-phase and quadrature (I and Q) channels to eliminate "blind phase" problems (See Paper 5.2). The availability of a larger number of delays allows DMTIs to be designed to have better performance than their analog counterparts. The performance is reflected in higher stability, better transient response and higher sub-clutter visibility or improvement factor.

The rapid advancement of digital components, developed mostly for the computer industry, soon changed the design philosophy of the DMTIs.

More complex designs were practical and the MTI designer could implement designs that approached theoretical limits. The DMTI became part of a larger unit called a signal processor that provided other radar processing functions such as pulse compression, adaptive constant-fasle-alarm-rate (CFAR) thresholding and video integration in addition to doppler processing.

A number of theoretical developments complemented the rapid development of digital hardware. The theoretical work of Emerson on optimum finite impulse response (FIR) MTI filters was extended and refined by Capon and others (See Papers 2.1, 2.2, 2.4 and 2.5). This procedure minimizes the interference (clutter and noise) output of the MTI under the constraint that all target doppler frequencies are equally likely.

Brennan and Reed (Paper 2.3) used a likelihood ratio statistical detection test to derive an expression that finds the optimum complex weights (amplitude and phase) for detecting a target of known doppler frequency in Gaussian interference (clutter and noise). The resulting Transversal filter can be implemented either digitally by using a set of in-phase and quadrature (I and Q) tapped delay lines (one for each doppler frequency of interest) or alternately by using a bank of doppler filters.

In practice the optimization procedures of Emerson and Brennan/Reed have served as theoretical limits which are useful for comparison purposes against more practical implementations. The implementation that has found wide acceptance because of its computational efficiency involves the use of the Fast Fourier Transform (FFT) algorithm.

The theoretical basis for using the FFT algorithm is described by McAulay in Paper 5.3. Using a likelihood ratio statistical detection test he shows that the optimum receiver structure can be approximated by a clutter filter followed by a narrow-band doppler filter bank. The practical approximation to this processor consists of a conventional MTI filter cascaded with a coherent filter bank implemented using the FFT algorithm. A significant advantage of this configuration is that the dynamic range at the input to the FFT narrow band processor is considerably reduced by the clutter rejection properties of the MTI filter thereby reducing the number of significant bits carried in the FFT processor. An additional advantage is that an improvement in signal-to-noise ratio (SNR) proportional to the number of pulses processed by the FFT is achieved.

A modern state-of-the-art signal processor for moving target detection (MTD) is described in Papers 5.4 and 5.5. The MTD processor was designed to improve the performance of Air-Traffic Control (ATC) radars to allow the use of data from these radars in automated air-traffic control systems.

The present ATC radars designed during the 1960's utilize coherent-on-receive MTI systems that employ receiver limiting for MTI residue equalization and 3-pulse staggered velocity shaped cancelers. The limitations of the present system include inadequate ground clutter suppression, degraded MTI performance due to clutter spreading induced by receiver limiting (See Paper 4.1), susceptability to second-time-around clutter caused by the coherent-on-receive operation, inadequate rejection of weather clutter and angel clutter (birds and insects), and rejection of tangential targets. The poor quality of the radar data is reflected by the primary reliance on the ATC radar beacon system as an input to automated ATC data processors.

The MTD processor, developed by Lincoln Laboratories of MIT, is a special purpose, hard-wired, digital signal processor that provides near optimum performance over the entire 50 n.mile coverage of an airport surveillance radar (ASR). This processor is illustrative of modern MTI design practice and demonstrates the excellent performance provided by a properly designed system.

Ground clutter suppression is accomplished by using a fully coherent system, with a wide dynamic range receiver, coupled into a near optimum digital MTI employing I and Q channels, consisting of a 3-pulse canceler, followed by a weighted 8-point FFT coherent filter. An MTI improvement factor of 52 dB was measured.

Super-clutter visability is provided by incorporating a digital clutter map which establishes the thresholds for the zero radial velocity cells. The clutter level in the ground clutter map continually adapts to a value based on the average of the previous 8 scans. The adaptive thresholding allows detection of large cross-section tangential targets (zero radial velocity) over clutter that otherwise might have fallen into the MTI's blind speed zone and been eliminated.

Weather and angel clutter is suppressed by a combination of the properties of the doppler filter bank formed by the FFT and a mean level threshold formed by averaging the signals in the same velocity filter averaged over one-half mile in range on either side of the cell of interest. With this method of thresholding the radar can detect aircraft with radial velocity components different from the rain even though their cross-sections are substantially less than the equivalent cross-section of the weather returns.

To eliminate blind speeds the pulse repetition rate is varied in blocks of 10 pulses (2 delay sections of 3 pulse canceler plus 8 delay sections of FFT). When used with a coherent transmitter, this technique eliminates second-time-around clutter as well as second-time-around stationary targets. The usual PRF staggering technique (See Papers 4.2 and 4.3) results in a different patch of second-time-around clutter being gated into the MTI on each PRF interval and hence does not cancel. This mode of operation is important where mountains or large cities provide large second-time-around returns.

The hardware necessary to implement the MTD consists of 900 integrated circuits and a disc memory to store the clutter map. During tests, using a modified FPS-18 radar against a small commercial aircraft, the MTD provided a blip-to-scan ratio of .99 and completely eliminated ground clutter returns.

The current trend in digital signal processors involve architectures that provide greater processing speeds coupled with high flexibility. These processors make extensive use of microprocessors and use highly parallel structures to provide reprogrammability coupled with the high speeds associated with multiple execution of parallel operations.

Some of the considerations involved in the design of a general purpose digital radar signal processor are discussed in Paper 5.6. This processor performs such diverse operations as pulse compression, matched filtering, digital filtering, integration, estimation and adaptive thresholding in addition to doppler processing. The basic operation performed in a radar digital processor is the FFT "butterfly" that consists of four real multiplies and six real additions. A figure of merit of radar digital signal processing hardware is the time to perform a basic "butterfly" operation (400 n-seconds for the processor described in Paper 5.6).

Charge Coupled Device (CCD) MTI

In contrast to the increasing sophistication of digital signal processing is another emerging MTI technology that is closely allied to analog signal processing. This type of MTI utilizes charge coupled devices (CCDs) to replace the ultrasonic delay lines of older type analog MTI's (See Paper 6.2). The charge coupled devices can be modeled by a large number of capacitors interconnected with lossless switches. The analog radar information is propagated by charge transfer through capacitive storage elements separated by switches which are commutated by a clock whose basic frequency corresponds to the Nyquist rate of the signal bandwidth.

The advantage of the CCD MTI is economy. However, the performance (50 dB cancellation ratio) provided by the unit described in Paper 6.2 is excellent considering the preliminary nature of the CCD devices used in the implementation. The performance is limited by a random pattern noise caused by utilization of the partial transfer mode of charge transfer from element-to-element. The performance should improve considerably as the state-of-the-art in CCD devices evolves. A promising application consists of using a CCD MTI as a clutter precanceler in front of a digital processor. The CCD MTI reduces the dynamic range at the input to the digital processor thereby considerably simplifying the A-D converter requirements by reducing the number of quantization bits.

Airborne MTI (AMTI)

In parallel with the rapid advancement in surface-based MTI techniques, a similar evolution occurred for airborne MTI (AMTI). The MTI problem for airborne radars is more complex than for surface based radars. The complication arises due to the clutter spectral spreading and translation effects caused by platform motion in addition to the aircraft-to-surface geometry whereby main beam and sidelobe clutter can be viewed at high grazing angles producing heavy clutter returns (See Paper 1.6). The "altitude line" clutter that is reflected, through the antenna's sidelobes, from the earth's surface directly below the aircraft can be particularly severe.

There are two general approaches for solving the AMTI clutter problem. The first is generically referred to as a Pulse Doppler radar. In this approach the radar's PRF is at least twice the doppler frequency corresponding to the sum of the aircraft's velocity and the target's closing velocity (typically 50 kHz to 300 kHz). This creates a clear doppler region, without competing clutter returns, where narrow band filters can be used to detect closing targets. The system is highly range ambiguous and special techniques are required to resolve the ambiguities. Pulse Doppler radars are used for airborne interceptor (AI) radars where "look down" in heavy clutter is required. It is also the technique used in the AWACS E-3A airborne radar.

The other approach for solving the AMTI radar clutter problem is to remove the effects of aircraft motion by antenna design and signal processing. Once the platform motion effects are removed, the same signal processing techniques as used in surface MTIs are applicable. This type of system is used in the E2-C airborne early warning (AEW) radar and is described in Paper 7.2.

A typical clutter spectrum for an AMTI without motion compensation is given in Paper 1.6 (pg. 2). The center of the spectrum is translated to a doppler frequency that corresponds to the closing velocity between the aircraft and the clutter patch in a direction perpendicular to the antenna aperture. The doppler frequency corresponding to the return from any clutter patch is a function of the relative closing velocity of the aircraft in that direction. Therefore, the clutter spectral shape is approximately proportional to the two-way antenna gain pattern. The clutter returns from antenna sidelobes have a differential velocity with respect to the center of the spectrum and hence fall within the pass band of the MTI filter. The spectral spread of main beam clutter is a function of the antenna's pointing angle with respect to the aircraft's ground track, and it increases as the antenna points away from the aircraft's ground track until a maximum is reached at the 90 degree point.

Paper 7.1 describes a simple single canceler non-coherent type AMTI. The performance of this type of MTI is severely limited by the clutter spectral spreading as the antenna points away from the aircraft's ground track. The build-up of residual clutter as the antenna points away from the aircraft's ground track produces a "butterfly" pattern when viewed on the radar's PPI and has been used to determine the aircraft's ground track heading.

In Paper 7.2 motion compensation techniques and design criteria for AMTI radars are described. Time Average Clutter Coherent Radar (TACCAR) is a technique to translate the center frequency of the clutter spectrum to a notch in the comb filter response of the MTI. This effectively removes the mean velocity of the clutter in a direction normal to the antenna aperture. However, the angle of arrival to the clutter is a function of the clutter's range, and complete correction requires a separate correction in each range cell. Approximately three corrections spread through the radar's range coverage provide adquate performance.

Displaced Phase Center Antenna (DPCA) corrects for the components of aircraft motion parallel to the antenna aperture. This correction is accomplished by electronically causing the phase center of the antenna to remain motionless over the period corresponding to the number of radar returns processed by the MTI; i.e., two pulses for a single canceler.

Performance of the motion compensated AMTI is proportional to the integrated antenna sidelobe level. This puts a premium on low antenna sidelobes that must be achieved when the antenna is installed within the aircraft. In general it is difficult to achieve the required low sidelobe performance in an airborne installation due to the proximity of the aircraft structure and radome in the near field of the antenna.

Adaptive Array AMTI Radar

Papers 7.3 and 7.4 discuss an advanced type of AMTI radar that adapts both its antenna pattern and MTI filter response to maximize the radar's probability of detection. This Adaptive Radar utilizes a phased array antenna to control the radar's antenna pattern to provide motion compensation while minimizing the sidelobe response in directions containing large clutter scatterers. In addition to antenna pattern control, the Adaptive Radar simultaneously adjusts the MTI filter response to reject mainbeam clutter.

The adaptive feature of the radar is achieved by an algorithm that utilizes an inverted interference covariance matrix. The matrix can be obtained by either of two recursive methods (Applebaum or Widrow algorithms) or by direct sample matrix inversion (SMI algorithm). The recursive methods require a modest number of computations but converge slowly, while the SMI method provides rapid convergence, but at the expense of a formidable computational load (approximately N^3 complex multiplications, where N is the total number of adaptive weights).

The Adaptive Radar is an attractive AMTI solution since it automatically adjusts its parameters to provide maximum performance. This implies that optimum motion compensation, low sidelobes in just the right directions, compensation for antenna installation problems, and maximum MTI clutter rejection are all achieved simultaneously. Even though Adaptive Radars have been analyzed extensively and the theoretical principles well established, an operational hardware test has not yet been accomplished.

Pulse-Burst MTI Radar

Each of the two general approaches to the AMTI problem (Pulse Doppler and low-PRF AMTI) has its respective advantage and disadvantages. Papers 3.1, 3.2, and 3.3 describe a class of Pulse-Burst radars that attempt to capitalize on the advantages of low and high PRF radars while minimizing their disadvantages. This class of radars appears to offer significant performance advantages over radars using the more conventional waveforms.

From a theoretical point of view, the Pulse-Burst waveform consists of a burst of relatively high PRF pulses that are repeated at a low PRF rate determined by the desired unambiguous range. The use of the Pulse-Burst waveform is advantageous in situations such as airborne radar where the doppler clutter spread is large, thus making the use of low-PRF designs impractical. In general, both amplitude and phase modulation is employed across the burst to optimize the design, resulting in a complex transmitter design.

The performance capability of a Pulse-Burst radar generally is determined by the number of pulses in the burst and the weighting applied. A practical design uses transmitter amplitude weighting (e.g., Chebyshev weighting) and a matched receiver in range extended clutter (clutter extent greater than the burst extent). If the range extent of the clutter is less than the burst extent, then a uniform burst and an optimum receiver (See Paper 2.3) provide almost optimum performance. The maximum signal-to-interference ratio (SIR) improvement of a Pulse-Burst radar in clutter is bounded by the clutter-to-noise ratio.

Reprinted Paper Summary

The reprinted papers are all referenced above in the general overview of the MTI field. A grouping by topic is provided to allow easy access to a particular subject. The papers generally are given in chronological order except where continuity is enhanced by changing the sequence. Editor's comments are interspersed throughout the reprinted papers to aid in their interpretation and to lend perspective. The topics covered by the reprinted papers are:

- **General** — General theory and analysis of analog delay line cancelers and range-gated filter MTIs.
- **Optimum MTI Filters** — Theory and properties associated with the design of optimum MTI filters.
- **Pulse-Burst MTI Radars** — Theory of Pulse-Burst MTI radars. Optimum MTI design using both transmitter waveform and receiver design.
- **Special MTI Effects** — Effects that cause MTI performance to degrade from the ideal. MTI signal limiting, PRF staggering and power spectral density of wind driven clutter.
- Digital MTI (DMTI) — Theory and design of digital MTIs, quantization effects, A-D converters, FFT MTI filters, clutter maps and general purpose digital signal processors.
- MTI Hardware — Description of modern Air Route Surveillance Radar (ARSR). Charge Coupled Device (CCD) MTI.
- Airborne MTI (AMTI) — Theory and design of airborne MTI radars. Motion compensation — TACCAR and DPCA. Theory of Adaptive Radar.

Glossary of MTI Terms

ADAPTIVE ARRAY MTI

An AMTI system using a phased array antenna where the antenna pattern is adaptively controlled to minimize interference, sidelobe clutter, and provide motion compensation while the MTI filter is adaptively controlled to minimize main lobe clutter. Important theoretical advantages of this type of AMTI are: 1) automatic platform motion compensation is provided by changing the array illumination function on a pulse-to-pulse basis; 2) both sidelobe jamming and clutter are minimized simultaneously and 3) antenna pattern compensation is provided for distortions caused by aircraft structure in the near field of the antenna (See Papers 7.3 and 7.4).

AIRBORNE MOVING TARGET INDICATOR (AMTI)

An airborne radar system designed to reject fixed targets and pass moving targets on the basis of their doppler shifts.

ALTITUDE LINE

In an airborne radar, the strong specular-like clutter return from the surface directly beneath the aircraft. This has been termed the "Altitude Line" because of its appearance on a B-scope as a line at a range equal to the airplane's altitude.

AMBIGUITY FUNCTION

The ambiguity function describes the complex envelope of the signal at the output of the radar receiver as a function of radar target range and radial velocity. This function is determined by the transmitter waveform and receiver filter characteristics and is useful in the design of transmitter waveforms and associated receivers. It provides, from visual inspection, a clear picture of the resolution, ambiguities, measurement precision and clutter rejection properties of a particular transmitter waveform-receiver filter design in both the range and doppler domains. By convention, the ambiguity function implies the use of a matched filter receiver while the cross-ambiguity function implies the use of a mismatched receiver. (See Paper 3.3). The ambiguity function is given by

$$\chi(\tau, f_d) = \int_{-\infty}^{\infty} u(x)\, u^*(\tau + x)\, e^{j2\pi f_d x}\, dx$$

where

τ = difference between time the signal actually arrives and time to which filter is matched

and

f_d = difference between received doppler frequency and doppler frequency to which filter is matched.

ANALOG-TO-DIGITAL (A/D) CONVERTER

A device that operates on an analog waveform to produce a quantized digital message consisting of a sequence of numbers each of which approximates a corresponding sample of the input waveform. The number of quantization levels associated with the A/D converter is expressed in bits (n bits provides 2^n comparison levels). The representation of the signal by a limited number of digits generates a quantization noise that limits the MTI cancellation ratio to approximately 6 dB per bit. The sampling rate should be at the Nyquist rate (sampling rate equal to twice the signal bandwidth) or greater to minimize sampling losses (See Papers 5.1 and 5.2).

ANGEL CLUTTER

Unwanted radar echoes primarily returned by birds and insects but also including returns from atmospheric disturbances such as those caused by fluctuations in the refractive index of the atmosphere. At S-band individual bird cross-sections range from 10^{-4} to 10^{-2} m^2. Typically there may be anywhere from one to several hundred birds in a radar resolution cell. Velocities can vary from 15 to 60 knots depending upon wind conditions. Individual insect cross-sections range from 10^{-5} to 10^{-3} m^2, generally increasing for higher frequencies. The radar cross-sections of both birds and insects have been found to exhibit a log-normal amplitude probability distribution function.

AREA MTI

A non-doppler type of MTI that processes radar

information on a frame-to-frame amplitude subtraction basis, rather than the usual pulse-to-pulse basis. Only returns that have changed position from one radar scan to the next are displayed thereby providing an absolute motion type of MTI. This type of MTI provides super-clutter visability as contrasted to the doppler type of MTI that provides subclutter visibility.

BACKSCATTER COEFFICIENT (CLUTTER REFLECTIVITY FACTOR)

For surface clutter, the dimensionless backscatter coefficient (σ^o) is defined as the average radar cross section of clutter per unit area of surface. For volume clutter, the backscatter coefficient (η) is defined as the average radar cross section of clutter per unit volume and has dimensions m^2/m^3.

BINOMIAL CANCELER

An MTI filter formed by n single cancelers in cascade. The filter can also be formed using a tapped delay line with k = n + 1 binomial weights (W_k)

$$W_k = (-1)^k \binom{n}{k} \qquad k = 0, 1, \ldots n$$

The power transfer function is given by

$$|H(\omega)|^2 = [2 \sin \frac{\omega T}{2}]^{2n}$$

where n is the number of delay sections.

BIPOLAR VIDEO

In a coherent MTI, the positive and negative video at the output of the phase detector that compares the moving target return against a transmitter cohered reference oscillator. The bipolar video is processed through the MTI filter and must be converted to unipolar video for display on the radar's PPI.

BLIND PHASES

A loss of target detectability that occurs in single channel video MTI systems due to the sampled data nature of pulsed radar. The input to the MTI filter can be visualized as a sine wave at the target's doppler frequency sampled at the pulse repetition frequency of the radar. If the number of target returns per beamwidth is small, the possibility exists that the sine wave can be sampled at a relatively low amplitude with respect to its peak. In an extreme case, the sine wave can be sampled at its zero point resulting in the name "blind phase." The detectability loss is a function of the system parameters but can approach 3 dB for a typical surveillance radar application. Use of in-phase and quadrature (I and Q) MTI channels avoids the loss as does processing with an IF canceler.

BLIND SPEEDS

Periodic nulls that occur in the MTI's velocity response due to the sampled data nature of pulsed radar. In uniform PRF MTI systems "blind speeds" occur in the velocity response at velocity nulls that coincide with multiples of the radar's PRF. PRF staggering is employed to fill in the nulls within the velocity range of interest. See Papers 1.1, 2.3, 4.2 and 4.3 for design information on PRF staggering. Block staggering (constant PRF within the block) is sometimes used to prevent noncancellation of second-time-around clutter (See Papers 5.4 and 5.5).

CANCELLATION RATIO

The ratio (usually expressed in dB) of the fixed-target voltage at the input of the MTI to that of the residue at the output of the MTI after cancellation. (The ratio must be adjusted for any latent gain or loss provided by the MTI.) The cancellation ratio is a measure of the imperfections in the MTI system due to pulse-time jitter, pulsewidth jitter, pulse-amplitude jitter, phase instabilities and MTI unbalances. The cancellation ratio must be greater than the desired improvement factor. The system improvement factor (I_s) is a function of the cancellation ratio (CR) and the ideal improvement factor (I_i) determined by considering the result of an ideal MTI filter acting on the clutter spectrum (See Paper 1.1).

$$I_s = \left[\frac{CR}{1 + CR}\left(\frac{1}{I_i} + \frac{1}{CR}\right)\right]^{-1}$$

The rule for combining cancellation ratios due to individual effects is

$$CR = \left[\frac{1}{CR_1} + \frac{1}{CR_2} + \cdots \frac{1}{CR_n}\right]^{-1}$$

CANCELER

An MTI filter that consists of a number of delay (radar interpulse interval) and subtraction circuits with possible feed-forward and feedback paths from one section to another to provide shaping of the MTI filter response. See also single, double, velocity shaped, binomial and FM cancelers.

CHAFF

A volumetric clutter that is composed of distributed metalized reflectors dispensed into the atmosphere to interfere with and confuse radar operation. The "chaff" usually consists of a large number of dipoles that are designed to resonate at the frequencies of the radars they are attempting to confuse. The properties of "chaff" are similar to that of weather clutter except the cross-sections can be many times larger. The mean velocity of "chaff" clutter and the spectral spreading caused by turbulence and wind shear must be considered in MTI design.

CHARGE COUPLED DEVICE (CCD) CANCELER

An analog type MTI canceler that utilizes charge coupled devices as delay elements. The CCDs consist of capacitive storage elements separated by switches. The switches coupling the capacitive elements alternately open and close causing the analog input signal to propagate through the CCD at a rate controlled by an input clock. Relatively high performance CCD MTI cancelers have been built. (See Paper 6.2), and rapidly evolving CCD technology shows promise of performance competitive with digital techniques. The advantage of CCD cancelers over digital MTIs is that relatively expensive and complex A/D converters are not required. The requirements on A/D converters in a digital doppler processor can be reduced by employing a CCD precanceler in front of the digital signal processor.

CLUTTER

Clutter is defined as a conglomeration of unwanted radar echoes. Surface or Area clutter consists of reflections from distributed surfaces intercepted by the radar antenna's main beam and sidelobes (called ground clutter and sea clutter as appropriate). Discrete clutter is returned from stationary objects such as water towers, buildings and other unwanted fixed targets and can be very large. Volumetric clutter is reflected from weather, chaff (distributed metal reflectors used to confuse radars), and other atmospheric disturbances. Angel clutter is primarily reflected from birds and insects and can be very disturbing. Clutter is characterized by its mean or median equivalent backscattering cross-section, amplitude probability density function and its power spectral density or equivalent autocorrelation function.

CLUTTER ATTENUATION (CA)

The ratio of the clutter power at the MTI filter input to that at the output (See Paper 1.6).

$$CA = \frac{\int_0^\infty S_c(\omega)\, d\omega}{\int_0^\infty S_c(\omega)\, |H(\omega)|^2\, d\omega}$$

where

$S_c(\omega)$ = Clutter power spectral density
$H(\omega)$ = MTI filter transfer function

The clutter output power is given by

$$P_c = \int_0^\infty |H(\omega)|^2\, S_c(\omega)\, d\omega$$

CLUTTER MAP

Storage of the average clutter backscatter in a radar resolution cell on a several scan basis so that it can be used to set a mean-level threshold for detection of relatively large moving targets (See Papers 5.4 and 5.5).

CLUTTER RESIDUE

The clutter power output of an MTI which is passed through the system after doppler filtering. The clutter residue spectrum is given by

$$S_R(\omega) = |H(\omega)|^2\, S_c(\omega)$$

where

$H(\omega)$ = MTI filter transfer function
$S_c(\omega)$ = Clutter power spectral density

The clutter output power is given by

$$P_c = \int_0^\infty |H(\omega)|^2\, S_c(\omega)\, d\omega$$

COHERENT MTI

An MTI system where the target return is compared against a motion corrected reference signal cohered to the transmitted signal and then doppler filtered to reject fixed targets and pass moving targets. Fully coherent systems use a master oscillator power amplifier (MOPA) to achieve coherent operation. Systems using noncoherent transmitters use a locked coherent oscillator (COHO) to remember the transmitter phase over a PRF period. Those systems are sometimes called coherent-on-receive MTI systems.

COHERENT-ON-RECEIVE MTI

An MTI using a locked coherent-oscillator (COHO) whose phase is locked to that of an noncoherent transmitter (e. g., magnetron) each time a radar transmitter pulse is generated. The COHO is used as a reference signal that is compared against the target return and then doppler filtered to reject fixed targets and pass moving targets. Second-time-around clutter is not cancelled in this type of system.

COHERENT-OSCILLATOR (COHO)

An IF reference oscillator that is cohered to the transmitter signal and then compared against target signals for subsequent doppler filtering to reject fixed targets and pass moving targets.

COMB FILTER

A filter whose transfer function is periodic in the frequency domain. A uniform PRF MTI filter has nulls at multiples of the radar's pulse repetition frequency (PRF).

COVARIANCE MATRIX

The covariance matrix provides a measure of the correlation between various samples of a stationary random process such as that associated with clutter. The covariance of two random samples (x_k, x_ℓ) is given by

$$\mathrm{Cov}\,[x_k, x_\ell] = E\,(x_k, x_\ell) - E(x_k)\,E(x_\ell)$$

The variance is given by

$$\mathrm{Var}\,[x_k] = \mathrm{Cov}\,[x_k, x_k]$$

The covariance matrix is given by

$$M_x = \begin{bmatrix} \mathrm{Var}\,(x_1) & \mathrm{Cov}\,(x_1, x_2) \ldots & \mathrm{Cov}\,(x_1, x_n) \\ \mathrm{Cov}\,(x_1, x_2) & \mathrm{Var}\,(x_2) & \mathrm{Cov}\,(x_2, x_n) \\ \cdot & \cdot & \cdot \\ \cdot & \cdot & \cdot \\ \mathrm{Cov}\,(x_1, x_n) & \mathrm{Cov}\,(x_2, x_n) \ldots & \mathrm{Var}\,(x_n) \end{bmatrix}$$

If the variance of each of the samples is σ_x^2 then the normalized covariance matrix is given by

$$\rho_x = \begin{bmatrix} 1 & R(x_1, x_2) \ldots & R(x_1, x_n) \\ R(x_1, x_2) & 1 & \cdot \\ \cdot & \cdot & \cdot \\ \cdot & \cdot & \cdot \\ R(x_1, x_n) & R\,(x_2, x_n) \ldots & 1 \end{bmatrix}$$

where

$$R(x_k, x_\ell) = \frac{\mathrm{Cov}\,(x_k, x_\ell)}{\sigma_x^2}$$

is the correlation coefficient of x_k and x_ℓ.

DIGITAL MTI (DMTI)

A digital implementation of any of the standard MTI or doppler filter configurations. The critical element of DMTIs is the associated analog-to-digital (A/D) converter (See Paper 5.1). The A/D converter generates quantization noise which limits the MTI cancellation ratio to approximately 6 dB per bit (See Paper 5.2). An appropriate number of bits must be carried within the DMTI arithmetic operations to prevent round-off errors from affecting MTI performance. See Papers 5.2, 5.4, 5.5 and 5.6 for design information relating to the design of both MTI cancelers and FFT digital filters. Theoretical consideration are given in Papers 1.1, 1.2, 2.1, 2.3, 5.1, 5.3 and 5.6.

DISCRETE CLUTTER (POINT CLUTTER, SPECULAR CLUTTER)

Clutter returned from spatially localized, dominant reflecting objects on the terrain. It includes the radar signals backscattered from both developed areas (i. e, cities and towns) and physically isolated structures such as water towers and buildings.

Discrete clutter is distinguished from distributed clutter in that it generally exhibits larger amplitudes, fluctuates more slowly, and cannot be described by Gaussian or quasi-Gaussian statistics.

The effects of discrete clutter must be considered for any MTI radar that operates in proximity to land. Cross sections can vary between 20 to 60 dBsm. Spectral spreading must be considered in AMTI radars due to platform motion.

DISPLACED PHASE CENTER ANTENNA (DPCA)

A motion compensation technique used in airborne radars that corrects for the component of platform motion parallel to the plane of the antenna aperture by physically or electronically displacing the phase center of the antenna by an amount that causes it to remain motionless in space for a period equal to the processing time of the MTI. Typically DPCA correction is obtained by using the sum-and-difference patterns of a monopulse antenna and is applied to the first canceler of an n-canceler MTI. This leads to adequate performance for clutter with narrow spectral spreads (i. e., ground clutter). More complete compensation requires array antennas with multiple elements and compensation in each canceler of an n-canceler MTI (See Paper 7.2).

DOPPLER BEAM SHARPENING

A method of reducing surface clutter returns in an airborne radar by using doppler filtering to narrow the antenna's beam width synthetically. Aircraft motion causes a unique doppler frequency to be associated with the clutter returned from each angular direction within the antenna's main beam pattern. This causes a spectral spreading of the doppler clutter return that is a function of the aircraft's velocity, antenna beam width and the antenna's pointing angle relative to the aircraft's velocity vector. Conversely the return from a target has a unique doppler frequency associated with its angular direction. If a narrow band doppler filter (bandwidth inversely proportional to dwell time) can be placed over the target, only the clutter returned from the angular directions associated with the filter's bandwidth compete with the target. Under favorable conditions a dramatic increase in signal-to-clutter ratio can be achieved for either stationary or slowly moving targets. The effect is equivalent to reducing the antenna's beam width — thus the descriptive term, doppler beam sharpening.

DOPPLER FILTER BANK (COHERENT INTEGRATION OF DOPPLER ECHOES - CIDE)

A bank of coherent contiguous narrow band doppler filters tuned to cover the expected band of target dopplers. When the filters are matched to the target spectrum, maximum coherent integration gain is achieved in a receiver noise background. The filters can be formed using either analog or digital techniques. Digitally implemented filters commonly use the FFT algorithm. Usually a mix of predetection (coherent) and post detection (noncoherent) integration is employed to reduce the total number of filters.

DOUBLE CANCELER

An MTI filter formed by two single cancelers in cascade. A practical advantage is that unbalances in each of the cancellation circuits is compensated for in the other circuit. The double canceler has a power transfer function equal to

$$|H(\omega)|^2 = 16 \sin^4 \frac{\omega T}{2}$$

where T is the radar's interpulse period.

FAST FOURIER TRANSFORM (FFT) DOPPLER FILTER

A bank of contiguous doppler filters formed by using the FFT algorithm to compute the discrete fourier transform of N complex radar target samples. This is equivalent to performing a spectral analysis of the N radar returns in a range resolution element. The basic operation in the FFT algorithm consists of a "butterfly," defined as (See Paper 5.6).

$$u(m+1) = u(m) + w^{\ell}\, v(m)$$
$$v(m+1) = u(m) - w^{\ell}\, v(m)$$

where

$w = \exp(-2j\pi/N)$, $u(\cdot)$ and $v(\cdot)$ are complex numbers and ℓ is an integer which varies with m. Four real multiplications and six real additions (subtractions) are required to implement a basic "butterfly" operation. To process N radar pulses $N/2 \log_2 N$ "butterfly" operations are required. Weighting of the input samples (called windowing) is usually employed to reduce the sidelobes of the FFT doppler filters. Popular weightings are Hamming, hanning, and Dolph-Chebyshev windows.

FEEDBACK CANCELER

An MTI filter of the velocity shaped canceler type where only feedback paths are used to shape the filter's response.

FM CANCELER

An MTI filter of the canceler type where signal information is passed through the direct and delay channels using frequency-modulation (FM). The subtraction of the direct and delayed channel is accomplished in a mixer or in a phase detector. The advantage of this type of analog MTI filter is that amplitude balance between the direct and delayed channels is accomplished automatically.

GROUND CLUTTER

Surface clutter associated with reflections from the ground. Its average magnitude is characterized by a dimensionless parameter, σ° — the backscattering cross section per unit area. The amplitude probability distribution can be approximated by a log-normal or Weibull distribution when the ground clutter is viewed at a grazing angle of less than 5 degrees. For higher grazing angles a Rayleigh distribution is applicable. Ground clutter has a narrow intrinsic spectral width determined by internal clutter motion induced by the wind. The spectral density is usually assumed to have a Gaussian shape. A spectral density with a cubic dependence has been measured (See Paper 4.4). and the actual shape must be considered when analyzing high performance MTI systems.

GROUND MOVING TARGET INDICATOR (GMTI)

An airborne radar system designed to reject fixed targets and pass moving ground targets on the basis of their doppler shifts.

IF CANCELER

An MTI filter of the canceler type where the subtraction of the delayed and direct channels is accomplished at an intermediate frequency (IF). This type of MTI filter provides cancellation before detection and is useful in applications where it is advantageous to remove clutter before other processing takes place (e.g., monopulse radar, digital processing). Also the radar's PRF must be a precise submultiple of the IF canceler's carrier frequency.

IMPROVEMENT FACTOR (REFERENCE GAIN)

The average increase in signal-to-clutter power ratio attributable to the MTI system for targets distributed uniformly over the radial velocity spectrum (See Papers 1.1, 2.1, 4.1 and 4.2). The improvement factor is given by

$$I = \frac{(P_s/P_c)_o}{(P_s/P_c)_i} ,$$

averaged over all possible target doppler frequencies.

INDUCED CLUTTER DOPPLER SPECTRUM

The spreading of the clutter spectrum induced by the radar due to antenna scanning, platform motion and transmitter frequency variations (See Paper 1.6).

INPHASE AND QUADRATURE CHANNELS (I and Q)

A two channel or homodyne type MTI system where the radar signal is mixed to zero carrier frequency (base-band) with a single sideband or quadrature mixer whose two video-output represents the in-phase (I) and quadrature (Q) components of the signal. The two channels (I and Q) are separately MTI doppler filtered and then combined by a sum-of-the-squares ($I^2 + Q^2$) power combiner. The processing of I and Q target data is equivalent to processing at IF and avoids the approximately 3 dB target detectability loss that results due to "blind phases."

INTERCLUTTER VISIBILITY (ICV)

The ability to resolve strong point clutter regions, possibly due to non-Gaussian clutter distributions, into isolated cells between which targets may be detected with a desired probability. High resolution radars provide ICV by resolving strong point targets thereby leaving weaker distributed clutter cells where strong targets can be detected. Low resolution radars tend to average strong point clutter into large resolution cells with distributed clutter and hence provide poor ICV.

INTRINSIC CLUTTER DOPPLER SPECTRUM

The clutter spectrum spreading due to the internal motion of the scatterers caused by wind effects (See Paper 1.6).

KALMUS FILTER

A directional sensitive type of MTI processor consisting of a correlator-integrator that suppresses clutter from vegetation and foliage that exhibits back-and-forth motion while emphasizing targets with unidirectional motion. The Kalmus filter consists of two quadrature MTI channels that are correlated in a phase detector and then integrated. The polarity of the correlator output depends on the target's direction, and clutter with back-and-forth motion produces alternate positive-and-negative outputs that are averaged to a low value in the integrator. The Kalmus filter is one form of a general type of balanced processor in which signals in the upper and lower sidebands of the doppler passband of a radar are subtracted to form a bipolar output whose polarity depends on the target direction.

MATCHED FILTER

A filter whose transfer function is the conjugate of the spectrum of the transmitted waveform, except for a phase shift that varies uniformly with frequency and causes a constant time delay. The matched filter provides the maximum output signal-to-noise ratio when the noise is white (uniform frequency spectrum). The matched filter can be modified for colored noise by using a prewhitening filter (See Paper 5.3 for application to optimum MTI systems).

MISMATCHED FILTER

A linear filter that maximizes the signal-to-interference ratio in the presence of clutter plus white noise. The optimum mismatched filter for Gaussian clutter is discussed in Paper 3.1 where it is shown that the maximum SIR improvement factor is equal to the clutter-to-noise ratio plus any coherent integration gain provided. Another approach, discussed in Paper 5.3, shows that a practical implementation of an optimum mismatched filter is an MTI clutter filter cascaded with a bank of doppler filters matched to the target spectrum.

MOVING TARGET DETECTOR (MTD)

A digital type of MTI that embodies a number of doppler and nondoppler signal processing techniques to achieve a high probability of detection on moving targets and a low false alarm probability in the presence of clutter (See Papers 5.4 and 5.5).

MOVING TARGET INDICATOR (MTI)

A radar system designed to reject fixed targets and pass mvoing targets on the basis of their doppler shifts.

MTI GAIN (SYSTEM GAIN)

MTI gain is defined as the normalized output of the MTI filter when excited with an input whose power spectral density is uniformly distributed over all target doppler frequencies of interest (See Paper 1.6). It provides a criteria for comparing target responses of various MTI filters.

$$\text{MTI Gain} = \frac{\int_0^{\Omega_{v\,max}} |H_n(\omega)|^2 \, d\omega}{\Omega_{v\,max}}$$

$\Omega_{v\,max}$ = Maximum doppler radian frequency of interest

$H_n(\omega)$ = Normalized MTI transfer function

MTI LIMITING

A limiting or nonlinearity introduced into the IF or RF section of an MTI to restrict the dynamic range of large clutter returns to a value consistent with the MTI's clutter attenuation capability. Ideally the limiting functions to reduce the rms value of the MTI's clutter residue to the receiver noise level thereby providing a constant display background. However, the limiting spreads the clutter spectrum significantly reducing the overall MTI performance (See Paper 4.1).

MULTIPLE - PRF

In a multiple - PRF MTI radar, the pulse-repetition-frequency is varied on a block-to-block (group of pulses) basis to extend the first blind speed zone to many times that of a single PRF system. The advantage of multiple-PRF systems over staggered PRF systems is that second-time-around clutter is cancelled in a fully coherent system (See Paper 5.4).

NONCOHERENT MTI

An MTI system where the amplitude fluctuations, caused by the beating of clutter and moving targets in the same resolution cell, is doppler filtered to detect moving targets. Clutter returns must be present in regions where target detection is required, resulting in this type of system being "blind in the clear." Popular form of MTI for moving platform radars since it eliminates the need for platform motion compensation.

NONRECURSIVE MTI FILTER (FIR MTI FILTER)

An MTI filter formed by combining the weighted ouputs of a tapped delay line. The weights can be either real or complex (phase and amplitude). The filter has a finite inpulse response (FIR) consistent with the number of MTI pulses processed. This type of MTI filter is less susceptable to clutter transients or asynchronous interference than a recursive or infinite inpulse response (IIR) type MTI. See Papers 2.1 through 2.5 for design information.

OPTIMUM SPEED TARGET

A target whose doppler frequency is one-half the radar's pulse repetition frequency.

PSEUDO-COHERENT MTI

An MTI system where clutter is coherently delayed by several range resolution cells and then used as a pseudo coherent reference for comparison against moving targets. As with noncoherent MTI, clutter must be present in regions where moving target detection is desired.

PULSE-BURST RADAR

The pulse-burst radar transmits burst of pulses, usually uniformly spaced, having a burst duration that is less than the range to the shortest range target, with the burst repeated at a relatively low burst repetition frequency consistent with the maximum range target. The pulses within each burst may be amplitude and phase modulated, depending upon the application. The high PRF for the pulses within the burst is usually selected to provide unambiguous doppler, while the low PRF for the burst block is selected to provide unambiguous range (See Papers 3.1, 3.2, and 3.3).

PULSE DOPPLER RADAR

Generally defined as an MTI radar whose pulse-repetition frequency is at least twice the highest target doppler frequency. This provides unambiguous doppler but usually results in a large number of range ambiguities. Pulse doppler radar is alternately defined as any pulse radar that detects targets using the doppler principle.

PULSE INTERFERENCE SEPARATOR AND BLANKER (PISAB)

An interference blanker, designed to work with MTI cancelers, that checks for synchronism on a pulse-to-pulse basis and rejects all return pulses that do not show the requisite synchronism. The unit uses an anticoincidence circuit to determine returns to be treated as interference and tags the interference pulses for subsequent blanking at the output of the MTI. This type of interference blanker reduces the susceptibility of velocity shaped cancelers to asynchronous interference by blanking the interference pulses that would normally circulate around the canceler for many interpulse periods.

QUANTIZATION NOISE

A noise-like effect created by the quantization of an analog signal into discrete levels, thereby causing an uncertainty as to the exact signal value. Clutter cannot be canceled to a level below this basic noise limitation that limits the MTI cancellation ratio to approximately 6 dB per bit.

RANGE-GATED FILTER (RGF) MTI

An MTI formed by range gating coherent video (possibly using I and Q channels) or noncoherent video from a single range resolution element into a conventional lumped constant audio filter. This allows the processing to be performed at audio frequencies where stable filters are easily built with good stop - and pass - band properties. A number of contiguous RGF MTIs are usually employed to cover target ranges of interest. Noncoherent integration can be provided by using an envelope detector and low-pass filter at the output of each RGF. See Papers 1.1, 1.6 and 4.4 for design information on RGF MTIs. Range-gated doppler filters can also be formed at IF by gating coherent target returns into a bank of doppler filters.

RECURSIVE MTI FILTER (IIR MTI FILTER)

An MTI filter formed by interconnecting a number of delay sections with feed forward and feedback paths. MTI filters of this type can be synthesized by translating designs developed using classical network theory to the MTI domain utilizing a bilinear transformation given in Paper 1.2. A canonical MTI filter configuration is also given in this paper. This type of filter is characterized by its excellent velocity transfer function but has the disadvantage that asynchronous interference or transients cause a disturbance that lasts for many radar interpulse periods. Design information is given in Papers 1.2, 1.3, 1.4 and 1.6.

SEA CLUTTER

Surface clutter associated with reflections from the sea. Its average magnitude is characterized by a parameter, σ^o — the backscattering cross section per unit area. For low resolution radars (Pulsewidth >0.5 μ sec.) and grazing angles greater than 5 degrees the amplitude probability density can be approximated by a Rayleigh distribution. For high resolution radars (Pulsewidth <0.5 μ sec) and low grazing angles (<5°) the distribution exhibits a much longer tail than associated with a Rayleigh distribution. The tail is associated with "sea spikes", a specular-like reflection from wave crests occuring during the formation of "white caps." Log-normal and Weibull distributions have been used to approximate the distribution on a scan-to-scan basis. The "sea spikes" are highly correlated (10 to 100 millisecs) and a time varying Rayleigh distribution has been used to model the sea clutter for slowly scanning radars. The smaller the radar's resolution patch (azimuth and range cell) and the greater the sea state, the greater the deviation from the Rayleigh distribution. The power spectral density is usually assumed to be Gaussian shaped with an rms velocity spread depending upon sea state. Sea clutter has a mean velocity dependent upon wind speed, radar polarization and direction of the sea.

SECOND-TIME-AROUND CLUTTER

Clutter returns from ranges beyond that associated with a basic PRF interval. These clutter returns are not rejected by coherent-on-receive or staggered - PRF MTI systems.

SECOND-TIME-AROUND TARGETS

Target returns from ranges beyond that associated with a basic PRF interval. Stationary second-time-around targets are not canceled by coherent-on-receive MTI systems.

SENSITIVITY TIME CONTROL (STC)

A dynamic receiver gain control whereby the receiver gain is programmed as a function of range to reduce sensitivity to close-in targets and clutter while maintaining maximum sensitivity for weak long range targets (See Paper 6.1). This technique effectively extends the dynamic range of the receiver, allowing operation within the linear region of the receiver, thereby preventing performance loss due to limiting as discussed in Paper 4.1.

SIGNAL-TO-CLUTTER RATIO (SCR)

The ratio of the target power to the average clutter power.

SIGNAL-TO-INTERFERENCE RATIO (SIR)

The ratio of the target power to the average clutter plus noise power.

SIGNAL-TO-NOISE RATIO (SNR)

The ratio of the target power to the average noise power.

SINGLE CANCELER

An MTI filter that is formed by delaying the radar return signal by a time corresponding to the radar's PRF interval and subtracting this delayed signal from the present signal. The single canceler has a power transfer function equal to

$$|H(\omega)|^2 = 4 \sin^2 \frac{\omega T}{2}$$

where T is the radar's interpulse period. This type of filter represents the optimum MTI receiver when only two-pulses are processed.

STABLE LOCAL OSCILLATOR (STALO)

A very stable reference oscillator used in an MTI system to translate the RF target signal to an IF target signal for comparison against a reference coherent oscillator (COHO) that is cohered to the transmitted pulse.

STAGGERED - PRF

In a staggered-PRF MTI radar the pulse-repetition-frequency is varied on a pulse-to-pulse basis to reduce the effects of nulls in the MTI's comb filter response. The MTI's first blind speed zone can be extended to many times that achieved with a single PRF radar. Optimum codes for PRF staggering are discussed in Paper 4.2. An alternate multiple PRF technique is discussed in Paper 5.4.

STEP SCAN

An antenna scanning technique whereby the antenna dwells in a single beam position for a length of time equal to the duration of the MTI processing time. This removes the antenna scanning modulation that spreads both the clutter and target spectrums and limits the performance of most ground and ship based MTIs. Step-scan is most appropriate for phased array radars that utilize electronic scan control. The transient introduced into the MTI filter using step-scan generally requires that the MTI have a short transient response such as that provided by nonrecursive MTI filters. The output of the MTI must be blanked during the transient settling period.

SUBCLUTTER VISIBILITY

The ratio by which the target signal power can be weaker than the clutter power and still be detected (See Paper 1.1). The above definition is subjective since the detection criterion is a function of the application. A more definitive definition is given in Paper 1.6. The SCV is defined as the ratio of clutter power at the input of the doppler filter to moving target power to produce equal doppler filter ouputs.

$$SCV = \frac{\int_0^\infty S_T(\omega-\omega_d)\,|H(\omega)|^2\,d\omega}{\int_0^\infty S_c(\omega)\,|H(\omega)|^2\,d\omega}; \quad P_c = P_T$$

where

$S_c(\omega)$ = Clutter power spectral density

$S_T(\omega-\omega_d)$ = Target power spectral density

$H(\omega)$ = MTI filter transfer function

The SCV equation is usually evaluated at ω_d equal to an optimum speed target; that is, a target whose doppler frequency is one-half the radar's pulse repetition frequency. However, it is more fundamental to evaluate the SCV at the target speeds of interest.

SUPER-CLUTTER VISIBILITY

Refers to the ability of a radar to detect targets in a resolution cell that are stronger than the background clutter. Operation of this type of system usually involves normalization of the clutter background or receiver threshold to prevent receiver or display saturation. Examples of this type of system include the logarithmic amplifier - FTC receiver, CFAR receivers, log-CFAR receivers and the mean-level threshold using a clutter map (See Papers 5.4 and 5.5). This type of system is sometimes used in conjunction with the MTI to permit "tangential targets" stronger than the clutter to be detected.

SURFACE ACOUSTIC WAVE (SAW) FILTERS

An analog type filter, usually operating at IF, that uses a Surface Acoustic Wave (SAW) device as the filter element. The SAW device converts electrical signals into surface acoustic waves that propagate approximately 10^5 slower than electrical waves thereby allowing significant signal delays to be accomplished. The surface acoustic wave can be sampled at any point along its propagation path, and appropriate shaping of sampling transducers allows phase and amplitude control at each sample point. SAW devices have been used as dispersive delay lines in pulse compression applications providing time bandwidths of 100. (SAW devices with time bandwidths up to 3600 are under development.) The same dispersive SAW delay lines can be used to provide MTI filtering by chirping the input signal before application to the matched pulse compression filter. Other applications such as velocity indicating coherent integrators (VICI) and transversal filters for MTI or array beamforming have been developed.

TANGENTIAL TARGET

A target moving approximately tangential to the radar so that its radial velocity component is too low to be within the MTI's filter passband.

TIME AVERAGE CLUTTER COHERENT RADAR (TACCAR)

A motion compensation technique used in airborne and shipboard radars to compensate for the average doppler velocity of the clutter parallel to the axis of the antenna beam pattern. Typically a gated phase-locked loop is locked to the mean frequency of a sampled portion of the clutter spectrum. The output of the phase-locked loop is used to translate the notch of the MTI comb filter to the mean sampled clutter frequency thereby compensating for the motion perpendicular to the antenna aperture. In airborne applications the angle of arrival to the clutter is a function of the clutter's range. Complete motion compensation would require a separate correction in each range cell. Approximately three corrections spread through the radar's range coverage provides adequate performance (See Paper 7.2).

TRANSVERSAL FILTER

An MTI filter that consists of a tapped delay line whose outpus are weighted and summed. The weights can be either real or complex. The tap spacings are selected to correspond to the transmitter interpulse code. For a uniform PRF radar the taps are uniformly spaced at an interval corresponding to the radar's PRF. The number of taps corresponds to the number of radar pulses processed by the MTI.

TWO-TONE MTI

An MTI system that simultaneously presents MTI and normal radar data. The MTI data is utilized in regions containing clutter, while normal data is used in regions with no clutter return or as a weak background against the MTI data.

ULTRASONIC DELAY LINE

A storage element used in analog type MTIs to delay target returns by a time equal to the radar's pulse repetition interval. Most analog type MTIs use quartz ultrasonic delay lines to achieve the long delays (several milliseconds for a long range radar) required in radar operation. In a quartz delay line the carrier signal is: a) converted into an acoustic wave by the input transducer; b) the acoustic wave is delayed by multiple passes through the quartz polygon; and then C) the signal is recovered by an output transducer that converts the acoustic wave into an electrical signal. The quartz delay line is capable of providing precise delays but has a temperature coefficient of approximately -100 PPM/° C. In MTI systems employing quartz delay lines the radar's pulse repetition interval is generated by recirculating a pilot pulse around the delay line. MTIs that process more than two pulses require precisely matched quartz delay lines.

VANDERMONDE MTI FILTER

A sub-optimum class of nonrecursive MTI filters which are useful for PRF staggering. These filters have the property of a maximally flat transfer-function about zero frequency (See Papers 4.2 and 4.3). The Vandermonde filter reduces to a binomial canceler for a uniform PRF MTI. The Vandermonde filter weights (W_k) are given by

$$W_k = -W_o \prod_{\substack{i=1 \\ i \neq k}}^{n-1} \frac{t_i}{t_i - t_k} \qquad k = 1, \ldots, n\text{-}1$$

where t_ℓ is the total delay to the ℓth delay line tap.

VECTOR MTI

An MTI filter that processes both the phase and amplitude information in the return associated with a moving target. Examples of this type of MTI are the IF canceler and dual channel systems that process inphase and quadrature (I and Q) MTI channels. This type of MTI radar avoids the detection loss due to "blind phases."

VELOCITY INDICATING COHERENT INTEGRATOR (VICI)

A coherent pulse doppler processing technique employing an IF recirculating coherent integrator that has a single sideband mixer within the feedback loop. The mixer is driven by an oscillator that sweeps through all possible target doppler frequencies within each range resolution element. When the target doppler and reference oscillator frequencies coincide, the integrator loop builds up, thus coherently integrating the target. The velocity of the target can be determined by noting the time within the range element when the coherent integration was effective.

VELOCITY SHAPED CANCELER (FEEDBACK CANCELER - IIR FILTER)

An MTI filter of the canceler type where the pass and stop bands are shaped through appropriate weighted feedback and feed-forward paths. A canonical form is given in Paper 1.2 in addition to a bilinear transformation that allows classical network theory to be applied to the synthesis of this type of MTI filter. Properties are given in Papers 1.2, 1.3, 1.4 and 1.6. A two-delay line configuration that has been used in several operational systems has a z-transform given by (See page 37 of Paper 1.6).

$$H(z) = \frac{(z-1)^2}{z^2 - (\beta_1 + \beta_2)\, z + \beta_2}$$

where feedback with weights β_1 and β_2 is taken from the output of the second delay line and applied to the inputs of the second and first canceler. This type of MTI filter has an infinite impulse response (IIR) and is susceptible to asynchronous interference.

WEATHER CLUTTER

Volumetric clutter associated with rain, snow, fog or clouds. Its average magnitude is characterized by a parameter, η — the backscattering cross section per unit volume. The amplitude probability density function of weather clutter usually follows a Rayleigh distribution. Weather clutter has a mean velocity proportional to the wind speed. Wind turbulence and shear can cause considerable spectral spreading.

Outline of MTI Theory

1. CLUTTER CROSS SECTIONAL AREA

1.1 Surface Clutter

$\sigma_e = \sigma^o R_\tau R \theta_{AZ} \sec \psi$

σ^o = Backscatter coefficient (m^2/m^2)

$R_\tau = (c\tau/2)$ = Transmitter pulse range (m)

τ = Transmitter pulse width(s)

R = Range to clutter (m)

θ_{AZ} = Azimuth beam width (radians)

ψ = Grazing angle (degrees)

For Large ψ ($\sigma_e' < \sigma_e$)

$\sigma_e' = \sigma^o R^2 \phi_{EL} \theta_{AZ} \csc \psi$

ϕ_{EL} = Elevation beam width (radians)

1.2 Volume Clutter

$\sigma_e = \eta(\pi/4) R^2 \theta_{AZ} \phi_{EL} R_\tau$

η = Backscatter coefficient (m^2/m^3)

η For Rain

$\eta_{|dB} \approx -93 + 40 \log (f_{GHz}/3) + 17 \log (r)$

f_{GHz} = Radar frequency (GHz)

r = Rainfall (mm/hr.)

2. CLUTTER SPECTRUM

2.1 Gaussian Shaped Spectrum

$$S(f) = \frac{P_c \exp(-f^2/2\sigma_f^2)}{\sqrt{2\pi}\ \sigma_f}$$

$\sigma_f = 2\sigma_v/\lambda$ = Standard deviation of spectrum (Hz)

σ_v = Standard deviation of velocity spectrum (m/s)

λ = Radar wavelength (m)

P_c = Clutter power (w)

2.1.1 Autocorrelation Function

$\rho(\tau) = P_c \exp(-\tau^2/2\sigma_\tau^2)$

$\sigma_\tau = 1/2\pi\sigma_f$ (s)

2.2 Antenna Scanning Modulation Spectrum — $S_\theta(\omega)$

2.2.1 Uniformly Illuminated Antenna

$$S_\theta(\omega) = (P_c/a)\,[1 - (\omega/a)]^2 \qquad \omega \leqslant a$$
$$= 0 \qquad \text{otherwise}$$
$$a = 5.6\,\dot{\theta}/\theta_{3dB} \qquad \text{(rad/s)}$$

$\dot{\theta}$ = Antenna scan speed (degrees/s)

θ_{3dB} = Antenna 3 dB beam width (degrees)

2.2.2 Gaussian Shaped Spectrum

$$\sigma_\theta = 1.4\,\dot{\theta}/\theta_{3dB} \qquad \text{(rad/s)}$$

2.3 Internal Clutter Motion Spectrum — $S_w(\omega)$

2.3.1 Gaussian Shaped Spectrum

$$\sigma_w = 12.6\,\sigma_v/\lambda \qquad \text{(rad/s)}$$

σ_v = Standard deviation of velocity spectrum (m/s)

λ = Radar wavelength (m)

σ_v **for Sea Clutter (Coherent Detection)**

$$\sigma_v = 0.051\,v_w \qquad \text{(m/s)}$$
$$\sigma_w = 0.643\,v_w/\lambda \qquad \text{(rad/s)}$$

v_w = Wind speed (knots)

λ = Radar wavelength (m)

σ_v **for Sea Clutter (Envelope Detection)**

$$\sigma_v = 0.072\,v_w \qquad \text{(m/s)}$$
$$\sigma_w = 0.909\,v_w/\lambda \qquad \text{(rad/s)}$$

σ_v **for Land Clutter (Nathanson)**

$$\sigma_v\big|_{dB} = 1.261\,v_w\big|_{dB} - 24.83$$
$$\sigma_w = 0.0414 \cdot v_w^{1.261}/\lambda \qquad \text{(rad/s)}$$

v_w = Wind speed (knots)

λ = Radar wavelength (m)

2.3.2 Cubic Spectrum (Fishbein — High Frequency Tails)

$$S(\omega) = \frac{P_o}{1 + |\omega/\omega_c|^3}$$

At X-Band

$$\omega_c = 8.36 \exp(.1356\,v_w)$$

v_w = Wind speed (knots)

$$P_o = 3\sqrt{3}\,P_c/2\,\omega_c$$

2.4 Platform Motion Spectrum — $S_{pm}(\omega)$

2.4.1 Uniformly Illuminated Antenna (Scan angle $\leqslant 30°$)

$$S_{pm}(\omega) = P_o\,\frac{\sin^4(\omega/a)}{(\omega/a)^4}$$

$$a = \frac{\omega_{ac}}{2.8}\left[\theta_{3dB}\cos\psi_o \sin\theta_o + \frac{\theta_{3dB}^2}{4}\cos\psi_o\cos\theta_o + \frac{R_\tau}{h}\cos\theta_o\,\frac{\sin^3\psi_o}{\cos\psi_o}\right]$$

ω_{ac} = Aircraft doppler radian frequency (rad/s)
θ_{3dB} = Azimuth 3 dB beam width (rad)
ψ_o = Antenna depression angle (degrees)
θ_o = Azimuth angle (degrees)
h = Aircraft height (m)
$R_\tau = (c\tau/2)$ = Transmitter pulse range (m)

2.4.2 Gaussian Shaped Spectrum (Scan angle $\leqslant 30°$)

$$\sigma_p = 0.3\,\omega_{ac}\left[\theta_{3dB}\cos\psi_o\sin\theta_o + \frac{\theta_{3dB}^2}{4}\cos\psi_o\cos\theta_o + \frac{R_\tau}{h}\cos\theta_o\,\frac{\sin^3\psi_o}{\cos\psi_o}\right] \quad \text{(rad/s)}$$

$$\sigma_p \approx 3.76\,\frac{v_{ac}\cos\psi_o\sin\theta_o}{D}$$

v_{ac} = Aircraft velocity (m/s)
D = Antenna azimuth aperture (m)

2.5 Transmitter Instability Spectrum Spreading — $S_f(\omega)$

2.5.1 Linear Frequency Drift

$$S_f(\omega) = P_o(1 - \beta\omega/2\pi f'_t); \quad |\omega| \leqslant 2\pi\,\frac{f'_t}{\beta}$$

$\beta = \frac{1}{\tau}$ = Transmitter bandwidth (Hz)

f'_t = Transmitter frequency drift (Hz/sec)

2.5.2 Gaussian Shaped Spectrum

$\sigma_f = 2.67\,f'_t/\beta$ (rad/s)

2.6 Combining Video Target and Clutter Spectrums

2.6.1 Gaussian Shaped Spectrum

$S_c(\omega) = (2\pi\sigma_c^2)^{-1/2}\exp(-\omega^2/2\sigma_c^2)$

$S_{tar}(\omega) = (2\pi\sigma_t^2)^{-1/2}\exp(-\omega^2/2\sigma_t^2)$

σ_c = Standard deviation of clutter spectrum (rad/s)
σ_t = Standard deviation of target spectrum (rad/s)

2.6.1.1 Ground Coherent MTI

$$\sigma_c^2 = \sigma_\theta^2 + \sigma_f^2 + \sigma_w^2$$
$$\sigma_t^2 = \sigma_\theta^2$$

2.6.1.2 Airborne Coherent MTI (Gaussian Shaped Antenna)

$$\sigma_c^2 = \sigma_\theta^2 + \sigma_f^2 + \sigma_w^2 + \sigma_{PM}^2$$

$$\sigma_t^2 = \sigma_\theta^2$$

2.6.1.3 Ground Noncoherent MTI

$$\sigma_c^2 = 2\sigma_\theta^2 + 2\sigma_f^2 + 2\sigma_w^2$$

$$\sigma_t^2 = 2\sigma_\theta^2 + \sigma_f^2 + \sigma_w^2$$

2.6.1.4 Airborne Noncoherent MTI (Gaussian Shaped Antenna)

$$\sigma_c^2 = 2\sigma_\theta^2 + 2\sigma_f^2 + 2\sigma_w^2 + 2\sigma_{pm}^2$$

$$\sigma_t^2 = 2\sigma_\theta^2 + \sigma_f^2 + \sigma_w^2 + \sigma_{pm}^2$$

3. GENERAL MTI PERFORMANCE

3.1 Clutter Attenuation (CA)

$$CA = \frac{P_{ci}}{P_{co}} = \frac{\int_0^\infty S_c(\omega)\,d\omega}{\int_0^\infty S_c(\omega)\,|H(\omega)|^2\,d\omega}$$

$S_c(\omega)$ = Clutter power spectral density
$H(\omega)$ = MTI filter transfer function

3.2 Improvement Factor (I)

$$I = \frac{(P_s/P_c)_o}{(P_s/P_c)_i} \quad ; \text{ averaged over all possible target doppler frequencies}$$

$$I = \frac{P_{ci}}{P_{co}} \cdot \overline{\left(\frac{P_{so}}{P_{si}}\right)} = CA \cdot \bar{G}$$

$\bar{G}$ = Average Power Gain of MTI

3.2.1 Overall System Improvement Factor

$$I_s = \left[\frac{CR}{1 + CR}\left(\frac{1}{I} + \frac{1}{CR}\right)\right]^{-1}$$

CR = Cancellation ratio of MTI

3.2.2 Combining Cancellation Ratios

$$CR = \left[\frac{1}{CR_1} + \frac{1}{CR_2} \cdots + \frac{1}{CR_n}\right]^{-1}$$

CR_n = Cancellation Ratio of individual effect

3.2.3 Improvement Factor (I′ for Clutter + White Noise)

$$I' = \frac{I}{1 + (P_N/P_c)I}$$

P_N = Noise power

P_c = Clutter power
I = Improvement factor for clutter only

4. MTI CANCELERS

4.1 Binomial Canceler

$$w_k = (-1)^k \binom{n}{k} \qquad k = 0, 1, \ldots, n$$

w_k = Weight of kth tap
n = Number of delay sections

4.1.1 Transfer Function

$$|H(\omega)|^2 = \left[2 \sin \frac{\omega T}{2}\right]^{2n}$$

T = 1/PRF = Radar interpulse period (s)

4.1.2 Improvement Factor (Gaussian Spectrum)

$$I_n = \frac{\sum_{j=0}^{n} w_j^2}{\sum_{j=0}^{n} \sum_{k=0}^{n} w_j w_k \rho_c(j-k)}$$

$$I_n = \frac{\sum_{j=0}^{n} \binom{n}{j}^2}{\sum_{j=0}^{n} \sum_{k=0}^{n} (-1)^{k+j} \binom{n}{j} \binom{n}{k} \rho_c(j-k)}$$

$\rho_c(i) = \exp(-i^2 \Omega^2/2)$; $\qquad \Omega = \sigma_c T = 2\pi\sigma_f T$

σ_c = Standard deviation of clutter spectrum (rad/s)

$$I_n = \frac{1}{1 - 2\frac{n}{n+1}\rho_c(1) + 2\frac{n}{n+1}\frac{n-1}{n+2}\rho_c(2) + \ldots}$$

4.1.3 Average Power Gain ($\bar{G}$)

$$\bar{G} = 1 + n^2 + \left[\frac{n(n-1)}{2!}\right]^2 + \left[\frac{n(n-1)(n-2)}{3!}\right]^2 + \ldots$$

4.1.4 Approximate Performance ($\sigma_c T \ll 1$)

$$CA = 1/[1 \cdot 3 \cdot 5 \ldots (2n-1)] \cdot (\sigma_c T)^{2n}$$
$$\bar{G} = 2^n \cdot [1 \cdot 3 \cdot 5 \ldots (2n-1)]/n!$$
$$I_n = CA \cdot \bar{G} = 2^n/(n!)\,(\sigma_c T)^{2n}$$

4.2 Single Canceler (Weights: 1, -1)

4.2.1 Transfer Function

$$|H(\omega)|^2 = 4 \sin^2 \frac{\omega T}{2}$$

4.2.2 Improvement Factor (Gaussian Spectrum)

$$I_1 = 1/\{1 - \exp[-(\sigma_c T)^2/2]\}$$

4.2.3 Clutter Attenuation (Gaussian Spectrum)

$$CA = 1/2\{1 - \exp[-(\sigma_c T)^2/2]\}$$

4.2.4 Average Power Gain ($\bar{G}$)

$$\bar{G} = 2$$

4.2.5 Scanning Limitation (Gaussian Spectrum)

4.2.5.1 Coherent MTI

$$CA = n_a^2/1.96$$

4.2.5.2 Noncoherent MTI

$$CA = n_a^2/3.92$$

n_a = Number of pulses per 3 dB beamwidth

4.3 Double Canceler (Weights: 1, -2, 1)

4.3.1 Transfer Function

$$|H(\omega)|^2 = 16 \sin^4 \frac{\omega T}{2}$$

4.3.2 Improvement Factor (Gaussian Spectrum)

$$I_2 = 1/\left\{1 - \frac{4}{3}\exp[-(\sigma_c T)^2/2] + \frac{1}{3}\exp[-(2\sigma_c T)^2/2]\right\}$$

4.3.3 Clutter Attenuation (Gaussian Spectrum)

$$CA = 1/2\{3 - 4\exp[-(\sigma_c T)^2/2] + \exp[-(2\sigma_c T)^2/2]\}$$

4.3.4 Average Power Gain ($\bar{G}$)

$$\bar{G} = 6$$

4.3.5 Scanning Limitation (Gaussian Spectrum)

4.3.5.1 Coherent MTI

$$CA = n_a^4/11.5$$

4.3.5.2 Noncoherent MTI

$$CA = n_a^4/46$$

n_a = Number of pulses per 3 dB beamwidth

4.4 Velocity Shaped Canceler (See Paper 1.6)

4.4.1 Transfer Function (Two Pole)

$$H(z) = \frac{(z-1)^2}{z^2 - (\beta_1 + \beta_2)z + \beta_2}$$

$$\text{Poles } z_{1,2} = \frac{\beta_1 + \beta_2}{2} \pm \frac{\sqrt{(\beta_1 + \beta_2)^2 - 4\beta_2}}{2}$$

$$\text{Butterworth } z_1 = \frac{\sqrt{2}\ \cot\frac{\phi_o}{2} - 1 + j}{\sqrt{2}\ \cot\frac{\phi_o}{2} + 1 - j}$$

$$z_2 = z_1^*$$

$$\phi_o = \omega_{co} T$$

ω_{co} = Transfer function 3 dB point (rad/s)

$T = 1/PRF$ = Radar interpulse period (s)

$$\beta_2 = |z_1|^2$$

$$\beta_1 = 2R_e z_1 - \beta_2$$

4.4.2 Determination of Pole Location in Z-Plane

4.4.2.1 Low-pass Pole (p) to MTI Z-Plane

$$z = \frac{p + \Omega}{p - \Omega}$$

$$p = \Omega \frac{z + 1}{z - 1}$$

$$\Omega = \omega_{co} \tan\frac{\phi_o}{2}\ ; \qquad \phi_o = \omega_{co} T$$

4.4.2.2 High-pass pole ($p^* = 1/p$) to MTI Z-Plane

$$z = \frac{\gamma + p^*}{\gamma - p^*}$$

$$p^* = \gamma \frac{z - 1}{z + 1}$$

$$\gamma = \omega_{co} \cot\frac{\phi_o}{2}\ ; \qquad \phi_o = \omega_{co} T$$

4.4.3 Asymptotic Filters (See Paper 1.6 — page 47)

4.4.3.1 Transfer Function

$$|H(\omega)|^2 = \left(\frac{\omega}{\omega_{co}}\right)^{2n} \qquad \omega \leqslant \omega_{co}$$

$$= 1 \qquad \omega \geqslant \omega_{co}$$

ω_{co} = Filter cut-off radian frequency (rad/s)

n = Number of delay lines or filter poles

4.4.3.2 Clutter Attenuation

$$CA = 1/[1 - \alpha_n \operatorname{erf}(x/\sqrt{2}) - \beta_n \exp(-x^2/2)]$$

$$\alpha_n = 1 - \frac{(2n-1)(2n-3)\ldots(1)}{x^{2n}}$$

$$\beta_n = \sqrt{\frac{2}{\pi}}\ \frac{x^n + (2n-1)x^{n-2} + (2n-1)(2n-3)x^{n-4} +}{x^{n+1}} \qquad \text{(use n terms)}$$

$x = \omega_{co}/\sigma_c$

σ_c = Standard deviation of clutter spectrum (rad/s)

4.4.3.3 Average Power Gain

$$\bar{G} = 1 - \frac{\omega_{co} T}{\pi} \cdot \frac{2n}{2n+1}$$

4.4.3.4 Improvement Factor

$$I = CA \cdot \bar{G}$$

$$I = \left[1 - \left(\frac{\omega_{co} T}{\pi}\right)\left(\frac{2n}{2n+1}\right)\right] \Big/ \left[1 - a_n \operatorname{erf} \frac{x}{\sqrt{2}} - \beta_n \exp\left(-\frac{x^2}{2}\right)\right]$$

T = Radar interpulse period (s)

4.5 Nonrecursive MTI Filters (See Papers 2.1 through 2.5)

4.5.1 Transfer Function ($z = e^{j\omega T}$)

$$H(z) = \frac{w_o z^n + w_1 z^{n-1} + \ldots + w_n}{z^n}$$

$$H(z) = \frac{(z-z_o)(z-z_1)(z-z_1^*)\ldots}{z^n} \quad ; \quad n \text{ odd}$$

$$H(z) = \frac{(z-z_1)(z-z_1^*)\ldots}{z^n} \quad ; \quad n \text{ even}$$

n = Number of delay sections

z_n = Zero of MTI z-plane transfer function

4.5.2 Emerson's Optimization Procedure (See Paper 1.1)

4.5.2.1 Optimization Problem

$$\text{Min } P_c = \sum_{j=0}^{n} \sum_{k=0}^{n} w_j w_k \rho_c[(j-k)T]$$

$$\text{Subject to } \sum_{i=0}^{n} w_i^2 = 1$$

w_i = n + 1 filter weights

$\rho_c(\cdot)$ = Elements of clutter covariance matrix

4.5.2.2 Solution Equation

$$(M_x - \lambda I)A = 0$$

M_x = Clutter Covariance Matrix
λ = Eigenvalues
A = Eigenvectors
I = Identity Matrix

4.5.2.3 Solutions (For curve See Fig. 2 — Paper 2.4)

$I = 1/\text{Min } P_c = 1/\lambda_{min}$

λ_{min} = Minimum Eigenvalue

A_{min} = Eigenvector for λ_{min} — gives weights

Zeros, when normalized, lie on unit circle

4.5.2.4 Transfer Functions

n-odd; weights anti-symmetrical

$$|H(\omega)| = \left| \sum_{i=0}^{(n-1)/2} (-1)^i \, 2|w_i| \sin\left(\frac{n}{2} - i\right) \omega T \right|$$

n-even; weights symmetrical

$$|H(\omega)| = \left| (-1)^{n/2} |w_{n/2}| + \sum_{i=0}^{(n-2)/2} (-1)^i \, 2|w_i| \cos\left(\frac{n}{2} - i\right)\omega T \right|$$

4.5.2.5 Optimum Solutions (See Paper 2.1)

$n = 1;\ w_o = 1;\ \ w_1 = -1$

$$I = 1/[1 - \rho_c(1)]$$

$$n = 2;\ w_o = w_2 = 1;\ w_1 = -\frac{\rho_c(2) + \sqrt{\rho_c^2(2) + 8\rho_c^2(1)}}{2\rho_c(1)}$$

$$I = 1/\left[1 - \frac{\rho_c(2)}{2}\left(\sqrt{1 + 8\left(\frac{\rho_c(1)}{\rho_c(2)}\right)^2} - 1\right)\right]$$

$n = 3;\ w_o = -w_3 = 1;\ w_2 = -w_1$

$$w_1 = -\frac{\sqrt{\frac{(\rho_c(1) - \rho_c(3))^2}{4} + (\rho_c(1) - \rho_c(2))^2} + \frac{\rho_c(1) - \rho_c(3)}{2}}{\rho_c(1) - \rho_c(2)}$$

$$I = 1/\left[1 - \frac{\rho_c(1) + \rho_c(3)}{2} - \sqrt{\frac{(\rho_c(1) - \rho_c(3))^2}{4} + (\rho_c(1) - \rho_c(2))^2}\right]$$

$\rho_c(i) = \exp[-i^2\Omega^2/2];\quad \Omega = \sigma_c T$

σ_c = Standard deviation of clutter spectrum (rad/s)

T = Radar interpulse period (s)

4.5.3 Transversal MTI Filters (General Theory)

4.5.3.1 Signal Power Output

$$P_s = W^T V V^{*T} W^* = W^T M_s W^*$$

W = Complex weight vector of transversal filter

V = Input signal voltage vector

M_s = Signal covariance matrix

4.5.3.1.1 Equally Likely Doppler Signals

$M_s = I$ = Identity Matrix

4.5.3.1.2 Specific Doppler Signal

$$V = [1 \; e^{j\omega_d T} \ldots\ldots e^{jn\omega_d T}]^T$$

$$M_s = \begin{bmatrix} 1 & e^{-j\omega_d T} & \cdots & e^{-jn\omega_d T} \\ e^{j\omega_d T} & 1 & \ddots & \vdots \\ e^{jn\omega_d T} & \cdots & \cdots & 1 \end{bmatrix}$$

n = Number of delay sections

ω_d = Dopper radian frequency

T = Radar interpulse period

4.5.3.1.3 Equally Likely Doppler Signals in Band (f = 1/NT)

$$m_{\alpha,\beta} = \exp\left[\frac{j2\pi k(\alpha-\beta)}{N}\right] \frac{\sin\left[\frac{\pi(\alpha-\beta)}{NT}\right]}{\left[\frac{\pi(\alpha-\beta)}{NT}\right]}$$

$m_{\alpha,\beta}$ = Element of signal covariance matrix

N = Number of sections in band, $f = 1/T$

k = Index of band segment, 0, 1, 2, . . . , N−1

4.5.3.2 Clutter Power Output

$$P_c = W^T M_x W^*$$

$$M_x = \begin{bmatrix} 1 & \rho_c(T) & \cdots & \rho_c(NT) \\ \rho_c^*(T) & 1 & \ddots & \vdots \\ \rho_c^*(NT) & \cdots & \cdots & 1 \end{bmatrix}$$

M_x = Clutter covariance matrix

4.5.3.2.1 Gaussian Shaped Clutter Spectrum

$$\rho_{\alpha,\beta} = \exp[-2\pi^2 \sigma_f^2 (\alpha-\beta)^2 T^2 - j2\pi\mu_c(\alpha-\beta)T]$$

$\rho_{a,\beta}$ = Element of covariance matrix
σ_f = Standard deviation of clutter spectrum (Hz)
μ_c = Mean Doppler clutter frequency (Hz)

4.5.3.2.2 Gaussian Shaped Clutter Spectrum and White Noise

$$\rho_{a,\beta} = P_c \exp[-2\pi^2 \sigma_f^2 (a-\beta)^2 T^2 - j2\pi\mu_c(a-\beta)T] + P_N \delta(a-\beta)$$

$P_c + P_N = 1$ for normalization
P_c = Clutter power
P_N = Noise power
$\delta(\cdot)$ = Dirac delta function

4.5.3.3 Improvement Factor

$$I = \frac{W^T M_s W^*}{W^T M_x W^*} = \frac{W^T V V^{*T} W^*}{W^T M_x W^*}$$

4.5.3.3.1 Optimum Improvement Factor (Gantmacher)

$$I_o = V^{*T} M_x^{-1} V$$

M_x^{-1} = Inverse covariance matrix
V = Specified input voltage vector

4.5.3.3.1.1 Optimum Weights (Gantmacher)

$$W = M_x^{-1} V^*$$

4.5.3.4 Solution for Markov Clutter Process

4.5.3.4.1 Gaussian Markov Process

$$S(\omega) = 2\gamma P_c/(\gamma^2 + \omega^2); R(\tau) = P_c \exp[-\gamma|\tau|]$$

$$m_{a,\beta} = \exp[-\gamma|a-\beta|T] = \rho[(a-\beta)T]$$

γ = 3 dB radian frequency of clutter (rad/s)

4.5.3.4.2 Inverse of Markov Covariance Matrix

$$M_x^{-1} = \frac{1}{1-\rho^2(T)} \begin{bmatrix} 1 & -\rho(T) & 0 \cdots\cdots 0 & 0 \\ -\rho(T) & 1+\rho^2(T) & -\rho(T) \cdots\cdots 0 & 0 \\ 0 & -\rho(T) & 1+\rho^2(T) \cdots 0 & 0 \\ \vdots & \vdots & \vdots \qquad\quad \vdots & \vdots \\ 0 & 0 & 0 \cdots\cdots 1+\rho^2(T) & -\rho(T) \\ 0 & 0 & 0 \cdots\cdots -\rho(T) & 1 \end{bmatrix}$$

4.5.3.4.3 Optimum Improvement Factor — Markov Clutter

$$I = \frac{(n+1) + (n-1)\rho^2(T) - 2n\rho(T)\cos\omega_d T}{1 - \rho^2(T)}$$

n = Number of delay sections

$\rho(T) = \exp(-\gamma T)$

4.5.3.4.4 Optimum Weights — Markov Clutter

$$w_o = \frac{1 - \rho(T)\cos\omega_d T}{1 - \rho^2(T)} + j\frac{\rho(T)\sin\omega_d T}{1 - \rho^2(T)}$$

$$w_i = \frac{[1 + \rho^2(T)]\cos i\omega_d T - \rho(T)[\cos(i-1)\omega_d T + \cos(i+1)\omega_d T]}{1 - \rho^2(T)}$$

$$+ j\frac{\rho(T)\sin(i+1)\omega_d T - [1 + \rho^2(T)]\sin i\omega_d T}{1 - \rho^2(T)}$$

$1 \leqslant i < n$

$$w_n = \frac{\cos n\omega_d T - \rho(T)\cos(n-1)\omega_d T}{1 - \rho^2(T)} + j\frac{\rho(T)\sin(n-1)\omega_d T - \sin(n\omega_d T)}{1 - \rho^2(T)}$$

4.5.4 Fast Fourier Transform (FFT) Digital Filters

4.5.4.1 Discrete Fourier Transform

$$F(n,k) = \sum_{n=0}^{N-1} f(n)e^{-j(2\pi/N)nk}$$

$f(n)$ = Signal sample at time point $n = 0, 1, \ldots, N-1$

N = Number of sample points

k = Output transform at point $k = 0, 1, \ldots, N-1$

4.5.4.2 Equivalent Weight of Transversal Filter

$$w_{n,k} = e^{-j(2\pi/N)nk}$$

k = Filter number

n = Delay line tap number

4.5.4.3 Transfer Function (Normalized)

$$|H(\omega)| = \left| \frac{1}{N} \frac{\sin[\pi(N\frac{f}{f_r} - k)]}{\sin[\frac{\pi}{N}(N\frac{f}{f_r} - k)]} \right|$$

N = Number of FFT points

k = Filter number $0, 1, 2, \ldots, N-1$

f_r = Radar pulse repetition frequency

4.5.4.4 Z-Plane Transfer Function

$$H(z) = \frac{z^N - 1}{z - e^{-j2\pi(k/N)}}$$

N zero's equally spaced about unit circle; with one pole calceling zero at point determined by filter number.

4.5.4.5 Post FFT Filter Weighting to Improve Sidelobes

$$U(n,k) = F(n,k) + \alpha[F(n,k+1) + F(n,k-1)]$$

$U(n,k)$ = Weighted FFT k^{th} filter output

$F(n,\cdot)$ = Unweighted k - 1, k, k + 1 filter output

a = Weighting = 0.5 for hanning weighting
0.426 for Hamming weighting

4.5.4.6 General Weighting Functions

$$a_n = a - (1 - a)\cos\frac{2\pi n}{N-1} \qquad n = 0, 1, \ldots, N-1$$

a = 0.5 hanning window

a = 0.54 Hamming window

4.5.4.7 Improvement Factor

4.5.4.7.1 General at Frequency (f_d) with Weighting

$$I = \frac{\left(\sum_{i=0}^{N-1} a_i\right)^2}{\sum_{i=0}^{N-1}\sum_{j=0}^{N-1} a_i a_j \cos\left(2\pi(i-j)\frac{k}{N}\right)\rho_c[(i-j)T]}$$

$a_{i,j}$ = Complex weights including windows

N = Number of FFT points

k = Filter number 0, 1, 2, . . . , N - 1

T = Radar interpulse period (s)

4.5.4.7.2 General at Frequency (f_d) — No Weighting

$$I = \frac{N^2}{N + \sum_{i=1}^{N-1} (N-i)\cos 2\pi\frac{k}{N} i \cdot \rho_c(i)}$$

$$\rho_c(i) = \exp[-i^2\Omega^2]; \qquad \Omega = \sigma_c T = 2\pi\sigma_f T$$

4.5.5 Cascade of MTI Canceler with FFT Digital Filter

4.5.5.1 Transfer Function (Binomial Canceler + FFT)

$$H(\omega) = \left| \sin^m \left(\frac{\pi f_d}{f_r}\right) \frac{\sin[\pi(N \frac{f_d}{f_r} - k)]}{\sin[\frac{\pi}{N}(N \frac{f_d}{f_r} - k)]} \right|$$

m = Number of canceler delay lines

N = Number of FFT points

k = Filter number 0, 1, 2, 3, . . . , N - 1

f_r = Radar pulse repetition rate (Hz)

4.5.5.2 Z-Transform of Transfer Function

$$H(z) = \sum_{i=0}^{m} (-1)^i \binom{m}{i} \sum_{\ell=i}^{i+N-1} e^{-j(2\pi/N)k(\ell-i)} z^{-\ell}$$

ℓ = Overall filter weight number 0, 1, 2, . . . n+m−1

4.5.5.3 Weights of Overall Filter

$$W_\ell = \sum_{i=0}^{m} (-1)^i \binom{m}{i} e^{-j(2\pi/N)k(\ell-i)}; \qquad \ell \geqslant i, \quad \ell - i < N$$

4.5.5.4 Improvement Factor for Single Canceler + N Point FFT

$$I = \frac{2N^2 \sin^2\left(\frac{a}{2}\right)}{N - (N-1)\cos a - \sum_{i=0}^{N-1} 2\rho_c(i)[\cos a(i-1) - \cos ai] - \sum_{i=1}^{N-2} (N-1-i)\rho_c(i)[\cos a(i-1) - 2\cos a(i) + \cos a(i+1)] - \rho_c(N)\cos a(N-1)}$$

$$a = 2\pi \frac{k}{N} \qquad k = 0, 1, 2, \ldots N-1$$

4.5.5.5 Improvement Factor for Double Canceler + N Point FFT

$$I = \frac{8N^2 \sin^4\left(\frac{a}{2}\right)}{\begin{aligned} &6 - 4\cos a + \left[1 - \frac{\delta(0)}{2}\right] \sum_{i=0}^{N-3} (N-2-i)[6 + 2\cos 2a - 8\cos a]\cos ia\, \rho_c(i) \\ &+ \sum_{i=1}^{N-2} [-4\cos(i+1)a + 10\cos ia - 8\cos(i-1)a + 2\cos(i-2)a]\rho_c(i) \\ &+ \sum_{i=1}^{N} [2\cos ia - 4\cos(i-1)a + 2\cos(i-2)a]\rho_c(i) + 4\cos a\, \rho(1) \\ &+ [4\cos(N-1)a - 4\cos(N-2)a + \cos(N-3)a]\rho_c(N-1) + \cos(N-2)a \cdot \rho_c(N+1) \end{aligned}}$$

$$a = 2\pi \frac{k}{N}; \qquad k = 0, 1, \ldots, N-1$$

$\cos(\cdot)\alpha, \; 0 \leq (\cdot) \leq N-1;$ zero otherwise

$$\left[\frac{\delta(0)}{2}\right] = \frac{1}{2}\;; \quad i = 0; \qquad \text{zero otherwise}$$

5. AMTI PLATFORM MOTION COMPENSATION

5.1 TACCAR

5.1.1 Improvement Factor for Single Loop (Andrews)

$$I'_n = I/L_n$$

I = Improvement factor of Binomial MTI without TACCAR

$$I \approx \frac{2^n}{n!}\left(\frac{f_r}{2\pi\sigma_f}\right)^{2n} \approx \frac{2^n}{n!}\left(\frac{D}{1.2\pi \sin\theta_a \cos\varphi_s V_p T}\right)^{2n}$$

$$L_n = \frac{\left(\frac{f_e}{\sigma_f}\right) + 2n \sum_{k=1}^{n} \binom{2n}{2k} \left(\frac{f_e}{\sigma_f}\right)^{2n-2k} [1 \cdot 3 \cdot 5 \ldots (2k-1)]}{[1 \cdot 3 \cdot 5 \cdot (2n-1)]}$$

$$f_e = \left| \frac{2v_p \cos\theta_a}{\lambda} \left(\cos\varphi_s - \frac{\cos\varphi_{max} + \cos\varphi_{min}}{2}\right) \right|$$

V_p = Aircraft velocity (m/s)
λ = Radar wavelength (m)
θ_a = Azimuth antenna angle (°)
φ_s = Elevation angle to clutter range (°)
φ_{max} = Elevation angle, maximum clutter range (°)
φ_{min} = Elevation angle, minimum clutter range (°)
n = Number of MTI delay stages
σ_f = Standard deviation of clutter spectrum (Hz)
D = Antenna azimuth aperture (m)
T = Pulse repetition period (s)
f_r = Pulse repetition rate (Hz)

5.2 DPCA

5.2.1 Transfer Function n-stage MTI with DPCA (Andrews)

$$|H_n(f)|^2 = \left(1 + \tan^2 \pi \frac{f'_d}{f_r}\right)\left\{ 2\sin\left[\pi\left(\frac{f-f'_d}{f_r}\right)\right]\right\}^2 \cdot \left[2\sin\pi\left(\frac{f}{f_r}\right)\right]^{2n-2}$$

$$f'_d \approx 2\,\frac{v_p \cos\varphi_s \sin\theta_a}{\lambda}\,\theta$$

f'_d = Differential Doppler shift after TACCAR correction
θ = Angle with respect to azimuth antenna axis (rad)

5.2.2 Improvement Factor (Andrews)

5.2.2.1 Single Canceler

$$I_1' = 2\left(\frac{f_r}{2\pi\sigma_f}\right)^2 \frac{\int_{-\theta_o}^{\theta_o} G^4(\theta)\,d\theta}{\int_{-\theta_o}^{\theta_o}\left[1+\tan^2\left(\pi\frac{f_d'}{f_r}\right) G^4(\theta)\,d\theta\right]}$$

$G^4(\theta)$ = Antenna two-way power gain

5.2.2.2 Double Canceler

$$I_2' = I_2 \frac{\int_{-\theta_o}^{\theta_o} G^4(\theta)\,d\theta}{\int_{-\theta_o}^{\theta_o} G^4(\theta)\left[1+\frac{2}{3}I_1\tan^2\left(\pi\frac{f_d'}{f_r}\right)\right]d\theta}$$

$$I_1 = 2\left(\frac{f_r}{2\pi\sigma_f}\right)^2 \qquad I_2 = 2\left(\frac{f_r}{2\pi\sigma_f}\right)^4$$

5.2.2.3 Uniformly Illuminated Antenna

$$G^4(\theta) = \left[\frac{\sin\left(\pi\frac{D}{\lambda}\sin\theta\right)}{\pi\frac{D}{\lambda}\sin\theta}\right]^4$$

D = Aperture length (m)
λ = Radar wavelength (m)

$$I_1' = I_1\frac{G_o}{G_o+G_1} \qquad I_2' = I_2\frac{G_o}{G_o+\frac{2}{3}I_1G_1}$$

$$G_o \approx \frac{\lambda}{\pi D}\int_{-\pi}^{\pi}\left(\frac{\sin x}{x}\right)^4 dx$$

$$G_1 \approx \frac{\lambda}{\pi D}\int_{-\pi}^{\pi}\tan^2\left(\frac{2v_pT\cos\varphi_s\sin\theta_a}{\lambda D}x\right)\left(\frac{\sin x}{x}\right)^4 dx$$

6. DOPPLER BEAM SHARPENING

6.1 Resolution

$$\delta_{(x)} = \frac{R\lambda\Delta f}{2\,v_p\sin a}$$

$\delta_{(x)}$ = Separation of azimuth points on surface (m)
R = Range to range resolution element (m)
v_p = Velocity of aircraft (m/s)

a = Angle between aircraft velocity vector and line-of-sight to point on surface (°)

Δf = Differential Doppler processor resolution (Hz)

$= 1/T_i$

T_i = Dwell time in resolution cell (s)

6.2 Effective Synthetic Aperture

$$D_e = v_p \sin a \cdot T_i$$

D_e = Effective azimuth aperture width (m)

6.3 Effective Azimuth Resolution

$$\theta_{Aze} = R \frac{\lambda}{2D_e} \qquad \text{(rad)}$$

6.4 Signal-to-Clutter Ratio

$$S/C \approx \frac{2\sigma_T v_p \sin a \, T_i}{R\lambda R_\tau \sigma^o L_c}$$

σ_T = Target cross-section (m^2)

v_p = Aircraft velocity (m/s)

$R_\tau = (c\tau/2)$ = Transmitter pulse range (m)

τ = Transmitter pulse width (s)

R = Range to clutter (m)

λ = Radar wavelength (m)

a = Angle between aircraft velocity vector and LOS to clutter patch (°)

L_c = Target detection losses

Clutter Models & Data

A clutter model assumption is implicit in the design or analysis of an MTI radar. The function of the clutter model is to represent the random clutter process in such a manner that system performance can be determined. Ideally, the model accurately reflects the clutter process in a realistic manner. Practically, the clutter data base is rarely adequate to achieve this objective, and the art of modeling the clutter involves a compromise between accurate representation and analytic convenience.

The simplest clutter model is the Gaussian model whereby the random clutter process is represented as a multidimensional Gaussian process. This model has been employed almost exclusively in MTI analyses. An attractive feature of this model is that the clutter process is completely specified by its mean value and covariance function. The Gaussian model assumption leads to an optimum MTI processor that is linear, and it is under this assumption that the term "optimum" is applied to MTI processors throughout the text and reprinted papers in this book.

The Gaussian clutter assumption generally leads to an optimistic evaluation of MTI performance. At present, analytical techniques are not available to allow the performance of MTI systems in the presence of non-Gaussian clutter to be determined. The principal method of attacking this type of problem is through a Monte Carlo computer simulation in which the actual MTI system is exercised by applying statistical samples of the postulated clutter and target processes.

In the following section a brief overview and several examples of the various clutter models are given. This material is intended to supplement and to aid in the interpretation of the extensive existing radar clutter data base available in the literature.

In formulating the clutter model both intrinsic and induced effects must be considered. Intrinsic effects are due to the inherent properties of the reflecting scatterers and are influenced primarily by the wind driven motion of the scatterers. Induced effects are functions of the radar system parameters and are primarily influenced by the antenna scanning, platform motion and transmitter instability characteristics of the radar. (See Outline of MTI Theory for formulas that apply to individual effects.)

1. The Gaussian (Rayleigh Envelope) Clutter Model

The Gaussian or Rayleigh envelope clutter model is applicable to distributed clutter that is returned from a spatially continuous distribution of scatterers with no subset of scatterers predominating. This type of clutter is usually associated with weather clutter, chaff, sea clutter observed with a low resolution radar (Pulse width> 0.5 μsec.) or with a high-resolution radar at high grazing angles ($\varphi > 5°$) and land clutter observed from high grazing angles ($\varphi > 5°$) over undeveloped terrain.

The random clutter process is represented at the clutter carrier frequency (f_c) as

$$c_t = x_t \cos \omega_c t - y_t \sin \omega_c t \quad (1)$$

where x_t and y_t are zero mean, identically distributed, low pass, independent normal processes with variance σ^2. The clutter process c_t is strict sense stationary and hence its characteristics are independent of the time origin.

The voltage envelope of c_t is given by

$$v_t = \sqrt{x_t^2 + y_t^2}, \quad (2)$$

and has a Rayleigh first probability density function

$$f_v(v) = \frac{v}{\sigma^2} \exp[-v^2/2\sigma^2] \; ; v \geqslant 0 \quad (3)$$

that is independent of time due to the stationarity, and is a function of only the clutter power

$$P_c = 2\sigma^2. \quad (4)$$

The power amplitude distribution function can be found by applying the transformation $P = v^2$

$$f_P(P) = \frac{1}{P_c} \exp[-P/P_c] \quad (5)$$

The clutter cross-sectional area (backscatter-coefficient x intercepted area) is proportional to the power envelope resulting in an exponential probability density function for the clutter cross-section (A)

$$f_p(A) = \frac{1}{\bar{A}} \exp[-A/\bar{A}] \tag{6}$$

$\bar{A}$ is the mean clutter cross-section and is given by

$$\bar{A} = \bar{\sigma}^o A_c \tag{7}$$

where $\bar{\sigma}^o$ is the mean backscattering coefficient, and A_c is the area intercepted by the radar cell. The power spectral density associated with the randomly fluctuating portion of the clutter cross-section is usually assumed to be Gaussian shaped in that it is the result of several independent effects. The power spectral density of the power envelope is given by

$$S_p(f) = \frac{P_c}{\sqrt{2\pi\sigma_f^2}} \exp[-f^2/2\sigma_f^2], \tag{8}$$

where P_c is the clutter power and σ_f is the standard deviation of the spectrum. The standard deviation of the spectrum (σ_f) is related to the rms velocity clutter spread (σ_v) by

$$\sigma_f = \frac{2\sigma_v}{\lambda}. \tag{9}$$

The autocorrelation function associated with the power envelope is given by

$$R_p(\tau) = P_c \exp[-\tau^2/2\sigma_\tau^2], \tag{10}$$

where

$$\sigma_\tau = 1/2\pi\sigma_f. \tag{11}$$

In many applications it is convenient to work with the normalized autocorrelation function,

$$\rho_p(\tau) = R_p(\tau)/P_c = \exp[-\tau^2/2\sigma_\tau^2] \tag{12}$$

Of interest in MTI work and other coherent processing systems is the power spectral density and auto correlation function of x_t and y_t. It is known (Uhlenbeck) that the relationship between the normalized autocorrelation functions of P_t and x_t is given by

$$\rho_p(\tau) = \rho_x^2(\tau) = \rho_y^2(\tau). \tag{13}$$

Thus the power spectral density of the power envelope can be found from the power spectral density of the coherent quadrature components by

$$S_p(\omega) = S_x(\omega) * S_x(\omega), \tag{14}$$

where * represents convolution. Hence for a Gaussian shaped spectrum all power spectral densitites and autocorrelations are Gaussian shaped, and the standard deviation of the power envelope can be found from those of the coherent quadrature components using

$$\sigma_p^2 = 2\sigma_x^2 = 2\sigma_y^2. \tag{15}$$

To summarize, for MTI work, the power spectral density or autocorrelation functions are given by equations (8), (10) and (12) using the appropriate standard deviation given by equation (15). When working with experimental or published data, care must be exercised to ascertain whether the standard deviations that are referenced refer to the coherent (σ_x) or noncoherent (σ_p) standard deviations.

In some MTI applications it is necessary to model clutter that has relative motion with respect to the radar. This is accomplished by modifying equation (8) by the mean clutter doppler shift ($\bar{f}_d$) resulting in

$$S_p(f) = \frac{P_c}{\sqrt{2\pi\sigma_f^2}} \exp[-(f - \bar{f}_d)^2/2\sigma_f^2]. \tag{16}$$

When the clutter process (1) is sampled by a pulsed radar the probability density of the samples is given by a multidimensional Gaussian distribution

$$f_{x_1 \cdots x_n}(x_1 \cdots x_n) = \frac{\exp[-\frac{1}{2} \mathbf{x}^T M_x^{-1} \mathbf{x}]}{(2\pi)^{n/2} |M_x|^{1/2}} \tag{17}$$

where $\mathbf{x} = (x_1, x_2 \cdots x_n)^T$, and the covariance matrix is given by

$$M_x = \sigma_x^2 \begin{bmatrix} 1 & \rho_{12} & \rho_{13} & \rho_{1n} \\ & \ddots & & \\ \rho_{21} & & \ddots & \rho_{2n} \\ & & & \\ \rho_{n1} & \cdots & \cdots & 1 \end{bmatrix} \tag{18}$$

where

$$\rho_{ij} = \exp[-(i-j)^2 \Omega^2/2]; \Omega = 2\pi\sigma_f T \tag{19}$$

for a Gaussian shaped distribution.

When the clutter is passed through a linear MTI the Gaussian form of the distribution is preserved. The probability density function at the output of the MTI can be found by using equation (17) with the covariance matrix modified by the effect of the MTI filter on the clutter process.

For analytical convenience the clutter process is sometimes assumed to be a Gaussian Markov process. This process has an exponential normalized covariance function given by

$$\rho_{\alpha\beta} = \exp\left[-\gamma|\alpha - \beta|\, T\right] \tag{20}$$

and a power spectral density given by

$$S(\omega) = 2\gamma\, P_c/(\gamma^2 + \omega^2). \tag{21}$$

The inverse covariance matrix (M_x^{-1}) has a particularly simple form:

$$M_x^{-1} = \frac{1}{1-\rho_{12}^2}\begin{bmatrix} 1 & -\rho_{12} & 0 \cdots\cdots & 0 \\ -\rho_{21} & 1+\rho_{12}^2 & -\rho_{12} \cdots\cdots 0 & 0 \\ 0 & -\rho_{21} & 1+\rho_{12}^2 \cdots\cdots 0 & 0 \\ 0 & 0 & 0 \cdots\cdots 1+\rho_{12}^2 & -\rho_{12} \\ 0 & 0 & 0 \cdots\cdots -\rho_{21} & 1 \end{bmatrix} \tag{22}$$

2. The Rician Clutter Model

This model is similar to the Gaussian model except that a predominant steady scatterer (S) is added to the distributed clutter. The first probability density function corresponding to equation (5) becomes

$$f_R(P) = \frac{1+m^2}{\overline{P}}\, e^{-m^2}\, e^{-\frac{P}{\overline{P}}(1+m^2)}\, I_0\left(2m\sqrt{(1+m^2)P/\overline{P}}\right), \tag{23}$$

where the ratio of the steady (S^2) to distributed (P_o) power ratio is given by

$$m^2 = S^2/P_o, \tag{24}$$

and the total power ($\overline{P}$) proportional to the cross-sectional area is given by

$$\overline{P} = S^2 + P_o. \tag{25}$$

For MTI work the clutter process at the carrier frequency is obtained by modifying equation (1) to

$$c_t = (S + x_t)\cos\omega_c t - y_t \sin\omega_c t. \tag{26}$$

The autocorrelation function for the in-phase (I) process and Gaussian shaped distributed clutter is

$$R(\tau) = P_o\,[m^2 + \exp(-\tau^2/2\sigma_t^2)]. \qquad (27)$$

The resulting power spectral density is

$$S(f) = m^2\,P_0\,\delta(f) + P_o \exp(-f^2/2\sigma_f^2)/\sqrt{2\pi\sigma_f^2}$$

and illustrates that MTIs operating in this type of clutter must be designed to provide a null at zero frequency.

3. The Log-Normal Clutter Model

The log-normal clutter model has been used to model high-resolution (pulse width $<.5\ \mu$ sec.) sea-clutter data, where the sea clutter is observed at grazing angles less than 5 degrees. In addition, ground clutter observed at low grazing angles has also been modeled using the log-normal model.

The log-normal distribution is highly skewed resulting in a relatively high probability of large clutter amplitudes. False alarm probabilities are controlled by raising the detection thresholds thereby causing a reduced probability of detection (P_d) as compared to that achieved in Rayleigh clutter.

The log-normal model generally tends to overestimate the dynamic range of the real clutter distribution, while the Rayleigh model tends to underestimate the dynamic range. Thus analyses with the log-normal model produce conservative results while optimistic results are produced with the Rayleigh model. A good design procedure is to design for the range of distributions between Rayleigh and log-normal.

The log-normal sea clutter model is associated with "sea spikes", a specular like reflection from wave crests occurring during the formation of "white caps." The log-normal model for ground clutter seems to be associated with large directive scatterers that, depending upon viewing aspect, produce bright ridges and shadowing resulting in a large clutter dynamic range.

The log-normal character of sea clutter is shown in Figure 1 for data taken with a Harbor Surveillance radar [1]. It is apparent that the data above the median can be closely approximated by straight lines indicating the close fit provided by the log-normal model.

The log-normal character of ground clutter data is apparent for clutter returned from cities as shown

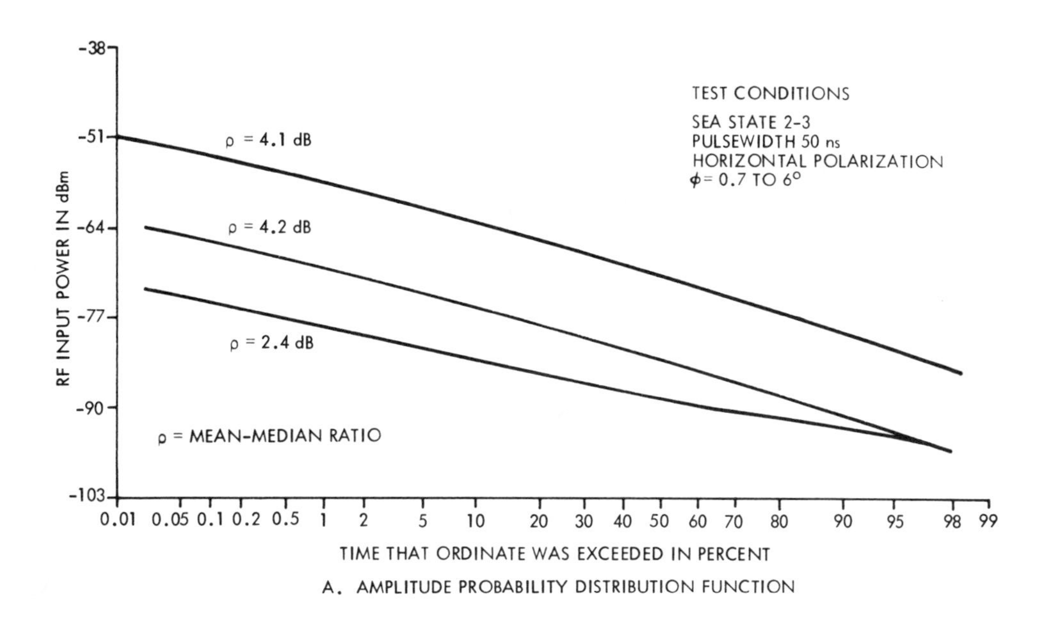

A. AMPLITUDE PROBABILITY DISTRIBUTION FUNCTION

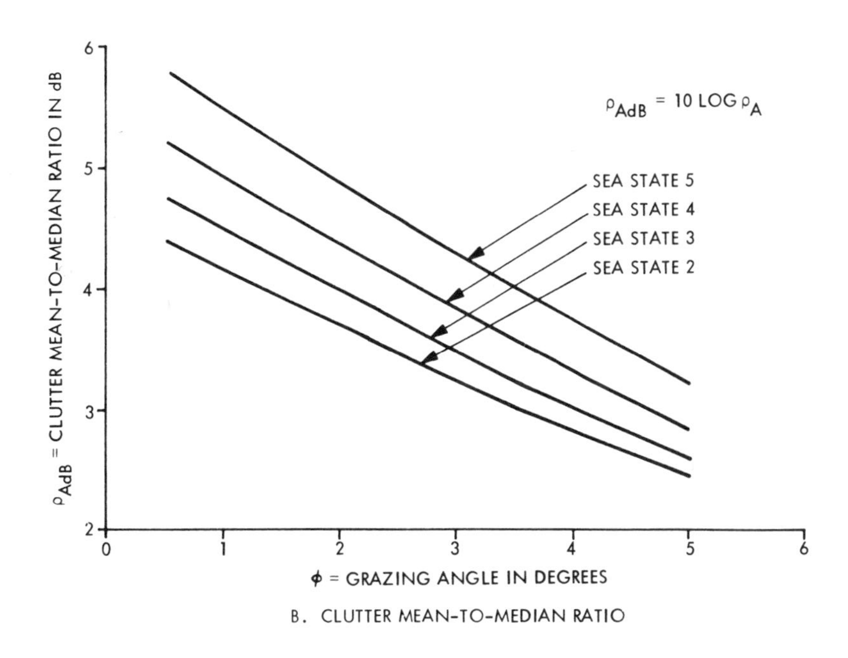

B. CLUTTER MEAN-TO-MEDIAN RATIO

Figure 1. Harbor Surveillance Radar, Sea Clutter Data

in Figure 2. These data were synthesized from high resolution synthetic aperture radar data.

The experimental log-normal data depicted in Figures 1 and 2 represent long-time averages of the clutter cross-section in a resolution cell. From an MTI standpoint the short-time characteristics of the clutter are important, and two alternative interpretations of the data have been proposed.

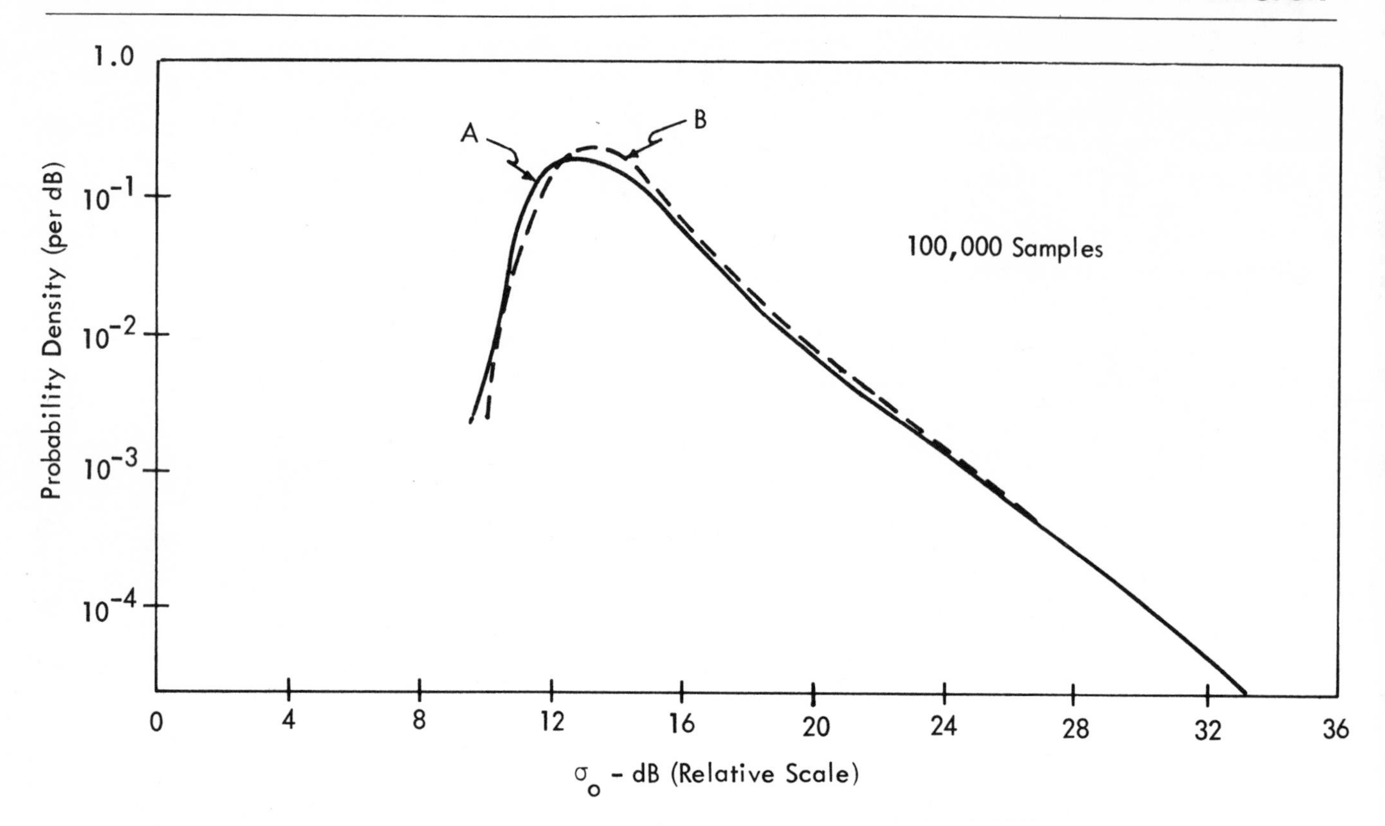

Figure 2. PDF of Two City Regions

The first interpretation represents the random clutter process as a wide-sense stationary log-normal process. This is obtained analytically by passing the multidimensional Gaussian process represented by equation (17) through an exponential nonlinearity. To achieve the desired stationarity the x_t and y_t processes of equation (1) must be uncorrelated. This interpretation has the advantage that it is a straightforward extension of the Gaussian clutter model, and many of the properties of this log-normal model can be deduced from corresponding properties of the Gaussian model.

The second interpretation models the clutter as a time-varying process whose parameters are stationary within a radar observation period over a specified spatial correlation distance. The conditional probability density function $f_v(v|\sigma^\circ)$ is Rayleigh or Rician over the spatial correlation distance associated with the clutter. The value of σ° associated with each spatial correlation block varies statistically in accordance with a log-normal distribution. Parameters of this model are discussed in a later section on the IIT clutter model.

Both models lead to the same approximate result if uncorrelated (either temporally or spatially) samples are taken from the distribution. This type of situation occurs in fast-scanning radars that sample the clutter on a scan-to-scan basis. For MTI systems the two interpretations can lead to different results, but guidelines are not currently available to indicate a clear choice.

Using the first interpretation the log-normal clutter model is formed by expressing equation (1) as

$$c_t = v_t \cos(\omega_c t + \varphi_c) \tag{29}$$

where v_t is the envelope of Equation (2) and φ_c is a uniformly distributed random variable. If the quadrature components $x_t = v_t \cos \varphi_c$ and $y_t = v_t \sin \varphi_c$ are zero mean, uncorrelated and have identical band limited spectral densities, then c_t is a wide sense stationary process. The first probability density function of the envelope is given by

$$f_v(v) = \frac{1}{\sqrt{2\pi}\sigma_v v} \exp\left[-\frac{\left(\ln \frac{v}{v_m}\right)^2}{2\sigma_v^2}\right] ; V \geqslant 0 \tag{30}$$

where σ_v is the standard deviation of the underlying normal distribution and v_m is the median value of the log-normal distribution.

The power envelope density function (proportional to the clutter cross-sectional area) is found by applying the transformation $A = v^2$ to equation (30) resulting in:

$$f_p(A) = \frac{1}{\sqrt{2\pi}\sigma_p A} \exp\left[-\frac{\left(\ln \frac{A}{A_m}\right)^2}{2\sigma_p^2}\right] ; A \geqslant 0 \tag{31}$$

where $\sigma_p = 2\sigma_v$ and A_m is the median value of the clutter cross-section. The mean value of the clutter cross-section is given by

$$\overline{A} = \rho_A A_m , \tag{32}$$

where ρ_A is the mean-to-median ratio:

$$\rho_A = \exp(\sigma_p^2/2) \tag{33}$$

The distribution function for the cross-sectional area is given by

$$F_p(A) = ½ \left[1 + \text{erf}\left(\frac{1}{\sqrt{2}\sigma_p} \ln \frac{A}{A_m}\right)\right] \tag{34}$$

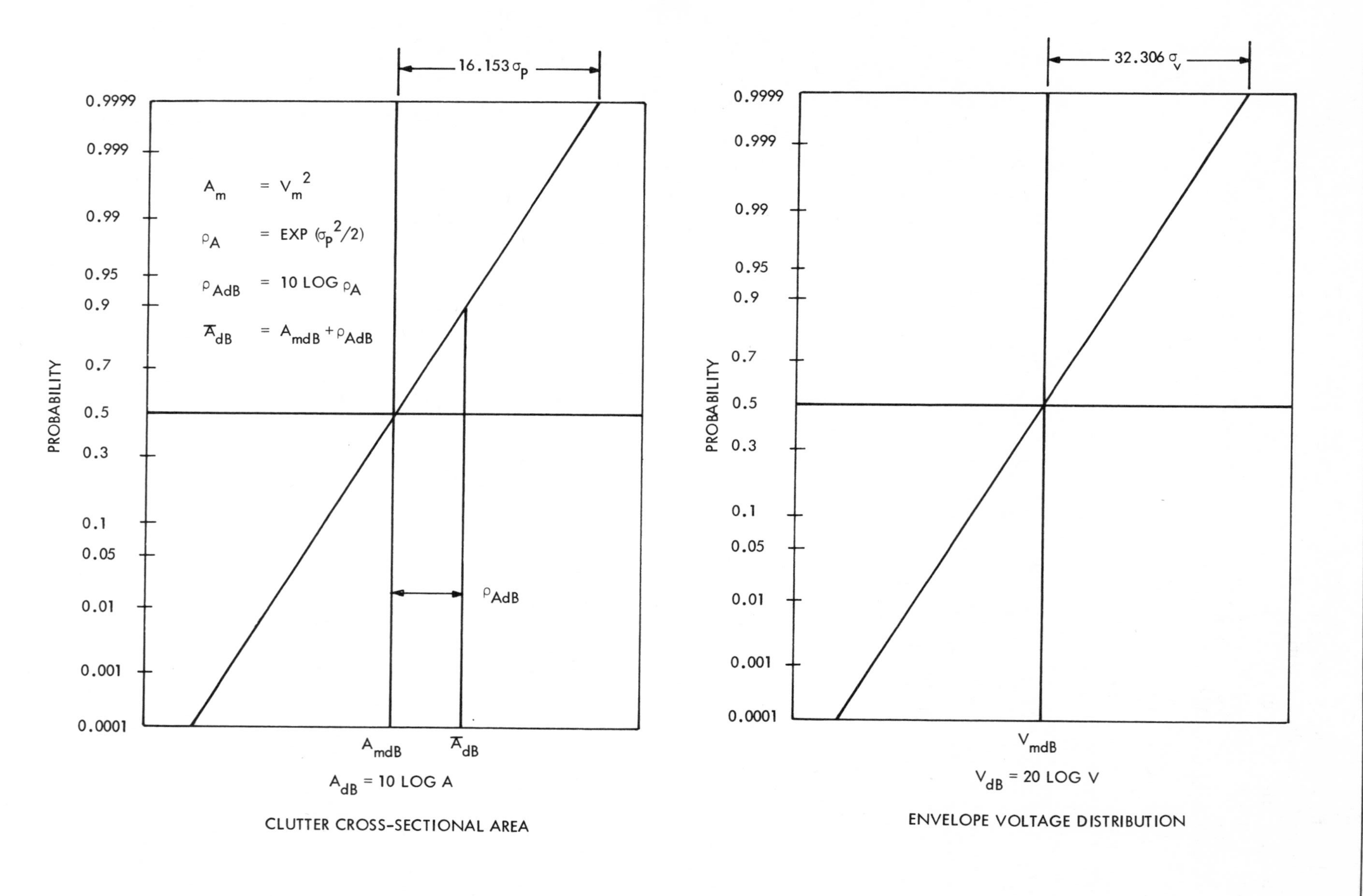

Figure 3. Log-Normal Clutter Model

On log-probability paper, this function plots as a straight line as depicted in Figure 3 ($A_{dB} = 10 \log A$). The straight line is easily determined for either the voltage envelope or cross-sectional area using the relationships provided in Figure 3 for the 0.5 and 0.9999 cumulative probability points. The mean or median cross-sectional area is determined from a knowledge of the clutter's backscattering coefficient.

The characteristics of log-normal clutter are often given in terms of its underlying normal distribution. From Figure 3 it can be seen that the log-normal clutter distribution can be expressed as

$$f_N(A_{dB}) = \frac{1}{\sqrt{2\pi}\sigma_{dB}} \exp\left[-\frac{A^2_{dB}}{2\sigma^2_{dB}}\right] \qquad (35)$$

where

$$A_{dB} = 10 \log_{10} A/A_m . \qquad (36)$$

Transforming equation (35) using equation (36) results in the log-normal distribution of equation (31) with

$$\sigma_p = 0.2303\, \sigma_{dB} . \qquad (37)$$

Equation (37) can be used to relate the log-normal clutter parameters in natural units (σ_p) to those in dB units (σ_{dB}).

Table 1 gives some reported values of σ_p for various types of clutter.

Terrain/Sea State	Frequency	φ	σ_p
Sea State 2 - 3	X	4.7°	1.382
Sea State 3	K_u	1 - 5°	1.440 - .960
Sea State 4	X	0.24°	1.548
Sea State 5*	K_u	0.50°	1.634
Land Clutter (Discrete)	S	low	3.916
Land Clutter (Distributed)	S	low	1.380
Land Clutter*	P-K_a	10° - 70°	.728 - 2.584
Rain Clutter	X and 95 GHz	—	.680
Radar Angels	—	—	1.352 - 1.620

* Extrapolated Values

Table 1. *Log-Normal Clutter Parameters*

For MTI work the correlation properties of the sample log-normal clutter model are important. A complete characterization of the log-normal model can be obtained by passing the multidimensional Gaussian distribution of equation (17) through an exponential nonlinearity. This is discussed in [2], and involves specification of the mean and covariance matrices associated with the multidimensional Gaussian distribution. The elements of the Gaussian covariance matrix (m_{xij}) are related to the elements of the log-normal covariance matrix ($m_{\ell n\ ij}$) by

$$m_{xij} = \ln\left[1 + \frac{m_{\ell n\ ij}}{\bar{x}_i\,\bar{x}_j}\right], \tag{38}$$

where $\bar{x}_{i,j}$ are the elements of the mean vectors of the log-normal clutter. Several relationships that are useful in converting from Gaussian (x_t) to log-normal (z_t) variables through an exponential nonlinearity, $A \exp(\cdot)$, are:

$$E(z_t) = A \exp\left[\frac{R_x(0)}{2}\right] \tag{39}$$

$$R_z(\tau) = A^2 \exp\left\{R_x(0) + R_x(\tau)\right\} \tag{40}$$

$$C_z(\tau) = A^2 \exp\left[R_x(0)\right]\left\{\exp\left[R_x(\tau)\right] - 1\right\} \tag{41}$$

where $R_x(\tau)$ is the ACF of the Gaussian variable, and $R_z(\tau)$ is the ACF of the log-normal variable while $C_z(\tau)$ is the covariance function of the log-normal variable.

4. The Weibull Clutter Model

The Weibull clutter model has properties that lie between the Rayleigh (a member of the Weibull family) and log-normal clutter models. The Weibull model has been used to model both land and sea clutter and offers the potential to accurately represent the real clutter distribution over a much wider range of conditions than either the log-normal or Rayleigh model.

Figure 4 shows experimental sea clutter distributions plotted on log-Rayleigh probability paper, while Figure 5 shows experimental land clutter data. The close fits provided by the straight lines indicates that the data can be closely modeled by Weibull statistics. In addition, the decreasing skewness with increasing grazing angle indicates that the distribution is approaching a Rayleigh distribution.

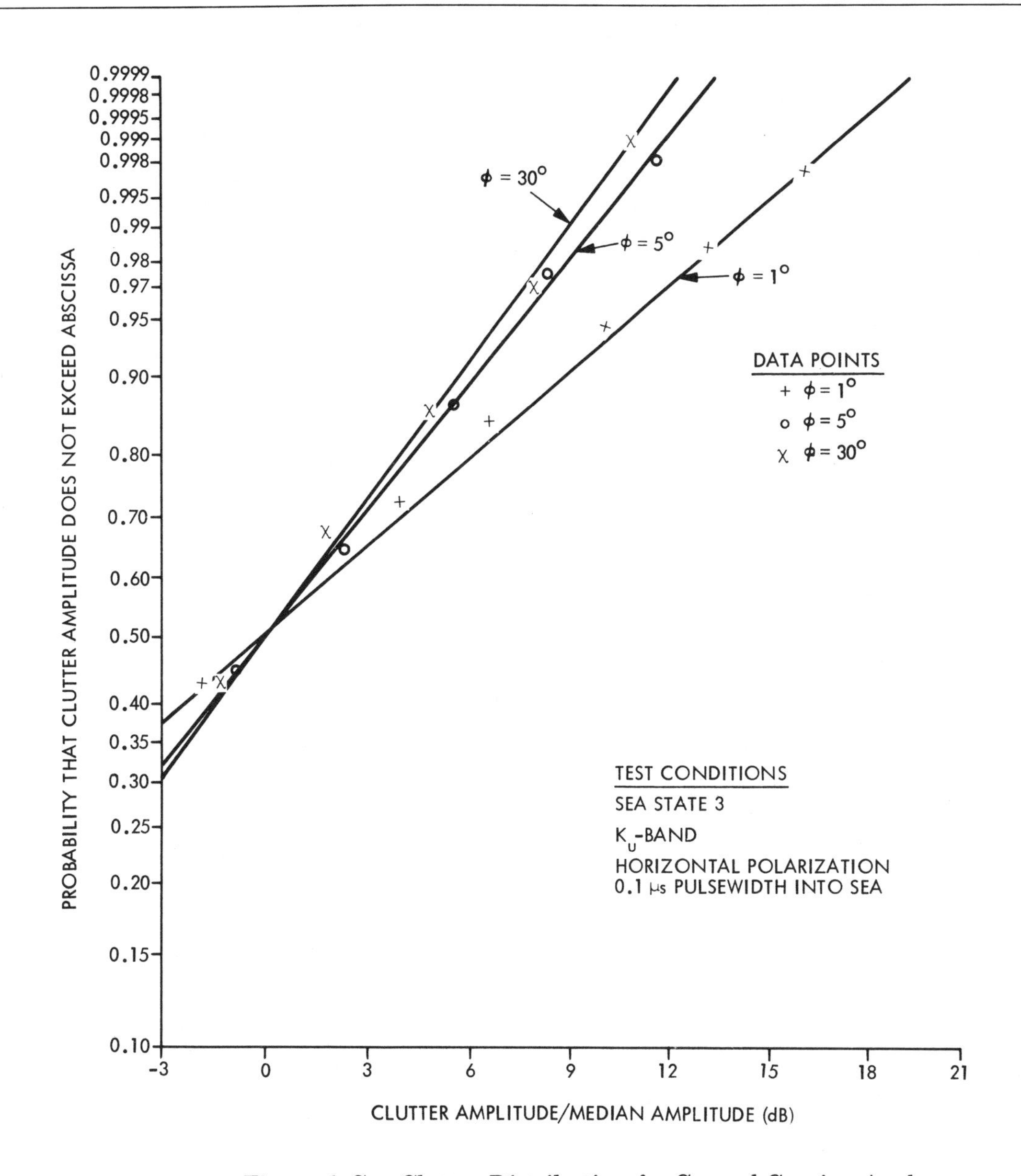

Figure 4. Sea Clutter Distribution for Several Grazing Angles

The two clutter model interpretations that apply to the log-normal model also apply to the Weibull clutter model. For MTI systems the two interpretations can lead to different results, but guidelines are not available at the present to indicate a clear choice.

In a Weibull clutter model the first probability distribution of the voltage output of an envelope detector normalized to the median v_m is given by

$$f_v(R) = a \cdot \ln 2 \cdot R^{a-1} \exp[-\ln 2 \cdot R^{\alpha}],\ R>0 \qquad (42)$$

where $R = v/v_m$, and α is a parameter that relates to the skewness of the distribution. This is related to the amplitude probability density function

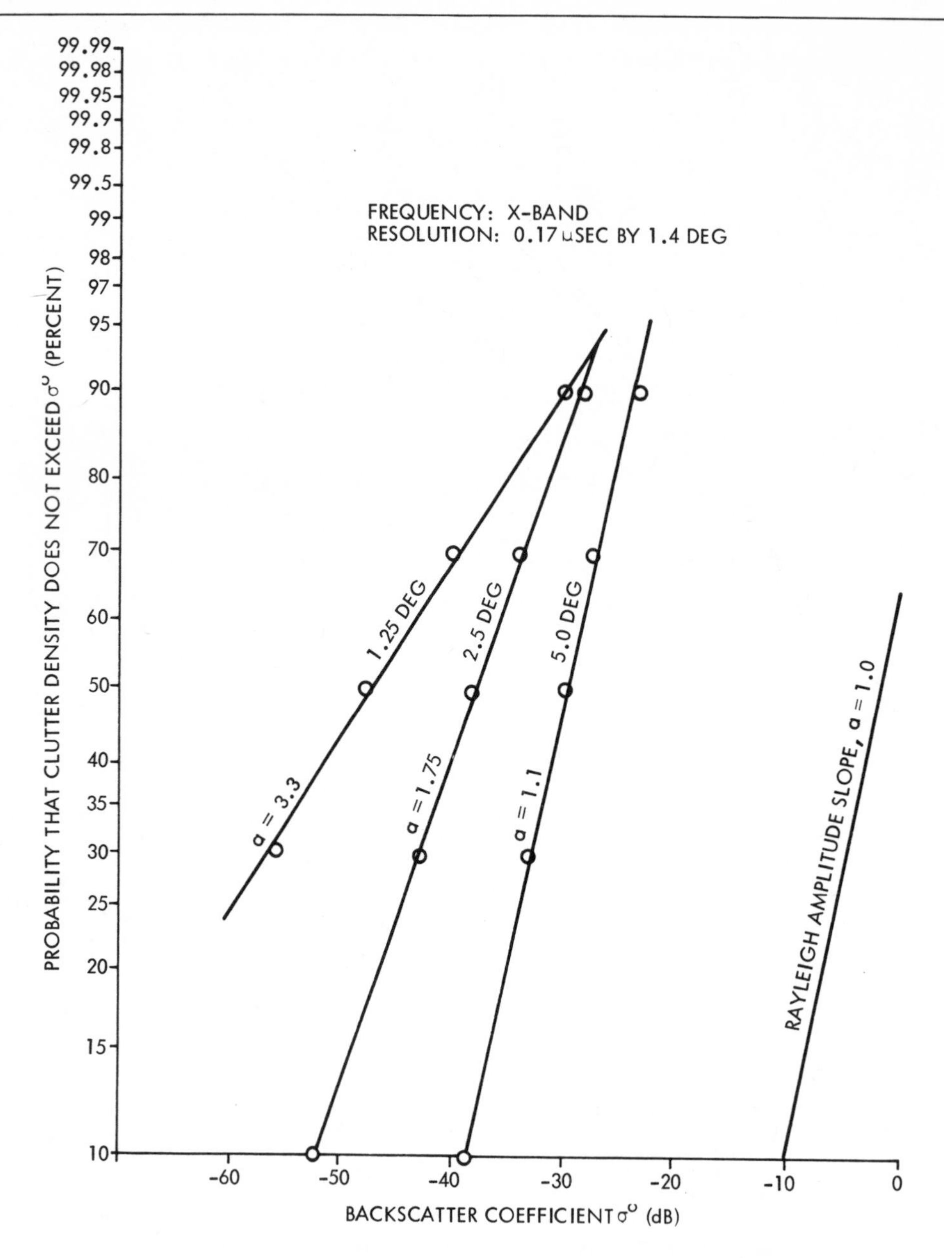

Figure 5. Ground Clutter Spatial Distributions for Cultivated Land

of the clutter cross-section A by the transformation $A = v^2$ resulting in

$$f_A(A_c) = \beta \cdot \ln 2 \cdot A_c^{\beta-1} \exp[-\ln 2 \cdot A_c^{\beta}],\ A_c > 0 \qquad (43)$$

where $A_c = A/A_m$, $\beta = a/2$ and A_m is the median of the clutter cross-sectional area. The mean of the clutter cross-section is given by

$$\overline{A} = A_m\, \Gamma(1 + 1/\beta) / (\ln 2)^{1/\beta}. \qquad (44)$$

The distribution function representing the clutter cross-sectional area on a dB scale is given by

$$A_{dB} = A_{mdB} + 1.592/\beta + (10/\beta) \log \left\{ \ln [1/(1-F_A(A))] \right\} \quad (45)$$

This function plots as a straight line on log-Rayleigh probability paper. The straight line is easily determined by plotting the 0.5 probability point as

$$A_{dB} = A_{mdB}, \quad (46)$$

and the 0.9999 probability point as

$$A_{dB} = A_{mdB} + 11.235/\beta. \quad (47)$$

The Weibull clutter distribution is sometimes defined in terms of a slope parameter (a) given by

$$a = 1/\beta \quad (48)$$

Some reported values of the Weibull model parameters are given in Table 2 below.

Terrain/ Sea State	Frequency	Beamwidth (degrees)	φ (degrees)	Pulsewidth (μ sec)	β
Rocky mountains	S	1.5	—	2	0.256
Wooded hills	L	1.7	≈ 0.5	3	0.313
Forest	X	1.4	0.7	0.17	0.253 - 0.266
Cultivated Land	X	1.4	0.7 -5	0.17	0.303 - 1
Sea State 1	X	0.5	4.7	0.02	0.726
Sea State 3	K_u	5	1-30	0.1	0.58 - 0.8915

Table 2. *Weibull Clutter Parameters*

The correlation properties of the sample Weibull clutter model are important for MTI work. The correlation matrix of the Weibull distribution can be related to the correlation matrix of a Gaussian distribution by the following set of transformations [3]. A Rayleigh vector **(v)** is formed from two zero mean independent Gaussian vectors x and y by

$$v_i = \sqrt{x_i^2 + y_i^2}. \quad (49)$$

The Rayleigh vector is transformed into a Weibull vector **(R)** by

$$r_i = v_i^{\,2/\alpha}. \quad (50)$$

Elements of the correlation matrix (m_{xij}) of the Gaussian vectors are related to elements of the correlation matrix of the Weibull vector (s_{ij}) by

$$s_{ij} = \left\{ \Gamma^2 (1 + 1/\alpha)/ [\Gamma(1 + 2/\alpha) - \Gamma^2 (1 + 1/\alpha)]\right\}.$$

$$[{}_2F_1 (-1/\alpha, -1/\alpha; 1; m^2_{xij}) - 1], \qquad (51)$$

where ${}_2F_1(\cdot)$ is the hypergeometric function.

5. The IIT Radar Clutter Model [4, 5]

Representing one of the more comprehensive clutter models, the IIT model was developed for use with overland airborne radars on the basis of extensive data analysis. The model provides for the spatial distribution of the clutter and does not explicitly address intrinsic or induced variations in the clutter return.

The IIT model treats radar ground clutter as consisting of two additive components consisting of 1) distributed clutter and 2) discrete clutter. The former is indicative of returns from natural terrains, while the latter corresponds to returns from physically isolated dominant man-made scatterers.

The complete model is intended to be used in computer simulations and is far too complex for hand calculations. In the following section a general description of the model is given to provide insight into clutter modeling and to provide a basis for approximations that might be used for simple calculations. The reader is referred to references [4 and 5] for a complete description of the model.

The distributed clutter backscattering cross-section per unit area ($\sigma°$) is a function of: 1) frequency, 2) polarization, 3) grazing angle, 4) terrain type and 5) environmental factors. The first four factors are deterministic in nature, while environmental factors are random in nature and take into account variations in grazing angle.

Figures 6 and 7 depict average backscattering coefficients for various terrains as a function of grazing angle. The data given in Figure 6 is independent of frequency and polarization since uncertainties in the data base are larger than variations due to these parameters. For desert terrain the variation with frequency is significant and hence is included in the model.

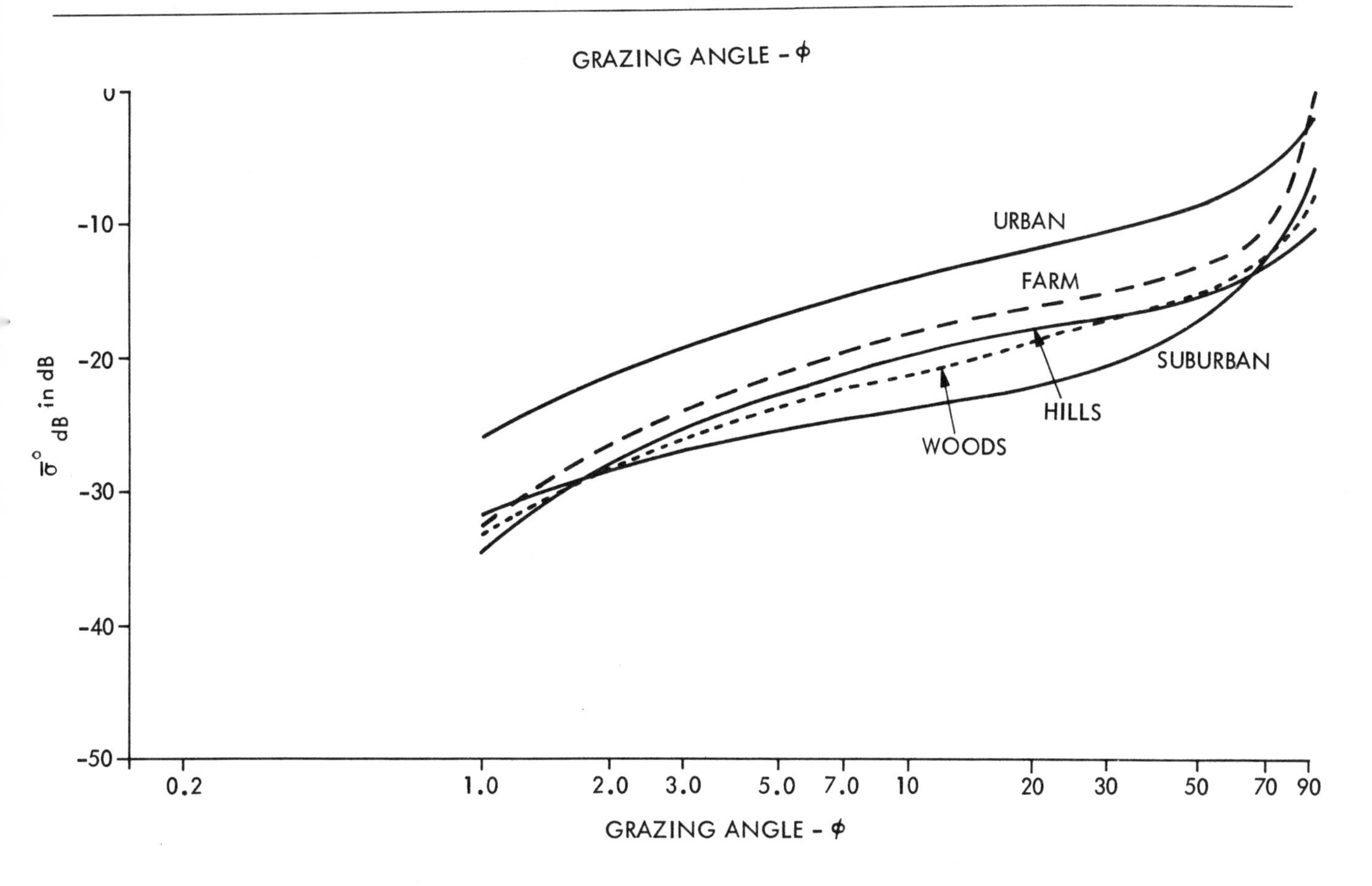

Figure 6. Mean dB Clutter Cross Section, $\overline{\sigma}^{\circ}$ dB, for Five Terrains

The instantaneous clutter power returned to the radar is proportional to the effective clutter cross-sectional area whose distribution is given by equation (6)

$$f_p(A) = \frac{1}{\overline{A}} \exp[-A/\overline{A}]. \tag{52}$$

$\overline{A}$ is a random variable whose value is a function of the spatial properties of the clutter distribution, and is given by

$$\overline{A} = \overline{\sigma}^{\circ}{}_{(x,y)} A_c , \tag{53}$$

where $\overline{\sigma}^{\circ}{}_{(x,y)}$ is the spatially dependent clutter backscatter coefficient, and A_c is the area intercepted by the radar resolution cell.

For any fixed set of spatial parameters (x_o, y_o) the value of the back scatter coefficient $\sigma^{\circ}(x_o, y_o)$ is selected from a log-normal distribution defined by

$$f_N(\sigma^{\circ}{}_{dB}) = \frac{1}{\sqrt{2\pi} s_o} \exp\left\{-1/2\left(\frac{\sigma^{\circ}{}_{dB} - \overline{\sigma}^{\circ}{}_{dB}}{s_o}\right)^2\right\} \tag{54}$$

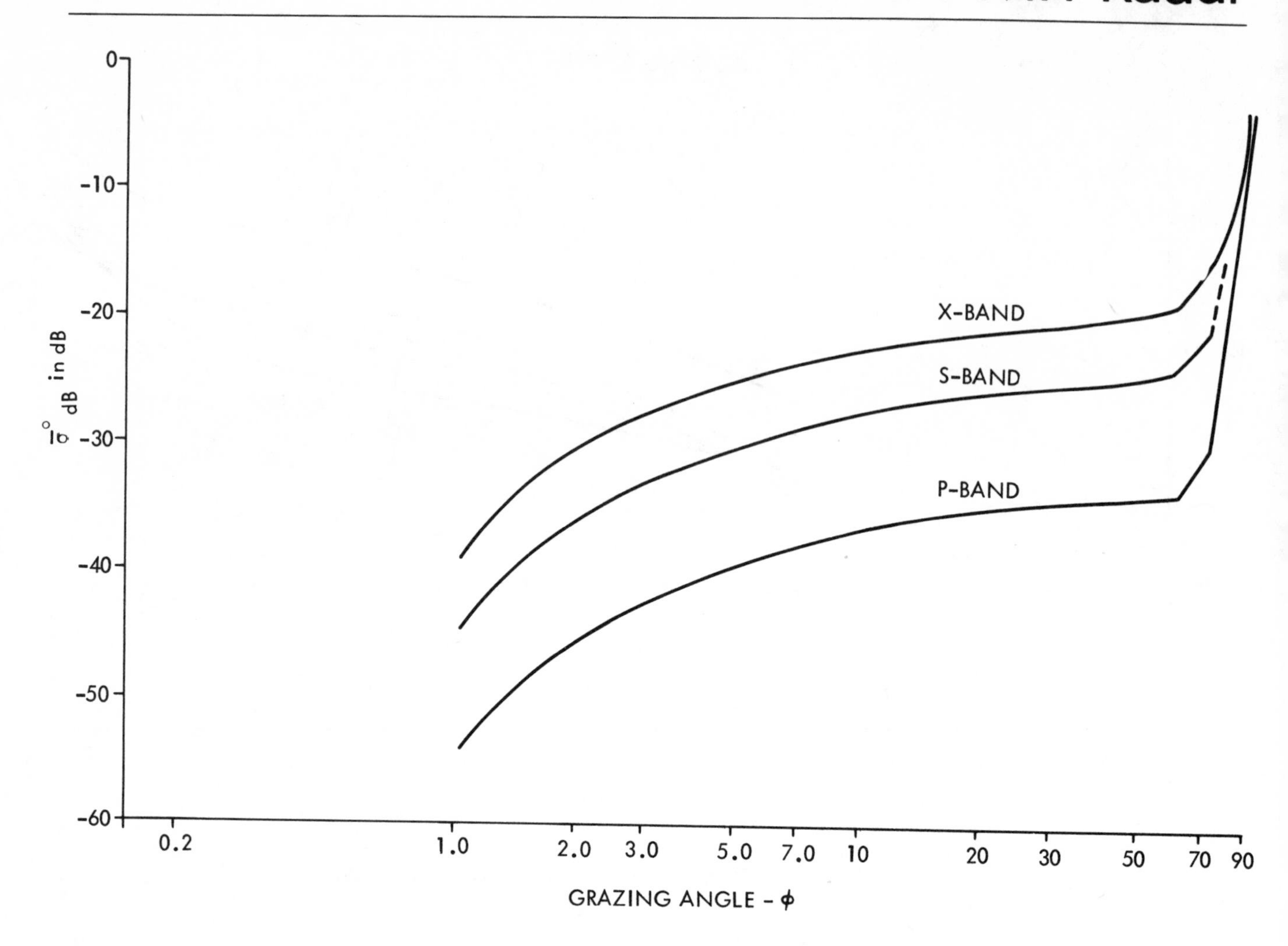

Figure 7. Mean dB Clutter Cross Section, $\bar{\sigma}^{\circ}$ dB, for Desert Terrain

where

$$\sigma^{\circ}_{dB} = 10 \log_{10} \sigma^{\circ} \tag{55}$$

and s_o is the standard deviation of σ°_{dB}. $\bar{\sigma}^{\circ}_{dB}$ is given in Figures 6 and 7, while s_o is given by

$$s_o = \sqrt{s_1^2 + s_2^2}\ , \tag{56}$$

where s_1 and s_2 are defined in Figures 8 and 9. Both $\bar{\sigma}^{\circ}_{dB}$ and s_o are functions of the grazing angle and terrain type.

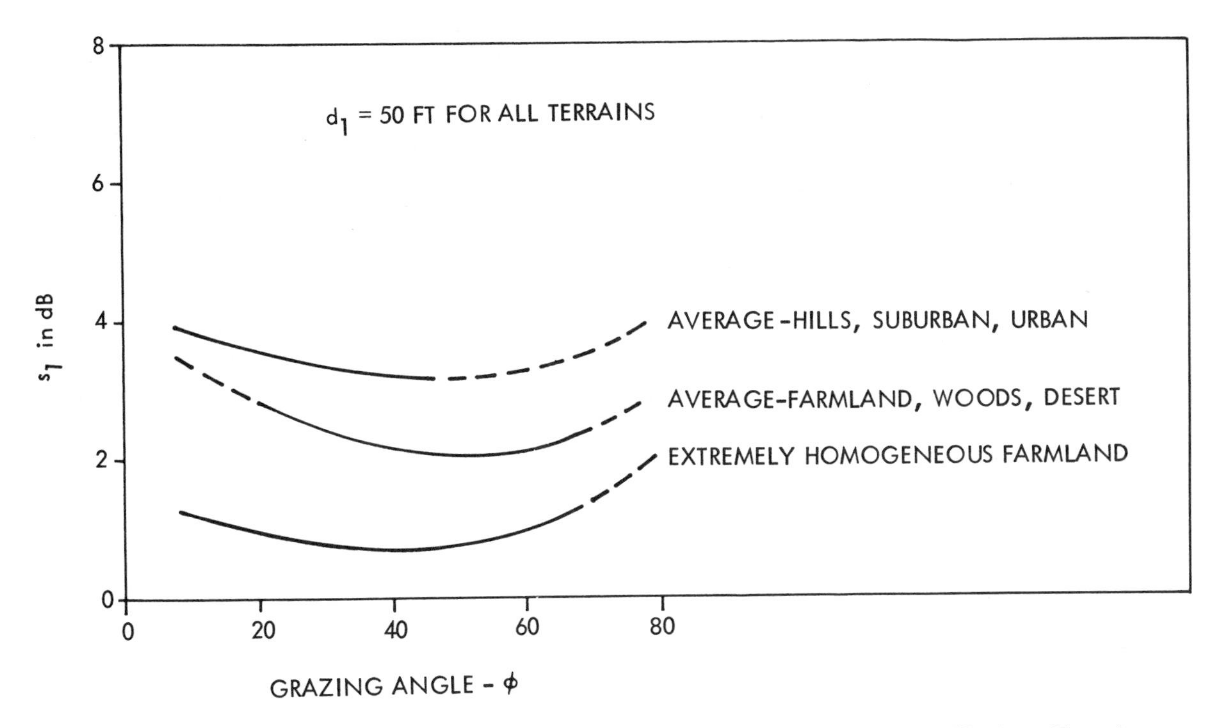

Figure 8. Standard Deviation, s_1, of dB Clutter Cross Section for Various Terrains (Short Range Spatial Variation)

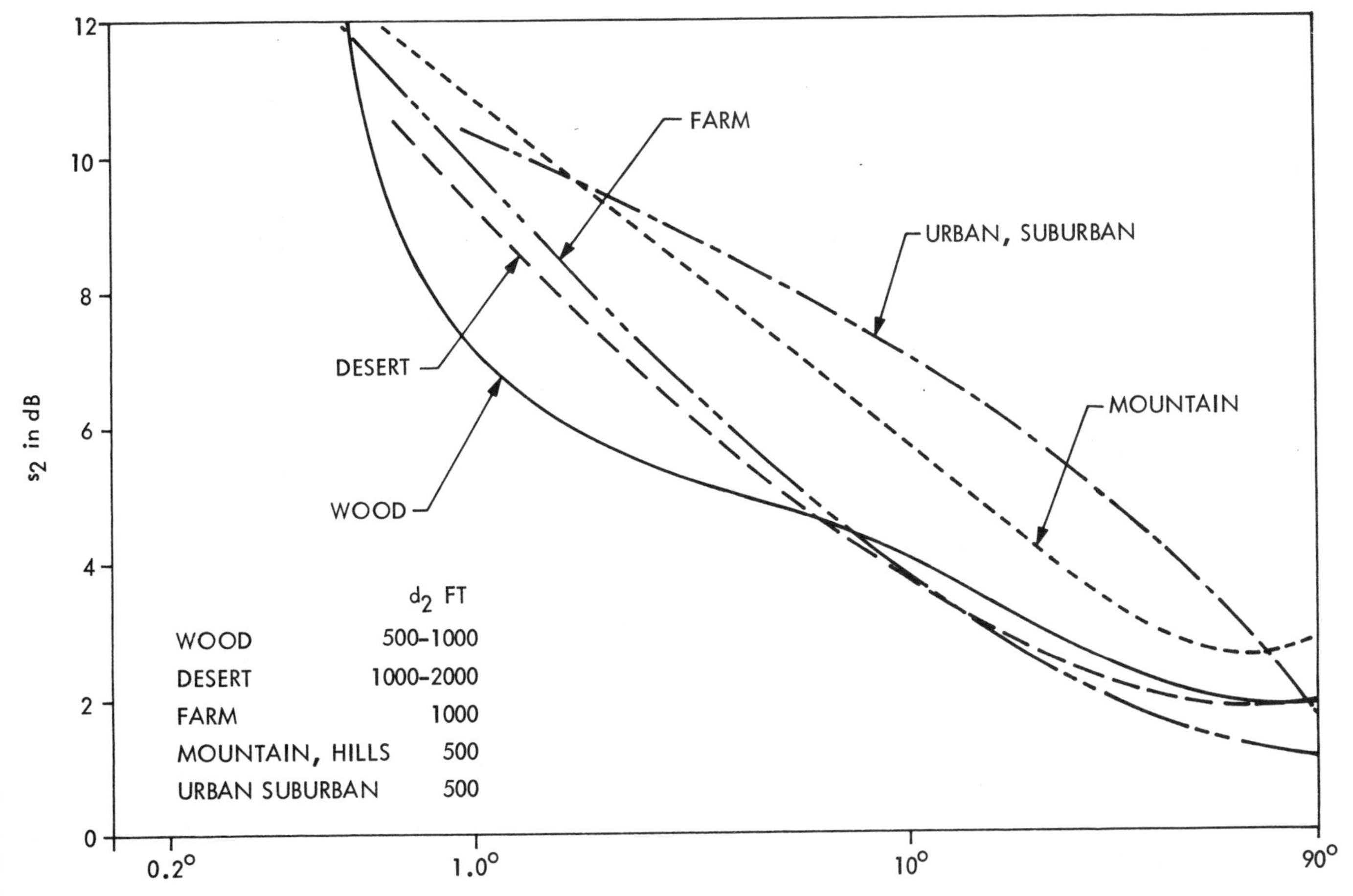

Figure 9. Standard Deviation, s_2, of dB Clutter Cross Section for Various Terrains (Long Range Spatial Variation)

As an intermediate step in the development of the model, consider that the terrain is broken up into a checkerboard-like pattern. Each cell of the checkerboard is defined by a correlation distance (d) over which the value of $\sigma^\circ(x_o, y_o)$ applies. The average value of the backscatter coefficient ($\overline{\sigma^\circ}(x, y)$) to be used in equation (53) is found by integrating over those cells of the checkerboard that are contained within the window formed by the radar's resolution cell. If high accuracy is desired, the radar's two-way antenna pattern is used as an amplitude weighting for one dimension of the window.

As the radar antenna is scanned, or when the radar is on a moving platform, new cells of the checkerboard enter the radar resolution window while old cells are discarded. This results in the radar inducing a temporal variation in the clutter statistics that accounts for the correlation properties of the clutter.

The next refinement of the model adds a triangular correlation function across each cell of the checkerboard. This spatial ACF is given by

$$r(\Delta x, \Delta y) = \frac{\exp\left\{(.2303\, s_o)^2 \rho_{12}^2(\Delta x, \Delta y)\right\} - 1}{\exp\left\{(.2303\, s_o)^2\right\} - 1}, \quad (57)$$

where

$$\rho_{12}^2(\Delta x, \Delta y) = \rho(\Delta x)\,\rho(\Delta y) \quad (58)$$

$$\rho(\Delta z) = 1 - \left|\frac{\Delta z}{d}\right|; \quad |\Delta z = \sqrt{(\Delta x)^2 + (\Delta y)^2}| \leqslant d$$

$$= 0; \quad |\Delta z| > d \quad (59)$$

and d is the correlation distance.

For computer simulation the model at any point in the terrain field can be generated by

$$\sigma^\circ_{dB}(x,y) = \overline{\sigma^\circ}_{dB} + s_o\, U_o(x, y), \quad (60)$$

where $U_o(x, y)$ is a correlated, zero mean, unit variance, sample function of a two dimensional Gaussian random process. Standard computer programs are available for generating $U_o(x, y)$. For this model values of $d = d_2$ should be used from Figure 9.

Analysis of Synthetic Aperture Radar (SAR) data [5] indicates that a further refinement of the ACF of the model is required to accurately depict the clutter return. The revised ACF depicted

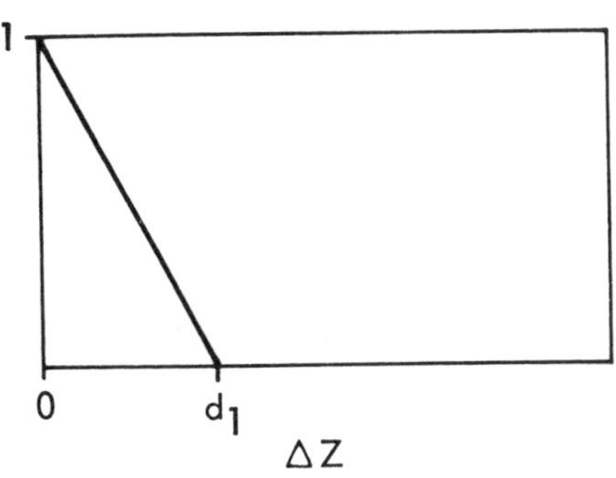

(A) COVARIANCE FUNCTION FOR U_1 (X,Y)

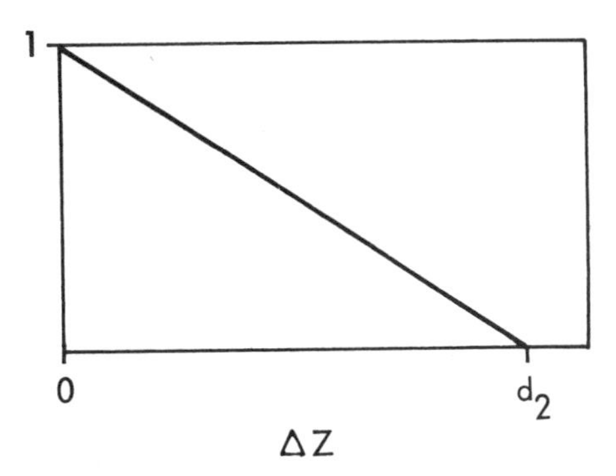

(B) COVARIANCE FUNCTION FOR U_2 (X,Y)

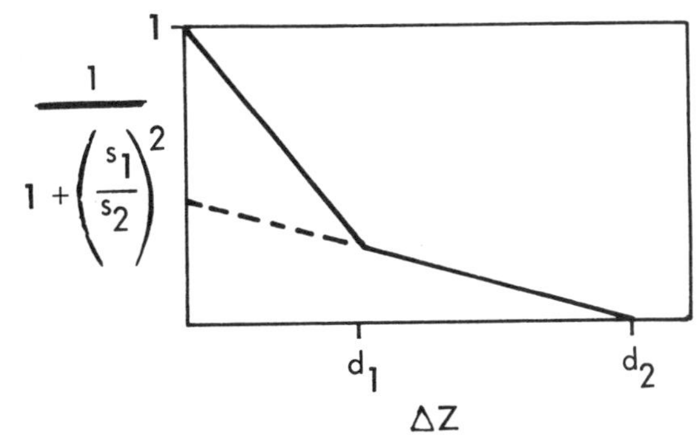

(C) NORMALIZED COVARIANCE FOR

$$\sigma^{o}_{dB}(X,Y) = \bar{\sigma}^{o}_{dB} + s_1\, u_1(X,Y) + s_2\, u_2(X,Y)$$

Figure 10. Covariance Functions for dB Values of $\sigma_0\,(x,y)$ in the Revised Model
$(\Delta z \triangleq \sqrt{\Delta x^2 + \Delta y^2})$

in Figure 10 is a two segment approximation to an exponential function. The first correlation distance (d_1) provides a rapid decorrelation while the second correlation distance (d_2) is much longer and represents a long term correlation. The value of d_2 is much longer for homogeneous terrain such as farms and woodland than it is for urban or mountainous terrain where sharp changes occur.

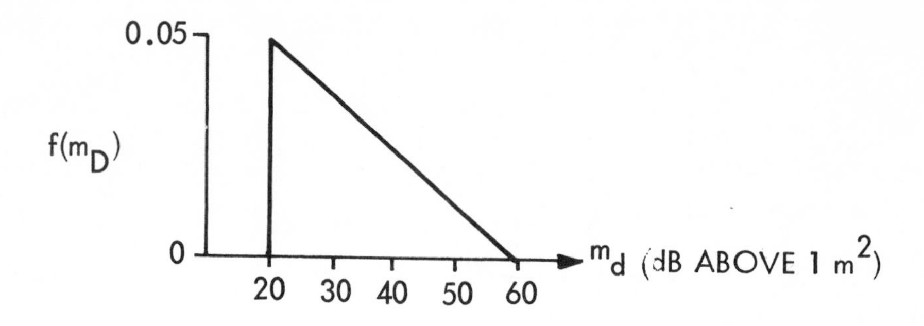

Figure 11. Distribution Law for Median Cross Section

For computer simulation equation (60) is modified to

$$\sigma^\circ_{dB}(x, y) = \overline{\sigma}^\circ_{dB} + s_1 U_1 (x, y) + s_2 U_2 (x, y), \quad (61)$$

where s_1 and s_2 are given in Figures 8 and 9 and U_1 (x, y) and U_2 (x, y) are sample functions from correlated, zero mean, unit variance two dimensional Gaussian processes.

The amplitude probability distribution of discrete clutter is given by a log-normal distribution defined by

$$f_N (A_{dB}) = \frac{1}{\sqrt{2\pi} s_d} \exp \left\{ -1/2 \left(\frac{A_{dB} - m_d}{s_d} \right)^2 \right\} \quad (62)$$

where

$$A_{dB} = 10 \log A_d \text{ (dBsm)} \quad (63)$$

and s_d is the standard deviation of A_{dB} while m_d is the median of A_d with respect to a cross-section of 1 m^2. The value of s_d is approximately equal to 5 dB. The value of m_d is given in Figure 11, and ranges from 10^2 to 10^6 m^2 with a triangular probability distribution.

6. Sea Clutter Model [6]

A number of the unique characteristics of sea clutter are illustrated in a sea clutter model developed by Georgia Tech for use with airborne radars. The backscattering coefficient (σ°) for sea clutter is dependent on 1) frequency, 2) grazing angle, 3) sea and wind aspect, 4) sea-state and wind speed and 5) polarization.

The variation with polarization is the most difficult to quantify. For moderate to low sea states, horizontal polarization has yielded consistently lower sea return than vertical polarization in the upwind and downwind directions. However, crosswind, vertical polarization may give a lower return. In rough seas the differences between the

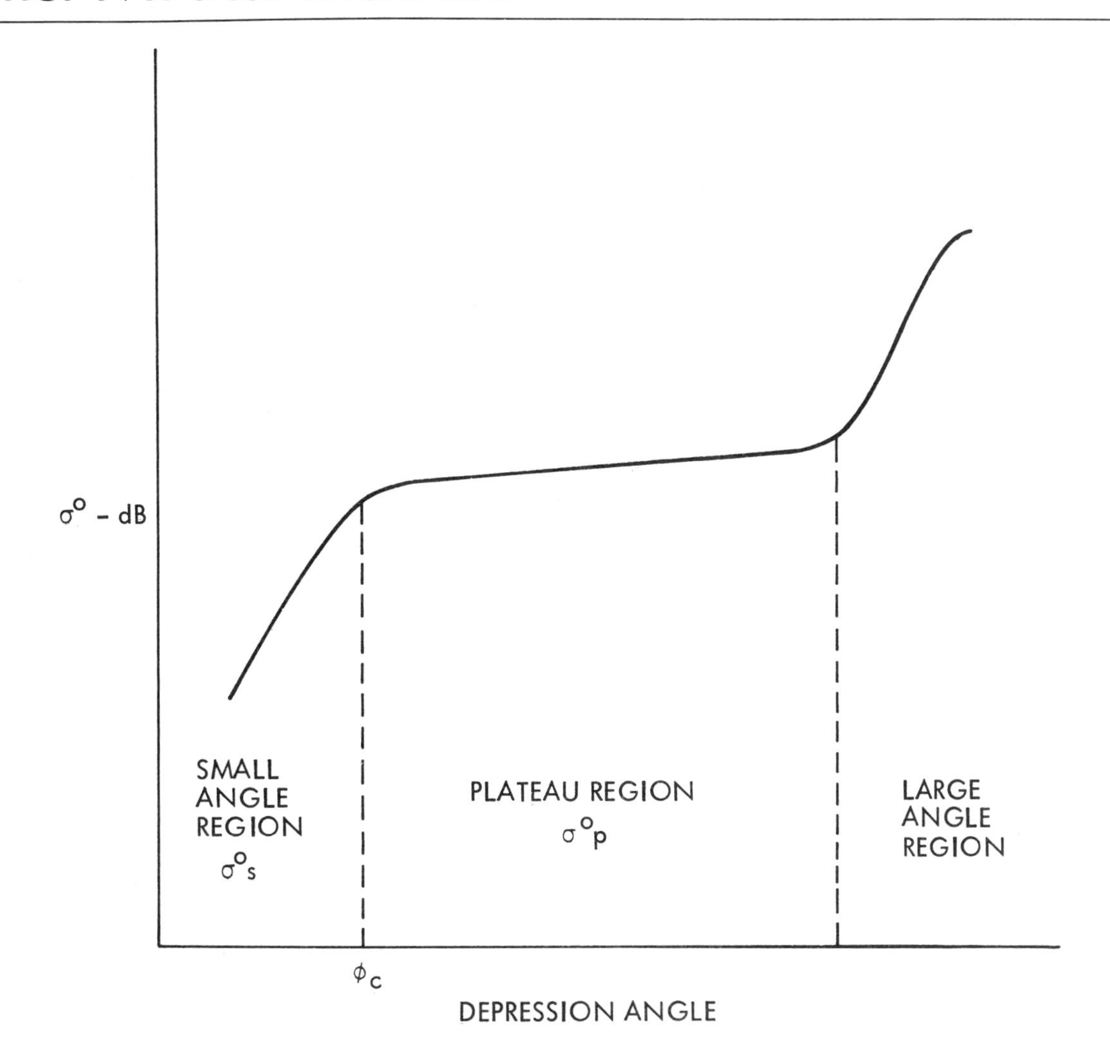

Figure 12. Radar Cross-Section per Unit Area of the Sea Surface as a Function of Depression Angle Showing the Three Principal Regions

two linear polarizations may diminish or reverse at large grazing angles. A disadvantage of horizontal polarization is the appearance of highly correlated "sea spikes", especially upwind. Because of this uncertainty a model for horizontal polarization is developed first, and then modified for vertical polarization.

The general shape of the curve for backscattering coefficient (σ°) as a function of grazing angle is shown in Figure 12. The three major regions of dependence are the small angle region, the plateau region and the large angle region.

The critical angle (φ_c) where the curves for the plateau and small-angle regions intersect is given by

$$\varphi_c = \frac{\lambda}{2.5 h_w} ,$$

where

φ_c = critical angle in radians

λ = wavelength in feet

h_w = peak-to-trough wave height in feet of the highest 1/10 waves.

An approximate relationship for h_w is given by

$$h_w = 1.05 \exp[0.495 s] \text{ (feet)}$$

where s is the Douglas sea state.

The variation of σ° in the plateau region is taken as proportional to $\varphi^{0.4}$ while the variation in the small angel regions is $\varphi^{5.0}$.

The aspect dependence is modeled as

$$\sigma^\circ_{dB}(\beta) = \sigma^\circ_{1dB} - 4 + 4 \cos\beta$$

where

β = angle between antenna boresight and upwind direction

σ°_{1dB} = backscatter coefficient for upwind direction

A frequency dependence of $\lambda^{-1/2}$ was chosen for the plateau region as best representing the data.

A calibration point is needed to quantify the model. This was taken as λ = 3 cm., $\varphi = 10°$, s = 4 and σ° = -42 dB. The value of σ°_p in the plateau region is then

$$\sigma^\circ_p = \frac{k\,\varphi^{0.4} \exp(0.792 s)}{\lambda^{1/2}},$$

where

s = Douglas sea state

φ = grazing angle in degrees

λ = wavelength in cm.

k = 1.83×10^{-6}

The value of σ°_s in the small angle region is given by

$$\sigma^\circ_s = \frac{\varphi^5 \exp(3.069 s)}{c\lambda^{5.1}}$$

where

c = 1.17×10^5

The value of σ°_L in the large angle region was not modeled, but the transition grazing angle appears to lie in the 30 to 60 degree region.

Analytical expressions were not obtained for vertical polarization. It was observed that grazing angle and frequency dependence should be the same as for horizontal polarization. The dependence on sea state is very complicated.

REFERENCES

[1] Schleher, D.C. "Harbor Surveillance Radar Detection Performance," *IEEE Journal of Oceanic Engineering, Vol. OE-2, No. 4,* Oct. 1977.

[2] Peebles, P., "The Generation of Correlated Log-Normal Clutter for Radar Simulations," *IEEE Trans. Aerosp. Electron. Syst., Vol. AES-7, No. 6,* Nov. 1971.

[3] Szajnowski, W., "The Generation of Correlated Weibull Clutter for Signal Detection Problems," *IEEE Trans. Aerosp. Electron. Syst.,* Vol. *AES-13, No. 5,* Sept. 1977.

[4] Greenstein, L., Brindley, A., and Carlson, R., "A Comprehensive Ground Clutter Model for Airborne Radars," IIT Research Institute, Sept. 1969.

[5] Kazel, S., Patten, J., and Dabkowski, J., Pipkin, R., and Brindley, A., "Extensions to the ORT Clutter Model," IIT Research Institute, June 1971.

[6] Rivers W., Zenner, S., and Dyer, F., "Airborne Search Radar Design Study," Georgia Institute of Technology, 20 Feb. 1969, *AD 685 911.*

MTI Bibliography

The significant literature relating to MTI Radars is extensive. The selected Bibliography provided below tabulates the significant MTI literature available in English books and journals. In addition, the Bibliography includes selected unclassified MTI literature available through the Defense Documentation Center.

A more complete Bibliography, including International MTI literature, is available in the January 1972, the May 1975 and the May 1977 issues of the *IEEE Transactions on Aerospace and Electronic Systems.*

The Bibliography covers the following topics:

1. Books on MTI
2. General
3. Optimum MTI Filters
4. Pulse-Burst MTI Radars
5. Special MTI Effects
6. Digital MTI
7. MTI Hardware
8. Airborne MTI
9. Pulse Doppler

1. Books on MTI

Ridenour, L. N., *Radar System Engineering*, N. Y.; McGraw-Hill Book Company, 1947, Chapter 16.

Povejsil, D. J., Raven, R. S. and Waterman, P., *Airborne Radar*, Princeton, N. J.; D. Van Nostrand, 1961, Chapters 3, 5 and 6.

Skolnik, M. I., *Introduction to Radar Systems*, N. Y.; McGraw-Hill Book Company, 1962, Chapters 3 and 4.

Wainstein, L. A., Zubakov, V. D., *Extraction of Signals from Noise*, Englewood Cliffs, N. J.; Prentice-Hall, 1962, Chapter 6.

Barton, D. K., *Radar System Analysis*, Dedham, MA; Artech House, 1976, Chapter 7.

Berkowitz, R. S. *Modern Radar*, N. Y.; John Wiley and Sons, 1965, Part VI, Chapters 1, 2, 3 and 4.

Nathanson, F. E., *Radar Design Principles: Signal Processing and the Environment*, N.Y.; McGraw-Hill Book Company, 1969, Chapters 8, 9, 10, 11 and 14.

Skolnik, M. I., *Radar Handbook*, N. Y.; McGraw-Hill Book Company, 1970, Chapters 16, 17, 18, 19 and 35.

Rabiner L., Gold B., *Theory and Application of Digital Signal Processing*, Englewood Cliffs, N.J.; Prentice-Hall, 1975, Chapter 13.

Brookner, E., *Radar Technology*, Dedham, Mass.; Artech House, 1977, Part II, Chapters 9, 10, 11.

2. General

Emerson, R. C.; "Some Pulsed Doppler MTI and AMTI Techniques," *RAND Corp. Report R-274*, March 1954 (Reprint Paper No. 1.1).

George, S. F. and Zamanakos, A. S.; "Comb Filters for Pulsed Radar Use," *Proc. 42, No. 7*, July 1954, pp. 1159-65.

Grisetti, R. S., Santa, M. M. and Kirkpatrick, G. M.; "Effect of Internal Fluctuations and Scanning on Clutter Attenuation," *Trans. ANE-2, No. 1*, March 1955, pp. 37-41.

White, W. D. and Ruvin, A. E.; "Recent Advances in the Synthesis of Comb Filters," *Conv. Rec. 57, Pt 2*, pp. 186-200 (Reprint Paper No.1.2)

Linden, D. A. and Steinberg, B. D.; "Synthesis of Delay Line Networks," *Trans. ANE-4, No. 1*, March 1957, pp. 34-39 (Reprint Paper No. 1.3)

Urkowitz, H.; "Analysis and Synthesis of Delay Line Periodic Filters," *Trans. CT-4, No. 2*, June 1957, pp. 41-53 (Reprint Paper No. 1.4)

Urkowitz, H.; "Analysis of Periodic Filters with Stationary Random Inputs," *Trans. CT-6, No. 4*, December 1959, pp. 330-34.

Fowler, C. A., Uzzo, A. P., Jr. and Ruvin, A. E.; "Signal Processing Techniques for Surveillance Radar Sets," *Trans. MIL-5, No. 2*, April 1961, pp. 103-08 (Reprint Paper No. 1.5)

Criag, S. E., Fishbein, W. and Rittenbach, O. E.; "Continuous-Wave Radar with High Range Resolution and Unambiguous Velocity Determination," *Trans. MIL-6, No. 2*, April 1962, pp. 153-61.

Fishbein, W. and Rittenbach, O. E.; "Combat Surveillance Correlation Radar," *AD 332 025, December, 1962.*

Ruina, J. P.; "The Fluctuation of Radar Ground Return Due To Antenna Scanning," *Trans. AP-11, No. 6*, November 1963, pp. 722-23c.

Schulkind, D.; "Performance Characteristics of Coherent, Phase-Only, MTI Cancellation," *Conv. Rec. 65, Pt. 4*, pp. 158-71.

Voles, R., "A Radar Phase Locking Technique," *Proc. 54, No. 12*, December, 1966, p. 1941c.

Barton, D.K.; "Radar Equations for Jamming and Clutter," *Trans. AES-3, No. 6*, November, 1967. pp. 340-55 (Sup).

Hall, W. M. and Ward, H. R.; "Signal-to-Noise Loss in Moving Target Indicator," *Proc. 56, No. 2*, February 1968, pp. 233-34c.

Shrader, W. W.; "An MTI Surveillance Radar Philosophy," *EASCON Rec. 1971*, pp. 172-74.

Hsiao, J. K.; "Comb Filter Design, Computer Synthesis and Analysis," *AD 900 333L*, May 1972, Rept. No. NRL-MR-2433.

Mark, J. W. and Woods, H. A.; "A Recursive Digital MTI Radar Filter," *Proc. 60, No. 6*, June 1972, pp. 728-29c.

Shrader, W. W. and Hansen, V. G.; "Improvement Factor of a Recursive MTI Radar Filter," *Proc. 60, No. 11*, November, 1972, pp. 1442-43c.

Fletcher, R. H., Jr.; "Recursive Digital Filter Design and Analysis with Applications to Radar Processing," *AD 766 689*, April, 1973.

Bagley, G. C.; "A Survey of Cancellation Versus Integration for Radar Clutter Rejection," *AD 787 623*, August, 1974, Rept. No. NRL-MR-2880.

Haykin, S. S. and Boulter, J. F.; "Performance of Recursive Digital Moving-Target Indicator Radar Filters with Butterworth Frequency Response," *Proc. IEE 121, No. 9*, September 1974, pp. 956-57.

Fishbein, W. and Rittenbach, O.E.; "Modulation Waveforms for Continuous Wave Radar," *Int. Rad. Conf. Rec. 1975*, pp. 192-97.

Dax, P. R.; "Eliminating Clutter in Radar Systems," *Microwaves 14, No. 4*, April, 1975, pp. 34-43.

Hsiao, J. K.; "Analysis of a Dual Frequency Moving Target Indication System," *Rad-Elect. Eng. 45, No. 7*, July, 1975, pp. 351-56.

Kroszczynski, J. J.; "Improvement Factor of the Phase-Sensitive Noncoherent MTI," *Trans. AES-11, No. 5*, September, 1975, pp. 899-902c.

Hawkes, C.D. and Haykin, S.S.; "Modelling of Clutter for Coherent Pulsed Radar," *Trans. IT-21, No. 6*, November, 1975, pp. 703-07.

Nourse, O. and Crossfield, M.; "Quad-Pulse Radar—A Frequency Agile Radar with Doppler Capability," *IEE Conf. Publ. No. 155 - Radar-77*, 25-28 October 1977, pp. 186-190.

Fancy, H.; "Multi-Filter MTI System," *IEE Conf. Publ. No. 155 - Radar-77*, 25-28 October, 1977, pp. 191-194.

Stoner, D.; "Adaptive MTI and Signal Processing Techniques for Long Range Chaff and Weather," *IEE Conf. Publ. No. 155-Radar-77*, 25-28 October, 1977, pp. 195-198.

Roberts J. et al.; "A New Approach to Pulse Doppler Processing," *IEE Conf. Publ. No. 155 - Radar-77*, 25-28 October, 1977, pp. 358-362.

Taylor, R., et al.; "Block Processing in Pulse Doppler Radar," *IEE Conf. Publ. No. 155 - Radar-77*, 25-28 October, 1977, pp. 373-378.

Wells, H.; "The Performance of Balanced Processing for Ground Clutter Suppression in Microwave Radars," *IEE Conf. Publ. No. 155 - Radar-77*, 25-28 October, 1977, pp. 379-383.

3. Optimum MTI Filters

Capon, J.; "Optimum Weighting Functions for the Detection of Sampled Signals in Noise," *Trans. IT-10, No. 2*, April, 1964, pp. 152-59 (Reprint Paper No. 2.1).

Kroszczynski, J.; "On the Optimum MTI Reception," *Trans. IT-11, No. 3*, July, 1965, pp. 451-52.

Brennan, L. E. and Reed, I. S.; "Optimum Processing of Unequally Spaced Radar Pulse Trains for Clutter Rejection," *Trans. AES-4, No. 3*, May, 1968, pp. 474-477 (Reprint Paper No. 2.3).

Benning, C. and Hunt, D.; "Coefficients for Feed-forward MTI Radar Filters," *Proc. 57, No. 10*, October, 1969, pp. 1788-89c.

Shrader, W. W. and Hansen, V. G.; Comments on "Coefficients for Feed-forward MTI Radar Filters," *Proc. 59, No. 1*, January, 1971, p. 101c.

Jacomini, O. J.; "Optimum Symmetrical Weighting Factors for a Video MTI Radar,"*Trans AES- 7, No. 1*, January, 1971, pp. 204-09.

Jacomini, O. J.; "Weighting Factor and Transmission Time Optimization in Video MTI Radar," *Trans AES-8, No. 4*, July, 1972, pp. 517-27.

Hsiao, J. K. and Cha, A.; "On the Optimization of Clutter Rejection of a Nonrecursive Moving Target Indication Filter," *AD 915 453L*, November, 1973, Rept. No. NRL-7622.

Kretschmer, F. F.; "MTI Weightings," *Trans AES-10, No. 1*, January, 1974, pp. 153-56c.

Hsiao, J. K.; "On the Optimization of MTI Clutter Rejection," *Trans. AES-10, No. 5*, September, 1974, pp. 622-629 (Reprint Paper No. 2.4).

Hsiao, J. K.; "MTI Improvement Factor and Clutter Spectra," *AD 8001 929L*, January, 1975, Rept. No. NRL-MR-2976.

Burlage, D., and Houts, R.; "Design Technique for Improved Bandwidth Moving Target Indicator Processors in Surface Radars," *AD A014 151*, June, 1975, Rept. No. RE-75-35.

Ewell, G.W. and Bush, A.M.: "Constrained Improvement MTI Radar Processors," *Trans. AES-11, No. 5*, September, 1975, pp. 768-80.

Hsiao, J. K.; "Clutter Rejection of a Radar Moving Target Indication Filter," *Journal of Appl. Sci. and Eng. 1, No. 2*, October, 1975, pp. 113-27.

Schleher, D. C. and Schulkind, D.; "Optimization of Digital MTI Using Quadratic Programming," *Int. Conf. on ASSP*, 1977, pp. 849-853.

Schleher, D. C. and Schulkind, D.; "Optimization of Nonrecursive Digital MTI," *IEE Conf. Publ. No. 155 - Radar - 77, 25-28* October, 1977, pp. 173-176 (Reprint Paper No. 2.5)

4. Pulse-Burst MTI Radars

Westerfield, E. C., Prager, R. H. and Stewart, J. L.; "Processing Gains Against Reverberation (Clutter) Using Matched Filters," *Trans. IT-6, No. 3*, June, 1960, pp. 342-49.

Rihaczek, A. W.; "Radar Resolution Properties of Pulse Trains," *Proc. 52, No. 2*, February, 1964, pp. 153-64.

Kelly, E. J. and Wishner, R. P.; "Matched-Filter Theory for High-Velocity, Accelerating Targets," *Trans. MIL-9, No. 1*, January, 1965, pp. 56-69.

Rihaczek, A. W.; "Radar Signal Design for Target Resolution," *Proc. 53, No. 2*, February, 1965, pp. 116-28.

Brookner, E.; "Optimum Clutter Rejection," *Trans. IT-11, No. 4*, October, 1965, pp. 597-99.

Rihaczek, A. W.; "Optimum Filters for Signal Detection in Clutter," *Trans. AES-1, No. 3*, December, 1965, pp. 297-98c.

Rihaczek, A.W.; "Doppler-Tolerant Signal Waveforms," *Proc. 54, No. 6*, June, 1966, pp. 849-57.

Rubin, W. L. and Kaiteris, C. P.; "Some Results Concerning Radar Waveform Design in Detecting Targets in Clutter," *Proc. 54, No. 11*, November 1966, pp. 1609-10c.

Rummler, W. D.; "Clutter Suppression by Complex Weighting of Coherent Pulse Trains," *Trans. AES-2 No. 6*, November, 1966, pp. 689-99.

Rihaczek, A.W.; "Radar Resolution of Moving Targets," *Trans. IT-13, No. 1, January 1967*, pp. 51-56.

Ares, M.; "Optimum Burst Waveforms for Detection of Targets in Uniform Range-Extended Clutter," *Trans. AES-3, No. 1*, January 1967, pp. 138-141c. (Reprint Paper No. 3.2)

Rihaczek, A. W. and Mitchell, R. L.; "Radar Waveforms for Suppression of Extended Clutter," *Trans. AES-3, No. 3*, May 1967, pp. 510-17.

DeLong, D. F., Jr. and Hofsetter, E. M.; "On the Design of Optimum Radar Waveforms for Clutter Rejection," *Trans. IT-13, No. 3*, July, 1967, pp. 454-63.

Rihaczek, A. W.; Delay-Doppler Ambiguity Function for Wideband Signals," *Trans. AES-3, No. 4*, July 1967, pp. 705-11.

Kibbler, G. O. T. H.; "CAVORT: A Radar Pulse Train Optimum Processor for Accelerating Targets," *Trans. AES-3, No. 5*, September, 1967, pp. 808-18.

Rummler, W. D.,; "A Technique for Improving the Clutter Performance of Coherent Pulse Train Signals," *Trans. AES-3, No. 6*, November 1967, pp. 898-906.

Thompson, J. S. and Titlebaum, E. L.; "The Design of Optimal Radar Waveforms for Clutter Rejection Using the Maximum Principle," *Trans. AES-3, No. 6 (Suppl)*, November 1967, pp. 581-89.

Stutt, C. A. and Spafford, L. J.; "A 'Best' Mismatched Filter Response for Radar Clutter Discrimination," *Trans. IT-14, No. 2*, March, 1968, pp. 280-87.

Urkowitz, H.; "Some High Velocity Clutter Effects in Matched and Mismatched Receivers," *Trans. AES-4, No. 3*, May 1968, pp. 481-85.

Mitchell, R. L. and Rihaczek, A. W.; "Matched-Filter Responses of the Linear FM Waveform," *Trans. AES-4, No. 3*, May 1968, pp. 417-32.

Spafford, L. J.; "Optimum Radar Signal Processing in Clutter," *Trans. IT-14, No. 5*, September 1968, pp. 734-43 (Reprint Paper No. 3.1).

Mitchell, R. L. and Rihaczek, A. W.; "Clutter Suppression Properties of Weighted Pulse Trains," *Trans. AES-4, No. 6*, November 1968, pp. 822-28 (Reprint Paper No. 3.3).

Rasmussen, C. P.; "Clutter Rejection Using a Coded Burst Waveform," *EASCON, 1968*, pp. 70-78.

DeLong, D. F., Jr. and Hofstetter, E. M.; "The Design of Clutter-Resistant Radar Waveforms with Limited Dynamic Range," *Trans. IT-15, No. 3*, May 1969, pp. 376-85.

Mosca, E.; "A Nonadaptive Processing of Radar Pulse Trains for Clutter Rejection," *Trans. AES-5, No. 6*, November 1969, pp. 951-58.

DeLong, D. F., Jr. and Hofstetter, E. M.; "Optimum Radar Signal-Filter Pairs in a Cluttered Environment," *Trans. IT-16, No. 1*, January 1970, pp. 89-90c.

DeLong, D. F., Jr.; "Design of Radar Signals and Receivers Subject to Implementation Errors," *Trans. IT-16, No. 6*, November 1970, pp. 707-11.

Zeoli, G. W.; "Some Results on Pulse-Burst Radar Design," *Trans. AES-7, No. 3*, May 1971, pp. 486-98.

Mitchell, R. L.; "Resolution in Doppler and Acceleration with Coherent Pulse Trains," *Trans. AES-7, No. 4*, July 1971, pp. 630-36.

Rihaczek, A. W. and Golden, R.M.; "Resolution Performance of Pulse Trains with Large Time-Bandwidth Products," *Trans. AES-7,No. 4*, July 1971, pp. 677-685.

McAulay, R. J. and Johnson, J. R.; "Optimum Mismatched Filter Design for Radar Ranging, Detection and Resolution," *Trans. IT-17, No. 6*, November 1971, pp. 696-701.

Rihaczek, A. W.; "Radar Waveform Selection — a Simplified Approach," *Trans. AES-7, No. 6*, Nov. 71, pp. 1093-99.

Rihaczek, A. W. and Golden, R. M.; "Clutter Performance of Coherent Pulse Trains for Targets with Range Acceleration," *Trans. AES-7, No.6*, November 1971, pp. 1093-99.

Ackroyd, M. H. and Ghani, F.; "Optimum Mismatched filters for Sidelobe Suppression," *Trans. AES-9, No. 2*, March 73, pp. 214-18.

Urkowitz, H., Geisler, J. D. and Ricciardi, N. A., Jr.; "The Effect of Weighting Upon Signal-to-Noise Ratio in Pulse Bursts," *Trans. AES-9, No. 4*, July 1973, pp. 486-94.

Mesiya, M. F. and McLane, P. J.; "Design of Optimal Radar Signal Subject to a Fixed Amplitude Constraint," *Trans. AES-9, No. 5*, September 1973, pp. 679-87.

Smith, F. W.; "Effect of Radar Amplitude and Phase Weighting Errors on Clutter Rejection by Burst Waveforms," *Trans AES-10, No. 1*, January 1974, pp. 160-62.

Cohen, A. I.; "An Algorithm for Designing Burst Waveforms with Quantized Transmitter Weight," *Trans. AES-11, No. 1*, January 1975, pp. 56-64.

Temes, C. L. and Thompson, W. E.; "Slope Coding of Coherent Pulse-Burst Waveforms," *Trans AES-11*, No. 2, March 1975, pp. 183-89.

Cohen, A. I.; "Ambiguity Removal When Using Burst Waveforms: Part 1: Coincidence Detection," *Trans. AES-11, No. 5*, September 1975, pp. 907-13c.

Kaveh, M. and Cooper, G. R.; "Average Ambiguity Function for a Randomly Staggered Pulse Sequence," *Trans. AES-12, No. 3*, May 1976, pp. 410-13.

Acampora, A. S.; "High Power Radar Implementation of Coherent Waveforms," *Trans. AES-12, No. 4*, July 1976, pp. 444-50.

Cohen, A. I.; "Ambiguity Removal When Using Burst Waveforms, Pt II: Nonlinear Processing," *Trans. AES-12, No. 6*, November 1976, pp. 805-11c.

5. Special MTI Effects

Raven, R. S.; "Requirements on Master Oscillators for Coherent Radar," *Proc. 54, No. 2*, February 1966, pp. 237-43.

Leeson, D. B. and Johnson, G. F.; "Short-Term Stability for a Doppler Radar: Requirements, Measurements, and Techniques," *Proc. 54, No. 2*, February 1966, pp. 244-48.

Shenoy, R. P.; "Performance of Doppler Radar in the Presence of Random Fading," *Trans. AES-2, No. 6*, November 1966, pp. 676-80.

Voles, R.; "A Radar Phase Locking Technique," *Proc. 54, No. 12*, December 1966, p. 1941c.

Fishbein, W., Graveline, S. W., Rittenbach, O. E.; "Clutter Attenuation Analysis," *AD 665 352*, March 1967, Rept. No. ECOM-2808, (Reprint Paper No. 18).

Kalmus, H. P.; "Doppler Wave Recognition With High Clutter Rejection," *Trans. AES-3, No. 6*, November 1967, pp. 334-39 (Sup).

Nathanson, F. E. and Reilly, J. P.; "Clutter Statistics Which Affect Radar Performance Analysis," *Trans. AES-3, No. 6*, November 1967, pp. 386-98 (Sup).

Ward, H. R. and Shrader, W. W.; "MTI Performance Degradation Caused By Limiting," *EASCON, 1968*, pp. 168-74.

Nishikawa, S.; "Nonlinear Processing of Radar Returns in MTI Systems," *AD 674 033*, March 1968, Rept. No. RM-5490-1-PR.

Grasso, G.; "Improvement Factor of a Nonlinear MTI in Point Clutter," *Trans. AES-4, No. 4*, July 1968, pp. 640-44c.

Rauch, S.; "Spectrum Distortion Due to Frequency Instabilities in a Two-way Coherent Doppler Radar," *Trans. AES-4, No. 5*, September 1968, pp. 802-05c.

Grasso, G. and Guarguaglini, P. F.; "Improvement Factor for Limited Coherent MTI," *Proc. 56, No. 11*, November 1968, pp. 2064-66c.

Barton, D. K. and Shrader, W. W.; "Interclutter Visibility in MTI Systems, EASCON, 1969, pp. 294-9

Grasso, G. and Guarguaglini, P. F.; "Clutter Residues of a Coherent MTI Radar Receiver," *Trans. AES-5, No* March 1969, pp. 195-204 (Reprint Paper No. 4.1).

Gray, M., *et. al.*; "Stability Measurement Problems and Techniques for Operational Airborne Pulse Doppler Radar," *Trans. AES-5, No. 4*, July 1969, pp. 632-37.

Roy, R. and Lowenschuss, O.; "Design of MTI Detection Filters With Nonuniform Interpulse Periods," *Trans. CT-17, No. 4*, November 1970, pp. 604-12.

Krosczynski, J.; "Efficiency of the Two-Frequency MTI Systems," *Radio Electron Eng. 41, No. 2*, February 1971, pp. 77-80.

Zeoli, G.W.; "IF Versus Video Limiting for Two-Channel Coherent Signal Processors," *Trans. IT-17, No. 5*, September 1971, pp. 579-86.

Johnson, S. M.; "Multiple Repetition Frequency Radar Coverage (Maximum No-Gap Cover)," *AD 759 155*, August 1972, Report No. R-969-PR

Thorsteinson, C. M., Haykin, S. S. and de Buda, R.; "Effect of Envelope Limiting in Pulse-Compression Moving-Target-Indicator Radar Systems," *Proc. IEE 119, No. 10*, October 1972, pp. 1463-64.

Fletcher, R. H., Jr. and Burlage, D. W.; "An Initialization Technique for Improved MTI Performance in Phased Array Radars," *Proc. 60, No. 12*, December 1972, pp. 1551-52.

Hansen, V. G., *et. al.*; "Adaptive Digital MTI Signal Processing," *EASCON Rec. 1973*, pp. 170-76.

Hsiao, J. K., Kretschmer, F. F., Jr.; "Design of a Staggered-PRF MTI Filter," *AD 908 181L*, January 1973, Report No. NRL-7545.

Fletcher, R.H., Jr.; "The State Variable Approach to the Analysis of Selecting Digital Moving Target Indicator Filters," *AD 757 611*, February 1973, Report No. RE-73-17.

Fletcher, R. H., Jr., and Burlage, D. W.; "Improved Moving - Target - Indicator Filtering for Phased Array Radars," *AD 759 901*, March 1973, Report No. *RE*-73-17

Roy, R. and Lowenschuss, O.; "A Velocity Adaptive MTI Filter," *Trans. AES-9, No. 2*, March 1973, pp. 324-26c.

McAulay, R. J.; "The Effect of Staggered PRFs on MTI Signal Detection," *Trans. AES-9, No. 4*, July 1973, pp. 615-618c.

Prinsen, P. J. A.; "A Class of High-Pass Digital MTI Filters With Non-uniform PRF," *Proc. 61, No. 8*, August 1973, pp. 1147-48c. (Reprint Paper No. 4.3).

Prinsen, P. J. A.; "Elimination of Blind Velocities of MTI Radar by Modulating the Interpulse Period," *Trans. AES-9, No. 5*, September 1973, pp. 714-24 (Reprint Paper No. 4.2).

Rihaczek, A. W.; "Choice of Burst Number for Blind-Speed Avoidance," *Trans. AES-9, No. 5*, September 1973, pp. 778-80c.

Voles, R.; "New Techniques for MTI Clutter-Locking," *IEE Conf. Publ. No. 105, Radar-Present and Future.*" London, October 23-25, 1973, pp. 274-79.

Fletcher, R. H., Jr. and Burlage, D.W.; "Improved MTI Performance for Phased Array Radars in Severe Clutter Environments," *IEE Conf. Publ. No. 105, Radar — Present and Future*, London October 23-25, 1973, pp. 280-85.

McQueen, J.G.; "The Design and Operational Simulation of a 4-Pulse-Per-Beamwidth MTI System," *IEE Conf. Publ. No. 105, Radar — Present and Future*, London, October 23-25, 1973, pp. 286-91.

Nitzberg, R.; "Limitation of MTI Improvement Factor Due to Oscillator Instability," *Electronic Letters 9, No. 20*, October 1973, pp. 470-71.

Voles, R.; "New Approach to MTI Clutter Locking," *Proc. IEE 120, No. 11*, November 1973, pp. 1383-90.

Rihaczek, A.W.; "A Systematic Approach to Blind speed Elimination," *Trans. AES-9, No. 6*, November 1973, pp. 940-47.

Voles, R.; "The Losses Due to the Error in Estimating the Velocity in Clutter-Locking MTI Systems," *Trans. AES-9, No. 6*, November 1973, pp. 950-53c.

Hsiao, J. K. and Krestschmer, F. F.; "Design of a Staggered — PRF Moving Target Indication Filter," *Rad-Elect. Eng. 43, No. 11*, November 1973, pp. 689-93.

Hughes, C. J. and Johnson, R.S.; "Calculated Performance of an Optimized Nonlinear MTI," *EASCON Rec. 1974*, pp. 206-13.

Wardrop, B.; "The Performance of MTI Systems When Used With PRF Stagger," *Marconi Rev. 37, No. 195*, 1974, pp. 217-32.

Lutte, N. P. and Thomas, H. W.; "Frequency-Domain Approaches to Moving-Target-Indicator Filters With Staggered Pulse-Repetition Frequencies," *Proc. IEE 121, No. 9*, September 1974, pp. 954-56.

Ekstrom, J. E.; "MTI Clutter Locking for Arbitrary Clutter Spectral Shapes," *Trans. AES-10, No. 6*, November 1974, pp. 872-74c.

Prinsen, P. J. A.; "Comparison of Some Classes of MTI-Filters for a 3-Pulse Canceller With Staggered PRF," *AD B002 407*, November 1974, Report No. PHL-1974-38.

Owen, P. L.; "The Effect of Random PRF Staggering on MTI Performance," *Int. Radar Conf. Rec. 1975*, pp. 73-78.

Urkowitz, H.; "Directional Velocity Sorting MTI With Staggered Pulse Spacing," *Int. Radar Conf. Rec. 1975*, pp. 91-96.

Sawai, A. and Irabu, T.; "Moving Clutter Rejection Using Doppler Information and its Range-to-Range Correlation Statistic," *Int. Radar Conf. Rec. 1975*, pp. 97-102.

Ward, H. R.; "Doppler Processor Rejection of Range Ambiguous Clutter," *Trans. AES-11, No. 4*, July 1975, pp. 519-22.

Voles, R.; "Effects of Noise in Clutter-Locking MTI System," *Proc. IEE 122, No. 11*, July 1975, pp. 689-92.

Acker, A. E.; "Eliminating Transmitted Clutter in Doppler Radar Systems," *Microwave Journal 18, No. 11*, November 1975, pp. 47-50.

Voles, R.; "MTI Clutter Locking When the Spectrum is Asymmetric," *Proc. IEE 122, No. 11*, November 1975, pp. 1169-72.

Thomas, H. W. and Abram, T. M.; "Stagger Period Selection for Moving-Target Radars," *Proc. IEE 123, No. 3*, March 1976, pp. 195-99.

Hsiao, J. K.; "MTI Optimization In a Multiple-Clutter Environment," *Trans. AES-21, No. 3*, May 1976, pp. 401-05.

McLane, P. J., Campbell, L. L. and Mesiya, M. F.; "Performance Degradation of MTI Circuits Due to Amplitude Limiting," *Trans. AES-12, No. 4*, July 1976, pp. 519-22.

Clectus, G. M.; "Properties of Staggered PRF Radar Spectral Components," *Trans. AES-12, No. 6*, November 1976, pp. 800-03c.

Hsiao, J. K.; "Statistical Properties of a Staggered PRF MTI System," *Int. Conf. on ASSP, 1977*, pp. 864-867.

Houts, R.; "Velocity Response Characteristics of MTI Radars Using Pulse or Block Stagger," *IEE Conf. Publ. No. 155, Radar — 77*, October 25-28, 1977, pp. 171-181.

6. Digital MTI

Brennan, L. E. and Reed, I.S.; "Quantization Noise in Digital Moving Target Indicator Systems," *Trans. AES-2, No. 6*, November 1966, pp. 655-58 (Reprint Paper No. 5.1).

Radar, C. M. and Gold, B.;"Digital Filter Techniques in The Frequency Domain," *Proc. 55, No. 2*, February 1967, pp. 149-71.

Cochran, W. T. *et. al.;* "What is the Fast Fourier Transform?," *Proc. 55, No. 10*, October 1967, pp. 1664-74.

Linder, R. A. and Kutz, G. H.; "Digital Moving Target Indicators," *Trans. AES-3, No. 6*, November 1967, pp. 374-85 (Sup). (Reprint Paper No. 5.2)

Groginsky, H. L. and Works, G. A.; "A Pipeline Fast Fourier Transform," *EASCON Rec. 1969*, pp. 22-29.

Nathanson, F. E. and Gray, G. A.; "Digital Processing of Radar Signals," *EASCON Rec. 1971*, pp. 208-15.

Hartt, J. K. and Sheats, L. F.; "Application of Pipeline FFT Technology in Radar Signal and Data Processing," *EASCON Rec., 1971*, pp. 216-21. (Reprinted: Barton, 1975, Vol. 3).

Halpern, H. M. and Perry, R. P.; "Digital Matched Filters Using Fast Fourier Transforms," *EASCON Rec. 1971*, pp. 222-23c.

Roecker, R.; "The Application of Digital Filters for Moving Target Indication," *Trans. AU-19, No. 1*, March 1971, pp. 72-77.

Inagaki, R. and Hagisawa, T.; "Digital MTI Processors," *NEC Res. and Dev., No. 23*, October 1971, pp. 23-29.

McAulay, R. J.; "A Theory for Optimal MTI Digital Signal Processing: Part 1, Receiver Synthesis," *AD 738 298*, February 1972, Rept. No. TN-1972-14-PT-1.

Echard, J. D. and Boorstyn, R. R.; "Digital Filtering for Radar Processing Application," *Trans. AU-20, No. 1*, March 1972, pp. 42-52.

Bergland, G. D. and Hunnicutt, C. F.; "Application of a Highly Parallel Processor to Radar Radar Data Processing," *Trans. AES-8, No. 2*, March 1972, pp. 161-67.

McAulay, R. J.; "A theory for Optimal Moving Target Indicator (MTI) Digital Signal Processing," Part II, Signal Design, *AD 750 747*, October 1972, Rept. No. TN-1972-14-PT-2.

McAulay, R. J.; "A Theory for Optimal Moving Target Indicator (MTI Digital Signal Processing," (Sup.1), *AD 751 923*, October 1972, Rept. No. TN-1972-14-Suppl-1. (Reprint Paper No. 5.3).

Ellis, J. G.; "Digital MTI, a New Tool for the Radar Used," *Marconi Rev. 36, No. 191*, 1973, pp. 237-48.

Andrews, G. A., Jr.; "Performance of Cascaded MTI and Coherent Integration Filters in a Clutter Environment," *AD 909 543L*, March 1973, Rept. No. NRL-7533.

Rosie, A. M. and O'Sullivan, B. A.; "Digital Filtering of Radar Signals," *IEE Conf. Publ. No. 105, Radar — Present and Future,*" London, October 23-25, 1973, pp. 292-96.

Muehe, C. E.; "Digital Signal Processor for Air Traffic Control Radars," *NEREM Rec,* 1974, Pt 4, pp. 73-82.

O'Donnell, R. M. *et. al.;* "Advanced Signal Processing for Airport Surveillance Radars," *EASCON Rec. 1974, pp. 71-71F (Reprint Paper No. 5.5).*

Peled, A. and Liu, B.; "Sone New Realizations of Dedicated Hardware Digital Signal Processors," *EASCON Rec.1974,* pp. 464-68.

Hsiao, J. K.; "FFT Doppler Filter Performance Computations," *AD 777 722,* March 1974, Rept. No. NRL-MR-2744.

Andrews, G. A. Jr.; "Optimum Radar Doppler Filtering Techniques," *AD 920 000L,* May 1974, Rept. No. NRL-7727.

Andrews, G. A., Jr.; "Comparison of Radar Doppler Filtering Techniques," *AD 923 915L,* October 1974, Rept. No. NRL-7811.

Hsiao, J. K.; "Performance Simulation of a Multiple-State MTI and Doppler Filter System," *AD B000 955L,* December 1974, Rept. No. NRL-7829.

Long, R. J.; "Clutter Suppression By Means of Digital MTI an Applied to Precision Approach Radar," *AD A004 475,* December, 1974, Rept. No. MTR-2851.

Benvenuit, P. and Guarguaglini, P. F.; "Improvement Factor Evaluation of Open Loop Adaptive Digital MTI," *Int. Rad. Conf. Rec. 1975,* pp. 52-56.

Hawkes, C. D. and Haykin, S. S.; "Adaptive Digital Filtering for Coherent MTI Radar," *Int. Rad. Conf. Rec. 1975,* pp. 57-62.

Dillard, G. M.; "Signal-to-Noise Ratio Loss in an MTI Cascaded With Coherent Integration Filters," *Int. Rad. Conf. Rec. 1975,* pp. 117-22.

Martinson, L.; "A Programmable Digital Processor for Airborne Radar," *Int. Rad. Conf. Rec. 1975,* pp. 186-91.

Rabiner, L. R., McElellan, J. H. and Parks, T. W.; "FIR Digital Filter Design Techniques Using Weighted Chebyshev Approximation," *Proc. 63, No. 4,* April 1975, pp. 596-610.

Drury, W. H.; "Improved MTI Radar Signal Processor," *AD A010 478,* April 1975, Rept. No. ATC-39.

Janardhanam, G. and Hegde, L. R.; "Solid State Digital Canceller for Moving Target Indicator," *Electro Technol 19, No. 2,* June 1975, pp. 35-40.

Bennett, R. and Holt, A. G. J.; "Design of Moving-Target-Indicator Filters Using a Quadrature Channel," *Electronic Letters 11, No. 24,* November 1975, pp. 595.

O'Donnell, R.M.; "Digital Moving Target Detector for Use with the FPS-20 Class of Radar," *AD B007 797L,* November 1975, Rept. No. TN-1975-58.

Muehe, C.E.; "Advances in Radar Signal Processing," *Electro Rec 1976,* pp. 25-3-1 to 9.

Collins, F.A. and Sicking, C.J.; "Properties of Low Precision Analog-to-Digital Converters," *Trans. AES-12, No. 5,* September 1976, pp. 643-46c.

Schleher, D.C. and Kozin, F.; "Radar Signal Processing Using Digital Nonlinear Filters," *Int. Conf. on ASSP,* 1977, pp. 854-858.

Castella, F., Miller, J., "Moving Target Detector Data Utilization Study," *IEE Conf. Publ. No. 155 — Radar-77,* 25-28 October 1977, pp. 182-185.

Kelly R., Charton, S., "A Programmable Digital Signal Processor Evaluated for Radar Applications," *IEE Conf. Publ. No. 155 — Radar-77,* 25-28 October 1977, pp. 20-24.

Muehe, C., *et. al.,* "The Parallel Microprogrammed Processor (PMP)," *IEE Conf. Publ. No. 155 — Radar-77,* 25-28 October 1977, pp. 97-100.

7. MTI Hardware

Fishbein, W.; "Moving Target Indication for Radar Set AN/MPQ-4," *AD 326 989,* May 1961.

Cappadona, W. and Kenneally, D.; "Microminiature MTI System for Surveillance Radars," *Conv. Rec. 66, Pt 4,* pp. 34-39.

Cappadona, W.L.; "Microminiaturized MTI Group AN/GPA (XW-2)," *AD 629 601,* March 1966, Rept. No. AIL-3453-1.

Graveline, S.W.; "Evolution of Multiple Range-gate Video Processors for Combat Surveillance Radar Systems," *AD 489 686,* July 1966, Rept. No. ECOM-2742.

Heck, J.C., Green, J.H.; "Lightweight Surveillance Radar AN/PPS-9," *AD 878 561L*, October 1970.

Amundsen, E.M.; "Ground Surveillance Radar Set AN/PPS-15," *AD 922 266L, Final Rept.* March 1972-November 1973, July 1974.

Patriarche, M.V.; "An Evaluation of Some Features of an Air Traffic Control Radar System in a Mountainous Environment," *AD 904 231*, April 1972, Rept. No. CRC-TN-635.

Butler, W.J., Puckette, C.M. and Barron, M.D.; "Implementation of a Moving-Target Indicator by Bucket-Brigade Circuits," *Electronic Letters 8, No. 22*, November 1972, pp. 543-44.

Maines, J.D. and Paige, E.G.S.; "Surface-Acoustic-Wave Components, Devices and Applications," *Proc. IEE 120, No. 10R*, 1973, pp. 1078-1110.

Lancaster, J.F.; "MICOM Experimental Array Radar Digital Signal Processor," *AD 764 227*, February 1973, Rept. No. RE-73-14.

Buss, D.D. and Brodersen, R.; "CCD Signal Processors," *AD B003 188L, Final Technical Rept.*, 1 July 1973 - 30, September 1974, Rept. No. TI-08-74-73.

Yoder, L.W.; TPN-19 "MTI Analysis," *AD 913 818L, Rept. No. MTR-2557*, August 1973.

Altman, J.L.; "Range Limitations of TPN-19 PAR Radar Caused by Rain," *AD 913 906L*, August 1973, Rept. No. MTR-2534.

Kailasam, T.K. and Venkatesan, M.; "A Solid State MTI Using RGF Canceller," *Electro Technol 17, No. 6*, November-December 1973, pp. 181-89.

Whateley, M.M. *et al.;* "Charge-Coupled Device Applications to Radar Signal Processing," *NAECON Rec*, 1974, pp. 284-91.

Buss, D.D. *et al.;* "Application of Charge-Coupled Devices to Radar Signal Processing," *NEREM Rec*, 1974, Pt 4, pp. 83-98.

Degan, L.J. and Stitts, J.R.; "Burst Waveform Processor Using Surface Wave Devices," *EASCON Rec*, 1974, pp. 214-20.

Bristol, T.W.; "Acoustic Surface-Wave-Device Applications," *Microwave J 17, No. 1*, January 1974, pp. 25-27.

Bounden, J.E. and Tomlinson, M.J.; "CCD Chebyshev Filter for Radar MTI Applications," *Electronic Letters 10, No. 7*, April 1974, pp. 89-90.

Holland, M.G. and Claiborne, L.T.; "Practical Surface Acoustic Wave Devices," *Proc. 62, No. 5*, May 1974, pp. 582-611.

Ludington, D.M. and Lobenstein, H.; "Discrete Signal Moving Target Indicator (MTI)," *AD B000 49IL*, October 1974.

Colin, J.M.; Racine: "An Unambiguous Pulsed Doppler Radar Process," *Int Rad Conf Rec* 1975, pp. 69-72.

Elliott, A.R., Haykin, S.S. and Hawkes, C.D.; "Hardware Implementation of a Recursive Digital Filter for MTI Radar," *Proc. IEE 122, No. 2*, February 1975, pp. 137-41.

Labitt, M., Giffin, D.; "Evaluation of the Applicability of Modern Signal Detection and Clutter Rejection Techniques To The AN/FPS-20A/67D/100 and AN/FPS-19 Classes of Air Force Radars," *AD B007 947L*, September 1975, Rept. No. TN-1975-48-Vol-1.

Sugie, M., Tomita, Y. and Hagisawa, T.; "Advanced Air Route Surveillance Radar With Various Unique Clutter Elimination Capabilities," *NEC Res and Dev (Japan) No. 39*, October 1975 pp. 71-83, (Reprint Paper No. 6.1).

Roberts, J.B.G.; "Radar Doppler Processing Using CCD and SAW Devices," *Electronic Letters 11, No. 22*, October 1975, pp. 525-26.

Butler, W.J. *et al.;* "Charge-Transfer Analog Memories for Radar and ECM Systems," *J Solid-State Circuits SC-11, No. 1*, February 1976, pp. 93-100.

Roberts, J.B.G. *et al.;* "A Processor for Pulse-Doppler Radar," *J. Solid-State Circuits SC 11, No. 1*, February 1976, pp. 100-04.

Maines, J.D. and Paige, E.C.S.; "Surface-Acoustic-Wave Devices for Signal Processing Applications," *Proc. 64, No. 5*, May 1976, pp. 639-52.

Scorer, M. and Watkons, C.; "An Advanced Man Portable Surveillance Radar," *IEE Conf. Publ. No. 155 — Radar-77*, 25-28 October 1977, pp. 29-32.

Marlow, P. and Sunderland, R.; "Design of High-Stability Transmitters for Pulse Doppler Radars and Measurement of Their Spectral Purity," *IEE Conf. Publ. No. 155 — Radar-77*, 25-28 October 1977, pp. 564-568.

8. Airborne MTI

George, T.S.; "Fluctuations of Ground Clutter Return in Airborne Radar Equipment," *Proc. IEE 99 IV*, April 1952, pp. 92-99.

Dickey, F.R., Jr.; "Theoretical Performance of Airborne Moving Target Indicators," *Trans. PGAE-8*, June 1953, pp. 12-23, (Reprint Paper No. 7.1).

Griffin, D.H.; Herlin, M.A.; "The IF TACCAR AMTI System," *AD 320 000,* December 1957, Rept. No. GR45 28.

Anderson, D.B.; "A Microwave Technique to Reduce Platform Motion and Scanning Noise in Airborne Moving-Target Radar," *Wesc Rec, 58, Pt 1,* pp. 202-11.

Solomon, K.; "A Double Delay and Subtraction Airborne Clutter Canceller," *MIL-E-CON,* June 16-18, 58, pp. 235-40.

Urkowitz, H.; "An Extension to the Theory of the Performance of Airborne Moving-Target Indicators," *Trans. ANE-5, No. 4,* December 1958, pp. 210-14.

Griffin, D.H., Labitt, M. and McGinn, J.W., Jr.; "Flight Test Summary," *AD 323 449,* May 1959, Rept. No. GR-45-30-4.

McCabe, K.E.; "A Broadband Airborne Moving Target Indicator," *MIL-E-CON,* September 9-11, 1963, pp. 94-96.

Longacre, A. and Mader, G.E.; "Continuous Drift Angle From a Search Radar," *Trans. ANE-10, No. 2,* June 1963, pp. 93-97.

Mallett, J.D.; "A Clutter Rejection Technique for an Airborne Radar Working Overland," *AD 353 062,* August 1964, Rept. No. RM4260PR.

Urkowitz, H.; "The Effect of Antenna Patterns on the Performance of Dual-Antenna Radar Airborne Moving Target Indicators," *Trans. ANE-11, No. 4,* December 1964, pp. 218-23.

Sharpe, C.B.; "Clutter Cancellation in Squinted-Beam AMTI," *AD 458 782,* March 1965, Rept. No. 6400-17-R.

Brennan, L.E.; "Optimum Signal Processing for AMTI Radars With Varied Interpulse Spacings," *AD 369 286L,* October 1965, Rept. No. RM-4754-PR.

Author Unknown; "Feasibility Study of a Multimode Radar Digital Signal Processor," *AD 380 221,* March 1967, Rept. No. 6452A.

Author Unknown; "E-2A Overland Program," *AD 384 752L,* May 1967, Grumman Aircraft Engineering Corp.

Bussgang, J.J.; "Approximate Results for Dual-Antenna AMTI Radar," *Trans. AES-3, No. 4,* July 1967, pp. 729-31c.

Widrow, B. *et al.;* "Adaptive Antenna Systems," *Proc. 55, No. 12,* December 1967, pp. 2143-59.

Clough, D.L.; "Maximizing Moving Target Indicator (MTI)," *AD 851 063,* March 1969, Rept. No. RADC-TR-69-13.

Sletten, C.J.; "Wide-Looking AMTI for Airborne Radars," *AD 862 179L,* June 1969, Rept. No. AFCRL-PSRP-382, AFCRL-69-0251.

Hofstetter, E.M., Weinstein, C.J. and Muehe, C.E.; "A Theory of Multiple Antenna AMTI Radar," *AD 724 076,* April 1971, Rept. No. TN-1971-21.

Andrews, G.A.; "Airborne Radar Motion Compensation Techniques, Evaluation of TACCAR," *AD 893 904L,* April 1972, Rept. No. NRL-7407.

Andrews, G.A.; "Airborne Radar Motion Compensation Techniques — Evaluation of DPCA," *AD 902 605L,* July 1972, Rept. No. NRL-7426.

Bottlik, I.P., Brennan, L.E. and Lank, G.W.; "Adaptive Filtering in AMTI Radar, *AD 904 853,* September 1972, Rept. No. TSC-PD-083-2.

O'Sullivan, M.R. and Pawula, R.F.; "Clutter Spectra of Low PRF AMTI Pulse-Doppler Radar," *Trans. AES-9, No. 2,* March 1973, pp. 93-103.

Brennan, L.E. and Reed, I.S.; "Theory of Adaptive Radar," *Trans. AES-9, No. 2,* March 1973 pp. 237-53, (Reprint Paper No. 7.3).

Hendrix, R.E.; "Overland Downlook Radar is Key Element of AWACS," *Westinghouse Eng. 33, No. 4,* July 1973, pp. 98-105.

Brennan, L.E.; "Adaptive AMTI Radar," *AD 771 385,* October 1973, Rept. No. TSC-PD-906-4.

Schindler, J.K. and Goggins, W.B.; "An Airborne Radar Technique for Moving-Target Detection, Location, and Tracking," *AD 777 214,* November 1973, Rept. No. AFCRL-TR-73-0719, AFCRL-PSRP-576.

Burgess, L.R. and Zeger, A.E.; "Adaptive Techniques for Radar Control," *AD 915 572L,* November 1973, Rept. No. GAC-2347-2592-17.

Andrews, G.A.; "Optimization of Radar Doppler Filters to Maximize Moving Target Detection," *NAECON Rec 1974,* pp. 279-83.

Zeger, A.E.; "An Adaptive MTI Radar Antenna Array," *NAECON Rec.* 1974, pp. 126-33.

Goggins, W.B., Sletten, C.J. and Holt, F.S.; "New Concepts in AMTI Radar," *Microwave J 17, No. 1,* January 1974, pp. 29-35.

Hsiao, J.K.; "A Digital Mean-Clutter-Doppler Compensation System," *AD 918 669L,* April 1974, Report No. NRL-MR-2772.

Brennan, L.E., Reed, I.S. and Swerling, P.; "Adaptive Arrays," *Microwave Journal 17, No. 5,* May 1974, pp. 43-46.

Reed, I.S., Mallett, J.D. and Brennan, L.E.; "Rapid Convergence Rate in Adaptive Arrays," *Trans. AES-10, No. 6,* November 1974, pp. 853-63, (Reprint Paper No. 7.4).

Goggins, W.B., Jr.; "Processing for Maximum Signal-To-Clutter in AMTI Radars," *AD A014 783,* November 1974, Rept. No. AFCRL-TR-74-0577.

Manoogian, D.V. and Ogar, G.; "The Generation of Non-Gaussian Airborne Clutter Spectra and the Implications Thereof on AMTI Performance," *Int. Rad. Conf. Rec.* 1975, pp. 79-84.

Heed, M. and Rizell, S.; "Influence of Video Detector Characteristic on Main Lobe Clutter Rejection in a Noncoherent Airborne Radar," *Int. Rad. Conf. Rec.* 1975, pp. 85-90.

Andrews, G.A.; "A Detection Philosophy for AMTI Radar," *Int. Rad. Conf. Rec.* 1975, pp. 111-16.

Muehe, C.E.; "Radar Techniques Program," *AD B003 152L,* January 1975.

Andrews, G.A.; "Platform Motion Compensation for AMTI Radar," *EASCON Rec. 1976,* pp. 64-A to 64-H.

Applebaum, S.P.; "Adaptive Arrays," *Trans. AP-24, No. 5,* September 1976, pp. 585-98.

Brennan, L.E., Mallett, J.D. and Reed, I.S.; "Adaptive Arrays in Airborne MTI Radar," *Trans. AP-24, No. 5,* September 1976, pp. 607-15.

Widrow, B. and McCool, J.M.; "A Comparison of Adaptive Algorithms Based on the Methods of Steepest Descent and Random Search," *Trans. AP-24, No. 5,* September 1976, 615-37.

Baird, C.A. and Rassweiler, G.G.; "Adaptive Sidelobe Nulling Using Digitally Controlled Phase-Shifters," *Trans. AP-24, No. 5,* September 1976, pp. 638-49.

Applebaum, S.P. and Chapman, D.J.; "Adaptive Arrays With Main Beam Constraints," *Trans. AP-24, No. 5,* September 1976, pp. 650-662.

Takao, K., Fujita, M. and Nishi, T.; "An Adaptive Antenna Array Under Directional Constraint," *Trans. AP-24, No. 5,* September 1976, pp. 662-69

White, W.D.; "Cascade Preprocessors for Adaptive Antennas," *Trans. AP-24, No. 5,* September 1976, pp. 670-84.

Chapman, D.J.; "Partial Adaptivity for the Large Array," *Trans. AP-24, No. 5,* September 1976, pp. 685-96.

Compton, R.T., Jr.; "An Experimental Four-Element Adaptive Array," *Trans. AP-24, No. 5,* September 1976, pp. 697-706.

9. Pulse Doppler

Barlow, E.J.; "Doppler Radar," *Proc. 37, No. 4,* April 1949, pp. 340-55.

Wimberly, F.T., and Lane, J.F., Jr.; "Navy's New Altimeter, the AN/APN-22," *Trans. ANE-1, No. 2,* June 1954, pp. 8-14.

Cohen, G.I. *et al.*; "Elimination of Ambiguities From High Pulse Repetition Rate Radars," *Proc. NEC 12,* 1956, pp. 271-81.

Cohn, G. *et al.*; "Suppression of False Range Indications in High Pulse Repetition Rate Radars," *Proc. NEC 13,* 1957, pp. 744-60.

Maguire, W.W.; "Application of Pulsed Doppler to Airborne Radar Systems," *NAECON 58,* pp. 291-96.

Potter, N.S.; "Range Ambiguity in High PRF Radar," *Conv. Rec., 60, Pt 8,* pp. 65-80.

Mooney, R.K. and Ralston, G.; "Performance in Clutter of Airborne Pulse, MTI, CW Doppler and Pulse Doppler Radar," *Conv. Rec., 61, Pt 5,* pp. 55-62.

Skillman, W.A. and Mooney, D.H.; "Multiple-High-PRF Ranging," *MIL-E-CON, 61,* pp. 37-40.

Coleman, S.D. and Hetrick. G.R.; "Ground Clutter and Its Calculation for Airborne Pulse Doppler Radar," *5th MIL-E-CON Rec.*, 1961, pp. 409-15.

Goetz, L.P., and Albright, J.D.; "Airborne Pulse-Doppler Radar," *Trans. MIL-5, No. 2,* April 1961, pp. 116-26.

Meltzer, S.A. and Thaler, S.; "Detection Range Predictions for Pulse Doppler Radar," *Proc. 49, No. 8,* August 1961, pp. 1299-1307.

Hetrick, G.R.; "Frequency Modulation Techniques Applied to Pulse Doppler Radar," *Conv. Rec.*, 1962, Pt 5, pp. 76-86.

Farrell, J.L. and Taylor, R.L.; "Doppler Radar Clutter," *Trans. ANE-11, No. 3,* September 1964, pp. 162-72.

Helgostam, L.F. and Ronnerstam, B.; "Ground Clutter Calculation for Airborne Doppler Radars," *Trans. MIL-9, Nos 3-4,* July-October 1965, pp. 294-97c.

Barna, A.; "Pseudorandom Frequency Modulation in Range-Doppler Radar," *Trans. AES-5, No. 2,* March 1969, pp. 221-29.

Gray, M. *et al.*; "Stability Measurement Problems and Techniques for Operational Airborne Pulse Doppler Radar," *Trans. AES-5, No. 4,* July 1969, pp. 632-37.

Friedlander, A.L. and Greenstein, L.J.; "A Generalized Clutter Computation Procedure for Airborne Pulse Doppler Radars," *Trans. AES-6, No. 1,* January 1970, pp. 51-61.

Hair, T. and Cross, M.G.; "Some Advances in Pulse Doppler Processing," *Int. Rad. Conf. Rec.* 1975, pp. 63-68.

Taylor, F.J.; "Adaptive Pulse Doppler Ambiguity Resolution," *Trans. AES-12, No. 2,* March 1976, pp. 98-103.

Ringel, M., "Computer-Calculated Ground Clutter for a Maneuvering Pulse Doppler Radar," *IEE Conf. Publ. No. 155 — Radar-77,* 25-28 October 1977, pp. 468-472.

Section 1 General Papers

EDITOR'S COMMENTS

Emerson's Paper

This classic report entitled "Some Pulsed Doppler MTI and AMTI Techniques" by R.C. Emerson of the Rand Corporation is one of the outstanding papers in the MTI field. Originally classified, it describes the MTI art through the early 1950's. More importantly, however, it provides a systematic analysis methodology that allows prediction of the overall performance of an MTI.

Chapter 1 introduces performance criteria that have found acceptance in the MTI field. The "reference gain" (also called improvement factor by a number of authors) is the normalized signal-to-clutter power ratio at the output of the MTI when the target's velocity is randomly distributed over all possible values.

In Chapter 2 optimization criteria are given that allow selection of the optimum weights for a finite impulse response (FIR) MTI based on the reference gain concept. PRF staggering to minimize the effects of MTI blind speeds is also discussed.

In Chapter 3, range-gated filter (RGF) MTIs are described and analyzed. This analog type of MTI was popular during the 60's since it allowed high MTI performance to be achieved utilizing the integrated circuits available at that time.

Chapter 4 describes noncoherent MTI systems that were extensively used during the 1950's and 1960's to detect moving targets from a moving platform. The attractiveness of this type of MTI is that clutter returned from the same range cell as the target is used as a reference thereby eliminating the need for a coherent reference corrected for platform motion. However, the performance potential for this type of MTI is less than that of a coherent system, and also it has the annoying characteristic that it is "blind in the clear."

Chapter 5 is devoted to the analysis of the complete MTI system that includes post detection integration. The balance between post detection integration and MTI filtering must be carefully considered in the design of any MTI system. The detected clutter residue at the output of the MTI filter tends to be highly correlated, and post detection integration is only weakly effective in enhancing the target-to-clutter ratio. The treatment of this subject given by Emerson still stands as one of the best available.

Chapter 6 describes the calculation of clutter spectra and system instabilities. In Chapter 7 the complete analysis is used to estimate the performance of a number of ground, shipboard and airborne radars operational in the early 1950's.

Velocity Shaped MTI Filters

Velocity shaped MTI filters are described in the next set of papers. This type of recursive MTI filter exhibits an infinite impulse response (IIR) thereby allowing precise control of the MTI's velocity response. MTI s of this type became popular in the early 1960's and are found in many currently operational radars. A disadvantage of this type of MTI is that nonsynchronous interference or clutter transients produce a disturbance consistent with the IIR. Paper 1.5 describes a method used to mitigate this effect but the added complexity often makes the FIR MTI filters described in the first paper preferable.

Paper 1.2, by White and Ruvin, describes a powerful bilinear transformation that allows the results of classical network theory to be applied to the design of velocity shaped MTI filters. Several examples of the design of Chebyshev and Elliptical filters are described. The MTI designs are implemented by applying feedback around double cancelers. By varying the feedback coefficients, velocity shaping is achieved, and can be used to compensate for the spectrum spreading induced by antenna rotation.

Paper 1.3, by Linden and Steinberg, describes the synthesis of MTI transfer functions using z-plane analysis. An important conclusion in this paper is that the number of delay sections required to implement a particular design is determined either by the number of z-plane poles located away from the origin or by the number of zeros, whichever is larger.

Paper 1.4, by Urkowitz, provides an in-depth discussion of the design of MTI filters and video integrators using z-plane analysis. Highlighting this paper is a catalog of z-plane transfer functions and pole-zero location for various canceler and integrator configurations.

Signal Processing Philosophy of the Early 60's

Paper 1.5, by Fowler, Uzzo and Ruvin, reflects the signal processing philosophy of the early 1960's. The processing techniques described in this paper were used in a series of Air Defense Radars designed at different frequencies to counter the bomber threat of that era.

Velocity shaped MTIs with adjustable pass bands provided SCVs of approximately 30 dB under a variety of clutter spectrum conditions induced by antenna scanning and internal clutter motion. A combination of coherent and noncoherent MTI provided performance against moving clutter such as rain clutter and chaff as well as against ground clutter. PRF staggering removed MTI blind zones.

The pulse interference separator and blanker (PISAB) was used to reject asynchronous interference thereby suppressing the undesirable effect that occurs in velocity shaped cancelers where a single interference pulse generates as many as 10 to 20 additional interference pulses.

Double loop FM sweep integrators that provide large feedback factors (≈ 0.99) are described as well as sidelobe blankers useful for discriminating against interference that enters through the radar's sidelobes.

MTI Analysis Theory

This paper, designated 1.6, develops a theory of MTI analysis that allows the overall capability of coherent and noncoherent shaped delay line and range gated filter (RGF) MTI systems to be determined. Both ground and airborne systems of the type utilized in the mid-1960's are discussed. The performance of several practical MTI radar systems are obtained by applying the performance curves provided in the paper.

In Sections I and II of the paper, the overall problem is approached by first considering the performance of an ideal implementation of the MTI system. The clutter is represented as a stochastic process with an associated spectral density and equivalent cross-sectional area. The clutter and point target signals are traced through coherent and noncoherent MTI systems considering the spectral spreading effects due to antenna scanning, platform motion, transmitter frequency stability and internal clutter motion. Equations for single, double, shaped-delay line filters and range gated filters are developed.

In Section III, performance criteria are defined and equations developed for computing MTI performance against these criteria. The application to several practical systems is discussed.

Section IV develops target and clutter characteristics, and provides these in tabulated form for both ground and airborne applications. The target and clutter spectra of ground and airborne MTI systems (coherent and noncoherent) is developed by matching Gaussian shaped spectrums to each of the individual effects and convolving to find the overall effect. Both intrinsic and induced effects are considered. In addition, the relationship for combining the effects of system instabilities and doppler filtering is derived.

Section V describes the characteristics of single, double and shaped delay line MTI filters using z-plane analysis. Synthesis techniques are discussed for delay line as well as RGF MTI filters.

Section VI discusses the calculation of the overall performance of the MTI system. The performance for shaped delay line and RGF MTI systems is developed from the viewpoint of the asymptotic filter that allows a determination of the number of delay lines or the number of filter sections in an RGF MTI to provide a given performance level. The overall system performance is calculated by considering the effect of system instabilities on the ideal implementation. Practical performance limitations due to system instabilities applicable to MTI systems of the mid-1960's are presented.

In Section VII, performance calculations are given for practical Combat Surveillance, Airborne and Air Traffic Control radars. The performance estimates obtained using the theory presented in this paper have been found to be in reasonable agreement with the measured performance of the system examined.

1.1 SOME PULSED DOPPLER MTI AND AMTI TECHNIQUES

R.C. Emerson

Rand Corporation Report R-274, March 1954, pp. 1-124

SUMMARY

An adequate air defense system depends to a large measure on an operationally reliable, integrated network of ground-based, airborne, and shipboard radars. Among other things, the reliability of these radars depends on their capabilities for detecting targets in the presence of ground clutter, sea return, and storm clouds, to say nothing of the various electronic countermeasures (ECM) possibilities. It is recognized that many of the existing sets scheduled for air defense service are severely limited by some of these forms of clutter, and consequently provisions are being made for their modification to provide moving target indication (MTI).

Until very recently MTI has almost always meant single-delay-line, two-pulse-cancellation equipment, even though it has long been known that superior clutter rejection can be achieved through the comparison of three or more pulses. This theoretically advantageous MTI circuitry has failed to appear in operational gear for at least two reasons: (1) the size and weight of the mercury delay lines and associated equipment, and (2) the more stringent design and manufacturing tolerances needed to guarantee electronic stability compatible with the high rejection ratios sought. With the development of fuzed quartz delay lines the size and weight obstacles are being removed. Similarly, as high-power klystrons become available, opening the way for crystal-controlled MOPA transmitter design, the problem of delivering stable equipment becomes less formidable, and when the stability problem is solved, it is probable that the so-called "range-gated filter" systems, which do not employ delay lines at all, will come into their own. Thus, in the years immediately ahead, one should expect to see radical changes in the whole character of MTI as it applies to air defense radar—the mercury delay lines giving way to multiple quartz lines for three-or-more-pulse comparison, and perhaps the abandonment of delay lines altogether.

This report presents a study of some of the aspects of MTI and AMTI (airborne MTI) design. The approach is characterized by mathematical analysis of the various relevant theoretical problems. The purpose is not to give a "bird's-eye view" of the state of the art, or to discuss the intricacies of present-day techniques, or to make recommendations for future systems. These subjects are not altogether avoided—the relevant hardware questions are dealt with as they arise, and the many recommendations for the development and adoption of advanced techniques, if not explicitly stated, are to be found between the lines.

The main purpose of this study is to provide under one cover a comprehensive method of analysis suitable for comparative evaluation of advanced systems. It is addressed primarily to technical people in the radar industry with the hope that interest will be stimulated in the acceleration of MTI and AMTI development.

The scope has been limited to systems which may find application in the defense network. This means that many important areas of MTI, such as bombing systems, the detection of slow-moving ground targets, antisubmarine systems, etc., have been neglected. Also omitted are various subjects including FM and pulsed-noise systems—in short, systems which are not likely to appear as interim developmental outgrowths of the AM-pulsed sets of today.

Specifically, this study provides a basis for estimating the performance of various pulse-comparison systems using either delay lines or range-gated filter techniques. It is divided into seven chapters. The first chapter begins with a discussion of various normal (i.e., non-MTI) anticlutter circuits, indicating the extent to which clutter background can be smoothed out and reduced in intensity without the use of doppler techniques. MTI and AMTI circuits, which distinguish moving targets on the basis of doppler shifts in signal frequency, are discussed in general terms to indicate the extent to which they are useful in suppressing clutter returns. Some of the basic mathematical concepts on which the later detailed analysis is based are stated. The remainder of Chap. 1 describes in condensed form the subject matter and basic concepts used in the other chapters of the report. A careful reading of this chapter will thus serve as a detailed introduction to the technical content of the study. Chapters 2 through 5 treat theoretical aspects of MTI and AMTI systems. Chapter 6 is devoted to analysis of clutter correlation and spectra and the effects of system-instability noise. Chapter 7 contains practical examples of the preceding theoretical discussions, in the form of numerical estimates of system performance presented graphically for a selection of specific air defense radars. The reader not interested in the theoretical chapters may prefer to skip directly to this chapter.

CONTENTS

CHAPTER 3

RANGE-GATED, LUMPED-CONSTANT FILTER SYSTEMS

SECTION

CHAPTER 4

NONCOHERENT SYSTEMS

CHAPTER 5

INTEGRATION AFTER DETECTION

SECTION

CHAPTER 6

CLUTTER AND SYSTEM NOISE

CHAPTER 7

PERFORMANCE ESTIMATES FOR SOME AIR DEFENSE RADARS

FIGURES

CHAPTER 1

CLUTTER SUPPRESSION AND MOVING-TARGET INDICATION

1.1. Introduction

The continually expanding requirement for high-performance radar detection and guidance equipment has stimulated the search for improved techniques for removing the echoes from storm clouds, rain, chaff, land masses, and the sea. The return from clutter can be many times larger than the echoes from targets within the clutter. In fact, unless special preventive circuitry is employed, the receiver may be saturated and detection of the actual targets will become impossible. This difficulty can be remedied to a certain extent by increasing the effective dynamic range of the receiver through the use of lin-log amplifiers and Instantaneous Automatic Volume Control (IAVC). The lin-log amplifier usually consists of a sequence of stages connected in such a way that weak signals derive amplification from all stages, while stronger signals which saturate the later stages are tapped off earlier at lower levels of amplification. The over-all effect is to reduce the sensitivity of the receiver in steps according to the strength of the echo. A similar effect is accomplished by IAVC, which provides a smooth gradation of sensitivity, depending on the signal strength. In addition to amplitude discrimination, the time constants of the IAVC circuit can frequently be chosen so that some "extended-target" discrimination takes place. This action is often desirable because of the greater spatial extension of most forms of clutter as compared with that of aircraft and other small targets.

More specifically for extended-target discrimination are the continuous differentiation circuits known as Fast Time Constant (FTC) and the discrete differencing circuits known as Detector Balanced Bias (DBB). The FTC type of anticlutter circuit is essentially an R-C high-pass filter having a (possibly variable) time constant of the order of a pulse width. It is useful for reducing the low-frequency components which are characteristic of extended clutter and low-frequency jamming signals. The DBB-type circuit employs a delay line of transit time comparable to a pulse width to provide a delayed balancing bias to the video detector. This tends to protect the video circuits from saturation by low-frequency extended clutter and cw jamming without serious loss of signal pulse shape.

The use of the optimum FTC adjustment and DBB, together with either IAVC or lin-log, will usually result in improved data for any given range. However, since the magnitude of all echoes depends strongly on range, it is likely that a good adjustment at one range will result in display saturation at shorter ranges and insufficient sensitivity at longer ranges. This can be corrected by employing Sensitivity Time Control (STC)—a circuit which provides an automatic control of the receiver sensitivity so that similar targets will be displayed with equal brightness, regardless of range.

It is sometimes possible by combining these circuits* to obtain a display in which the background of clutter approaches uniform grain and brightness, having the appearance of random system noise. When this is accomplished it represents the best that can be done by means of gross compensations. Beyond this point further improvements in clutter rejection must be based on the relative statistics of the echoes.

1.2. Clutter Suppression by Means of MTI

One respect in which clutter masses differ from actual targets is that they move at different relative speeds. In MTI and AMTI systems, the velocity characteristics conveyed by the doppler shift are used to distinguish the desired from the undesired returns. This is accomplished in principle by employing filter networks (or their equivalent), which annul the returns from fixed targets bearing no doppler shift while passing the frequency-shifted echoes from moving targets. It is, therefore, apparent that clutter elimination by these means is feasible to the extent that the energy spectrum returned by the clutter mass is distinguishable from that of the target. Actually, this is seldom the case. Either because of random fluctuations in scatter cross section, or because of random motions within the clutter mass,† the power spectrum occupies a continuous range of frequencies, and those components in the vicinity of the target frequencies are indistinguishable from target energy. Thus, depending on the width of the clutter spectrum in relation to the range of target velocities to be accepted, a residue of uncancelable clutter persists.

Various other factors contribute to the residual clutter. Briefly, these are "wind noise," caused by the unified motion of the clutter mass; "scanning noise," due to modulation of the clutter return by the scanning antenna; "platform-motion noise," occurring in airborne or shipboard radars and caused by the translation of the antenna; and "instability noise," which is due to imperfections in the radar system. These factors are discussed in more detail in Chap. 6.

With ideal adjustment, the proper combination of IAVC (or lin-log), STC, and MTI should produce a display on which the residual clutter is uniformly presented and of as low an intensity as is consistent with good moving-target visibility. In other words, the average clutter power residue should be kept small and constant, while the signal-to-residual-clutter power ratio should be as high as possible. To the extent that the MTI unit brings about an improvement in this signal-to-clutter ratio, it will contribute to the quality of the display.

1.3. Various Performance Criteria

The ultimate measure of system performance, i.e., the capability of the equipment to detect targets submerged in the clutter, does not depend on power considerations alone. A more detailed description of the output probability density function is required before the probability of detection for such targets can be determined precisely. The output power residue, which is the variance of the output voltage around its mean value, is but one factor characterizing this distribution. In principle, by computing a sufficient number of "higher moments" the distribution can be determined accurately. It is then possible to find the detection probability for targets of arbitrary intensity against backgrounds of various types of clutter and subsequently to calculate the "subclutter visibility factor," which is the ratio by which the signal power can be weaker than the clutter power and still be detected.

The probability of detection for normal (not MTI) receivers where the desired signals are embedded in thermal ("white") noise has been determined by Marcum[9] and others.[10] Since white noise is uncorrelated pulse to pulse, the corresponding analysis can be based on well-known techniques for integrating independent statistics. This is not the case for MTI receivers, because the clutter is definitely correlated from pulse to pulse (the cancellation of permanent echoes depends on this), and even if the clutter were not so correlated, independence would be destroyed by the pulse-combining action of the MTI circuits. Consequently, these results are not directly applicable to MTI systems.

The corresponding analysis for MTI is more involved. Two general classes of systems are presently soluble. They are

1. Coherent systems with postdetector integration, and
2. Noncoherent systems without postdetector integration.

In either case, the (single) nonlinear element must be a quadratic detector or mixer. The procedure is based on a rather general treatment by M. Kac and A. J. F. Siegert[11] for receivers with square-law detectors. Their work provides a method of obtaining the characteristic function for the first output probability density function for systems using arbitrarily specified predetector and postdetector filters. The method presents two major difficulties, however:

1. An integral equation must be solved to obtain the characteristic function explicitly, and
2. The characteristic function must then be inverted to get the output probability density function.

An adaptation of the method of Kac and Siegert is available[12] which directly provides the cumulants (a class of semi-invariants) for the output probability density function. These cumulants are then used in forming approximations to the density function via Gram-Charlier or Laguerre series, etc. Thus, the probability of detection and subsequently the subclutter visibility factor can, in principle, be computed.

The probability-of-detection calculations are tedious. First, the probability density functions for the system output must be determined with clutter alone and with signal plus clutter. (These distributions depend on the clutter correlation as well as the signal-to-clutter power ratio, and so constitute a two-parameter family of distributions.) Next, a threshold is established for the purpose of deciding whether or not a signal is present (output voltages exceeding this level indicate a signal present). The threshold level is then set so that false alarms are sufficiently infrequent. Finally, the probability of detection is computed by integrating the output density function for signal plus clutter from the threshold level to infinity. This is an extremely laborious process, and since relatively

*Discussions of the various anticlutter circuits will be found in Refs. 1, 2, 3, and 4, p. 123.
†These produce what will be called "internal-motion noise." See Refs. 5, 6, 7, and 8.

persuasive arguments can be given favoring simpler criteria,* the probability of detection will not be used as a performance criterion in this work.

Using the Neyman-Pearson criterion,† it can be shown that the "optimum receiver" for detecting a weak sinusoidal signal of unknown frequency and phase in the presence of correlated Gaussian noise consists of a *linear* process followed by a squaring device and then an integrator. Since this is essentially the problem of the MTI receiver, the following conclusions may be drawn:

1. The optimum MTI receiver is linear for weak signals up to the video detector.
2. The optimum video system for weak signals consists of a square-law detector followed by an integrator.

Lin-log and IAVC receivers essentially satisfy point (1) if both the moving target and the clutter signals are weak. For strong clutter and weak target signals, these systems will still be "linear" if the clutter fluctuations do not materially alter the i-f amplification factor from pulse to pulse. For strong, violently fluctuating clutter, the lin-log and IAVC systems should be less satisfactory than the ideal linear systems; but, of course, they should be preferred to the pseudolinear systems which saturate on strong clutter returns.‡

With regard to point (2), almost every detector behaves as a square-law detector for weak signals.[15][16] If the radar equipment provides for video integration, then the system performance can be estimated on the basis of the integrated output statistics. This will be done in Chap. 5. For receivers without provision for video integration, performance can be estimated on the basis of i-f output power alone. The particular criterion to be used in these cases is the system "gain," defined as the ratio of the output signal-to-clutter power ratio to the input signal-to-clutter power ratio.** This ratio is measured around the MTI cancellation unit and consequently indicates the improvement in the signal-to-clutter ratio brought about by and creditable to the MTI.

1.4. MTI Gain, Reference Gain, and Enhancement Factor

In the next chapters of this report, formulas will be derived for the gain, $G(v)$, as a function of target velocity, and it will be found that it can be expressed in factored

*At least for the purpose of deciding which of two competitive systems is preferable.

†The Neyman-Pearson criterion is a basis for testing statistical hypotheses based on joint minimization of the probabilities of error. In this case the hypotheses are (1) that the observed sample is due to clutter alone, and (2) that the sample is due to target plus clutter. The errors are (type I) a detection called when no target is present (a false alarm); and (type II) a detection not called when a target is present. The joint minimization involves finding the decision processes which "take up the slack" in the error probabilities—processes for which a reduction in the probability of one type of error necessarily forces an increase in the probability of the other type of error. For all detection problems of this type where the noise is due to natural causes (this excludes ECM), the optimal decision functions are realized by making a clear-cut decision (i.e., mixed strategies employing the toss of a coin, die, etc., are not warranted) between the hypotheses solely on the basis of the size of the processed sample. The optimal process is then determined by the structure of the problem.

‡Systems employing i-f limiting and phase detection (see, for example, Refs. 13 and 14) are not specifically treated in this report.

**These quantities are precisely defined in Sec. 1.10.1.

form as a product; i.e.,

$$G(v) = \overline{G} \cdot E(v), \tag{1.1}$$

where $\overline{G}$, a quantity to be called the "reference gain," is independent of target velocity, and $E(v)$, the "velocity enhancement factor," is independent of the clutter characteristics.

The quantity $\overline{G}$ is a convenient reference level above or below which the actual gain, $G(v)$, for a particular target velocity can be expressed. $\overline{G}$ depends only on the clutter characteristics and the particular type of cancellation mechanism employed. It is numerically equal to the average of $G(v)$ over all velocities, v, and so can be construed as the "expected" gain in the probabilistic sense that a target whose velocity is unknown is equally likely to be moving at all velocities. Alternatively, $\overline{G}$ can be interpreted as the ratio of clutter attenuation in the cancellation mechanism after the irrelevant latent amplifications (or losses) have been normalized. It is desirable for $\overline{G}$ to be as large as possible.

$E(v)$, on the other hand, being independent of the clutter characteristics, reflects the relative sensitivity of the system to targets of various speeds. For the conventional two-pulse-comparison, single-delay-line system, $E(v)$ has the form $1 - \cos 4\pi(v\tau/\lambda)$ and exhibits the well-known "blind speeds" when v is a multiple of $\lambda/2\tau$. By employing more delay lines and/or interpulse periods, or by using range-gated lumped-constant filter systems (which will be discussed in Chap. 3), the enhancement factor may be altered so as to remove these blind speeds (or "dead zones," as they are sometimes called), or it may be otherwise reshaped. See, for example, Fig. 1.1.

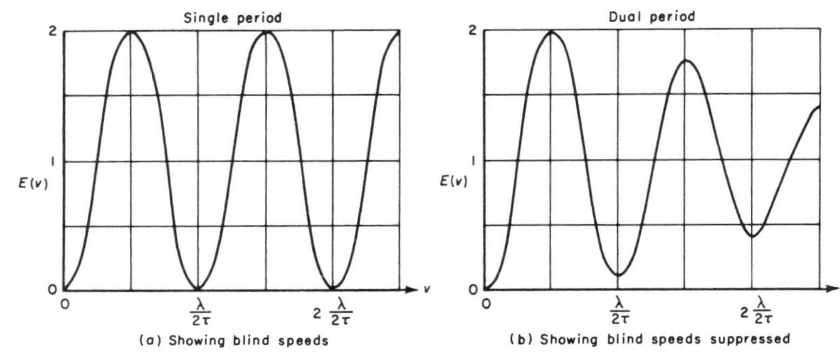

Fig. 1.1—Enhancement curves for delay-line systems

Since both $\overline{G}$ and $E(v)$ are affected by the type of cancellation mechanism employed, they cannot be manipulated independently.* Nevertheless, it is convenient to associate

*In some cases $E(v)$ can be materially altered without significantly changing $\overline{G}$. For example, a system using a single interpulse period can be changed into one having two periods by slightly

$E(v)$ with the pass-band characteristics for the moving targets and $\overline{G}$ with the clutter cancellation efficiency.

1.5. Specific Systems To Be Treated

Chapters 2 and 3 are concerned with determining the theoretical gain for a variety of MTI cancellation systems. The systems to be investigated are grouped as follows:

1.5.1. Finite-pulse-group Systems*

This group includes the conventional single-delay-line system, together with its generalizations to more than one delay line and/or interpulse period, as follows:

- Two-pulse comparison, single interpulse period using a single delay line;
- Two-pulse comparison, double interpulse period using one long and one short delay line;
- Three-or-more-pulse comparison, multiple interpulse period using many separate delay lines or fewer delay lines with feedback loops, etc.

The above systems will be considered in Chap. 2. Chapter 3 will deal with range-gated systems.

1.5.2. Range-gated Lumped-constant Filter Systems

These systems do not employ delay lines at all. They are designed on the basis of lumped-constant networks which exhibit the desired target velocity enhancement characteristics. Since these "optimal" filters are usually not range-preserving, the systems must be range gated into separate channels, one for each range resolution interval desired, and must each be provided with a separate filter. These systems allow somewhat more flexibility of design than do the delay-line systems, since the enhancement characteristics may be controlled (up to a point); and, if satisfactory range-gating circuitry can be developed, this group of systems should show great promise for future MTI. This group includes

- Single interpulse period,
- Double interpulse period,
- Three or more interpulse periods.

1.6. Coherence and Noncoherence

Any of the systems in Secs. 1.5.1 and 1.5.2 may be employed with radars operating on either the "coherent" or the "noncoherent" principle.

1.6.1. The Coherent Principle

Coherent reception is accomplished by mixing the i-f signals with a strong local i-f oscillator which has been forced by phase-locking techniques into phase and frequency

lengthening alternate pulse spacings. In this way some of the dead zones can be removed with no appreciable loss in $\overline{G}$.

*Many of these systems are discussed in Ref. 10.

agreement with the returns from fixed targets. The resultant of the mixing is called the "bipolar" video, which, for fixed targets, consists of a sequence of pulses of constant amplitude and polarity, and which, for moving targets, consists of a sequence of pulses whose amplitude and polarity alternate at the frequency of the doppler shift. The mixer is operated in the presence of a strong signal (the coho) and is, therefore, essentially a linear device. It needs no special attention in the analysis. This is not the case for noncoherent operation.

1.6.2. The Noncoherent Principle

Noncoherent detection is accomplished by beating the received i-f signals with themselves rather than with a coherent oscillator. The required doppler beats are produced between the moving-target signal and the clutter (which must be present for detection). These systems are usually used for AMTI where the precise ground speed is not available for coherent reference.

The production of the necessary doppler components in the noncoherent detector relies on certain essential nonlinearities, and therefore this type of detection cannot be analyzed by means of conventional linear circuit theory.

1.7. Effect of Nonlinearities

In addition to the nonlinearities essential to noncoherent detection, there will be other sources of signal distortion. These may be intentionally introduced, as is the case with the lin-log or IAVC process, or they may be unavoidable, as are vacuum-tube imperfections, etc. In order to ascertain the effects of these nonlinearities, a quadratic mixer is introduced into the analysis in Chap. 4, which deals specifically with noncoherent systems. The results, however, indicating only minor sensitivity to this type of distortion, suggest a rather general insensitivity to other *similar* nonlinearities.

1.8. Integration after Detection

Chapter 5 deals with the subject of video integration. The analysis is applicable to coherent radars in which the output of the MTI system is rectified (with a quadratic detector) and then integrated either with delay lines or narrow-band filters. A measure of integration efficiency based on the "Deflection Criterion" of Lawson and Uhlenbeck[2] is determined for correlated and uncorrelated clutter, with and without MTI. The outstanding result of this chapter is that the integration efficiency for correlated clutter, which usually decreases when i-f (coherent) integration is employed, does not decrease so rapidly with MTI. This result supports the contention that higher detection probabilities can be achieved for certain ground-based MTI radars by augmenting the "scope" and "eye" integration through the use of electronic integration, such as delay lines with feedback, etc.

1.9. Numerical Evaluation of Set Performance

Chapters 2 through 5 deal with the development of general formulas for computing the reference gain, enhancement factors, and integration efficiencies. In these formulas the fluctuation characteristics of the clutter appear implicitly in terms of the "autocorrelation function" and "power spectrum." Some of these quantities are determined explicitly in Chap. 6, which deals with the internal-motion, wind, scanning, and system-instability noises. In addition, there is a discussion of "cancellation ratio," its measurement and effect on system performance, etc. These quantities are brought together in Chap. 7, where specific performance estimates are given in graphical form for a collection of ground-based, shipboard, and airborne radars.

1.10. Autocorrelation Functions and Power Spectra

The power output from an MTI cancellation unit can be determined in either of two ways. One method is to measure the time average of the square of the output voltage; this can be accomplished with a power-registering device in the laboratory. Alternatively, one may theoretically compute the *expected* value of the power in any particular output pulse as governed by the probability distribution for the fluctuation statistics. The former measure, at least for stationary or "pulsed stationary" processes,* is independent of time; but the latter—the "ensemble" estimate of the pulse power—depends on time, for the estimate can be made during an interval between pulses, in which case the power is zero; or, if made at the time of pulse arrival, it may depend on the position of the antenna, etc.

The treatment of delay-line systems (see Chap. 2) will draw exclusively on the theory of the ensemble autocorrelation function. It is defined by

$$\psi(t_1, t_2) = \overline{E(t_1)E(t_2)}, \tag{1.2}$$

where $E(t)$ is the voltage time series applied to the cancellation unit, and the bar indicates the average with respect to the probability distribution. For pulsed series, $\psi(t_1, t_2)$ will be evaluated for t_1 and t_2 corresponding to pulse arrival epochs. In particular, if $t_1 = t_2$, then

$$\psi(t_1, t_2) = \overline{E(t_1)^2} \tag{1.3}$$

is the ensemble estimate of the power in the pulse occurring at time t_1. For pulsed stationary series this quantity is independent of which pulse epoch is involved. Furthermore, as long as t_1 and t_2 correspond to pulse epochs, then $\psi(t_1, t_2)$ will depend only on the time difference $\tau = t_2 - t_1$. One may, therefore, write

$$\psi(\tau) = \overline{E(t_1)E(t_1 + \tau)}. \tag{1.4}$$

The restriction that t_1 and t_2 correspond to pulse epochs may be lifted by interpreting $\psi(\tau)$ as the autocorrelation function for the underlying stationary process accounting for the clutter fluctuations. For this analysis, the underlying process includes internal-motion, wind, scanning, platform-motion, and system-instability noise. Some of these sources are

*A stationary time series is one for which the amplitude statistics are the same for all time. White thermionic noise is an example of an uncorrelated stationary series. The correlated clutter returns received with a nonscanning cw system constitute what is essentially a stationary time series. A "pulsed-stationary" series is a stationary time series that has been "gated" into pulses. Examples are gated "white noise" and the clutter returns from a fixed, nonscanning, pulsed radar.

less statistical than others; e.g., the effects of scanning and platform motion are precisely determined by the direction and position of the antenna with respect to the clutter. This fact is employed in "step-scanning" (see Sec. 6.4.2) to eliminate scanning noise. Similar attempts have been made to eliminate platform-motion noise (see Sec. 6.5.2). Ordinarily, the deterministic character of these noises is not utilized, and an uncanceled power residue similar to that caused by statistical decorrelation results. For this reason, scanning and platform-motion noise can be replaced with stationary noise having the same autocorrelation functions and included as a part of the underlying statistical source. This is done in Chap. 6, where the respective autocorrelation functions are discussed.

The analysis of range-gated filter systems in Chap. 3 draws on the theory of time-average autocorrelation functions and power spectra. The time-average autocorrelation function is defined by

$$\phi(\tau) = \lim_{T\to\infty} \frac{1}{T} \int_0^T E(t)E(t + \tau)\, dt, \tag{1.5}$$

where the limit is presumed to exist. If $E(t)$ represents the underlying stationary process, then $\phi(\tau) = \psi(\tau)$—this is a well-known property of stationary time series* which permits the interchange of time and ensemble averaging. If, however, $E(t)$ is pulsed-stationary, one must put $\phi(\tau) = \theta(\tau)\psi(\tau)$, where $\theta(\tau)$ is a pulsed function of τ which depends on the duty cycle, uniformity of pulse spacings, etc.

A fundamental property of independent statistics is that "the mean of the product equals the product of the means." This fact will be used frequently. In particular, it can be used to determine the function $\theta(\tau)$. To this end let the received pulsed-stationary series be represented by

$$E_{ps}(t) = E_s(t)P(t), \tag{1.6}$$

where $E_s(t)$ is the underlying stationary process and $P(t)$ is the transmitted pulse envelope.† The time-average autocorrelation function of $E_{ps}(\tau)$ is the mean (time-average) value of the product $[E_s(t)E_s(t + \tau)] \times [P(t)P(t + \tau)]$ where each of the bracketed terms is statistically independent. Accordingly, it can be shown that

$$\phi_{ps}(\tau) = \phi_s(\tau) \times \phi_p(\tau). \tag{1.7}$$

Since $\phi_s(\tau) = \psi(\tau)$, and since

$$\phi(\tau) = \psi(\tau)\theta(\tau), \tag{1.8}$$

it follows that

$$\theta(\tau) = \phi_p(\tau) = \lim_{T\to\infty} \frac{1}{T} \int_0^T P(t)P(t + \tau)\, dt; \tag{1.9}$$

i.e., $\theta(\tau)$ is the autocorrelation function for the original pulse envelope.

*Or, more properly, ergodic time series.

†It is assumed here that the transmitter is perfectly stable.

The functions $\phi(\tau)$, $\psi(\tau)$, and $\theta(\tau)$ reflect power levels which, being irrelevant to considerations of system gain, integration efficiency, etc., may be eliminated by normalization. This is done by dividing the respective functions by their value for $\tau = 0$. Thus,

$$\pi(\tau) = \frac{\phi(\tau)}{\phi(0)}, \tag{1.10}$$

$$\rho(\tau) = \frac{\psi(\tau)}{\psi(0)}, \tag{1.11}$$

$$\sigma(\tau) = \frac{\theta(\tau)}{\theta(0)}. \tag{1.12}$$

Equation (1.8) now takes the form

$$\pi(\tau) = \rho(\tau)\sigma(\tau). \tag{1.13}$$

(Note that $\pi(0) = \rho(0) = \sigma(0) = 1$.) It follows from their definitions that these autocorrelation functions are real-valued symmetric functions of τ. Accordingly, their Fourier transforms are real-valued symmetric functions of frequency.

To call attention to the fact that $\rho(\tau)$ is the normalized autocorrelation function of the underlying clutter, it will be indicated with the subscript, $\rho_c(\tau)$. Its Fourier transform, denoted $W_c(f)$, is called the unit-power spectrum for the clutter. Similarly, the Fourier transform of $\sigma(\tau)$ is the unit-power spectrum for the transmitted pulse chain. These spectra combine by convolution to form the unit-power spectrum for the received fluctuating pulses. This is illustrated in Fig. 1.2.

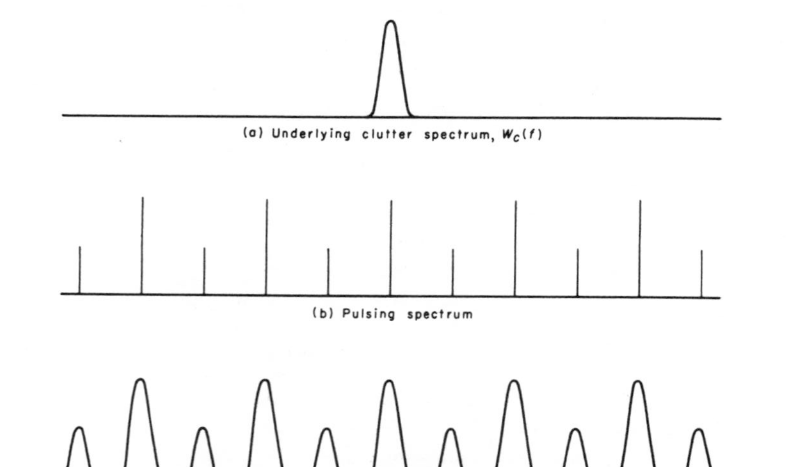

Fig. 1.2—Comparison of cw and pulsed clutter spectra

1.11. General Definition of System Gain

The uncanceled power residue at the output terminals of the cancellation unit depends on the type of cancellation system used, the input signal and clutter power levels, and their respective correlation functions or power spectra. If a moving-target signal is present, this residue will be denoted $R(C, S, v)$, where C is the clutter power present at the input terminals of the cancellation unit, S is the input power for the target signal, and v is the target velocity. In the absence of a moving-target signal the residue will be designated $R(C, 0, —)$. The incremental increase in output power residue that can be credited to the presence of the moving target is

$$R(C, S, v) - R(C, 0, —), \tag{1.14}$$

and it will be called the output *signal* power. Since the output *clutter* power is $R(C, 0, —)$, the output signal-to-clutter power ratio, designated X_o, is given by the ratio

$$X_o = \frac{R(C, S, v) - R(C, 0, —)}{R(C, 0, —)}. \tag{1.15}$$

The input signal-to-clutter power ratio, designated X_i, is simply S/C. In terms of these quantities, the gain, $G(v)$, is defined by

$$G(v) = \frac{X_o}{X_i} = \frac{C}{S} \cdot \frac{R(C, S, v) - R(C, 0, —)}{R(C, 0, —)}. \tag{1.16}$$

This quantity is evaluated in Chap. 2 for a variety of specific delay-line systems. In addition, a general formula is developed for N-pulse comparison which on passage to the limit ($N \to \infty$) provides the basic gain formula for the range-gated filter systems of Chap. 3.

Chapters 4 and 5 are in the nature of appendixes to Chaps. 2 and 3 in that they provide modifications to account for noncoherent operation and video integration.

CHAPTER 2

FINITE-PULSE-GROUP (DELAY-LINE) SYSTEMS

2.1. Introduction

The finite-pulse-group systems to be treated in this section belong to a class of techniques which, employing continuous delay memory, accomplish the MTI filtering through the comparison of a finite number of successively received pulses. A sufficient number of delay lines are provided to store each pulse upon arrival until the last pulse of the group to be compared is received. At this time the delay lines are "read out," and the stored pulses are weighted and combined with the last pulse received to yield the system output. Thus, if the respective pulse amplitudes are denoted $p_1, p_2, \cdots, p_N$, then the output pulse amplitude is $\alpha_1 p_1 + \alpha_2 p_2 + \cdots + \alpha_N p_N$, where the various α_j are the weighting factors selected for the particular MTI characteristics desired. This output pulse occurs at the time of arrival of the last pulse in the group, p_N. At this time the system being cleared is ready to go into the storage phase with the next N pulses, $p_{N+1}, p_{N+2}, \cdots, p_{2N}$, yielding another output pulse when p_{2N} is received. Depending on the regularity of the interpulse periods, this system may or may not be capable of furnishing useful output pulses in the intervals between the terminal pulses, p_N, p_{2N}, p_{3N}, etc.

If the transmitted pulses are uniformly spaced, any reasonable delay-line arrangement will match, in regard to delay times, any of the successive overlapping pulse groups $(p_1, \cdots, p_N)$, $(p_2, \cdots, p_{N+1})$, $(p_3, \cdots, p_{N+2})$, etc., as well as any other, with the result that a useful output pulse is available at the time of arrival of each pulse transmitted. If, however, the pulses are not uniformly spaced, this will not be the case; and the system output, being improperly formed for MTI in certain of the range frames, will have to be discarded.

In principle, this can be remedied by employing N different delay-line systems, each being specifically matched for a group of the type $P_j, P_{j+1}, \cdots, p_{N+j-1}$. In this way the successive range frames can be displayed without interruption, the $N - 1 + j$th frame being furnished by the jth system. Because of the large number of delay lines that would be required, however, such a procedure is impractical if N is greater than 2 or 3. This, coupled with the fact that essentially all information is eventually presented even with the interrupted system, renders this procedure unnecessarily extravagant, with the following possible qualification:

For small N, the generalized N-system discussed above can be used to eliminate the blind speeds associated with uniform pulse spacing. In this regard, use is made of the fact that each of the N different delay-line systems can be made to exhibit a different blind speed associated with the particular set of interpulse spacings in its pulse group, thereby ensuring display from some of the systems when others are dead. For example, let N be 2 and let the pulses be transmitted with alternate spacings τ_1 and τ_2. Let a delay line of length τ_1 be used in one system to delay and subtract the pulses spaced by τ_1, and let a delay line of length τ_2 be used in the other to delay and subtract the alternate pulse pairs spaced by τ_2. The output for the first system, if displayed during the proper range frames, will exhibit blind speeds at multiples of $v_1 = \lambda/2\tau_1$. Similarly, the output from the second system, displayed on alternate range frames, will have blind speeds divisible by $v_2 = \lambda/2\tau_2$. By choosing τ_1 and τ_2 different but nearly the same, all blind speeds for the composite display will be eliminated for velocities ranging all the way out to the least common multiple of v_1 and v_2. Clearly, such a system has utility in situations where the blind speeds are objectionable. The practicality of the system is further enhanced by the fact that only one "long" delay line is needed, the function of the other line being performed by the same line to which is alternately adjoined a "short" differential line of length $\tau_1 - \tau_2$. This system will be analyzed along with the elementary single-delay-line, uniform-interpulse-period systems and other more complicated multiple-delay-line systems in later paragraphs of this chapter.

2.2. Single-delay-line, Two-pulse Comparison

The MTI kits for many of the present-day radars incorporate a single delay line to provide storage so that successive pairs of bipolar video pulses may be subtracted. The theory of operation is that the returns from fixed targets, being of constant amplitude from pulse to pulse, will successively cancel, whereas the doppler-modulated pulses from moving targets will not. Actually, since the clutter mass does not represent a truly fixed target, and since scanning, platform-motion, and system-instability noise may be present, power output residues exist.

Let the pulse chain at the input to the cancellation unit be designated $E_i(t)$ volts. The output voltage is then given by

$$E_o(t) = E_i(t) - E_i(t - \tau). \tag{2.1}$$

The output power residue is given by

$$\begin{aligned}\overline{E_o(t)^2} &= \overline{[E_i(t) - E_i(t - \tau)]^2} \\ &= 2[\psi_i(0) - \psi_i(\tau)],\end{aligned} \tag{2.2}$$

where the bar indicates the ensemble average. Actually $\overline{E_o(t)^2}$ depends on time, being zero during the interval between pulses. Equation (2.2) gives the correct value of $\overline{E_o(t)^2}$ only at the time of pulse arrival.

The input pulse chain is now expressed as the sum of a target-signal voltage, $S(t)$, and a clutter voltage, $C(t)$. These are each pulse chains representing the respective returns from the same range interval. Thus,

$$E_i(t) = S(t) + C(t). \tag{2.3}$$

The input autocorrelation function, $\psi_i(\tau)$, is thus expressed

$$\begin{aligned}\psi_i(\tau) &= \overline{E_i(t)E\,(t + \tau)} \\ &= \overline{[S(t) + C(t)][S(t + \tau) + C(t + \tau)]},\end{aligned} \tag{2.4}$$

which, by virtue of the statistical independence of $S(t)$ and $C(t)$, reduces to

$$\begin{aligned}\psi_i(\tau) &= \overline{S(t)S(t + \tau)} + \overline{C(t)C(t + \tau)} \\ &= \psi_s(\tau) + \psi_c(\tau).\end{aligned} \tag{2.5}$$

To evaluate the autocorrelation function, $\psi_s(\tau)$, for the moving-target return, only the beat signals between the target i-f signal and the coherent oscillator need be considered. These may be expressed as a pulse chain modulated by the function

$$S(t) = \sqrt{2S}\, f(t) \cos(\omega_d t + \phi), \tag{2.6}$$

where S is the signal *power* returned. $f(t)$ accounts for fluctuations of the echoing high lights of the target about the mean target velocity as well as the fluctuations in illumination due to scanning. The cosine term describes the doppler modulation. The frequency of this "butterfly" modulation, ω_d, is given by the formula

$$\omega_d = 4\pi \frac{v}{\lambda}, \tag{2.7}$$

where v is the target mean velocity and λ is the radar wavelength.

The phase angle, ϕ, in Eq. (2.6) is indeterminate, and hence $S(t)$ constitutes an ensemble of target returns, the specific members of which are identified by ϕ. The autocorrelation function, $\psi_s(\tau)$, for this ensemble is therefore

$$\begin{aligned}\psi_s(\tau) &= 2S\overline{f(t)f(t + \tau)}\;\overline{\cos[\omega_d t + \phi]\cos[\omega_d(t + \tau) + \phi]} \\ &= S\rho_f(\tau)\cos\omega_d\tau.\end{aligned} \tag{2.8}$$

The function $\rho_f(\tau)$ is the autocorrelation faction for the target fluctuation. It will be discussed in Sec. 2.2.1.

The autocorrelation function, $\psi_c(\tau)$, for the clutter return is simply

$$\psi_c(\tau) = C\rho_c(\tau), \tag{2.9}$$

where C is the clutter power and $\rho_c(\tau)$ is the normalized autocorrelation function.

The autocorrelation function $\psi_i(\tau)$ for the input pulse chain can now be expressed as

$$\psi_i(\tau) = S\rho_f(\tau)\cos\omega_d\tau + C\rho_c(\tau), \tag{2.10}$$

and the output power residue is

$$\begin{aligned}R(C, S, v) &= \overline{E_o(t)^2} \\ &= 2\{S[1 - \rho_f(\tau)\cos\omega_d\tau] + C[1 - \rho_c(\tau)]\}.\end{aligned} \tag{2.11}$$

Consequently, the gain is given by (see Sec. 1.11)

$$G(v) = \frac{1 - \rho_f(\tau) \cos \omega_d \tau}{1 - \rho_c(\tau)}. \tag{2.12}$$

This is factored into

$$G(v) = \overline{G}E(v), \tag{2.13}$$

where

$$\overline{G} = \frac{1}{1 - \rho_c(\tau)} \tag{2.14}$$

is the reference gain, and

$$E(v) = 1 - \rho_f(\tau) \cos \omega_d \tau$$
$$= 1 - \rho_f(\tau) \cos 4\pi \frac{v\tau}{\lambda} \tag{2.15}$$

is the velocity enhancement function.

2.2.1. Effect of Target Fluctuation

If $\rho_f(\tau) = 1$—i.e., if there were no system or scanning noises, and if the target were a perfectly steady reflector—$E(v)$ would exhibit the familiar blind speeds. However, because of propeller modulation, vibration of reflecting surfaces, scanning, etc., the return will not generally be steady, and $\rho_f(\tau)$ will be somewhat less than unity. Consequently, the dead zones will not be complete and some visibility will be possible at these speeds, depending on the severity of the fluctuation (see Fig. 2.1). However, this phenomenon is of only secondary importance as far as delay-line system design is concerned, and, since the reference gain is totally unaffected by it, $\rho_f(\tau)$ will be neglected in the subsequent development in this chapter.

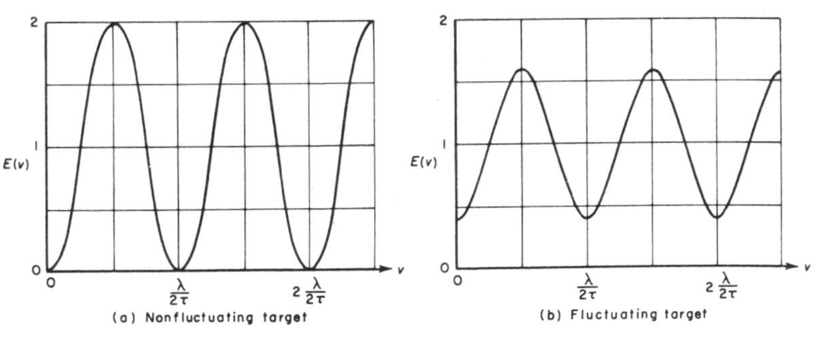

Fig. 2.1—Effect of target fluctuation on single-delay-line enhancement

2.3. Two-pulse-comparison, Dual-period System

Since target fluctuation cannot be depended on to render all moving targets visible, it is of interest to consider a process which will remove the nulls in the desired pass-band. This can be accomplished by altering the successive interpulse periods before transmission. By means of a short "differential" delay line in the receiver, which is switched in and out of operation on alternate range frames, a uniform pulse spacing is restored. The pulses are then applied to the single-delay-line cancellation unit previously discussed (see Sec. 2.2).

Let the transmitted interpulse spacings be denoted τ_1 and τ_2, and consider any instant when two pulses spaced by τ_1 are subtracted. The residue at this time is given by

$$\overline{E_o^1(t)^2} = 2\{S[1 - \cos \omega_d \tau_1] + C[1 - \rho_c(\tau_1)]\}. \tag{2.16}$$

Similarly, at any instant of comparison between pulses spaced by τ_2 the residue will be

$$\overline{E_o^2(t)^2} = 2\{S[1 - \cos \omega_d \tau_2] + C[1 - \rho_c(\tau_2)]\}. \tag{2.17}$$

These two energies are alternately sampled and presented to the radar display, and the average of the two will be used for computation. That is,

$$\overline{E_o(t)^2} = 2\left\{S\left[1 - \frac{1}{2}(\cos \omega_d \tau_1 + \cos \omega_d \tau_2)\right] + C\left[1 - \frac{1}{2}[\rho_c(\tau_1) + \rho_c(\tau_2)]\right]\right\}. \tag{2.18}$$

From this, the enhancement function is found to be

$$E(v) = 1 - \frac{1}{2}[\cos \omega_d \tau_1 + \cos \omega_d \tau_2]; \tag{2.19}$$

and the reference gain,

$$\overline{G} = \frac{1}{1 - \frac{1}{2}[\rho_c(\tau_1) + \rho_c(\tau_2)]}. \tag{2.20}$$

Thus, if the pulse spacings are only slightly different, the reference gain will be slightly higher than would have been the case had the longest spacing been used uniformly, and will be slightly lower than if the shortest spacing had been used. Actually, in order to maintain the highest possible interpulse correlation, the shortest interpulse spacing consistent with the maximum unambiguous range requirement should be used. This means that slight staggering of the pulse spacings will result in a small net loss in gain.

The velocity enhancement curve, shown in Fig. 2.2, is the average of two periodic functions, one having blind speeds at multiples of $\lambda/2\tau_1$ and the other at multiples of $\lambda/2\tau_2$. Thus, unless its velocity is a multiple of each, the target will be visible.

2.4. Three-or-more-pulse Comparison, Multiple Periods

The two preceding cases (Secs. 2.2 and 2.3) form the basis for a generalized system which will now be described. Let the radar transmit a sequence of N pulses at

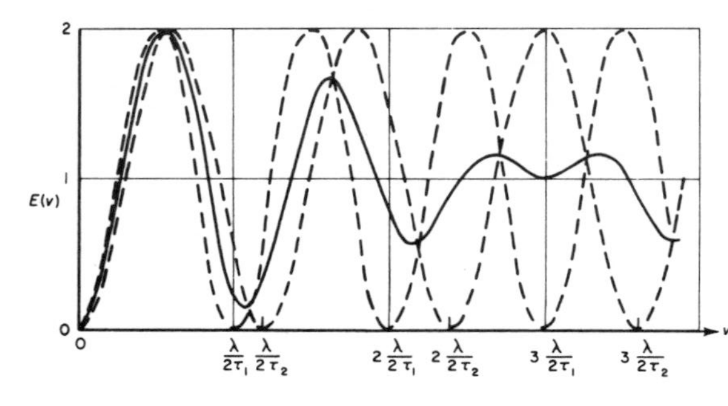

Fig. 2.2—Enhancement function for a dual-period system (showing $E(v)$ as the average of two single-period functions)

times $t_1, t_2, \cdots, t_N$, and thereafter repeat with pulse spacings $t_{N+1} - t_N = t_2 - t_1$, $t_{N+2} - t_{N+1} = t_3 - t_2$, and so on. The transmission will consist of an infinite chain of pulses with a fundamental repetition period $t_{j+N} - t_j$, and various shorter interpulse periods $t_{j+1} - t_j$. The receiver is equipped with a bank of N cancellation systems designated $\Delta_1, \Delta_2, \cdots, \Delta_N$, each of which, operating independently of the others, consists of enough delay lines (with or without feedback), amplifiers, etc., to perform N-pulse comparison on a separate subset of N successively received pulses. The output of the lth system, Δ_l, occurs only at the time of arrival of the lth pulse and at multiples of $t_N - t_1$ thereafter. This output consists of a weighted combination of the lth pulse return and the $N - 1$ preceding returns. Thus, the output voltage can be expressed as

$$E_o(t_l) = \sum_{j=0}^{N-1} a_j^l E_i(t_{l-j}), \tag{2.21}$$

where in this expression the t_l represents the time of arrival rather than transmission of the lth pulse, and the a_j^l are the various weight factors associated with Δ_l.

The output of the various systems will be presented to the display consecutively, so that Eq. (2.21) will indicate the voltage presented to the display at time t_l. The display will merge the N presentations, so that the energy displayed will be the sum or average of the $\overline{E_0(t_l)^2}$ over the N values of l.

The output power residue for the lth system is easily found to be

$$\overline{E_o(t_l)^2} = 2\Big\{S \sum_{j=0}^{N-1} \sum_{k=0}^{N-1} a_j^l a_k^l \cos \omega_d (t_{l-j} - t_{l-k}) + C \sum_{j=0}^{N-1} \sum_{k=0}^{N-1} a_j^l a_k^l \rho_c (t_{l-j} - t_{l-k})\Big\}. \tag{2.22}$$

This is summed (or averaged) over l and converted into gain. The resulting formula for the reference gain is

$$\overline{G} = \frac{\sum_{l=1}^{N} \sum_{j=0}^{N-1} (a_j^l)^2}{\sum_{l=1}^{N} \sum_{j=0}^{N-1} \sum_{k=0}^{N-1} a_j^l a_k^l \rho_c (t_{l-j} - t_{l-k})}, \tag{2.23}$$

and, for the enhancement function,

$$E(\omega_d) = \frac{\sum_{l=1}^{N} \sum_{j=0}^{N-1} \sum_{k=0}^{N-1} a_j^l a_k^l \cos [\omega_d (t_{l-j} - t_{l-k})]}{\sum_{l=1}^{N} \sum_{j=0}^{N-1} (a_j^l)^2}. \tag{2.24}$$

These formulas can perhaps be best understood by studying them in parts, i.e., by considering separately the contributions from each cancellation system, Δ_l. The combined performance will be an appropriately normalized average for them all. To this end, the single N-pulse cancellation unit system will be considered next.

2.4.1. Single *N*-pulse Cancellation System, Multiple Periods

Almost every practical system is included under the category of the single N-pulse system (i.e., only one Δ_l). In this case the gain formula is

$$G(v) = \frac{\sum_{j=0}^{N-1} \sum_{k=0}^{N-1} a_j a_k \cos [\omega_d (t_j - t_k)]}{\sum_{j=0}^{N-1} \sum_{k=0}^{N-1} a_j a_k \rho_c (t_j - t_k)}, \tag{2.25}$$

where $t_j - t_k$ represents the delay between the jth and the kth pulses.

The reference gain in this case is given by

$$\overline{G} = \frac{\sum_{j=0}^{N-1} a_j^2}{\sum_{j=0}^{N-1} \sum_{k=0}^{N-1} a_j a_k \rho_c (t_j - t_k)}. \tag{2.26}$$

The single-delay-line, two-pulse-comparison system discussed earlier (Sec. 2.2) is the special case obtained by setting $N = 2$ and $a_1 = -a_0$.*

The sum $\sum_j a_j^2$ is called the latent gain, G_L, for the system. It acts as a normalizing factor so that the characteristic amplification or loss for the comparison unit will not appear in the reference gain. The latent gain is equal to the power gain of the system for white noise. This can be seen as follows: Let the input noise be denoted $N_i(t)$, voltage; then the output noise, $N_o(t)$, is

*It will be shown later that this choice of weight factors is the best possible choice for the two-pulse, single-period system in the sense of maximizing the reference gain.

$$N_o(t) = \sum_j a_j N_i(t - t_j), \tag{2.27}$$

and the output noise power is

$$\overline{N_o(t)^2} = \sum_j \sum_k a_j a_k \overline{N_i(t - t_j) N_i(t - t_k)}$$
$$= \overline{N_i(t)^2} \sum_j a_j^2. \tag{2.28}$$

Hence,

$$\frac{\overline{N_o(t)^2}}{\overline{N_i(t)^2}} = G_L. \tag{2.29}$$

2.4.1.1. Effect of Changing the Interpulse Spacing

It is of interest to examine the dependence of this reference gain on the interpulse spacing, τ. For reasons of bandwidth limitations it will be presumed that $\rho_c(\tau)$ behaves quadratically with τ near the origin. Consequently, $\rho_c(\tau)$ is written

$$\rho_c(\tau) = 1 - \alpha\tau^2 + \beta|\tau|^3 + \gamma\tau^4 + \cdots, \tag{2.30}$$

where the coefficients α, β, γ, etc., depend on the clutter characteristics, i-f bandwidth, etc. On introducing this formula for $\rho_c(\tau)$ into Eq. (2.14) one obtains

$$\overline{G} = \frac{1}{\alpha\tau^2 - \beta|\tau|^3 - \gamma\tau^4 - \cdots} \approx \frac{1}{\alpha\tau^2} \quad \text{for small } \tau. \tag{2.31}$$

This indicates that for two-pulse comparison the reference gain varies as the square of $1/\tau$ (i.e., as the square of the prf).

For three-pulse comparison, the weight factors are chosen* to satisfy

$$a_0 : a_1 : a_2 = (t_2 - t_1) : (t_0 - t_2) : (t_1 - t_0)$$
$$= \mu : -(1 + \mu) : 1, \tag{2.32}$$

where $t_1 - t_0$ is designated by τ and $t_2 - t_1$ by $\mu\tau$. The reference gain becomes

$$\overline{G} = \frac{1 + \mu + \mu^2}{1 + \mu + \mu^2 - \mu(1 + \mu)\rho_c(\tau) - (1 + \mu)\rho_c(\mu\tau) + \mu\rho_c[(1 + \mu)\tau]}. \tag{2.33}$$

Substituting the Taylor series, Eq. (2.30), for $\rho_c(\tau)$ into this formula gives

$$\overline{G} = \frac{1 + \mu + \mu^2}{2\mu^2(1 + \mu)\beta|\tau|^3 + 3\mu^2(1 + \mu)^2\gamma\tau^4 + \cdots}, \tag{2.34}$$

*This choice of weight factors will be shown later to be very nearly the best possible in the sense of maximizing $\overline{G}$ for three-pulse comparison.

which indicates that the reference gain rises as the cube of $1/\tau$ if $\beta \neq 0$, and rises as the fourth power if $\beta = 0$.*

The choice of weight factors in the last example was made in order to annul the coefficient of τ^2 in the denominator of $\overline{G}$. This is not necessarily the best possible choice, as will be shown.

2.4.1.2. Best Possible Choice of Weight Factors

The expression for the reference gain (Eq. 2.25) for the N-pulse-comparison, single-cancellation-unit system is the reciprocal of a normalized quadratic form. Its maximum value is therefore assumed when the weight numbers $(a_0, a_1, \cdots, a_{N-1})$ form an eigenvector for the matrix $\rho_c(t_j - t_k)$, and specifically when that eigenvector corresponds to the smallest eigenvalue.

For the two-pulse-comparison system, the correlation matrix possesses two eigenvectors, (a, a) and $(a, -a)$, which are independent of the nature of the clutter. The former, (a, a), corresponds to the larger eigenvalue $1 + \rho_c(\tau)$ and leads to the minimum value of $\overline{G}$. This corresponds to the addition of adjacent pulses, and hence the returns from fixed targets are enhanced rather than canceled. The latter, $(a, -a)$, leads to the smaller eigenvalue $1 - \rho_c(\tau)$, and this maximizes the gain.

For three-pulse comparison, the correlation matrix possesses three eigenvectors which depend on the relative correlations. These eigenvectors will not be explicitly determined, for it can be shown from spectral considerations that the maximum gain will be only slightly greater than that given by Eq. (2.33).

As was stated earlier (see Sec. 2.4.1.1), the weight functions used (Eq. 2.32) were selected to annul the term in τ^2 in the reference-gain formula (Eq. 2.33). The power spectrum, $E(f)$, for the cancellation unit (i.e., the enhancement function) consequently vanishes at $f = 0$ and departs from the origin as f^4. Now, depending on the frequency spread of the clutter, a slight improvement may result from relaxing the requirement that $E(0) = 0$ and, instead, establishing a pair of nulls situated symmetrically about the origin and well within the clutter band. In this way, the power spectrum $E(f)$, being small on a larger interval of frequencies, may accept a smaller total clutter energy. The best adjustment of this kind will, of course, correspond to the eigenvector combination of weights.† It will be different for different types of clutter.

Figure 2.3 is an illustration of the particular case of three-pulse comparison with equal pulse spacings. The clutter power spectrum is Gaussian,

$$W_c(f) = \frac{1}{\sqrt{2\pi\sigma^2}} e^{-f^2/2\sigma^2}, \tag{2.35}$$

*In Chap. 6 it will be shown that, for an ideal radar system (no instability noise), the cubic factor, β, will vanish unless platform-motion noise is present. Consequently, for ground-based radars, the fourth-power law is more appropriate.

†For a similar development of these ideas, see Ref. 17—the Eleventh Interim Report.

with standard frequency deviation, σ. Curve (a) is the near-optimum enhancement curve obtained with the "binomial" combination

$$a_0 : a_1 : a_2 = 1 : -2 : 1 . \tag{2.36}$$

Curve (b) is the best possible enhancement curve. It corresponds to the weight numbers

$$a_0 : a_1 : a_2 = 1 : -2\left(\frac{2\alpha + \beta + \sqrt{\beta^2 + 8\alpha^2}}{4\alpha - \beta + \sqrt{\beta^2 + 8\alpha^2}}\right) : 1 , \tag{2.37}$$

where $\alpha = \rho_c(\tau)$ and $\beta = \rho_c(2\tau)$. From this last equation it is apparent that, if $\alpha = \beta$, then the best possible choice of the a_j reduces to the binomial choice, Eq. (2.36).

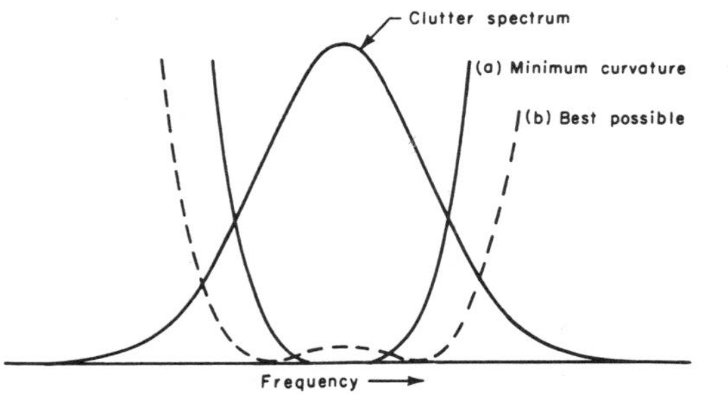

Fig. 2.3—Comparison of weight-factor selections (three-pulse-comparison, dual-period system)

Case $\alpha = \beta$, or $\rho_c(\tau) = \rho_c(2\tau)$, corresponds to the superposition of a truly fixed target echo upon white noise, or, equivalently, the combination of a fixed ground target plus a moving target whose velocity is equally likely to be any velocity between two dead zones. This result generalizes: for any fixed number of delay lines, the maximum reference gain against completely correlated clutter with white noise is achieved with the binomial weights

$$a_j = (-)^j \binom{N}{j} . \tag{2.38}$$

A comparison of the reference gains for the optimum and the binomial weights for three-pulse comparison and Gaussian clutter appears in Fig. 2.4. It can be seen that the difference is less than 2 db for all values of the product $\sigma\tau$.

2.4.1.3. Enhancement Functions

Turning now to the general formula, Eq. (2.24), the enhancement function for the

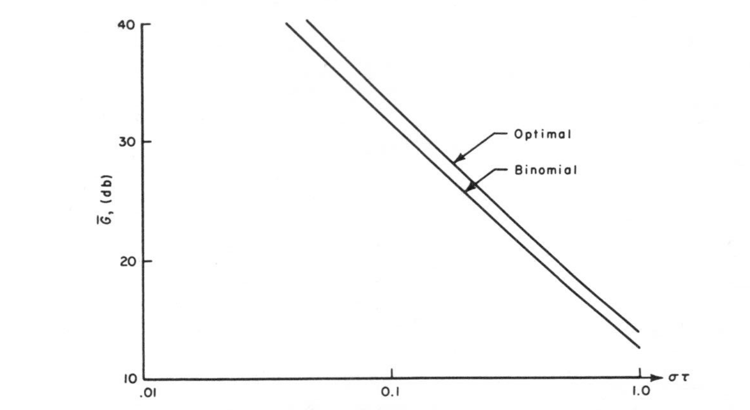

Fig. 2.4—Comparison of optimal and binomial weights (three-pulse-comparison, equal-interpulse-period system)

single N-pulse cancellation system is

$$E(\omega_d) = \frac{\sum_{j=0}^{N-1} \sum_{k=0}^{N-1} a_j a_k \cos \omega_d (t_j - t_k)}{\sum_{j=0}^{N-1} (a_j)^2} , \tag{2.39}$$

where $t_j - t_k$ represents the delay between the jth and the kth pulse.

For two-pulse comparison, this reduces to

$$E(\omega_d) = 1 - \cos \omega_d \tau , \tag{2.40}$$

which exhibits the dead zones for f equal to multiples of $1/\tau$.

For three-pulse comparison, the enhancement function is

$$E(\omega_d) = \frac{1 + \mu + \mu^2 - \mu(1 + \mu) \cos \omega_d \tau - (1 + \mu) \cos \omega_d \mu \tau + \mu \cos \omega_d (1 + \mu)\tau}{1 + \mu + \mu^2} , \tag{2.41}$$

where μ and τ are as defined in Sec. 2.4.1.1. If μ is set equal to unity (i.e., if the pulse spacings are made equal), this formula reduces to

$$E(\omega_d) = 1 - \frac{4}{3} \cos \omega_d \tau + \frac{1}{3} \cos 2\omega_d \tau . \tag{2.42}$$

This system is equivalent to two single-delay-line systems (Sec. 2.2) in cascade, and the dead zones at multiples of $1/\tau$ cycles are still present. However, if μ is made slightly greater than unity, complete dead zones do not occur until frequencies are reached corresponding to multiples of the least common multiple of $1/\tau$ and $1/\mu\tau$. In this way few or many of the dead zones may be removed. It should be stressed, however, that an attempt to remove a great many dead zones by choosing $\mu - 1$ small and perhaps irrational will be met with more severe oscillations in the pass-band of $E(v)$, with the first dip which occurs between $1/\tau$ and $1/\mu\tau$ doppler cycles becoming deeper and deeper and approaching a complete null with $\mu = 1$. Thus, the choice of μ involves a compromise between the frequency with which the complete nulls occur on the one hand and the depth of the incomplete dips on the other. This is illustrated in Fig. 2.5, where the velocity enhancement curves corresponding to various ratios for interpulse spacing are shown. (The numerical values of the enhancement function are plotted in decibels, 0 db corresponding to $E(v) = 1$.)

2.4.2. Multiple N-pulse Cancellation Systems

The more complex arrangements which were discussed in the introduction to this chapter, i.e., those employing more than one cancellation system, Δ_l, can now be evaluated separately in terms of the performance of each system. Furthermore, if the output data are merged onto a single display, then the over-all enhancement function will be a suitably weighted average of the separate enhancement functions, and the over-all reference gain will be the reciprocal of the weighted average of the reciprocals of the separate reference gains. In this case the weights are determined by the various proportions with which the data are mixed.

2.5. Reference Gain for Gaussianly Correlated Clutter

When the blind speeds are to be suppressed by using multiple interpulse periods, it will usually be desirable to have these periods only slightly different. Under these conditions the reference gain will be essentially that of the equal-period system, with period τ equal to the average. That is,

$$\overline{G} \approx \frac{\sum_{j=1}^{N} a_j^2}{\sum_{j=1}^{N}\sum_{k=1}^{N} a_j a_k \rho_c[(j-k)\tau]}. \tag{2.43}$$

For purposes of evaluation the autocorrelation function will be presumed Gaussian, i.e.,

$$\rho_c(x) = \exp\left[-(2\pi)^2\frac{\sigma^2 x^2}{2}\right], \tag{2.44}$$

where the parameter σ measures the decorrelation. Specifically, σ is equal to the standard frequency deviation of the unit-power spectrum for the clutter.

Using the argument of Sec. 2.4.1.2 that the best possible choice of weight numbers, a_j, will not produce appreciably higher gain than will the binomial combination, the latter will be used for computation:

$$a_j = (-)^j\binom{N}{j}. \tag{2.45}$$

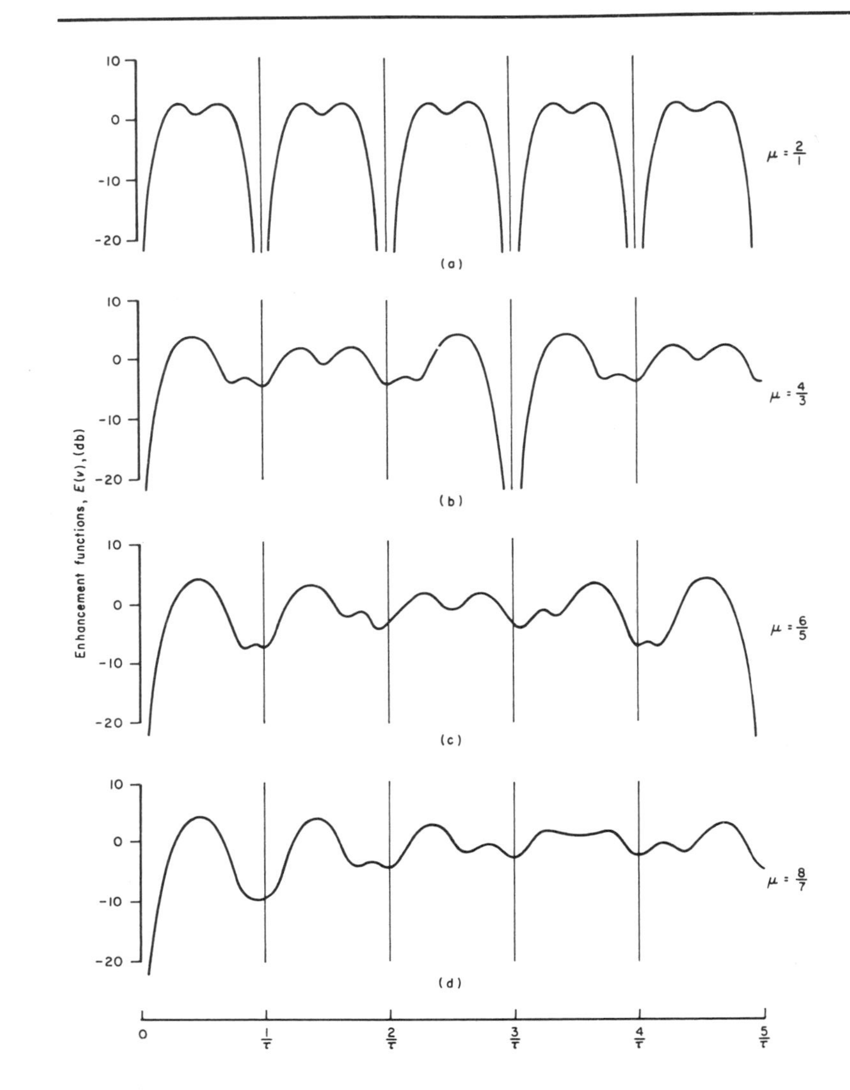

Fig. 2.5—Enhancement functions for three-pulse-comparison, dual-period system (showing effect of changing the ratio, μ, of interpulse spacings)

For these a_j, Eq. (2.43) can be written

$$\overline{G} = \frac{1}{1 - 2\dfrac{N}{N+1}\rho_c(\sigma\tau) + 2\dfrac{N}{N+1}\dfrac{N-1}{N+2}\rho_c(2\sigma\tau)\cdots}. \tag{2.46}$$

This has been evaluated for selected values of σ and N. The results appear in Fig. 2.6.

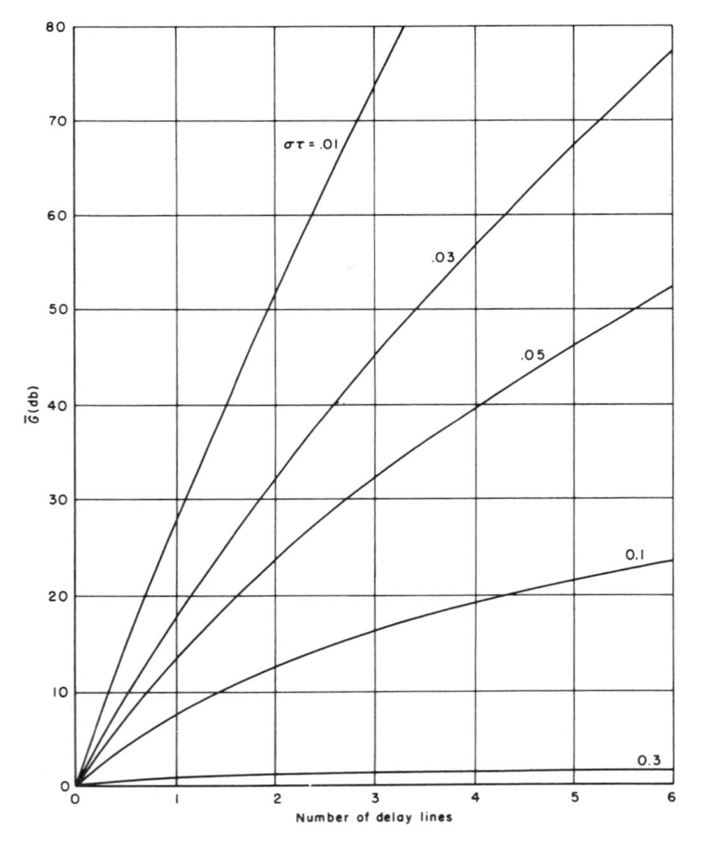

Fig. 2.6—Reference gain vs number of delay lines (Gaussian clutter spectra)

CHAPTER 3

RANGE-GATED, LUMPED-CONSTANT FILTER SYSTEMS

3.1. Introduction

If the receiver can be gated into separate range channels, it becomes possible to dispense with the distributed-constant filters, such as delay lines, and deal only with the lumped-constant, nonperiodic type. In particular, these filters may be inducto-capacitive, magnetostrictive, piezoelectric, etc., and whatever suits the purpose as to Q, size and weight, cost, mechanical rigidity, and electrical dependability. There is the requirement, however, that there be as many separate filter units as there are range resolution intervals desired. This is an imposing restriction, for it can be seen that if, say, 1-mi range resolution is desired out to 100 mi, then up to 100 separate filter blocks are required, and each of these blocks may comprise many elementary filter units. It has been conjectured that as many elementary filtering units are needed for the range-gated system as would be needed to approximate the delay lines, if the range resolution and velocity enhancement characteristics are to be the same. Nevertheless, since the simplest velocity enhancement curves obtainable with delay lines are not necessarily well shaped, one might expect, by using the gating principle, to obtain more desirable curves with fewer elementary filters.

3.2. Possible Methods of Range Gating

The range-gating mechanism consists of a pair of selector switches (or their equivalent) which complete a sequence each range frame. The first switch is employed to divert the bipolar video output into the successive velocity filters, and the second is used to reassemble the respective filter outputs for amplification and display. (This second switch would not be required if all range data could be displayed simultaneously.)

It is essential for satisfactory MTI performance that these switches be electrically stable and quiet. Timing stability is important in order that identical ranges and range intervals may be sampled each frame. In addition, it is necessary that the "contact noise" be small relative to the clutter energy involved. These requirements, together with the necessarily high speed of operation, almost preclude switching by mechanical means. On the other hand, electronic switching by means of conventional range-gating techniques may be feasible. A reduction in the number of vacuum tubes needed might be brought about through the use of transistors; and, rather than a separate range trigger generator for each channel, fewer generators might be used in combination together with coincidence circuits, etc. Alternatively, cathode-ray principles may be employed for switching, and in this case the timing circuitry may be simpler.

Delay lines may also be used to provide the gating impulses. In this case either a bank of delay lines or a single, long, "tapped" delay line may be used to delay the transmitted pulse envelope. Alternatively, the transmitted pulses may be mixed with the Stable Local Oscillator (stalo) and the resulting i-f pulses delayed to provide "coho pulses" for each respective range cell. In this case the separate coherent oscillator and phase-locking circuitry are not needed, and the individual range gates assume the form of coherent mixers.

Since the relative merits of these schemes must be determined empirically, the analysis of the gated systems will be based on the assumption that adequate gates are available. To the extent that the actual gates are unstable or noisy, the theoretical results will have to be degraded by an equivalent cancellation ratio.

3.3. System Gain

As far as theoretical manipulation is concerned, any of the range-gated systems of Sec. 1.5.2 may be considered as being a generalization from or extension of a finite-pulse-group system to an infinite-pulse-group system. The various weight factors, a_j (see Sec. 2.4), then form an infinite sequence and may be identified with the impulse time-response function, $f(t)$, for the filter. Thus,

$$a_j = f(t_j), \tag{3.1}$$

where the t_j represent the times of arrival of the successive pulses (or the times when the gate is open). In terms of this notation, the system gain, $G(v)$, is given by the formula

$$G(v) = \frac{\sum_{j=0}^{\infty}\sum_{k=0}^{\infty} f(t_j)f(t_k)\rho_s(t_j - t_k)}{\sum_{j=0}^{\infty}\sum_{k=0}^{\infty} f(t_j)f(t_k)\rho_c(t_j - t_k)}, \tag{3.2}$$

where ρ_s and ρ_c are, respectively, the normalized autocorrelation functions for the signal and clutter. These correlation functions are related to the unit-power spectra $W_s(f)$ and $W_c(f)$ by the Wiener-Khintchine formulas

$$\rho_{s,c}(x) = \int_{-\infty}^{\infty} e^{i\omega x} W_{s,c}(f)\, df. \tag{3.3}$$

The integrals above for ρ_s and ρ_c are introduced into Eq. (3.2), yielding

$$G(v) = \frac{\int_{-\infty}^{\infty} H(f)W_s(f)\, df}{\int_{-\infty}^{\infty} H(f)W_c(f)\, df}, \tag{3.4}$$

where

$$H(f) = \left| \sum_{j=0}^{\infty} e^{i\omega t_j} f(t_j) \right|^2 \tag{3.5}$$

is the spectrum function for the gate-filter combination. It is necessarily a symmetric function; i.e., $H(-f) = H(f)$. This function will be discussed in more detail in Sec. 3.3.3.

3.3.1. Nonfluctuating Target

If the target echoes were perfectly stable (except for the doppler modulation), then the autocorrelation function $\rho_s(\tau)$ would be

$$\rho_s(\tau) = \cos \omega_d \tau, \tag{3.6}$$

and the unit power spectrum would be

$$W_s(f) = \frac{1}{2}[\delta(f = -f_d) + \delta(f = +f_d)], \tag{3.7}$$

where $\delta(x = y)$ is the unit impulse function at $x = y$, and f_d is the doppler-modulation frequency.

Since $H(f)$ is symmetric, the system gain reduces to

$$G(v) = \frac{H(f_d)}{\int_{-\infty}^{\infty} H(f)W_c(f)\, df}. \tag{3.8}$$

The reference gain, $\overline{G}$, is given by the average of $G(v)$ over v. Letting $\overline{H}$ denote the average of $H(f_d)$ over f_d, then

$$\overline{G} = \frac{\overline{H}}{\int_{-\infty}^{\infty} H(f)W_c(f)\, df}, \tag{3.9}$$

and the enhancement function is simply

$$E(v) = \frac{H(f_d)}{\overline{H}}. \tag{3.10}$$

Thus, for a nonfluctuating target signal, the velocity enhancement function is simply $H(f)$ normalized to unit average.

The power gain of the system for white noise is the latent gain, G_L. This is equal to $\overline{H}$. To show this, let $N_i(t)$ be the input white noise voltage. The output noise voltage is

$$N_o(t) = \sum_j f(t_j)N(t - t_j). \tag{3.11}$$

The output power is then

$$\overline{N_o(t)^2} = \sum_j \sum_k f(t_j)f(t_k)\overline{N_i(t - t_j)N_i(t - t_k)}$$
$$= \overline{N_i(t)^2} \sum_j f(t_j)^2. \tag{3.12}$$

Hence,

$$G_L = \sum_j f(t_j)^2. \tag{3.13}$$

Now, from Eq. (3.5),

$$H(f) = \sum_j \sum_k \exp\{i\omega(t_j - t_k)\} f(t_j) f(t_k). \tag{3.14}$$

This is averaged over f to give

$$\overline{H} = \sum_j \sum_k f(t_j) f(t_k) \overline{\exp\{i\omega(t_j - t_k)\}}$$

$$= \sum_j f(t_j)^2. \tag{3.15}$$

3.3.2. Fluctuating Target

In actual practice, the target echoes will fluctuate. Consequently, the autocorrelation function for the video pulse modulation must be written

$$\rho_s(\tau) = \rho_f(\tau) \cos \omega_d \tau, \tag{3.16}$$

where $\rho_f(\tau)$ accounts for the additional decorrelation. Letting $W_f(f)$ denote the Fourier transform of $\rho_f(\tau)$, $W_s(f)$ takes the form

$$W_s(f) = \frac{1}{2}[W_f(f + f_d) + W_f(f - f_d)]. \tag{3.17}$$

The fluctuation spectrum, $W_f(f)$, for aircraft targets should exhibit low-frequency energy in amounts proportional to the scatter cross section of the airframe, arising from the relatively slow pitch, yaw, and roll motions; and, in addition, if the aircraft is propeller-driven, higher frequency components will be present in proportion to the propeller cross section. Depending on the type of airplane, the propeller tip velocity may be comparable to or may exceed the aircraft cruising speed, so that this part of the fluctuation energy may be spread over the entire useful velocity spectrum. Furthermore, if the propeller modulation constitutes a sufficiently large proportion of the total return, this spectral smear may effectively "fill in" the dead zones and otherwise smooth out the enhancement function.

At the present time, data are being analyzed by the Naval Research Laboratory in an attempt to determine the functions $\rho_f(\tau)$ and $W_f(f)$ for certain types of aircraft targets.[18] When $W_f(f)$ is sufficiently well determined, it may be introduced into the analysis to yield the system gain, etc., for specific types of aircraft targets.

The effect of target fluctuation on the velocity enhancement curve is illustrated in Fig. 3.1.

3.3.3. The Gated-filter Response Function, *H(f)*

In order to determine the spectrum function, $H(f)$, Eq. (3.5), for the gate-filter combination, it is necessary to make certain assumptions concerning the gate. It will simplify matters to suppose that, when it goes off, the gate goes *completely* off, is per-

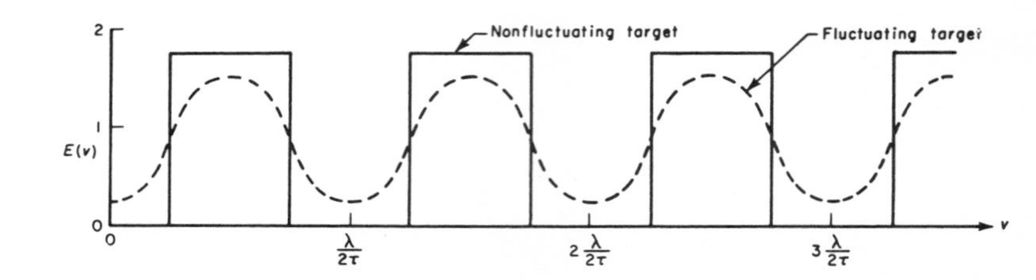

Fig. 3.1—Effect of target fluctuation on the velocity enhancement function (single-interpulse-period, rectangular-filter system)

fectly stable in time, and has an infinitely short (pulse) width.* In this case the gate may be viewed as an input modulating voltage consisting of a sequence of impulses, $\delta(t = t_j)$, occurring at the arrival times for the returns from the specified range. Denoting this chain of impulses by $g(t)$, the sum on the right-hand side of Eq. (3.5) can then be replaced by an integral, thus:

$$H(f) = \left| \int_{-\infty}^{\infty} e^{i\omega t} f(t) g(t)\, dt \right|^2$$

$$= \left| \int_{-\infty}^{\infty} F^*(g) G^*(f - g)\, dg \right|^2, \tag{3.18}$$

where $F^*(f)$ is the *voltage* spectrum of the ungated filter, and $G^*(f)$ is the *voltage* spectrum of the gating impulses.

Presuming a fundamental repetition frequency, f_g, for the gate, the spectrum $G^*(f)$ will consist of discrete lines of diverse intensities spaced by f_g. (Except for a uniform phase shift per cycle corresponding to the range delay, $G^*(f)$ should be identical with the spectrum of the transmitted modulation.) The positive frequency pass-band of $F^*(f)$ will now be confined between two positive frequency lines of $G^*(f)$ in such a way as to reject the fixed-target energy coincident with $G^*(f)$ and to accept some of the frequency-shifted energy from moving targets. (The negative frequency pass-band will then be similarly confined between corresponding negative frequency lines of $G^*(f)$.) Under these circumstances, since no frequency cross products arise, the expression for $H(f)$, Eq. (3.18), may be replaced with

$$H(f) = \int_{-\infty}^{\infty} F(g) G(f - g)\, dg, \tag{3.19}$$

*This last assumption makes for compatibility with the delay-line treatment, wherein it was assumed that the delay lines were perfectly range preserving. This implies an infinite delay-line bandwidth—an ideal which for all practical purposes is achieved with bandwidths of the order of the reciprocal pulse width. Similarly, the infinitely narrow gate is sufficiently approximated by gate widths of the order of the pulse width. In either case, the errors of approximation should be small and of comparable magnitude.

where $F(f)$ and $G(f)$ are the respective *power* spectra for the ungated filter and the gate.

Let the intensity of the jth line of $G(f)$ be denoted g_j; then $H(f)$ can finally be represented as

$$H(f) = \sum_{j=-\infty}^{\infty} g_j F(f - jf_g). \tag{3.20}$$

Thus, as a result of the gating operation, the frequency pass characteristics of the ungated filter become manifest between each pair of gate harmonics, so that the spectrum, $H(f)$, possesses an infinite range of pass frequencies similar to the delay lines. A graphical presentation of a single-period, rectangular-filter system is shown in Fig. 3.2 for $F(f)$, $G(f)$, and $H(f)$.

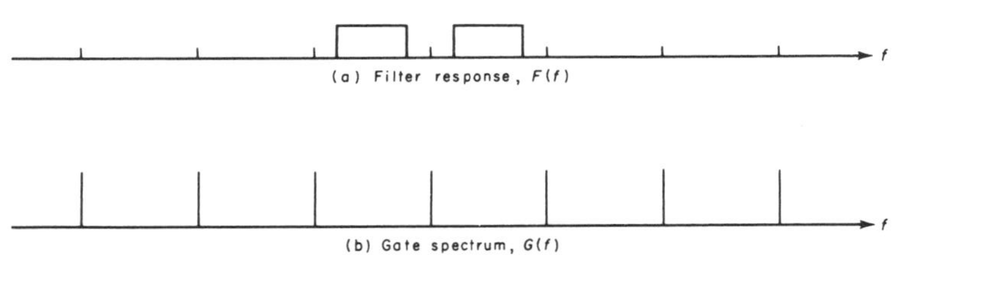

Fig. 3.2—F(f), G(f), and H(f) for single-period, rectangular-filter system

Since both $F(f)$ and $G(f)$ are necessarily symmetric functions, i.e., $F(-f) = F(f)$, and $g_{-j} = g_j$, it follows that $H(f)$ as expressed in Eq. (3.20) is also symmetric. The notation of Eq. (3.20) tends to hide an important consequence of this symmetry—that the positive frequency pass characteristic $F(f > 0)$, which need not be symmetric about its mid-band frequency, must be accompanied in the determination of $H(f)$ by its mirror image, $F(f < 0)$. Under ordinary circumstances, since $F(-f)$ always equals $F(f)$, the redundant $F(f < 0)$ may be disregarded in describing the filter power response. If it is desired to deal only with positive frequencies, then it may be assumed that the superheterodyne detector splits each doppler-shifted spectral component into a video *pair*, creating the impression that the entire r-f spectrum has been "folded" about the carrier and translated to d-c, and that the negative frequency components have been deleted. All members of the folded spectrum are then presented to $F(f > 0)$. On the other hand, if the negative frequency components are preserved as in the preceding analysis, an unfolded video spectrum results, which must be presented to both $F(f < 0)$ and $F(f > 0)$. Either procedure, when properly carried out, will lead to the correct results. The latter point of view has been adopted in this analysis because the mathematical notation is more compact.

3.4. Single-interpulse-period Systems

Single-interpulse-period systems may be useful under circumstances which make multiple periods difficult to obtain (e.g., MTI modification kits for existing radars), and for applications in which the dead zones are not objectionable. The narrow "velocity gate" is an example of the latter and will be discussed in Sec. 3.4.2.

3.4.1. Broad-band Rectangular Filter

The analysis will proceed by considering a broad-band rectangular filter, whose positive frequency pass-band is centered at half the prf (see Fig. 3.3), i.e.,

$$F(f) = \begin{cases} \dfrac{1}{2B\tau} & \text{if} \left[\dfrac{1}{2\tau} - \dfrac{B}{2}\right] < |f| < \left[\dfrac{1}{2\tau} + \dfrac{B}{2}\right], \\ 0 & \text{otherwise}. \end{cases} \tag{3.21}$$

Here τ is the interpulse period, and B is the "total bandwidth," presumed to be less than $1/\tau$. The gate spectrum, $G(f)$, now consists of an infinite system of lines of uniform intensity, g_i (which will be chosen equal to unity), of uniform spacing, $1/\tau$, and situated at the multiples of the prf. Consequently, $H(f)$ becomes a periodic extension of $2F(f)$ in the interval $(0, 1/\tau)$ to all the intervals between gate harmonics. Under these conditions the value of $\overline{H}$ is unity, and, see Eqs. (3.10) and (3.9),

$$E(v) = H(f_d), \tag{3.22}$$

and

$$\overline{G} = \left\{ \int_{-\infty}^{\infty} H(f) W_c(f)\, df \right\}^{-1}. \tag{3.23}$$

From Eq. (3.3) it can be seen that the total power in $W_c(f)$ is $\rho(0) = 1$. This power is clustered about d-c and will be presumed, for purposes of evaluation, to be

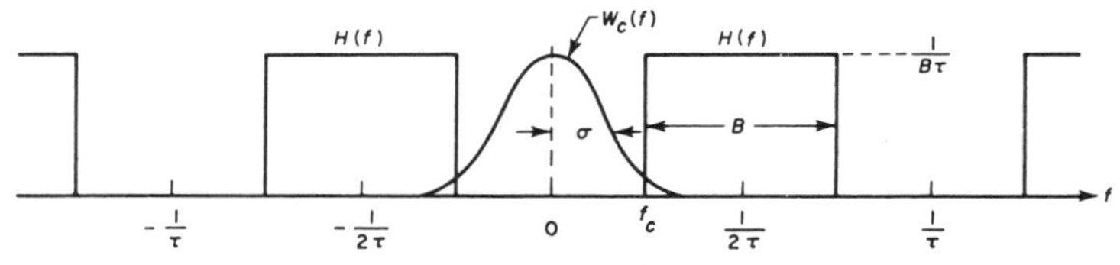

Fig. 3.3—Gaussian clutter spectrum and gated-filter response (single-interpulse-period, broad-band rectangular filter)

Gaussianly distributed according to

$$W_c(f) = \frac{1}{\sqrt{2\pi\sigma^2}} e^{-f^2/2\sigma^2}, \tag{3.24}$$

where the standard deviation parameter, σ, is small compared with $1/\tau$.* In this case, the reference gain reduces to

$$\overline{G} = \frac{B\tau}{\operatorname{erf}\left(\frac{1 + B\tau}{2\sqrt{2}\,\sigma\tau}\right) - \operatorname{erf}\left(\frac{1 - B\tau}{2\sqrt{2}\,\sigma\tau}\right)}, \tag{3.25}$$

where

$$\operatorname{erf}(x) = \frac{2}{\sqrt{\pi}} \int_0^x e^{-g^2}\, dg. \tag{3.26}$$

It is assumed here that the only significant contribution to the clutter residue comes from the pass-bands of $H(f)$ in the intervals $(-1/\tau, 0)$ and $(0, 1/\tau)$. This is consistent with the assumption that $\sigma\tau << 1$.

If the standard deviation parameter for the clutter, σ, is small compared with the rejection band for the filter, i.e., $\sigma\tau << (1 - B\tau)$, and if the bandwidth is large enough so that the clutter is significant only near the edges of the pass-band, then $\overline{G}$ can be approximated (using the asymptotic formula for the error integral) by

$$\overline{G} \sim \sqrt{\frac{\pi}{2}}\, \frac{B\tau(1 - B\tau)}{2\sigma\tau} \exp\left[\frac{1}{2}\left(\frac{1 - B\tau}{2\sigma\tau}\right)^2\right]. \tag{3.27}$$

Recalling that the enhancement function has the value $1/B\tau$ in the pass-band, the actual gain, $G(v)$, for moving targets in the pass-band is

$$G(\text{pass-band}) = \frac{\overline{G}}{B\tau} \sim \sqrt{\frac{\pi}{2}} \left[\frac{1 - B\tau}{2\sigma\tau}\right] \exp\left[\frac{1}{2}\left(\frac{1 - B\tau}{2\sigma\tau}\right)^2\right]. \tag{3.28}$$

This is shown in Fig. 3.4. The independent variable, $(1 - B\tau)/(2\sigma\tau)$, is the ratio of the rejection bandwidth of the filter to the "width," 2σ, of the clutter spectrum.

3.4.2. Narrow-band Rectangular Filter (Velocity Gate)

If a narrow-band filter is centered at $1/2\tau$, it can be specified by Eq. (3.21) with B small. In this case the target velocities, $\overline{v}$, corresponding to the pass-band, will be multiples of $\lambda/4\tau$, i.e., midway between the blind speeds. Since it is desirable to "see" other velocities, the filter will be centered at an arbitrary positive frequency, $f^* < 1/\tau$. In

*This is essential, because otherwise the corresponding "pulsed" spectrum, which consists of Gaussianly distributed clusters of energy about each repetition rate harmonic, would exhibit severe overlapping. This would destroy the utility of the Gaussian approximation.

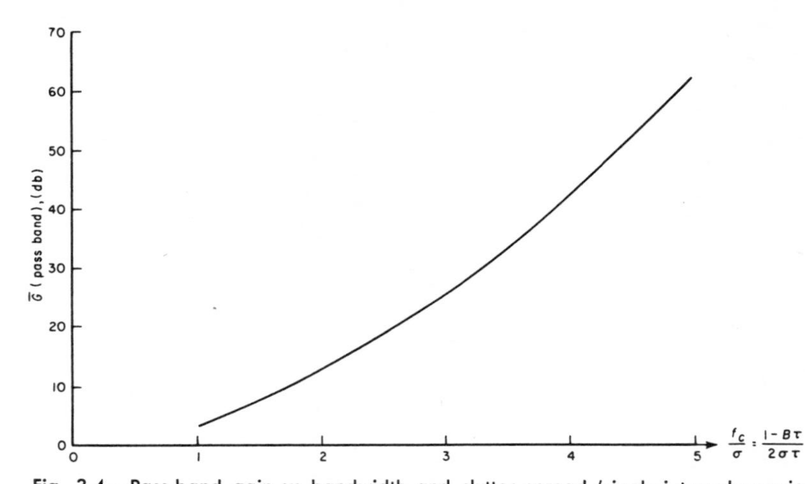

Fig. 3.4—Pass-band gain vs bandwidth and clutter spread (single-interpulse-period, broad-band rectangular filter)

this case the filter specification is

$$F(f) = \begin{cases} \frac{1}{2B\tau} & \text{if} \left[f^* - \frac{B}{2}\right] < |f| < \left[f^* + \frac{B}{2}\right], \\ 0 & \text{otherwise}. \end{cases} \tag{3.29}$$

This response function has a double pass-band (see Fig. 3.5), one around f^* and the other around $-f^*$. (It is presumed that these pass-bands do not overlap, i.e., $B < 2f^*$.)

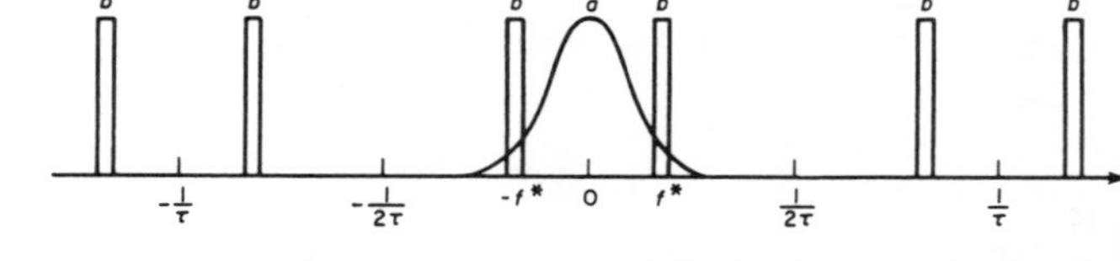

Fig. 3.5—Gaussian clutter spectrum (a) and gated-filter response (b) (single-interpulse-period, narrow "velocity-gate" filter)

As before, $H(f)$ is obtained by disposing $F(f)$ about each prf harmonic, but in this case twin pass-bands occur in each interval between the harmonics—the twins being situated $1/2\tau - f^*$ cycles above and below the mid-interval frequencies. The average, $\overline{H}$, is again unity, so that $E(v) = H(f_d)$.

If the Gaussian distribution, Eq. (3.24), is taken for the clutter, the reference gain reduces to

$$\overline{G} = \frac{2B\tau}{\operatorname{erf}\left(\frac{2f^*\tau + B\tau}{2\sqrt{2}\,\sigma\tau}\right) - \operatorname{erf}\left(\frac{2f^*\tau - B\tau}{2\sqrt{2}\,\sigma\tau}\right) + \operatorname{erf}\left(\frac{1 - 2f^*\tau + B\tau}{2\sqrt{2}\,\sigma\tau}\right) - \operatorname{erf}\left(\frac{1 - 2f^*\tau - B\tau}{2\sqrt{2}\,\sigma\tau}\right)}, \quad (3.30)$$

which can be approximated for narrow bandwidths, i.e., $B << \sigma$, by

$$G \approx \sqrt{\frac{\pi}{2}} \frac{2\sigma\tau}{\exp\left[-\frac{1}{2}\left(\frac{f^*\tau}{\sigma\tau}\right)^2\right] + \exp\left[-\frac{1}{2}\left(\frac{1 - f^*\tau}{\sigma\tau}\right)^2\right]}. \quad (3.31)$$

In terms of the pass-band velocity, $\bar{v}$, this becomes

$$\overline{G} \approx \sqrt{\frac{\pi}{2}}\,\sigma\tau \frac{2\exp\left[\frac{1}{2}\left(\frac{2\bar{v}}{\sigma\lambda}\right)^2\right]}{1 + \exp\left[\frac{4\bar{v}\tau - \lambda}{2\lambda\sigma^2\tau^2}\right]}, \quad (3.32)$$

in which, of course, $\bar{v}$ must be kept smaller than half the blind speed. For higher target speeds, the velocity must be reduced by multiples of the blind speed, and the residue, $\bar{v}$, must be used for computation.

To obtain the pass-band gain, the value of $\overline{G}$ must be multiplied by the enhancement factor, which is $1/2B\tau$. Thus, for targets in the velocity gate, the gain is

$$G(\text{pass-band}) = \sqrt{\frac{\pi}{2}} \frac{\sigma\tau}{2B\tau} \frac{2\exp\left[\frac{1}{2}\left(\frac{1}{\sigma\tau}\right)^2\left(\frac{2\bar{v}\tau}{\lambda}\right)^2\right]}{1 + \exp\left[\left(\frac{1}{\sigma\tau}\right)^2\left(\frac{2\bar{v}\tau}{\lambda} - \frac{1}{2}\right)\right]}, \quad (3.33)$$

which is illustrated graphically in Fig. 3.6.

This pass-band gain can be made indefinitely large by reducing the bandwidth. In reality, however, the gain will be limited by the target fluctuation, the limiting value being the ratio of the target power density to the clutter power density at the filter frequency. That is,

$$\max_{B} G(\text{pass-band}) = \frac{W_s(f^*)}{W_c(f^*)}, \quad (3.34)$$

which can be infinite only if no clutter energy is present at f^* or if the target energy is infinitely concentrated there. The former is conceivable, but the latter would imply absolute target stability over an infinite observation time.

3.4.3. Effect of Changing σ

In the case of the broad-band filter, when the clutter width, σ, is increased, the reference gain decreases, as should be expected. However, this is not the case for the narrow-

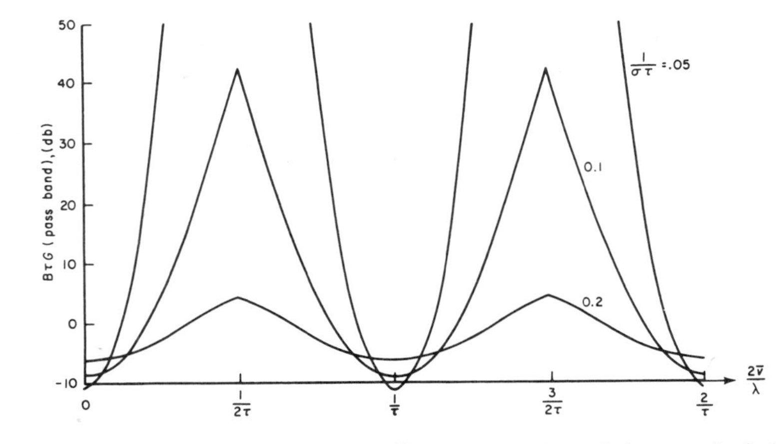

Fig. 3.6—Pass-band gain vs gate velocity, $\bar{v}$ (single-interpulse-period, narrow "velocity-gate" filter)

band filter. Inspection of Eq. (3.32) reveals that $\overline{G}$ will rise on increasing σ if the gate velocity is sufficiently small. This is explained by the fact that the clutter-energy *density* (which is what the narrow filter "sees") diminishes in the neighborhood of low velocities as the total clutter energy is spread out over a larger band of frequencies. Of course, σ is limited, for this analysis, to be small compared with $1/\tau$ (see Sec. 3.4.1).

Figure 3.7 shows the dependence of G on σ for the two cases.

3.5. Dual-interpulse-period System

The blind speeds inherent in the single-period system can be substantially eliminated by using dual-period transmission. Of course, the range gate is alternately adjusted to correspond with these periods, so that data from the same range are sampled each frame.

Let the interpulse periods be τ_1 and τ_2. Then the gate spectrum, $G(f)$, will consist of discrete lines uniformly spaced by $f_g = 1/(\tau_1 + \tau_2)$ and of intensities, g_j, proportional to

$$g_j \sim 1 + \cos 2\pi j \frac{\tau_1}{\tau_1 + \tau_2}. \quad (3.35)$$

If τ_1 and τ_2 are commensurate, then the g_j will exhibit a periodic pattern repeating every f_r cycles, where f_r is the least common multiple of $1/\tau_1$ and $1/\tau_2$, denoted $(1/\tau_1, 1/\tau_2)$.

If τ_1 and τ_2 are so chosen that they are in the ratio of two odd integers, then the spectral lines, g_j, will vanish whenever j is an odd multiple of j_0 defined by

$$j_0 = \frac{\tau_1 + \tau_2}{2}\left(\frac{1}{\tau_1}, \frac{1}{\tau_2}\right). \quad (3.36)$$

The corresponding frequencies are odd multiples of $f_r/2$.

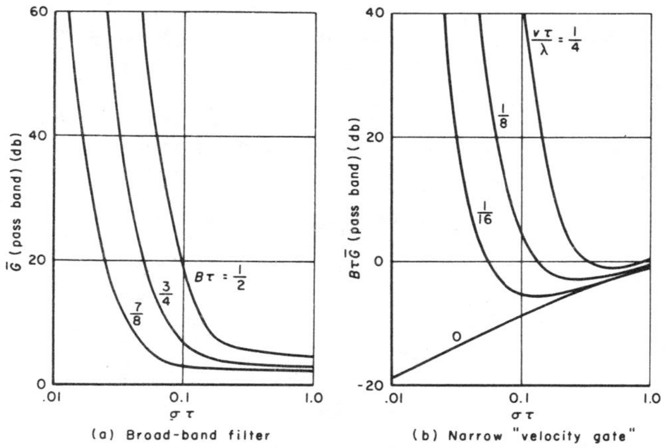

Fig. 3.7—Effect on pass-band gain of changing clutter spread, σ (single-interpulse-period, rectangular-filter systems)

A rectangular filter of bandwidth $B = 1/(\tau_1 + \tau_2)$ is now introduced into the dead interval between g_{j_0-1} and g_{j_0+1} (this refers to the positive frequency pass-band; the negative frequency pass-band will automatically be in the interval g_{-j_0-1} to g_{-j_0+1}). This filter is specified by

$$F(f) = \begin{cases} \dfrac{1}{2} \text{ if } \dfrac{1}{2}\left[\left(\dfrac{1}{\tau_1}, \dfrac{1}{\tau_2}\right) - \dfrac{1}{\tau_1 + \tau_2}\right] < |f| < \dfrac{1}{2}\left[\left(\dfrac{1}{\tau_1}, \dfrac{1}{\tau_2}\right) + \dfrac{1}{\tau_1 + \tau_2}\right], \\ 0 \text{ otherwise.} \end{cases} \tag{3.37}$$

If the g_j are chosen to be equal to the right-hand side of Eq. (3.35), then the gated-filter response function $H(f)$ becomes

$$H(f) = 1 - \cos 2\pi j \frac{\tau_1}{\tau_1 + \tau_2}, \tag{3.38}$$

where

$$\frac{j - \frac{1}{2}}{\tau_1 + \tau_2} < f \leq \frac{j + \frac{1}{2}}{\tau_1 + \tau_2}. \tag{3.39}$$

This function is zero in the interval $\{-1/[2(\tau_1 + \tau_2)] < f \leq 1/[2(\tau_1 + \tau_2)]\}$ around d-c. It then takes a step to a new value for the adjacent intervals, etc., stepping higher and lower from one interval to the next until it returns to zero in the interval around f_r (see Fig. 3.8).

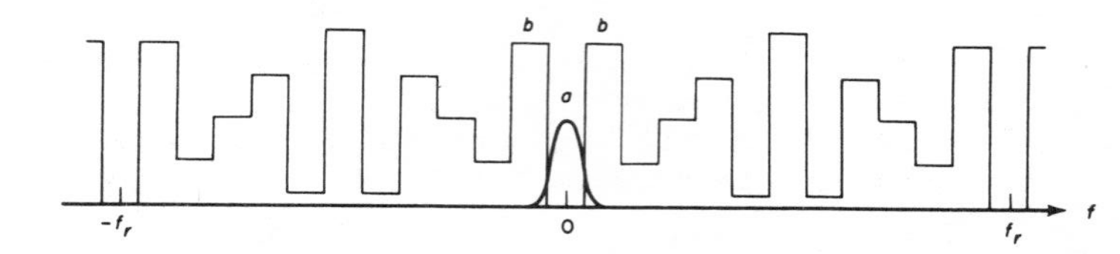

Fig. 3.8—Gaussian clutter spectrum (a) and gated-filter response (b) (dual-interpulse-period, $\mu = 7:5$, rectangular filter)

The average value, $\overline{H}$, is again unity, and hence $E(v) = H(f_d)$ and

$$\overline{G} = \left\{ \int_{-\infty}^{\infty} H(f) W_c(f)\, df \right\}^{-1}. \tag{3.40}$$

For purposes of evaluation, it will be presumed that the only significant contribution to the clutter residue comes from the two intervals neighboring the null interval around d-c, in which case

$$\overline{G} = \left\{ 2\left[1 - \cos 2\pi \frac{\tau_1}{\tau_1 + \tau_2}\right] \times \int_{1/[2(\tau_1+\tau_2)]}^{\infty} \frac{1}{\sqrt{2\pi\sigma^2}} \exp\left(-\frac{f^2}{2\sigma^2}\right) df \right\}^{-1}$$

$$= \frac{1}{\left[1 - \cos 2\pi \frac{\tau_1}{\tau_1 + \tau_2}\right]\left[1 - \operatorname{erf}\left(\frac{1}{2\sqrt{2}\,\sigma(\tau_1 + \tau_2)}\right)\right]}. \tag{3.41}$$

This appears in Fig. 3.9 for selected values of the ratio $\mu = \tau_1/\tau_2$.

As was mentioned previously in Sec. 3.5, in order that this system may operate properly it is essential that the interpulse periods bear the ratio of two odd integers. Any departure from this condition will result in a failure to nullify the spectrum line at $f_r/2$, which in turn will present a strong fixed-target (i.e., clutter) component to the filter mid-frequency. This vulnerability to interpulse spacing instability may or may not be serious, but it can be eliminated by simply alternating the polarity of the successive pulses in the output of the gate. This will shift the null frequency $f_r/2$ down to d-c, where it will remain regardless of the spacings τ_1 and τ_2. Variations in τ_1 and τ_2 will then affect only the range of visible velocities, which is of relatively minor importance. Merely shifting the coho by $f_r/2$ will not accomplish this.

An additional point in favor of alternating the pulse polarity is that then any interpulse periods can be used; i.e., they need not be in the ratio of odd integers.

3.6. More than Two Interpulse Periods

Having materially eliminated the blind speeds through the use of two interpulse periods, there is little to be gained by using more than two periods. True, a greater range

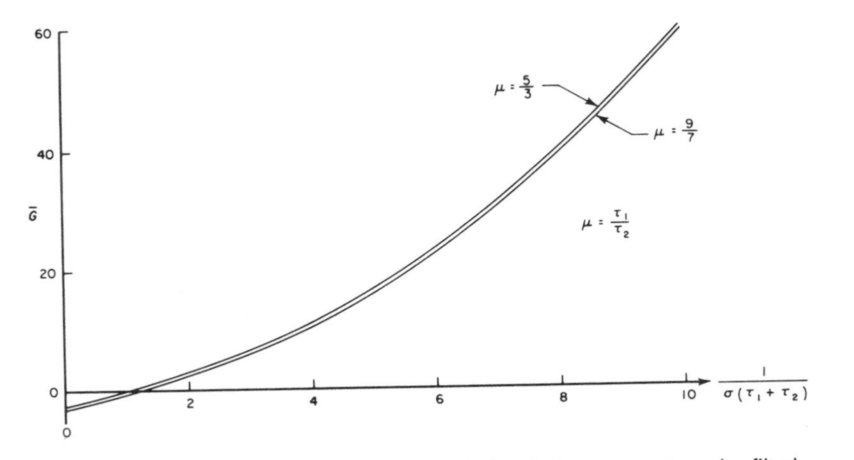

Fig. 3.9—Reference gain vs clutter spread (dual-period-system, rectangular filter)

of visible velocities with less violent oscillations in the pass-band may be possible with the use of more periods, but when the smoothing effect of target fluctuation is taken into account these refinements may be unnecessary.

With more interpulse periods, however, it is possible to realize a somewhat higher "velocity cut-off" than is obtainable with a dual-period system. This will be illustrated by consideration of the gate spectrum for the general N-period system. As was mentioned in Sec. 3.5, by alternating the polarity of successive pulses in the dual-period system, the d-c component of the gate spectrum may be annulled. Similarly, by appropriate modulation of the gate sensitivity for the N-period system, a group of at most $N - 1$ adjacent lines of $G(f)$ may be annulled. (There are infinitely many such groups, recurring every f_r cycles, where $f_r = (1/\tau_1, 1/\tau_2, \cdots, 1/\tau_N)$ is the least common multiple of the $1/\tau_j$.

Denoting the successive modulating factors by α_k, the nth spectral line, g_n, of $G(f)$ is proportional to

$$g_n \approx \left| \sum_{k=1}^{N} \alpha_k \exp\left\{ 2\pi i n \left(\frac{\tau_1 + \tau_2 + \cdots + \tau_k}{\tau_1 + \tau_2 + \cdots + \tau_N} \right) \right\} \right|^2, \tag{3.42}$$

and the first $N - 1$ of these lines can be annulled by selecting the α_k so that

$$\sum_{k=1}^{N} \alpha_k \exp\left\{ 2\pi i n \left(\frac{\tau_1 + \tau_2 + \cdots + \tau_k}{\tau_1 + \tau_2 + \cdots + \tau_N} \right) \right\} = 0 \quad (\text{for } n = 0, 1, \cdots, N-2). \tag{3.43}$$

The required α_k may be complex, i.e.,

$$\alpha_k = a_k e^{i\phi_k}; \tag{3.44}$$

but this can be achieved in principle by modulating the gate sensitivity by the real factor a_k and shifting the coho frequency from pulse to pulse so that its instantaneous phase is ϕ_k.

If a rectangular filter of bandwidth $B = 1/(\tau_1 + \tau_2 + \cdots + \tau_N)$ is now introduced in the center of the annulled group of lines, it will be blind to all doppler frequency shifts whose absolute magnitude is less than

$$f_c = \frac{N-1}{2} \frac{1}{\tau_1 + \tau_2 + \cdots + \tau_N}. \tag{3.45}$$

Beyond this point, and all the way out until doppler shifts in the vicinity of f_r occur, the filter bandwidth is just wide enough to accept energy from some one of the non-annulled lines.

The unambiguous range, r_u, for the radar is defined by

$$r_u = \frac{c\tau}{2}, \tag{3.46}$$

where c is the velocity of light and τ is the shortest interpulse period. Let

$$\tau_1 + \tau_2 + \cdots + \tau_N = N\tau\mu, \tag{3.47}$$

where μ is a factor somewhat larger than unity to account for the fact that all but one of the interpulse periods are larger than τ. The cut-off frequency, f_c, can then be expressed by

$$f_c = \frac{N-1}{2\mu N} \frac{c}{2r_u}. \tag{3.48}$$

Consequently, the cut-off velocity, v_c, is given by

$$v_c = \frac{N-1}{\mu N} \frac{c^2}{8f_0 r_u}, \tag{3.49}$$

where f_0 is the carrier frequency. Since $\mu > 1$,

$$v_c < \frac{N-1}{N} \frac{c^2}{8f_0 r_u}, \tag{3.50}$$

indicating that, as more interpulse periods are employed, the cut-off-velocity limit rises toward the upper bound $c^2/8f_0 r_u$, which is just double that for the dual-period system.

The limiting form of the inequality (3.50) as $N \to \infty$ can be written

$$f_0 < \frac{1170}{v_c r_u}, \tag{3.51}$$

where f_0 is in megacycles per second, v_c is in hundreds of knots, and r_u is in hundreds of nautical miles. This inequality expresses the theoretical upper limit on the carrier frequency for extremely complex systems of this type, when it is required that clutter out to v_c hundreds of knots be rejected and that targets out to r_u hundreds of nautical miles be detected without ambiguity. (Of course, the right-hand side should be multiplied by the factor $(N-1)/N\mu$ for any specific N-period system, and an additional degradation factor should be included to account for the fact that the filter will not be "rectangular.")

3.6.1. Limiting Reference Gain for Infinitely Many Periods

If the number of interpulse periods, N, is even, and if the α_k are chosen so that

$$\alpha_k = (-1)^k + \varepsilon_k, \tag{3.52}$$

where ε_k is a small ($|\varepsilon_k| \sim 0$) complex number related to the nonuniformity of the pulse spacings, then the $N-1$ gate lines symmetrically disposed about d-c can be annulled. The corresponding situation ensues for N odd if

$$\alpha_k = \omega^k + \varepsilon_k, \tag{3.53}$$

where ω is a complex cube root of unity.

Assuming now that N is infinitely large, that the various τ_k are incommensurate, and that all the gate harmonics in the interval $-f_c$ to $+f_c$ have been annulled, the non-annulled harmonics will be dense and will oscillate violently outside the interval. The filter bandwidth can now be infinitely narrow; consequently, $H(f) = G(f)$, i.e.,

$$H(f) = \begin{cases} 0 \text{ for } |f| < f_c, \\ \text{violently oscillating function for } |f| \geq f_c. \end{cases} \tag{3.54}$$

$H(f)$ must now be multiplied by $W_c(f)$ and integrated to determine the reference gain. In the process of integration, however, the oscillations of $H(f)$ are effectively smoothed out, and therefore one may as well replace $H(f)$ by the step function

$$H'(f) = \begin{cases} 0 \text{ for } |f| < f_c, \\ 1 \text{ for } |f| \geq f_c. \end{cases} \tag{3.55}$$

(The value unity is used above so that $H(f)$ will have unit average.) Thus, the reference gain reduces to

$$\overline{G} = \left\{ 2 \int_{f_c}^{\infty} W_c(f)\, df \right\}^{-1}. \tag{3.56}$$

For the Gaussian clutter spectrum, Eq. (3.24), used previously, this becomes

$$\overline{G} = \left\{ 1 - \operatorname{erf}\left(\frac{f_c}{\sigma\sqrt{2}} \right) \right\}^{-1} \leq \left\{ 1 - \operatorname{erf}\left(\frac{1}{2\sqrt{2}\,\sigma\tau} \right) \right\}^{-1}, \tag{3.57}$$

where τ is the shortest interpulse period. This is shown graphically in Fig. 3.10.

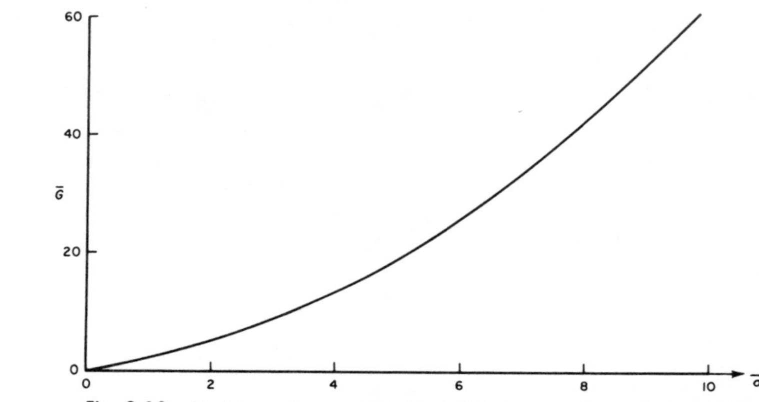

Fig. 3.10—Limiting reference gain for infinitely many interpulse periods

3.7. Reference Gain for Practical Filters

Rectangular filters were used in the preceding analysis mainly to illustrate the general nature of the enhancement curves, the sensitivity of the reference gain to the filter bandwidth, clutter spectral spread, etc. In cases where the clutter spectrum is fairly broad with respect to the filter bandwidth (i.e., low reference gain), or when the actual filter is very nearly rectangular, the appropriate gain formulas of Secs. 3.4, 3.5, and 3.6 will be useful. However, in cases where the cut-off frequency for the filter is far out on the tail of the clutter spectrum, the reference gain will be extremely sensitive to the "slope" of the filter characteristic, and the rectangular filter approximation may be very bad. Of course, the exact value for $\overline{G}$ can be found from Eq. (3.9) when the filter characteristic is specified, but this formulation is not particularly well suited for computation with nonrectangular filters. For these practical filters, it is much more convenient to use the delay-line formulation, Eq. (2.25), and to this end a correspondence is set up between the filter as characterized by its "slope" in the clutter region and an equivalent number of delay lines.

In any particular system, it will be supposed that the filters have been designed so that complete nulls occur at the two adjacent prf harmonics, and that sufficient rejection has been provided at all other harmonics so that their residues may be neglected. Under these circumstances, $H(f)$, see Eq. (3.20), will tend to zero as $f \to 0$ in a manner dictated by the "slope" of $F(f)$ near the nulls. The "slope" of $H(f)$ as $f \to 0$ will be measured in decibels per decade, as follows:

$$\text{"Slope of } H(f)\text{"} = \lim_{f \to 0} 10 \log_{10} \frac{H(10f)}{H(f)}. \tag{3.58}$$

In general, a single tuned circuit can provide 20 db per decade at its null, for it can be shown that in the neighborhood of the null frequency, f_0,

$$F(f) \sim \frac{(f - f_0)^2}{Q^2}. \tag{3.59}$$

Similarly, N single tuned circuits in cascade can produce $20N$ db per decade at a common null. To obtain a specified number of decibels per decade for $H(f)$, one must therefore provide that number of decibels per decade at each of two nulls. Beyond this point additional filters may be required to shape the pass-band of $H(f)$, i.e., the enhancement curve, between the nulls and to provide the necessary rejection for the extraneous prf harmonics. However, as far as the reference gain is concerned, the number of decibels per decade for $H(f)$ is the important parameter.

A single delay line is similarly capable of 20 db per decade. This follows from the shape of the enhancement curve

$$E(f) \sim 1 - \cos 2\pi\left(f - \frac{n}{\tau}\right) \rightarrow 2\pi^2\left(f - \frac{n}{\tau}\right)^2 \tau^2 \qquad \text{as } f \rightarrow \frac{n}{\tau}, \tag{3.60}$$

and, in general, N delay lines in cascade, i.e., the binomial system of Sec. 2.5, can produce $20N$ db per decade. Thus, the correspondence between filter characteristics and number of delay lines is established.

The set of curves corresponding to Fig. 2.6 for use with these range-gated filter systems appears in Fig. 3.11. In this case the ordinate is labeled in decibels per decade and is therefore applicable to filters in general.

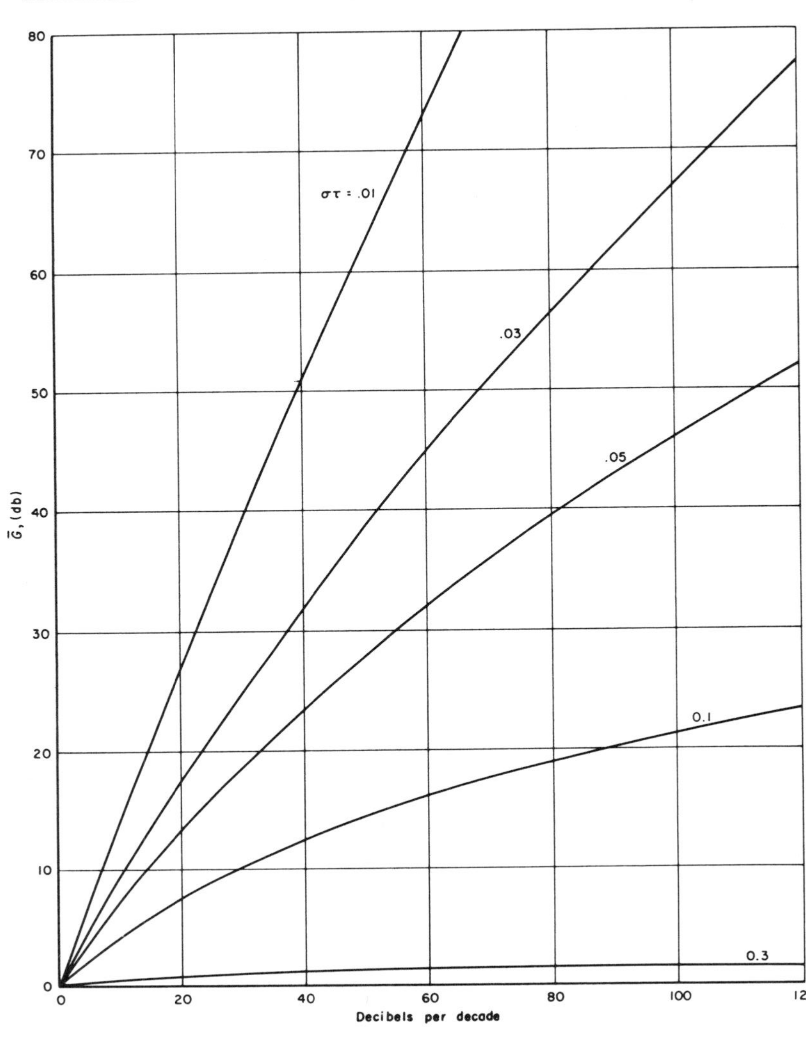

Fig. 3.11—Reference gain vs number of decibels per decade (Gaussian clutter spectra)

CHAPTER 4

NONCOHERENT SYSTEMS

4.1. Introduction

In principle, the coherent MTI systems outlined in Chaps. 2 and 3 may be successfully employed whenever the mean radial component of the clutter velocity relative to the radar is known. This is the case for ground clutter with ground-based sets. For airborne equipment, however, such a mode of operation hinges on the ability of the radar operator (or automatic equipment) to manipulate the frequency of the coherent oscillator one way or the other around the normal intermediate frequency by the amount required to compensate for the doppler shift associated with the radar platform velocity. The resulting bipolar video signals are then normal in the sense that targets moving at the same relative speed as the clutter appear as "fixed targets," and those moving at relatively different speeds exhibit the usual "butterfly" modulation.

In certain airborne applications, such as high-speed air-interception (AI) radars, accurate determination of ground speed, mean chaff velocity, etc., is not feasible, and noncoherent operation becomes necessary. Noncoherent radars are essentially the same as coherent radars, except that the stabilized local oscillator (stalo), locking circuits, and coho are eliminated, and the undesired (clutter) return is used for the doppler reference. These modifications present two major problems—consideration of the fact that clutter must be present for the detection of moving targets, and determination of the effects that the nonlinearities essential to noncoherent detection have on system performance.

4.2. Detection of Targets in the Clear

The video signals resulting from the noncoherent detection of clutter alone are identical with those in a normal radar without MTI; i.e., the voltages are unipolar and undulate at a relatively slow rate, pulse to pulse, as determined by the clutter fluctuations. Similarly, targets in the clear give rise to normal unipolar video pulses varying at a rate associated with the target fluctuation. It is not until both target and clutter appear together that doppler "beats" arise. Since, in the absence of these beats, the doppler filter (or delay lines) will reject each equally well,* some remedy must be found, e.g., such as using a switch to short-circuit the filters when the clutter is not present.

For airborne early-warning (AEW) radars operating over water, such a switch could be based on target range, the filters being bypassed for all ranges beyond the clutter horizon. The system would be fairly simple to design and could eliminate all but the

*Actually, neither is completely rejected, and each leaves a residue according to the echo instability. For example, conditions may arise in which propeller-driven aircraft are readily detected, whereas jet fighters are not. In any case detection will be degraded to some extent if this fault is not eliminated.

borderline sea return. It would not, however, provide MTI for the volume of sky beyond the horizon—a region in which storm clouds and chaff could be a menace.

An alternative would be a switch based on echo extension, such as that used in the AN/APA-78(XA-1) proposal[19] and more recently in the clutter-gated system built by Naval Research Laboratory.[20] These systems provide MTI for all returns of long duration, as compared with a pulse width and no MTI for the others. The presence of the clutter mass, which usually measures many pulse widths in depth, activates the MTI in that area. On the other hand, the targets in the clear that measure only fractions of a pulse width are detected, without regard to their velocities, with the MTI turned off. This type of system would have some advantages over that described in the previous paragraph, in that it would provide cancellation for storm clouds and chaff sown in dense blanket, and that it would not be limited to use over water. However, it could be confused by small fixed undesired targets in the clear, such as chaff laid in small clusters, etc.

An indication of the quality of result obtainable with the NRL system is provided by Fig. 4.1. The photographs of the PPI presentation were taken using the SR-3 radar, which is similar to the AN/SPS-6. Figure 4.1(a) shows the normal 20-mi presentation from the Anacostia site. The large cloud-like signal returns between 10 and 20 mi are from a thunderstorm, and those in the 0- to 7-mi region are from hills and from the city of Washington, D.C. Figure 4.1(b) shows the same environment, but with the clutter switch in operation.

A somewhat more sophisticated approach to the problem would be to use either a two-color or a two-tone display system for the simultaneous presentation of both the normal and the MTI data. Perhaps such systems would represent the ultimate, in that all the information would be displayed and the experienced observer might then be capable of finer judgments than those that could be "built in" with automatic switches. The two-color principle is probably the more attractive, but it must await the development of multicolor display tubes suitable for military use and capable of sufficient brightness for daylight viewing. A two-tone system could be built with existing equipment, however, and tests could be run to determine the effects of operator fatigue, display degradation due to reduced dynamic range, etc.

As the enemy ECM threat developed, a shift to noncoherent ground-based radars might be indicated. The advantages of such a change would be

- Reduced vulnerability to chaff sown in high winds;
- Reduced vulnerability to the steady-wind component of storm-cloud noise;
- Generally superior performance, because the equipment is simpler and easier to maintain.

The prerequisite, of course, would be the development of a really satisfactory clutter switch or other device, such as the two-color or two-tone simultaneous display.

4.3. Theoretical Analysis

The following sections will provide the mathematical context by which to evaluate the typical noncoherent mode of operation:

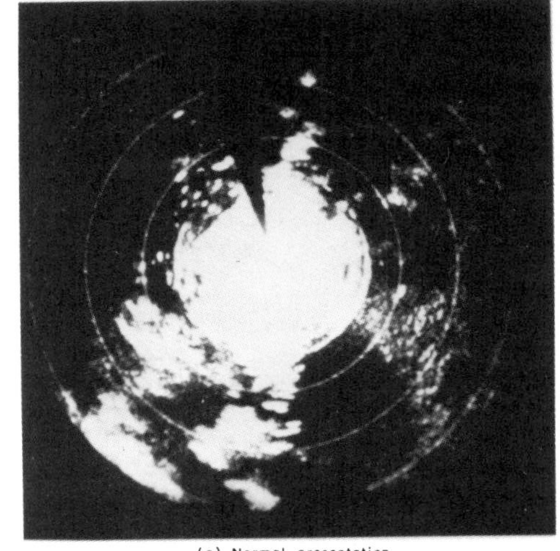

(a) Normal presentation

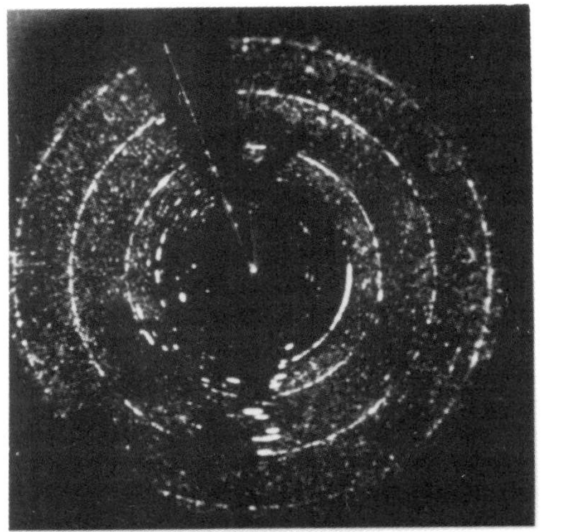

(b) Clutter-gated presentation

Fig. 4.1—Comparison of normal- and clutter-gated display (noncoherent radar type SR-3: 20-mi range, 1-μs pulse, clutter gate set for 2.5 μs) (courtesy of Naval Research Laboratory)

4.3.1. Simulation of the Signal and Clutter Voltages

Essential to the analysis of noncoherent MTI is the nonlinear i-f video mixer. In the subsequent treatment, this will be presumed to be a square-law detector having the property

$$e_o(t) = e_i(t)^2, \tag{4.1}$$

where $e_i(t)$ is the instantaneous i-f output voltage applied to the detector, and $e_o(t)$ represents the detected video input. The usual low-pass filtering associated with such a mixer is accomplished in the analysis by neglecting the high-frequency residue.

The end product of the mathematics is the instantaneous system gain, $G(v)$, and it is obtained by performing ensemble rather than time averages. Consequently, the pulsed nature of the returns may be neglected and the analysis carried through on a cw basis.

The i-f clutter voltages are presumed to have the Gaussian amplitude density given by

$$P'[C(t)] = \frac{1}{\sqrt{2\pi C}} \exp\left[-\frac{C(t)^2}{2C}\right], \tag{4.2}$$

where $C(t)$ is the instantaneous i-f clutter voltage, C the average clutter power, and $P'[C(t)]$ the corresponding probability density. Their autocorrelation functions are designated $\psi_c(\tau)$, unnormalized, and $\rho_c(\tau)$, normalized, and their unit-power spectrum is designated $W_c(f)$. These voltages are simulated in the analysis by passing "white (i.e., uncorrelated Gaussian) noise" through an accessory filter whose power spectrum is $W_c(f)$.

The signal voltages are represented by a steady sinusoid at the doppler frequency, f_d, corresponding to a nonfluctuating target moving with velocity $v = \lambda f_d/2$. For reasons that will become apparent later, this signal voltage is applied to the accessory "clutter-shaping" filter, and the input power is adjusted to compensate for the resultant attenuation, $W_c(f_d)$. Thus, at the input to this filter the voltages are

$$C'(t) + \sqrt{\frac{2S}{W_c(f_d)}} \cos(\omega_d t + \phi'), \tag{4.3}$$

where $C'(t)$ is the *uncorrelated* noise of power C, S is the actual (uncompensated) signal power, and ϕ' is an irrelevant phase angle. At the output of the accessory filter then appears

$$C(t) + \sqrt{2S} \cos(\omega_d t + \phi), \tag{4.4}$$

where $C(t)$ has the desired autocorrelation, $\psi_c(\tau)$, and the signal is in the standard form.

4.3.2. Detection and Cancellation

In order ultimately to derive the system-gain formula corresponding to Eq. (2.25), the treatment will be directed along the lines of the finite-pulse-group (i.e., delay-line) systems. The corresponding theory for range-gated-filter systems will then extend directly from this formula, as was the case in Chap. 3.

The simulated signal-plus-clutter voltages, Expression (4.4), are now applied to the square-law detector. In turn, the detected video voltages are applied to a delay-line cancellation network specified by the weight numbers a_j. The impulse time response function for this network is represented by the expression

$$f(t) = \sum_{j=1}^{N} a_j \delta(t = t_j), \tag{4.5}$$

where the $\delta(t = t_j)$ is a unit impulse function at time t_j. The over-all system takes the form shown in Fig. 4.2.

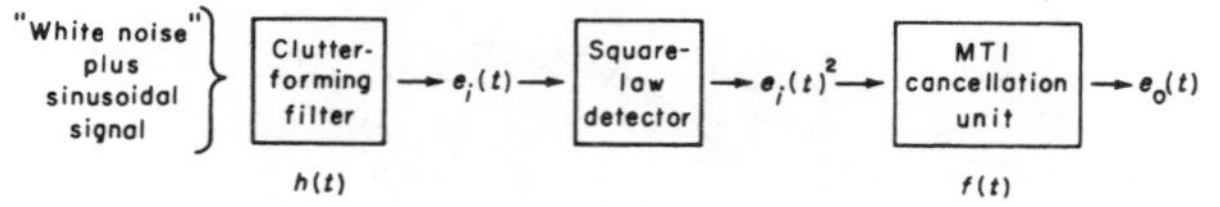

Fig. 4.2—Block diagram of noncoherent system

In Fig. 4.2, $h(t)$ is the impulse time response function for the clutter-forming network.

4.3.3. Output Statistics by the Method of Kac and Siegert

In order to find the system gain, $G(v)$, the output power residues must be computed. Based on a method of Kac and Siegert,[11] Ref. 12 provides the following formula for the second output cumulant (i.e., the power output) for the system above:

$$K_2(t) = (2C)^2\left\{\frac{1}{2}\int_{-\infty}^{\infty} g^2(u,u)\,du + \frac{1}{C}\iint_{-\infty}^{\infty} S(t-u)g^2(u,v)S(t-v)\,du\,dv\right\}$$

$$= 2C^2\left\{\int_{-\infty}^{\infty} g^2(u,u)\,du + \frac{2S}{CW_c(f_d)}\iint_{-\infty}^{\infty} [\cos\omega_d(u-v) + \cos\omega_d(2t-u-v)]g^2(u,v)\,du\,dv\right\}, \tag{4.6}$$

where

$$g^2(u,v) = \int_{-\infty}^{\infty} g(u,x)g(x,v)\,dx, \tag{4.7}$$

and the system "kernel" $g(u,v)$ is given by

$$g(u,v) = \int_{-\infty}^{\infty} h(u-z)f(z)h(v-z)\,dz. \tag{4.8}$$

where $h(z)$ denotes the impulse time response of the clutter-shaping filter.

These formulas simplify as follows: The terms involving $\cos\omega_d(2t-u-v)$ in

Eq. (4.6) represent the high-frequency residue of detection. They are neglected. Next, $f(t)$ from Eq. (4.5) is introduced into the integral (4.8), yielding

$$g(u, v) = \sum_{j=1}^{N} a_j h(u - t_j) h(v - t_j). \tag{4.9}$$

On performing the indicated integration, the second iterate, Eq. (4.7), of $g(u, v)$ is found:

$$g^2(u, v) = \sum_{j=1}^{N} \sum_{k=1}^{N} a_j a_k h(u - t_j) h(v - t_k) \int_{-\infty}^{\infty} h(x - t_j) h(x - t_k)\, dx$$

$$= \sum_{j=1}^{N} \sum_{k=1}^{N} a_j a_k h(u - t_j) h(v - t_k) \rho_c(t_j - t_k). \tag{4.10}$$

The trace of $g^2(u, v)$ is now evaluated, i.e.,

$$\int_{-\infty}^{\infty} g^2(u, u)\, du = \sum_{j=1}^{N} \sum_{k=1}^{N} a_j a_k \rho_c(t_j - t_k)^2. \tag{4.11}$$

This gives the clutter component in Eq. (4.6).

The signal component is

$$\int_{-\infty}^{\infty} \cos \omega_d(u - v) g^2(u, v)\, du\, dv = \sum_{j=1}^{N} \sum_{k=1}^{N} \left\{ a_j a_k \rho_c(t_j - t_k) \times \iint_{-\infty}^{\infty} h(u - t_j) h(v - t_k) \cos \omega_d(u - v)\, du\, dv \right\}. \tag{4.12}$$

The cosine is now replaced with its exponential equivalent, and the double integral reduces to

$$\iint_{-\infty}^{\infty} h(u - t_j) h(v - t_k) \cos \omega_d(u - v)\, du\, dv = W_c(f_d) \cos \omega_d(t_j - t_k). \tag{4.13}$$

Thus, the power output is

$$K_2 = 2C^2 \left\{ \sum_{j=1}^{N} \sum_{k=1}^{N} a_j a_k \rho_c(t_j - t_k)^2 + \frac{2S}{C} \sum_{j=1}^{N} \sum_{k=1}^{N} a_j a_k \rho_c(t_j - t_k) \cos \omega_d(t_j - t_k) \right\}. \tag{4.14}$$

4.3.4. System Gain

The system gain was originally defined by Eq. (1.16). Substituting K_2, Eq. (4.14), for $R(C, S, v)$ yields the formula for $G(v)$ for the noncoherent system:

$$G(v) = 2 \frac{\sum_{j=1}^{N} \sum_{k=1}^{N} a_j a_k \rho_c(t_j - t_k) \cos \omega_d(t_j - t_k)}{\sum_{j=1}^{N} \sum_{k=1}^{N} a_j a_k \rho_c(t_j - t_k)^2}. \tag{4.15}$$

The above formula corresponds to Eq. (2.25) for the coherent system and differs from it in the following respects:

- The clutter autocorrelation function $\rho_c(\tau)$ appears *squared* in the denominator. This is directly traceable to the generation of clutter-clutter cross products in the mixer. The corresponding terms arising from frequency addition among the various clutter pairs, each being of higher frequency than its constituents, contribute to the weakening of the correlation.
- The signal autocorrelation function, $\cos \omega_d \tau$, is multiplied by $\rho_c(\tau)$. This is traceable to the signal-clutter cross products weakening the signal correlation.
- $G(v)$ is multiplied by the factor 2. This is accounted for by the fact that the signal-clutter cross products, absent in the coherent mixer, are identified with the signal and contribute to the signal output power.

4.4. Comparison of Coherent and Noncoherent Gain

The gain for delay-line systems can now be obtained by using the appropriate formulas in Chap. 2, where in place of $\rho_c(\tau)$, $\rho_c(\tau)^2$ is used; in place of $\rho_s(\tau)$, the product $\rho_c(\tau) \times \rho_s(\tau)$ is used;* and $G(v)$ is doubled. This will be carried out for the single-delay-line, single-interpulse-period system in Sec. 4.4.1, below, and it will be seen that the results differ only negligibly from those for the coherent system.

4.4.1. Single-interpulse-period, Single-delay-line System

The reference gain for this system is

$$\bar{G} = \frac{2}{1 - \rho_c(\tau)^2}. \tag{4.16}$$

In order to compare this with the coherent system, $\rho_c(\tau)$ is expanded in the Taylor series, Eq. (2.31). Thus,

$$\bar{G} = \frac{2}{1 - \{1 - \alpha\tau^2 + \beta |\tau|^3 + \cdots\}^2} = \frac{1}{\alpha\tau^2 - \beta|\tau|^3 - \dfrac{\alpha^2 + 2\gamma}{2}\tau^4 - \cdots}. \tag{4.17}$$

*This rule applies only to nonfluctuating targets, i.e., $\rho_s(\tau) = \cos \omega_d \tau$.

This differs from the coherent gain formula, Eq. (2.32), only in the coefficients of τ^4 and higher. For small τ, i.e., for high $\overline{G}$, this difference is negligible.

The corresponding enhancement function is

$$E(v) = 1 - \rho_c(\tau) \cos \omega_d \tau. \tag{4.18}$$

This shows that $\rho_c(\tau)$ enters now as did $\rho_f(\tau)$, the target-fluctuation correlation function, in the coherent analysis. Consequently, depending on the extent of clutter decorrelation, the peaks of $E(v)$ will be reduced, and the dead zones will be made somewhat less severe.

4.4.2. Range-gated Filter Systems

The rules set down in Sec. 4.4 for adjusting the gain formulas for noncoherent operation can be stated more appropriately for range-gated systems in terms of the signal spectrum and the clutter spectrum, $W_c(f)$. They are as follows: Instead of $W_c(f)$, use $W_c^{\mathrm{II}}(f)$, where

$$W_c^{\mathrm{II}}(f) = \int_{-\infty}^{\infty} W_c(y) W_c(f - y)\, dy \tag{4.19}$$

is the autoconvolution of $W_c(f)$. Instead of the spectrum for a steady sinusoidal signal, use $W_c(f - f_d)$; and, finally, double the resulting gain.

For the Gaussian clutter spectrum used in Chap. 3, this results in σ multiplied by $\sqrt{2}$, and in the enhancement curves being smeared as they would be for a fluctuating target with spectrum $W_f(f) = W_c(f)$. Depending on which system is being considered, this may raise or lower the reference gain, but it will usually amount to only a few decibels one way or the other.

CHAPTER 5

INTEGRATION AFTER DETECTION

5.1. Introduction

When a sequence of signal-plus-noise pulses is returned from a target, the detectability of the target can usually be increased by adding or averaging the successive returns. This process, called integration, capitalizes on the relative stability of the target echoes as opposed to the randomly fluctuating noise and tends to sharpen the contrast between them as more pulses are included. If the noise is completely uncorrelated pulse to pulse and is Gaussianly distributed in the i-f (i.e., white noise), and if the target echoes are perfectly stable, then it is well known that the signal detectability increases as if the signal-to-noise ratio had been increased by $\sqrt{N}$, where N is the number of samples integrated in the video.*

The subject of uncorrelated integration has been exhaustively treated by Marcum[9] and others,[10] and their analysis is directly applicable to normal (i.e., non-MTI) radars. In MTI radars, however, the successive noise (i.e., clutter) pulses are not independent—either in the i-f, where correlation is essential to the MTI operation, or in the video output, where additional correlation has been introduced by the pulse-combining action of the cancellation networks. Nevertheless, the independent sample theory and the square-root rule are occasionally applied to MTI calculations. In these cases, rather than all hits per target, N, a smaller number, ηN, with $\eta \sim 1/2$ for two-pulse comparison, etc., is used for the number of "independent hits." This type of modification is in the right direction, insofar as it leads to a degradation in integration efficiency, and a judicious guess with respect to η may be adequate for the purposes at hand.

The correct value for η, however, must depend not only on the type of cancellation mechanism, but on the clutter characteristics as well. For example, if the clutter mass is barren ground terrain, the clutter echoes may be strongly correlated pulse to pulse (perhaps even more strongly than the target echoes), and in this case no benefit whatever can be obtained from integration. Here the equivalent number of independent samples is 1, so that $\eta \sim 1/N$. In contrast with this situation is that of violently blowing rain. In this case the clutter echoes may be essentially independent, and, if two-pulse comparison is used, as many as $2/3\, N$ independent samples may be efficiently integrated.

*Marcum[9] has shown that, with the "best possible detector," i.e., a threshold device with an automatic alarm for calling detections, the detectability corresponds to a signal-to-noise ratio somewhat greater than $\sqrt{N}$. He suggests a law of the type N^{θ}, where $\theta = 1$ for $N = 1$ and diminishes toward about $3/4$ for $N = 1000$. The discrepancy between his N^{θ} law and the $\sqrt{N}$ law, which seems to fit observed data, is attributed to the various inefficiencies associated with visual integration and integration on the display tube.

The purpose of this section is to clarify the dependence of η on both $\rho(\tau)$ and the type of cancellation network. This will be accomplished by means of the Kac-Siegert theory as adapted in Ref. 12, and a formula for η will be derived, based on a square-law video-to-integrator detector. The analysis will be restricted to coherent systems—the methods not being applicable to noncoherent operation.*

5.2. Methods of Integrating Pulses

A certain amount of signal "integration" takes place in the mind and eyes of the operator when he perceives a target and tracks it through the background. For the most part this is scan-to-scan integration, as evidenced by the fact that an operator can usually track a receding target out to a range which is much greater than the acquisition range for oncoming targets. Some pulse-to-pulse integration occurs visually if, as on a PPI, the hits per target are distributed in azimuth so that target echoes exhibit a characteristic arc. This identifying arc then tends to stand out from the random clutter background. These aids to detection, however, are not classed as pulse addition per se.

More in the nature of integration is the process by which pulses are stored and added on the face of the display tube. Efficient integration by this means requires enough dynamic range in the phosphor so that the superposed pulses contribute equally to the final brightness. Some phosphors are better than others in this respect, the P1 (better than most) being capable of integrating efficiently up to about 30 or so weak pulses. The number of pulses presented to the phosphor, of course, being the number of pips per spot diameter, depends on the scan rate, the prf, the pulse width, and the beam resolution of the cathode-ray tube (CRT). Since the choice of these parameters may be forced by other considerations, and since more hits per target may be available than can be efficiently integrated on the display tube, it is desirable to consider other methods of integration which can be employed prior to the display.

For purposes of bandwidth compression, transmission, and assembly of data from many radars onto one central display, a recently developed method for integration prior to transmission is available.† It employs a CRT on which the radar video data are presented with a circular J-type scan. The successive range frames are overlaid on the same circle in such a way that pulses from the same range are superposed on the same spot. In this way integration takes place to within the capabilities of the phosphor. The integrated picture is then read out at a slower rate by means of a rotating photoelectric cell. Tests of this equipment have indicated that the P1 phosphor is somewhat better than the tested P4 and P7, and that it is capable of integrating the 30 or so pulses mentioned earlier. Storage tubes might also be used in this connection.

*A noncoherent system with integration involves two essential nonlinearities—one for the i-f-to-video mixing, and the other between the MTI filter and the integrator. No general formula for the output statistics for such a system is known to the author. However, the various moments can be computed by brute force when the output voltage is determined. This is an extremely tedious method and is not suitable for general analysis.

†This is the "Rafax" flying spot scanner engineered by Haller, Raymond & Brown, Inc. (see Ref. 21).

Similar to the CRT and storage-tube methods, which integrate without loss of range information, are the delay lines, which, when used with sufficient feedback, are in principle capable of integrating a large number of pulses. However, in practice delay lines are fundamentally limited by their latent bandwidth. For example, if without feedback the bandwidth is B, then with feedback corresponding to N-pulse integration, the effective bandwidth is reduced to approximately $B/\sqrt{N}$.* This means that if the pulse width, μ, is to be preserved after integration, the delay line must have a latent bandwidth of the order of $\sqrt{N}/\mu$. This may or may not be severe, depending on the pulse width, and future developments in delay-line transducers may open the way to high-efficiency integration by these means.

The delay-line feedback schemes are appropriate to the finite-pulse-group MTI system of Chap. 2. Their complexity and vulnerability to bandwidth deterioration, etc., are principally outgrowths of the requirement for range preservation. Such is not the case for the range-gated MTI systems, and so integration for these systems comes more easily. It is accomplished in each range channel separately by rectifying the MTI filter output and inserting a low-pass filter of bandwidth $B \sim 1/N\tau$. The outputs of the respective integration filters are then rectified (if desired) and assembled for display.

5.3. Detectability Criterion

In order to assess the theoretical merits of integration, a more sophisticated criterion than system gain is needed. This is due to the fact that the integrating detector converts a certain amount of a-c power into d-c voltage, which is then compounded by the integrator. The system gain, on the other hand, is a measure of a-c power ratios and is consequently insensitive to the shifts in the d-c output levels which carry the bulk of the integrated information.

A suitable criterion which does not embark on the complicated theory of threshold detection is the deflection criterion of Lawson and Uhlenbeck.[2] According to this criterion, a detection is called if the mean d-c integrated output increases in the presence of a target signal by an amount comparable to the standard deviation of the output noise voltages. That is,

$$C_d = \frac{K_1^{S+N} - K_1^N}{\sqrt{K_2^N}} \sim 1 \tag{5.1}$$

is sufficient for detection.† (Here K_1^{S+N} is the mean d-c integrated output voltage for signal plus clutter, and K_2^N is the variance of the integrated noise.) It is realized that this

*This assumes a Gaussian-shaped pass-band, which may be too pessimistic. More rectangular-shaped characteristics suffer less drastically.

†The deflection criterion actually constitutes a variation of the threshold detector in the following sense: When applied to reading a meter, the mean deflection K_1^{S+N} is estimated by visually integrating or averaging the output voltages for a period of time. This averaged estimate is then required to exceed the threshold $\xi = K_1^N + \sqrt{K_2^N}$. The probability, P_f, of incorrectly calling a detection, i.e., the false-alarm probability, is consequently specified by ξ. Other false-alarm probabilities can be achieved by merely requiring $C_d \sim \gamma$ where $\gamma > 1$ for fewer and $\gamma < 1$ for more frequent false alarms.

criterion is somewhat pessimistic (see Ref. 9 and the footnote on page 55, above) as compared with the ideal detection criterion, but certain intangible losses almost always present in operational gear tend to justify some initial pessimism. Besides, the more elaborate criteria simply do not lead to tractable results.

5.4. Theoretical Analysis

The following sections deal with the mathematical analysis leading to general formulas for estimating integration efficiency both with and without MTI:

5:4.1. Simulation of the Signal and Clutter Voltages

As was the case in treating the noncoherent systems in Chap. 4, the clutter voltages are presumed Gaussian in amplitude density and are simulated by passing white noise through a spectrum-shaping filter, $W_c(f)$. Similarly, the signal is presumed perfectly stable. It is sent through the clutter-shaping filter and then compensated for the resulting attenuation, $W_c(f_d)$. At the output of this filter appear the voltages (see Sec. 4.3.1):

$$C(t) + \sqrt{2S}\cos(\omega_d t + \phi), \tag{5.2}$$

where $C(t)$ is the normally distributed clutter voltage possessing autocorrelation function $\psi_c(\tau)$, and the signal is in standard form.

5.4.2. The Coherent Mixer and MTI Filter

The simulated signal-plus-clutter voltages are now applied to the coherent mixer. This mixer is a nonlinear device which, by virtue of the strong injected coho signal, behaves linearly as far as the relatively weaker signal and clutter voltages are concerned. Consequently, the distorting effects of this mixer can be neglected. Regarding the frequency conversion to video, this has been accounted for by expressing the voltages, Expression (5.2), as they would appear around d-c after conversion. The signal-plus-clutter voltages are now applied directly to the MTI filters.

As was the case with the noncoherent system analysis (see Sec. 4.3.2), the MTI filter will be considered as being the delay-line type. The results are then extendable to range-gated systems, as before.

The impulse time-response function for the delay-line network is designated*

$$f(t) = \sum_{j=1}^{\infty} a_j \delta(t = t_j). \tag{5.3}$$

5.4.3. Detection and Integration

The delay-line residues are now rectified by means of the square-law detector specified earlier (see Sec. 4.3.1), and the output voltages are applied to an integrating filter. This filter is characterized by the impulse time-response function $k(t)$, given by

*Only those a_j corresponding to the set of pulses to be compared are presumed different from zero. In this way the summation may be construed to encompass all j from 1 to ∞.

$$k(t) = \sum_{l=1}^{\infty} b_l \delta(t = t_l), \tag{5.4}$$

where the b_l are associated with the impulse time-response function, $k'(t)$, for a low-pass integrating filter. Specifically,

$$b_l = k'(t_l). \tag{5.5}$$

Alternatively, if delay lines are used for integration, the b_l are associated with the various weight factors according to which the successive pulses are combined in the integrator. The latter interpretation applies to integrators for delay-line MTI, whereas the former applies to the range-gated system.

The entire system is depicted in Fig. 5.1. The impulse time-response function for the clutter-forming filter is denoted by $h(t)$.

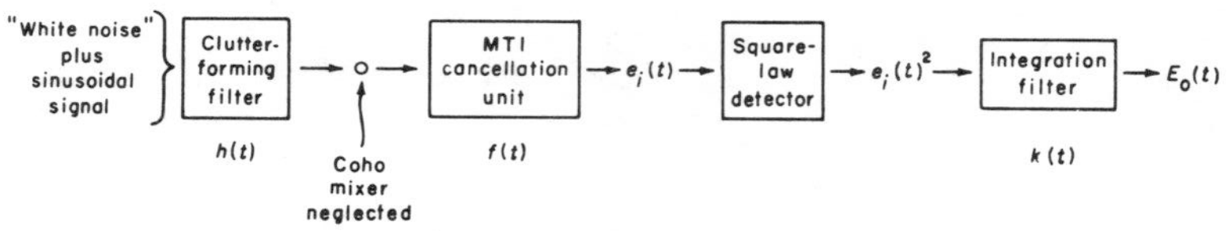

"White noise" plus sinusoidal signal → Clutter-forming filter $h(t)$ → ○ (Coho mixer neglected) → MTI cancellation unit $f(t)$ → $e_i(t)$ → Square-law detector → $e_i(t)^2$ → Integration filter $k(t)$ → $E_o(t)$

Fig. 5.1—Block diagram of video integration model

5.4.4. Output Statistics

On applying the method of Kac and Siegert referred to earlier (see Sec. 4.3.3), the shift in the mean output voltage creditable to signal is found to be

$$K_1^{S+N}(t) - K_1^N(t) = \iint_{-\infty}^{\infty} S(t-u)g(u,v)S(t-v)\,du\,dv$$

$$= \frac{S}{W_c(f_d)} \iint_{-\infty}^{\infty} [\cos\omega_d(u-v)$$

$$+ \cos\omega_d(2t-u-v)]g(u,v)\,du\,dv, \tag{5.6}$$

where the system kernel, $g(u, v)$ is given by

$$g(u,v) = \sum_j \sum_k \sum_l a_j a_k b_l h(u - t_j - t_l) h(v - t_k - t_l). \tag{5.7}$$

The integrals are evaluated first by neglecting the high-frequency terms involving $\cos\omega_d(2t-u-v)$ and then expressing $\cos\omega_d(u-v)$ in exponential form. Thus,

$$K_1^{S+N} - K_1^N = S\Big(\sum_l b_l\Big) \sum_j \sum_k a_j a_k \cos\omega_d(t_j - t_k). \tag{5.8}$$

Similarly, the output power for noise alone is given by

$$K_2 = 2C^2 \int_{-\infty}^{\infty} g^2(u, u)\, du. \tag{5.9}$$

The kernel $g(u, v)$ is now iterated, using Eq. (4.7), yielding $g^2(u, v)$, which is then integrated to give

$$K_2 = 2C^2 \sum_l \sum_m b_l b_m \Big\{ \sum_j \sum_k a_j a_k \rho_c[t_j - t_k + (t_l - t_m)] \Big\}^2. \tag{5.10}$$

The deflection function, Eq. (5.1), reduces to

$$C_d = \frac{X_i}{\sqrt{2}} \frac{(\sum_l b_l) \sum_j \sum_k a_j a_k \cos \omega_d (t_j - t_k)}{\sqrt{\sum_l \sum_m b_l b_m \{\sum_j \sum_k a_j a_k \rho_c [t_j - t_k + (t_l - t_m)]\}^2}}, \tag{5.11}$$

where X_i is the input signal-to-clutter power ratio.

The spectral counterpart of this formula is

$$C_d = \frac{X_i}{\sqrt{2}} \frac{H(f_d)\sqrt{K(0)}}{\sqrt{\iint_{-\infty}^{\infty} W_c(f_1) H(f_1) K(f_1 + f_2) H(f_2) W_c(f_2)\, df_1\, df_2}}, \tag{5.12}$$

where, as usual, $W_c(f)$ is the unit-power spectrum for the clutter, $H(f)$ is the power spectrum for the gated MTI filter, and

$$K(f) = \Big| \sum_l e^{i\omega t_l} b_l \Big|^2 \tag{5.13}$$

is the power spectrum for the gated integrating filter.

If delay lines are employed for both the MTI filter and the integration filter, then the first formulation, Eq. (5.11), is directly applicable. However, the second expression, Eq. (5.12), may be used, and in this case $H(f)$ and $K(f)$ are the actual power spectra for the delay-line units, i.e., gating is unnecessary. If delay lines are to be used for MTI, together with a low-pass filter for integration, then range gates must be employed in the integrator to preserve the range information. In this case $H(f)$ is the actual delay-line power spectrum, and $K(f)$ is the gated spectrum for the low-pass filter. Similarly, if the entire system is range gated, then both $H(f)$ and $K(f)$ should be gated spectra. (If the clutter spectrum is reasonably confined, occupying only a small portion of the interval between prf harmonics, i.e., $\sigma\tau << 1$, then this gating operation is relatively unimportant as far as computing C_d is concerned.* This will become apparent in the next paragraph.)

*With one exception—$H(f_d)$ in the numerator should be the gated spectrum so that the effects of velocity enhancement will be accurately portrayed.

A rearrangement of the terms in the denominator of Eq. (5.12) makes possible a simple interpretation of the double integral, as follows:

$$C_d = \frac{X_i}{\sqrt{2}} \frac{H(f_d)\sqrt{K(0)}}{\sqrt{\int_{-\infty}^{\infty} K(f) L(f)\, df}}, \tag{5.14}$$

where

$$L(f) = \int_{-\infty}^{\infty} \{W_c(g) H(g)\}\{W_c(f - g) H(f - g)\}\, dg \tag{5.15}$$

is the convolved power spectrum for the MTI filter power residues. The procedure is illustrated in Fig. 5.2.

The denominator in Eq. (5.14) is the square root of the total power in $K(f) \times L(f)$, i.e., the square root of the area under curve (f) in Fig. 5.2. A relatively broad clutter spectrum is indicated in curve (a), resulting in a considerable power density beyond the main groupings in curve (c). This will not be the case if the clutter is reasonably correlated over the interpulse period, and the side groups will possess correspondingly less energy. The convolved spectrum, curve (d), appears at the output of the square-law detector.* The spectrum also exhibits a large amount of energy at the prf and higher frequencies which would not be present for a narrower clutter spectrum. The integration filter now accentuates $L(f)$ in the vicinity of d-c and the prf harmonics, with the result shown in curve (f). Thus, had the clutter spectrum been realistically narrow, the high-frequency components could have been negligible, and the ungated spectra could have been used throughout.

In Eq. (5.14), if $H(f)$ is normalized so that $\overline{H(f)} = 1$, and if $K(f)$ is normalized so that $K(0) = 1$, then C_d reduces to

$$C_d = \frac{X_i}{\sqrt{2}} \frac{E(v)}{\sqrt{\int_{-\infty}^{\infty} K(f) L(f)\, df}}. \tag{5.16}$$

This is often a more convenient form.

5.4.5. Signal-to-clutter-ratio Improvement Factor

Rather than to deal with the effective number of uncorrelated pulses integrated, ηN, it is more convenient to use the signal-to-clutter-ratio improvement factor $\gamma = \sqrt{\eta N}$. This factor describes the efficiency of the video integration when integration of the

*This spectrum is unlike the power spectrum for Gaussian noise in that it does not tell the whole story about the detected voltages. There is, in fact, an elaborate interphase correlation over frequency which is not accounted for in $L(f)$, and which materially influences the statistics of the rectified voltages. (For Gaussian noise the phases are uncorrelated, so that the spectrum of the power intensities is all that needs to be specified. Rectified noise, however, is not Gaussian, and the phase correlation influences the amplitude distribution—e.g., the output voltage is nonnegative, etc.). If the probability of detection is sought, then these phase correlations must be accounted for.

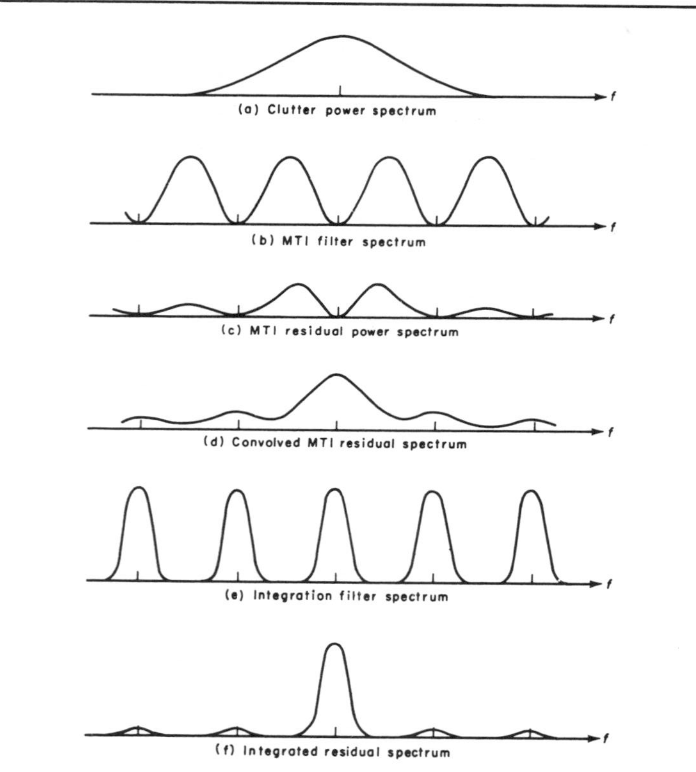

Fig. 5.2—Clutter spectra in transition through MTI filter, square-law detector, and integration filter

correlated clutter has previously taken place in the i-f.* Specifically, γ is defined by

$$\gamma = \frac{C_d(N \text{ correlated pulses integrated})}{C_d(1 \text{ pulse})}. \tag{5.17}$$

Notice that, regardless of the influence of the MTI network and the clutter correlation, $\gamma = 1$ for no video integration. It will generally be larger than unity if pulses are integrated after detection, but its size will depend on the initial clutter correlation and the amount of additional correlation or decorrelation that has been introduced by the MTI

*MTI filtering can be interpreted as frequency-sensitive i-f integration, the sensitivity as a function of doppler frequency being proportional to $E(v)$. Clutter in this sense is integrated poorly (i.e., rejected), and the moving target echoes in the pass-band of $E(v)$ are integrated more or less effectively.

filter. In general, the more correlated the clutter, and the more i-f integration there is, the smaller γ will become for any fixed N. In the limit, for completely correlated clutter, $\gamma = 1$ for all N. At the opposite extreme, for "white" clutter, "infinite" i-f bandwidth, and no MTI, γ attains its largest value, $\sqrt{N}$.

C_d, for no video integration (i.e., $N = 1$), can be evaluated from Eq. (5.11) by setting $b_1 = 1$ and $b_l = 0$ for $l \neq 1$. Thus,

$$C_d(1 \text{ pulse}) = \frac{X_i}{\sqrt{2}} \frac{\sum_j \sum_k a_j a_k \cos \omega_d (t_j - t_k)}{\sum_j \sum_k a_j a_k \rho_c (t_j - t_k)}. \tag{5.18}$$

In terms of the system gain, see Eq. (2.25), this is

$$C_d(1 \text{ pulse}) = \frac{X_i}{\sqrt{2}} G(v). \tag{5.19}$$

Consequently, except for the factor $\sqrt{2}$, which can be associated with a false-alarm time, the deflection function is simply the product of the input signal-to-clutter ratio and the system gain.

With video integration, the deflection function is (by definition of γ) equal to

$$C_d = \frac{X_i}{\sqrt{2}} G(v) \cdot \gamma. \tag{5.20}$$

The video integration factor, γ, can thus be written

$$\gamma = \frac{\sum_l b_l \sum_j \sum_k a_j a_k \rho_c (t_j - t_k)}{\sqrt{\sum_l \sum_m b_l b_m \{\sum_j \sum_k a_j a_k \rho_c [t_j - t_k + (t_l - t_m)]\}^2}}. \tag{5.21}$$

Alternatively, in terms of the spectra,

$$\gamma = \left\{ \frac{K(0) \int_{-\infty}^{\infty} L(f)\, df}{\int_{-\infty}^{\infty} L(f) K(f)\, df} \right\}^{1/2}. \tag{5.22}$$

5.5. Integration of Clutter without MTI

Apart from any MTI considerations, the generality of the treatment so far permits a comparison of the relative merits of i-f and video integration when used separately or in combination. This will be discussed in Sec. 5.5.1, where white noise will be treated. In Sec. 5.5.2 the video integration of Gaussianly correlated clutter will be discussed. These topics serve as an introduction to the video integration of correlated noise with MTI.

5.5.1. Video vs I-F Integration of White Noise (Ideal Integrator)

The formula, Eq. (5.21), will now be evaluated for white noise with an ideal i-f integrator and an ideal video integrator. The white noise is specified by setting $\rho_c(0) = 1$ and $\rho_c(\tau > 0) = 0$. The i-f integrator is specified by the numbers $a_j = 1$ for $j = 1, 2, \cdots, M$, and otherwise $a_j = 0$. It can consist of $M - 1$ delay lines with feed-forward circuits connected in cascade in such a way as to add successive sets of M pulses. The video integrator can consist of a similar set of $N - 1$ delay lines for integrating N pulses. It will be characterized by $b_l = 1$ for $l = 1, 2, \cdots, N$, and otherwise $b_l = 0$.

Under these conditions, Eq. (5.21) becomes

$$\gamma = \frac{NM}{\sqrt{\sum_{l=1}^{N} \sum_{m=1}^{N} \{(M - |l - m|) \vee 0\}^2}}, \tag{5.23}$$

where the symbol $a \vee b$ means "the greater of a and b." Depending on whether N is larger or smaller than M, γ assumes different forms. They are

$$\gamma(M < N) = \left\{ \frac{N}{1 + \frac{M-1}{6MN}[4MN - 2N - M - M^2]} \right\}^{1/2}, \tag{5.24}$$

$$\gamma(M > N) = \left\{ \frac{N}{1 + \frac{N-1}{6M^2}[6M^2 - 4M - 4MN + N + N^2]} \right\}^{1/2}, \tag{5.25}$$

$$\gamma(M = N) = \frac{N\sqrt{2}}{\sqrt{N^2 + 1}}. \tag{5.26}$$

On examination of Fig. 5.3 the degrading effect on γ of integration prior to detection

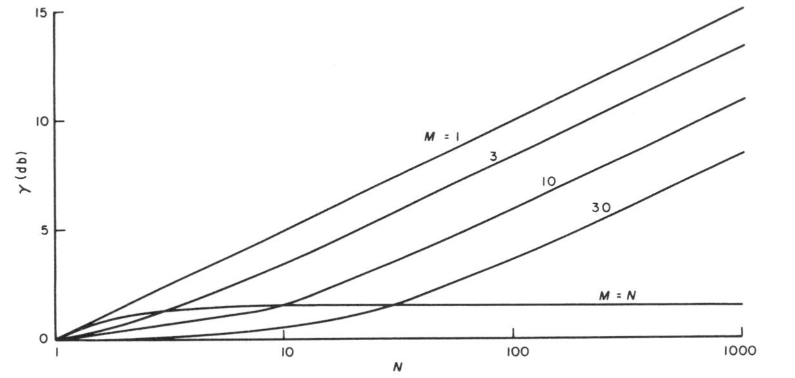

Fig. 5.3—Signal-to-noise-ratio improvement factor for mixed integration without MTI (*N* video and *M* i-f pulses integrated with delay lines)

becomes apparent.* This degradation is expressed in terms of the equivalent fraction, η, of independent (i.e., not previously integrated) pulses shown in Fig. 5.4.

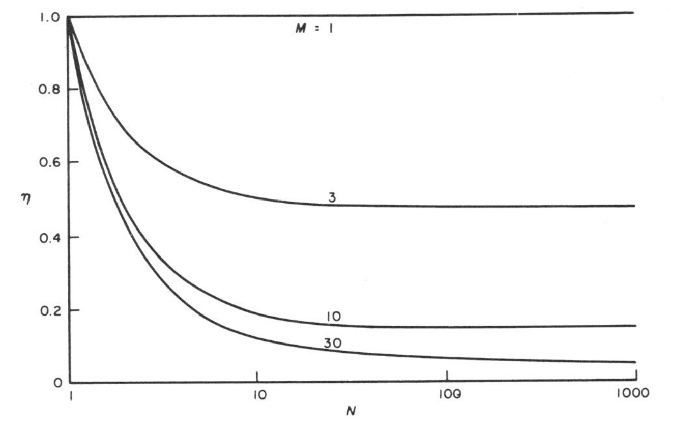

Fig. 5.4—Equivalent fraction of independent pulses for mixed integration without MTI (*N* video and *M* i-f pulses integrated with delay lines)

5.5.2. Integration of Gaussianly Correlated Clutter (Low-pass Gaussian Integration Filter)

In these paragraphs the spectral formula for γ, Eq. (5.22), will be evaluated. The Gaussianly correlated clutter will be used here as in Eq. (3.24), and no MTI filter will be employed (i.e., $H(f) = 1$). Consequently,

$$W'_c(f)H(f) = W'_c(f) = \frac{1}{\sqrt{2\pi\sigma^2}} e^{-f^2/2\sigma^2}. \tag{5.27}$$

This is easily convolved, yielding

$$L(f) = \frac{1}{\sqrt{4\pi\sigma^2}} e^{-f^2/4\sigma^2}, \tag{5.28}$$

for which

$$\int_{-\infty}^{\infty} L(f)\, df = 1. \tag{5.29}$$

*This is not to say that i-f integration is not advantageous; on the contrary, it is more beneficial than video integration. But this improvement appears in the enhancement factor. It amounts to an additional factor of M multiplying γ on arriving at C_d.

For video integration a low-pass Gaussian filter* of half-bandwidth δ will be used. Its power spectrum is

$$K(f) = e^{-f^2/2\delta^2}. \tag{5.30}$$

The product $L(f)K(f)$ is now easily integrated

$$\int_{-\infty}^{\infty} L(f)K(f)\,df = \sqrt{\frac{\delta^2}{2\sigma^2 + \delta^2}}, \tag{5.31}$$

so that, finally,

$$\gamma = \sqrt[4]{1 + 2\left\{\frac{\sigma\tau}{\delta\tau}\right\}^2}, \tag{5.32}$$

where, as before, $\sigma\tau < 1$.† This is plotted in Fig. 5.5.

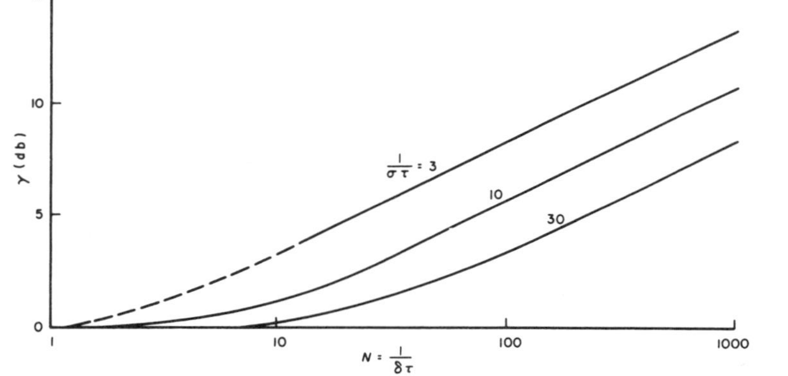

Fig. 5.5—Signal-to-clutter-ratio improvement factor for video integration of correlated clutter without MTI (clutter Gaussianly correlated with spectral width 2σ; Gaussian integration filter of bandwidth δ)

If the quantity $1/\delta\tau$ is identified with N, the number of correlated pulses integrated in the video, then the fraction η, effectively integrated, becomes

$$\eta = \frac{1}{N}\sqrt[4]{1 + 2N^2(\sigma\tau)^2}. \tag{5.33}$$

This appears in Fig. 5.6.

*Such a filter is actually unrealizable. It is useful, however, for purposes of evaluation and helps provide a feeling for the general dependence on bandwidths, etc.

†In these formulas τ is the interpulse period for a single-period system. If a multiple-period system is used and the various periods are nearly equal, then τ can be any one of the periods.

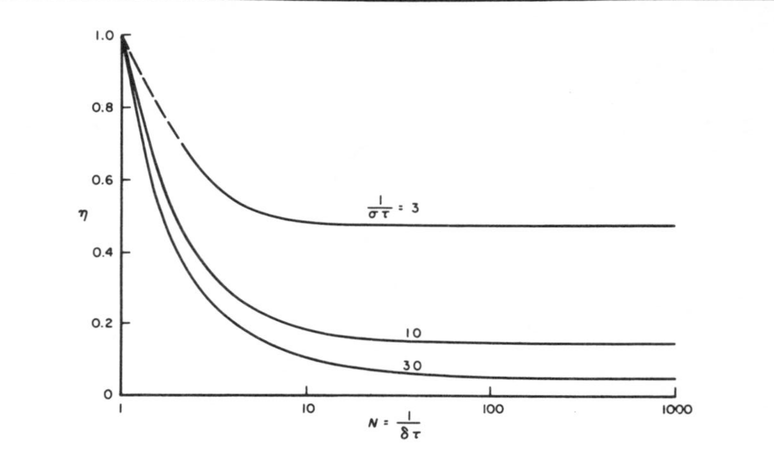

Fig. 5.6—Equivalent fraction of independent pulses for video integration of correlated clutter without MTI (clutter Gaussianly correlated with spectral width 2σ; Gaussian integration filter of bandwidth δ)

5.6. Integration of Correlated Clutter with MTI Filter

When the MTI is included, the evaluation of the formulas for γ becomes much more difficult. In order to simplify the computation, the work will be confined to single-interpulse-period systems. This will not incur very much loss in generality because, for multiple-period systems having periods almost equal, $H(f)$ near the origin will be very nearly the same as for single-period systems, and, if the clutter is reasonably correlated over the interpulse period, these differences can be neglected.

Sections 5.6.1 and 5.6.2, below, will be devoted to the "ideal" N-pulse integrator in connection with two- and three-pulse-comparison delay-line MTI. The sections following these will deal with range-gated MTI using a broad-band rectangular filter (see Sec. 5.6.3.1) and a narrow-band velocity gate (see Sec. 5.6.3.2). For each of these gated systems a low-pass narrow-band filter will be used for video integration.

In all these cases the clutter correlation will be presumed Gaussian.

5.6.1. Two-pulse-comparison, Single-period Ideal Integrator

For two-pulse comparison, the weight numbers a_j are chosen to be $a_1 = 1$, $a_2 = -1$, and otherwise $a_j = 0$. The ideal integrator is characterized by $b_j = 1$ for $j = 1, 2, \cdots, N$, and otherwise $b_j = 0$. And, for the single-interpulse period, τ, the $t_j - t_k = (j - k)\tau$.

Equation (5.21) is now used for computing γ. It becomes

$$\gamma = \frac{A}{\sqrt{B}}, \tag{5.34}$$

where

$$A = 2N[1 - \rho_c(\tau)], \tag{5.35}$$

and

$$B = 4N[1 - \rho_c(\tau)]^2 + 2(N - 1)[1 - 2\rho_c(\tau) + \rho_c(2\tau)]^2 + 2(N - 2)[\rho_c(\tau) - 2\rho_c(2\tau) + \rho_c(3\tau)]^2 + \cdots. \tag{5.36}$$

This has been evaluated for selected values of N and σ, using the Gaussian correlation function

$$\rho_c(\tau) = \exp[-(2\pi\sigma)^2(\tau^2/2)], \tag{5.37}$$

which corresponds to the Gaussian power spectrum, Eq. (3.24), used earlier.

For large values of σ, this can be computed by hand. Specifically, for $\sigma = \infty$,

$$\gamma(\sigma = \infty) = \frac{2N}{\sqrt{6N - 2}}. \tag{5.38}$$

It will be noted that this same value is obtained from Eq. (5.24) by setting $M = 2$. This is as it should be, since, for "white" clutter, pulse subtraction and pulse addition are statistically indistinguishable in the i-f.

For large values of N, γ was computed by means of IBM machinery. The general results appear in Fig. 5.7.

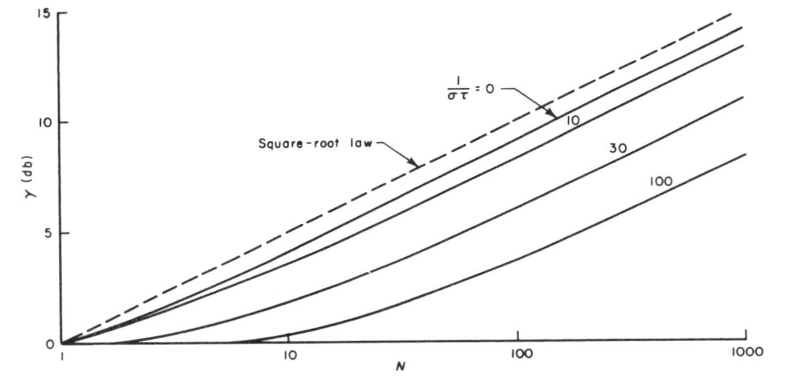

Fig. 5.7—Signal-to-clutter-ratio improvement factor for video integration of N pulses with delay lines (two-pulse-comparison, single-interpulse-period MTI)

5.6.2. Three-pulse-comparison, Single-period Ideal Integrator

For three-pulse-comparison MTI, the binomial combination of weights will be used,

i.e., $a_1 = 1$, $a_2 = -2$, $a_3 = 1$; otherwise $a_j = 0$. This choice corresponds to the selection of Sec. 2.4.1.1 for the single-period system.

Equation (5.21) now reduces to

$$\gamma = \frac{A}{\sqrt{B}}, \tag{5.39}$$

where

$$A = 2N[3 - 4\rho(\tau) + \rho(2\tau)], \tag{5.40}$$

and

$$B = 4N[3 - 4\rho(\tau) + \rho(2\tau)]^2 + 2(N - 1)[-4 + 7\rho(\tau) - 4\rho(2\tau) + \rho(3\tau)]^2 + 2(N - 2)[1 - 4\rho(\tau) + 6\rho(2\tau) - 4\rho(3\tau) + \rho(4\tau)]^2 + \cdots. \tag{5.41}$$

In the limiting case for white clutter,

$$\gamma(\sigma = \infty) = \frac{6N}{\sqrt{70N - 36}} \quad \text{for } N \geq 2. \tag{5.42}$$

By virtue of the nonuniformity of the weight numbers, this formula does not correspond to any of those obtained in Sec. 5.5.1 for the i-f integration of white noise. It would correspond, however, to that case in which three i-f samples were added with the weights 1, 2, 1.

The general formula, Eq. (5.39), has been evaluated for selected values of N and σ for the Gaussian correlation function, Eq. (5.35), by means of IBM machinery. The results are given in Fig. 5.8.

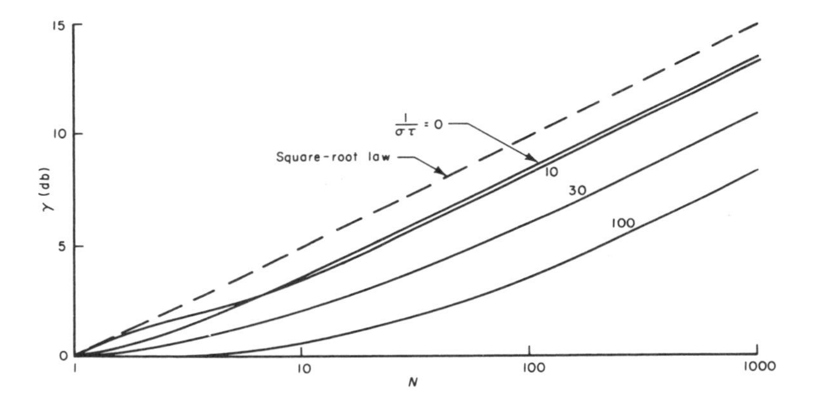

Fig. 5.8—Signal-to-clutter-ratio improvement factor for video integration of N pulses with delay lines (three-pulse-binomial-comparison, single-interpulse-period MTI)

5.6.3. Range-gated MTI Systems with Narrow-band Video Integration

For a narrow-band video integrator, the spectral formula for γ, Eq. (5.22), may be approximated by

$$\gamma \sim \frac{1}{\sqrt{\delta}} \cdot \frac{\int_{-\infty}^{\infty} W_c(f)H(f)\, df}{\sqrt{\int_{-\infty}^{\infty} \{W_c(f)H(f)\}^2\, df}}, \tag{5.43}$$

where δ is the video bandwidth ($-\delta/2$ to $\delta/2$). This formula is deduced from the identities

$$\int_{-\infty}^{\infty} L(f)\, df = \left\{\int_{-\infty}^{\infty} W_c(f)H(f)\, df\right\}^2 \tag{5.44}$$

and

$$L(0) = \int_{-\infty}^{\infty} \{W_c(f)H(f)\}^2\, df. \tag{5.45}$$

The integral ($-\infty$ to ∞) of $L(f)K(f)$ is then approximated, for $\delta << \sigma$, by

$$\int_{-\infty}^{\infty} L(f)K(f)\, df \sim \delta K(0)L(0). \tag{5.46}$$

Equation (5.43) follows immediately.

5.6.3.1. Broad-band Rectangular MTI Filter

The broad-band MTI system of Sec. 3.4.1 will now be considered. The following expression will be used for $H(f)$:

$$H(f) = \begin{cases} 0 & \text{for } |f| < \dfrac{1}{2\tau} - \dfrac{B}{2} \\ \dfrac{1}{B\tau} & \text{otherwise.} \end{cases} \tag{5.47}$$

This neglects the periodic recurrence of pass intervals, but for small σ it is unimportant (see discussion, Sec. 5.4.4).

Equation (5.43) for γ now reduces to

$$\gamma = \sqrt[4]{4\pi} \cdot \sqrt{\frac{\sigma\tau}{\delta\tau}} \cdot \chi, \tag{5.48}$$

where

$$\chi = \frac{1 - \operatorname{erf}\left(\dfrac{1 - B\tau}{2\sqrt{2}\,\sigma\tau}\right)}{\sqrt{1 - \operatorname{erf}\left(\dfrac{1 - B\tau}{2\sigma\tau}\right)}}. \tag{5.49}$$

This result shows that the integration efficiency, γ, again depends on the ratio of the video bandwidth, δ, to the spread of clutter frequencies, σ. And, as before, the square-root rule applies. The quantity γ/χ appears in Fig. 5.9.

The factor χ describes the degrading effect on γ of the MTI filtering. It is plotted in Fig. 5.10. As the cut-off frequency, f_c, for the broad-band MTI filter is adjusted so as to reject more and more of the clutter energy, the filter residue becomes correspondingly more correlated. This results in the video integration process becoming less effective. At

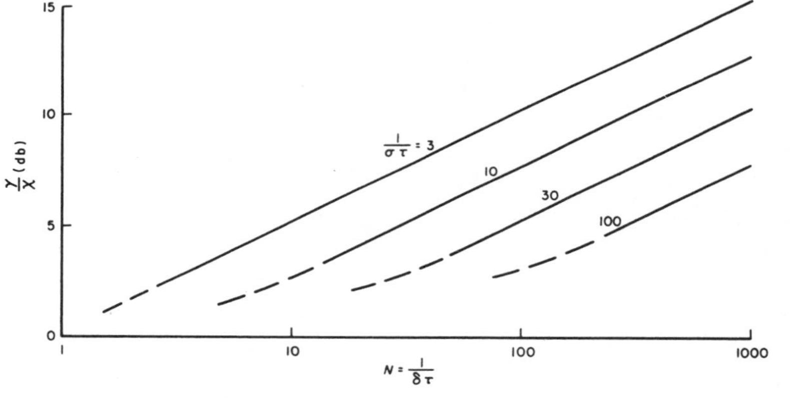

Fig. 5.9—Relative signal-to-clutter-ratio improvement factor for narrow-band video integration filter and range-gated broad-band rectangular MTI filter (clutter Gaussianly correlated with spectral width 2σ; integration bandwidth δ)

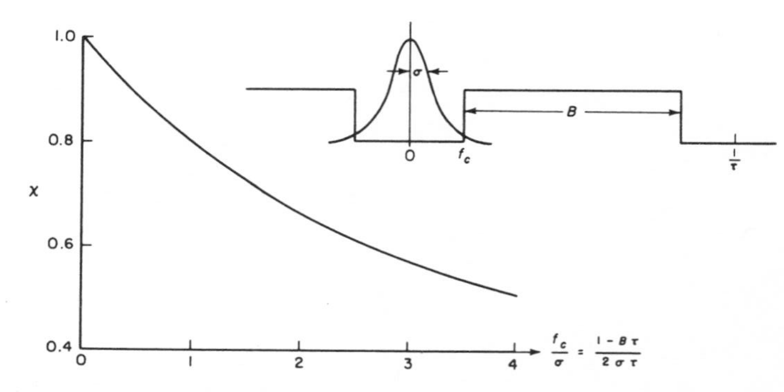

Fig. 5.10—Signal-to-clutter-ratio degradation factor due to MTI filtering (for use with Fig. 5.9)

the other extreme, when $f_c = 0$, no MTI filtering takes place, and, as was the case in Sec. 5.5.2, γ then varies as the square root of σ/δ without degradation. Of course, the restrictions $\delta\tau << \sigma\tau << 1$ still apply.

5.6.3.2. Narrow-band Velocity-gate MTI Filter

The narrow-band rectangular filter system of Sec. 3.4.2 will now be considered. For this system the function

$$H(f) = \begin{cases} \frac{1}{2B\tau} & \text{for} \mid |f| - f^* \mid < \frac{B}{2} \\ 0 & \text{otherwise} \end{cases} \tag{5.50}$$

will be used, which neglects the recurring pass-band pairs beyond the clutter region.

In this case γ becomes

$$\gamma = \sqrt{2\frac{B\tau}{\delta\tau}}, \tag{5.51}$$

which appears in Fig. 5.11.

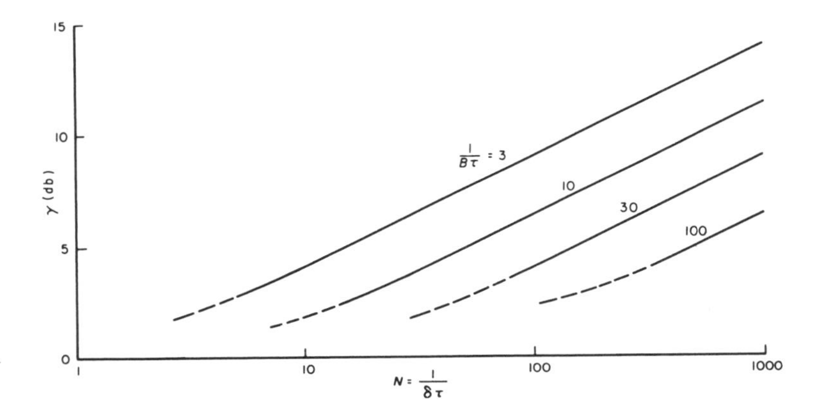

Fig. 5.11—Signal-to-clutter-ratio improvement factor for narrow-band video integration filter and range-gated narrow-band "velocity-gate" MTI (video integration bandwidth δ; velocity-gate bandwidth B)

Notice that γ previously depended on the ratio σ/δ, whereas now it depends on the ratio B/δ. This is as it should be; for, as far as the integration process is concerned, the clutter might as well have spread an amount B instead of the larger amount σ, the energy outside this band being entirely rejected by the velocity gate.

5.7. Conclusions

In this section an attempt has been made to study the effects of postdetector integration when used in conjunction with MTI equipment. Although analysis based on probability of detection would be more desirable, a compromise was necessary because of the intractabilities of such an approach. The alternative procedure selected leads to the integration efficiency factor, γ, which indicates the amount by which the signal-to-noise ratio can be smaller with video integration than without for the same degree of performance, as measured by the deflection criterion of Lawson and Uhlenbeck. This factor, γ, as well as a related quantity, η, which is the "equivalent fraction of *independent* pulses integrated," has been evaluated for a selection of special cases, and the results appear in graphical form in the figures.

Although the principal results of this section are numerical, there are some general conclusions that deserve special mention. In particular, the analysis of video integration for correlated pulses reveals a characteristic of i-f integration which is not well known. Hitherto, it has generally been agreed that the signal-to-noise-ratio benefit for i-f (or *coherent*) integration accrues in direct proportion to the number of pulses integrated, whereas for video integration the square-root or N^θ rule (see footnote, page 55) applies. That this is not the case when a mixture of i-f and video integration is used is illustrated in Fig. 5.12. This is a cross plot of Fig. 5.3 applicable to delay-line integrators in both the

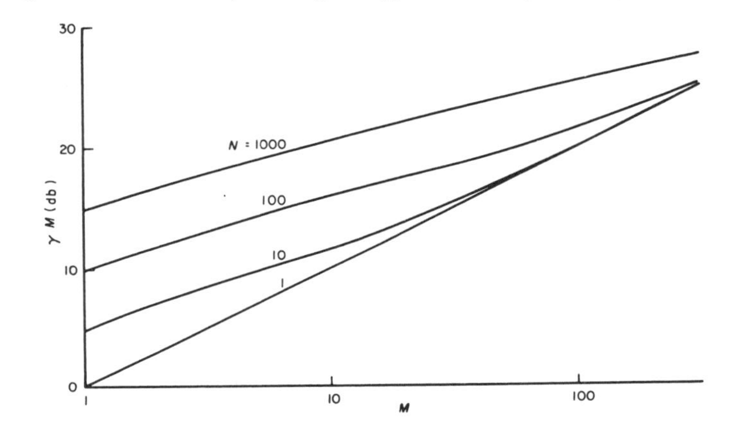

Fig. 5.12—Over-all signal-to-clutter-ratio improvement factor for mixed integration without MTI (N video and M i-f pulses integrated with delay lines)

i-f and the video.* It will be noted that when $M < N$, the performance criterion very

*The ordinate $M\gamma$, which equals $2C_d/X_i$, see Eq. (5.20), is used here rather than γ in order to include the i-f integration factor, $G(v)$, which in this case is equal to M, the number of pulses integrated in the i-f.

nearly follows a square-root law as a function of M. On the other hand, for $M > N$ the rule approximates the well-known linear law. The explanation for this behavior lies in the reduced efficiency of the video integrator arising out of the correlations introduced by the i-f integrator.

Another interesting result is that a comparison of the data for integration with MTI with those for integration without MTI shows that there are cases in which the introduction of MTI enhances the performance of the postdetection integrator. At first this may seem peculiar, since MTI produces a kind of i-f integration (see footnote, page 62) and it has been shown that the video integration efficiency is usually degraded by prior integration in the i-f. However, the somewhat enigmatic situation occurs only with the two- and three-pulse-comparison system considered.

The explanation is found on examining the autocorrelation functions for the clutter at the output of the MTI circuits. For two-pulse comparison, this autocorrelation function is proportional to the second difference of ρ_c, and for three-pulse comparison it is proportional to the fourth difference. For the Gaussian case, these higher differences result in generally reduced correlation over a range of values of σT, and it is this decorrelating action of the MTI filter which is responsible for the slightly greater video integration efficiency.

CHAPTER 6

CLUTTER AND SYSTEM NOISE

6.1. Introduction

The characteristic fluctuations of any particular patch of clutter return arise from the combined influence of several statistically independent factors. First among these is the relative motion of the reflecting particles constituting the clutter mass. This "internal motion"—the extent of which depends on sea state and/or atmospheric conditions, such as wind velocity and turbulence as well as the mobility of the particles—accounts for the basic randomness of the return.

In addition to the internal-motion fluctuations, there will, in general, be variations in the echo intensity due to scanning. The severity of this "scanning noise" depends on the ratio of the scanning rate to the beam width, i.e., it depends on the time on target, becoming more severe as this interval is reduced. Thus, depending on the information rate of the radar system, the scanning noise may be negligible or it may dominate the basic internal fluctuations.

A third factor affecting the clutter statistics is the so-called motion noise. This is a phenomenon encountered in airborne or shipborne installations where the radar platform undergoes a translational motion with respect to the clutter mass. Motion noise comes about because the motion of the antenna imparts to the various clutter particles an additional gradation of relative radial velocities. This results in an increased spread of the clutter power spectrum, the extent of which depends directly on the planar translation velocity of the antenna and inversely on the antenna size. Thus, for large platform velocities and small antennas sighted well off the radar path, the motion noise can be severe—in fact, for high-speed AI radars it is the most important single factor affecting MTI performance.

The fourth factor, system-instability noise, is included here for completeness, although it might properly be discussed separately as a distinct phenomenon on a par with clutter. These noises arise because of basic equipment imperfections, such as magnetron interpulse amplitude and frequency variations, repetition-frequency and pulse-width jitter, stalo and coho instabilities, pulse distortion, power-line ripple, etc. In general, radars for use with MTI should be designed to have enough stability so that the operational MTI gain reasonably approaches (say, within 4 or 5 db) the theoretical gain. This means that, with more elaborate cancellation schemes for obtaining better clutter rejection, more stringent requirements for stability are imposed on the equipment. This is presently not a serious problem with the conventional (properly designed and adjusted) two-pulse delay-line systems, but it should be kept in mind when evaluating and comparing the more complicated arrangements.

6.2. Internal Motion

Direct measurements of internal-motion noise have been made by H. Goldstein[22] using A-scope photographic techniques. The data obtained were processed and converted to probability distributions and autocorrelation functions. The power spectra for certain representative types of clutter are given by E. J. Barlow.[8] Both investigators suggest the Gaussian form

$$W_{im}(f) = \frac{\lambda}{2v\sqrt{2\pi}} \exp\left(-\frac{\lambda^2 f^2}{8v^2}\right) \tag{6.1}$$

for the power spectrum, and

$$\rho_{im}(\tau) = \exp\left(-8\pi^2 \frac{v^2\tau^2}{\lambda^2}\right) \tag{6.2}$$

for the normalized autocorrelation function. The parameter, v, is the standard deviation of the particle velocities about the mean velocity of the clutter mass. If the radar wavelength, λ, is measured in meters, the frequency, f, in cycles per second, and the time interval, τ, in seconds, the values for the parameter, v, corresponding to the representative clutter types of Barlow are given below:

Description of Clutter Mass	v (knots)
Sparsely wooded hills, calm day	.01
Heavily wooded hills, 20-mph wind	.50
Sea echo, windy day	1.8
Window "jamming"	2.0
Rain clouds	3.9

Graphs of the normalized autocorrelation functions and power spectra, Eqs. (6.1) and (6.2), appear in Figs. 6.1 and 6.2. These curves should be interpreted as idealizations in the sense that they represent the decorrelating effect of internal motion only. A perfectly stable transmitter and receiver is assumed, having no thermionic noise. The presence of any poorly-correlated noise (as would be the case in the best of controlled measuring experiments) would destroy the gently quadratic nature of the curves at the origin. Experiments of this type were conducted by the Admiralty Signal Establishment,[6] from which is was concluded that $\rho_{im}(\tau)$ for sea return behaves linearly for small τ. The disagreement between these results and those of Goldstein and Barlow could be due in part to system noise. This cannot be decided definitely, however, because their interpulse period (2 ms) is too long (for the wavelengths of 3 and 10 cm used) to obtain a good description of $\rho_{im}(\tau)$ near the origin.

The Philco Corporation[23] has conducted tests to determine the probability distributions and correlation functions for various types of land clutter as seen from the air. Their results also indicate nonquadratic behavior of $\rho_{im}(\tau)$ near the origin, but this is consistent with their theoretical evaluation, which takes account of system noise. They

estimate that from 8 to 10 per cent of the transmitted power was of low MTI grade, the effect of which, if deleted from their measured data, would lead to results in agreement with those of Goldstein and Barlow.

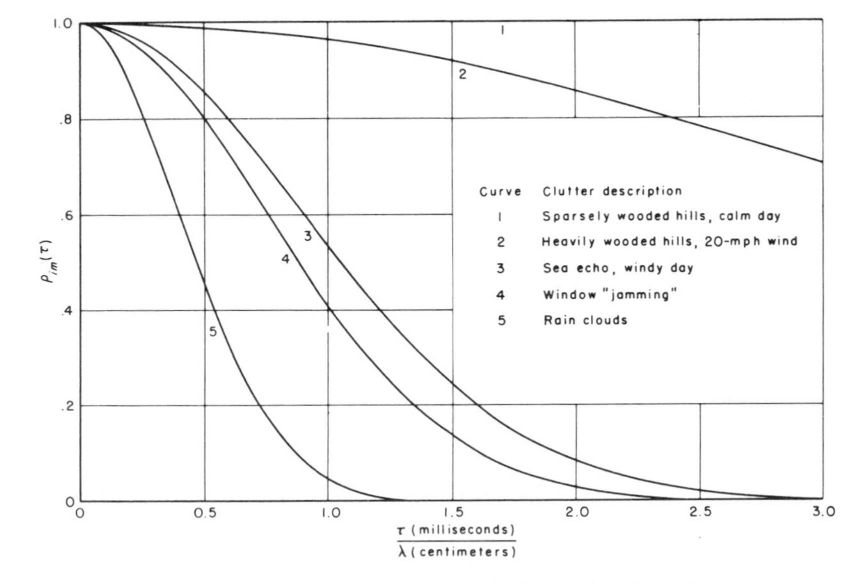

Fig. 6.1—Autocorrelation functions for internal-motion noise

There is some feeling that $\rho_{im}(\tau)$ should not be linear in τ for small τ. The argument is based on the fact that the highest frequencies in the power spectrum, $W_{im}(f)$, are finite; i.e., $W_{im}(f) > 0$ only in some finite range $-v/2\lambda \leq f \leq v/2\lambda$, where v is the highest particle velocity in the clutter mass. Furthermore, $W_{im}(f)$ is symmetric, i.e., $W_{im}(-f) = W_{im}(f)$. Consequently,

$$\alpha = \int_{-\infty}^{\infty} f W_{im}(f)\, df \text{ exists and } = 0. \tag{6.3}$$

On differentiating the Wiener-Khintchine relation

$$\rho_{im}(\tau) = \int_{-\infty}^{\infty} e^{i\omega\tau} W_{im}(f)\, df \tag{6.4}$$

with respect to τ and setting $\tau = 0$, one obtains

$$\rho'_{im}(0) = 2\pi i\alpha = 0, \tag{6.5}$$

so that $\rho_{im}(\tau)$ is at least quadratic in τ near the origin. Conversely, since $\rho_{im}(-\tau) = \rho_{im}(\tau)$, if

$$\lim_{\tau \to +0} \rho'_{im}(\tau) \neq 0, \tag{6.6}$$

then $\rho'_{im}(0)$, and consequently the integral α, does not exist. This failure to converge can only be due to energy in $W_{im}(f)$ at unbounded frequencies.

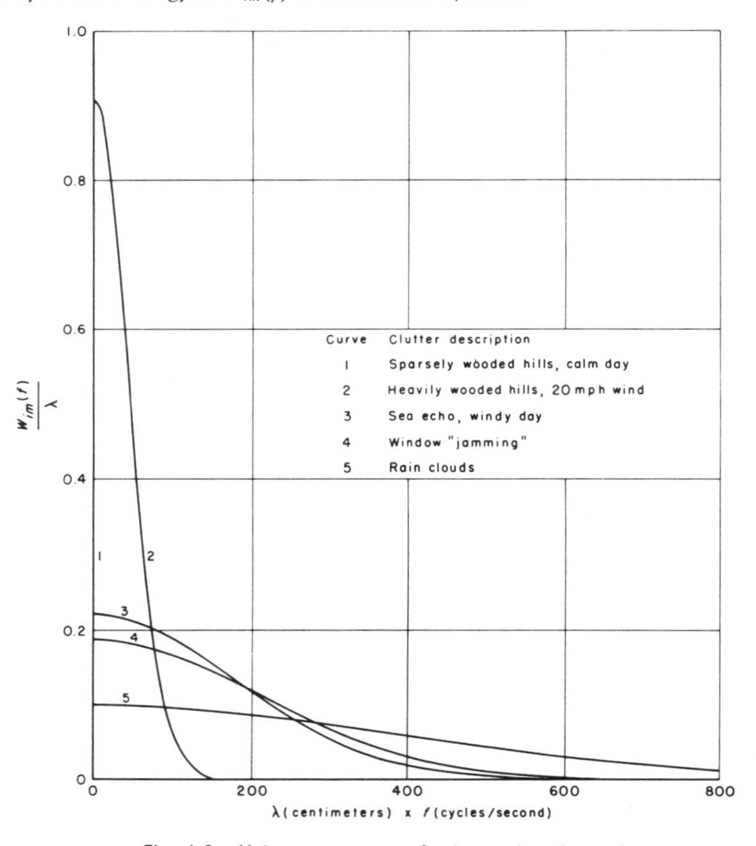

Fig. 6.2—Unit-power spectra for internal-motion noise

The preceding argument concerning the bounded-frequency domain of $W_{im}(f)$ is not consistent with the Gaussian form given by Barlow and Goldstein. This is due to the fact that it is only an approximation to the measured data. The approximation can no doubt

be improved by truncating the Gaussian curve at the frequencies $\pm v/2\lambda$. Whether or not this is worth while depends on the type of MTI being considered. For two- and three-pulse delay-line systems the error can probably be neglected. For certain range-gated filter systems, however, such as those using near-rectangular filters, the reference gain may be entirely determined by the clutter energy in the "tails" of the spectrum, and in these cases the computations may not be very reliable.

6.3. Wind Noise

The decorrelating effect of particle motion discussed in Sec. 6.2, above, describes the effect of particle motion relative to the center of gravity of the clutter mass. This agitation may be primarily due to wind, in which case the standard deviation parameter, v, will have to be chosen accordingly. For ground clutter, the center of gravity of the clutter mass will be fixed, regardless of the wind velocity. However, for airborne clutter, such as rain clouds and chaff, the center of gravity will drift at the wind velocity, v_w, with respect to the ground. If the velocity is steady, this drift will produce a doppler shift of the entire clutter spectrum away from the central ground-clutter frequency by an amount

$$f_w = \frac{2v_w}{\lambda}. \tag{6.7}$$

For ground-based radars, this shift will result in an additional decorrelation of the clutter returns. The extent of this degradation is expressed by the factor $\rho_w(\tau)$, which is the normalized autocorrelation function for the return from a single point reflector moving with radial velocity v_w. It is given by

$$\rho_w(\tau) = \cos 4\pi \frac{v_w \tau}{\lambda}. \tag{6.8}$$

The corresponding power spectrum is

$$W_w(f) = \frac{1}{2}\left\{ \delta\left(f = -\frac{2v_w}{\lambda}\right) + \delta\left(f = \frac{2v_w}{\lambda}\right)\right\}. \tag{6.9}$$

The composite autocorrelation function for the drifting clutter is then given by the product,* $\rho_{im}(\tau) \cdot \rho_w(\tau)$, and the power spectrum is given by the convolution of $W_{im}(f)$ with $W_w(f)$.

6.3.1. Elimination of Wind Noise

In noncoherent systems wind decorrelation will not be present. Similarly, it can be removed in coherent systems if the coho can be adjusted (as a function of beam position) so as to "zero beat" on the center of the returned clutter spectrum.

6.4. Scanning Noise

Another type of noise affecting MTI performance is the so-called scanning noise,

*The product rule for independent phenomena will be discussed in detail in Sec. 6.6.

Curve Description
1 Rectangular
2 Elliptical
3 Cosine
4 Triangular
−0.5a −0.4a −0.3a −0.2a −0.1a 0 0.1a 0.2a 0.3a 0.4a 0.5a
Displacement from apperture center, x

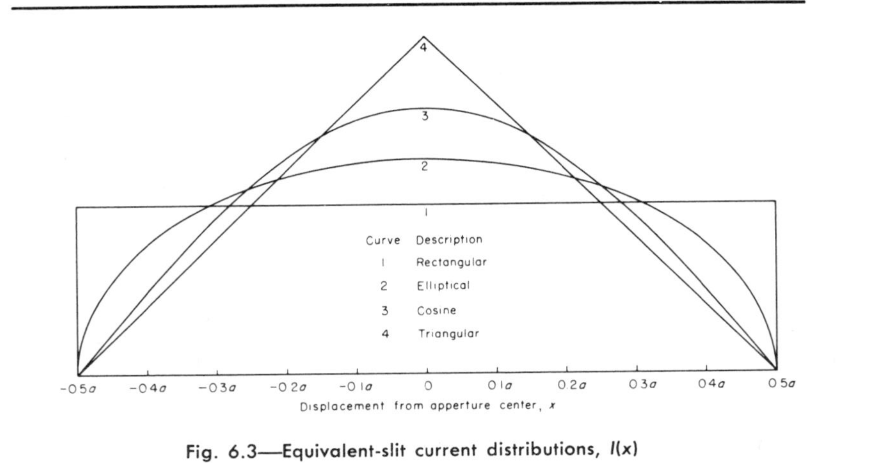

Fig. 6.3—Equivalent-slit current distributions, $I(x)$

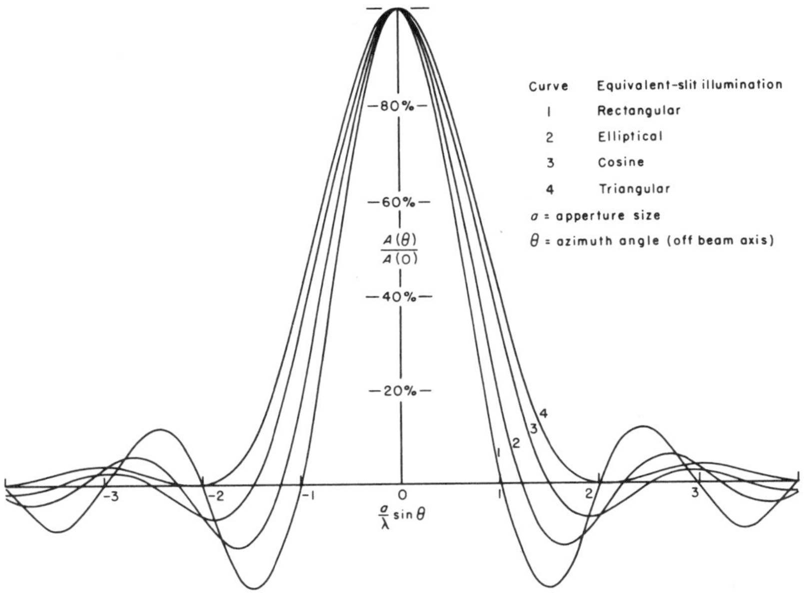

Curve Equivalent-slit illumination
1 Rectangular
2 Elliptical
3 Cosine
4 Triangular
a = apperture size
θ = azimuth angle (off beam axis)
−80%
−60%
A(θ)/A(0)
−40%
−20%
−3 −2 −1 0 1 2 3
a/λ sin θ

Fig. 6.4—Typical horizontal one-way voltage patterns

The resulting scanning spectra, $W_a(f)$, and autocorrelation functions, $\rho_a(\tau)$, have been determined analytically where possible and otherwise by numerical integration. They appear in Figs. 6.5 and 6.6, respectively.

Curve Equivalent-slit illumination
1 Rectangular
2 Elliptical
3 Cosine
4 Triangular
--- Gaussian representative (see Chap 7)
θ' = scan rate (radians per unit τ)
ρa(τ)
a/λ θ' τ

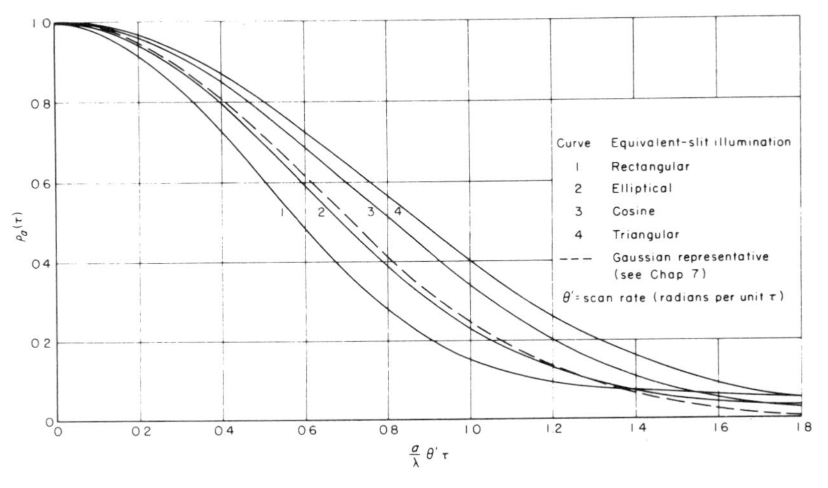

Fig. 6.5—Autocorrelation functions for scanning noise

Curve Equivalent-slit illumination
1 Rectangular
2 Elliptical
3 Cosine
4 Triangular
--- Gaussian representative (see Chap 7)
θ' = scan rate (radians per unit τ)
a/λ θ' x Wa(f)
λf/aθ'

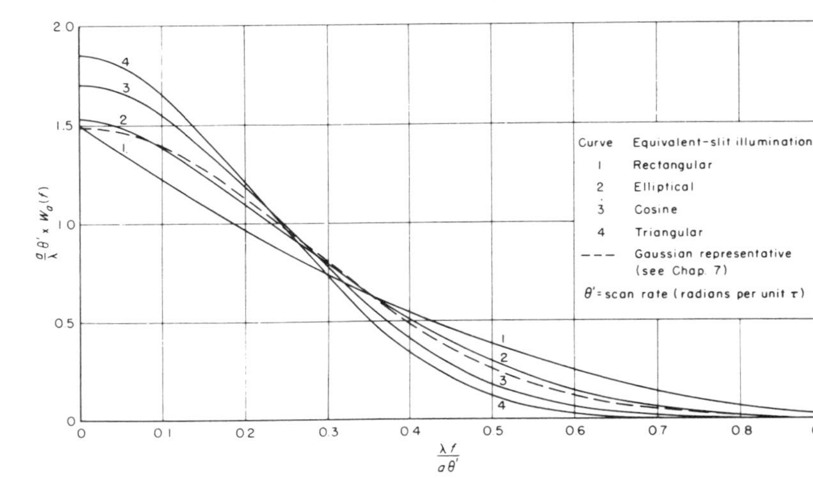

Fig. 6.6—Unit-power spectra for scanning noise

which is caused by the angular motion of the radar antenna. The theory of scanning noise is fairly well established, and discussion of it will be found in Refs. 1 and 10. This theory will be briefly outlined below in order to determine the "scanning spectrum," $W_a(f)$, and the associated normalized autocorrelation function, $\rho_a(\tau)$.

Suppose that the same antenna is used for transmitting and receiving. Let its one-way voltage radiation pattern in the plane of the scanning motion be denoted by $A(\theta)$, where θ is a radian measure relative to some reference direction. Suppose also that the antenna scans continuously at the rate of θ' radians per unit time. If a fixed particle is situated at a certain azimuth angle, θ_0, then the voltage return from this particle will be proportional to $A^2(\theta' t - \theta_0)$. Since the angle, θ_0, must be considered unknown in advance of detection,* $A^2(\theta' t - \theta_0)$ constitutes an ensemble of wave forms indexed by θ_0. Assuming uniform likelihood for θ_0 in the interval 0 to 2π, the autocorrelation function for the ensemble is

$$\psi_a(\tau) = \frac{1}{2\pi}\int_0^{2\pi} A^2[\theta' t - \theta_0]A^2[\theta'(t+\tau) - \theta_0]\, d\theta_0, \tag{6.10}$$

and, in normalized form,

$$\rho_a(\tau) = \frac{\int_0^{2\pi} A^2[\theta' t - \theta_0]A^2[\theta'(t+\tau) - \theta_0]\, d\theta_0}{\int_0^{2\pi} A^4[\theta' t - \theta_0]\, d\theta_0}. \tag{6.11}$$

These functions appear to depend on t as well as on τ. However, this is not the case. Since $A(\theta)$ is periodic (with period 2π), the integration limits may be extended to $(-\infty$ to $+\infty)$ and $\theta' t$ may be absorbed in the integration variable. Thus,

$$\rho_a(\tau) = \frac{\int_{-\infty}^{\infty} A^2(\theta_0)A^2(\theta_0 + \theta'\tau)\, d\theta_0}{\int_{-\infty}^{\infty} A^4(\theta_0)\, d\theta_0}, \tag{6.12}$$

which is independent of time. An alternative form for $\rho(\tau)$ is

$$\rho_a(\tau) = \frac{\int_{-\infty}^{\infty} A^2[\theta' t]A^2[\theta'(t+\tau)]\, dt}{\int_{-\infty}^{\infty} A^4(\theta' t)\, dt}. \tag{6.13}$$

The associated power spectrum is found, by means of the Wiener-Khintchine formulas, to be

$$W_a(f) = \frac{\left|\int_{-\infty}^{\infty} e^{i\omega t}A^2(\theta' t)\, dt\right|^2}{\int_{-\infty}^{\infty}\left|\int_{-\infty}^{\infty} e^{i\omega t}A^2(\theta' t)\, dt\right|^2 df}. \tag{6.14}$$

This formula may be simplified by introducing the current distribution, $I(x)$, for the linear array which would produce the beam pattern, $A(\theta)$; i.e.,

$$I(x) = \int_0^{2\pi} \exp\left(2\pi i \frac{x}{\lambda}\theta\right)A(\theta)\, d\theta. \tag{6.15}$$

*The discussion does not apply to "step-scanning," which is discussed in Sec. 6.4.2.

(This is an approximate formula based on the assumption of narrow beamwidths.) The autoconvolution of $I(x)$ is denoted $I^{II}(x)$. It is given by

$$I^{II}(x) = \int I(x-y)I(y)\, dy, \tag{6.16}$$

and, in terms of $A(\theta)$,

$$I^{II}(x) = \int_0^{2\pi} \exp\left(2\pi i \frac{x}{\lambda}\theta\right)A^2(\theta)\, d\theta. \tag{6.17}$$

Consequently, Eq. (6.14) for $W_a(f)$ becomes

$$W_a(f) = \frac{|I^{II}(\lambda f/\theta')|^2}{\int |I^{II}(\lambda f/\theta')|^2\, df}. \tag{6.18}$$

Hence, the scanning spectrum is obtained by convolving and squaring the equivalent current distribution, $I(x)$, and then scaling the frequency according to θ'/λ and normalizing to unit power.

It can now be seen that the highest frequencies present in the scanning spectrum are associated with the velocity of the edges of the antenna aperture. Thus, if the aperture measures a feet from tip to tip, then $I(x)$ truncates at $x = a/2$, and $I^{II}(x)$ truncates at $x = a$. From Eq. (6.18) it follows that $W_a(f)$ truncates at $f = a\theta'/\lambda$. This frequency corresponds to a doppler velocity $(a/2)\theta'$ which is precisely the tip velocity of the aperture.

6.4.1. Calculations for Typical Antennas

Four elementary antenna-current distributions, $I(x)$, have been selected for comparison. They are listed below:

Case	Type	$I(x)$
A	Rectangle	1
B	Ellipse	$\frac{1}{a}\sqrt{a^2 - 4x^2}$
C	Cosine	$\cos \pi \frac{x}{a}$
D	Triangle	$\frac{1}{a}\{a - 2\|x\|\}$

The entries for $I(x)$ in the table above are for $|x| \leq a/2$, where a is the aperture diameter. For $|x| > a/2$, $I(x)$ is supposed to be zero. Graphs of these functions are shown in Fig. 6.3. The corresponding horizontal one-way voltage radiation patterns appear in Fig. 6.4.

6.4.2. Elimination of Scanning Noise

A method for eliminating scanning noise is discussed in Ref. 24. The procedure is called "step-scanning," and it employs a principle analogous to that of the motion-picture projector, which presents the successive picture frames stepwise rather than continuously so that the motion of the film will not be displayed. Similarly, the step-scanning antenna is held (electrically) fixed through the prescribed time interval, Δt, assigned to each direction. In this way the reflected echoes, rather than being modulated by $A^2(\theta' t)$, are modulated by a rectangular function, the truncation points of which are known. If a two-pulse-comparison, delay-line system is used, the returned pulses from fixed targets will then cancel, with the exception of the first and last pulses. These uncanceled pulses can then be eliminated by blanking the display during the range frames immediately following the antenna steps. Similarly, if N pulses are received in the interval, Δt, between steps, then up to N-pulse comparison may be employed in the MTI system, the display being blanked whenever a step occurs within the set of pulses being compared.

If the MTI system employs range gates and filters, a similar procedure can be used in conjunction with step-scanning to eliminate the scanning noise. The procedure is to sample the system output only at such times (toward the end of the interval, Δt) as the stepping transients from fixed targets vanish. If the combined integration time for the MTI and/or video filter is longer than Δt, then it may be necessary to quench the filtering networks so that only those pulses received in Δt are integrated. Otherwise the filters will have to be carefully designed so that the stepping transients from successive Δt intervals will vanish together. If the integration time is less than Δt, this may not be necessary—but then the target detectability for the radar might be improved by extending the integration time to Δt.

6.5. Platform-motion Noise

In the case of airborne radars where the antenna platform translates at a relatively high speed with respect to the clutter mass, an additional noise is introduced into the system. As to origin, this platform-motion noise is distinct from the internal-motion, wind, and scanning noises previously described; however, it has elements of similarity with each. As fixed particles drift through the antenna beam pattern they are scanned, and a certain amount of scanning noise results. This noise will be greatest when the antenna is sighted at 90° to the ground track, and will be worse for narrow- than for broad-beam patterns. If the antenna is sighted forward or aft, the scanning noise will be small, but in this case the fixed particles will drift through the range resolution intervals, producing what might be called "ranging noise." Both the scanning noise and the ranging noise can be accounted for in the analysis by appropriately limiting the integration time.

The unified relative motion of the fixed clutter mass can be decomposed into a translation and a rotation, each producing a characteristic noise. The translation noise is the same as wind noise and so is unimportant for noncoherent systems (and coherent systems in which the coho frequency can be adjusted as a function of sighting angle and ground speed). The rotation noise, on the other hand, can be very serious.* Let the antenna be

*This rotation noise is discussed in Ref. 1 under "Beating due to Finite Pulse Packet."

sighted at an angle, γ, with respect to the ground track (see Fig. 6.7). Let the aircraft velocity be v, and consider a fixed particle at p, in the rth range-resolution interval making an angle, θ, with the beam axis. To the moving observer, the clutter mass will appear to rotate about the point, 0, at an angular rate, $\tilde{\omega}$, given by

$$\tilde{\omega} = \frac{v \sin \gamma}{r}. \tag{6.19}$$

Consequently, the particle at p will exhibit a relative radial velocity, v_p, which for small beamwidths is approximately

$$\begin{aligned} v_p &= \tilde{\omega} r \theta \\ &= v\theta \sin \gamma. \end{aligned} \tag{6.20}$$

The antenna one-way voltage pattern is designated by $A(\theta)$ and will be considered to have arisen from a linear-array current distribution, $I(x)$. The voltages returned from a reflector at p will be proportional to $A^2(\theta)$, and they will be spectrally distributed around the doppler-shift frequency corresponding to v_p, i.e.,

$$f = \frac{2v}{\lambda}\,\theta \sin \gamma. \tag{6.21}$$

Thus, neglecting the truncating effect of the range resolution, the unit-power spectrum $W_{pm}(f)$ is given by

$$W_{pm}(f) = \frac{A^4\left(\dfrac{\lambda f}{2v \sin \gamma}\right)}{\int A^4\left(\dfrac{\lambda f}{2v \sin \gamma}\right) df}. \tag{6.22}$$

The effect of truncation is to further smear this spectrum. For high-resolution radars, such as tracking equipment in missiles, etc., this may be serious, but in such equipment it may conceivably be eliminated by employing moving range gates. For equipment of lower

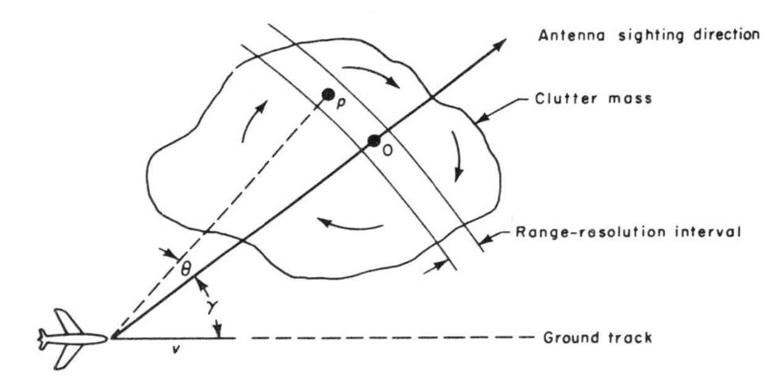

Fig. 6.7—Illustration of relative rotation of clutter mass due to platform motion

resolution capability, the range noise will be less severe, and, fortunately, it will diminish as the sighting angle, γ, increases—i.e., in contrast to the rotation noise, which is most severe for $\gamma = 90°$. Since both the scanning and ranging noises can be accounted for by integrating the correct number of hits per target, the platform-motion noise will be presumed to be entirely due to rotation. The corresponding normalized autocorrelation function is

$$\rho_{pm}(\tau) = \frac{I^{IV}(2v\tau \sin \gamma)}{I^{IV}(0)}, \tag{6.23}$$

where $I^{IV}(x)$ is the twice-autoconvolved current distribution, $I(x)$.

6.5.1. Calculation for Typical Antennas

Generalized curves for the power spectrum, $W_{pm}(f)$, and autocorrelation function, $\rho_{pm}(x)$, obtainable with the four representative antenna current distributions are shown in Figs. 6.8 and 6.9. Some of these curves have been obtained analytically; the others were found by numerical means.

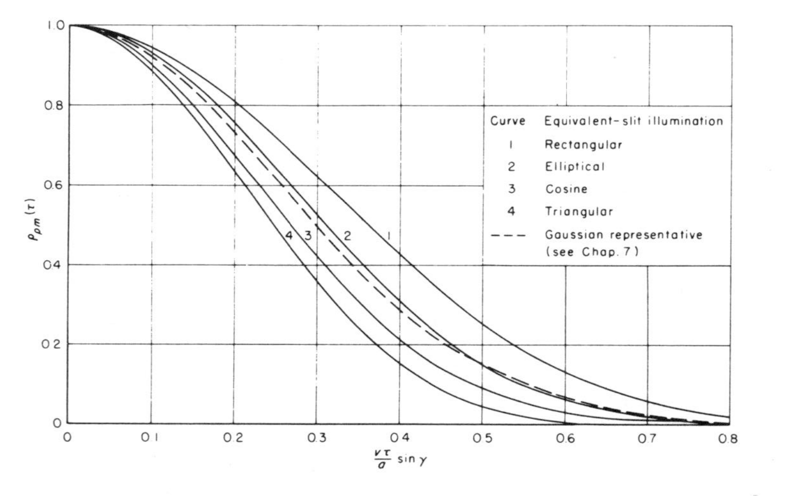

Fig. 6.8—Autocorrelation functions for platform-motion noise

6.5.2. Elimination of Platform-motion Noise

Investigations have been conducted by the Philco Corporation[25] to determine the feasibility of the so-called pick-a-back antenna for eliminating platform-motion noise. The scheme employs two identical antennas located one ahead of the other on the airplane and spaced by the distance the airplane travels in one interpulse period. The first pulse in every other pair of pulses is transmitted and received on the lead antenna, the second pulses using the antenna farther back. In this way two consecutive samples of clutter are taken from the same point in space. These pulses are then subtracted in a two-pulse-comparison MTI system.

There are many difficulties connected with this procedure for eliminating platform-motion noise. Chief among them are

- The required spacing between the antennas is very small for many aircraft speeds and interpulse periods commonly used, and so special antennas must be designed that do not overlap physically and that can still scan.
- Extremely tight manufacturing tolerances are required to achieve the necessary beam-pattern duplication.
- In order to operate successfully over a range of aircraft speeds, either the antennas must be designed for adjustable spacing—which worsens the manufacturing tolerances by an order of magnitude—or the interpulse period must be flexible. On the other hand, a variable interpulse period will rule out the use of fixed delay lines.

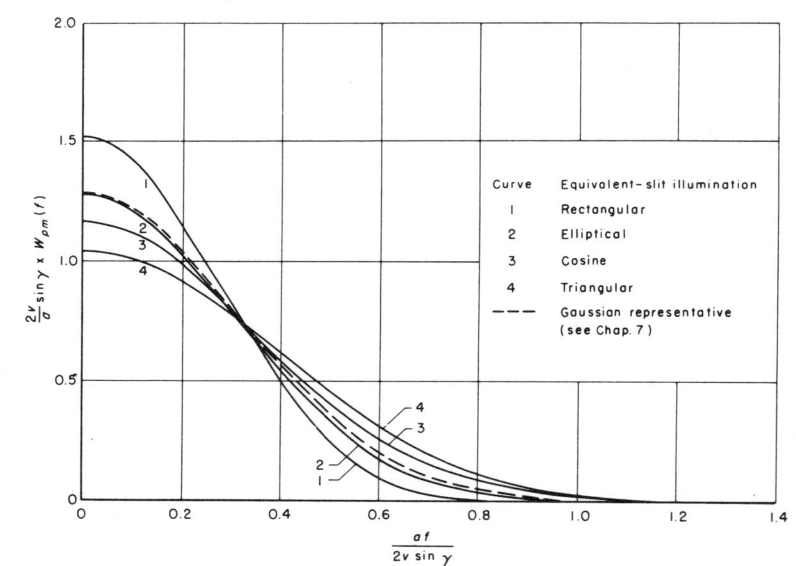

Fig. 6.9—Unit-power spectra for platform-motion noise

6.6. Composition Rule for Determining the Clutter Correlation Function and Power Spectrum

In Secs. 6.2 through 6.5, autocorrelation functions and power spectra were found for internal-motion, wind, scanning, and platform-motion noise, each on the premise that the others were absent. If the specific system is not scanning, or if there is no platform motion, etc., then, in these particulars, the signals are considered to be completely correlated. Thus, in general, one may consider that all four of these noise sources are acting to some degree as specified by ρ_{im}, ρ_w, ρ_a, and ρ_{pm}. The composite autocorrelation function, ρ_c, and power spectrum, W_c, needed for evaluating the formulas derived in Chaps. 2, 3, 4, and 5 will now be determined for the combination of noise sources.

The four noise sources may be considered statistically independent for the following reasons: The internal-motion and wind noises are certainly independent of the scanning and platform-motion noises, inasmuch as the latter are generated at the radar and the former rise out of conditions at the clutter mass. The internal-motion and wind noises are independent because, by definition, the wind noise is due to the unified motion of the entire clutter mass and the internal motion refers to fluctuations within the mass relative to the moving (or fixed) center of gravity. Although the standard-deviation parameter, v, for the particle velocities about their mean (wind) velocity may very well be affected by the presence of wind (the wind may contribute turbulent energy to the particles), the wind can in no way bias the selection of particle velocities from their distribution about the mean. Similarly, the scanning and platform-motion noises, as defined, are independent.

Each noise is now represented as a separate modulation on the correlated carrier, i.e.,

$$\text{Received voltage} = n_{im}(t)n_w(t)n_a(t)n_{pm}(t) \times \text{carrier}. \tag{6.24}$$

The result of demodulation is then the product of the noises. The autocorrelation function, $\psi_c(\tau)$, is consequently

$$\psi_c(\tau) = \overline{n_{im}(t)n_{im}(t+\tau) \cdot n_w(t)n_w(t+\tau) \cdots}, \tag{6.25}$$

which, by virtue of the statistical independence, splits into

$$\psi_c(\tau) = \overline{n_{im}(t)n_{im}(t+\tau)} \times \overline{n_w(t)n_w(t+\tau)} \times \cdots \tag{6.26}$$

$$= \psi_{im}(\tau) \cdot \psi_w(\tau) \cdot \psi_a(\tau) \cdot \psi_{pm}(\tau); \tag{6.27}$$

and, in normalized form,

$$\rho_c(\tau) = \rho_{im}(\tau) \cdot \rho_w(\tau) \cdot \rho_a(\tau) \cdot \rho_{pm}(\tau). \tag{6.28}$$

Thus, the composite normalized autocorrelation function is simply equal to the product of the respective normalized autocorrelation functions.

Accordingly, the unit-power spectrum, $W_c(f)$, is given by

$$W_c(f) = W_{im}(f) * W_w(f) * W_a(f) * W_{pm}(f), \tag{6.29}$$

where ($*$) indicates the convolution procedure.

6.7. System-instability Noise

There are many possible causes for system instability, both in the transmitter and in the receiver. These instabilities are due to the inevitable imperfections in the installed and

operating equipment, and they give rise to what will be called "system-instability noise." Some of the sources of this noise are controllable up to a point: in the design stage by calling for intrinsically stable components, providing automatic stabilizing circuitry, etc.; in manufacture, by tightening critical tolerances, etc.; and in the field, by ensuring careful maintenance. Such procedures, however, become progressively more expensive; and a compromise is usually reached, with the result that a certain amount of instability noise still remains to degrade the MTI performance.

A commonly used yardstick for measuring instability noise and its degrading effect on MTI performance is the *cancellation ratio*. This quantity (denoted *CR*) is the ratio of attenuation creditable to the MTI cancellation unit for truly fixed target returns when the latent amplifications or losses have been normalized. It will be shown in the sequel that the *CR* is equal to the ratio of the average power in the stable part of the voltages to the average power in the instability noise. This is the case—regardless of the number of delay lines employed, or whether delay lines or range-gated filters are used. Of course, one might expect to have more instabilities and correspondingly lower *CR* with the more complicated arrangements, but the reduced *CR* is due to the increased power in the noise and not to any changes in the enhancement curves, filter bandwidths, etc.

The most outstanding sources of instability noise* are—

- pulse-time jitter,
- pulse-width jitter,
- pulse-amplitude jitter,
- pulse distortion,
- i-f phase and gain variations,
- frequency modulation within the pulse,
- coho, stalo, and locking-circuit instability,
- gating noise.

These noises, although quite dissimilar as to sources, have a great deal in common. In the first place, they are all representable as small perturbations on otherwise stable pulses, and so are more or less correlated with those pulses. In the second place, their action time is small, i.e., of the order of the pulse width. And, finally, the successive perturbations are essentially independent, pulse to pulse. These common features result in autocorrelation functions which may be "split" into two parts: the one, being periodic, is almost identical in shape with the fixed-target autocorrelation function; the other, the aperiodic part, representing correlations only over the duration of the pulse, corresponds to noise which is *almost* white. Accordingly, then, the *CR* is expressible as the unperturbated average power divided by the power in the continuous part of the perturbated spectrum. This last is the instability-noise power, and it must be determined for each type of instability. This will be done in Sec. 6.8, where the *CR* for many of the noise sources listed above will be determined.

*Thermionic noise is omitted here in that it is not relevant to the MTI rejection of clutter as such.

6.7.1. Autocorrelation Functions and Power Spectra

The autocorrelation function, $\psi_{s+n}(\tau)$, for the sum of fixed-target signal voltage, $s(t)$, and instability-noise voltage, $n(t)$, will now be found. The signal function, $s(t)$, will be considered a completely determined infinite chain of pulses, whereas the noise, $n(t)$, will be a statistical function which is zero everywhere except in the vicinity of the pulses. During these intervals of activity, $n(t)$ will be determined to within one parameter by the general nature of the perturbation, and the parameter will be a random variable, independent pulse to pulse and distributed according to the statistics of the noise source.

The infinite pulse chain $s(t) + n(t)$ can be split into statistically independent parts, as follows:

$$s(t) + n(t) = [s(t) + \bar{n}(t)] + [n(t) - \bar{n}(t)], \tag{6.30}$$

where $\bar{n}(t)$ is the ensemble expectation of $n(t)$. Since $s(t) + n(t)$ is presumed stationary, pulse to pulse, the expectation $\overline{s(t) + n(t)} = s(t) + \bar{n}(t)$ is a chain of identical pulses which for small noise levels is almost identical with $s(t)$. The quantity $n(t) - \bar{n}(t)$ is the fluctuating part of the noise. It has mean zero. The autocorrelation functions for the sum can now be expressed as

$$\psi_{s+n}(\tau) = \overline{\{s(t) + n(t)\}\{s(t+\tau) + n(t+\tau)\}}, \tag{6.31}$$

which, when expanded and averaged, reduces, by virtue of the independence of $s + \bar{n}$ and $n - \bar{n}$, to

$$\psi_{s+n}(\tau) = \psi_{s+\bar{n}}(\tau) + \psi_{n-\bar{n}}(\tau). \tag{6.32}$$

Since $s + \bar{n}$ is a steady chain of identical pulses, $\psi_{s+\bar{n}}(\tau)$ will be numerically the same for all pulse pairs. On the other hand, since $n - \bar{n}$ is uncorrelated pulse to pulse, $\psi_{n-\bar{n}}(\tau)$ will vanish unless the same pulses are being matched, i.e., unless τ is of the order of the pulse width or less. Thus, $\psi_{s+n}(\tau)$ can be represented as the sum of a periodic function, $\psi_{s+\bar{n}}(\tau)$, which is essentially the same as $\psi_s(\tau)$, and an aperiodic function, $\psi_{n-\bar{n}}(\tau)$, which corresponds to noise that is essentially flat over the i-f bandwidth.

As far as the MTI system is concerned, the sum $s + \bar{n}$ is as good a "stable" signal as is s. In some circumstances it may be difficult to measure the fixed-target average power in the laboratory without, in fact, measuring $\overline{(s + \bar{n})^2}$. On the other hand, except when the noise is generated within the cancellation unit, $\overline{(s + \bar{n})^2}$ is readily obtained using a low-pass power-measuring instrument. For this reason $\overline{(s + \bar{n})^2}$ will be used rather than s^2 for the fixed-target input power in determining the CR. Similarly, since it is the fluctuating part of the perturbation, i.e., $n - \bar{n}$, which is active in degrading the MTI performance, the quantity $\overline{(n - \bar{n})^2}$ will be used for the average power in the instability noise rather than $\overline{n^2}$.

The general formula

$$CR = \frac{\overline{(s + \bar{n})^2}}{\overline{(n - \bar{n})^2}} \tag{6.33}$$

will be derived in the next sections. Also will be derived the general formula

$$\bar{G}_{\text{actual}} = \frac{1}{\dfrac{1}{\bar{G}_{\text{ideal}}} + \dfrac{1}{CR}}, \tag{6.34}$$

which expresses the degrading effect of low CR on the system performance. Both of the above formulas apply to all MTI systems considered.

6.7.2. The Cancellation Ratio

The cancellation ratio is customarily used to indicate the degree of cancellation for fixed-target returns. It is sometimes determined by presenting a completely correlated signal to the MTI unit and measuring the power residue in the output. The input power is then divided by this residue, and the ratio, after accounting for the latent gain or loss (see Secs. 2.4.1 and 3.3.1), is the cancellation ratio. This measurement gives the CR for the MTI unit alone, i.e., it does not include transmitter or receiver noises. A more complete measurement would involve directing the radar on a truly fixed target; but such a target is hard to find, and great care must be exercised in performing the experiment so that the results will be meaningful. There are various compromises that can be made in this regard, and through indirect measurements the over-all CR can often be tied down fairly accurately.

For the N-pulse cancellation system, the output voltage can be expressed

$$E_o(t) = \sum_{j=1}^{N} a_j[s(t - t_j) + n(t - t_j)], \tag{6.35}$$

where $s(t)$ and $n(t)$ are defined in the preceding paragraphs. The average output power residue is

$$\begin{aligned}\overline{E_o(t)^2} &= \sum_{j=1}^{N}\sum_{k=1}^{N} a_j a_k \psi_{s+n}(t_j - t_k) \\ &= \sum_{j=1}^{N}\sum_{k=1}^{N} a_j a_k \psi_{s+\bar{n}}(t_j - t_k) + \sum_{j=1}^{N}\sum_{k=1}^{N} a_j a_k \psi_{n-\bar{n}}(t_j - t_k).\end{aligned} \tag{6.36}$$

The first double sum above vanishes because the a_j are chosen so that fixed targets cancel. The second double sum involves $\psi_{n-\bar{n}}(\tau)$, which vanishes except when $j = k$. Consequently,

$$\overline{E_o(t)^2} = \psi_{n-\bar{n}}(0) \sum_{j=1}^{N} a_j^2. \tag{6.37}$$

The latent amplification factor (see Sec. 2.4.1) is $\sum_{j=1}^{N} a_j^2$. This is normalized out, and the result is divided into the input fixed-target power $\psi_{s+\bar{n}}(0)$, yielding the cancellation ratio

$$CR = \frac{\psi_{s+\bar{n}}(0)}{\psi_{n-\bar{n}}(0)}. \tag{6.38}$$

For a range-gated filter system characterized by the response function, $H(f)$, the output power residue is

$$\overline{E_o(t)^2} = \int H(f)[\psi_{s+\bar{n}}(0)W_{s+\bar{n}}(f) + \psi_{n-\bar{n}}(0)W_{n-\bar{n}}(f)]\, df, \tag{6.39}$$

where $W_{s+\bar{n}}(f)$ and $W_{n-\bar{n}}(f)$ are the unit-power spectra corresponding to the *normalized* autocorrelation functions, $\rho_{s+\bar{n}}(\tau)$ and $\rho_{n-\bar{n}}(\tau)$, respectively. The power spectrum, $W_{s+\bar{n}}(f)$, corresponds to the fixed-target signals, $s + \bar{n}$, and therefore, by the design of $H(f)$ for fixed-target rejection, the product $H(f)W_{s+\bar{n}}(f) = 0$. This leaves

$$\overline{E_o(t)^2} = \psi_{n-\bar{n}}(0) \int H(f)W_{n-\bar{n}}(f)\, df. \tag{6.40}$$

Since $W_{n-\bar{n}}(f)$ is essentially flat with unit area, the integral above equals $\overline{H}$, which, according to Sec. 3.3.1, equals the latent amplification factor for the system. Accordingly, the integral is normalized out, leaving $\overline{E_o(t)^2} = \psi_{n-\bar{n}}(0)$, which, when divided into the input fixed-target power, yields

$$CR = \frac{\psi_{s+\bar{n}}(0)}{\psi_{n-\bar{n}}(0)}. \tag{6.41}$$

Since the CR is a power ratio, either ensemble averages or time averages may be used. To be specific,

$$\psi_{s+\bar{n}}(0) = \frac{1}{DC}\,\phi_{s+\bar{n}}(0), \tag{6.42}$$

and

$$\psi_{n-\bar{n}}(0) = \frac{1}{DC}\,\phi_{n-\bar{n}}(0), \tag{6.43}$$

where DC is the average duty cycle (fraction of time on) and the ϕ's are the time-average autocorrelation functions (see Sec. 1.10). Consequently,

$$CR = \frac{\phi_{s+\bar{n}}(0)}{\phi_{n-\bar{n}}(0)}. \tag{6.44}$$

This formula will be used for calculation in Sec. 6.8.

6.7.3. Effect of Cancellation Ratio on System Performance

The actual reference gain can now be expressed in terms of the measured cancellation ratio and the reference gain theoretically computed on the basis of $CR = \infty$. For this

purpose it is convenient to suppose that all the system noises arise in the transmitter.* The transmitted signal is denoted $s(t) + n(t)$, where, as before, $s(t)$ is the intended transmission and $n(t)$ is the perturbation due to instability. The corresponding autocorrelation function, $\psi_{s+n}(\tau)$, splits into $\psi_{s+\bar{n}}(\tau) + \psi_{n-\bar{n}}(\tau)$. The transmitted signals are then reflected by the clutter mass, the statistics of which are independent of the system noise, $n - \bar{n}$. Consequently, the autocorrelation function of the received signals is proportional to the product

$$\rho_c(\tau) \times \psi_{s+n}(\tau) = \rho_c(\tau)\psi_{s+\bar{n}}(\tau) + \rho_c(\tau)\psi_{n-\bar{n}}(\tau), \tag{6.45}$$

where $\rho_c(\tau)$ is the normalized autocorrelation function for the clutter. Under these conditions the reference gain, Eq. (2.26), for an N-pulse-comparison system may be written

$$\overline{G}_{\text{actual}} = \frac{\psi_{s+n}(0) \sum_{j=1}^{N} a_j^2}{\sum_{j=1}^{N}\sum_{k=1}^{N} a_j a_k \rho_c(t_j - t_k)\psi_{s+n}(t_j - t_k)}. \tag{6.46}$$

The reciprocal of $\overline{G}_{\text{actual}}$ is now split into two parts, using Eq. (6.32), i.e.,

$$\frac{1}{\overline{G}_{\text{actual}}} = \frac{\sum_{j=1}^{N}\sum_{k=1}^{N} a_j a_k \rho_c(t_j - t_k)\psi_{s+\bar{n}}(t_j - t_k)}{\psi_{s+n}(0) \sum_{j=1}^{N} a_j^2} + \frac{\sum_{j=1}^{N}\sum_{k=1}^{N} a_j a_k \rho_c(t_j - t_k)\psi_{n-\bar{n}}(t_j - t_k)}{\psi_{s+n}(0) \sum_{j=1}^{N} a_j^2}. \tag{6.47}$$

Recalling that $\psi_{s+\bar{n}}(t_j - t_k)$, which corresponds to the steady component, $s + \bar{n}$, is the same for all pulse pairs, it can be replaced by $\psi_{s+\bar{n}}(0)$ and brought outside the double sum. Furthermore, $\psi_{n-\bar{n}}(t_j - t_k)$ is zero unless $j = k$, in which case it equals $\psi_{n-\bar{n}}(0)$. Thus, in the second double sum, $\rho_c(t_j - t_k)$ may be discarded. These changes yield

$$\frac{1}{\overline{G}_{\text{actual}}} = \frac{\psi_{s+\bar{n}}(0)}{\psi_{s+n}(0)}\left\{\frac{1}{\overline{G}_{\text{ideal}}} + \frac{1}{CR}\right\} = \frac{CR}{1 + CR}\left\{\frac{1}{\overline{G}_{\text{ideal}}} + \frac{1}{CR}\right\}. \tag{6.48}$$

Finally, for sufficiently large CR (15 db or so), the initial factor is essentially unity, and

*This assumption, although not entirely realistic, is not impractical, for many of the predominant instabilities are in the transmitter, and the only error involved by making this assumption comes from the false clutter modulation of the noises actually arising in the receiver. These may be neglected.

the resulting formula is

$$\overline{G}_{\text{actual}} = \frac{1}{\dfrac{1}{\overline{G}_{\text{ideal}}} + \dfrac{1}{CR}}. \tag{6.49}$$

This function is plotted versus $\overline{G}_{\text{ideal}}$ for various cancellation ratios in Fig. 6.10. On examining these curves, it will be seen that, for the actual reference gain to be within,

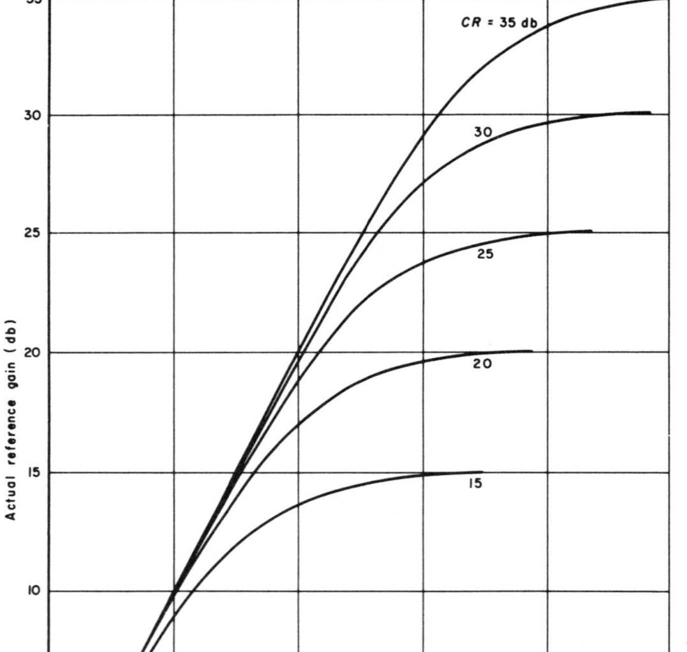

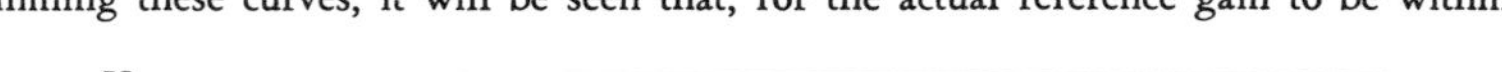
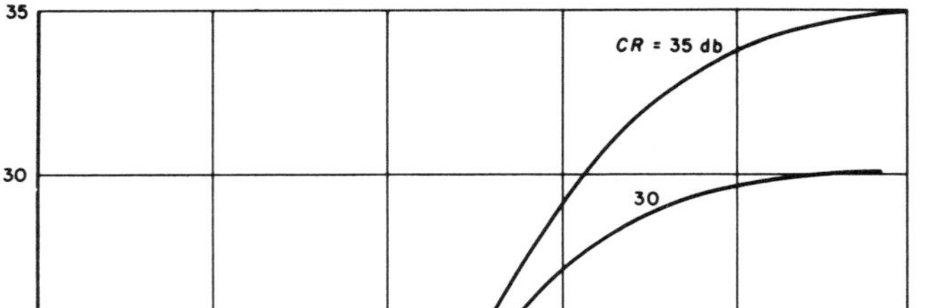
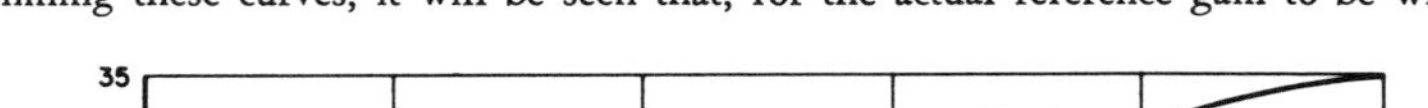
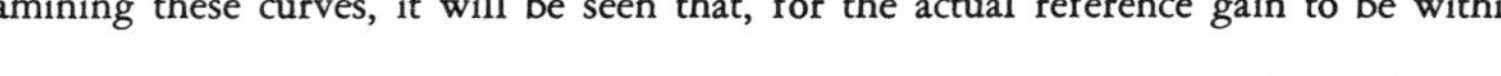

Fig. 6.10—Degradation due to system-instability noise

say, 3 db of the ideal, it is necessary for the CR to be at least as high as $\overline{G}_{\text{ideal}}$. For high-gain delay-line-type systems the consequences are obvious. As more delay lines are incorporated, more noise sources are introduced to force the CR down to meet the rising $\overline{G}_{\text{ideal}}$, until a point is reached where they are comparable. Beyond this point, additional circuit complexity will result in worse rather than in better performance.

Since the range-gated filter systems are viewed here as extensions of finite-pulse-group systems to infinite-pulse-group systems, the same formula, Eq. (6.49), holds. This is difficult to see from the spectral point of view because of certain difficulties associated with the spectrum of white noise of unit energy. However, by giving the instability noise a rectangular spectrum of unit energy and width, W, and letting $W \to \infty$, the formula can be derived from Eq. (3.9).

6.8. Cancellation Ratios for Various Types of System Instability

The cancellation ratios for some of the isolated system instabilities listed in Sec. 6.7 will now be determined. It will be assumed that the pulses are of the Gaussian form

$$p_j(t) = a \exp\left[-\frac{(t - t_j)^2}{2\mu^2}\right], \tag{6.50}$$

where $p_j(t)$ is the jth pulse (voltage), a is the pulse amplitude (peak), t_j is the pulse epoch, and μ is the pulse width. The statistical parameters, a, t_j, and μ, will be studied separately, and they will be assumed to have normal distributions about their intended values.

The cancellation ratios will be determined by calculating the fixed-target power, $\phi_{s+\bar{n}}(0)$, and the noise power, $\phi_{n-\bar{n}}(0)$. The quotient, Eq. (6.44), will then yield the CR.

6.8.1. Pulse-time Jitter

Let Δ denote the random error in t_j, and let the probability density for Δ be the Gaussian form with mean zero and variance $\overline{\Delta^2}$; thus,

$$P'(\Delta) = \frac{1}{\sqrt{2\pi\overline{\Delta^2}}} \exp\left[-\frac{\Delta^2}{2\overline{\Delta^2}}\right]. \tag{6.51}$$

The noise voltage, $n(t)$, is equal to the perturbed pulse less the intended signal, i.e.,

$$n(t) = a\left\{\exp\left[-\frac{(t - t_j - \Delta)^2}{2\mu^2}\right] - \exp\left[-\frac{(t - t_j)^2}{2\mu^2}\right]\right\}. \tag{6.52}$$

The expected noise pulse, $\bar{n}(t)$, is the ensemble average of $n(t)$, i.e.,

$$\bar{n}(t) = \int_{-\infty}^{\infty} n(t)P'(\Delta)\, d\Delta$$

$$= a\left\{\sqrt{\frac{\mu^2}{\mu^2 + \overline{\Delta^2}}} \exp\left[-\frac{(t - t_j)^2}{2(\mu^2 + \overline{\Delta^2})}\right] - \exp\left[-\frac{(t - t_j)^2}{2\mu^2}\right]\right\}. \tag{6.53}$$

The intended signal, $s(t)$, equals the unperturbed pulse, $p_j(t)$; consequently,

$$s(t) + \bar{n}(t) = a\sqrt{\frac{\mu^2}{\mu^2 + \overline{\Delta^2}}} \exp\left[-\frac{(t - t_j)^2}{2(\mu^2 + \overline{\Delta^2})}\right]. \quad (6.54)$$

This is squared and averaged over time, thus:

$$\phi_{s+\bar{n}}(0) = \frac{a^2\mu\sqrt{\pi}}{\tau}\sqrt{\frac{\mu^2}{\mu^2 + \overline{\Delta^2}}}. \quad (6.55)$$

The fluctuating part of the noise is

$$n(t) - \bar{n}(t) = a\left\{\exp\left[-\frac{(t - t_j - \Delta)^2}{2\mu^2}\right] - \sqrt{\frac{\mu^2}{\mu^2 + \overline{\Delta^2}}}\exp\left[-\frac{(t - t_j)^2}{2(\mu^2 + \overline{\Delta^2})}\right]\right\}. \quad (6.56)$$

This is squared and averaged over one interpulse period and then averaged over the ensemble to obtain the time-average noise power:

$$\phi_{n-\bar{n}}(0) = \overline{\frac{1}{\tau}\int_{-T/2}^{T/2} [n(t) - \bar{n}(t)]^2\, dt}$$

$$\approx \overline{\frac{1}{\tau}\int_{-\infty}^{\infty} [n(t) - \bar{n}(t)]^2\, dt}. \quad (6.57)$$

Accordingly,

$$\frac{1}{\tau}\int_{-\infty}^{\infty} [n(t) - \bar{n}(t)]^2\, dt = \frac{a^2\mu}{\tau}\sqrt{\pi}\left\{1 + \sqrt{\frac{\mu^2}{\mu^2 + \Delta^2}} - 2\sqrt{\frac{2\mu^2}{2\mu^2 + \overline{\Delta^2}}}\exp - \frac{1}{2}\frac{\Delta^2}{2\mu^2 + \overline{\Delta^2}}\right\}, \quad (6.58)$$

and

$$\phi_{n-\bar{n}}(0) = \frac{a^2\mu}{\tau}\sqrt{\pi}\left(1 - \sqrt{\frac{\mu^2}{\mu^2 + \overline{\Delta^2}}}\right). \quad (6.59)$$

The cancellation ratio is now determined:

$$CR(\text{time jitter}) = \left(\sqrt{1 + \frac{\overline{\Delta^2}}{\mu^2}} - 1\right)^{-1}. \quad (6.60)$$

This is shown in Fig. 6.11.

6.8.2. Pulse-width Jitter

Let Δ denote the random pulse-width error. The perturbed signal pulse is

$$s(t) + n(t) = a\exp\left[-\frac{t^2}{2(\mu + \Delta)^2}\right] \quad (\text{here } t_j = 0). \quad (6.61)$$

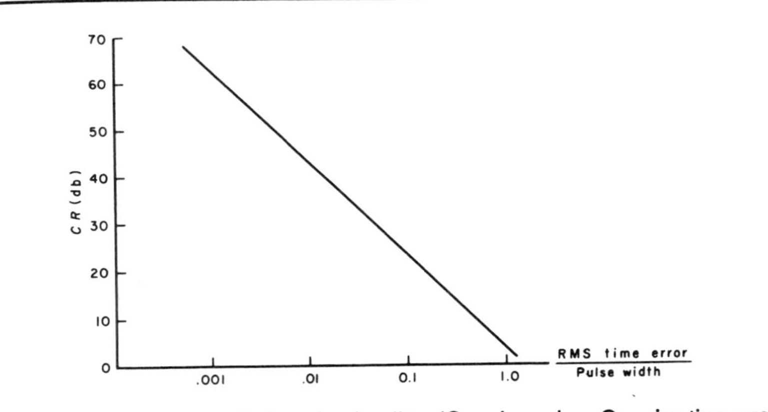

Fig. 6.11—Cancellation ratio for pulse-time jitter (Gaussian pulses; Gaussian time error)

This is averaged with respect to the normal distribution, Eq. (6.51), then squared and represented as a double integral, thus:

$$[s(t) + \bar{n}(t)]^2 = \frac{a^2}{2\pi\overline{\Delta^2}}\iint_{-\infty}^{\infty} \exp\left[-\frac{t^2}{2(\mu + \Delta_1)^2} + \frac{t^2}{2(\mu + \Delta_2)^2} + \frac{\Delta_1^2 + \Delta_2^2}{2\overline{\Delta^2}}\right] d\Delta_1\, d\Delta_2. \quad (6.62)$$

The time average is

$$\phi_{s+\bar{n}}(0) = \frac{1}{\tau}\frac{a^2}{\overline{\Delta^2}\sqrt{2\pi}}\iint_{-\infty}^{\infty}\sqrt{\frac{(\mu + \Delta_1)^2(\mu + \Delta_2)^2}{(\mu + \Delta_1)^2 + (\mu + \Delta_2)^2}}\exp\left(-\frac{\Delta_1^2 + \Delta_2^2}{2\overline{\Delta^2}}\right) d\Delta_1\, d\Delta_2. \quad (6.63)$$

The integration variables are changed as follows: $\Delta_1 = x\sqrt{\overline{\Delta^2}}$, $\Delta_2 = y\sqrt{\overline{\Delta^2}}$. This yields

$$\phi_{s+\bar{n}}(0) = \frac{a^2}{\tau\sqrt{2\pi}}\iint_{-\infty}^{\infty} F(x, y, \overline{\Delta^2})\exp\left[-\frac{x^2 + y^2}{2}\right] dx\, dy, \quad (6.64)$$

where

$$F(x, y, \overline{\Delta^2}) = \frac{(\mu + x\sqrt{\overline{\Delta^2}})(\mu + y\sqrt{\overline{\Delta^2}})}{\sqrt{(\mu + x\sqrt{\overline{\Delta^2}})^2 + (\mu + y\sqrt{\overline{\Delta^2}})^2}}. \quad (6.65)$$

This function is now expanded in powers of $(\overline{\Delta^2})^{1/2}$, i.e.,

$$F(x, y, \overline{\Delta^2}) = \frac{1}{\sqrt{2}}\left\{\mu + \frac{x + y}{2}(\overline{\Delta^2})^{1/2} - \frac{3}{8}\frac{(x - y)^2}{\mu}(\overline{\Delta^2}) + \cdots\right\}. \quad (6.66)$$

This is used in the double integral, Eq. (6.64). The result is

$$\phi_{s+n}^{-}(0) = \frac{a^2\mu}{\tau}\sqrt{\pi}\left\{1 - \frac{3}{4}\frac{(\overline{\Delta^2})}{\mu^2} + \cdots\right\}. \tag{6.67}$$

The fluctuating part of the noise is

$$n(t) - \bar{n}(t) = a\left\{\exp\left[-\frac{t^2}{2(\mu+\Delta)^2}\right] - \frac{1}{\sqrt{2\pi\overline{\Delta^2}}}\int_{-\infty}^{\infty}\exp\left[-\frac{t^2}{2(\mu+\Delta)^2} - \frac{\Delta^2}{2\overline{\Delta^2}}\right]d\Delta\right\}. \tag{6.68}$$

This is squared and averaged over the interpulse period and then averaged over the ensemble. The result is

$$\phi_{n-n}^{-}(0) = \frac{a^2\mu}{\tau}\sqrt{\pi} - \phi_{s+n}^{-}(0). \tag{6.69}$$

The cancellation ratio is, therefore,

$$CR(\text{pulse-width jitter}) = \frac{\phi_{s+n}^{-}(0)}{\phi_{n-n}^{-}(0)} \approx \frac{4}{3}\frac{\mu^2}{\overline{\Delta^2}}, \tag{6.70}$$

as shown in Fig. 6.12.

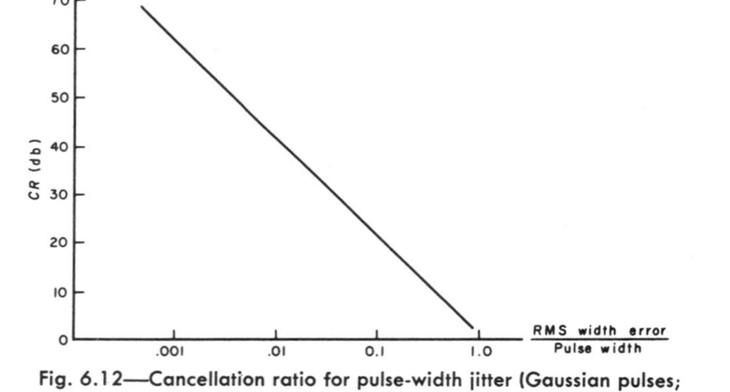

Fig. 6.12—Cancellation ratio for pulse-width jitter (Gaussian pulses; Gaussian width error)

6.8.3. Pulse-amplitude Jitter

Letting Δ denote the pulse-amplitude error, the perturbed signal is

$$s(t) + n(t) = (a+\Delta)e^{-t^2/2\mu^2} \qquad (\text{here } t_j = 0). \tag{6.71}$$

In this case $\bar{n}(t) = 0$, so that

$$\phi_{s+n}^{-}(0) = \phi_s(0) = \frac{a^2\mu}{\tau}\sqrt{\pi}. \tag{6.72}$$

The fluctuating part of the noise is

$$n(t) - \bar{n}(t) = n(t) = \Delta e^{-t^2/2\mu^2}. \tag{6.73}$$

This is squared, averaged over the interpulse period, and then averaged over the ensemble. The result is

$$\phi_{n-n}^{-}(0) = \frac{\mu}{\tau}\sqrt{\pi}\,\overline{\Delta^2}. \tag{6.74}$$

The cancellation ratio becomes

$$CR(\text{amplitude jitter}) = \frac{a^2}{\overline{\Delta^2}}. \tag{6.75}$$

This is illustrated in Fig. 6.13.

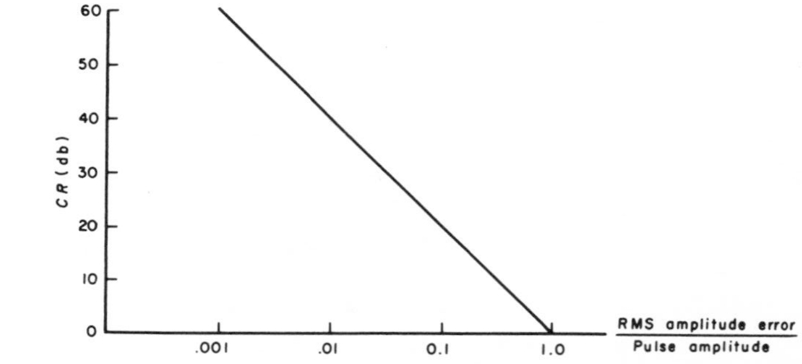

Fig. 6.13—Cancellation ratio for pulse-amplitude jitter (Gaussian pulses; Gaussian amplitude error)

6.8.4. Phase Instabilities

Variations in pulse amplitude can arise as a result of phase instability either in the coho, stalo, or locking circuitry, or through frequency modulation within the transmitted pulse. Let the phase error be Δ radians; then, after beating with the coherent oscillator, the pulse amplitude will be

$$\Omega = a\cos\Delta, \tag{6.76}$$

where a is the intended pulse amplitude. If Δ is normally distributed with mean zero and variance $\overline{\Delta^2}$, then the first and second moments of Ω will be

$$\overline{\Omega} = \frac{1}{\sqrt{2\pi\overline{\Delta^2}}} \int_{-\infty}^{\infty} \exp\left[-\frac{\Delta^2}{2\overline{\Delta^2}}\right] \cdot a \cos \Delta \, d\Delta$$
$$= a e^{-\frac{1}{2}\overline{\Delta^2}}, \tag{6.77}$$

and

$$\overline{\Omega^2} = \frac{1}{\sqrt{2\pi\overline{\Delta^2}}} \int_{-\infty}^{\infty} \exp\left[-\frac{\Delta^2}{2\overline{\Delta^2}}\right] \cdot a^2 \cos^2 \Delta \, d\Delta$$
$$= \frac{a^2}{2}\{1 + e^{-2\overline{\Delta^2}}\}. \tag{6.78}$$

For the Gaussian pulse shape, the expected signal-plus-noise pulse is

$$s(t) + \overline{n}(t) = \overline{\Omega} e^{-t^2/2\mu^2} \qquad (\text{here } t_j = 0). \tag{6.79}$$

This is squared and averaged over the interpulse period, yielding

$$\phi_{s+\overline{n}}(0) = \frac{(\overline{\Omega})^2 \mu}{\tau} \sqrt{\pi}. \tag{6.80}$$

The fluctuating part of the noise is

$$n(t) - \overline{n}(t) = (\Omega - \overline{\Omega}) e^{-t^2/2\mu^2}. \tag{6.81}$$

This is squared, averaged over the interpulse period, and then averaged over the ensemble, yielding

$$\phi_{n-\overline{n}}(0) = \{\overline{\Omega^2} - (\overline{\Omega})^2\} \frac{\mu}{\tau} \sqrt{\pi}. \tag{6.82}$$

The cancellation ratio is

$$CR(\text{phase instabilities}) = \frac{(\overline{\Omega})^2}{\overline{\Omega^2} - (\overline{\Omega})^2}$$
$$= \frac{1}{\cosh \overline{\Delta^2} - 1}. \tag{6.83}$$

This is shown graphically in Fig. 6.14.

6.8.5. Conservative Rule for Combining Cancellation Ratios

On the premise that the various system-instability noises are independent and that they add in power, it is possible to determine an upper bound for the over-all cancellation ratio. The aggregate noise power is expressed

$$\phi_{n-\overline{n}}(0) = \phi_{n_1-\overline{n}_1}(0) + \phi_{n_2+\overline{n}_2}(0) + \cdots, \tag{6.84}$$

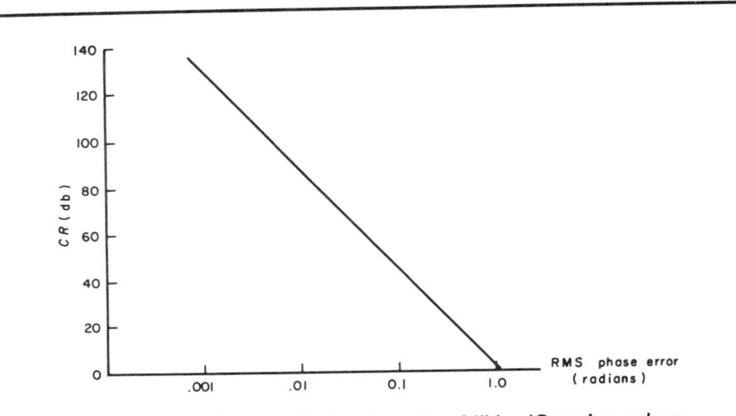

Fig. 6.14—Cancellation ratio for phase instabilities (Gaussian pulses; Gaussian phase errors)

where the subscripts denote the different noise sources. The steady signal power is approximated by

$$\phi_{s+\overline{n}}(0) \approx \phi_s(0) \approx \phi_{s+n_j}(0) \qquad (\text{all } j), \tag{6.85}$$

which is the same for all sources. This results in an over-all cancellation ratio given by

$$CR = \frac{1}{\frac{1}{CR_1} + \frac{1}{CR_2} + \cdots}. \tag{6.86}$$

CHAPTER 7

PERFORMANCE ESTIMATES FOR SOME AIR DEFENSE RADARS

7.1. Introduction

A question frequently raised in connection with MTI and AMTI is, How much suppression is needed to effectively eliminate the clutter? To arrive at a quantitative answer, something must first be known about the intensity of the returns that are most likely to be troublesome. Theoretical and experimental investigations in this direction are listed in Refs. 26 and 27.*

To determine the magnitude of the return from any particular patch of clutter, one must know the transmitted power and pulse width, the range to the clutter patch, the angle of incidence, and the reflection properties of the illuminated region at the transmitted wavelength. These factors vary over a wide range of conditions, depending on the type of radar (if airborne, shipboard, or ground based), the weather (if the echo is from clouds, chaff, or the sea), the type of terrain (if the echo is from land masses), etc. Consequently there is no general rule for determining a priori the seriousness of the clutter problem.

Nevertheless, experience with single-delay-line ground-based MTI sets has indicated that from 15- to 20-db rejection ratio will usually eliminate most of the objectionable clutter from neighboring terrain. The situation is similar for moderate sea return and shipboard MTI. Fortunately this amount of rejection is typical of that afforded by single delay lines with calm and mildly fluctuating clutter. The more severely fluctuating targets, such as chaff and storm clouds, usually produce weaker echoes. This compensates somewhat for the lower rejection capability due to the fluctuation.

With AMTI, however, the most unfavorable combination of circumstances exists. Here, because of the higher probability of specular reflections, the returns from land and sea will be somewhat larger, on the average; and at the same time, due to the motion of the radar platform, the echoes are caused to fluctuate even more violently than would otherwise be the case. Both of these effects contribute to a larger clutter residue.† Unfortunately, a backlog of experience similar to that of ground-based and shipboard MTI has not yet been accumulated for AMTI.

There are many instances where single-delay-line MTI will be adequate, at least for

*Particular attention is directed to the reports of General Precision Laboratory, Inc., where many of the earlier references are discussed.

†This is most applicable to the interception AMTI, where high platform velocity is the rule and direct ground illumination in search of low-flying targets is a possibility. These factors are less severe in the case of AMTI for the airborne early warning (AEW) over water because of the lower cruising speed and the generally lower clutter intensities.

the present; however, it is not clear to what extent this will be true as the enemy ECM capability increases. It is therefore of interest to examine the utility of the more complex cancellation system as compared with that of the single-delay-line systems presently employed. To illustrate the relative advantages, performance estimates are given for three-pulse as well as the standard two-pulse comparison. (Three-pulse comparison was selected because it is achievable with the minimum modification of existing systems, i.e., through the addition of an extra delay-line unit or through the employment of recirculation principles with the one delay line.)

The radar sets considered are as follows:

	Figure
Ground-based surveillance	
AN/CPS-6B, AN/FPS-10	7.4
AN/FPS-7(XW-)	7.5
AN/FPS-8, AN/MPS-11	7.6
AN/FPS-3, AN/MPS-7	7.7
AN/TPS-1D	7.8
AN/CPN-4 (search section)	7.9
AN/CPN-18A	7.10
ASR-1	7.11
Shipboard surveillance	
AN/SPS-6B	7.12
AN/SPS-13	7.13
Airborne early warning	
AN/APS-20B	7.14
Air interception	
AN/APG-33 (type)	7.15

7.2. General Remarks

The clutter cancellation capability is measured in terms of the reference gain, $\bar{G}$, defined in Sec. 1.4. This is plotted versus clutter-particle velocity, v, a measure of the clutter fluctuation. The bench marks at $v = 0.01$, 0.50, 1.8, 2.0, and 3.9 knots correspond to Barlow's typical clutter types (see table on page 76). These are intended to indicate only very general areas of reference, since there is a great deal of overlap among the categories.

In order to simplify the calculations as much as possible, Gaussian autocorrelation functions and power spectra are used in this chapter. Accordingly, the scanning autocorrelation functions corresponding to the different antenna current distributions (see Figs. 6.3, 6.4, and 6.5) are approximated by the one normal curve

$$\rho_a(\tau) = \exp\left[-1.39\left(\frac{a\theta'\tau}{\lambda}\right)^2\right]. \tag{7.1}$$

To illustrate the degree of approximation, this is shown by the dashed curve on Fig. 6.5. The corresponding unit-power spectrum is

$$W_a(f) = 1.5\frac{\lambda}{a\theta'}\exp\left[-7.12\left(\frac{\lambda f}{a\theta'}\right)^2\right], \tag{7.2}$$

which is shown by the dashed curve on Fig. 6.6.

Similarly, the autocorrelation functions for platform motion have been condensed into

$$\rho_{pm}(\tau) = \exp\left[-7.7\left(\frac{v\tau}{a}\sin\gamma\right)^2\right], \tag{7.3}$$

and the corresponding unit-power spectrum is given by

$$W_{pm}(f) = .64\frac{a}{v\sin\gamma}\exp\left[-1.29\left(\frac{af}{v\sin\gamma}\right)^2\right]. \tag{7.4}$$

These appear as dashed curves on Figs. 6.8 and 6.9.

In all cases the decorrelating effect of wind and system noise has been neglected. To account for system imperfections, an appropriate cancellation ratio may be introduced through the formula

$$\bar{G}_{\text{actual}} = \frac{1}{\dfrac{1}{\bar{G}_{\text{ideal}}} + \dfrac{1}{CR}}, \tag{7.5}$$

or Fig. 6.10 may be used.

The internal-motion autocorrelation function, $\rho_{im}(\tau)$, is used as given in Sec. 6.2. This function, together with $\rho_a(\tau)$ and $\rho_{pm}(\tau)$, as given above, gives rise to the composite autocorrelation function (see Sec. 6.6),

$$\rho_c(\tau) = \exp(-2\pi^2\sigma_c^2\tau^2), \tag{7.6}$$

and to the composite unit-power spectrum,

$$W_c(f) = \frac{1}{\sqrt{2\pi\sigma_c^2}}\exp\left[-\frac{f^2}{2\sigma_c^2}\right], \tag{7.7}$$

where

$$\sigma_c^2 = \sigma_{im}^2 + \sigma_a^2 + \sigma_{pm}^2, \tag{7.8}$$

and

$$\sigma_{im} = \frac{2v}{\lambda}, \tag{7.9}$$

$$\sigma_a = .266\frac{a\theta'}{\lambda}, \tag{7.10}$$

and

$$\sigma_{pm} = .625\frac{v\sin\gamma}{a}. \tag{7.11}$$

The parameters are as follows:

Symbol	Definition	Units
ν	RMS clutter-particle velocity	length/time
λ	Radar wavelength	length
a	Antenna diameter	length
θ'	Scan rate	radians/time
v	Radar platform velocity	length/time
γ	Antenna sighting angle (off ground track)	....

Nomographs for determining σ_{im}, σ_a, and σ_{pm} appear in Figs. 7.1, 7.2, and 7.3. Since the system gain is substantially the same for multiple- and single-interpulse-period operation, all calculations have been based on the single-interpulse period. The reference-gain formulas used are as follows:

	Single Delay Line	Double Delay Line
Coherent.......................	$\{1 - \rho_c(\tau)\}^{-1}$	$3\{3 - 4\rho_c(\tau) + \rho_c(2\tau)\}^{-1}$
Noncoherent....................	$2\{1 - \rho_c(\tau)^2\}^{-1}$	$6\{3 - 4\rho_c(\tau)^2 + \rho_c(2\tau)^2\}^{-1}$

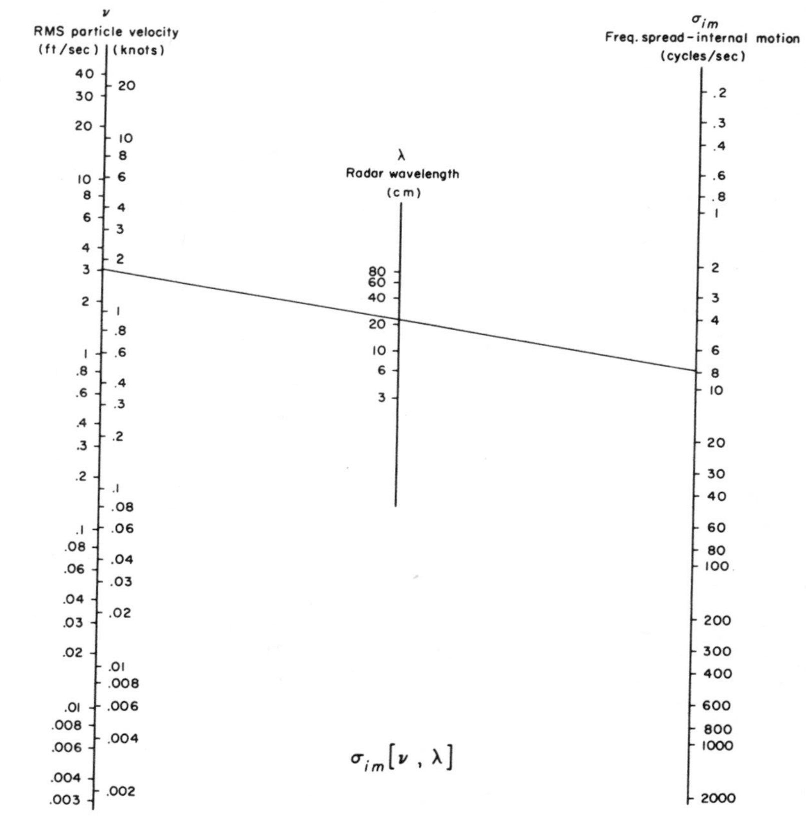

Fig. 7.1—Nomograph for determining frequency spread due to internal-motion noise

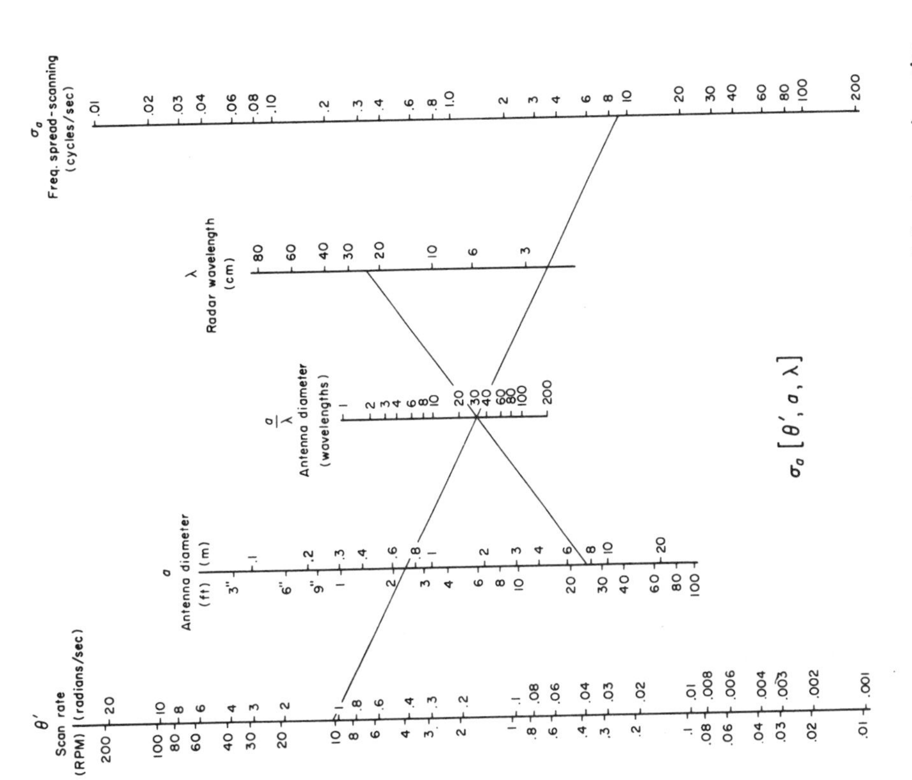

Fig. 7.2—Nomograph for determining frequency spread due to scanning noise

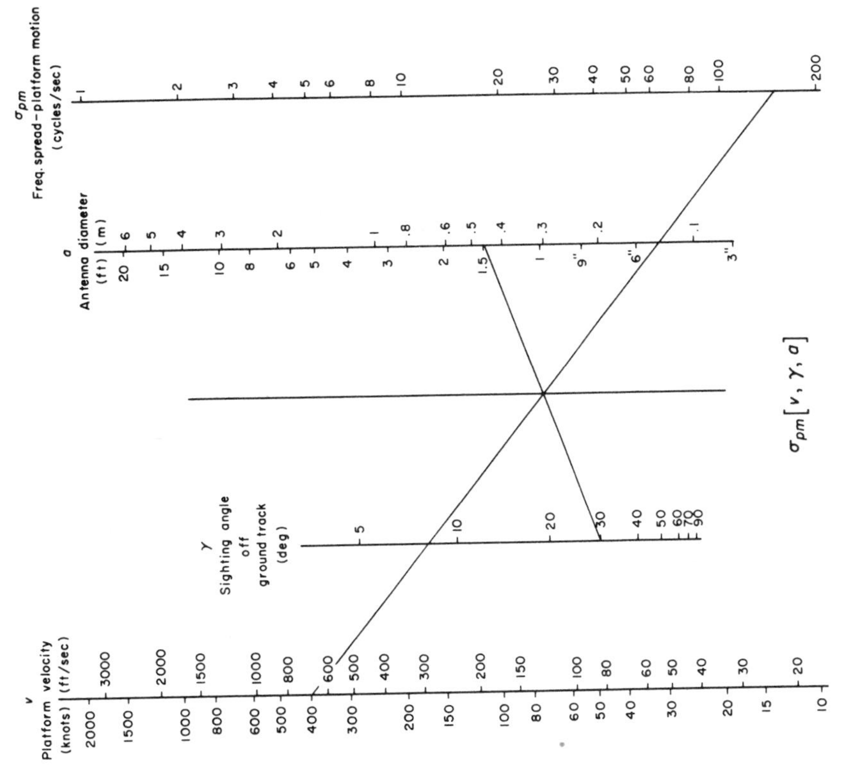

Fig. 7.3—Nomograph for determining frequency spread due to platform-motion noise

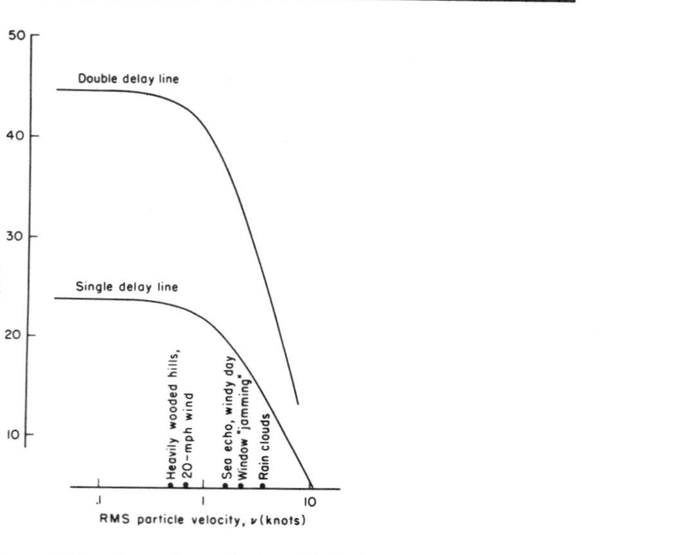

Fig. 7.4—Reference gain vs clutter fluctuation velocity: AN/CPS-6B and AN/FPS-10

Function*

The AN/CPS-6B is a ground-based radar suitable for early-warning and ground-control-intercept (GCI) application. It employs a V-beam type of antenna and is capable of providing elevation as well as range and azimuth information.

Facts Relevant to the MTI

PRF	300 pps (normal) 600 pps (MTI)
Frequency	2700–3019 Mc
Wavelength	$\lambda = 10$ cm
Antenna diameter	
Vertical beam	$a = 25$ ft (width)
Slant beam	$a = 32$ ft (length)
Scan rate	$\theta' = 6$ rpm

Remarks

The present AN/CPS-6B equipment includes a single delay line for two-pulse-comparison MTI. Radar set AN/FPS-10 is identical with the AN/CPS-6B except for reduced display facilities.

*Information concerning these radars was obtained from Ref. 28.

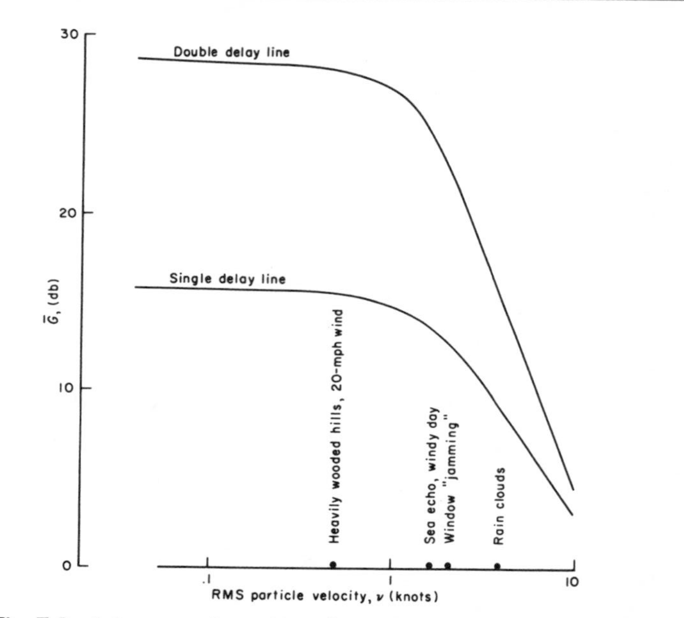

Fig. 7.5—Reference gain vs clutter fluctuation velocity: AN/FPS-7(XW-)

Function*

The AN/FPS-7(XW-) is a ground-based version of the shipboard radar AN/SPS-2. It is a high-powered (10 Mw), long-range search set providing, in addition to range and azimuth information, target elevation by means of interpolation among seven receiving beams.

Facts Relevant to the MTI

PRF	244 pps
Frequency	1250–1350 Mc
Wavelength	$\lambda = 10$ cm
Antenna diameter	$a = 40$ ft (width)
Scan rate	$\theta' = 6$ rpm

Remarks

This set will provide for the selection of either coherent or clutter-gated noncoherent MTI. The MTI will employ a single delay line for two-pulse comparison.

*Information concerning this radar was obtained from Ref. 28.

Fig. 7.6—Reference gain vs clutter fluctuation velocity: AN/FPS-8 and AN/MPS-11

*Function**

The AN/FPS-8 (fixed version) and the AN/MPS-11 (mobile version) are ground-based search radars providing range and azimuth information. A separate height finder is required when these sets are used for GCI.

Facts Relevant to the MTI

PRF	360 pps
Frequency	1280–1350 Mc
Wavelength	$\lambda = 23$ cm
Antenna diameter	$a = 25$ ft (width)
Scan rate	$\theta' = 0–10$ rpm

Remarks

Both sets include a single delay line for two-pulse-comparison MTI. Figure 7.6 is based on a scan rate of 10 rpm.

*Information concerning these radars was obtained from Ref. 28.

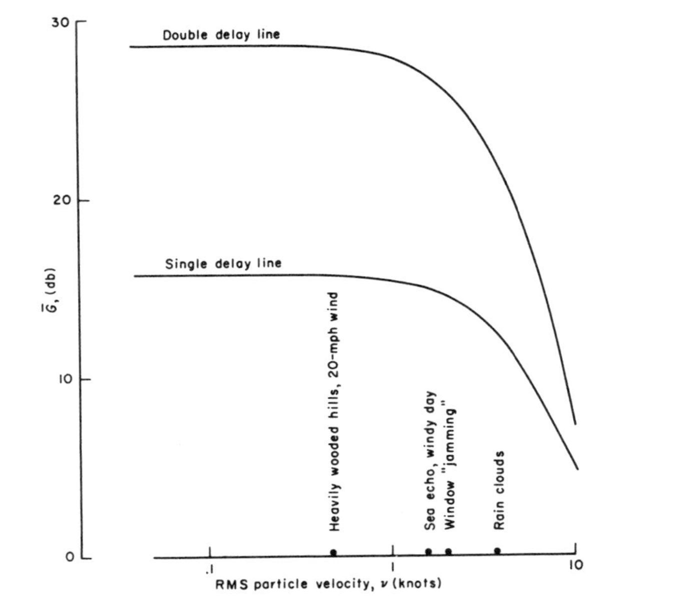

Fig. 7.7—Reference gain vs clutter fluctuation velocity: AN/FPS-3 and AN/MPS-7

*Function**

The AN/FPS-3 (fixed version) and AN/MPS-7 (mobile version) are ground-based search radars providing range and azimuth information. For use in GCI, additional height-finding equipment such as the AN/FPS-6 is required.

Facts Relevant to the MTI

PRF	200 pps (normal) 400 pps (MTI)
Frequency	1220–1350 Mc
Wavelength	$\lambda = 23$ cm
Antenna diameter	$a = 40$ ft (width)
Scan rate	$\theta' = 3–10$ rpm

Remarks

The AN/FPS-3 includes a single delay line for two-pulse-comparison MTI. Figure 7.7 is based on a scan rate of 10 rpm.

*Information concerning these radars was obtained from Ref. 28.

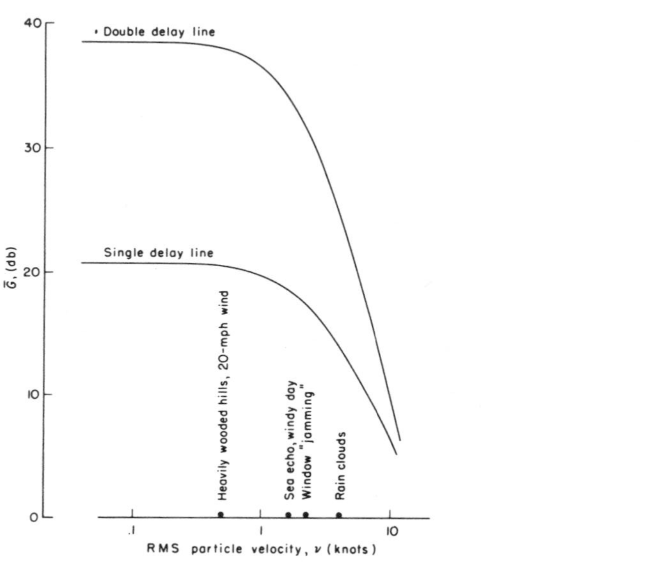

Fig. 7.8—Reference gain vs clutter fluctuation velocity: AN/TPS-1D

Function*

The AN/TPS-1D is a lightweight, transportable, ground-based search radar providing range and azimuth information. Its primary function is envisioned to be that of a gap-filler in an early-warning network; however, it has some GCI capability when used in conjunction with a separate height finder.

Facts Relevant to the MTI

PRF . . . . 400 pps
Frequency . . . . 1220–1350 Mc
Wavelength . . . . $\lambda = 23$ cm
Antenna diameter . . . . $a = 15$ ft (width)
Scan rate . . . . $\theta' = 0$–15 rpm

Remarks

A single delay line for two-pulse comparison has been incorporated as an integral part of this system.

Figure 7.8 is based on a scan rate of 15 rpm.

*Information concerning this radar was obtained from Ref. 28.

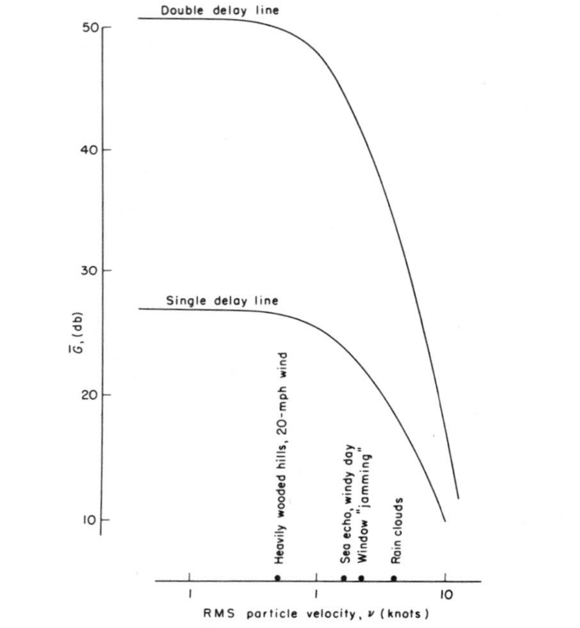

Fig. 7.9—Reference gain vs clutter fluctuation velocity: AN/CPN-4 (search section)

Function*

The AN/CPN-4 is a combination S-band search-surveillance and X-band precision landing system for airport GCA. The search section of this equipment is a developmental outgrowth of the ASR-1.

Facts Relevant to the MTI

PRF . . . . 1500 pps
Frequency . . . . 2780–2820 Mc
Wavelength . . . . $\lambda = 10$ cm
Antenna diameter . . . . $a = 9$ ft (width)
Scan rate . . . . $\theta' = 20$ rpm

Remarks

This radar is equipped with a single mercury delay line for two-pulse-comparison MTI.

*Information concerning this radar was obtained from Ref. 29.

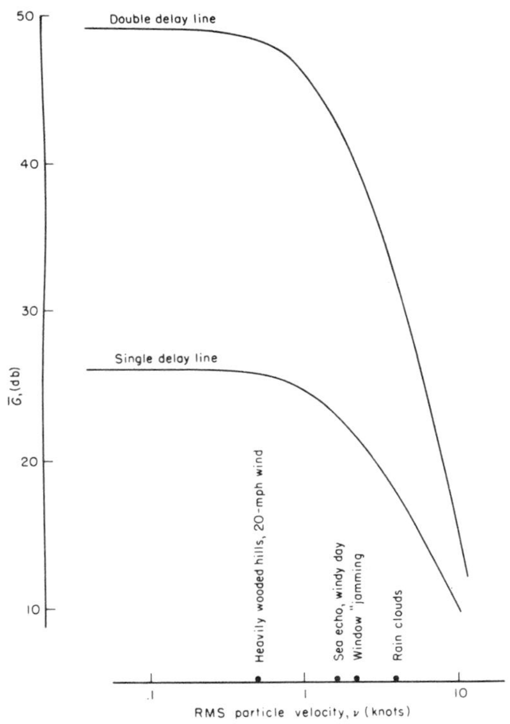

Fig. 7.10—Reference gain vs clutter fluctuation velocity: AN/CPN-18A

*Function**

The AN/CPN-18A is a lightweight, air-transportable, ground-based search radar primarily intended for air-traffic-control surveillance and for use in conjunction with precision GCA equipment.

Facts Relevant to the MTI

PRF	1500 pps
Frequency	2700–2900 Mc
Wavelength	$\lambda = 10$ cm
Antenna diameter	$a = 20$ ft (width)
Scan rate	$\theta' = 10$ rpm

*Information concerning this radar was obtained from Ref. 28.

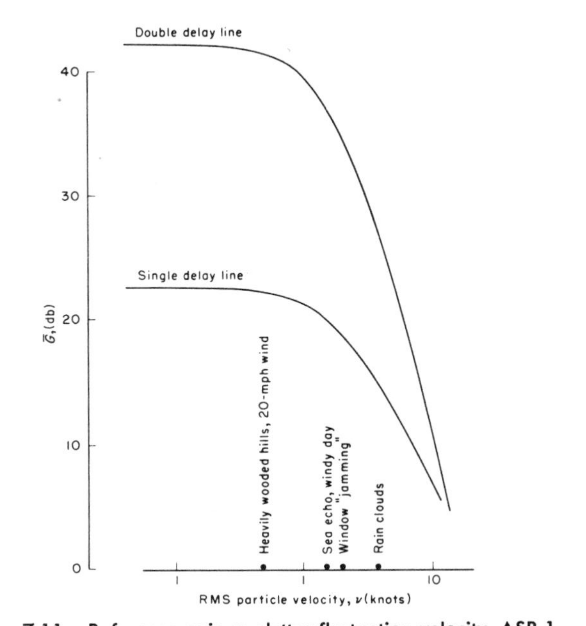

Fig. 7.11—Reference gain vs clutter fluctuation velocity: ASR-1

*Function**

The ASR-1 is a fixed-installation, ground-based airport-surveillance radar primarily intended for air traffic control and for use in conjunction with precision GCA equipment.

Facts Relevant to the MTI

PRF	1000 pps
Frequency	2780–2820 Mc
Wavelength	$\lambda = 10$ cm
Antenna diameter	$a = 7\frac{1}{2}$ ft (width)
Scan rate	$\theta' = 26$ rpm

Remarks

This set includes a single mercury-delay-line, two-pulse-comparison MTI section.

*Information concerning this radar was obtained from Ref. 30.

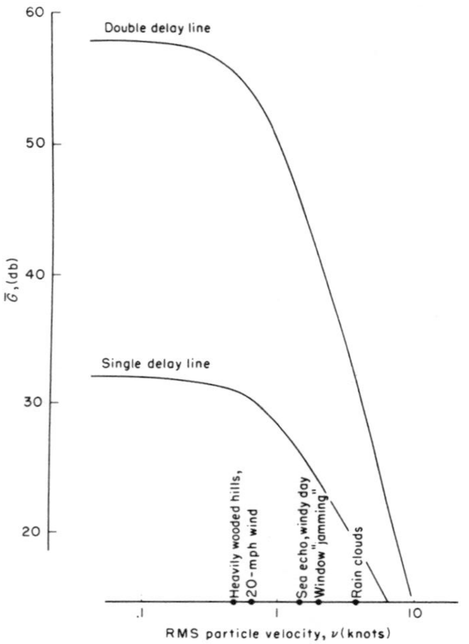

Fig. 7.12—Reference gain vs clutter fluctuation velocity: AN/SPS-6B

*Function**

The AN/SPS-6B is a shipboard air-search radar for vessels of the destroyer class and larger. Separate height-finding equipment is needed if this set is used for over-water GCI.

Facts Relevant to the MTI

PRF	150 pps (normal) 600 pps (MTI)
Frequency	1300 ± 50 Mc
Wavelength	$\lambda = 23$ cm
Antenna diameter	$a = 18$ ft
Scan rate	$\theta' = 5$ rpm

Remarks

This set is equipped with single-delay-line noncoherent MTI.

*Information concerning this radar was obtained from Ref. 31.

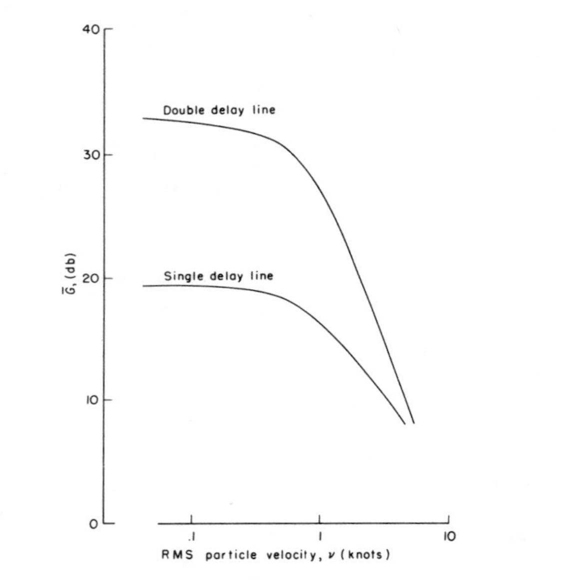

Fig. 7.13—Reference gain vs clutter fluctuation velocity: AN/SPS-13

*Function**

The AN/SPS-13 is a shipboard radar, being designed for early warning, air traffic control, and control of close-support aircraft for beachhead ground troops. This radar provides simultaneous height information by means of seven vertically stacked beams.

Facts Relevant to the MTI

PRF	400 pps
Frequency	2800 ± 30 Mc
Wavelength	$\lambda = 10$ cm
Antenna diameter	$a = 20$ ft (width)
Scan rate	$\theta' = 3$ or 6 rpm

Remarks

This radar is a conceptual descendant of the AN/SPS-2 (Air Force version, AN/FPS-7 (XW-)) multiple-beam search-and-control radar. It will employ a single-delay-line, clutter-gated noncoherent MTI.

Figure 7.13 is based on a scan rate of 6 rpm.

*Information concerning this radar was obtained from Ref. 32.

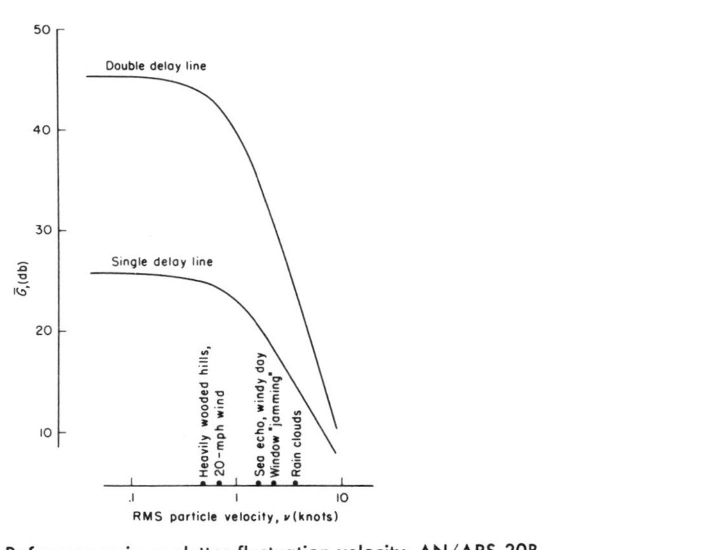

Fig. 7.14—Reference gain vs clutter fluctuation velocity: AN/APS-20B

*Function**

The AN/APS-20B is a long-range, AEW search-and-surveillance radar with facilities for interceptor control. It is suitable for installation in the PO-2W- or RC-121-C-type aircraft, or it may be used for ground-based search and control (e.g., in connection with the AN/CPS-6B to provide additional beam coverage).

Facts Relevant to the MTI

PRF	300 pps (normal) 900 pps (MTI)
Frequency	2700–2900 Mc
Wavelength	$\lambda = 10$ cm
Antenna diameter	$a = 17$ ft (width)
Scan rate	$\theta' = 6$ rpm

Remarks

As presently produced, no MTI is included.

Figure 7.14 is based on the AEW cruising speed (platform motion) of 180 knots and a sighting angle (γ) of 30° off the ground track.

*Information concerning this radar was obtained from Ref. 31.

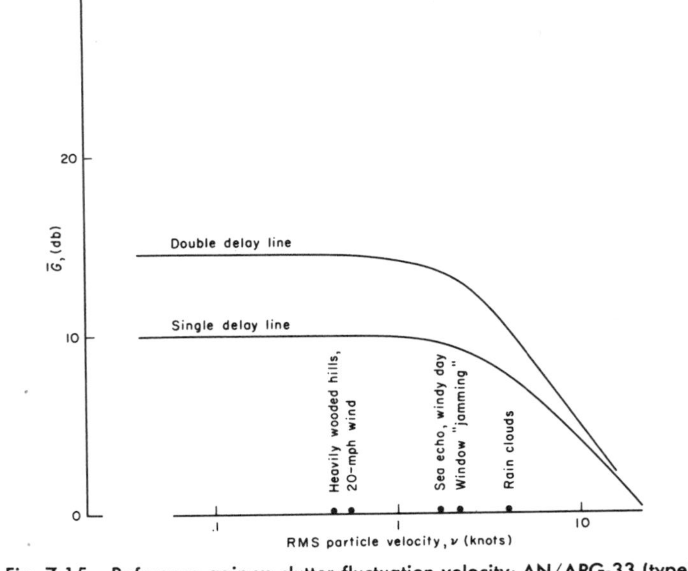

Fig. 7.15—Reference gain vs clutter fluctuation velocity: AN/APG-33 (type)

*Function**

The AN/APG-33 is the airborne-interception-radar part of the E-1 fire-control system.

Facts Relevant to the MTI

PRF	2000 pps
Frequency	10,000 Mc
Wavelength	$\lambda = 3$ cm
Antenna diameter	$a = 1.5$ ft (width)
Scan rate	$\theta' = 100° \times 50°$ in 3 sec

Remarks

As presently produced, this radar has no MTI.

Figure 7.15 is based on the interceptor speed (platform motion) of 400 knots and a sighting angle (γ) of 30° off the ground track.

*Information concerning this radar was obtained from Ref. 33.

REFERENCES

1. Ridenour, L. N. (ed.), *Radar Systems Engineering*, Radiation Laboratory Series, Vol. 1, McGraw-Hill Book Company, Inc., New York, 1947.
2. Lawson, J. L., and G. E. Uhlenbeck (eds.), *Threshold Signals*, Radiation Laboratory Series, Vol. 24, McGraw-Hill Book Company, Inc., New York, 1950.
3. *Anticlutter Circuit for AEW*, Radiation Laboratory, MIT, Report 5-52, August 1, 1945 (Confidential).
4. Hobbs, Charles F., *A Discussion of Some Anticlutter Circuits*, AMC, Cambridge Field Station, Report 4–5, January 15, 1947.
5. Goldstein, H., *The Effects of Clutter Fluctuations on MTI*, Radiation Laboratory, MIT, Report 700, December 27, 1945 (Restricted).
6. Ross, A. W., *An Analysis of Present Information on the Scattering of Radar Signals by the Sea*, Part II, "The Pulse-to-pulse Fluctuation of Radar Signals," Admiralty Signal Establishment Extension, Witley, England, XRD/46/3, October 7, 1946.
7. Siegert, A. J. F., and F. W. Martin, *Fluctuation in the Return Signals from Random Scatterers (Window, Rain, Sea Echo)*, Radiation Laboratory, MIT, Report 773, January 24, 1946.
8. Barlow, Edward J., "Doppler Radar," *Proc. I.R.E.*, Vol. 37, No. 4, April, 1949, pp. 340–354.
9. Marcum, J. I., *A Statistical Theory of Target Detection by Pulsed Radar, Mathematical Appendix*, The RAND Corporation, Research Memorandum RM-753, July 1, 1948.
10. *Radar Systems Analysis*, Final Engineering Report on Doppler Systems Study, AMC–Watson Laboratories Contract No. W28-099 AC-115, Sperry Gyroscope Company, June, 1948 (Secret).
11. Kac, M., and A. J. F. Siegert, "On the Theory of Noise in Radio Receivers with Square Law Detectors," *J. Appl. Phys.*, Vol. 18, No. 4, April, 1947, pp. 383–397.
12. Emerson, R. C., "First Probability Densities for Receivers with Square Law Detectors," *J. Appl. Phys.*, Vol. 24, No. 9, September, 1953, pp. 1168–1176.
13. Middleton, D., "Rectification of a Sinusoidally Modulated Carrier in the Presence of Noise," *Proc. I.R.E.*, Vol. 36, No. 12, December, 1948, pp. 1467–1477.
14. Fubini, E. G., and D. C. Johnson, "Signal-to-noise Ratio in A.M. Modulated Receivers," *Proc. I.R.E.*, Vol. 36, No. 12, December, 1948, pp. 1461–1466.
15. *Quarterly Progress Report, Division 3—Communications and Components*, Lincoln Laboratory, MIT, September 30, 1952 (Secret).
16. *Quarterly Progress Report, Division 3—Communications and Components*, Lincoln Laboratory, MIT, October 1, 1952 (Secret).
17. *Airborne Detection of Moving Ground Targets*, Interim Engineering Reports 1–12, Contract AF-33(038)-12473, H-1008, Philco Corporation (Confidential).
18. *Quantitative Measurements of Radar Echoes from Aircraft*, Series of Memoranda and Reports, NRL.
19. *Evaluation of Airborne Moving Target Indication (AMTI) Kit (Moving Target Indicator Group AN/APA-78(XA-1))*, Engineering Memorandum Report, Electronic Subdivision, Engineering Division, AMC, Wright-Patterson AFB, March 1, 1950 (Confidential).
20. Ellis, R. E., *A Clutter-gated Noncoherent MTI and Video Integrator*, NRL Report 4106, February 16, 1953 (Confidential).

21. Harrington, J. V., *Tests of the Rafax Scanner*, Memorandum to G. E. Valley, Lincoln Laboratory, MIT, March 26, 1952.
22. Kerr, D. E., *Propagation of Short Radio Waves*, Radiation Laboratory Series, Vol. 13, McGraw-Hill Book Company, Inc., New York, 1951.
23. *Research on Radar Terrain Characteristics*, Final Engineering Reports, Vols. 1, 2, and 3, Contract AF-33(038)-9219, CD 954, Philco Corporation (Confidential).
24. *A Step-scanning Frequency-coded Pulse Radar*, Final Engineering Report on Long-range L-band Search Radar, AMC–Watson Laboratories Contract W28-099 AC-454, Sperry Gyroscope Company, July, 1950 (Secret).
25. *Airborne Radar Detection of Moving Ground Targets*, Summary Engineering Reports, Vols. 1, 2, and 3, Contract AF-33(038)-12473, H-1008, Philco Corporation (Confidential).
26. *A Bibliography of Radar Reflection Characteristics*, Vols. 1 and 2, Georgia Inst. of Tech., 1952 (Confidential).
27. *Automatic Ground Position Indicator*, Interim Engineering Report A11-32, Contract W33-038-ac-14192, General Precision Laboratory, Inc., May 20, 1949 (Confidential).
28. *Ground Radar Sets, Systems and Related Components*, RADC Technical Report 52–22, November 1, 1952 (Secret).
29. *Survey of USAF Ground Primary Radars for Air Navigation and Traffic Control*, 3151st Electronics Group, Watson Laboratories Report ENRPE-2, May 15, 1950 (Secret).
30. *Instruction Book for Airport Surveillance Radar Model ASR-1*, CAA–U.S. Dept. of Commerce Contract Cca-26345, Gilfillan Bros., Inc.
31. *U.S. Navy Radar Systems Survey*, 2d ed., NRL Report 4128, April 24, 1953 (Secret).
32. *Proposed Military Characteristics, Radar Set AN/SPS-13*, Applied Physics Laboratory, Johns Hopkins University, TSO-M-231, February 28, 1951 (Confidential).
33. Barlow, E. J., and R. C. Emerson, *Airborne Interception Radar Detection Capabilities*, The RAND Corporation, Research Memorandum RM-290, November 22, 1949 (Secret).

References Not Specifically Cited

Barlow, E. J., and J. F. Digby (eds.), *Air Defense Study*, The RAND Corporation, Report R-227, October 15, 1951 (Secret).

Graham, W. B., J. D. Mallett, and R. C. Emerson, *The Capabilities and Limitations of Some MTI Radar Systems*, The RAND Corporation, Research Memorandum RM-527, February 1, 1951 (Secret).

Final Engineering Report on the MTI Measurement Problem, Vol. 1 and Appendixes, Contract AF28(099)-89, Laboratory for Electronics, July 1, 1950 (Restricted).

Reich, E., and P. Swerling, "The Detection of a Sine Wave in Gaussian Noise," *J. Appl. Phys.*, Vol. 24, No. 3, March, 1953, pp. 289–296.

Campbell, G. A., and R. M. Foster, *Fourier Integrals for Practical Applications*, Bell Telephone Monograph B-584, 1931.

Rice, S. O., *Mathematical Analysis of Random Noise*, Bell Telephone Monograph B-1589, 1944.

1.2

RECENT ADVANCES IN THE SYNTHESIS OF COMB FILTERS

W.D. White and A.E. Ruvin

IRE National Convention Record, 1957, pp. 186-199.

INTRODUCTION

The term "Comb Filter" is used to describe a filter whose transfer function is periodic in the frequency domain. While a strict periodicity would imply the ability to pass signals to infinite frequency, practical comb filters can be adequately represented by the cascade connection of a conventional low pass filter and an ideal comb filter in which strict periodicity is observed at least throughout the useful range of frequencies.

MTI Filters

The simplest form of comb filter in common use is the conventional cancellation circuit of an MTI radar. Such a circuit is shown schematically in Figure 1 and its frequency response shown in Figure 2 is of the form

$$\left|\frac{E_o}{E_i}\right| = \left|\sin\frac{\omega T}{2}\right| = \left|\sin\pi\frac{f}{f_r}\right| \qquad (1)$$

This type of filter provides MTI action by virtue of the fact that in the coherent video signal, the spectrum of a fixed target is a comb shaped spectrum with the peaks of the comb occurring at the harmonics of the pulse repetition frequency. If the delay time T is equal to the reciprocal of the repetition frequency, then these spectrum peaks will fall at the zero response points and will be cancelled. In the case of moving targets, the spectrum is shifted by virtue of the Doppler effect and the peaks will not fall on the zero response points. The moving target signal will thus be passed with less attenuation.

Complete cancellation of a fixed target will occur only when the spectrum is a true line spectrum where all of the energy is confined to the harmonics of the repetition frequency. If the radar in question is scanning or if the fixed targets are not truly fixed but include such things as trees waving in the wind, the spectrum lines will have a finite width and some of the energy will be passed through the filter. As radars are called upon to yield faster data rates and greater ranges, this problem becomes more serious and it becomes desirable to provide more elaborate control over the frequency response shape. An example of this requirement is shown in Figure 3 where the spectrum of a typical fixed target is shown superimposed on a shaped frequency response. The response of a simple canceller is shown dotted for comparison and the superiority of the shaped response is clearly evident.

Video Integrators

Another application of comb filters to radar signals is their use for video integration. It is well known that if several successive sweeps of radar video are added together linearly, the signal which is consistent from sweep to sweep will add linearly but the noise which is random from sweep to sweep will add in a root-sum-square manner and the net result is an enhancement in signal to noise ratio. One method of achieving this integration is to use a positive feedback of a delayed signal as shown in Figure 4. Each pulse recirculates around the feedback loop being added to the successive pulses as they arrive. If stability is to be maintained, it is necessary for the pulses to be attenuated somewhat each time they complete the loop. This type of integrator may be thought of as a comb filter in which the multiple pass bands are tailored to fit around the signal spectrum. This effect is illustrated in Figure 5.

As in the case of MTI filters, increasing demands on the radar performance are causing a requirement for improved control of the integrator response. Here the requirement is not only a need for improved signal to noise enhancement but also a requirement for the preservation of angular accuracy. That an integrator can cause a loss of angular accuracy is due to the asymmetrical nature of its impulse response and the wide dynamic range encountered with signals. This effect is illustrated in Figure 6. The response of the filter to a single pulse shown in Figure 6a, is a string of pulses of exponentially decaying amplitude. As the antenna of a scanning radar sweeps through a target, the signal received is a string of pulses whose envelope traces out the shape of the antenna pattern as shown in Figure 6b. When this signal is then applied to the input of the integrator, the output is of the form shown in Figure 6c. It is seen that the center of gravity of the output signal is delayed somewhat relative to the input. This delay which is equivalent to an angular shift of the target is not in itself objectionable so long as it can be compensated. It is objectionable however, if a strong signal causes a different apparent shift than does a weak signal. This will be the case if any threshold device or other nonlinear operation is used in determining the angular position of the target. This situation could be greatly improved by providing an integrator with a more symmetrical response function.

MATHEMATICAL PRELIMINARIES

Having established the need for more sophisticated synthesis procedures in connection with comb filters let us now review some of the mathematical techniques which are useful in this regard.

Generating Functions

Comb filters containing a finite number of delay elements are conveniently analyzed through the use of generating functions or, as they are sometimes called, Z transforms. These transforms or generating functions have received considerable attention in the literature recently (see references 1, 2, and 3), particularly as they are useful for obtaining approximate solutions to conventional problems in continuous circuits. Here, however, our use of them is limited to the application for which they are directly suited, namely the analysis of discrete time series. The generating functions are related to the Laplace and Fourier transforms as we shall see.

As may be shown the impulse response of a comb filter is a series of impulses regularly spaced in time but of varying amplitude. In other words, the response can be written

$$h(t) = \sum_{0}^{\infty} h_k \, \delta(t - kT) \tag{2}$$

where $\delta(t)$ is the Dirac delta function. The transfer function is the Laplace transform of the impulse response or

$$H(p) = \int_{0}^{\infty} h(t) e^{-pt} \, dt = \sum_{k=0}^{\infty} h_k e^{-kpT} \tag{3}$$

Now if we make the transformation of variable,

$$z = e^{pT} \tag{4}$$

we obtain

$$H(p) = H(z) = \sum_{k=0}^{\infty} h_k z^{-k} \tag{5}$$

$H(z)$ is the generating function which represents the filter and the time sequence can be recovered by expanding $H(z)$ in a power series in descending powers of Z.

Without going into the detailed proofs, we will state that if we have an input signal of the form

$$f(t) = \sum_{k=0}^{\infty} f_k p(t - kT) \tag{6}$$

where $p(t)$ is the form of a single input pulse, then the output signal will be of the form

$$g(t) = \sum_{k=0}^{\infty} g_k p(t - kT) \tag{7}$$

The amplitudes of the output pulses g_k can be obtained from a series expansion of the output generating function $G(Z)$ i.e.

$$G(z) = \sum_{k=0}^{\infty} g_k z^{-k} \tag{8}$$

And the output generating function $G(Z)$ is related to $F(Z)$ the generating function of the input wave by the equation

$$G(z) = H(z)F(z) = H(z) \sum_{0}^{\infty} f_k z^{-k} \tag{9}$$

When $G(Z)$ is a rational function of Z, the expansion to recover the coefficient g_k can normally be accomplished by means of a long division process. Where this is not convenient, the following formula may be used

$$g_k = \frac{1}{2\pi j} \oint_c G(z) z^{k-1} dz \tag{10}$$

where $j = \sqrt{-1}$
and the contour of integration includes the origin and all the singularities of $G(Z)$.

The great advantage of the generating function over the more conventional Laplace transform is that whereas the Laplace transform $H(p)$ is periodic having an infinite number of singularities and is therefore transcendental, $H(z)$ is not periodic and can usually be expressed as a rational function of z. The periodicity involved in the transformation of variable just matches the periodicity of $H(p)$ leaving $H(z)$ non-periodic. The transformation may be regarded as a mapping process in which a strip of the p plane lying between the limits $p = -\omega_r/2$ and $p = \omega_r/2$ is mapped into the entire z plane, that portion of the p plane lying to the left of the imaginary axis becoming the interior of the unit circle and that portion to the right of the imaginary axis becoming the exterior of the unit circle. The imaginary axis itself is transformed into the unit circle. Zero frequency and the harmonics of the repetition frequency are represented by the point $z = 1$. If $z = e^{j\phi}$ is a point on the unit circle, it corresponds to a frequency $j\omega$ such that $\omega T = \phi$.

Bilinear Transformation

Of value to the synthesis procedure we shall present is another transformation of variable in which the unit circle of the z plane is transformed into the imaginary axis of the p plane although in this case, one trip around the unit circle of the z plane corresponds to a traverse of the entire imaginary axis of the p plane. It is this transformation that enables us to draw on the wealth of information available from the art of synthesis of conventional filters. Two forms of this transformation are useful. The first form is applicable when we wish to make use of a low pass filter design in the synthesis of a video integrator or a high pass filter design in the synthesis of an MTI canceller. To avoid confusion with (4) we will use the sympol p* instead of p. The first form of the bilinear transformation is

$$Z = \frac{\Omega + P^*}{\Omega - P^*} \tag{11a}$$

and its converse is

$$p^* = \Omega \frac{Z-1}{Z+1} \tag{11b}$$

When we have a low pass filter design that we wish to use as the basis for the synthesis of a canceller or a high pass filter design that we wish to use as the basis for the synthesis of an integrating filter we may use the second form

$$Z = \frac{p^* + \Omega}{p^* - \Omega} \tag{12a}$$

and the converse of this transformation is

$$p^* = \Omega \frac{Z+1}{Z-1} \tag{12b}$$

In both cases, Ω is a scale adjusting parameter which enables us to place the cutoff frequencies where we want them.

Microwave engineers will note a similarity between these relations and the formulae for impedance in terms of reflection coefficient. The significance of this similarity will be apparent when we show how the calculations can be performed on a Smith chart.

APPLICATIONS

Simple Integrater

We now illustrate the application of generating functions by considering the analysis of a simple video integrator as shown in Figure 4. The transfer function of a delay line whose time delay is T is e^{-kT} or $1/z$. We may thus write the circuit equation in terms of generating functions as follows:

$$E_0(z) = E_1(z) + \gamma E_0(z) \frac{1}{z} \tag{13}$$

or solving for E_o.

$$E_0(z) = E_i(z) \frac{1}{1-\gamma/z} \tag{14}$$

and the transfer function is

$$H(z) = \frac{E_0(z)}{E_i(z)} = \frac{1}{1-\gamma/z} = \frac{Z}{Z-\gamma} \tag{15}$$

The frequency response is obtained by substituting $Z = e^{j\phi}$ and obtaining the absolute value of the resulting expression. The curve of Figure 5 is for $\Sigma = 0.9$. Suppose now that we apply to this integrator a semi-infinite string of pulses

$$f(t) = \sum_0^{\infty} p(t-kT) \tag{16}$$

The generating function of the input is then

$$F(Z) = \sum_0^{\infty} Z^{-k} = \frac{1}{1-Z^{-1}} = \frac{Z}{Z-1} \tag{17}$$

and the generating function of the output is given by

$$G(Z) = H(Z)F(Z) = \frac{Z^2}{(Z-1)(Z-\gamma)} \tag{18}$$

This function may be expanded by long division yielding the series

$$G(Z) = 1 + (1+\gamma)Z^{-1} + (1+\gamma+\gamma^2)Z^{-2} + \tag{19}$$

or alternatively the coefficient g_k may be obtained directly from

$$g_K = \frac{1}{2\pi j} \oint_c G(Z) Z^{K-1} dZ \tag{20}$$

$$= \frac{1}{2\pi j} \oint_c \frac{Z^{K+1}}{(Z-1)(Z-\gamma)} dz \tag{21}$$

This integral may be evaluated by summing the residues of the simple poles located at $Z = 1$ and $Z = \gamma$ yielding

$$g_k = \frac{1}{1-\gamma} + \frac{\gamma^{K+1}}{\gamma - 1} = \frac{1-\gamma^{K+1}}{1-\gamma} \tag{22}$$

Canonical Circuit Configuration

Consider the circuit shown in Figure 7. The circuit equations can be written

$$v_1(t) = v_0(t) + \beta_1 v_1(t-t_0) + \beta_2 v_1(t-2t_0) + \cdots + \beta_n v_1(t-nt_0) \quad (23)$$

$$v_2(t) = a_0 v_1(t) + a_1 v_1(t-t_0) + a_2 v_1(t-2t_0) + \cdots + a_n v_1(t-nt_0) \quad (24)$$

or in terms of generating functions

$$V_1(Z) = V_0(Z) + \frac{\beta_1}{Z} V_1(Z) + \frac{\beta_2}{Z^2} V_1(Z) + \cdots + \frac{\beta_n}{Z^n} V_1(Z) \quad (25)$$

$$V_2(Z) = a_0 V_1(Z) + \frac{a_1}{Z} V_1(Z) + \frac{a_2}{Z^2} V_1(Z) + \cdots + \frac{a_n}{Z^n} V_1(Z) \quad (26)$$

Eliminating $V_1(z)$ from these two equations and solving for $V_2(z)$, we obtain

$$V_2(Z) = V_0(Z) \frac{a_0 + \frac{a_1}{Z} + \frac{a_2}{Z^2} + \cdots + \frac{a_n}{Z^n}}{1 - \frac{\beta_1}{Z} - \frac{\beta_2}{Z^2} - \cdots - \frac{\beta_n}{Z^n}} \quad (27)$$

and the transfer function is

$$H(Z) = \frac{V_2(Z)}{V_0(Z)} = \frac{Z^n a_0 + Z^{n-1} a_1 + \cdots + a_n}{Z^n - Z^{n-1}\beta_1 - \cdots - \beta_n} \quad (28)$$

It will be noted that since the coefficients of the numerator and denominator polynomials are related directly to the controllable parameters of the circuit, $H(z)$ is a completely general rational fraction. In other words, any rational fraction in z can be synthesized by a network of the general configuration shown in Figure 7. Because of this property, this configuration is called the canonical configuration.

Reduced Configurations

Given a general rational fraction such as (28) in which the coefficients are all real, the roots of the numerator and denominator polynomials will be either real or in the form of conjugate complex pairs. The significance of this is that the numerator and denominator polynomials may be factored into component factors none of which is of higher order than quadratic. In other words, we may break the overall configuration into cascaded sections, no section having more than two delay elements with no loss of generality. This means that no feedback or feed forward path need span more than two delay elements.

It so happens that in some cases, greater freedom from circuit drifts and instabilities can be achieved by using something other than the canonical configuration. A case in point is the synthesis of the transfer function

$$H(Z) = \left(1 - \frac{1}{Z}\right)^2 = 1 - \frac{2}{Z} + \frac{1}{Z^2} \quad (29)$$

Either of the configurations in Figure 8 may be used to implement this function. The upper of these two configurations consists of two simple cancellers in cascade and is called double cancellation. The lower of the two forms the simultaneous sum of the direct signal and two delayed signals and is commonly called "three pulse comparison". So long as the circuits are perfectly adjusted, their performance is identical but let us examine what happens when through tube aging or other causes the circuits drift out of adjustment. The most significant feature of this particular circuit is the double zero at $z = 1$ or the ability to cancel steady strings of pulses. If one of the two single cancellers in Figure 8a drifts out of adjustment, the other is still capable of cancelling a steady string and if both are out of adjustment by a small amount, the residual which gets through is the product of two small quantities or a very small quantity. In Figure 8b on the other hand, any drift of the pulse amplitudes from their correct value results in a first order lack of cancellation. It sometimes happens also that a revision of the canonical configuration is desirable which will reduce the dynamic range of the signals encountered in one part of the circuit.

SYNTHESIS PROCEDURE

Illustrative Example

The synthesis procedure can probably best be outlined by means of an example. Suppose for an MTI application we desire a filter having maximum attenuation in the neighborhood of $\omega = 0, \pm\omega_r, \pm 2\omega_r$, and a relatively flat (say 1 db ripple) transmission in the pass band which includes those frequencies for which $\emptyset_0 < \omega T < 2\pi - \emptyset_0$. Phrasing the problem another way, the magnitude of $H(z)$ should be constant (within 1 db) for values of z on the unit circle $Ze^{j\emptyset}$ where $\emptyset_0 < \emptyset < 2\pi - \emptyset_0$ and should be minimized in the neighborhood $z = 1$. Through the use of equation 12b this problem may be transformed into the analogous problem of designing a conventional filter whose response is flat (within 1 db) for a pass band

$$-\Omega \cot\left(\frac{\emptyset_0}{2}\right) < W^* < \Omega \cot\left(\frac{\emptyset_0}{2}\right) \quad (30)$$

and having maximum attenuation in the neighborhood of W = OO. If we set

$$\Omega = \tan\left(\frac{\emptyset_0}{2}\right) \quad (31)$$

the band limits are simplified to

$$-1 < W^* < 1 \tag{32}$$

The solution of this problem is well known (ref.4) and the locations of the poles are readily calculated. If there are N poles, the location of the Kth pole is given by the formula

$$P_k^* = \sinh\gamma\cos\phi_k + j\cosh\gamma\sin\phi_k \tag{33}$$

where

$$\phi_k = \frac{\pi}{2}\,\frac{n+1-2k}{n} \qquad k = 1, 2, 3, --- n \tag{34}$$

$\tanh(n\gamma) = \frac{V_v}{V_p}$ the voltage ratio between peaks and valleys of the pass band

For the case of 1 db ripple and N = 3 for example, the pole locations are

$$\begin{aligned} P_1^* &= -.2471 + j\,.9659 \\ P_2^* &= -.4942 \\ P_3^* &= -.2471 - j\,.9659 \end{aligned} \tag{35}$$

These pole locations are shown in Figure 9 and the resulting frequency response is shown in Figure 10. The pole locations may now be transformed to the z plane by the use of equation 12a. The results for the case where $\phi_0 = 45^\circ$ are

$$\begin{aligned} Z_1 &= .8374\, e^{j\alpha} \\ Z_2 &= .0881 \\ Z_3 &= .8374\, e^{-j\alpha} \end{aligned} \tag{36}$$

where

$$\alpha = 44.2^\circ \tag{37}$$

In the transformed problem, the zeros of transmission were all located at this point transforms to z = 1. We may therefore write the transfer function in the z plane as

$$\begin{aligned} H(Z) &= \frac{(Z-1)^3}{(Z-Z_1)(Z-Z_2)(Z-Z_3)} \\ &= \frac{(Z-1)^3}{(Z-.0881)(Z^2-1.2001Z+.7012)} \end{aligned} \tag{38}$$

The synthesis is conveniently accomplished by dividing the filter into two cascaded portions, the first having the transfer function

$$H_1(Z) = \frac{Z-1}{Z-.0881} \tag{39}$$

and the second

$$H_2(Z) = \frac{(Z-1)^2}{Z^2-1.2001\,Z+.7012} \tag{40}$$

The circuit configuration corresponding to the implementation of these two sections in the canonical form is shown in Figure 11 and the response characteristic is shown in Figure 12. The response of a simple cancelling filter is shown dotted for comparison.

Use of Smith Chart

The transformation from the p* plane to the z plane can be conveniently accomplished through the use of the Smith chart familiar to microwave engineers. This chart is a map of the relation between reflection coefficient and impedance. If we let R be the complex reflection coefficient and z be the impedance, this relation may be expressed.

$$R = \frac{z-z_0}{z+z_0} \tag{41}$$

where z_0 is the characteristic impedance. If we now substitute $-p^*$ for z and Ω for z_0 we obtain

$$R = \frac{-p^*-\Omega}{+p^*-\Omega} = \frac{p^*+\Omega}{p^*-\Omega} \tag{42}$$

which if we substitute z for R becomes identical with equation 12a. On the other hand, if we substitute -z for R, we obtain

$$Z = -\frac{p+\Omega}{p-\Omega} = \frac{\Omega+p}{\Omega-p} \tag{43}$$

which is identical with equation 11a. Since multiplying by -1 is the same as multiplying by $e^{j\pi}$, this effect is conveniently accomplished by rotating the chart 180°.

We now illustrate the use of the chart by plotting the pole locations of the previous problem. First we compute

$$\begin{aligned} -\frac{p_1^*}{\Omega} &= -\frac{p_1^*}{\tan\phi_0/2} = 0.597 - j\,2.332 \\ -\frac{p_2^*}{\Omega} &= -\frac{p_2^*}{\tan\phi_0/2} = 1.194 \\ -\frac{p_3^*}{\Omega} &= -\frac{p_3^*}{\tan\phi_0/2} = 0.597 + j\,2.332 \end{aligned} \tag{44}$$

These points are plotted on Figure 13 along with the triple zero at p* = 00. The z plane coordinates of these poles may now be determined by measuring the length and direction.

It is apparent that the same technique may be applied to any type of known low pass filter whether it is a Butterworth, Tschebycheff, or Bessel filter or one of the filters based on elliptic function transformation which has equal ripple in the rejection band as well as in the pass band (ref. 5, 6). Figure 14 is an illustration of this latter type of filter. The pass has the same width and ripple as the filter shown in Figure 12 but the rejection band is of the equal ripple type with a minimum attenuation of 34 db.

Alternative Configurations for Two Element Network

As was indicated previously the double cancellation network is preferable to the canonical form for the implementation of the function $(1-1/z)^2$. We now generalize on this concept somewhat. Consider the network of Figure 15. It may be seen that this configuration is similar to that of the double cancellation circuit of Figure 8a with the addition of feedback paths. Writing the circuit equations we have

$$\begin{aligned} E_2(Z) &= E_1(Z) + \beta_2 E_5(Z) \\ E_3(Z) &= (1-\tfrac{1}{Z})E_2(Z) \\ E_4(Z) &= E_3(Z) + \beta_1 E_5(Z) \\ E_5(Z) &= \tfrac{1}{Z} E_4(Z) \\ E_6(Z) &= \left\{1-\tfrac{1}{Z}\right\} E_4(Z) \end{aligned} \tag{45}$$

Solving these equations for $E_6(z)$ we obtain

$$E_6(Z) = E_1(Z)\,\frac{(Z-1)^2}{Z^2-(\beta_1+\beta_2)Z+\beta_2} \tag{46}$$

so that the transfer function is

$$H(Z) = \frac{(Z-1)^2}{Z^2-(\beta_1+\beta_2)Z+\beta_2} \tag{47}$$

This formula gives us the means for relating the circuit parameters to the pole locations for this type of configuration. As in the configuration of Figure 8a if one of the two cancellation circuits drifts out of adjustment, the other will still cancel a steady string of pulses and if both are slightly out of adjustment, the residual will be the product of two small quantities.

Configuration for Separated Zeroes

It may be desired in some cases such as for the filter of Figure 14 to provide two separate and distinct zeroes on the unit circle. The preferred configuration for this application appears to be the one shown in Figure 16. This configuration has the advantage of being easier to adjust than would be a canonical configuration having the same characteristic. Each of the two cancelation circuits can be adjusted for minimum output on a steady signal with the α_1 branch open. α_1 then controls the separation of the zeroes without displacing them from the unit circle. The transfer function for this configuration is

$$H(Z) = \frac{Z^2 - Z(2-\alpha_1) + 1}{Z^2 - Z(\beta_1+\beta_2) + \beta_2} \tag{48}$$

Pole Loci Curves

In the simple two pole case the peak to valley ratio of a Tschebycheff low pass filter is determined by the angle between the negative real axis and a line between the origin and one of the poles. The relation between this angle θ and the peak to valley ratio is given by

$$\tan\theta = \coth\gamma \tag{49}$$

where $\tan 2\gamma = V_v/V_p$ = peak to valley ratio (50)

The following tables gives some typical values

Vv/Vp	db	θ
1.0	0	45°
0.891	1	58.5°
0.707	3	67.5°
0.500	6	75°

(51)

These values are conveniently transferred to the z plane through the use of the impedance-phase angle plot as shown in Figure 17.

Video Integrators

The simple video integrator shown in Figure 4 has two deficiencies. The frequency response does not closely match the signal spectrum as it should for optimum performance and the single pulse response is high asymmetrical leading to angular inaccuracies in certain types of systems. Both of these deficiencies can be alleviated through the use of more elaborate configurations. While the theory is by no means worked out in detail to the point of providing designs for optimum integrators, we will illustrate the type of improvement obtainable through the use of a two section integrator.

Consider the case where it is desired to pass more or less uniformly the band of frequencies from $\omega T = -\phi_0$ to $\omega T = \phi_0$ and to provide maximum attenuation to frequencies in the neighborhood of $\omega T = \pi$. Through the use of equation 4, we see that this means uniform response on the unit circle from $Z = e^{-j\phi_0}$ to $Z = e^{j\phi_0}$ and maximum attenuation in the neighborhood of $Z = -1$. This requirement is further translated into the p* plane through equation 11a to be a problem of providing uniform response from $\omega^* = -\Omega \tan(\emptyset_0/2)$ to $\omega^* = \Omega \tan(\emptyset_0/2)$. If we let $\Omega = \cot(\emptyset_0/2)$ the problem then resolves to the design of a low pass filter with the pass band located so that

$$-1 < \omega^* < 1 \tag{52}$$

As before, the solution to this problem is well known. The pole locations for a Tschebycheff filter with 1 db ripple in the pass band and two poles are

$$\begin{aligned} P_1 &= -.5489 + j\,.8951 \\ P_2 &= -.5489 - j\,.8951 \end{aligned} \tag{53}$$

Let us now assume $\emptyset_c = 9.2^0$ and compute

$$\begin{aligned} -\frac{p_1}{\Omega} &= -p_1 \tan\frac{\emptyset_0}{2} = .0441 - j\,.0719 \\ -\frac{p_2}{\Omega} &= -p_2 \tan\frac{\emptyset_0}{2} = .0441 + j\,.0719 \end{aligned} \tag{54}$$

These points are shown plotted in Figure 18. If we now measure these points carefully, we find the radius is 0.914 units and the angle is 8-1/2°. The two zeroes are of course located at $\omega^* = \infty$ or at $Z = -1$ We can thus write for the transfer function

$$H(Z) = \frac{(Z+1)^2}{Z^2 - 2Z(.914)\cos 8\frac{1}{2}^0 + (.914)^2}$$

$$= \frac{Z^2 + 2Z + 1}{Z^2 - 1.808Z + .835} \tag{55}$$

The frequency response is plotted in Figure 19 and the single pulse response is plotted in Figure 20. It will be noted that although the single pulse response is still asymmetrical it is not so much so as is the simple integrator whose response is plotted in Figure 6. With 5 or 6 delay lines it is believed possible to achieve a quite symmetrical response although this has not been done as yet.

EXPERIMENTAL VERIFICATION

Figure 21 presents calculated frequency responses of a two pole MTI filter adjusted to give 30 db cancellation of fixed targets at various rotation rates on a particular radar. The corresponding pulses-per-beamwidth figures for the different rotation rates are approximately as follows.

RPM	P/BW
3.3	17.3
5.0	11.5
6.6	8.6
10.0	5.8

Also shown for comparison purposes are the responses of a simple single cancellation circuit and a cascade connection of two single cancellation circuits to provide double cancellation.

Figure 22 is a series of scope photographs obtained through the use of a sweeping oscillator that sweeps through two adjacent harmonics of the repetition frequency. In each case, the response of the single cancellation circuit is superimposed for reference. As may be seen, the correllation with the theory is good. It is believed that the slight asymmetries noticeable in the scope photos are due to small timing errors in the feedback loops.

ACKNOWLEDGEMENT

The authors would like to acknowledge the contribution of S. F. Hall who computed the responses shown in Figure 20 and was responsible for the experimental verification. This development was sponsored by the Rome Air Development Center under Contract #AF 30(602-381) with AIL. Our thanks are due engineer Al Feiner and William T. Pope who had the foresight to push the development.

REFERENCES

1. H. M. James, N. B. Nichols, & R. S. Phillips "Theory of Servomechanisms" MIT Radiation Laboratory Series, Vol. 25, McGraw Hill Book Company, 1947, Chapter 5
2. J. R. Ragazzini & A. R. Bergen, "A Mathematical Technique for the Analysis of Linear Systems", Proc. IRE, Vol. 42, No. 11, Nov. 1954, p. 1645
3. S. Thaler & R. Boxer, "An Operational Calculus for Numerical Analysis", Convention Record of the IRE, Volume IV-1956, Part 2-Circuit Theory, p. 100
4. Louis Weinberg, "Modern Synthesis Network Design from Tables", Electronic Design Sept. 15, 1956, Oct. 1, 1956, Oct. 15, 1956
5. George L. Matthaei, "Filter Transfer Function Synthesis", Proc. IRE, Vol. 41, No. 3, Mar. 1953, p. 377
6. George L. Matthaei, "Conformal Mappings for Filter Transfer Function Synthesis", Proc. IRE, Vol. 41, No. 11, Nov. 1953, p. 1658

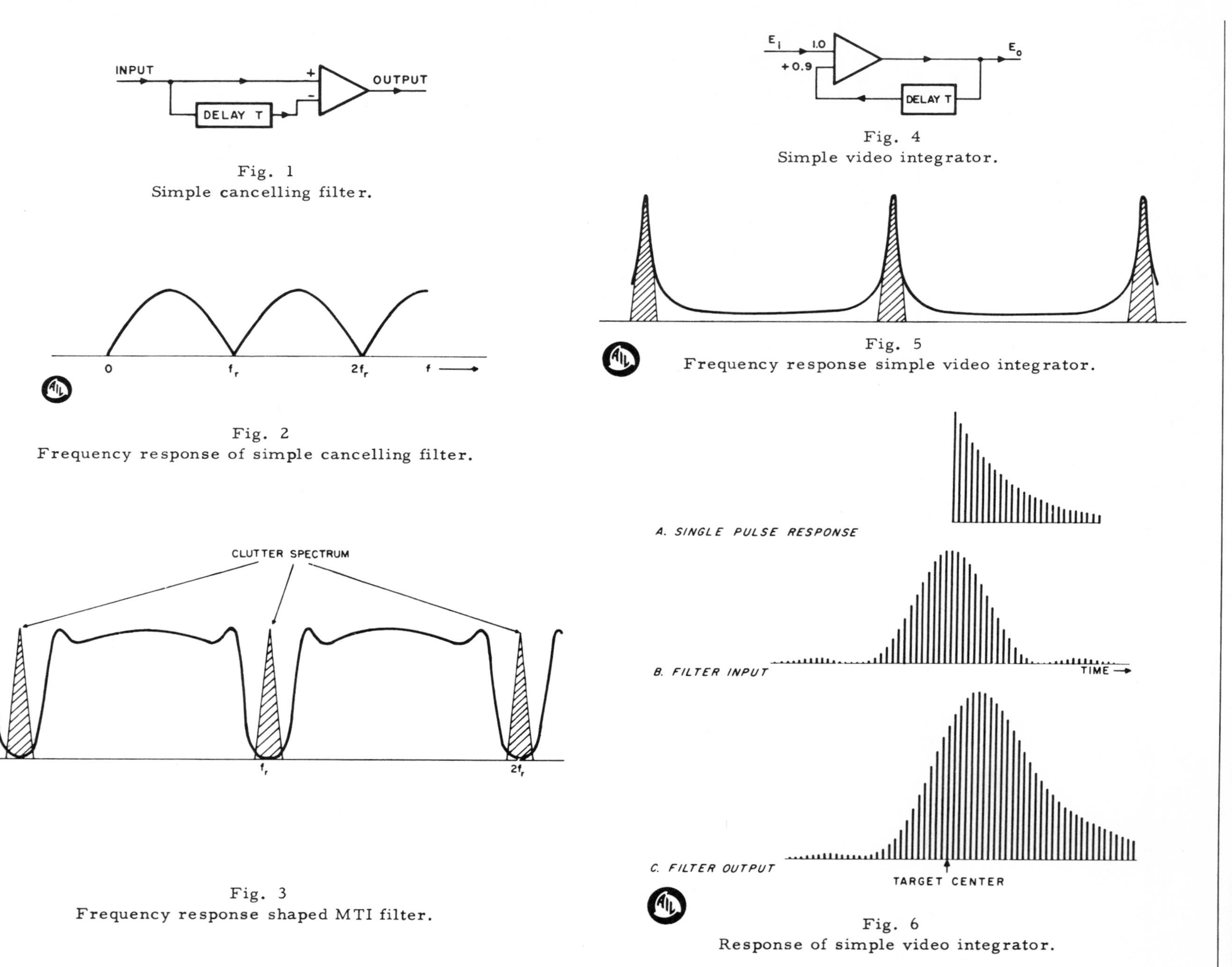

Fig. 1
Simple cancelling filter.

Fig. 2
Frequency response of simple cancelling filter.

Fig. 3
Frequency response shaped MTI filter.

Fig. 4
Simple video integrator.

Fig. 5
Frequency response simple video integrator.

Fig. 6
Response of simple video integrator.

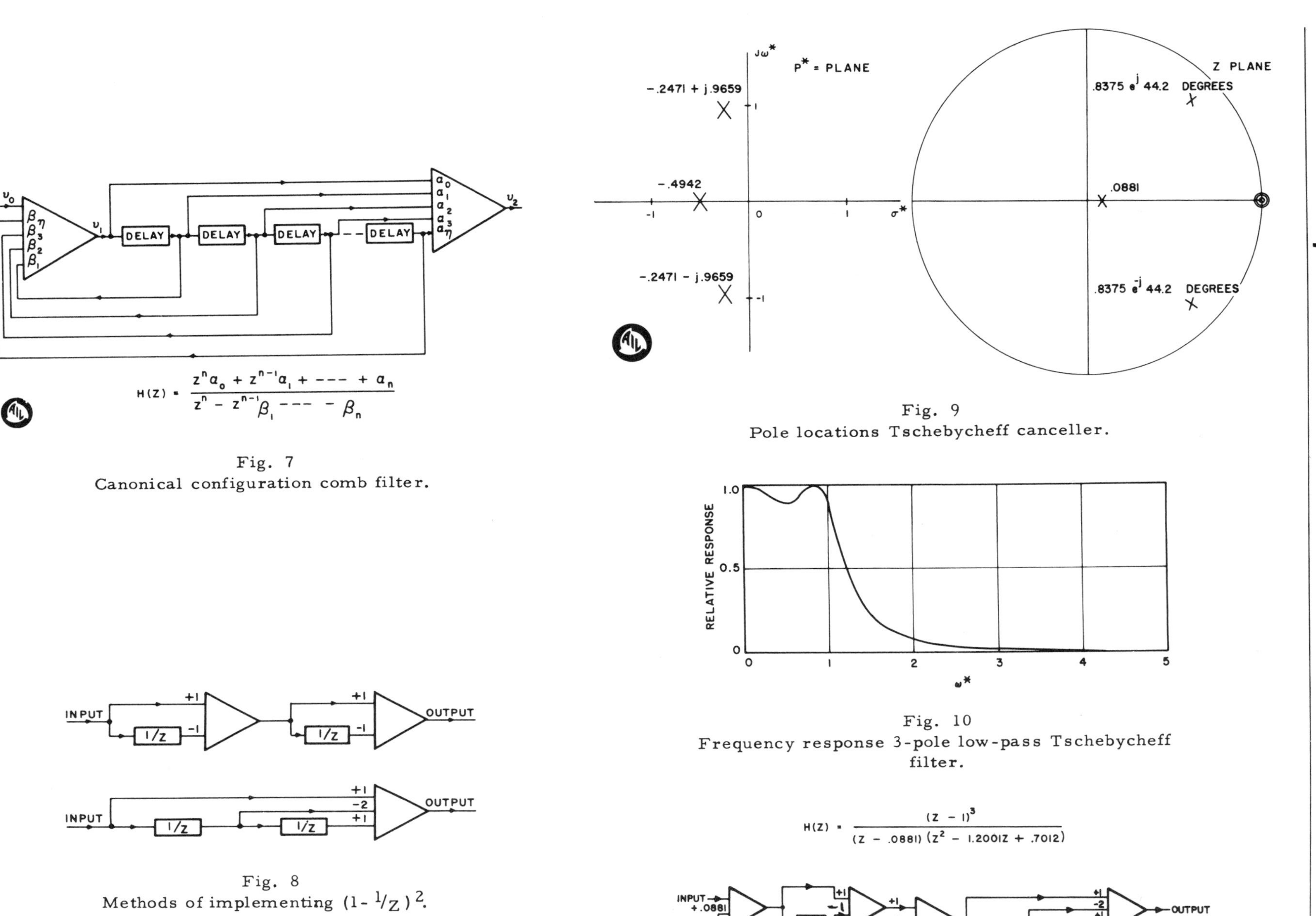

$$H(Z) = \frac{z^n a_0 + z^{n-1} a_1 + --- + a_n}{z^n - z^{n-1}\beta_1 --- - \beta_n}$$

Fig. 7
Canonical configuration comb filter.

Fig. 8
Methods of implementing $(1 - {}^1/_Z)^2$.

Fig. 9
Pole locations Tschebycheff canceller.

Fig. 10
Frequency response 3-pole low-pass Tschebycheff filter.

$$H(Z) = \frac{(Z-1)^3}{(Z-.0881)(Z^2-1.2001Z+.7012)}$$

Fig. 11
Implementation of 3-pole Tschebycheff cancellation filter.

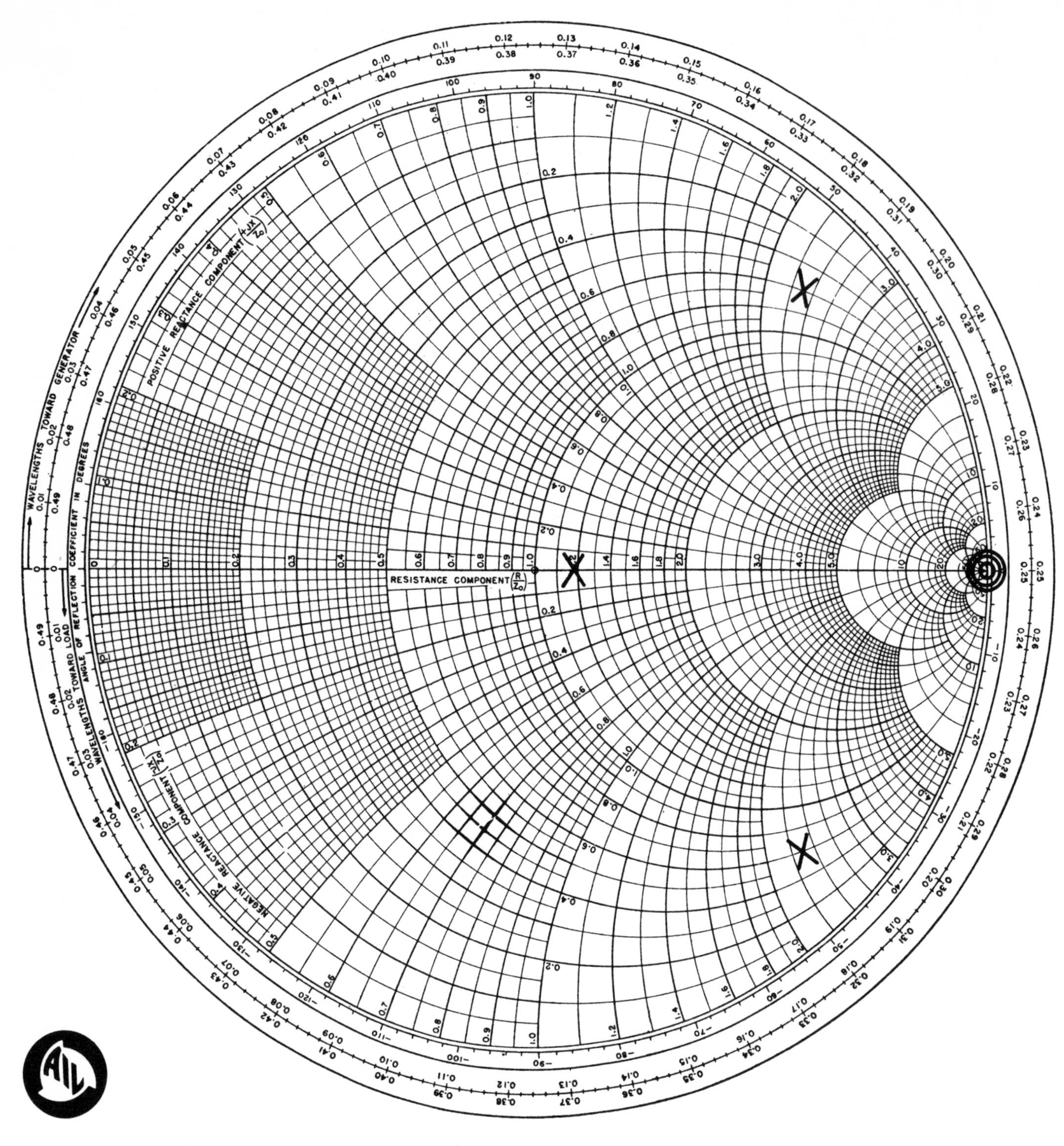

Fig. 13
Pole locations Tschebycheff canceller.

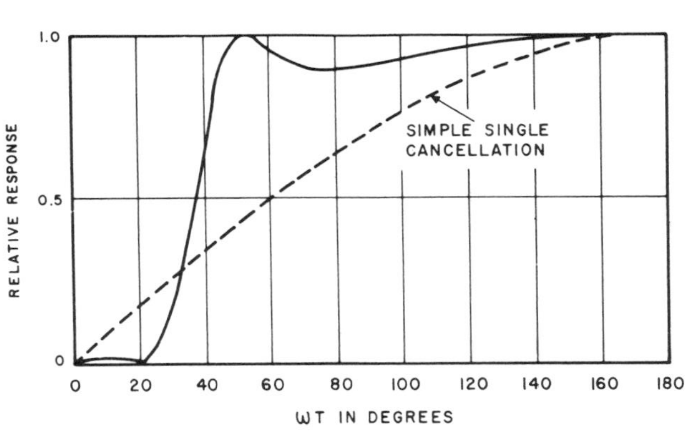

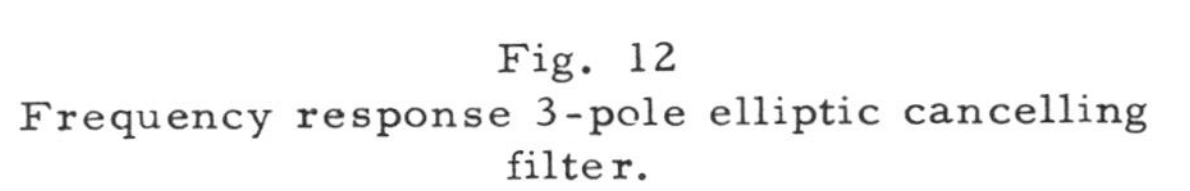

Fig. 12
Frequency response 3-pole elliptic cancelling filter.

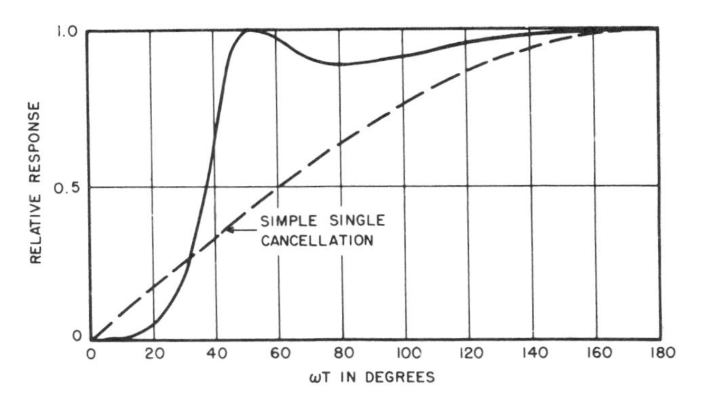

Fig. 14
Frequency response 3-pole Tschebycheff cancelling filter.

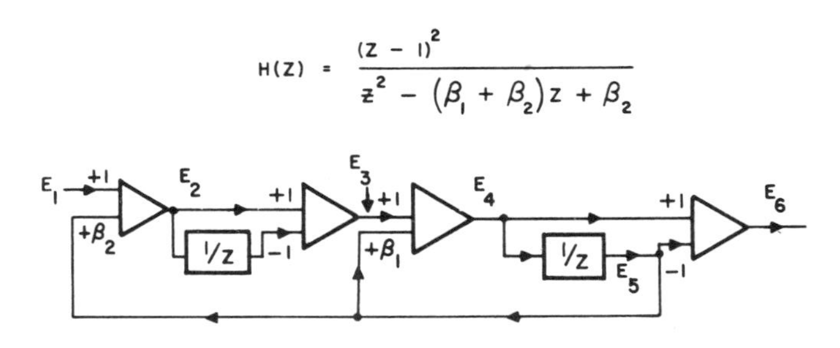

Fig. 15
Two pole filter with double cancellation

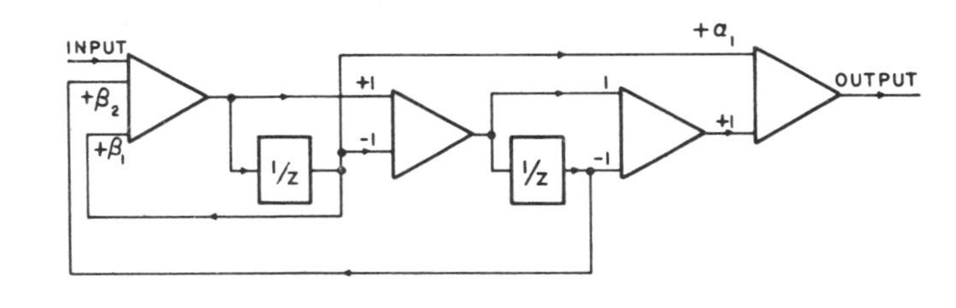

Fig. 16
Two-section filter with separated zeroes.

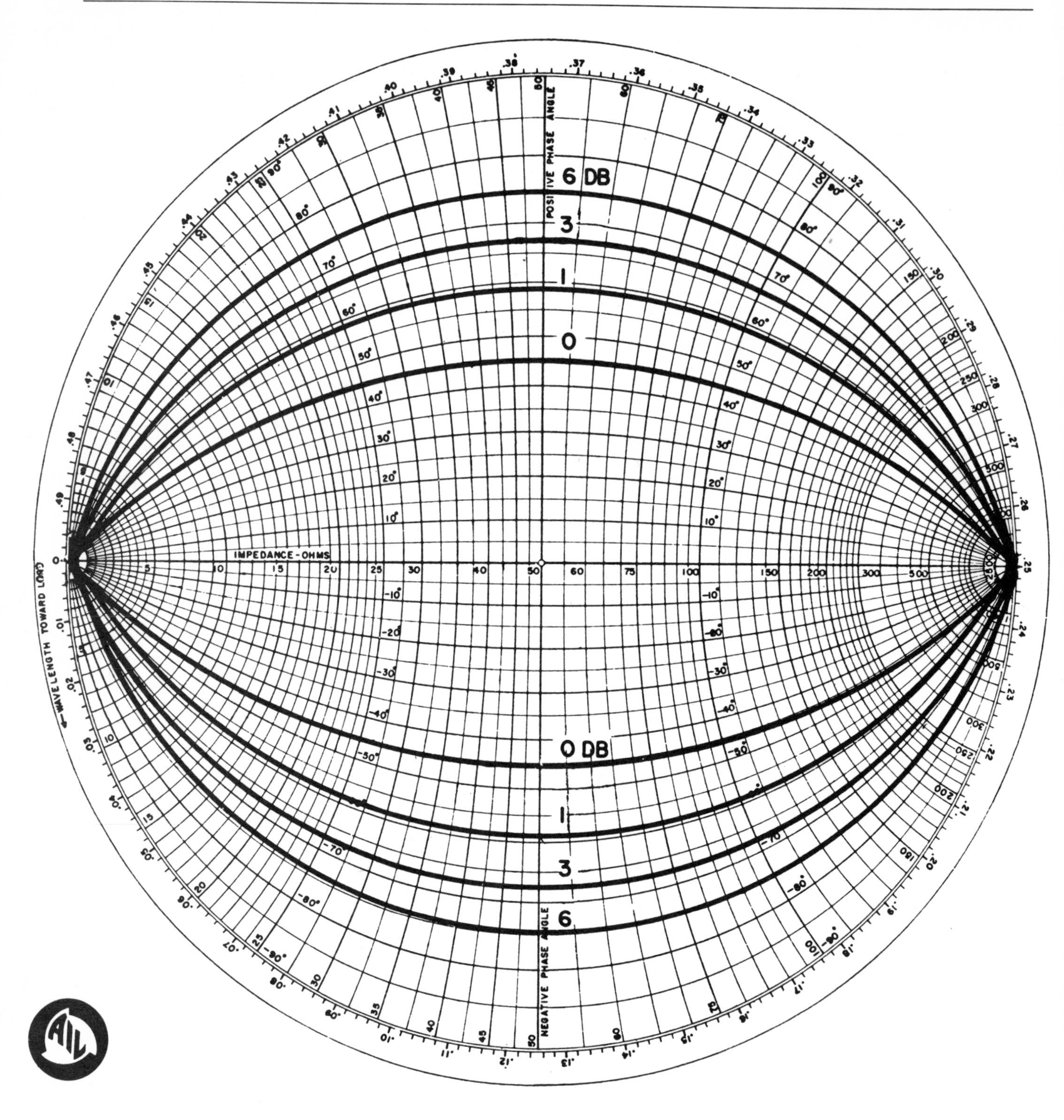

Fig. 17
Pole loci (two-pole Tschebycheff filters).

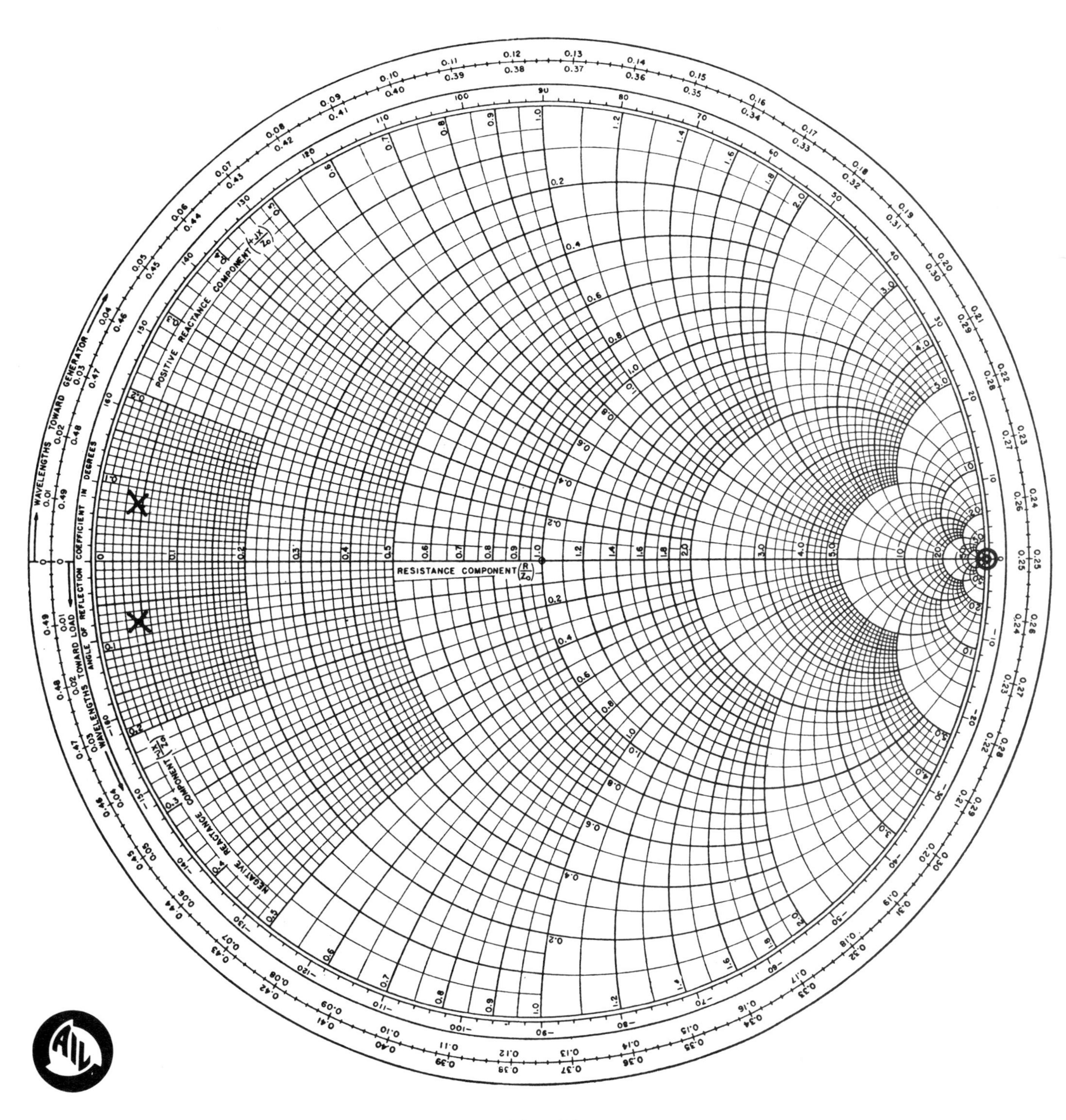

Fig. 18
Pole and zero locations, two-pole Tschebycheff integrator.

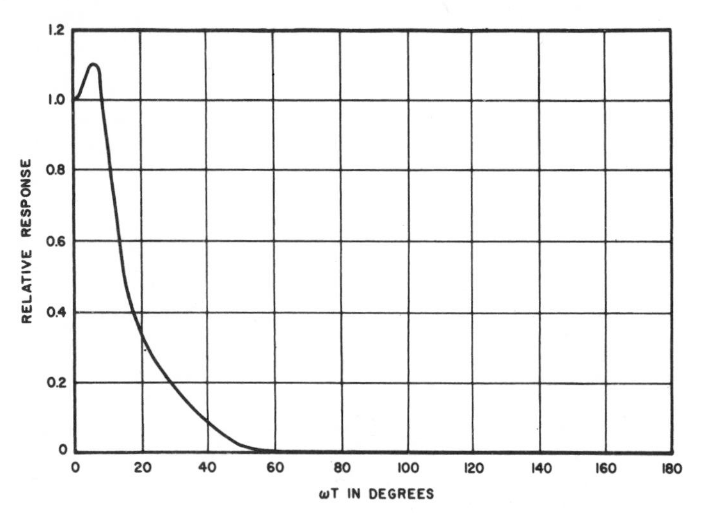

Fig. 19
Response of double delay line integrator
($Z_{1,\,2}$ = 0.9 _ jo. 135).

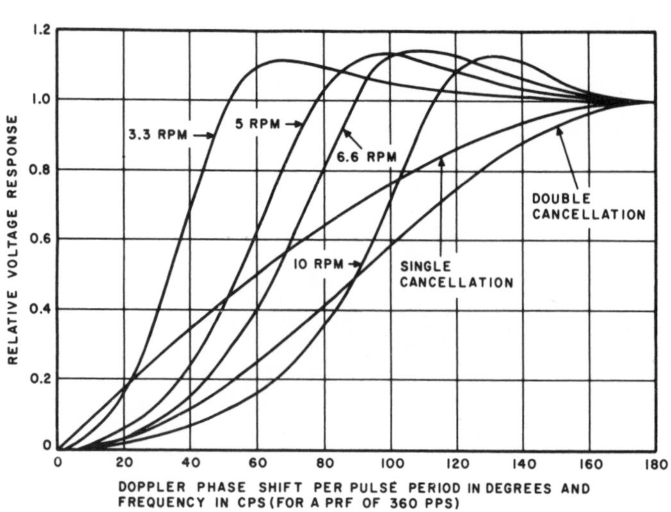

Fig. 21
Calculated sweep responses for various velocity-response shapes.

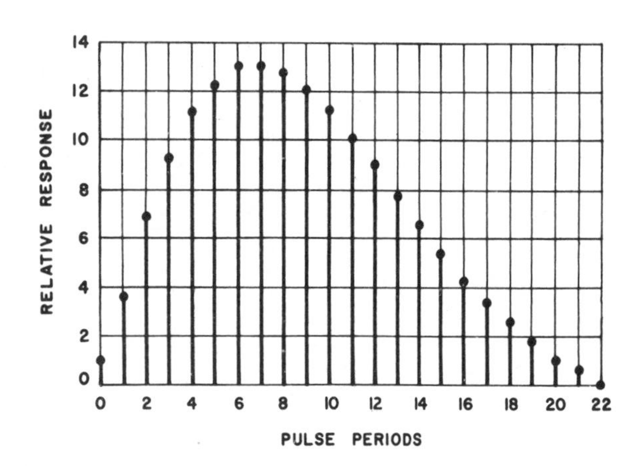

Fig. 20
Response of double delay integrator to a single pulse.

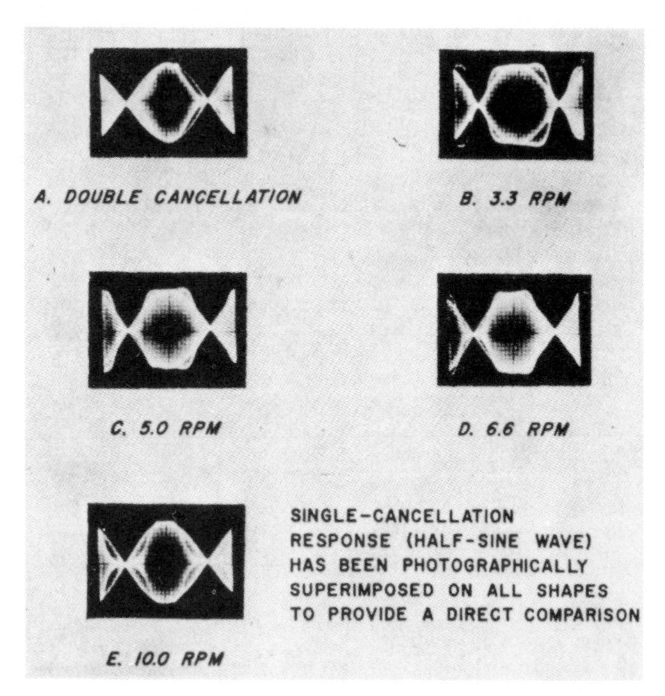

Fig. 22
Sweep presentations of various velocity-response shapes.

1.3
SYNTHESIS OF DELAY LINE NETWORKS
D.A. Linden and B.D. Steinberg
IRE Transactions on Aeronautical and Navigational Electronics, Vol. ANE-4, No. 1, March 1957, pp. 34-39.

Synthesis of Delay Line Networks*

D. A. LINDEN† AND B. D. STEINBERG‡

Summary—**Radar video information may be processed on a continuous-trace (as opposed to range-gated) basis by using delay elements whose delay times are integral multiples of the repetition period τ. Networks consisting of such lines and of linear amplifiers and adders yield periodic pass bands. Their properties may be analyzed most conveniently by the transformation $z=e^{p\tau}$, where p is the complex frequency variable.**

A systematic synthesis procedure is presented which realizes a prescribed delay-line transfer function as a cascade of elementary building blocks. Flow-graph techniques may be used to obtain different embodiments corresponding to the same transfer function.

It is shown that the required number of delay elements is equal to the number of poles of the z-transfer-function which are located away from the origin, or to the number of zeros, whichever is larger.

INTRODUCTION

IN THE PROCESSING of video radar return traces, one may distinguish between two basic methods. In *range-gated* systems, properly positioned samples of each return trace are selected so as to yield a succession of pulses representing successive returns from a selected range element. This amplitude-modulated string of pulses may then be subjected to suitable processing by conventional video networks. Alternately, *continuous-trace* processing makes use of delay elements capable of delaying the entire return trace by multiples of the interpulse period τ. A composite output trace may then be obtained by suitable weighting of past return traces. The necessary delays may be achieved by use of mercury or—more recently employed—quartz ultrasonic delay lines.

This paper is concerned with continuous-trace processing of video radar traces, and deals with the analysis and synthesis of networks consisting of delay elements[1] with delay periods equal to a basic repetition period τ. These networks may also comprise adders and linear amplifiers. All elements are unidirectional.

The analysis of delay-line networks is greatly simplified by a suitable transformation of variables.[2-4] This method of analysis will be reviewed, and the application of flowgraph techniques to the manipulation of delay-line networks will be discussed.

It will be shown that within the limits of realizability, arbitrary transfer functions may be synthesized as cascades of elementary building blocks containing one or two delay lines. This synthesis technique will then be used to establish a relation between the required number of delay elements and the degree of the transfer function.

POLE-ZERO REPRESENTATION

Networks of the type specified above result in periodic pass-bands, and their pole-zero configurations in the complex frequency plane ($p=\sigma+j\omega$) are repetitive

* Manuscript received by the PGANE, June 21, 1956; revised manuscript received, December 26, 1956.
† Philco Corporation, Philadelphia, Pa.
‡ General Atronics Corp., Bala-Cynwyd, Pa. Formerly with Philco Corp., Philadelphia, Pa.

[1] Separate paths through the same delay line are considered as separate delay elements.

[2] P. I. Richards, "Resistor-transmission-line circuits," PROC. IRE, vol. 36, pp. 217–220; February, 1948. (Driving-point impedances of networks containing lumped resistances and lossless transmission lines of commensurable lengths.)

[3] "Final Engineering Report on Investigation of MTI Systems," Airborne Instruments Laboratory, Inc., Appendix F; July, 1952. (The work is credited to W. D. White. Application to radar filters.)

[4] H. M. James, N. B. Nichols, and R. S. Phillips, "Theory of Servomechanisms," Rad. Lab. Ser., McGraw-Hill Book Co., Inc., New York, N. Y., vol. 25, Ch. 5; 1947. (Application of the transformation to more general sampled-data systems.)

structures.[5] The periodicity of these structures (*i.e.*, the periodicity of their frequency characteristics) is given by $f_r = 1/\tau$.

The transformation

$$z = e^{p\tau}$$

leads to considerable simplification and ease of visualization. This transformation maps the zero to $2\pi/\tau$ interval along the $j\omega$ axis into the unit circle in the manner shown in Fig. 1. Successive $2\pi/\tau$ intervals are mapped into the same path; it is thus seen that the mapping places in evidence one of the (identical) periodic intervals along the frequency axis. The left half-plane strip defined by

$$\sigma < 0; \qquad 0 < \omega < \frac{2\pi}{\tau}$$

is mapped into the interior of the unit circle and the corresponding right half-plane strip into its exterior. Since the poles of any stable network are restricted to the left half-plane, including its boundary, it follows that the poles of the transfer function of a delay line network must lie within or on the unit circle of the z plane.

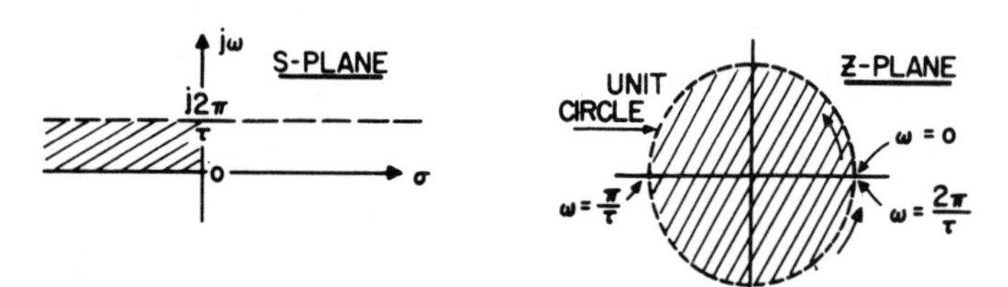

Fig. 1—Properties of the transformation $z = e^{pt}$.

The transfer function of a delay line network is a rational function of z with real coefficients, *i.e.*,

$$T(z) = \frac{a_n z^n + a_{n-1} z^{n-1} + \cdots + a_0}{b_m z^m + b_{m-1} z^{m-1} + \cdots + b_0}.$$

It follows that the z-plane poles and zeros of the transfer function must occur either on the real axis or in complex conjugate pairs. It should be noted that the locations of the zeros are not otherwise restricted. There is also no restriction on the multiplicity of poles or zeros.

Since the transfer function is a rational function of z, it may be written as

$$T(z) = K \frac{(z - z_{01})(z - z_{02}) \cdots (z - z_{0n})}{(z - z_{p1})(z - z_{p2}) \cdots (z - z_{pm})},$$

and may thus be visualized as the ratio of two complex vector products. The numerator is given by the product of all vectors emanating from the zeros and terminating at the observation point; the denominator consists of a similar product of pole vectors. It is thus clear that a pole or zero at the origin of the z plane has no effect on the frequency response, *i.e.*, on $\left| T(z) \right|_{|z|=1}$. Two networks differing only in a pole or zero at the origin may therefore be considered equivalent.

ANALYSIS

As an example of the manner in which the analysis of delay line networks proceeds, the simple delay and subtraction circuit of Fig. 2(a) will be considered. The steady state transfer function of a delay line is given by

$E_i \longrightarrow$ [delay = τ] $\longrightarrow E_o \qquad \frac{E_o}{E_i} = e^{-p\tau} = \frac{1}{z}$.

It follows that the output of the delay and subtraction system is

$$E_0 = \left(\frac{1}{z} - 1\right) E_i,$$

so that its transfer function becomes

$$T(z) = \frac{E_0}{E_i} = \frac{1 - z}{z}.$$

The corresponding z-plane diagram is shown in Fig. 2 (b).

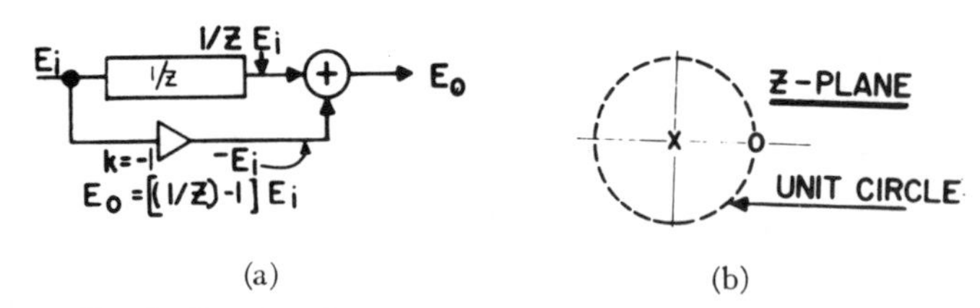

(a) (b)

Fig. 2—Simple delay and subtraction system. (a) Delay and subtraction system; (b) Pole-zero diagram.

Another example is the network of Fig. 3(a), which is designed to add successive radar traces with an exponential weighting factor. In this case, the open-loop transfer function is simply that of a single delay element, *i.e.*, $1/z$. The closed loop transfer function may then be written immediately by use of the conventional feedback formula as

$$T(z) = \frac{\frac{1}{z}}{\left(1 - k\frac{1}{z}\right)} = \frac{1}{z - k}.$$

The z-plane diagram is shown in Fig. 3(b).

[5] This fact may easily be established by noting that the impulse response of such networks must be of the form

$$\sum_{n=0}^{\infty} a_n \delta(t - n\tau).$$

Fourier transformation yields the corresponding system response function

$$T(i\omega) = \sum_{n=0}^{\infty} a_n e^{-j\omega n\tau}$$

Fig. 3—Single delay feedback network. (a) Block diagram; (b) pole diagram.

In this connection it should be pointed out that the numerator of a transfer function $T(z)$ cannot exceed in degree that of the denominator. If it did, the function could be written in the form

$$T(z) = T'(z) + Az + Bz^2 + \cdots$$

where $T'(z)$ has a numerator and denominator of equal degree. A transfer function of the type $T'(z)$ is always realizable. If, in addition, $T(z)$ were realizable, it would be possible to obtain a negative delay function by subtracting the outputs of $T'(z)$ and $T(z)$. One concludes that the numerator degree of a transfer function $T(z)$ may not exceed that of the denominator.

Application of Flow-Graph Techniques

Because delay line networks consist of unilateral branches, flow-graph techniques appear to be ideally suited for obtaining equivalent networks without the necessity of intervening analytical steps. As an example, consider the cascade of the networks of Fig. 2(a) and Fig. 3(a). The original circuit is shown in Fig. 4(a). The transformation to the equivalent circuit of Fig. 4(b) is achieved by noting that the following flowgraph situations are identical. The block labeled T may have any

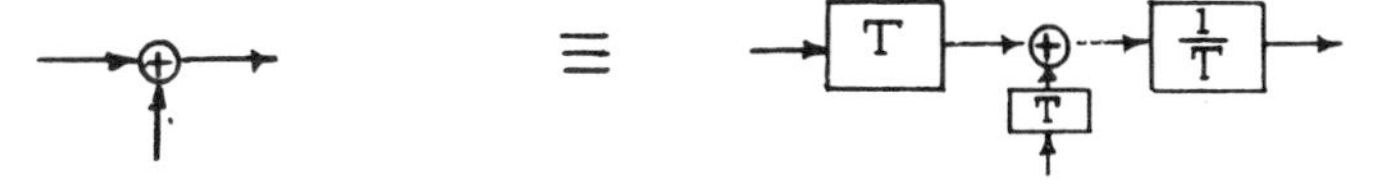

transfer function whatsoever. If one identifies T with the, physically not realizable, function z, Fig. 4(b) results. A further simplification, which will eliminate the physically nonrealizable z block, results from the observation that the function delivered by the second feed-forward loop is simply $-E_1$. It may thus be replaced by a feed-forward loop emanating from the output side of the first adder as shown in Fig. 4(c). The final $1/z$ block does not affect the frequency response, and may therefore be ignored. Fig. 4(d) shows the final network.

As a second illustration of the technique, the circuit of Fig. 4(d), redrawn as Fig. 5(a), will be modified; in particular, it will be shown that the take-off point of the feedback loop may be moved to the right of the last adder without affecting the frequency characteristics of the network. If the feed-forward loop were to end on adder A_2 instead of A_3, only one error would result. The feedback loop would add an extra signal $-kE_1$ into the first adder. To compensate for this error, a compensat-

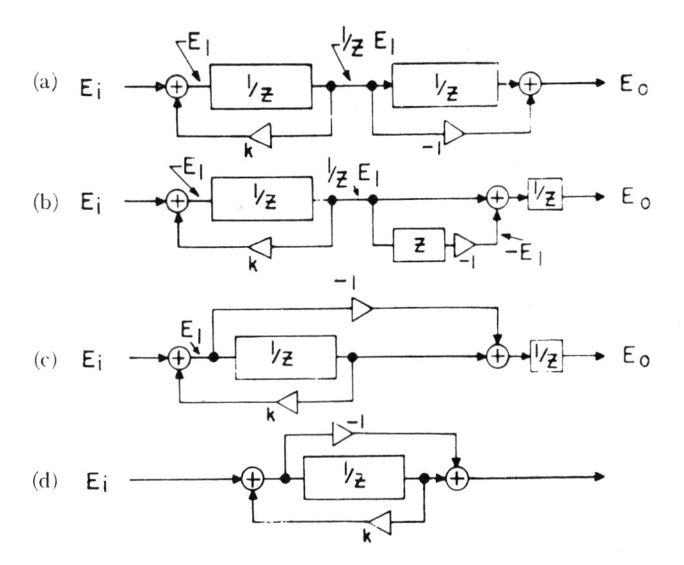

Fig. 4—Flow-graph technique applied to cascade of two networks.

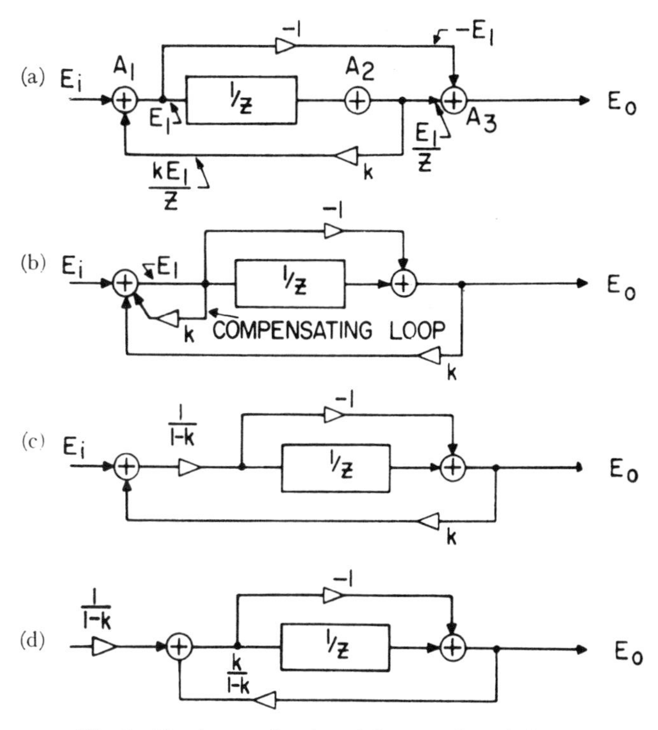

Fig. 5—Further application of flow-graph technique.

ing loop is inserted, as shown in Fig. 5(b). Use is now made of the following equivalents to obtain Fig. 5(c).

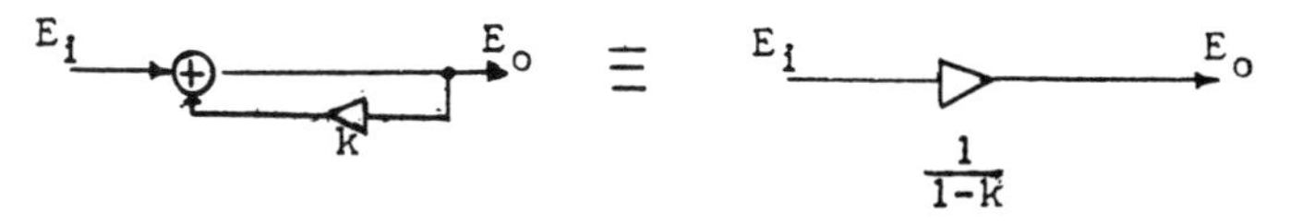

This may further be transformed, and the resultant network is indicated in Fig. 5(d).

Basic Building Blocks

For the time being, it will be assumed that a complex structure is to be built up of a cascade of building blocks each of which consist of, at most, two delay lines, and realizes as many poles and zeros as is possible. The most general two-delay elements building block is shown in Fig. 6. The transfer function of this filter is

$$T(z) = \frac{E_0}{E_i} = (k_1k_5 + k_2)\frac{z^2 + m_1z + m_2}{z^2 + m_3z + m_4}$$

where the m's are functions of the gain factors, as shown in Fig. 6.

Examination of the expressions for the pole and zero locations shows that some of the signal paths of Fig. 6 are redundant. The choice[6] $k_5 = k_6 = 0$ does not sacrifice any generality, and leads to the basic two-element building block shown in Fig. 7. Its transfer function is

$$T(z) = k_2 \frac{z^2 + \frac{k_1}{k_2}z + \frac{1}{k_2}}{z^2 - k_3z - k_4}.$$

It should be noted that the pole locations are controlled solely by the two feedback loops, while the two feed-forward loops govern only the zero locations.

By setting $k_2 = 0$, one may realize a transfer function consisting of two poles and one zero. Similarly, when $k_1 = k_2 = 0$, two poles and no zeros are obtained.

In order to realize a section with only one pole, the network shown in Fig. 8 may be used. Its transfer function is given by

$$T(z) = k_1 \frac{z + \frac{1}{k_f}}{z - k_b}$$

and both pole and zero are seen to be restricted to the real axis. The zero may be eliminated by choosing $k_f = 0$. The pole cannot be eliminated since a network having a number of zeros in excess of poles is not realizable. However, the frequency dependence generated by the pole may be completely nullified by setting $k_b = 0$, thus generating a pole at the origin of the z plane.

Synthesis Procedure

Given a specified transfer function in the form of a rational function of z,

$$T(z) = K\frac{z^n + a_{n-1}z^{n-1} + \cdots + a_0}{z^m + b_{m-1}z^{m-1} + \cdots + b_0},$$

the latter may be written in terms of a product of factors. Each individual factor corresponds to one of the sections discussed in the preceding paragraph; it contains either two poles, which may be real or complex conjugate, or it contains a single real pole. The zeros may be assigned among these factors in an arbitrary fashion, the only restriction being that complex conjugates must be kept together and must be associated with a factor containing two poles. It follows that in this particular synthesis procedure, the number of delay elements is equal to the number of poles which are located away from the origin, or the number of zeros, whichever is larger. Conversely, n arbitrary zeros and

[6] While it is possible to eliminate other pairs of signal paths, the above choice is the only one which has the property that the feedback factors involved in the location of poles do not affect the zeros, and vice versa.

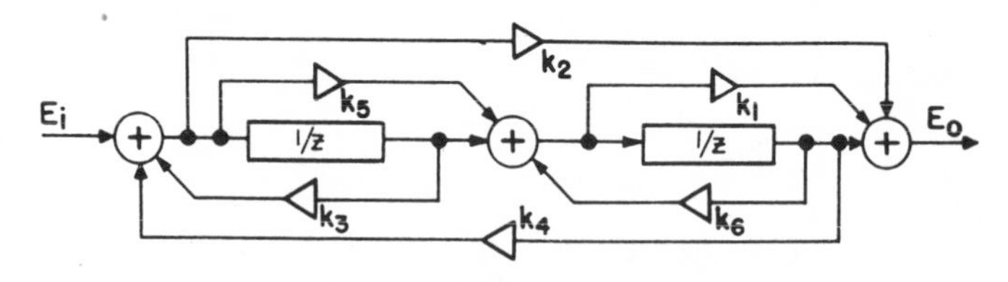

$$T(z) = \frac{E_0}{E_i} = (k_1k_5 + k_2)\frac{z^2 + M_1z + M_2}{z^2 + M_3z + M_4}$$

WHERE $M_1 \equiv \frac{k_1 + k_5 - k_2k_6}{k_1k_5 + k_2}$ $\quad M_2 \equiv \frac{1}{k_1k_5 + k_2}$

$M_3 \equiv -(k_3 + k_4k_5 + k_6)$ $\quad M_4 \equiv k_3k_6 - k_4$

ZEROS AT $1/2[-M_1 \pm \sqrt{M_1^2 - 4M_2}]$

POLES AT $1/2[-M_3 \pm \sqrt{M_3^2 - 4M_4}]$

Fig. 6—Most general 2-element building block.

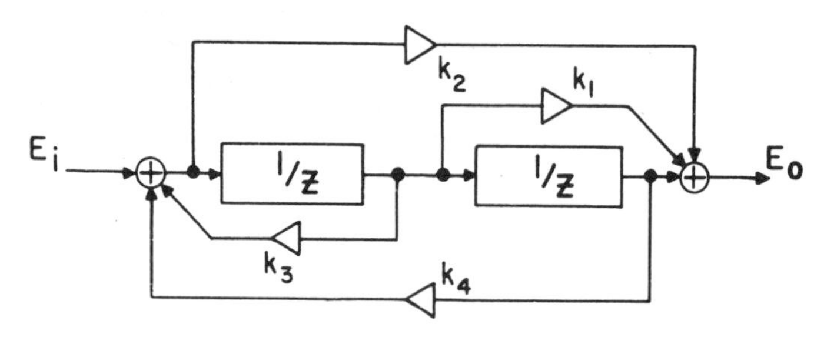

$$T(z) = k_2 \frac{z^2 + \frac{k_1}{k_2}z + \frac{1}{k_2}}{z^2 - k_3z - k_4}$$

ZEROS AT $1/2k_2\left[-k_1 \pm \sqrt{k_1^2 - 4k_2}\right]$

POLES AT $1/2\left[k_3 \pm \sqrt{k_3^2 + 4k_4}\right]$

Fig. 7—Basic 2-element building block.

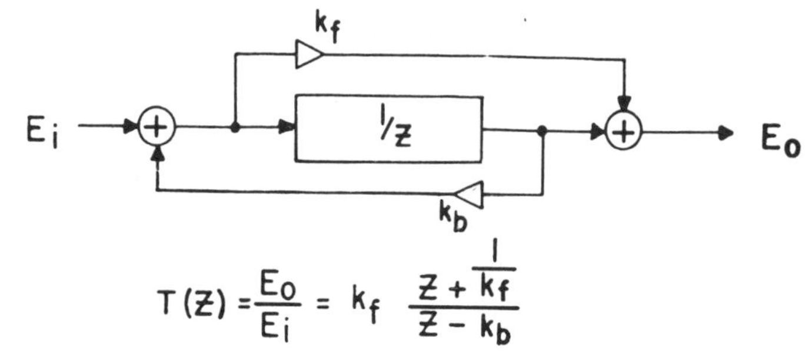

$$T(z) = \frac{E_0}{E_i} = k_f \frac{z + \frac{1}{k_f}}{z - k_b}$$

Fig. 8—One-element building block.

n arbitrary poles[7] may be synthesized with n delay elements.

The generality of the preceding synthesis procedure will be proven by noting its structural constraints and by showing that these do not affect the relation between the number of elements and the degree of the transfer function. These constraints are as follows:

1) All sections are in cascade; *i.e.*, the output is not obtained by realizing a number of subsections whose outputs are combined in one final adder.

2) All delay elements appear on the "main line" of the structure; *i.e.*, a continuous path may be traced from input to output in such a fashion that it is unidirectional and goes through all of the delay lines. In particular, no delay elements appear in any of the feedback or feed-forward loops.

3) No feed-forward or feedback loops span more than two delay elements; *i.e.*, no loops are added to those which exist within the building blocks.

4) The main line of the network does not contain any amplifiers[8] but passes only through delay elements and adders.

It is easy to see that the last restriction is a trivial one. Consider a section of the delay network as shown in Fig. 9(a), which is completely equivalent to the configuration of Fig. 9(b). In an entirely similar matter, the amplifiers may be absorbed into the feedback loops and displaced forward until finally there remains a structure whose main line is entirely free of amplifiers with the possible exception of one final output amplifier. The latter may, for the purposes of this analysis, be disregarded.

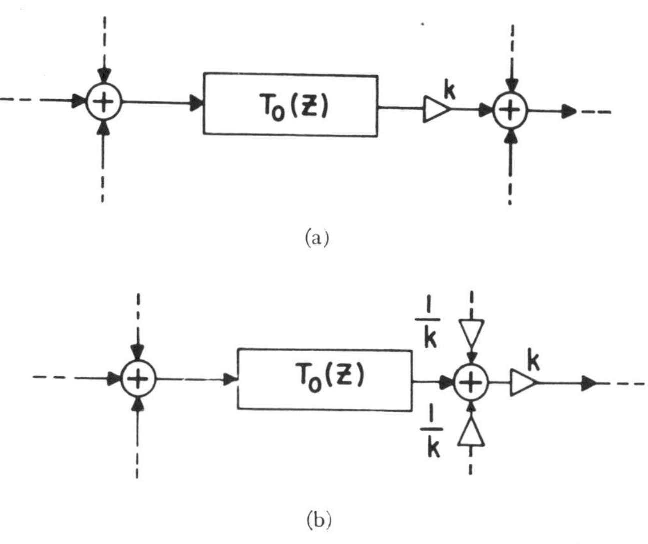

Fig. 9—Elimination of amplifiers from main line of networks. (a) Network with amplifier in main line; (b) equivalent network.

The above synthesis method results in a maximum of n arbitrary poles and n arbitrary zeros for n delay elements. Subsequent paragraphs show that removal of the remaining constraints does not permit additional poles and zeros.

The third constraint may be shown to be nonrestrictive in the sense that it does not reduce the possible degree of the transfer function. In order to demonstrate this fact, the network of Fig. 10 will be considered.

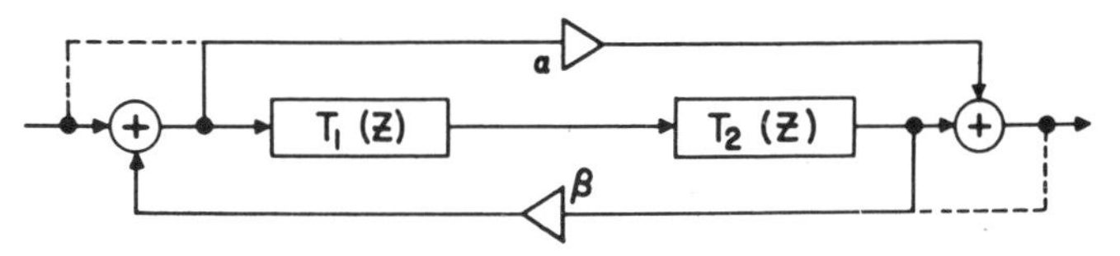

Fig. 10—Delay networks with additional feed-forward and feedback loops.

Terms $T_1(z)$ and $T_2(z)$ are the transfer functions of two cascaded networks having n_1 and n_2 delay elements, respectively. Each network is made up of building blocks of the type shown in Fig. 7; *i.e.*, its transfer function has as many poles and zeros as there are delay elements. The over-all transfer function is given by

$$T(z) = \frac{T_1(z)T_2(z) + \alpha}{1 - \beta T_1(z)T_2(z)},$$

and it is readily seen that $T(z)$ has at most (n_1+n_2) zeros and (n_1+n_2) poles. Thus, adding additional loops to a cascade of building blocks of the type of Fig. 7 has not increased the possible number of poles and zeros.[9]

In the building block synthesis, each additional delay element adds, at most, one pole and one zero to the transfer function. The question arises whether it would be possible to utilize additional delay elements more efficiently by inserting them in places other than the main line. In order to show that this is not the case, the circuit of Fig. 11 will be considered. Terms T_1, T_2, T_α,

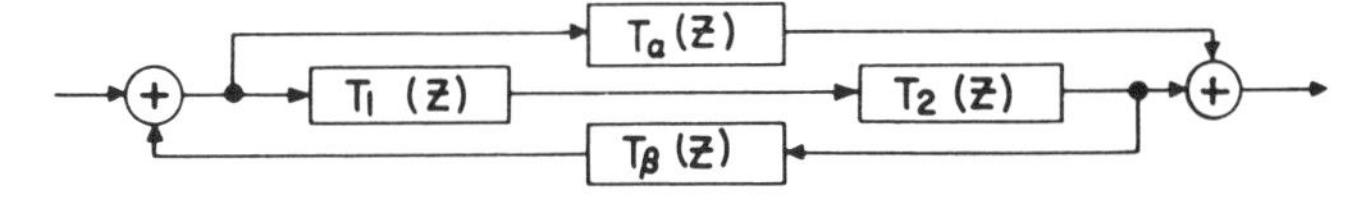

Fig. 11—Delay elements in feed-forward and feedback loops.

and T_β are all assumed to have one zero and one pole per delay element. The over-all transfer function is

$$T(z) = \frac{T_1(z)T_2(z) + T_\alpha(z)}{1 - T_1(z)T_2(z)T_\beta(z)}.$$

[7] Within the restrictions imposed by stability considerations.

[8] There is, of course, no implication that such an arrangement is practically feasible or desirable; the restriction is included solely for analytical convenience and can readily be modified by the application of simple block-diagram manipulation.

[9] The same conclusion is reached if the α and β loops are connected differently; *e.g.*, as indicated by the broken lines in Fig. 10.

Examination of this expression shows that $T(z)$ has at most $(n_1+n_2+n_\alpha+n_\beta)$ poles and zeros. Thus, the $(n_\alpha+n_\beta)$ delay elements which were added in the feedback and feed-forward loops contributed at most $(n_\alpha+n_\beta)$ poles and zeros. An equal increase in the degree of the transfer function could have been obtained, however, by adding the appropriate number of building block sections in cascade.

Finally, there remains the task of showing that constraint (1) does not affect the number of poles and zeros which can be obtained by means of a given number of delay elements. For this purpose, one need only consider a network whose output consists of the sum of the outputs of a number of networks, each of which has one pole and one zero per delay element. The over-all transfer function is

$$T(z) = \sum T_k(z),$$

and contains, at most, $\sum n_k$ poles and zeros where n_k is the number of delay elements in the kth parallel network. It is clear that an equal number of poles and zeros could have been obtained by employing the n_k delay elements in a cascade realization.

It may now be concluded that the transfer function of a delay network containing n delay elements has at most n poles and n zeros.

Conclusion

A systematic synthesis method has been presented for the realization of periodic transfer functions by means of networks consisting of delay lines, adders, and linear amplifiers. It has been shown that the required number of delay channels is equal to the number of z-plane poles located away from the origin, or to the number of zeros, whichever is larger.

The synthesis is accomplished as a cascade of "building blocks." Flow-graph techniques may be used for manipulating block diagrams in order to obtain more advantageous configurations having the same transfer function.

1.4

ANALYSIS AND SYNTHESIS OF DELAY LINE PERIODIC FILTERS

H. Urkowitz

IRE Transactions on Circuit Theory, Vol. CT-4, No. 2, June 1957, pp. 41-53.

PREFACE

WHEN my colleagues and I first became aware of the applications of z-transform techniques as applied to MTI filters, it seemed to us that here was an immensely useful and rather undeveloped method for the study of all types of periodic filters characterized by having one or more delay elements. This paper is the result of the awakening of such interest. It does not represent simply my own research but rather reflects the results of many lively discussions with B. D. Steinberg, D. A. Linden, and N. C. Randall of the Philco Research Division. To these men, my sincere thanks for providing the stimulating atmosphere of lively and intensive discussions.

I. INTRODUCTION

A periodic filter is one whose frequency characteristic is periodic when plotted as a function of real frequency, $\omega = 2\pi f$. Such filters are sometimes used in pulsed radar systems to process the video output of the radar receiver. Examples of such filters are cancellers for detecting moving targets and video sweep integrators for improving signal-to-noise ratio. Continuous processing is possible if the periodic filter employs delay elements (delay lines) capable of delaying or storing an entire radar trace for one interpulse period. The frequency period is, of course, the reciprocal of the delay time, τ.

The essential element in all such filters is a delay line which simply delays for a time τ whatever its input may be. The delay line is indicated symbolically in Fig. 1.

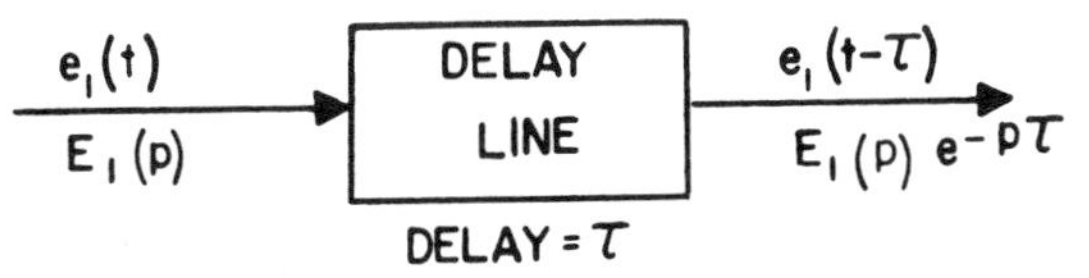

Fig. 1—Representation of delay line.

If the input were a unit impulse, $\delta(t)$, the output would be $\delta(t - \tau)$. Thus, the frequency characteristic of the delay line is $e^{-p\tau}$. It is assumed that the frequency response of the delay line is so broad that there is no shape distortion of input voltages.

II. EXAMPLES OF PERIODIC FILTERS

A. Canceller

A block diagram of a video canceller is shown in Fig. 2. Since $E_0 = E_1 - E_1e^{-p\tau}$, the transfer function is

$$T(p) = \frac{E_0}{E_1} = 1 - e^{-p\tau}. \tag{1}$$

* Manuscript received by the PGCT, August 5, 1956.
† Philco Corp., Philadelphia, Pa.

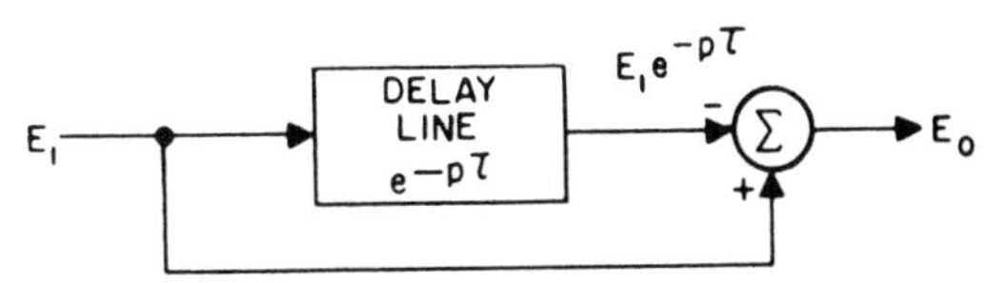

Fig. 2—Block diagram of a canceller.

The response to a unit impulse, or delta function, $\delta(t)$, (sometimes called the characteristic transient) is given by the inverse Laplace transform of $T(p)$.

$$h(t) = \mathcal{L}^{-1}[T(p)] = \delta(t) - \delta(t - \tau). \tag{2}$$

This is shown in Fig. 3.

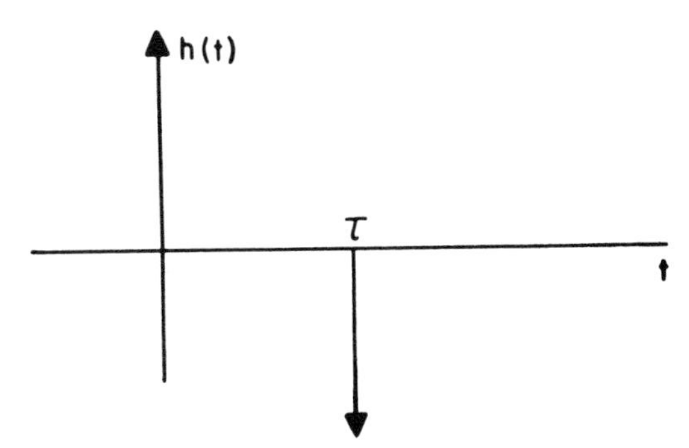

Fig. 3—Characteristic transient of a canceller.

We can find amplitude (frequency) characteristic by finding $|T(j\omega)|$. We would get the familiar form

$$|T(j\omega)| = 2\left|\sin\frac{\omega}{2}\right|. \tag{3}$$

This is plotted in Fig. 4.

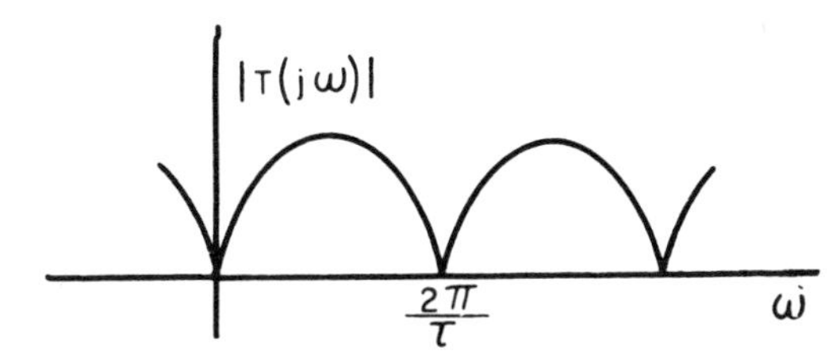

Fig. 4—Frequency characteristic of canceller.

The p-plane pole-zero (P-Z) diagram can be found from (1). There are no poles, and $T(p)$ is zero when $p = 2\pi n/\tau$, n = any positive or negative integer, including zero. The P-Z diagram is shown in Fig. 5. All zeros are simple.

B. Video Sweep Integrator

A block diagram of a sweep integrator is shown in Fig. 6. The transfer function is given by

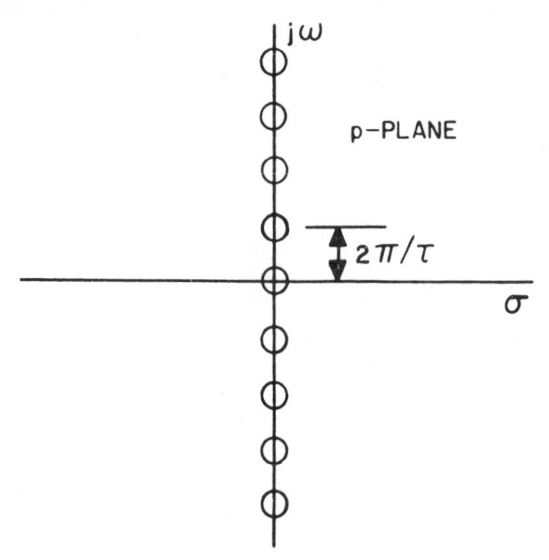

Fig. 5—Pole-zero diagram of canceller.

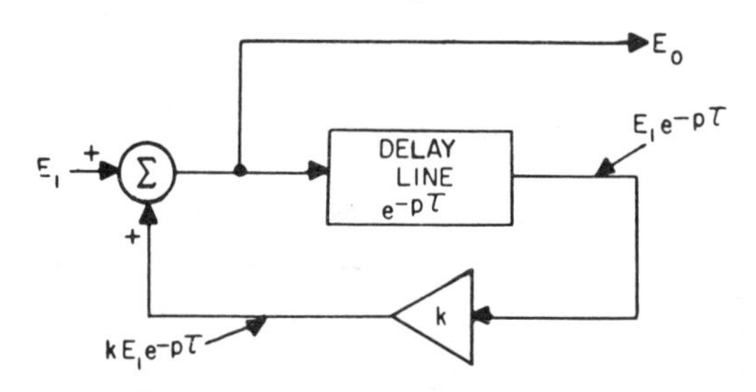

Fig. 6—Block diagram of sweep integrator.

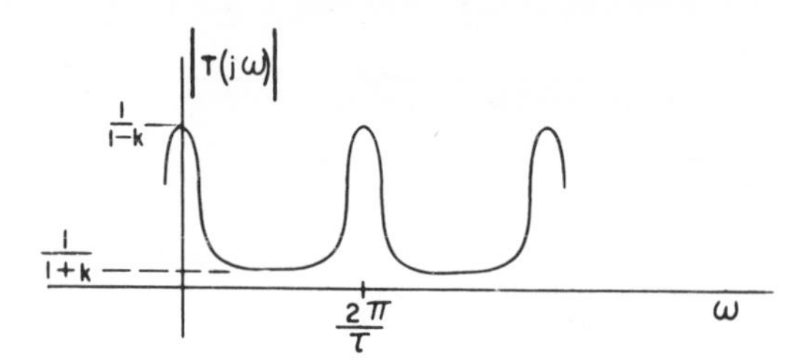

Fig. 7—Frequency characteristic of sweep integrator.

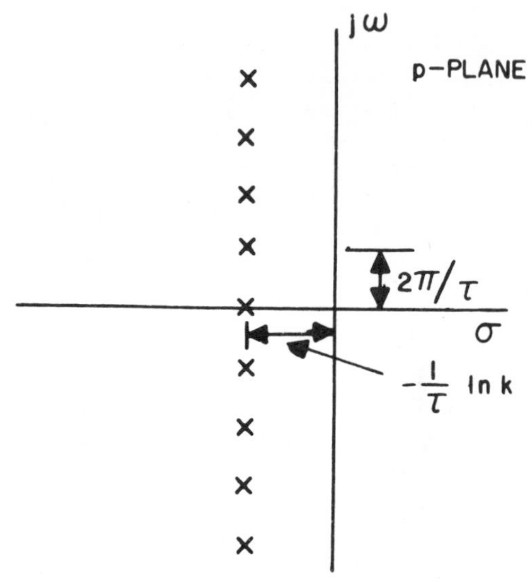

Fig. 8—*P-Z* diagram for sweep integrator.

$$T(p) = \frac{E_0}{E_1} = \frac{1}{1 - ke^{-p\tau}}. \tag{4}$$

It is found that

$$|T(j\omega)| = \frac{1}{\sqrt{(1 - k\cos\omega\tau)^2 + k^2\sin^2\omega\tau}} \tag{5}$$

which is plotted in Fig. 7. The feedback factor, k, must be less than unity; it determines the "width" and height of each "tooth" of the response shown in Fig. 7. $T(p)$ has no zeros but has poles wherever

$$1 - ke^{-p\tau} = 0. \tag{6}$$

Roots of (6) occur for

$$e^{-p\tau} = \frac{1}{k} \tag{7}$$

or

$$\exp(\sigma + j\omega)\tau = k. \tag{8}$$

Since the right-hand side is real and positive, this equation is satisfied for

$$\omega = \frac{2\pi n}{\tau}.$$

n = positive or negative integer including zero.

All poles are simple. The *P-Z* diagram is shown in Fig. 8.

The characteristic transient can be obtained from physical considerations or from (4). Eq. (4) can be written as

$$T(p) = \sum_{n=0}^{\infty} k^n e^{-np\tau} \tag{9}$$

since (4) will be recognized as the sum of a geometric series. Then

$$h(t) = \mathfrak{L}^{-1}[T(p)] = \sum_{n=0}^{\infty} k^n\, \delta(t - n\tau). \tag{10}$$

$h(t)$ is shown in Fig. 9 opposite.

III. The z Transform

The z transform was introduced by Hurewicz[1] for the study of sampling servomechanisms, and has been used extensively for the analysis and design of such systems.[2-4]

[1] H. M. James, N. B. Nichols, and R. S. Phillips, "Theory of Servomechanisms," M. I. T. Rad. Lab. Ser., McGraw-Hill Book Co., Inc., New York, N. Y., ch. 5; 1947.

[2] J. R. Raggazzini and L. A. Zadeh, "The analysis of sampled data systems," *Trans. AIEE*, vol. 71, part II, pp. 225-232; 1952.

[3] J. G. Truxal, "Automatic Feedback Control System Synthesis," McGraw-Hill Book Co., Inc., New York, N. Y., ch. 9; 1955.

[4] After the manuscript was submitted it was brought to the author's attention that a general system which will synthesize pulsed transfer functions which consist of ratios of polynomials in z was first proposed by R. H. Barker, "The Theory of Pulse Monitored Servos and Their Use for Prediction," Rep. No. 1046, Signals Res. and Dev. Est., Christchurch, Hants, England; November, 1950.

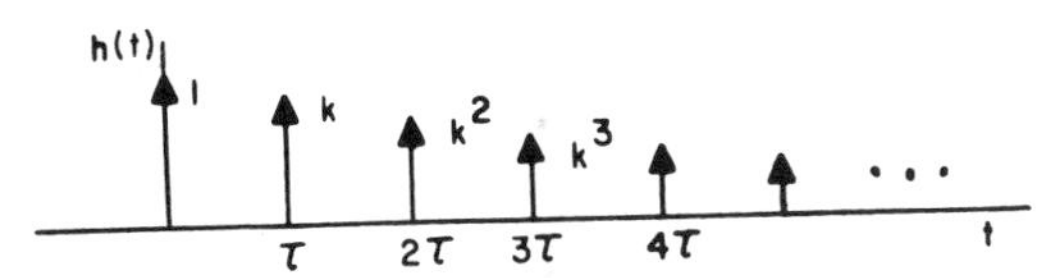

Fig. 9—Characteristic transient of sweep integrator.

z-plane algebra has been applied to the study of cancellation circuits in radar by White and his colleagues at Airborne Instruments Laboratory.[5,6]

The transformation is as follows:

$$z = e^{p\tau} \tag{11}$$

This transformation transforms the left half p plane into the interior of the unit circle in the z plane, with the $j\omega$ axis going over into the boundary of the circle. This is illustrated in Fig. 10. Actually, the transformation is multiple valued, so that a strip in p plane is transformed into the unit circle represented by one traverse around the contour $|z| = 1$. This is illustrated in Fig. 11.

The condition for stability in the z plane is that poles of the z-transfer function must be inside the unit circle. Zeros may be anywhere.

In general, a transfer function expressed as a function of z will be of the form of a rational fraction:

$$T(z) = \frac{a_n z^n + a_{n-1} z^{n-1} + \cdots + a_0}{b_m z^m + b_{m-1} z^{m-1} + \cdots + b_0}. \tag{12}$$

The poles and zeros may be placed in evidence by finding the roots of the denominator and numerator of $T(z)$, and writing it in the form

$$T(z) = K\left[\frac{(z - z_{01})(z - z_{02}) \cdots (z - z_{0n})}{(z - z_{p1})(z - z_{p2}) \cdots (z - z_{pm})}\right]. \tag{13}$$

Poles and zeros must either be real or occur in conjugate complex pairs. To find $|T(j\omega)|$ from the z-plane P-Z diagram of $T(z)$, we note in (13) that both the numerator and denominator consist of the product of several complex vectors, so that the magnitude of $T(z)$, at any value of z, is given by the product of the vector lengths from the zeros to the point in question divided by the product of the vector lengths from the poles to the same point. That is

$$|T(j\omega)| = |T(z)|\Big|_{z=e^{j\omega\tau}}$$

$$= \frac{\text{product of vector lengths from zeros to point on unit circle}}{\text{product of vector lengths from poles to point on unit circle}}. \tag{14}$$

This is illustrated in Fig. 12. Note that a pole or zero at

[5] W. D. White, "The mathematics of difference equations and the concept of the z-plane," unpublished lecture notes of seminar lectures at Airborne Instr. Lab.; 1951.

[6] W. J. Dauksher, C. A. Fowler, *et al.*, "Investigation of MTI Systems," Final Eng. Rep., Contract No. AF 28(099)-260, Airborne Instr. Lab.; July, 1952. See Appendix F, "Shaping of velocity response in an MTI."

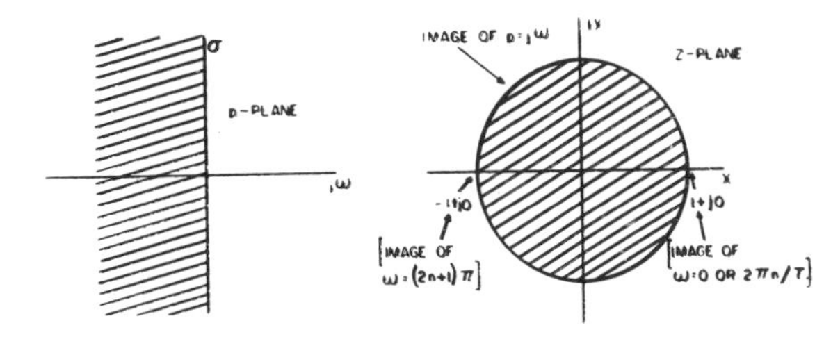

Fig. 10—The transformation $z = e^{p\tau}$.

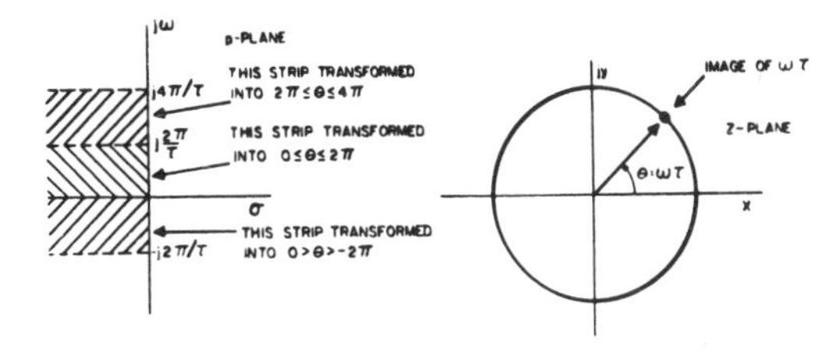

Fig. 11—Further illustration of z transform.

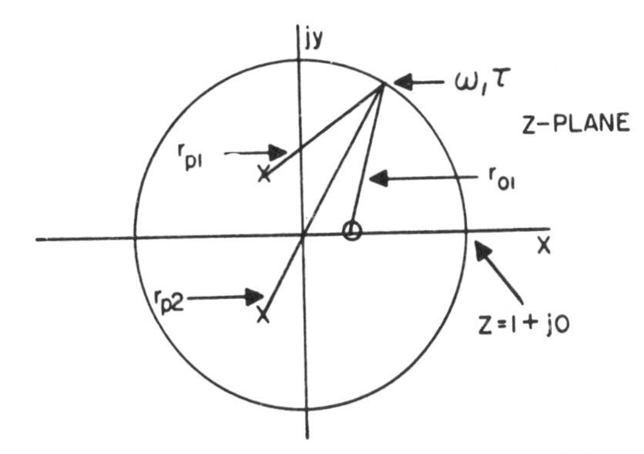

Fig. 12—Finding the frequency characteristic from P-Z diagram.

the origin has no effect on $|T(j\omega)|$, since the corresponding vector length is unity.

IV. z-Plane P-Z Diagram of Some Periodic Filters

A. Canceller

From (1), we can write

$$T(z) = 1 - z^{-1} = \frac{z-1}{z}. \tag{15}$$

This has a pole at the origin and a zero at $z = 1$, as shown in Fig. 13. The expression for $|T(j\omega)|$, (or rather, $|T(j\omega\tau)|$), is easily obtained by a graphical construction on the diagram of Fig. 13. To show this, we use the construction of Fig. 14. From this figure

$$r = \frac{b}{\sin\epsilon}, \quad \epsilon = \frac{\pi - \omega\tau}{2}, \quad b = \sin\omega\tau$$

$$\therefore r = \frac{\sin\omega\tau}{\sin\left(\frac{\pi}{2} - \frac{\omega\tau}{2}\right)} = \frac{2\sin\frac{\omega\tau}{2}\cos\frac{\omega\tau}{2}}{\cos\frac{\omega\tau}{2}} \tag{16}$$

$$\therefore r = 2\sin\frac{\omega\tau}{2}.$$

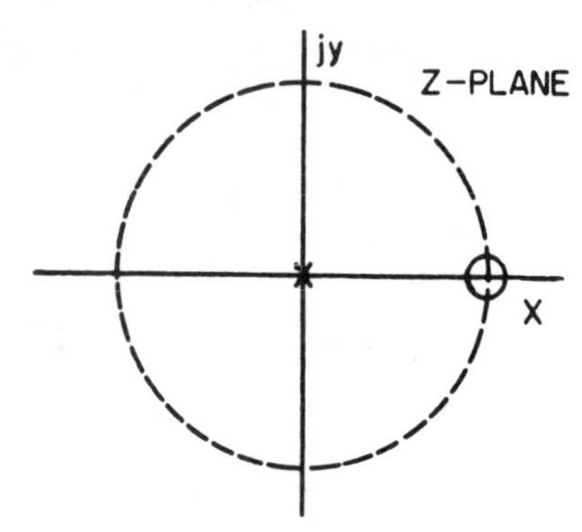

Fig. 13—z-plane diagram of a canceller.

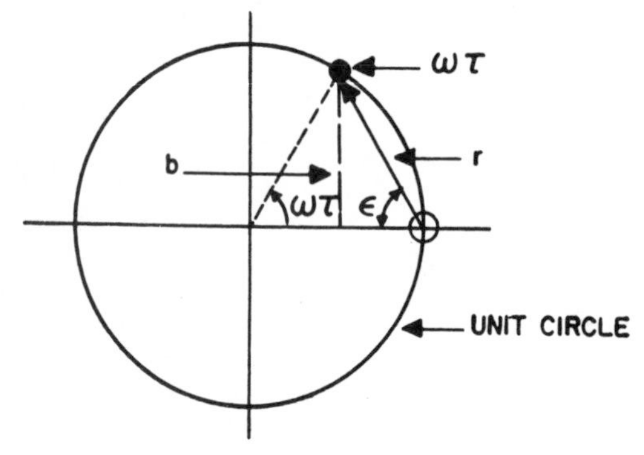

Fig. 14—Figure used to derive (17).

Since r has the same value whatever the sign of $\omega\tau$,

$$|\,T(j\omega)\,| = r = 2\left|\sin\frac{\omega\tau}{2}\right|. \tag{17}$$

B. Video Sweep Integrator

Eq. (4) gives us

$$T(z) = \frac{1}{1 - k/z} = \frac{z}{z - k}. \tag{18}$$

The P-Z diagram is shown in Fig. 15.

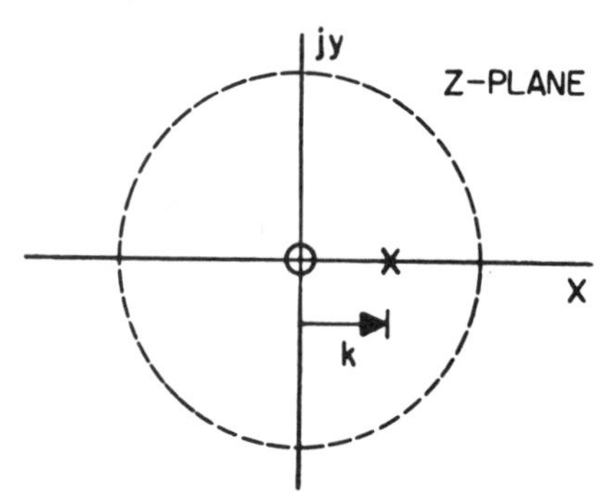

Fig. 15—P-Z diagram of a video sweep integrator.

C. Double Canceller

A block diagram of a double canceller is shown in Fig. 16. Its transfer function is

$$T(z) = \frac{(z-1)^2}{z^2}.$$

The P-Z diagram is shown in Fig. 17. Note that both the zero and pole are of double order.

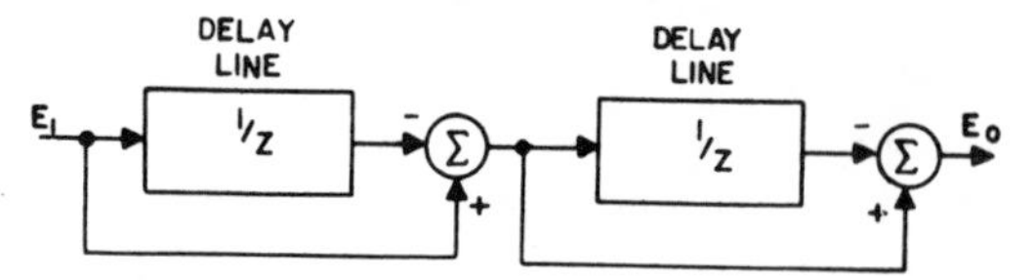

Fig. 16—Double canceller.

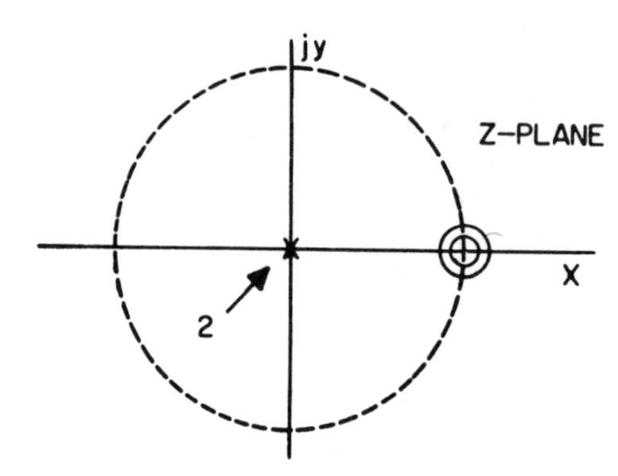

Fig. 17—P-Z diagram of double canceller.

In Fig. 16, two separate delay lines are shown. Actually only one delay line need be used if the two signals are multiplexed.

D. Double Loop Integrator

A block diagram of a double loop sweep integrator is shown in Fig. 18. While this may appear more complicated than the diagram of Fig. 6, Fig. 18 actually allows a simpler and more reliable type of construction.

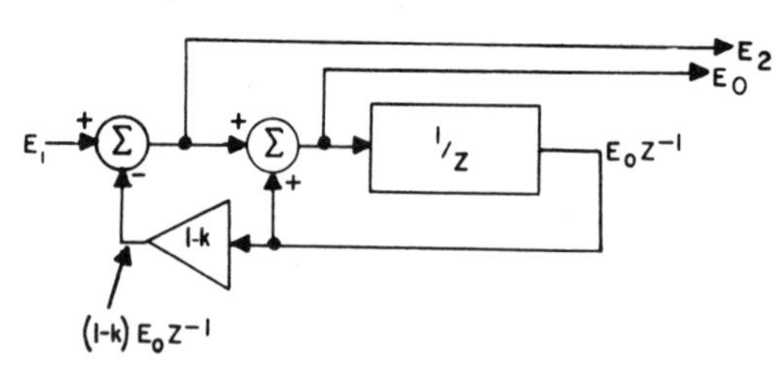

Fig. 18—Double loop integrator.

Turning to Fig. 18,

$$\begin{aligned} E_2 &= E_1 - (1-k)E_0 z^{-1} \\ E_0 &= E_2 + E_0 z^{-1} \\ &= E_1 - (1-k)E_0 z^{-1} + E_0 z^{-1} \end{aligned} \tag{20}$$

$$E_0 = E_1 + kE_0 z^{-1}$$

$$\therefore E_0(1 - kz^{-1}) = E_1 \tag{21}$$

$$\therefore T(z) = \frac{E_0}{E_1} = \frac{1}{1 - kz^{-1}} = \frac{z}{z-k}. \tag{22}$$

Comparison with (18) shows that E_0 gives an integrated output.

Now, let us consider E_2. If we substitute E_0 from (21) into (20) we get

$$E_2 = E_0 - (1 - k)z^{-1}\left(\frac{E_1}{1 - kz^{-1}}\right) \qquad (23)$$

$$E_2 = E_1\left[\frac{1 - kz^{-1} - (1 - k)z^{-1}}{1 - kz^{-1}}\right]$$

$$= E_1\left(\frac{1 - z^{-1}}{1 - kz^{-1}}\right) = E_1\left(\frac{z - 1}{z - k}\right)$$

$$T_2(z) = \frac{E_2}{E_1} = \frac{z - 1}{z - k}. \qquad (24)$$

The z-plane P-Z diagram is shown in Fig. 19. Note that there is a zero at $z = 1$. This corresponds to a zero at every multiple of $\omega = 2\pi/\tau$ in the p plane, a property

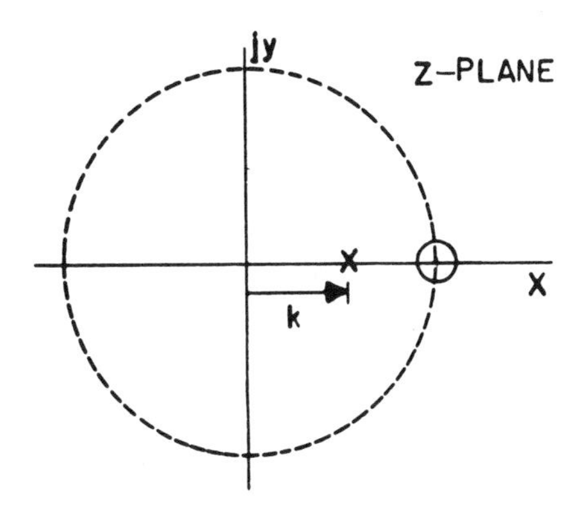

Fig. 19—P-Z diagram for alternate output of double loop integrator.

which is characteristic of a canceller. The position of the pole is movable by variation of k, the feedback factor. This suggests that the circuit of Fig. 18, with the output taken at E_2, is a velocity shaped canceller, with the shaping dependent on the value of k. The effect of k is illustrated in Fig. 20. The circuit may properly be called a response-shaped canceller. The P-Z diagram of Fig. 19, with k adjustable, represents the best that can be done to shape the response of a canceller with one delay line.

Block diagrams of periodic filters together with their z-plane pole-zero diagrams are given in Table II in Appendix II.[7]

V. Frequency Response Shaping

The preceding discussion suggests that it is possible to obtain almost any periodic frequency response if one wishes to work hard enough. The problem of response shaping can be formulated in terms of specific questions, provided it has already been decided just what shape is desired.

[7] The author is indebted to B. D. Steinberg, D. A. Linden, and N. C. Randall of the Res. Div., Philco Corp. for some of the diagrams in Appendix II.

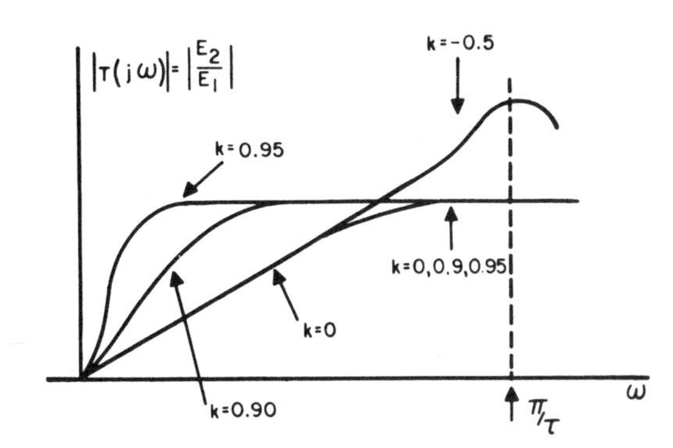

Fig. 20—Response shaping with changing feedback factor (see Fig. 18).

1) How many z-plane poles and zeros are required to approach the desired shape to within a given tolerance and what is their geometric arrangement?
2) What is the minimum number of delay lines needed to achieve the pole-zero pattern?
3) How shall the various paths around the delay line be placed, and with what weighting?
4) For a given number of delay elements, how close can one come to the desired shape?

The answer to the first question is closely related to the approximation problem of ordinary network synthesis. The people at Airborne Instruments Laboratory[5,6] have attempted to supply an answer through the use of a further transformation. However, their method appears to be incomplete. It is the author's opinion that a satisfactory method remains to be found. The second question can be answered almost by intuition. The number of delay lines required is equal to the degree of the largest degree polynomial in the z transform, whether it is the numerator or the denominator. For example, suppose that $T(z)$ is written as the ratio of two polynomials in z, with positive exponents, such that the highest power represented is z^3. This means that there are two signals somewhere in the system such that one is obtained from the other after passage through three delay elements.

A more formal demonstration of this fact has been given by Linden and Steinberg.[8] We have already noted that one zero and one pole can be obtained with a single delay line. We shall also show that two poles and two zeros with the poles and zeros either complex or real can be obtained from two delay lines. Then, it is not difficult to see that a cascade of such sections can produce a number of poles and a number of zeros equal to the number of delay lines. These sections can be termed "basic building blocks." Is there another arrangement, other than cascade, which will produce more zeros and/or poles? If building blocks are not cascaded they must be arranged in feedforward or feedback paths. In Fig. 21, such an arrangement is illustrated. The over-all transfer function is given by

[8] D. A. Linden and B. D. Steinberg, "Synthesis of delay line networks," IRE Trans., vol. ANE-4, pp. 34–39; March, 1957

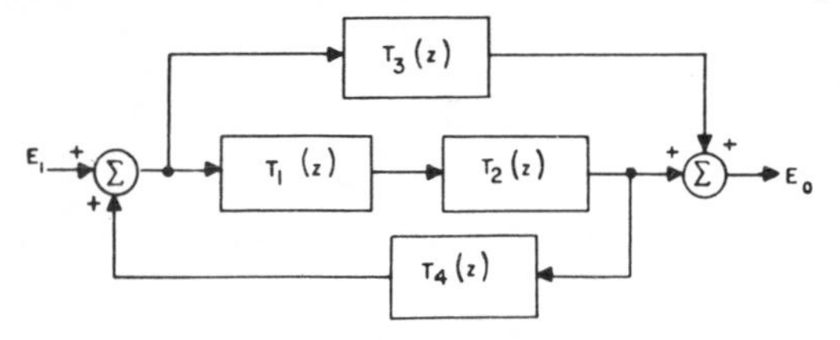

Fig. 21—A general arrangement for component delay line networks. (T_1 has N_1 poles and N_1 zeros, etc.)

$$T(z) = \frac{E_0}{E_1} = \frac{T_1(z)T_2(z) + T_3(z)}{1 - T_1(z)T_2(z)T_4(z)}. \quad (25)$$

Examination of (25) shows that $T(z)$ has at most $N_1 + N_2 + N_3 + N_4$ poles and zeros. But, since each of the component networks has a number of delay lines equal to the number of its poles and zeros, we see that the arrangement did not result in a number of poles and zeros greater than the number of delay lines. However, we know that to get N poles and/or N zeros we can cascade basic building blocks so that the total number of delay lines will be N. Therefore, we conclude that for N poles and/or N zeros in the z plane, just N delay lines are required.

The answer to the third question as to how the various paths should be arranged is the problem of synthesis which is covered in the next section.

VI. Synthesis

The problem of synthesis can be stated as follows: Given a z-plane transfer function:

$$T(z) = \frac{a_n z^n + a_{n-1} z^{n-1} + \cdots a_0}{b_m z^m + b_{m-1} z^{m-1} + \cdots b_0}, \quad (26)$$

draw a block diagram of a delay line network containing m or n (whichever is larger) delay lines, having this transfer function. We shall do this by synthesizing first the general one delay element circuit, and then synthesizing the general two delay element circuit. These can then be considered as the basic building blocks for obtaining any $T(z)$. The two delay-element building block can give either real or complex poles and zeros.

Problem 1

To synthesize

$$T(z) = \frac{z - z_{01}}{z} = 1 - z_{01}z^{-1}. \quad (27)$$

Now

$$E_0 = E_1(1 - z_{01}z^{-1}). \quad (28)$$

For this simple expression, it is seen that all we need do is combine E_1 with E_1 delayed and weighted with z_{01}. The circuit is shown in Fig. 22.

Problem 2

To synthesize

$$T(z) = \frac{z}{z - z_{p1}} = \frac{1}{1 - z_{p1}z^{-1}}. \quad (29)$$

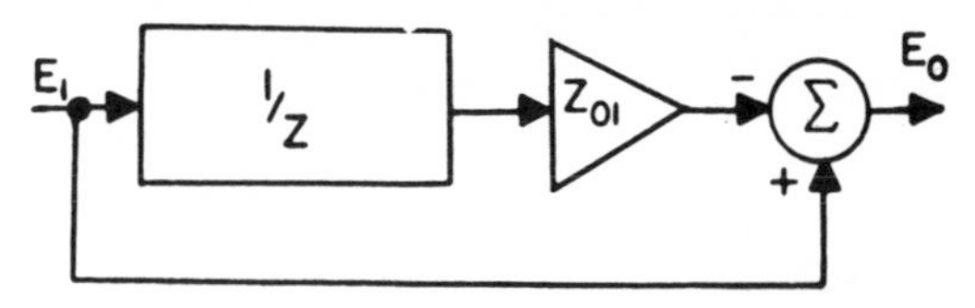

Fig. 22—Realization of $T(z) = (z - z_{01})/z$.

We know the answer to this problem already; it is the sweep integrator of Fig. 15. However, it is instructive to derive the circuit in a formal way. To do this, we note that the sum of signals flowing into a combining network is equal to the sum of the signals flowing out. This is illustrated in Fig. 23. From (29), we can write

$$E_0 = E_1\left(\frac{1}{1 - z_{p1}z^{-1}}\right) \quad (30)$$

or

$$E_1 = E_0(1 - z_{p1}/z). \quad (31)$$

Fig. 23 shows that (31) can be represented by Fig. 24. The quantity $E_0 z_{p1}/z$ can be obtained by delaying E_0 and multiplying it by z_{p1}. The result is shown in Fig. 25.

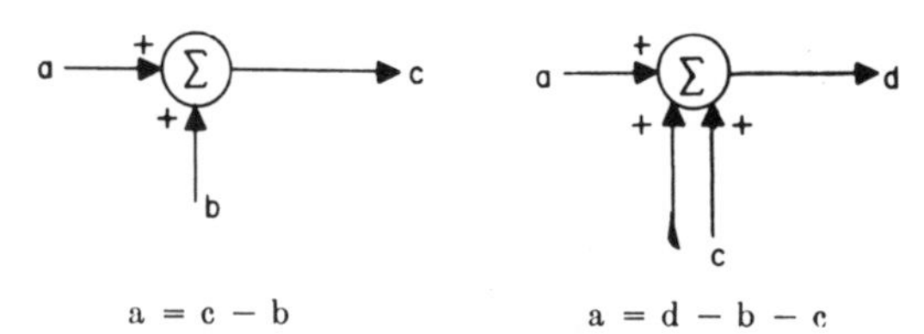

a = c − b   a = d − b − c

Fig. 23—Signals in and out of combining networks.

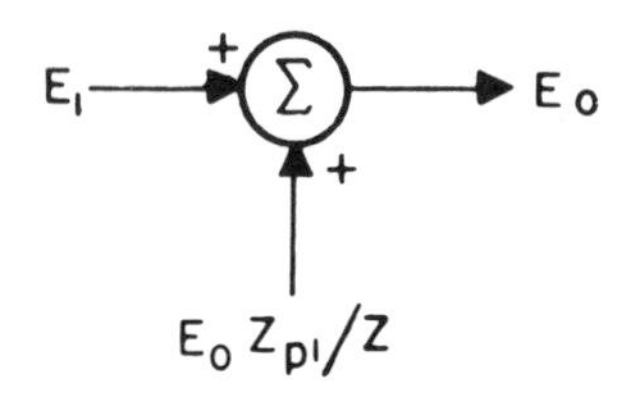

Fig. 24—Representation of (31).

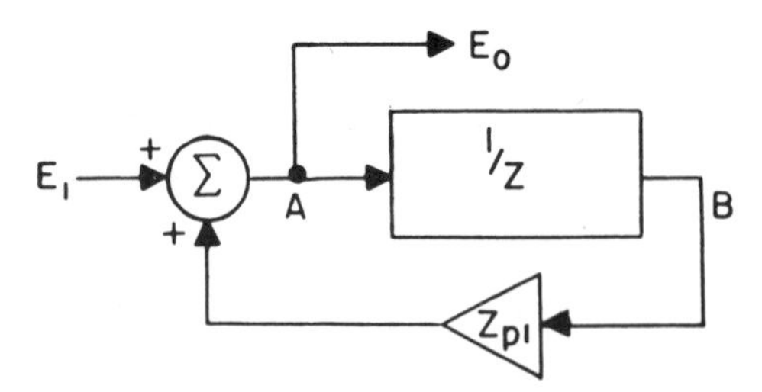

Fig. 25—Realization of $T(z) = z/(z - z_{p1})$.

Problem 3

To synthesize

$$T(z) = \frac{1}{z - z_{p1}} \quad (32)$$

$$E_0 = E_1\left(\frac{1}{z - z_{p1}}\right) = E_1 z^{-1}\left(\frac{z}{z - z_{p1}}\right). \quad (33)$$

In this case, E_0 is the same as in the previous example,

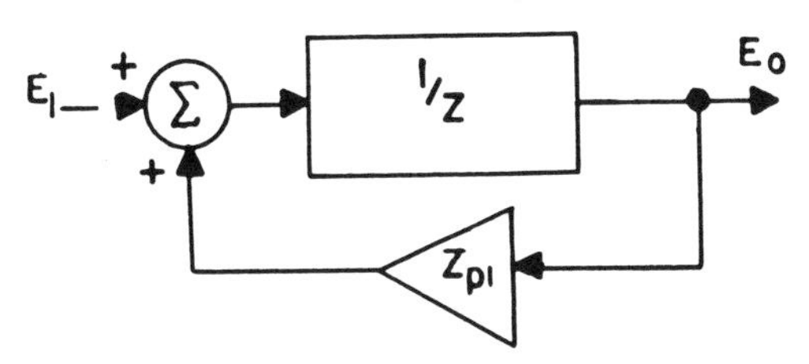

Fig. 26—Realization of $T(z) = 1/(z - z_{p1})$.

except that it is delayed once. Thus, in Fig. 25, if the output is taken from point B, we will have a transfer function given by (32). The block diagram is shown in Fig. 26.

Problem 4

To synthesize

$$T(z) = \frac{z^2}{(z - z_{p1})(z - z_{p2})} = \frac{1}{(1 - z_{p1}/z)(1 - z_{p2}/z)}$$
$$= \frac{1}{1 - (z_{p1} + z_{p2})/z + z_{p1}z_{p2}/z^2}. \tag{34}$$

Then

$$E_1 = E_0[1 - (z_{p1} + z_{p2})/z + z_{p1}z_{p2}/z^2]. \tag{35}$$

Eq. (35) is a description of the combining network of Fig. 27. The inputs for the combining network are obtained quite simply from delay lines and weighting amplifiers, and the result is shown in Fig. 28. It is worthwhile

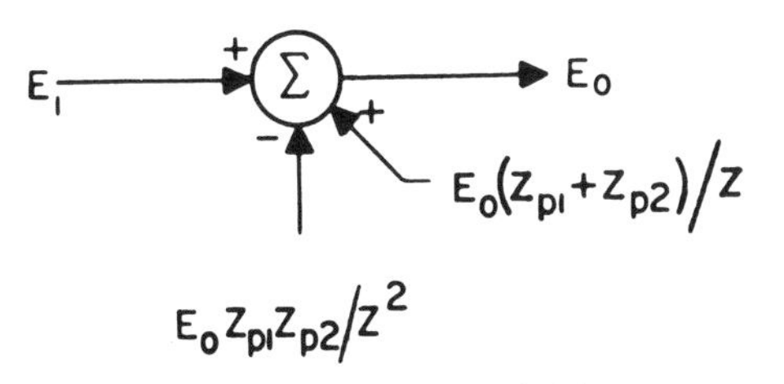

Fig. 27—Representation of (35).

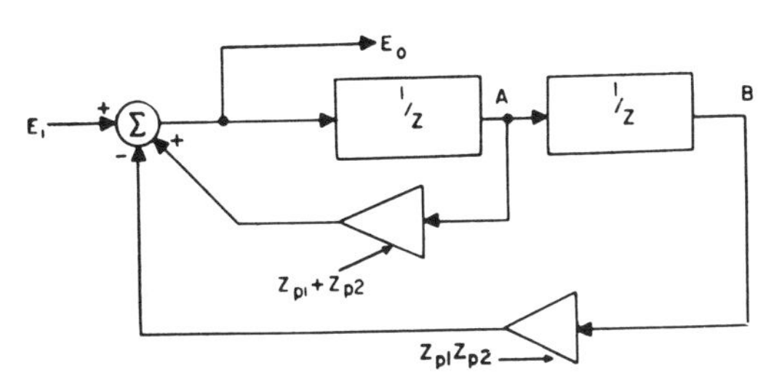

Fig. 28—Realization of $T(z) = z^2/[(z - z_{p1})(z - z_{p2})]$.

to note that $z_{p1} + z_{p2}$ is real, as is $z_{p1}z_{p2}$. In Fig. 28, if the output is taken at A, we would have E_0z^{-1} and at B we would have E_0/z^2. Therefore we can get both

$$T(z) = \frac{z}{(z - z_{p1})(z - z_{p2})} \tag{36}$$

and

$$T(z) = \frac{1}{(z - z_{p1})(z - z_{p2})}. \tag{37}$$

Problem 5

To synthesize

$$T(z) = \frac{(z - z_{01})(z - z_{02})}{(z - z_{p1})(z - z_{p2})} = \frac{1 - (z_{01} + z_{02})/z + z_{01}z_{02}/z^2}{1 - (z_{p1} + z_{p2})/z + z_{p1}z_{p2}/z^2}.$$

From (38), we can write

$$E_0 = E_1\left[\frac{1}{1 - (z_{p1} + z_{p2})/z + z_{p1}z_{p2}/z^2} - \frac{(z_{01} + z_{02})/z}{1 - (z_{p1} + z_{p2})/z + z_{p1}z_{p2}/z^2} + \frac{z_{01}z_{02}/z^2}{1 - (z_{p1} + z_{p2})/z + z_{p1}z_{p2}/z^2}\right]. \tag{39}$$

Thus, the output is the sum of three terms; the first is the same as E_0 of Fig. 28, the second is obtained by delaying E_0 once (point A of Fig. 28) and weighting with $-(z_{01} + z_{02})$, and the third is obtained by delaying E_0 twice (point B of Fig. 28) and weighting with $z_{01}z_{02}$. Then these three quantities are combined. The result is shown in Fig. 29.

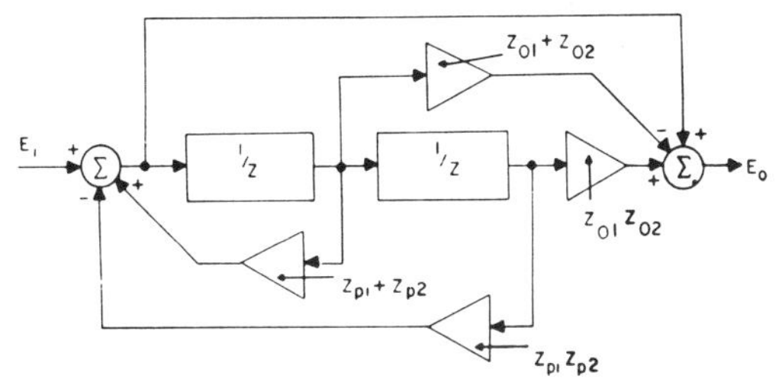

Fig. 29—Realization of $T(z) = (z - z_{01})(z - z_{02})/[(z - z_{p1})(z - z_{p2})]$.

The forms

$$T(z) = \frac{z(z - z_{01})}{(z - z_{p1})(z - z_{p2})} \tag{40}$$

and

$$T(z) = \frac{z - z_{01}}{(z - z_{p1})(z - z_{p2})} \tag{41}$$

are easily obtained from Fig. 29 by making $z_{02} = 0$ and taking the output at E_0 for (40) and after the first delay line for (41).

VII. Functions Which Cannot Be Synthesized with Delay Lines

A transfer function which requires a time advance cannot be synthesized with delay lines. That is, polynomials (not rational fractions) in z with positive exponents cannot be so synthesized. Furthermore a rational fraction of z whose numerator is of higher degree than the denominator cannot be synthesized.

However, if one's interest lies only in $|T(j\omega)|$, then one can multiply numerator or denominator by z raised to the appropriate power to make the numerator and denominator of the same degree. So, if $|T(j\omega)|$ is the only thing of interest, then any rational fraction with poles inside the unit circle can be synthesized.

VIII. Alternative Forms

The form shown in Fig. 29 is probably the simplest block diagram that will give two poles and two zeros. However, it should not be inferred that an actual network of this form will be the simplest to construct and to keep stable. It would be helpful to have means for obtaining a number of alternative block diagrams from a known block diagram. Steinberg and Linden[8] have applied flow graph techniques[9-12] to the problem and have shown how alternatives may be derived. In fact, their derivation of the basic two delay element building block makes use of flow graph techniques. The designer of a periodic filter would start with a cascade of building blocks to achieve a particular z-transfer function, and then, if the block diagram looks unfavorable for implementation, would apply flow graph techniques to find alternatives.

One interesting set of alternative forms may be found without using flow graph techniques. This alternative form may be found from the partial fraction expansion of $T(z)$. Since the degree of the numerator cannot exceed that of the denominator, we can always write, except for a multiplying constant,

$$T(z) = 1 + \frac{A_1}{z - z_{p1}} + \frac{A_2}{z - z_{p2}} + \cdots. \quad (42)$$

In (42), the A's are the residues of $T(z)$ at the poles. If z_{p1} and z_{p2} are conjugate complex, A_1 and A_2 are conjugate complex.

Example—real poles:

$$T(z) = 1 + \frac{A_1}{z^2} + \frac{A_2}{z - z_{p2}}. \quad (43)$$

Three delay lines are necessary. The block diagram is shown in Fig. 30.

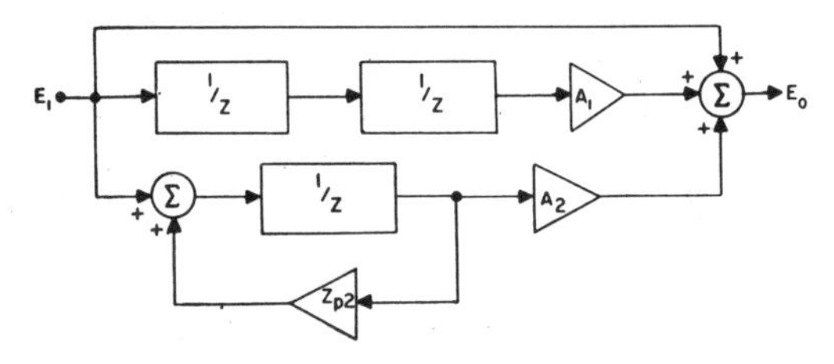

Fig. 30—Realization of $T(z) = 1 + A_1/z^2 + A_2/(z - z_{p2})$.

Example—complex poles:

$$T(z) = 1 + \frac{A_1}{z^2} + \frac{A_2}{z - z_{p2}} + \frac{A_2^*}{z - z_{p2}^*}. \quad (44)$$

We can write $T(z)$ in the form

$$T(z) = 1 + \frac{A_1}{z^2} + \frac{A_2(z - z_{p2}^*) + A_2^*(z - z_{p2})}{z^2 - (z_{p2} + z_{p2}^*)z + z_{p2}z_{p2}^*}. \quad (45)$$

Now, let

A_r = real part of A_2
A_i = imaginary part of A_2
z_r = real part of z_{p2}
z_i = imaginary part of z_{p2}
$|z_{p2}|$ = magnitude of z_{p2}.

Then, after some algebraic manipulation, we get

$$T(z) = 1 + \frac{A_1}{z^2} + \frac{2(A_r z - A_r z_r - A_i z_i)}{z^2 - (2z_r)z + z_{p2}^2}. \quad (46)$$

Then, by using (36) and (37) in conjunction with Fig. 28, we get Fig. 31.

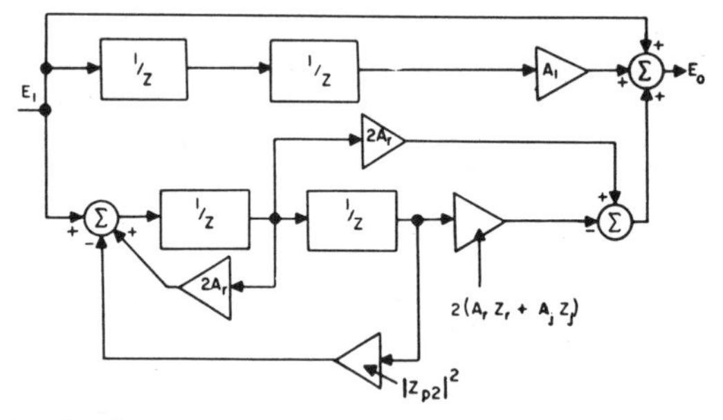

Fig. 31—Realization of $T(z) = 1 + A_1/z^2 + A_2/(z - z_{p2}) + A_2/(z - z_{p2})$.

Another interesting synthesis is the expansion of $T(z)$ in a power series. If we have

$$T_1(z) = A_0 + A_1 z + A_2 z^2 + A_3 z^3 + \cdots \quad (47)$$

we know that this cannot be synthesized. However, if we are interested only in $|T_1(j\omega)|$ we can approximate $|T_1(j\omega)|$ (or $|T_1(j\omega)|$ delayed) with a finite number of terms. Let us consider

$$T_1(z) = A_0 + A_1 z + A_2 z^2 + A_3 z^3. \quad (48)$$

Then, we can form

$$T_2(z) = z^{-3}T_1(z)$$
$$= A_0 z^{-3} + A_1 z^{-2} + A_2 z^{-1} + A_3 \quad (49)$$

and $T_2(z) = T_1(z)$ delayed by a time 3τ. $T_2(z)$ is achieved by the structure shown in Fig. 32. The extension to any finite power series with real coefficients is obvious.

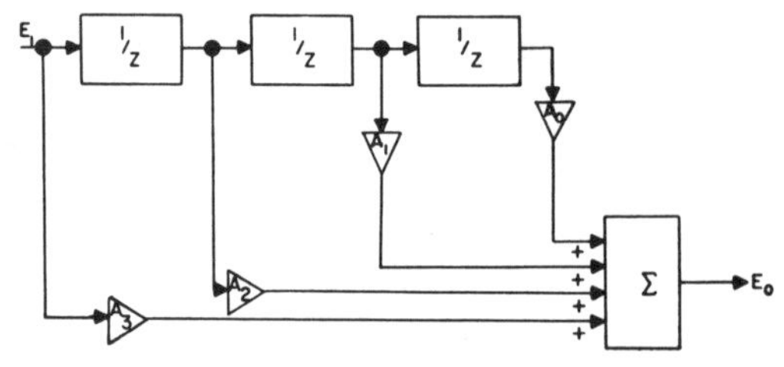

Fig. 32—Realization of $T(z) = (A_0 + A_2 + A_{2}z^2 + A_{3}z^3)/z^3$.

[8] S. J. Mason, "Feedback theory—some properties of signal flow graphs," Proc. IRE, vol. 41, pp. 1144-1156; September, 1953.
[10] J. G. Truxal, "Automatic Feedback Control System Synthesis," McGraw-Hill Book Co., Inc., New York, N.Y., ch. 2; 1955.
[11] T. M. Stout, "A block diagram approach to network analysis," *Trans. AIEE*, vol. 71, part II, pp. 225-260; 1952.
[12] T. M. Stout, "Block diagram solutions for vacuum tube circuits," *Trans. AIEE*, vol. 72, part I, pp. 561-567; 1953.

VIII. Approximating a Periodic Frequency Characteristic with a Power Series in z

If $T(j\omega)$ (or $|T(j\omega)|$) is periodic, it can be represented by a trigonometric series as follows:

$$T(j\omega) = \sum_{n=-\infty}^{\infty} A_n e^{-jn\omega\tau}. \tag{50}$$

This suggests that we may write $T(p)$ as

$$T(p) = \sum_{n=-\infty}^{\infty} A_n e^{-np\tau} \tag{51}$$

or

$$T(z) = \sum_{n=-\infty}^{\infty} A_n z^{-n}. \tag{52}$$

For a finite approximation we can write

$$T_n(z) = \sum_{n=-N}^{N} A_n z^{-n}. \tag{53}$$

Multiplication by any negative power of z simply represents a delay. Then, the series in (53) can be made to have all negative powers of z by multiplying by z^{-N}. Then

$$T_N(z) = \sum_{i=0}^{2N} A_{i-N} z^{-i}. \tag{54}$$

In order that all the A's be real, $T(j\omega)$ should be an even function of ω. Of course, only $|T(j\omega)|$ is even. Therefore, the series method of approximation can be used only to approximate $|T_N(j\omega)|$. With real coefficients, the structure of Fig. 32 can be used.

The coefficients, A_n, in (50) may be Fourier coefficients. These will be obtained if it required that the finite approximation fit the desired function with minimum mean square error. Other coefficients are possible if the criterion of goodness of fit is different from minimum mean square error.

Appendix I

Some Properties of Periodic Filters

The frequency characteristic of a periodic filter has a period $2\pi j/\tau$. This suggests that we can write the transfer function as follows:

$$T(p) = \sum_{n=0}^{\infty} A_n e^{-pn\tau}. \tag{55}$$

The characteristic transient (response to a delta function) is obtained by finding the inverse Laplace transform of $T(p)$. It is found that

$$h(t) = \sum_{n=0}^{\infty} A_n\, \delta(t - n\tau). \tag{56}$$

The type of time function which is of interest to us is the pulsed or sampled function. This type is conveniently expressed as a sequence of delta functions spaced τ in time. The input time function, then, can be expressed by

$$e_1(t) = \sum_{m=0}^{\infty} b_m\, \delta(t - m\tau) \tag{57}$$

and

$$E_1(p) = \sum_{m=0}^{\infty} b_m e^{-pm\tau} \tag{58}$$

or

$$E_1(z) = \sum_{m=0}^{\infty} b_m z^{-m}.$$

The output transform is

$$E_0(p) = E_1(p)T(p). \tag{59}$$

Equivalently,

$$E_0(z) = E_1(z)T(z). \tag{60}$$

Once $E_0(z)$ has been found, the various b_m of (58) must be found to get $e_0(t)$. What is really of interest is the envelope of the sequence of output delta functions. This process is materially aided by a table of sampled time functions and their transforms. A sequence of samples is considered equivalent to a continuous time function which is the envelope of the sequence of functions. For example, the unit step function is considered equivalent to a sequence of delta functions each of unit area. Tables of z transforms have been given in some of the references[2,3] and also in Table I on the next page.

A few examples will illustrate the use of the table.

Example 1

Sweep integrator with feedback factor unity and uniform pulses as input, output taken after delay line: here

$$T(z) = \frac{1}{z-1}$$

and

$$E_1(z) = \frac{z}{z-1}.$$

Then

$$E_0(z) = T(z)E_1(z) = \frac{z}{(z-1)^2}.$$

Entry 4) in Table I shows that the envelope of $e_0(t)$ is the ramp function. This problem and its solution are shown in Fig. 33.

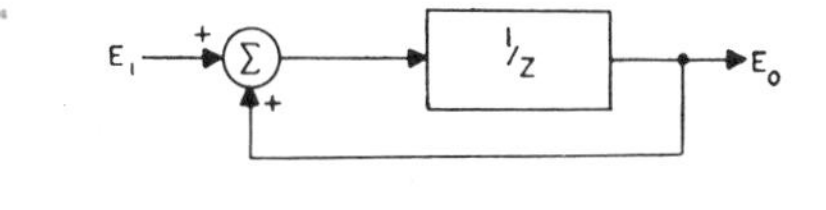

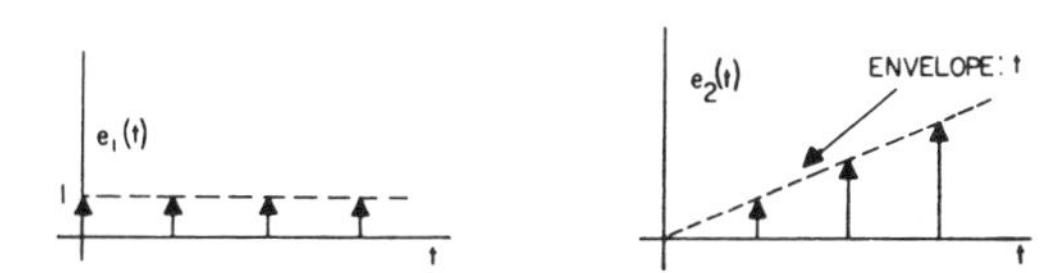

Fig. 33—Illustration of example 1.

Example 2

Sweep integrator with feedback factor k and uniform pulses as the input, output taken before delay line: here

TABLE I

z TRANSFORMS

	Time Function	Envelope	Laplace Transform	z Transform
1)	$\delta(t)$ (delta function)	$\delta(t)$	1	1
2)	$\sum_{n=0}^{\infty} \delta(t - n\tau)$, (uniform pulses)	$u(t)$ (step function)	$\frac{1}{1 - e^{-p\tau}}$	$\frac{z}{z - 1}$
3)	$\sum_{n=0}^{\infty} k^n \delta(t - n\tau)$, $k < 1$ (exponentially decaying pulses)	$e^{-\alpha t}$, $t > 0$; $\alpha = -\frac{1}{\tau} \ln k$	$\frac{1}{1 - ke^{-p\tau}}$	$\frac{z}{z - k}$
4)	$\sum_{n=0}^{\infty} n\delta(t - n\tau)$ (linearly increasing pulses)	t, for $t > 0$ (ramp function)	$\frac{e^{-p\tau}}{(1 - e^{-p\tau})^2}$	$\frac{z}{(z - 1)^2}$
5)	$\sum_{n=0}^{\infty} \frac{1 - k^{n+1}}{1 - k} \delta(t - n\tau)$	$\frac{1}{1 - k}(1 - ke^{-\alpha t})$, $t > 0$	$\frac{1}{(1 - e^{-p\tau})(1 - ke^{-p\tau})}$	$\frac{z^2}{(z - 1)(z - k)}$
6)	$\sum_{n=0}^{\infty} (-1)^n \delta(t - n\tau)$ (alternate positive and negative pulses)	$\cos \frac{\pi t}{\tau}$; $t > 0$	$\frac{1}{1 + e^{-p\tau}}$	$\frac{z}{z + 1}$

$$T(z) = \frac{z}{z - k}$$

and

$$E_1(z) = \frac{z}{z - 1}.$$

Then

$$E_0(z) = \frac{z^2}{(z - 1)(z - k)}.$$

Item 5) of Table I shows that

$$e_0(t) = \sum_{n=0}^{\infty} \frac{1 - k^{n+1}}{1 - k} \delta(t - n\tau)$$

and the envelope is $[1/(1-k)]\ (1-ke^{-\alpha t})$, with $\alpha = -1/\tau \ln k$. The problem and its solution are shown in Fig. 34.

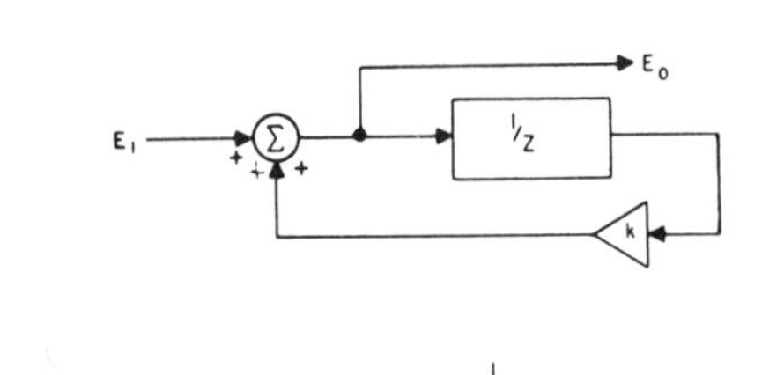

Fig. 34—Illustration of example 2.

Example 3

Canceller with a train of uniform pulses as input: here

$$T(z) = \frac{z - 1}{z}$$

and

$$E_1(z) = \frac{z}{z - 1}.$$

Then

$$E_0(z) = 1.$$

Thus, the output is a single delta function at $t = 0$. This is illustrated in Fig. 35.

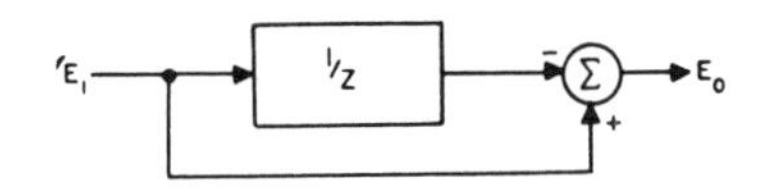

Fig. 35—Illustration of example 3.

Example 4

Canceller with alternating pulses as input:

$$T(z) = \frac{z-1}{z}$$

and

$$E_1(z) = \frac{z}{z+1}.$$

Then

$$E_0(z) = \frac{z-1}{z+1}.$$

Now

$$E_0(z) = \frac{z-1}{z+1} - 1 + 1 = -\frac{2}{z+1} + 1$$

$$= -z^{-1}\left(\frac{2z}{z+1}\right) + 1$$

$$= -z^{-1}[2E_1(z)] + 1.$$

Thus, the solution consists of two terms: the first term is the input reversed in sign, doubled in amplitude, and delayed one period; the second term is a delta function at $t = 0$. This example is illustrated in Fig. 36.

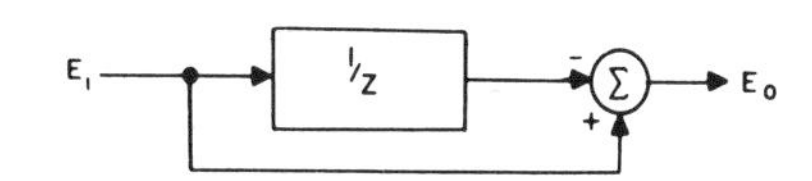

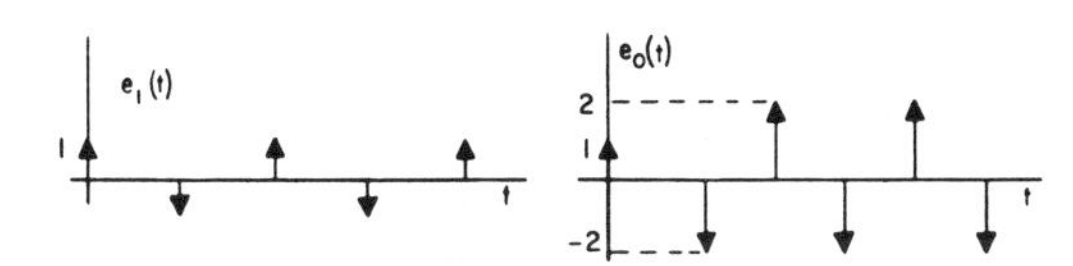

Fig. 36 -Illustration of example 4.

APPENDIX II

TABLE II

Periodic Filters

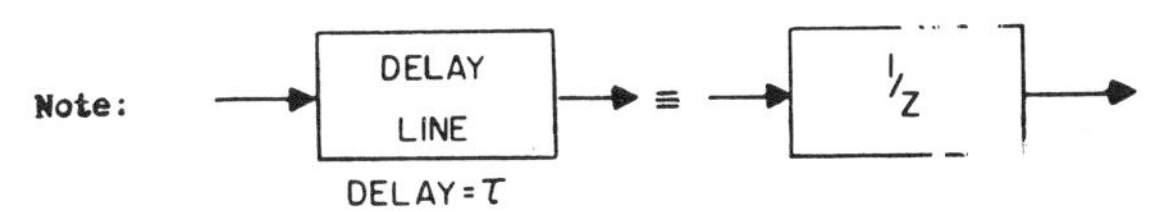

BLOCK DIAGRAM	T(z)	P-Z DIAGRAM
1) Single Canceller E₁, 1/z, Σ, E₀	$\frac{z-1}{z}$	jy, Z PLANE, X, UNIT CIRCLE
2) Double Canceller E₁, 1/z, Σ, 1/z, Σ, E₀	$\left(\frac{z-1}{z}\right)$	2

TABLE II (*Cont'd*)

BLOCK DIAGRAM	T(z)	P-Z DIAGRAM
3) Video Sweep Integrator 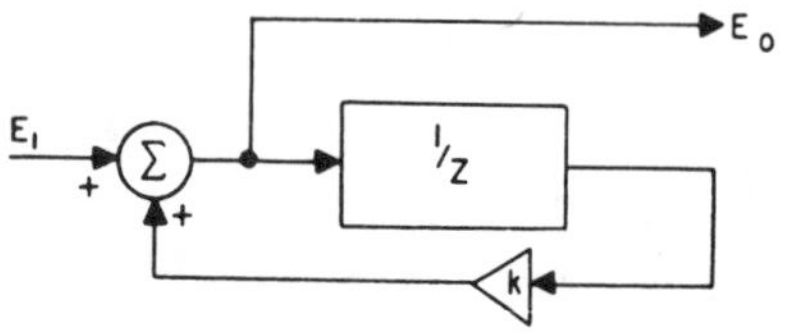 E_I + Σ + 1/z k E_O	$\frac{z}{z - k}$	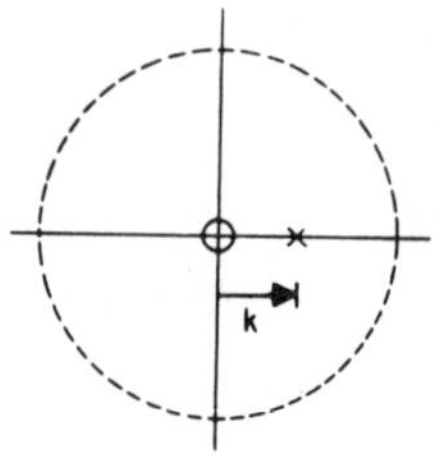 k
4) Double Loop Sweep Integrator 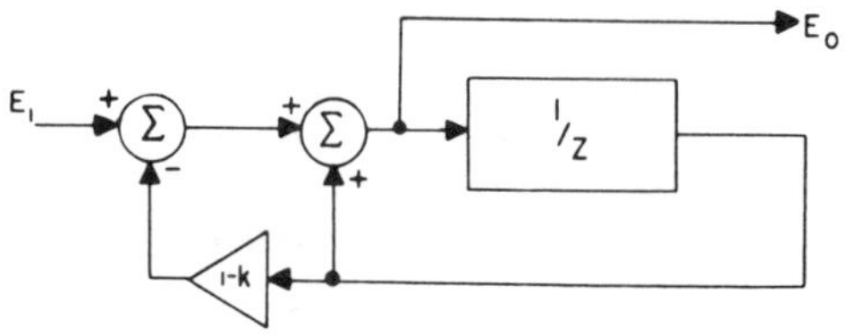 E_I + Σ − + Σ + 1/z 1-k E_O	same as 3	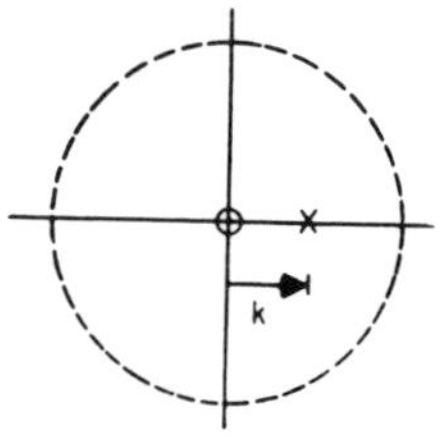 k
5) Response Shaped Single Canceller 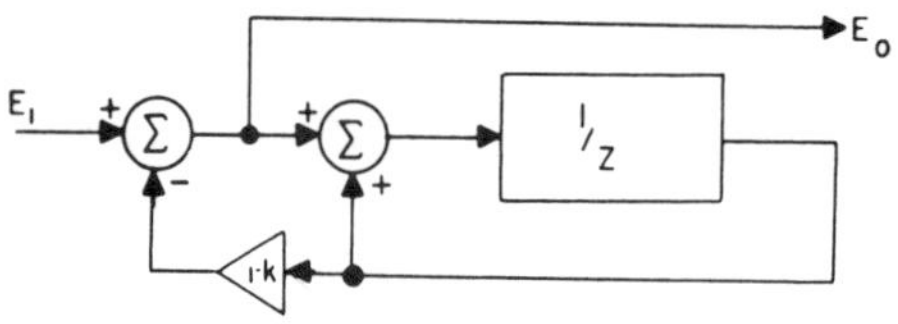 E_I + Σ − + Σ + 1/z 1-k E_O	$\frac{z - 1}{z - k}$	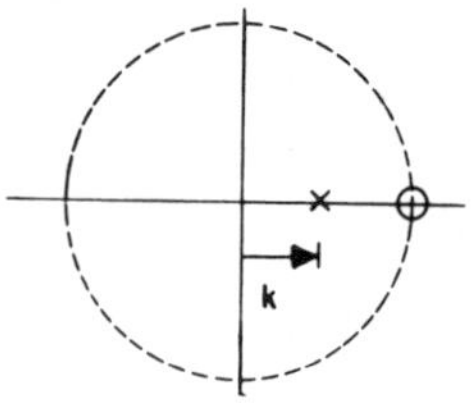 k
6) Response Shaped Double Canceller 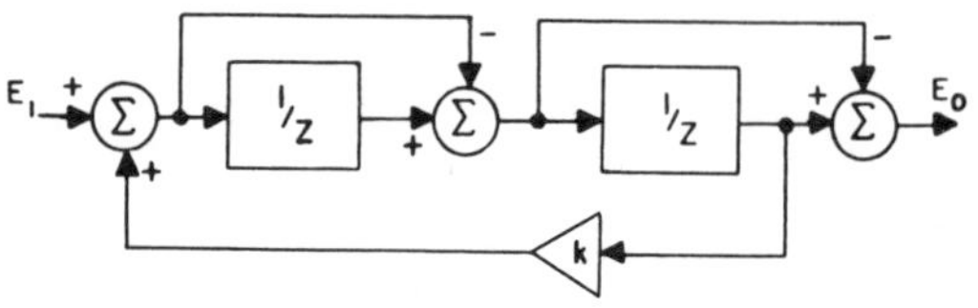 E_I + Σ + 1/z − + Σ 1/z − + Σ k E_O	$\frac{(z-1)^2}{(z-z_1)(z-z_2)}$ $z_1,\ z_2 = \frac{k \pm \sqrt{k^2-4k}}{2}$	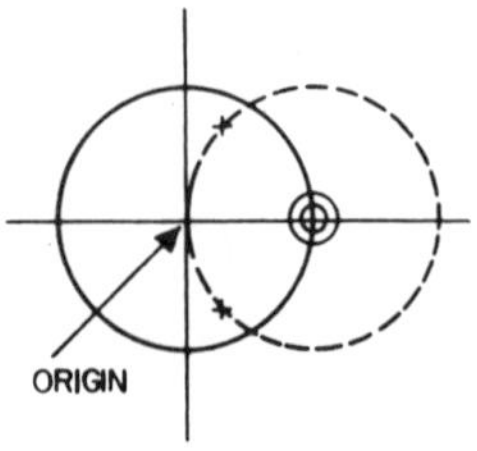 ORIGIN

TABLE II (*Cont'd*)

BLOCK DIAGRAM	T(z)	P-Z DIAGRAM
7) Frequency (or Velocity) Rejection 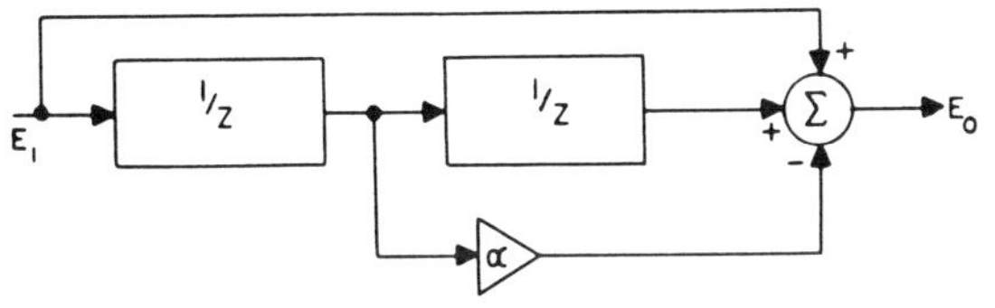	$\frac{(z-z_1)(z-z_2)}{z^2}$ $z_1, z_2 = \frac{\alpha \pm \sqrt{\alpha^2-4}}{2}$	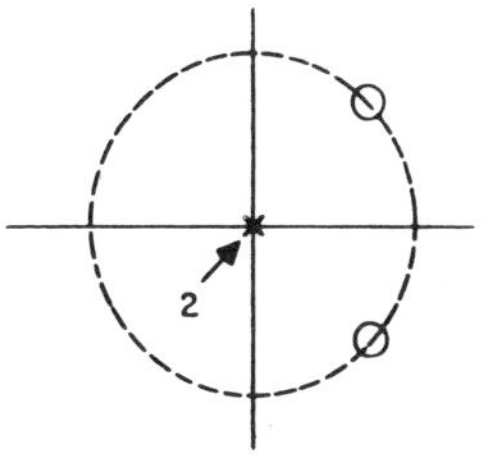 2
8) General Double Delay Filter 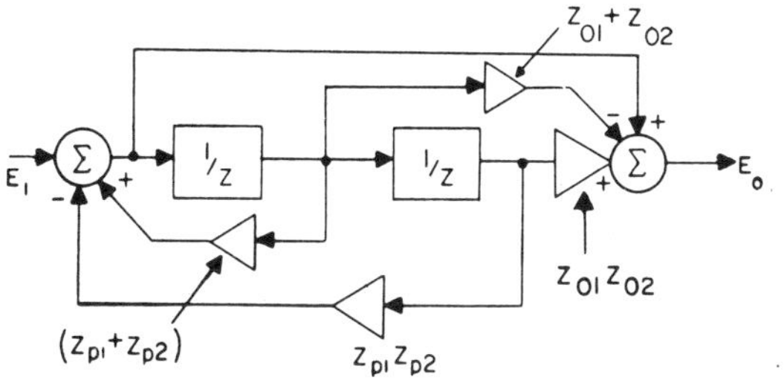	$\frac{(z-z_{01})(z-z_{02})}{(z-z_{p1})(z-z_{p2})}$	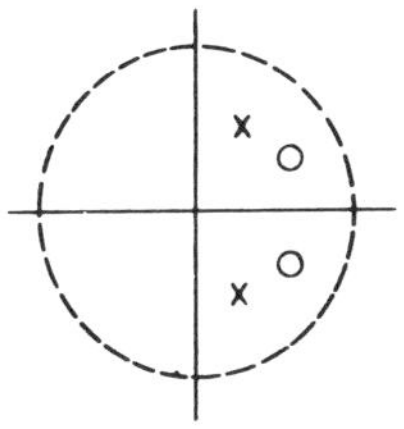 Poles and zeros real or conjugate complex pairs. Poles inside circle.

Reprinted from the IRE TRANSACTIONS OF THE PROFESSIONAL GROUP ON *CIRCUIT THEORY*
Volume CT-4, Number 2, June, 1957
PRINTED IN THE U.S.A.

1.5 SIGNAL PROCESSING TECHNIQUES FOR SURVEILLANCE RADAR SETS

C.A. Fowler, A.P. Uzzo and A.E. Ruvin

IRE Transactions on Military Electronics, Vol. MIL-5, No. 2, April 1961, pp. 103-108.

***Summary*—One of the major recent advances in radar technology has been in the processing of the received signals. Several techniques have been developed to enhance the desired signals (aircraft) relative to ground clutter, sea clutter, rain, interference from other radars, and active countermeasures.**

The following techniques are discussed: MTI, Sweep Integration, and Blanking and Switching. A signal processing system utilizing these techniques is described.

INTRODUCTION

OVER the past decade, one of the major advances in radar technology has been in the processing of radar signals. Technique developments have been principally associated with the relief of effects due to ECM and ground clutter and improvement in target detection capability. Additional effort has also gone into the development of automatic and semi-automatic devices that select the best signal processing techniques and output signals for use in particular areas and in specific environments with emphasis on the multiple processing of the radar video.

Signal processing techniques can be divided into two broad classes—techniques that are essentially linear and attempt to separate the desired signals from interference or noise on the basis of some selective property of the system, and techniques that are based on blanking of the radar output in the presence of undesired signals. Both approaches have areas of utility, but it is preferable to use enhancement of the desired signals prior to the use of blanking.

Radar returns of interest can be characterized as having a known frequency, spectrum, pulse length and repetition period.

MTI

MTI (Moving Target Indicator) is a radar technique used to separate moving targets from fixed targets. The signal received from a moving target is essentially a line spectrum shifted from that of fixed targets by the Doppler effect. In an MTI, the composite signal of fixed clutter and moving targets is processed to filter out the fixed target spectral components and to pass the others.

The MTI, developed during World War II for surveillance radar,[1] was a coherent, uniform PRF system utilizing a simple single canceler as a comb filter (Fig. 1). The velocity (frequency) response of this comb filter is shown in Fig. 2. This canceler has nulls (blind speeds) at multiples of the radar PRF (fr), and thus completely cancels an ideal fixed target with a true line spectrum. However, the scanning of the radar antenna causes the spectrum lines to broaden as shown in Fig. 2. The spectral width is proportional to the tip speed of the radar antenna.

A more ideal response shape would be less sensitive near the nulls and have a more uniform response away from them. A response of this type can be obtained by the use of a velocity-response-shaping-canceler,[2] which consists of two single cancelers in cascade with feedback, as shown in Fig. 3. The shape of the response is controlled by the feedback factors. Fig. 4 shows a velocity

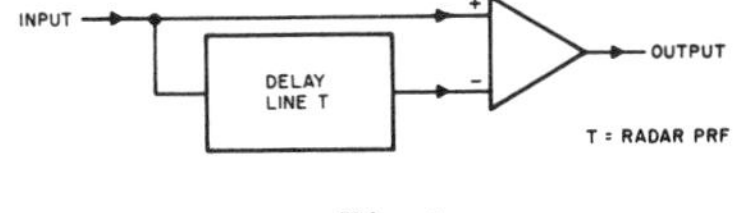

Fig. 1.

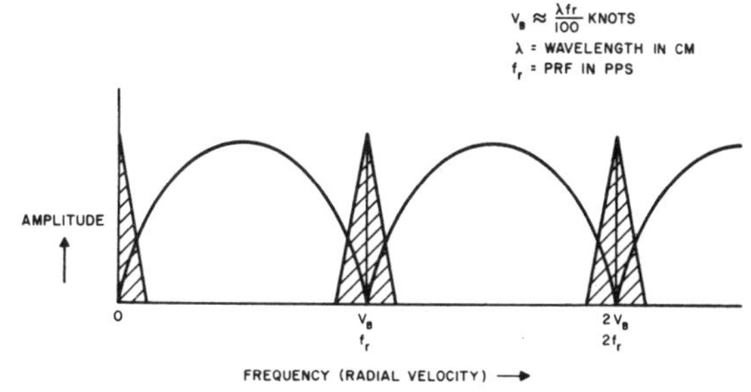

Fig. 2.

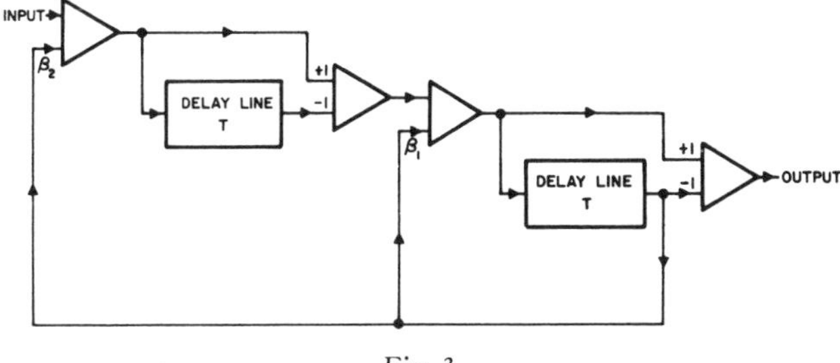

Fig. 3.

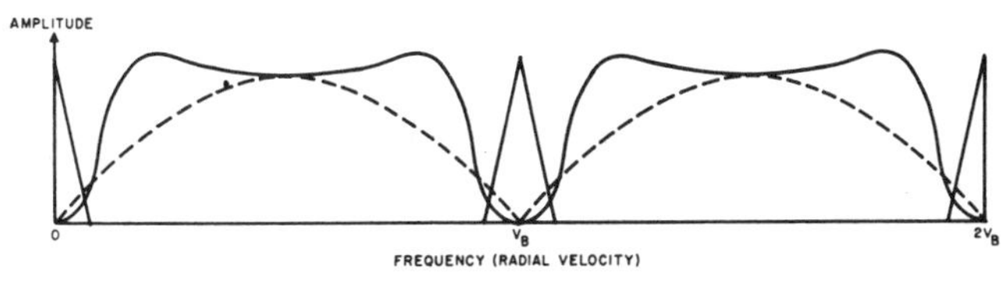

Fig. 4.

* Received by the PGMIL, January 18, 1961.

† Airborne Instruments Lab., Div. Cutler-Hammer, Inc., Melville, N. Y.

[1] L. N. Ridenour, "Radar Systems Engineering," M.I.T. Rad. Lab. Ser., McGraw-Hill Book Co., Inc., New York, N. Y., vol. 1, ch. 16; 1947.

[2] W. D. White and A. E. Ruvin, "Recent advances in the synthesis of comb filters," 1957 IRE NATIONAL CONVENTION RECORD, pt. 2, pp. 186–200.

response typically available with this approach. (The response of the single canceler is also shown for reference.)

With this technique the scanning limitation can be readily removed for nearly all surveillance radars. The exact response selected depends upon the scanning clutter and the performance desired. Fig. 5 shows four typical responses each adjusted to give 30-db cancellation of fixed targets for the designated number of PBW (pulses per beamwidth). In many radars the parameters are such that only a modest amount of shaping is needed to reduce the scanning fluctuations. It may still be desirable, however, to have the more severe shapes available to reduce clutter, having additional fluctuation such as rain return or sea clutter. For example, the rain return on a 400 PRF, *L*-band radar is reduced 30 db with the 5.8 PBW shape compared with 15 db obtained from a single canceler.

This example assumed that the fluctuation spectrum of the rain is centered about the blind speeds. Therefore, to be valid, the group velocity of the rain due to wind must be removed. This can be done by the use of noncoherent MTI. This type of MTI uses the clutter (or rain) as the reference signal for moving target detection. Therefore, in the case of a target in rain clutter, it is the radial component of the relative motion of the target to the rain that produces the Doppler beat. (Noncoherent MTI does not discriminate between group velocity of rain and sea clutter and platform motion, and therefore can be readily used on moving platforms such as aircraft and ships to provide MTI capability against clutter.)

The combination of noncoherent MTI with velocity response shaping is therefore a powerful tool in dealing with targets such as rain and sea clutter.

Since noncoherent MTI utilizes the beat between target and clutter, it is "blind" to targets not over clutter and loses sensitivity on targets over weak clutter. It is therefore desirable to use it only in those regions where it is necessary (*e.g.*, rain clutter areas) and to use either coherent MTI or normal video everywhere else in the coverage area. Means for accomplishing this are discussed later.

When severe response shapes are used to suppress extraneous targets, the blind speed zones are widened. This can be alleviated by using a nonuniform interpulse interval—a staggered PRF. Staggered PRF is accomplished by delaying every other transmitted pulse by an amount, ΔT. The ratio of the two unequal interpulse intervals is called the stagger ratio. Upon reception, the signals are destaggered using the same ΔT delay element that generated the original stagger. Fig. 6 shows the timing diagram of a stagger/destagger system.

Fig. 7 shows the over-all velocity response of a 5/7 stagger combined with a velocity-shaping canceler. The gross effect of the stagger is to multiply the first blind speed by the factor $T/\Delta T$. However, the larger this factor is, the deeper the holes in the response curve.

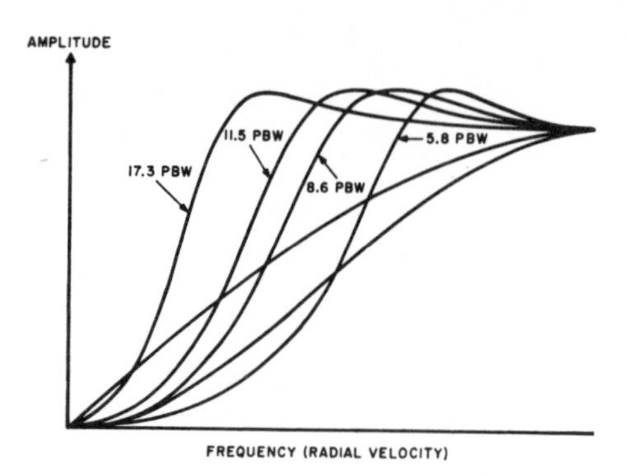

Fig. 5.

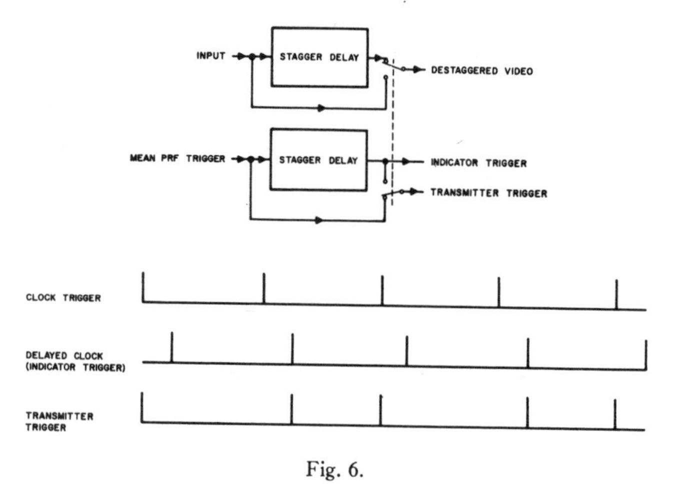

Fig. 6.

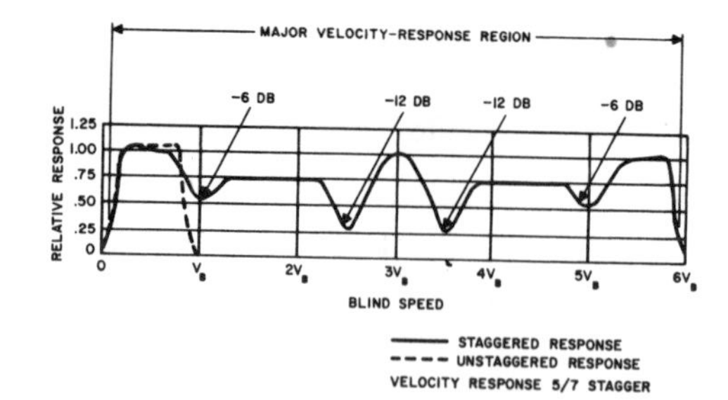

Fig. 7.

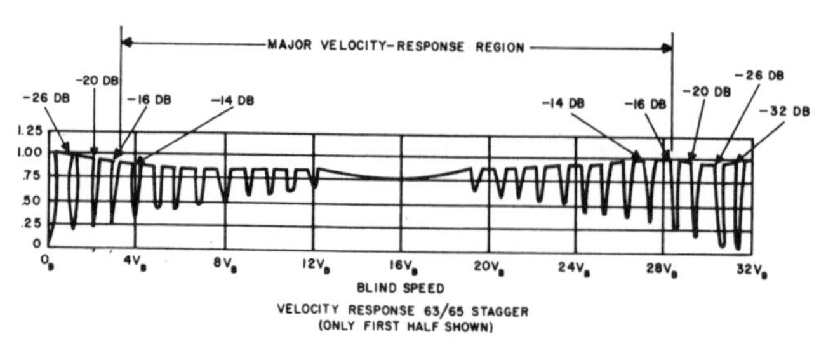

Fig. 8.

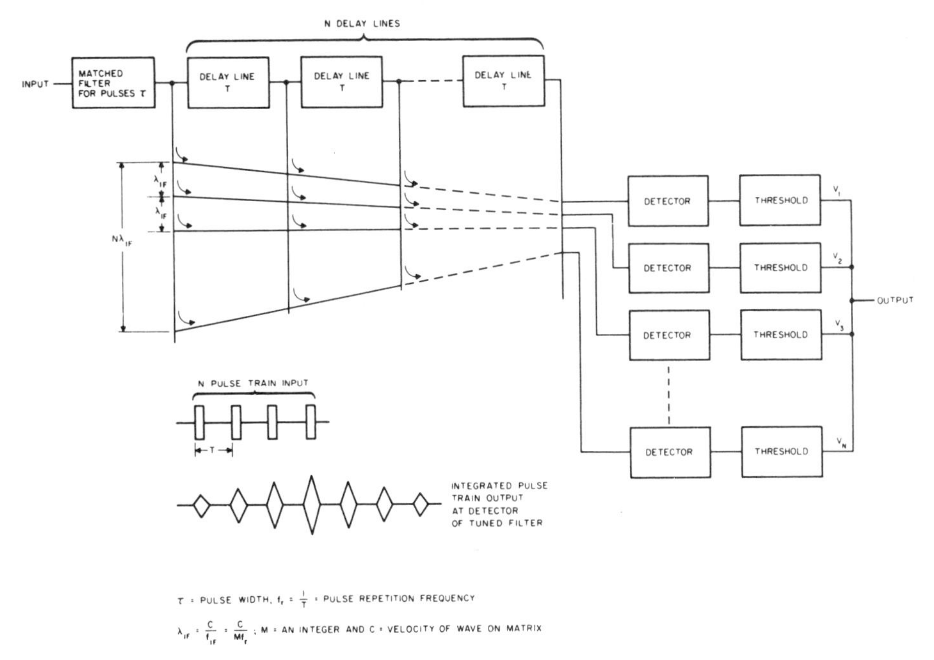

Fig. 9.

This is evident in Fig. 8, which shows the response obtained with a 63/65 stagger ratio. The higher stagger ratios are useful in optimizing the response in a particular velocity region.

INTEGRATORS

The threshold detection for a group of weak pulses in a background of random noise can be greatly enhanced by integration.[2] If the integration of the pulses takes place prior to envelope detection, the process is called coherent integration. When the phasing of N successive pulses has been corrected for target motion, the enhancement of the signal power will be N times the noise power. (The signal voltages add linearly while the noise voltages add in an rms manner.) If the integration takes place after detection, the process is called noncoherent integration.

A hypothetical example of a coherent integration bank of filters for the N pulse train is shown in Fig. 9. It is assumed that the individual pulses are optimally filtered by means of a matched IF filter. The concept of N equal pulses from a point target implies a rectangular beam shape. A more realistic example is a uniformly illuminated rectangular aperture resulting in a sin X/X one-way beam pattern or a $\sin^2 X/X^2$ round trip scanning modulation. The signal voltage spectral shape of the line is triangular, and it is not physically realizable for an integrator to have a finite pass band required for a matched filter response. In practice, an optimum 3-db bandwidth for an integrator is approximately fr/2(PBW).

Sweep integrators have been designed around a single period delay line with a positive feedback factor K slightly less than unity enhancing the harmonics of the

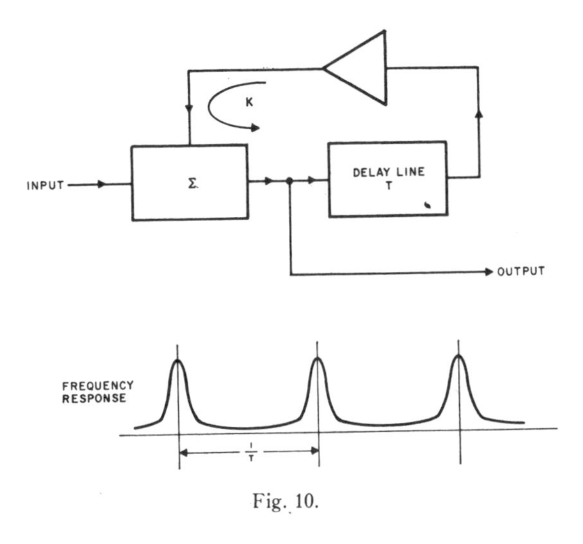

Fig. 10.

pulse repetition frequency. Fig. 10 shows the basic block diagram and frequency response of such an integrator. The 3-db bandwidth of such an integrator is approximately $(1-K)\mathrm{fr}/\pi$. The response of such an integrator for an infinite train of optimum Doppler pulses is $1/1-K$, and the noise response is $1/\sqrt{1-K^2}$. The response of the integrator to a finite train of equal level pulses is $1-K^N/1-K$. The signal enhancement of an N pulse train is

$$(1 - K^N)\left(\frac{1 + K}{1 - K}\right)^{1/2}.$$

The feedback sweep integrator exponentially weights the pulse in the train with maximum weighting of the latest pulse. A bank of delay line sweep integrators is required for coherent integration or else the single delay line integrator may be multiplexed either in time or in frequency if sufficient bandwidth is available.

If Doppler resolution is not required, noncoherent (post-detection) integration may be used. The losses compared with coherent integration are not excessive for scanning radars (less than 100 pulses per beamwidth).[3] Post-detection integration losses of 1.2 db, 3 db, 4.1 db, 8 db and 12.3 db exist for 10, 50, 100, 1000, and 10,000 pulses respectively for a square law detection probability of 0.99 and a false alarm probability of 10^{-10}. A signal-to-noise ratio at the detector input of unity (0 db) results in a signal-to-noise ratio after detection of -4 db. The post-detection integration of 100 pulses results in a 20-db enhancement giving the $+16$-db signal-to-noise ratio required for 0.99 detection probability. Detector suppression results in a square law relation between output and input signal-to-noise ratios for weak signal-to-noise levels.

A mixture of coherent and noncoherent integration can be employed to reduce the total number of integrators required. For example a bank of 10 coherent integrators followed by a bank of 10 noncoherent integrators (total of 20) is nearly as effective from a detection standpoint as 100 coherent integrators. In addition, the target acceleration in some cases results in a Doppler which precludes the use of the very narrow filters required for fully coherent integration. The coherent integration must be adequate to enhance the signal-to-noise ratio to a level of at least 0 db at the detector.

Sweep integrators are useful for pulse and low duty interference reduction. The signal dynamic range into the integrator is restricted by means of a limiter so that the response to random pulses in the integration time can be restricted to a level below the detection threshold.

The use of an integrator ahead of a PPI does not result in an increase in detection range of a normal search radar because the PPI phosphor is itself a good integrator for this case. An integrator ahead of a PPI, however, results in an improved contrast that is more pleasing to the operator. A black-faced indicator is required in order to avoid any integrated noise painting. Integrated noise, if presented, has the blip dimensions similar to targets. The use of a black-faced tube also allows the easy recognition of target trails.

The stability problems associated with high feedback integrators are serious. An interesting technique (devised by the Philco Corporation) for achieving stability makes use of a double loop. One loop is oscillating (unity feedback) and the other has a small negative feedback factor (β). The result is a high positive feedback factor, $K=1-\beta$. The system shown in Fig. 11 uses frequency modulation of a deviable oscillator. A stable buildup of $1/1-K=1/\beta$ appears at the integrated video output for an infinite unit pulse train.

[3] J. I. Marcum and P. Swerling, "Study of target detection by pulsed radar," IRE Trans. on Information Theory, vol. IT-6, pp. 59–144; April, 1960.

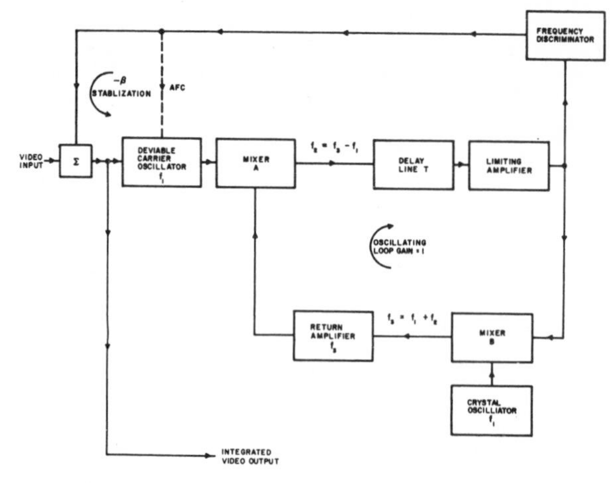

Fig. 11.

Blanking and Switching Techniques

There is a class of video processing devices that can be characterized by the common property of desensitization of the radar output or selection of an alternate signal source whenever activated by an anomaly in the radar video signal: one such device, the pulsewidth discriminator, serves a variety of useful purposes as part of a video processing system. Fig. 12 shows a block diagram of a typical pulsewidth discriminator. The video input is delayed by an amount equal to the desired pulsewidth criterion; that is, by some definition (usually about two pulsewidths) that is chosen as the basis for deciding between desired radar returns and interference or ground clutter, etc. The delayed and undelayed signals are then compared in a coincidence gate that produces an output whenever the input signal exceeds the desired criterion. The output signal from the coincidence gate is equal to the input signal width less a width equal to the criteria and this is added on by the pulse stretcher.

The output signal from the pulsewidth discriminator is usually applied to a video switch. The pulsewidth discriminator can be used as a simple clutter suppressor, blanking out the areas where clutter is sensed, as defined as being targets of excessive pulsewidth. If MTI is substituted in the areas where clutter is sensed, a clutter-gated MTI system is obtained wherein MTI is presented only in the regions of clutter and normal video outside these areas. Since MTI is more susceptible to certain forms of ECM and interference, the pulsewidth discriminator can be used to advantage as a clutter gate to confine MTI operation to the minimum required area. Furthermore, when noncoherent MTI (which is blind in areas of no clutter) is used, clutter gating is a necessity.

A "capture gate" may be used to suppress isolated fixed point targets, which are not gated by simple pulsewidth discriminator. If a PPI picture is examined, it will be found that, almost invariably, heavy clutter areas are followed by several isolated fixed point targets. The capture gate (shown in Fig. 12 within the

1.6

MTI PERFORMANCE ANALYSIS
D.C. Schleher
Internal Report for Airborne Instruments Laboratory, Cutler-Hammer, Inc. 1966, pp. 1-64.

MTI PERFORMANCE ANALYSIS

D. Curtis Schleher

AIRBORNE INSTRUMENTS LABORATORY
A DIVISION OF CUTLER-HAMMER, INC.
Deer Park, New York 11729

TABLE OF CONTENTS

LIST OF ILLUSTRATIONS

LIST OF TABLES

ABSTRACT

A unified procedure is developed that allows the overall system performance of both coherent and noncoherent MTI systems to be determined. Performance curves for delay line and range-gated doppler filters are given.

Three examples are given to illustrate the performance analysis: (1) a noncoherent, range-gated filter, ground combat-surveillance radar; (2) a noncoherent, variable cutoff range-gated filter, airborne surveillance radar; and (3) a coherent, shaped delay-line filter, ground airport surveillance radar.

MTI PERFORMANCE ANALYSIS

I. INTRODUCTION

This paper analyzes the performance of delay-line and range-gated-filter coherent and noncoherent MTI systems. Both ground and airborne situations are considered.

The analysis stresses overall performance of the MTI system and allows the capability of the complete MTI system to be assessed.

The theory is developed by representing the clutter as a stochastic process with an associated spectral density. A simple point target is assumed. The predetection spectral distribution of the target and the clutter is determined by considering the spreading effects induced by (1) antenna scanning, (2) platform motion, (3) internal motion among the scatterers forming the clutter, (4) transmitter instability, and (5) system instabilities.

The coherent detection process translates the predetection spectral distribution to a video frequency. The noncoherent detection process results in an interaction between the clutter with itself and with the target spectrum. This effect widens both the clutter and target spectrums.

The MTI filter functions to extract the target from the clutter and is usually placed after the detector. (Several coherent systems have been built which filter before detection.) Single and double delay-line MTI filters have fixed shapes determined by the radar's pulse repetition frequency. These simple MTI filters have been extensively used and provide good performance under favorable target and clutter situations.

This situation is depicted in Fig. 1 for an airborne MTI (AMTI) with its antenna pointed along the aircraft's ground track. Fig. 2 shows the same system with its antenna pointed away from the aircraft's ground track and illustrates an unfavorable clutter situation for the

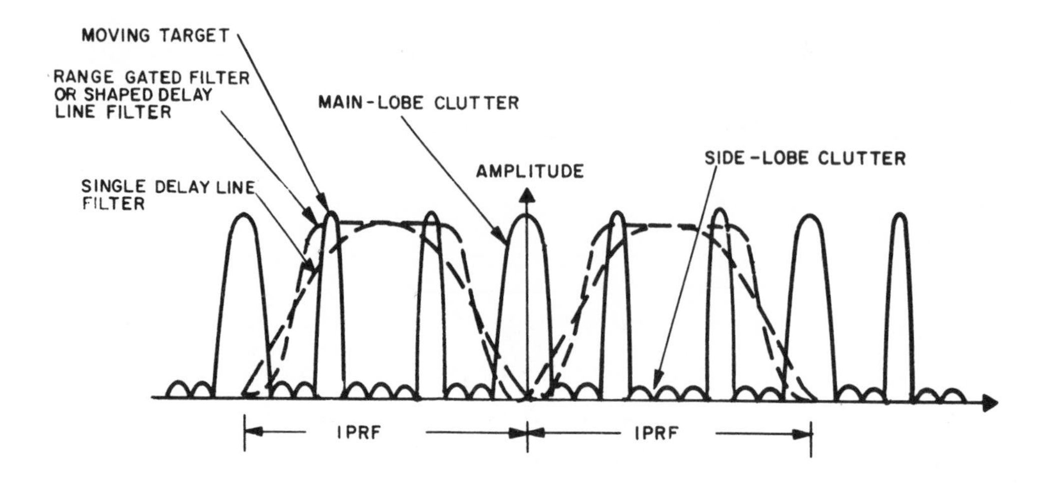

Figure 1. MTI Filter and Clutter Spectrum (Antenna Pointing Along Ground Track)

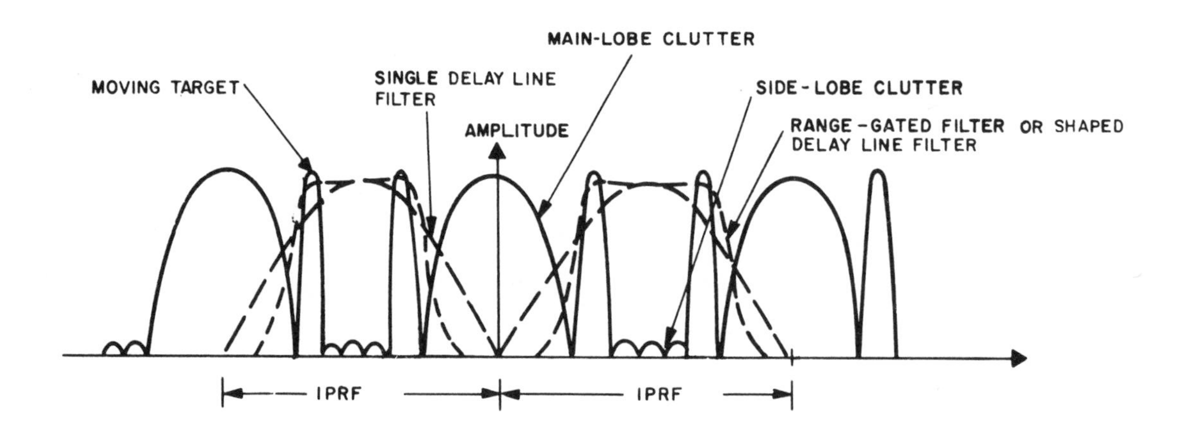

Figure 2. MTI Filter and Clutter Spectrum (Antenna Pointing Away from Ground Track)

single delay-line filter. Shaped delay-line and range-gated filters allow adjustment of the pass and stop bands as illustrated in Figs. 1 and 2, and hence presents the capability of providing good performance over a variety of clutter and target situations.

MTI systems can be analyzed in either the time or frequency domains. References 1 through 4 illustrate both types of analyses and also summarize most of the available unclassified literature on MTI systems. Most of the available literature emphasize the analysis of single and double delay-line filters and does not include all the effects necessary to compute system performance. (Reference 3 contains some material and references on shaped delay-line filters.) Furthermore the theory has evolved over many years and represents many disjointed viewpoints, making it difficult to apply to the analysis of modern MTI systems.

A fundamental viewpoint can be obtained by considering the interrelationships between inputs and outputs of the linear filter given in Fig. 3. The mean-squared output of the filter when excited with a stochastic process, $u_{(t)}$, is given by:

$$\overline{v_{(t)}^2} = \lim_{T \to \infty} \frac{1}{2T} \int_{-T}^{T} \left| v_{(t)}^2 \right| dt = \frac{1}{2\pi} \int_{-\infty}^{\infty} S_{v(\omega)} \, d\omega \quad (1)$$

where $S_{v(\omega)}$ is the spectral density of the output.

The impulse responses and transfer functions of several MTI filters are given in Fig. 4. The output of the single delay-line filter is found by inspection or by applying the convolution relationship to obtain:

$$v_{(t)} = u_{(t)} - u_{(t-T)} \quad (2)$$

where T is the radar interpulse period. For random inputs equation 2 is substituted in equation 1 and ensemble averaged, resulting in

$$\overline{v^2_{(t)}} = 2\left[R_{u(0)} - R_{u(T)}\right] \quad (3)$$

where $R_{u(T)}$ is the auto correlation function (ACF) of the input. For nonrandom inputs, the values of the input displaced in time by T are differenced, squared, and time averaged. Therefore, to analyze a single delay-line filter, it is sufficient to know the value of the auto correlation function at times 0 and T, or the difference between the input signal displaced in time by T. For a double delay-line filter the values displaced by T and by 2 T must be known. For a shaped delay-line filter or a range-gated filter, values displaced by n T (n = 0, 1, 2, 3, - - -) must be known. Hence the complexity of analysis in the time domain varies directly with the complexity of the MTI filter.

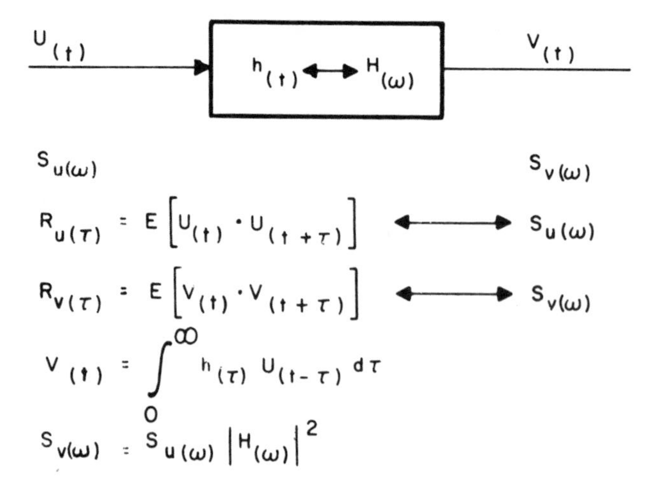

Figure 3. Linear Filter

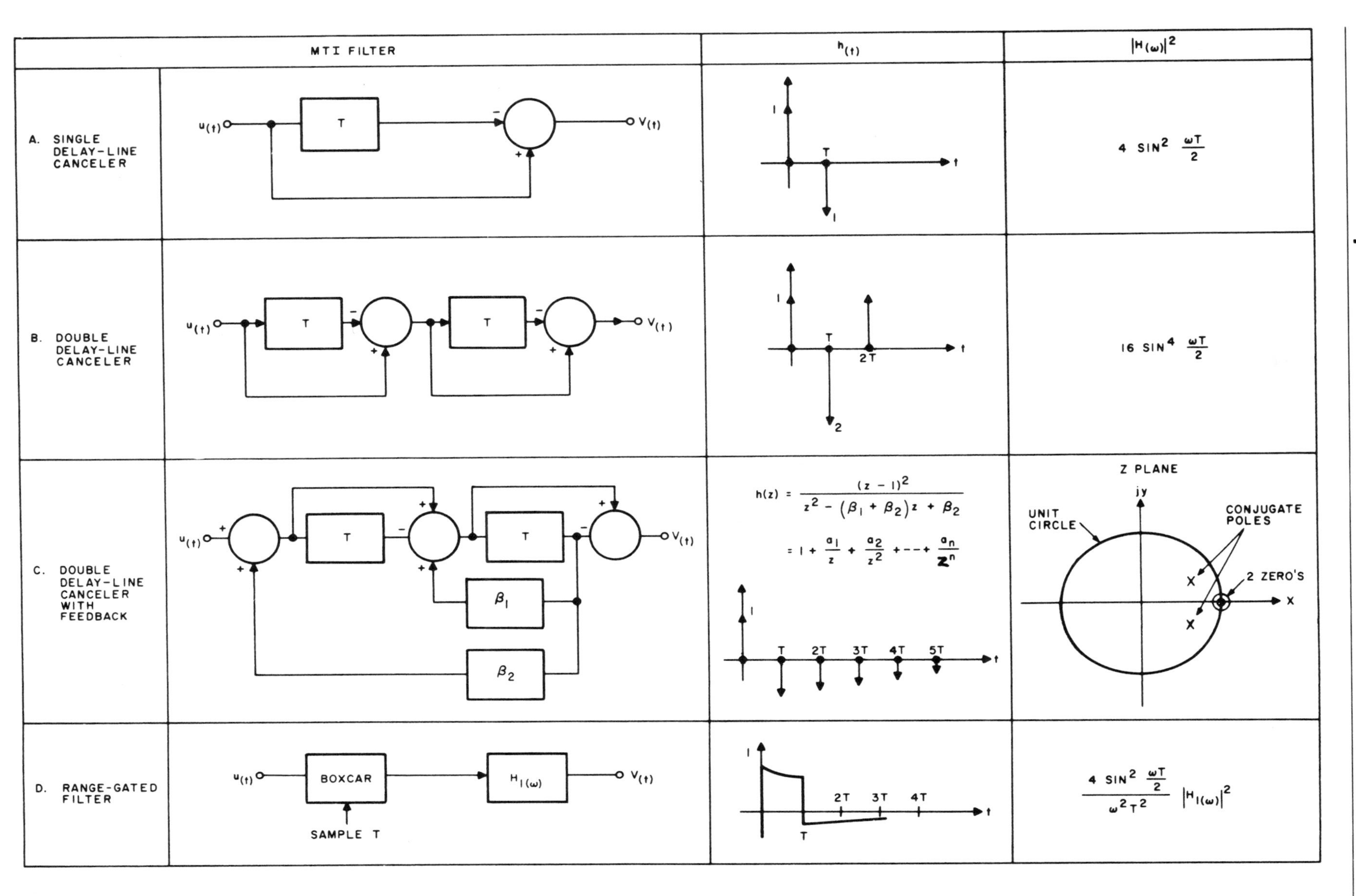

Figure 4. MTI Filters

Alternately the mean-squared output is given by:

$$\overline{v^2_{(t)}} = \frac{1}{2\pi} \int_{-\infty}^{\infty} S_{u(\omega)} \left| H_{(\omega)} \right|^2 d\omega \tag{4}$$

where $S_{u(\omega)}$ is the spectral density of the input and $H_{(\omega)}$ is the voltage transfer function of the filter. With this approach, all the necessary signal and filter information is contained in a single function and the analysis is concentrated on finding $S_{u(\omega)}$ and $H_{(\omega)}$ in a suitable form to evaluate equation 4. The procedure employed to find $S_{u(\omega)}$ is (1) to evaluate the spectrum of individual effects, (2) to match a gaussian curve to the spectrum, and (3) to convolve the gaussian spectrums of the individual effects (add variances) to obtain the overall spectrum. To evaluate the performance of single and double delay-line filters, the actual transfer function can be used. For shaped delay-line filters and range-gated filters, an initial evaluation can be made by an asymptotic approximation to the desired filter. More complete analysis can be made by using the exact filter characteristic.

II. COHERENT AND NONCOHERENT MTI SYSTEMS

A. Clutter-Signal Representation

The clutter is assumed to consist of many elemental stationary scatterers randomly oriented with respect to each other. When these scatterers are illuminated by the radar, the return consists of many randomly phased independent components whose sum forms a clutter vector as shown in Fig. 5. Resolving the clutter vector into orthogonal components and applying the central limit theorem results in orthogonal vectors with gaussian amplitude distributions. The clutter vector has a rayleigh amplitude distribution and a uniformly distributed phase distribution. (Reference 5 presents detailed discussion of this topic.) Accordingly, the clutter return has all the characteristics of band limited gaussian noise and may be represented as a stochastic process.

$$c_{(t)} = x_{(t)} \cos \omega_c t - y_{(t)} \sin \omega_c t \tag{5}$$

where $x_{(t)}$ and $y_{(t)}$ are low pass $\left[S_{x(\omega)} = 0, \ \omega \geq \Omega\right]$, independent $\left\{E\left[x_{(t)} \, y_{(t)}\right] = 0\right\}$, zero mean $\left\{E\left[x_{(t)}\right] = E\left[y_{(t)}\right] = 0\right\}$ processes with identical spectral densities. (Reference 6 presents an excellent treatment of this representation.) It is interesting to note that this representation is valid with as little as six contributing phasors.

The target is assumed to be a point reflector moving with respect to the elemental clutter scatterers. The combined clutter-target signal return is given by:

$$v_{(t)} = \sqrt{2}\, S_{(t)} \cos\left[\left(\omega_c + \omega_d\right) t\right] + x_{(t)} \cos \omega_c t - y_{(t)} \sin \omega_c t \tag{6}$$

where ω_d is the doppler frequency of the target with respect to the clutter; the clutter power $P_c = E\left[x^2_{(t)}\right] = E\left[y^2_{(t)}\right]$ and the signal power $P_s = E\left[s^2_{(t)}\right]$. In this representation provision is made for radar motion by taking a doppler shifted carrier frequency, ω_c, rather than the radar transmitter frequency, ω_t. (When there is no relative motion between clutter and radar $\omega_c = \omega_t$.)

Equation 6 can be rewritten as:

$$v_{(t)} = r_{(t)} \cos\left[\omega_c t + \phi_{(t)}\right] \quad (7)$$

where

$$r_{(t)} = \left\{\left[x_{(t)} + \sqrt{2}\, S_{(t)} \cos \omega_d t\right]^2 + \left[y_{(t)} + \sqrt{2}\, S_{(t)} \sin \omega_d t\right]^2\right\}^{1/2} \quad (8)$$

is the envelope of the process. This equation is shown in vector form in Fig. 5. Examination of this figure shows that the moving target information is contained in both the amplitude and phase characteristics of the IF signal. Coherent MTI systems extract the phase information, while noncoherent MTI systems use the amplitude information.

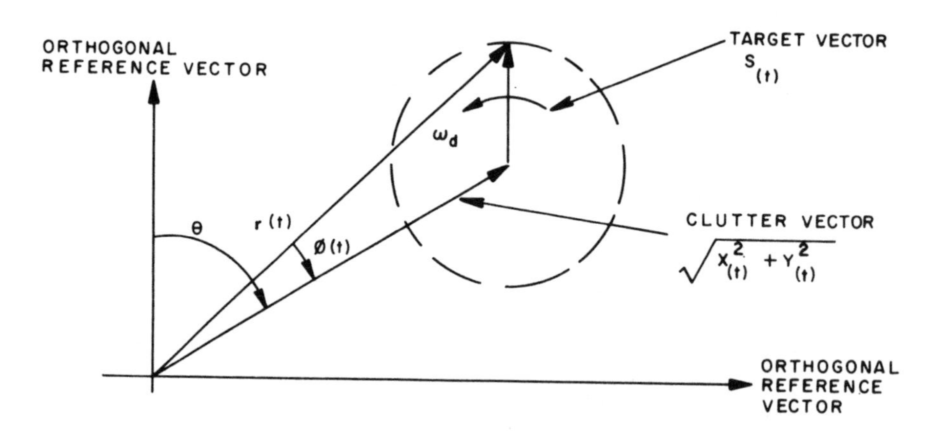

Figure 5. Vector Diagram of IF Envelope Signals

B. Coherent MTI Systems

In a coherent MTI system the target return is compared with a motion-corrected transmitter reference signal. The motion correction shifts the frequency of the reference signal to compensate for the average doppler frequency of the ground clutter at the same range as the target. (For stationary MTI systems no motion correction is necessary.) The comparison is usually accomplished in a phase detector whose output is applied to a doppler filter that rejects fixed targets and passes moving targets.

A block diagram of this type of receiver is shown in Fig. 6. The phase detector performs a synchronous detection and its output can be shown to be:

$$p_{(t)} = \sqrt{2}\, S_{(t)} \cos \omega_d t + x_{(t)} \tag{9}$$

The ACF of this process is given by:

$$R_{p(\tau)} = R_{S(\tau)} \cos \omega_d \tau + R_{x(\tau)} \tag{10}$$

The spectral density of the phase detector output is found by taking the Fourier transform of the ACF:

$$S_{P(\omega)} = \underbrace{S_{x(\omega)}}_{\text{Clutter Spectrum}} + \underbrace{\frac{1}{2}\left[S_{S(\omega + \omega_d)} + S_{S(\omega - \omega_d)}\right]}_{\text{Target Spectrum}} \tag{11}$$

The output of the coherent MTI system is found by passing equation 11 through a doppler filter whose voltage transfer function is $H_{(\omega)}$.

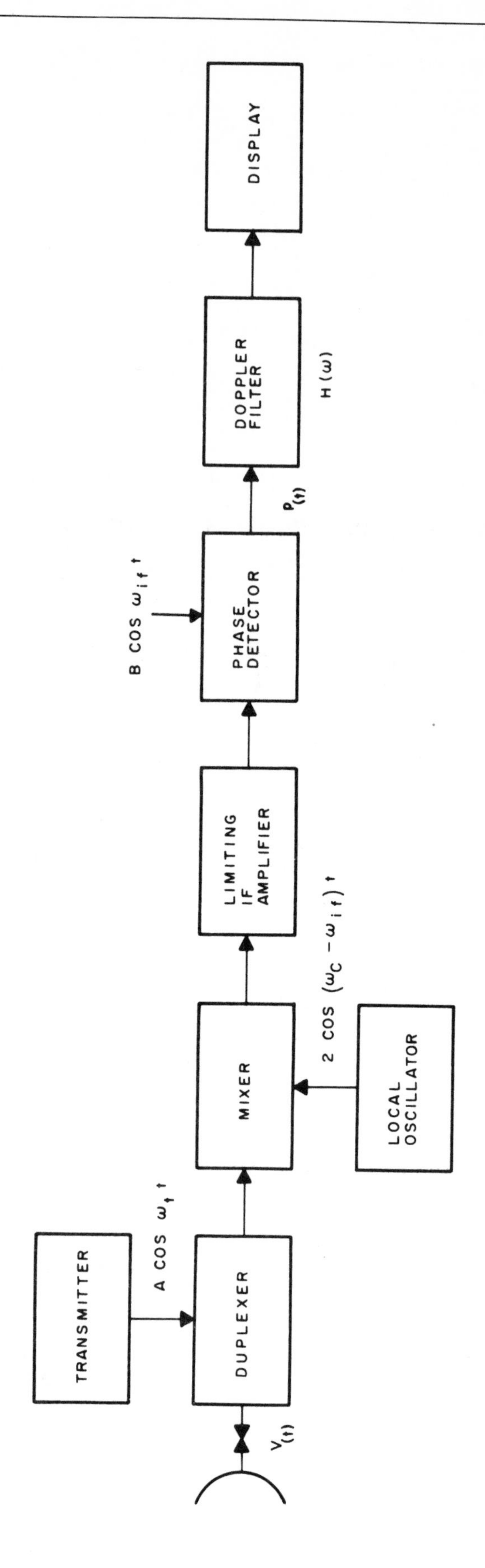

Figure 6. Coherent MTI System

Equation 11 is made applicable to a pulse radar by sampling the voltage waveform by a sequence of equally spaced pulses separated by a time T, where T is the period between transmitter pulses: (T = 1/PRF). Reference 7 gives the transformation:

$$S^*_{P(\omega)} = \sum_{-\infty}^{\infty} S_P\left(\omega + \frac{2\pi n}{T}\right) \tag{12}$$

for a series of inpulses sampling the voltage waveform. The resulting spectrum is shown in Fig. 1. To be completely accurate, this function should be weighted by the Fourier transform of the transmitter pulse. However, all functions are ambiguous about one-half the radar PRF and only this interval need be considered. Calculations can be made on the one-sided spectrum where the spectrum is given by:

$$S_{P(\omega)} = 2\,S_{x(\omega)} + S_S\left(\omega - \omega_d\right);\ 0 \leq \omega \leq 2\pi \cdot \frac{PRF}{2} \tag{13}$$

The factor 2 appears before the clutter spectrum because only one-half of the spectrum appears in the region of interest while the whole target spectrum is included in this region.

C. Noncoherent MTI Systems

Noncoherent MTI systems use the clutter return at the target range as a reference signal. As shown in Fig. 5, the beat between the fixed ground clutter and the moving target induces an amplitude modulation on the radar return. This return is processed through a wide dynamic range receiver that preserves the amplitude information, and is applied to a doppler filter that extracts the moving target information. This type of system has been extensively used where the radar is moving with respect to the ground clutter since it eliminates the need for motion compensation in the MTI receiver.

The noncoherent MTI system is shown in Fig. 7. Translation of the signal to IF frequencies is conventional with the exception that care must be exercised to prevent any extraneous amplitude modulation from being induced on the signal.

Typically the wide dynamic range IF amplifier has a linear-logarithmic characteristic. Alternately sensitivity time control (STC) can be employed, separately, or in conjunction with the lin-log characteristic to compress the dynamic range of the system.

The lin-log characteristic serves to compress the dynamic range of typically 80 to 100 dB to about 40 to 50 dB.

The slope of the logarithmic characteristic is matched to the rayleigh amplitude probability distribution of the clutter so that the rms value of the fluctuation output of the receiver is constant across the logarithmic region. The transition between the linear and logarithmic regions is determined by the clutter attenuation capability of the doppler filter following the linear logarithmic receiver. For example, if the doppler filter provides a CA of 30 dB, then the transition is set at 30 dB above the rms receiver noise level. This will ensure that the rms value of the clutter approximately equals the rms value of the noise level at the output of the doppler filter throughout the full dynamic range of the lin-log receiver. A clutter attenuation (CA) of less than 30 dB will allow strong clutter returns to exceed the receiver noise level and report as targets.

The performance of the noncoherent MTI system can be analyzed by assuming square-law detection. Reference 8 gives the spectrum of linear and square-law detector outputs when excited with narrow-band random noise processes. The spectrums are similar in the region of interest and the square-law process will be used because of its mathematical simplicity. The receiver process is:

$$z_{(t)} = i^2_{(t)} = r^2_{(t)} \cos^2 \left[\omega_{if} t + \phi_{(t)} \right] \qquad (14)$$

where the envelope characteristic $r^2_{(t)}$ is given by the square of equation 8:

$$r^2_{(t)} = \left[\sqrt{2}\, S_{(t)} \cos \omega_d t + x_{(t)}\right]^2 + \left[\sqrt{2}\, S_{(t)} \sin \omega_d t + y_{(t)}\right]^2 \qquad (15)$$

The auto correlation function is given by:

$$R_{z(\tau)} = \underbrace{8\, P_s\, P_c + 4\, P_c^2}_{\text{DC Terms}} + \underbrace{8\, R_{x(\tau)}\, R_{s(\tau)} \cos \omega_d \tau + 4R^2_{s(\tau)}}_{\text{Signal Terms}} + \underbrace{4\, R^2_{x(\tau)}}_{\text{Clutter Term}} \qquad (16)$$

The dc-suppressed spectrum for small signal-to-clutter ratios is given by:

$$S_{z(\omega)} = \underbrace{4\, S_{x(\omega)} * \left[S_s\left(\omega - \omega_d\right) + S_s\left(\omega + \omega_d\right)\right]}_{\text{Signal Terms}} + \underbrace{4\, S_{x(\omega)} * S_{x(\omega)}}_{\text{Clutter Terms}} \qquad (17)$$

where * represents convolution:

$$S_{1(\omega)} * S_{2(\omega)} = \frac{1}{2\pi} \int_{-\infty}^{\infty} S_{1(\omega)}\, S_{2(\omega - y)}\, dy$$

The signal term results from the beat between the target and the clutter, while the clutter term results from the beat between the clutter with itself. The target has a spectral width variance equal to the sum of target and clutter spectral width variances. The clutter spectral width variance is twice that of the coherent case.

Spreading of the target and clutter spectrums results in a lower performance capability for a noncoherent system when compared with an equivalent coherent system.

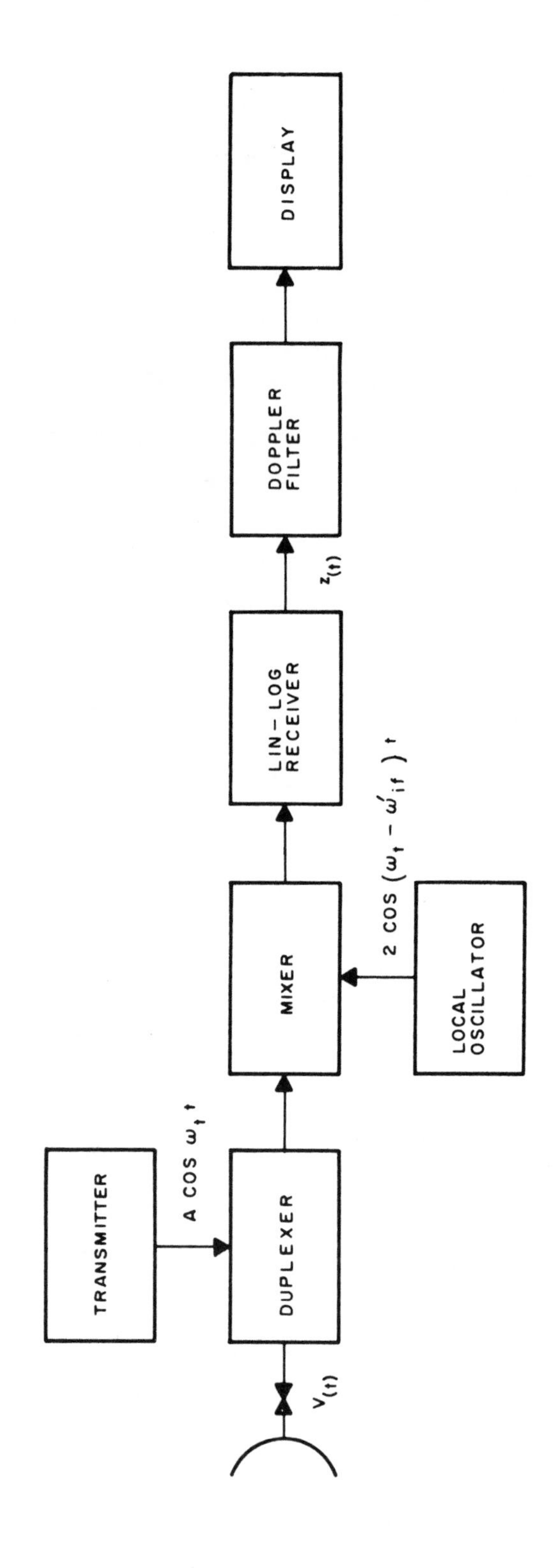

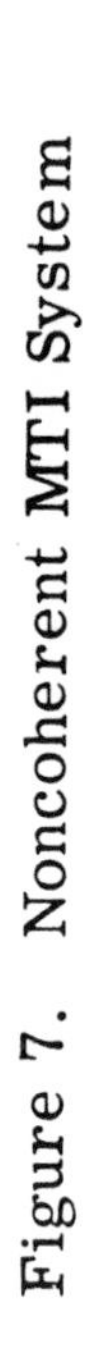

Figure 7. Noncoherent MTI System

Equation 17 can be made applicable to a pulse radar by applying equation 12 resulting in a spectrum that is similar to that given in Figs. 1 and 2. For calculation purposes, the one-sided spectrum:

$$S_{z(\omega)} = 8\,S_{x(\omega)} * S_{s}\left(\omega - \omega_{d}\right) + 8\,S_{x(\omega)} * S_{x(\omega)}; \quad 0 \leq \omega \leq 2\pi\,\mathrm{PRF}/2 \tag{18}$$

can be used. It is interesting to note that the assumption of a square-law detector results in an apparent video gain of 3 dB in the target-to-clutter power ratio for small target-to-clutter ratios.

III. MTI PERFORMANCE CRITERIA

Several performances indexes are in common use.

Clutter attenuation (CA) is found by comparing the power in the clutter before and after doppler filtering.

$$CA = \frac{\int_0^\infty S_c(\omega)\, d\omega}{\int_0^\infty S_c(\omega)\,|H(\omega)|^2\, d\omega} \tag{19}$$

Subclutter visibility (SCV) is defined as the ratio at the input of the doppler filter of clutter power to moving target power to produce equal doppler filter outputs:

$$SCV = \frac{\int_0^\infty S_T(\omega - \omega_d)\,|H(\omega)|^2\, d\omega}{\int_0^\infty S_c(\omega)\,|H(\omega)|^2\, d\omega}\,; \quad P_c = P_T \tag{20}$$

By convention, equation 20 is usually evaluated at ω_d equal to an optimum speed target--that is, a target whose doppler frequency is one-half the radar's PRF. However, it is more fundamental to evaluate equation 20 at the target speeds of interest.

MTI gain is defined as the output of the doppler filter when excited with a process whose spectral density corresponding to all velocities of interest is uniformly distributed.

$$\text{MTI Gain} = \frac{\int_0^{\Omega_{vmax}} |H_n(\omega)|^2 \, d\omega}{\Omega_{vmax}} \qquad (21)$$

where Ω_{vmax} is the maximum doppler frequency of interest, and $|H_n(\omega)|^2$ is the normalized power transfer function.

Clutter attenuation and subclutter visibility are equivalent criteria when target and clutter spectral shapes are similar and the response of the doppler filter is unity for the targets under consideration. For single and double delay-line filters, this occurs for only optimum speed targets, while for shaped delay-line filters and range-gated filters equivalence may occur for a wide range of target velocities.

MTI gain provides a criteria for comparing target responses of various doppler filters.

In general, the performance criteria that best represents the desired objective of the system should be selected. For example, consider an MTI surveillance requirement where an observer is to locate moving targets by examining a PPI or B-type display. The MTI system must reduce the ground clutter to a level consistent with the radar's noise level to keep the false alarms at a tolerable level, while allowing the detection of weak targets. For this type of situation, the clutter attenuation and MTI gain (the product CA $\times$ MTI gain might be used) are appropriate indexes; the first relating to the false alarm performance; the second to the ability to detect a variety of targets. As a second example, consider a combat surveillance radar whose function is to detect personnel whose velocities are 1 to 5 knots. The performance criteria of interest here is subclutter visibility evaluated at 1 to 5 knot velocities.

IV. CLUTTER AND TARGET CHARACTERISTICS

A. Clutter Amplitude

The clutter geometry for an airborne situation is shown in Fig. 8. (This figure can be applied to a ground situation by letting h equal zero.) If we assume the target is located at point T on Fig. 8, then the clutter that competes with the target return is back scattered from a concentric ring about the radar whose width is equal to the projection on the ground of two equally spaced range contours that are separated by a radar range corresponding to the radar's transmitter pulse width (150 meters per μs). In Fig. 8, the portion of the concentric

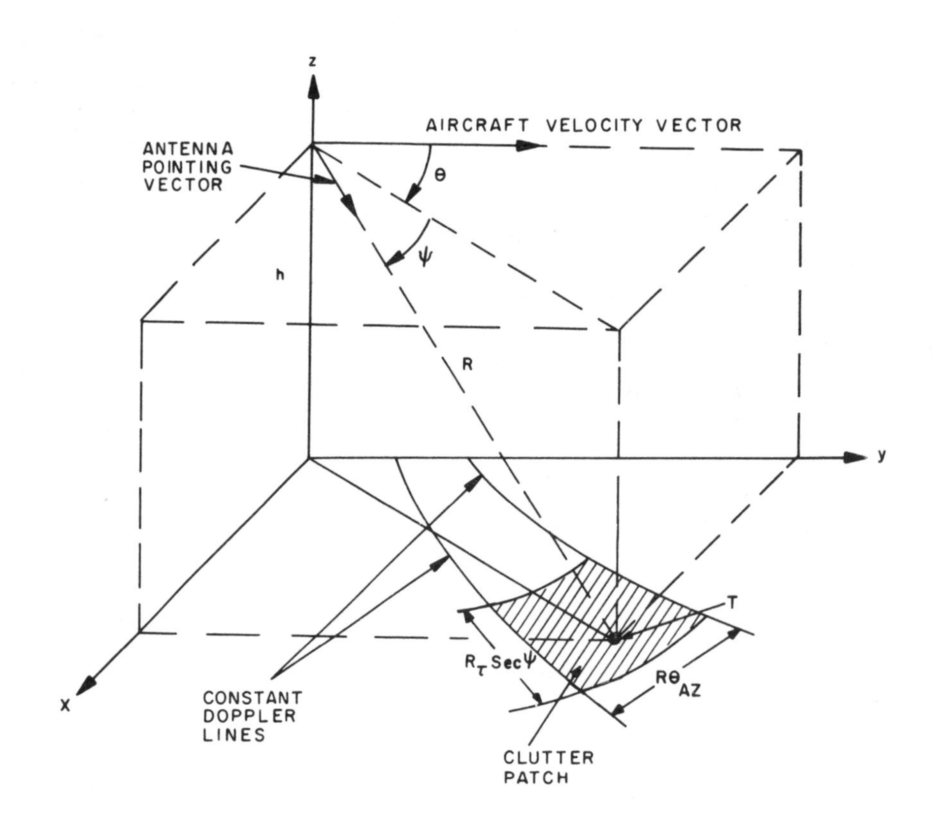

Figure 8. Airborne Clutter Geometry

ring illuminated by the radar's main azimuthal beamwidth is shaded and the major component of the clutter return is back scattered from this region. If the radar antenna has low side lobes, the return from the concentric ring outside the shaded area can be neglected. However, in many airborne installations, it is difficult to achieve low side lobes, and the return from the full concentric ring must be considered.

The magnitude of the clutter can be found by computing an equivalent cross sectional area that is inserted into the radar equation to compute the clutter power. This equivalent cross sectional area is given by:

$$\sigma_e = \sigma_o R_\tau \cdot R \cdot \theta_{az} \sec \psi \tag{22}$$

where σ_o is the back scattering coefficient of the illuminated terrain; R_τ is the radar range corresponding to the transmitter pulse width (150 meters per μs), θ_{az} is the radar's azimuthal beamwidth in radians, and ψ is the angle to the clutter patch as defined in Fig. 8.

Reference 9 gives values of σ_o for various terrains, frequencies, polarizations, and depression angles. Fig. 9 plots σ_o versus depression angle for several frequencies and terrains.

B. Clutter and Target Spectrums

The spectral distributions of the clutter and the target ultimately determines the performance of the MTI system. The IF clutter spectrum is spread by: (1) antenna scanning modulation, (2) radar platform motion, (3) transmitter frequency drift, (4) internal clutter motion, and (5) system instabilities. The IF target spectrum is spread by: (1) antenna scanning modulation, and (2) system instabilities.

1. Antenna Scanning Modulation

As the radar antenna scans by a point target, it induces an amplitude modulation on the target return.

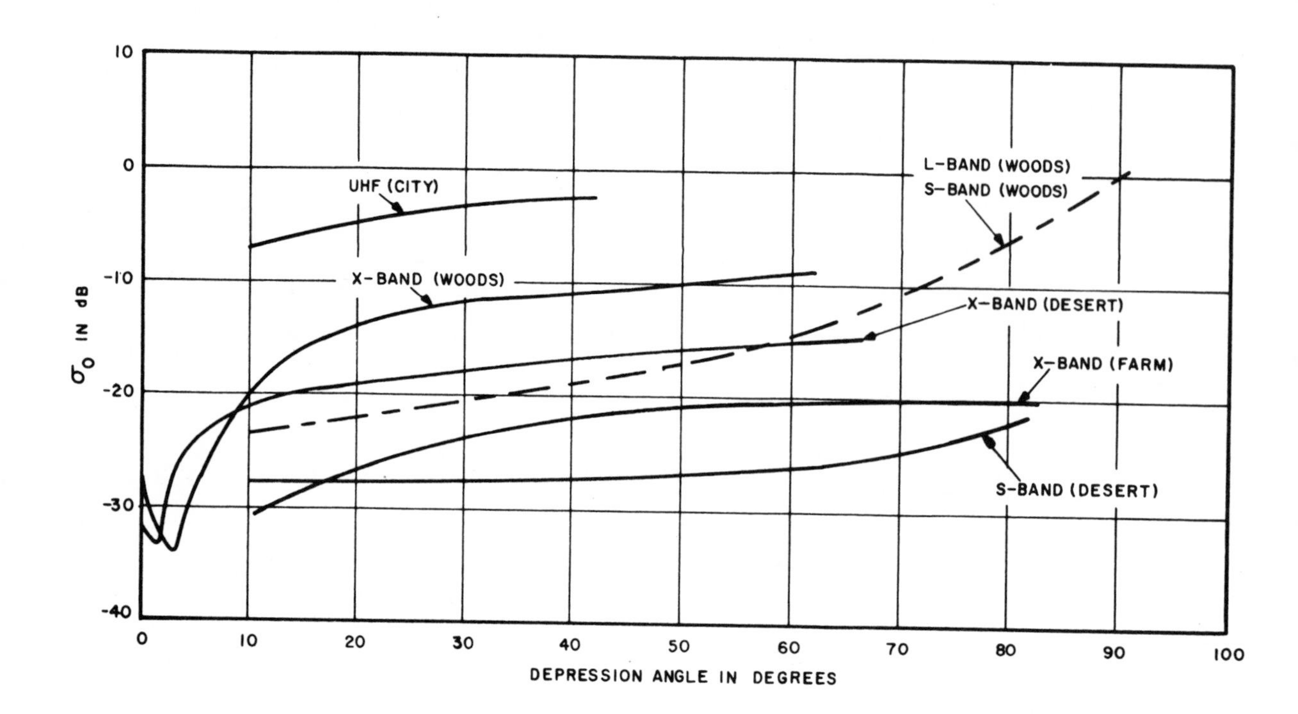

Figure 9. Backscattering Coefficient (σ_0) for Various Terrains and Frequencies

A uniformly illuminated antenna scanning at a constant speed, $\dot{\theta}$, and with a 3-dB beamwidth of θ_{3dB}, induces an amplitude modulation on the target given by:

$$V_{\theta(t)} = V_m \frac{\sin^2 \frac{\alpha}{2} t}{\left(\frac{\alpha}{2} t\right)^2} \tag{23}$$

where

$$\alpha = 5.6 \frac{\dot{\theta}}{\theta_{3dB}} = 5.6 \frac{D\dot{\theta}}{\lambda} \text{ (rad/s)} \tag{24}$$

where

D = antenna diameter

λ = radar wavelength both in consistent units

$\dot{\theta}$ = antenna scan rate in rad/s.

The spectral density is given by:

$$S_\theta(\omega) = V_m^2 \left(1 - \frac{\omega}{\alpha}\right)^2 \quad \omega \leq \alpha$$
$$= 0 \quad \omega \geq \alpha \tag{25}$$

Matching a gaussian spectrum at the half-power point results in a spectrum given by:

$$S_\theta(\omega) = \frac{P_{t,c}}{\sqrt{2\pi\sigma_\theta^2}} \epsilon^{-\omega^2/2\sigma_\theta^2} \tag{26}$$

where

$$\sigma_\theta = 1.4 \frac{\dot{\theta}}{\theta_{3dB}} = 1.4 \frac{D\dot{\theta}}{\lambda} \text{ rad/s} \tag{27}$$

Equation 25 is applicable to both target and clutter spectrums. The spectrum shapes are similar since the clutter can be viewed as a collection of independent scatterers each with a spectrum given by equation 25.

Spectrums generated by various antenna illuminations (reference 10) and the gaussian approximation are shown in Fig. 10. It is interesting to note that the spectrums generated by real antenna illuminations are truncated at frequencies that can be shown to correspond to the doppler frequency associated with the tip speed of the scanning antenna.

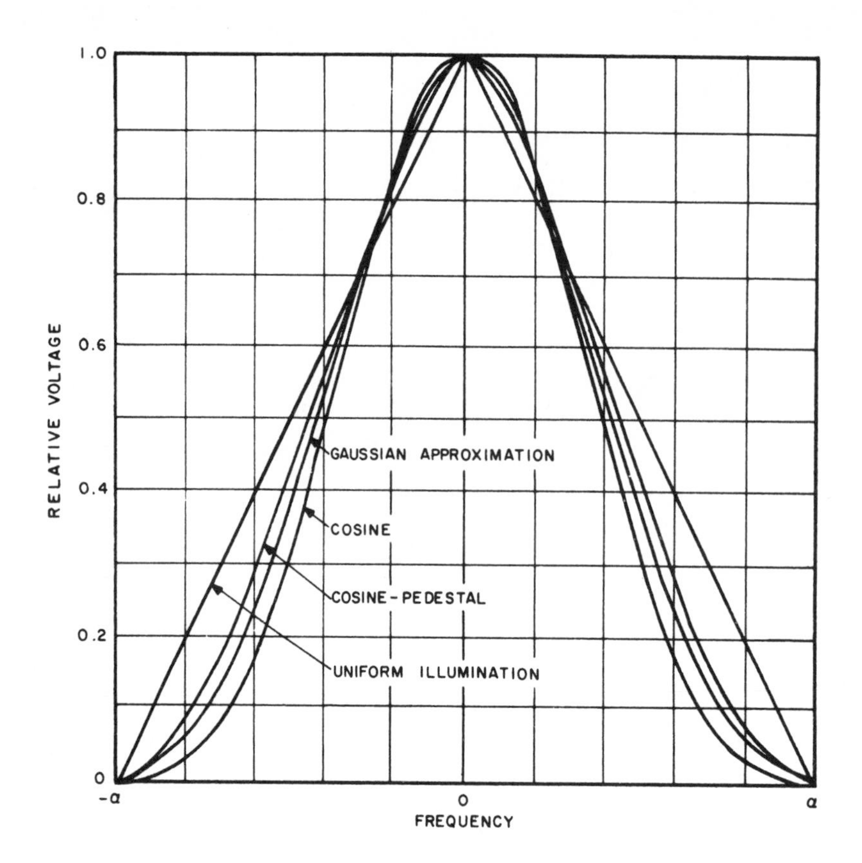

Figure 10. Voltage Spectrum due to Scanning Antenna for Various Illumination Functions

2. Platform Motion Spectral Density

Fig. 8 shows the geometrical relationship between the clutter patch, the aircraft velocity vector, and the antenna pointing vector. The doppler frequency corresponding to the closing velocity between the aircraft and the center of the clutter patch is given by:

$$f_c = f_{ac} \cos \psi_o \cos \theta_o \tag{28}$$

where f_{ac} is the doppler frequency corresponding to the aircraft velocity and ψ_o, θ_o are respectively the depression angle and azimuth angle to the center of the clutter patch as defined in Fig. 8.

The differential doppler frequency with respect to the center of the clutter patch is given by:

$$\Delta f_d = f_{ac} \, \Delta\theta \cos \psi_o \sin \theta_o{}^* \tag{29}$$

where $\Delta\theta$ is the azimuth angle with respect to the antenna boresight. Equation 29 is valid for small deviations from the antenna boresight (approximately 30 degrees) and indicates that the spectral spread is a linear function of $\Delta\theta$.

The voltage spectral density due to platform motion is deduced from equation 29 to be:

$$V_{p(\omega)} = V_m \frac{\sin^2 \frac{\omega}{a}}{\left(\frac{\omega}{a}\right)^2} \tag{30}$$

where a is given by:

$$a = \frac{\omega_{ac} \, \theta_{3dB} \cos \psi_o \sin \theta_o{}^{**}}{2.8} \tag{31}$$

---

* This expression is useful for most practical calculations. A more accurate expression (within 4 percent for $\Delta\theta > \pi/6$ and $\frac{R_\tau}{h} \frac{\sin^2 \psi}{\cos \psi} \geq \frac{\pi}{6}$) is: $\Delta f_d = f_{ac} \left[\Delta\theta \cos \psi_o \sin \theta_o + \frac{(\Delta\theta)^2}{2} \cos \psi_o \cos \theta_o + \frac{1}{2} \cos \theta_o \frac{R_\tau}{h} \frac{\sin^3 \psi_o}{\cos \psi_o} \right]$.

** The more accurate expression is: $a = \frac{\omega_{ac}}{2.8} \left[\theta_{3dB} \cos \psi_o \sin \theta_o + \frac{(\theta_{3dB})^2}{4} \cos \psi_o \cos \theta_o + \frac{R_\tau}{h} \cos \theta_o \frac{\sin^3 \psi_o}{\cos \psi_o} \right]$.

where ω_{ac} is the aircraft doppler radian frequency, θ_{3dB} is the azimuth 3-db beamwidth (radians) and ψ_o and θ_o are angles defined in equation 29. Equation 30 assumes a uniformly illuminated antenna aperture. The spectral density is given by:

$$S_{pm(\omega)} = V_m^2 \frac{\sin^4 (\omega/a)}{(\omega/a)^4} \tag{32}$$

with a as defined in equation 31.

The width of $S_{pm(\omega)}$ is a function of the antenna's pointing angle (θ_o and ψ_o), beamwidth (θ_{3dB}), and the aircraft's velocity vector.

The spreading due to this effect usually predominates over the spreading due to other effects when the antenna is pointed away from the aircraft's ground track.

Matching a gaussian spectrum at the half power point results in a spectrum given by:

$$S_{pm(\omega)} = \frac{P_c}{\sqrt{2\pi\sigma_p^2}} \epsilon^{-\omega^2/2\sigma_p^2} \tag{33}$$

where

$$\sigma_p = 0.3\ \omega_{ac}\ \theta_{3dB} \cos\psi_o \sin\theta_o = 3.76 \frac{V_{ac} \cos\psi_o \sin\theta_o}{D} \tag{34}$$

where V_{ac} is the aircraft's velocity and D is the antenna diameter in consistent units.*

---

* The more accurate expression is given by: $\sigma_p = 0.3\ \omega_{ac}\left[\theta_{3dB} \cos\psi_o \sin\theta_o + \frac{(\theta_{3dB})^2}{4} \cos\psi_o \cos\theta_o + \frac{R_T}{h} \cos\theta_o \frac{\sin^3\psi_o}{\cos\psi_o}\right]$.

The spectral densities for uniform antenna illumination and the gaussian approximation are plotted in Fig. 11. Performance calculations or systems employing range-gated or shaped delay-line doppler filters sometimes require a more precise represetation of the antenna side-lobe structure than that given by equations 32 or 33. This can be seen from Fig. 1 and is a result of the antenna side-lobes illuminating the ground away from the antenna's ground track, thereby creating doppler components in the pass band of the doppler filter in accordance with equation 29. A representation that allows an arbitrary side-lobe structure to be considered is given by:

$$S_{pm(\omega)} = V_m^2 \left[K_1^2 \frac{\sin^4 (\omega/a)}{\left(\frac{\omega}{a}\right)^4} P_{\pi a(\omega)} + \frac{\sin^4 \frac{\omega}{a}}{\left(\frac{\omega}{a}\right)^4} \right] \tag{35}$$

where

$$P_{\pi a} = 1 \qquad |\omega| \leq \pi a$$

$$= 0 \qquad \text{elsewhere}$$

and K_1 is an antenna side-lobe factor given in Fig. 12.

3. Transmitter Instability Spectrum Spreading

Transmitter frequency instability causes the clutter to decorrelate on a pulse-to-pulse basis thereby spreading the clutter spectrum.

To examine this effect, the clutter patch of Fig. 8 is sliced in range at intervals separated by a radar wavelength (a 1-μs pulse at 10 GHz contains 10,000 slices) as shown in Fig. 13. When the scatterers located in each slice are illuminated by the radar, they return in-phase signals. If the radar's transmitter frequency changes by an amount that causes a change of one in the number of wavelengths within one radar

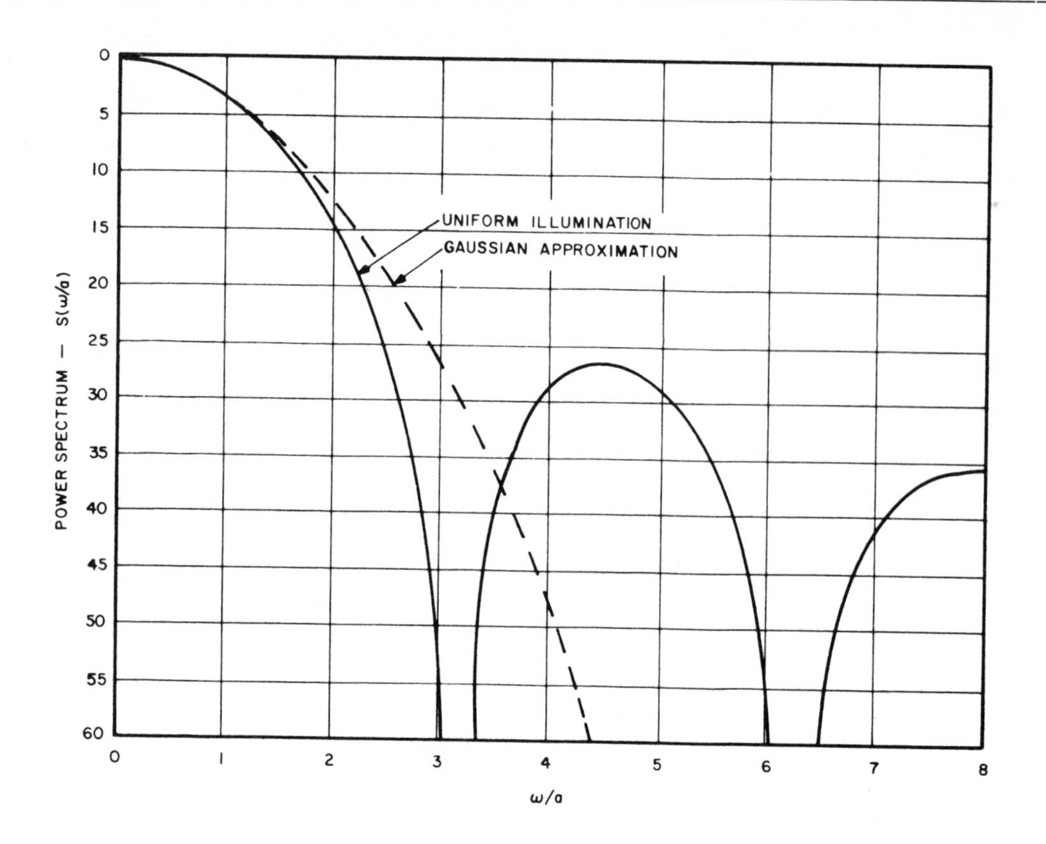

Figure 11. Spectral Density due to Platform Motion

pulse width, then the slices return a signal that contains all possible phases. Hence, the signal returns at the two frequencies are completely decorrelated.

A complete description of this process is given in reference 11. The auto-correlation function is given by:

$$R_{f(\tau)} = \frac{\sin^2 \pi \frac{f'_t}{B} \tau}{\left(\pi \frac{f'_t}{B} \tau\right)^2} \tag{36}$$

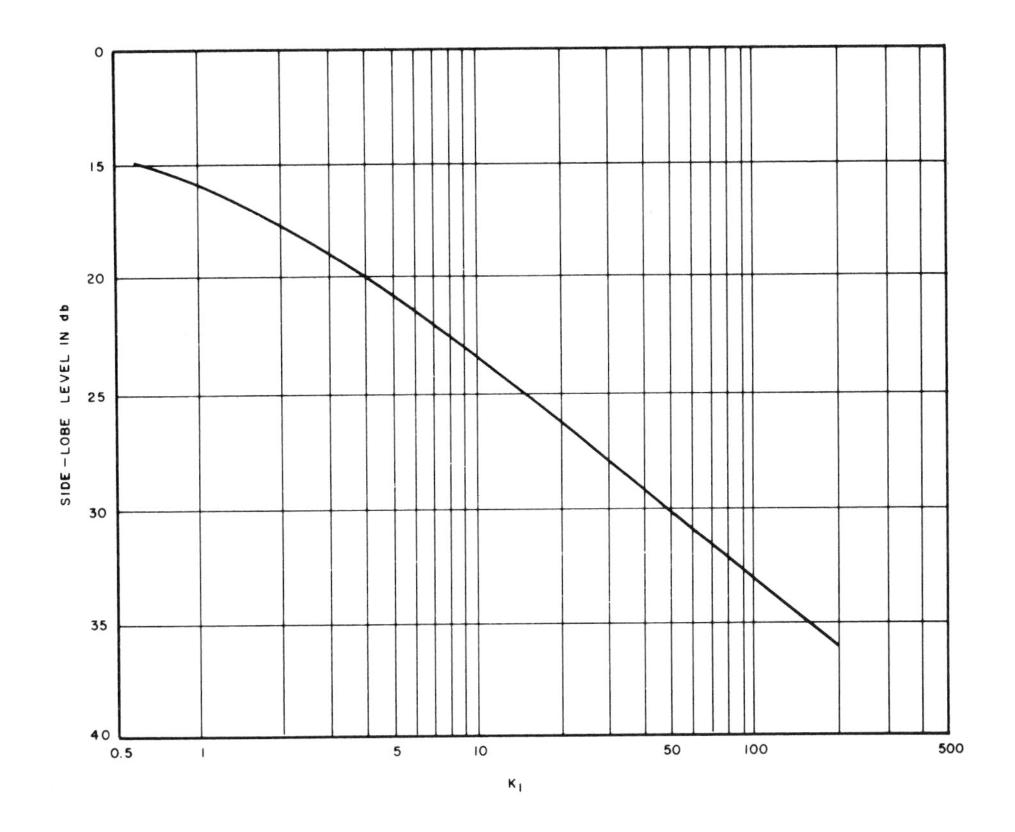

Figure 12. Antenna Side-Lobe Factor

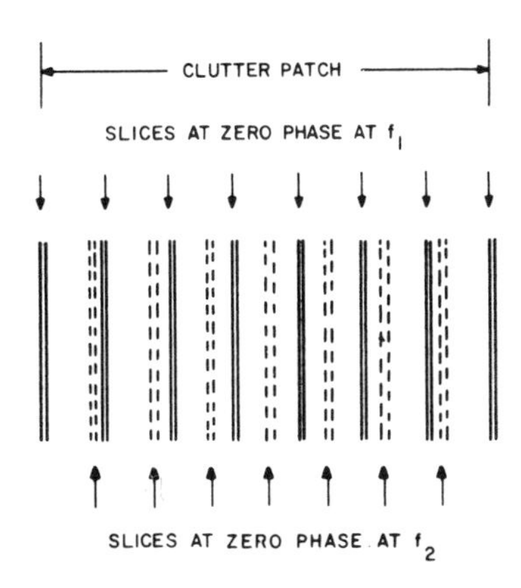

Figure 13. Phase of Clutter Slice for Frequencies f_1 and f_2

where f'_t is the transmitter drift rate and B is the transmitter pulse bandwidth in consistent units. The spectral density is given by:

$$S_{f(\omega)} = V_m^2 \cdot \left(1 - \frac{B\omega}{2\pi f'_t}\right), \quad |\omega| \leq 2\pi \frac{f'_t}{B} \tag{37}$$

$$= 0$$

Matching a gaussian spectrum at the half-power point gives:

$$S_{f(\omega)} = \frac{P_c}{\sqrt{2\pi\sigma_f^2}} \epsilon^{-\omega^2/2\sigma_f^2} \tag{38}$$

where

$$\sigma_f = 2.67 \frac{f'_t}{B} \text{ (rad/s)} \tag{39}$$

Transmitter frequency instability also causes additional effects in a coherent system. These effects are covered under system instabilities.

4. Internal Clutter Motion

The scatterers forming the clutter return often move in the wind causing a spreading of the clutter spectrum. This effect is well covered in references 1, 2, and 3 and exhibits a gaussian spectral density with a standard deviation given by:

$$\sigma_\omega = 12.6 \frac{v_{rms}}{\lambda} \tag{40}$$

where v_{rms} is the apparent rms velocity spread of the scatterers and λ is the radar wavelength in consistent units. Values of v_{rms} for wooded hills from reference 2 are given in Table I.

Table I. Apparent Clutter Velocity for Wooded Hills

Wind Speed (knots)	v_{rms} (ft/sec)
10	0.13
25	0.38
40	1.06

5. System Instabilities

The maximum performance of MTI systems is normally limited by imperfections in the radar system that are termed system instabilities. System instabilities associated with various parts of the MTI system are given in Table II.

Table II. System Instabilities

Subsystem	Coherent	Noncoherent
Transmitter	Frequency instability, firing jitter	Transmitter AM, firing jitter
Receiver	STALO drfit, COHO lock and drift	AM due to LO
Power supply	Ripple	Ripple
Delay-line doppler filter	Timing jitter, pulse width jitter, cancellation ratio	Timing jitter, pulse width jitter, cancellation ratio
Range-gated filter	Pulse width jitter, cancellation ratio	Pulse width jitter, cancellation ratio

The various system instabilities and the spectral spreading effects act independently on the radar signal. Independent effects can be represented by independent spectral densities:

$$S_{t(\omega)} = \sum_{n=1}^{N} S_{n(\omega)} \tag{41}$$

The CA for independent effects can be found by substituting equation 41 into equation 19.

$$CA = \frac{\int_{0}^{\infty} \sum_{n=1}^{N} S_{n(\omega)} \, d\omega}{\int_{0}^{\infty} \sum_{n=1}^{N} S_{n(\omega)} \, |H(\omega)|^2 \, d\omega} \tag{42}$$

Interchanging summation and integration, and noting that the numerator always equals the clutter power, provides a relationship for combining the CA due to independent effects

$$CA = \frac{1}{\frac{1}{[CA]_1} + \frac{1}{[CA]_2} + \cdots + \frac{1}{[CA]_N}} \tag{43}$$

Fig. 14 plots equation 43 and allows the CA for two effects to be found.

The CA due to system instabilities is best found by measuring the performance of the MTI system by searchlighting on a hard target such as a water tower. Alternately, if the system is in the paper design stage, typical performance as given in Section VI can be used as a design guide, or equation 43 can be evaluated by considering each effect. In either event, the CA due to system instabilities is found.

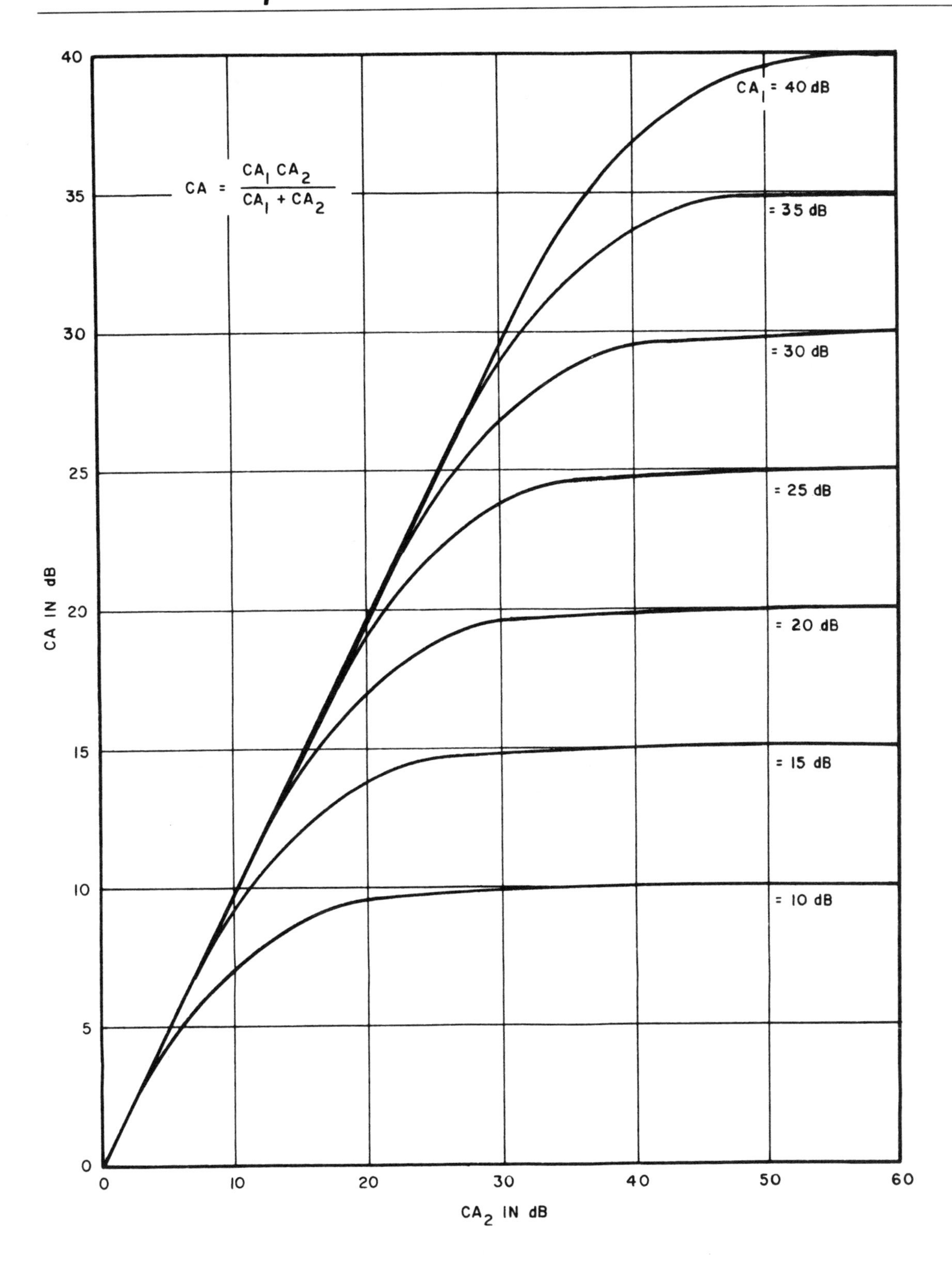

Figure 14. Combining Clutter Attenuations

The next step depends on whether the MTI system is being synthesized or analyzed. Synthesis implies a knowledge of the overall requirement on CA. In this event, Fig. 14 is entered, and the CA required of the doppler filter acting on the clutter spectrum is determined. From this information, the shape of the doppler filter is synthesized to meet the system requirement. If analysis is being performed, the shape of the doppler filter is known and the CA due to its acting on the clutter spectrum is determined. Fig. 14 is entered and the overall CA is determined.

6. Video Clutter and Target Spectrum

The video target and clutter spectrums can be found by combining the effects given by equations 26, 33, 38, and 40 and applying equation 13 for coherent MTI systems and equation 18 for noncoherent MTI systems.

To combine the effects, it is assumed that they are independent and disturb the return in a multiplicative way.

$$V_{(t)} = a_{(t)}\, b_{(t)}\, c_{(t)}\, d_{(t)} \tag{44}$$

where $a_{(t)}$, $b_{(t)}$, $c_{(t)}$, and $d_{(t)}$ are the various effects. The ACF of the envelope is given by:

$$R_{abcd\,(\tau)} = R_{a(\tau)} \cdot R_{b(\tau)} \cdot R_{c(\tau)} \cdot R_{d(\tau)} \tag{45}$$

The spectrum is found from equation 45 to be

$$S_{abcd(\omega)} = S_{a(\omega)} * S_{b(\omega)} * S_{c(\omega)} * S_{d(\omega)} \tag{46}$$

where * represents convolution. Convolving gaussian functions results in a gaussian function whose variance is the sum of the individual variances. Hence, the combined IF spectrum is given by:

$$S_{t(\omega)} = \frac{P_c}{\sqrt{2\pi\sigma_t^2}} \epsilon^{-\omega^2/2\sigma_t^2} \tag{47}$$

where

$$\sigma_t^2 = \sigma_\theta^2 + \sigma_{pm}^2 + \sigma_f^2 + \sigma_\omega^2 \tag{48}$$

Equation 48 must be evaluated for both the target and the clutter. In a coherent MTI system, equation 48 is translated to video, while in a noncoherent system, the clutter spectrum variance is given by

$$\sigma_c^2 = 2\sigma_t^2 \tag{49}$$

and the target spectrum variance is given by

$$\sigma^2_{tar} = \sigma_t^2 \Big|_{target} + \sigma_t^2 \Big|_{clutter} \tag{50}$$

Table III lists the video target and clutter variances that apply to airborne and ground coherent and noncoherent MTI systems.

Table III. Video Target and Clutter Spectrum

		Coherent MTI	Noncoherent MTI
Ground:	Target	σ_θ^2	$2\sigma_\theta^2 + \sigma_f^2 + \sigma_\omega^2$
	Clutter	$\sigma_\theta^2 + \sigma_f^2 + \sigma_\omega^2$	$2\sigma_\theta^2 + 2\sigma_f^2 + 2\sigma_\omega^2$
Airborne:	Target	σ_θ^2	$2\sigma_\theta^2 + \sigma_f^2 + \sigma_\omega^2 + \sigma_{pm}^2$
	Clutter	$\sigma_\theta^2 + \sigma_f^2 + \sigma_\omega^2 + \sigma_{pm}^2$	$2\sigma_\theta^2 + 2\sigma_f^2 + 2\sigma_\omega^2 + 2\sigma_{pm}^2$

$$S_{tar(\omega)} = \frac{1}{\sqrt{2\pi\sigma_{tar}^2}}\,\epsilon^{-\omega^2/2\sigma_{tar}^2} \qquad S_{c(\omega)} = \frac{1}{\sqrt{2\pi\sigma_c^2}}\,\epsilon^{-\omega^2/2\sigma_c^2}$$

$$\sigma_\theta = 1.4\,\frac{D\dot{\theta}}{\lambda} \qquad \sigma_p = 3.76\,\frac{V_{ac}\cos\psi_o\sin\theta_o}{D} \qquad \sigma_f = 2.67\,\frac{f'_t}{B}$$

$$\sigma_\omega = 12.6\,\frac{v_{rms}}{\lambda}$$

V. MTI DOPPLER FILTERS

The characteristics of single and double delay-line doppler filters, shaped delay-line filters and range-gated doppler filters will be considered. The performance of these filters is determined by the filter's transfer function and its cancellation ratio (CR). The transfer function is found for an ideal implementation of the filter, while the cancellation ratio is a measure of the deviation of the filter from the ideal.

The cancellation ratio is measured by applying a maximum level video pulse at the radar's PRF to the doppler filter and noting the output. The PRF is halfed and the input video pulse is attenuated until the output is the same as noted with the full PRF. The attenuator setting is the cancellation ratio.

A. Single Delay-Line Filter

The simplest implementation of an MTI filter is the delay and subtract circuit shown in Fig. 4. The most common delay element is a quartz delay line, although digital shift registers, magnetostrictive delay lines, and storage tubes have been used. Both AM and FM techniques (subtraction by mixing frequencies) have been used to process signals through the delay element.

The transfer characteristic is given by:

$$|H(\omega)|^2 = 4 \sin^2 \frac{\omega T}{2} \quad (51)$$

The cancellation ratio is usually low due to amplitude imbalance between direct and delay channels, dynamic range problems, and difficulties in generating a stable trigger for the radar. It is difficult to maintain a field system having a CR of better than 30 dB for an AM system and 35 dB for an FM system.

B. Double Delay-Line Filter

The double delay-line filter is simply two single delay-line filters in series. Its principal advantage over the single delay-line system is a superior cancellation ratio. This occurs because the canceled output of the first filter is considerably reduced in the second filter. Cancellation ratios of 40 dB are attainable with double delay-line filters.

The transfer function of this filter is given by:

$$|H(\omega)|^2 = 16 \sin^4 \frac{\omega T}{2} \tag{52}$$

C. Shaped Delay-Line Filters

Shaped delay-line filters allow adjustment of the filters pass and stop bands and hence can provide better performance than the inflexible filter shapes provided by single and double delay-line filters.

The shaped delay-line filter shown in Fig. 4 is only one of many possible arrangements. (Refer to references 3 and 12 for other configurations.) However, this configuration is advantageous in that it provides a cancellation ratio equivalent to that of a double delay-line filter and has a performance in terms of transfer functions as good as any other two delay-line configurations. This configuration has been used in many operational equipments.

Shaped delay-line filters are best analyzed by transforming the pole-zero pattern from the P-plane into the Z-plane. The Z-transformation is given by:

$$Z = \epsilon^{pT} \tag{53}$$

and maps the left half P-plane into the unit circle in the Z-plane. This transformation is effective because it allows one period of the filter's transfer characteristic to be examined by a single rotation about the unit circle.

The voltage transfer function in the Z-plane of the shaped delay-line filter of Fig. 4 is:

$$H(z) = \frac{(z - 1)^2}{z^2 - (\beta_1 + \beta_2) z + \beta_2} \tag{54}$$

Fig. 15 shows the pole-zero pattern of the shaped delay-line filter of Fig. 4. The amplitude of the voltage transfer function of the filter is found by the following procedure.

1. Locate the point on the unit circle intersected by a vector from the origin that makes an angle with the real axis of 360-degree frequency of interest/PRF.
2. Form vectors from the poles and zeros to the point on the unit circle.

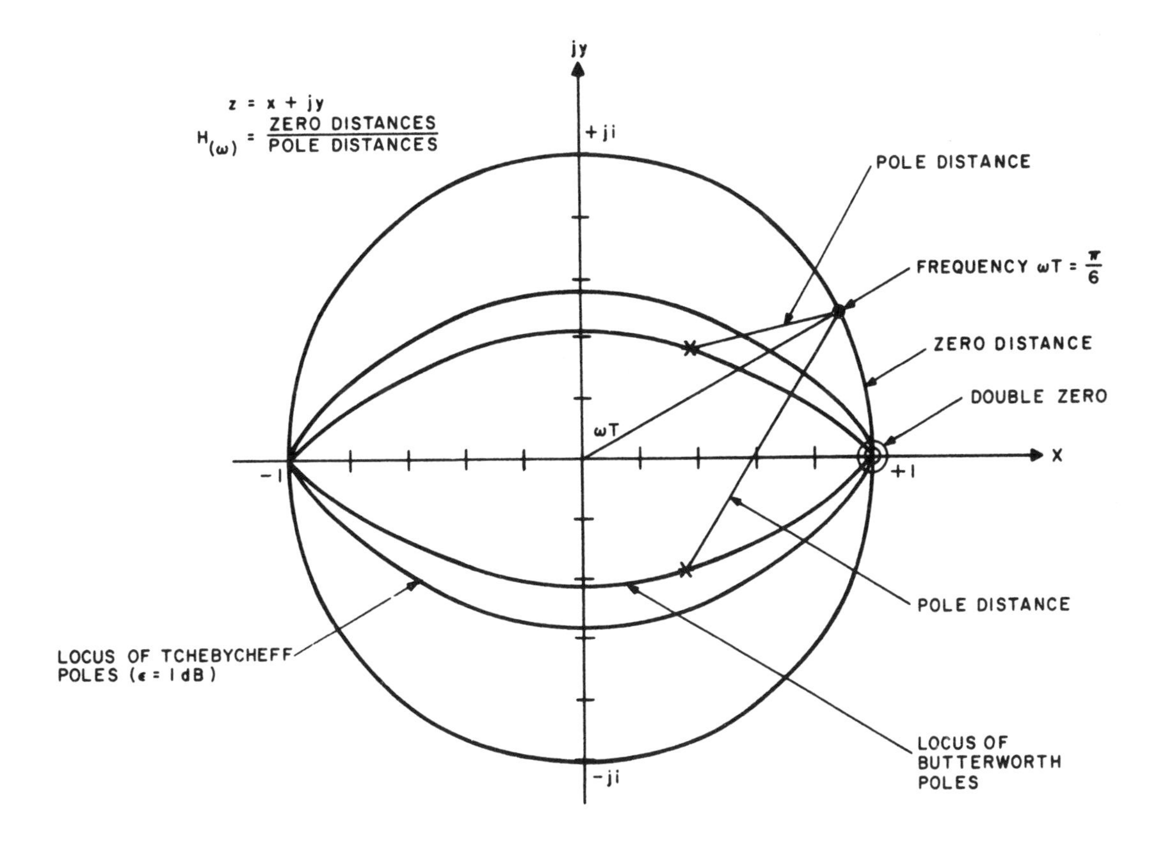

Figure 15. Pole-Zero Pattern of Shaped Delay-Line Filter

3. Find the product of the magnitudes of the two pole vectors. Repeat for the zero vectors.
4. Find the response at the frequency of interest by dividing the zero vector product by the pole vector product.

Determination of the pole-zero pattern to synthesize a particular transfer function is found by employing a transformation given in reference 12.

$$Z = \frac{\gamma + p^*}{\gamma - p^*} \tag{55}$$

where γ is a scale factor. This transformation maps the unit circle in the Z-plane into the left half P* plane and allows conventional filter theory to be applied to the Z-plane.

Reference 13 treats the Butterworth and Tchebycheff approximations to the ideal high-pass filter. Fig. 16 gives the Z-plane pole-zero locations of Butterworth and Tchebycheff filter functions for various cutoff frequencies. Fig. 17 gives the Butterworth and Tchebycheff amplitude functions found from the Z-plane pole-zero patterns of Fig. 16.

The feedback factors, β_1 and β_2, can be found from the pole locations (z_1 and z_1^*) in the Z-plane:

$$\beta_2 = |z_1|^2 \tag{56}$$

and

$$\beta_1 = 2 R_e z_1 - \beta_2 \tag{57}$$

D. Range-Gated Filters

An MTI filter that provides the greatest flexibility in terms of its transfer function and an excellent cancellation ratio is the range-gated filter (RGF).

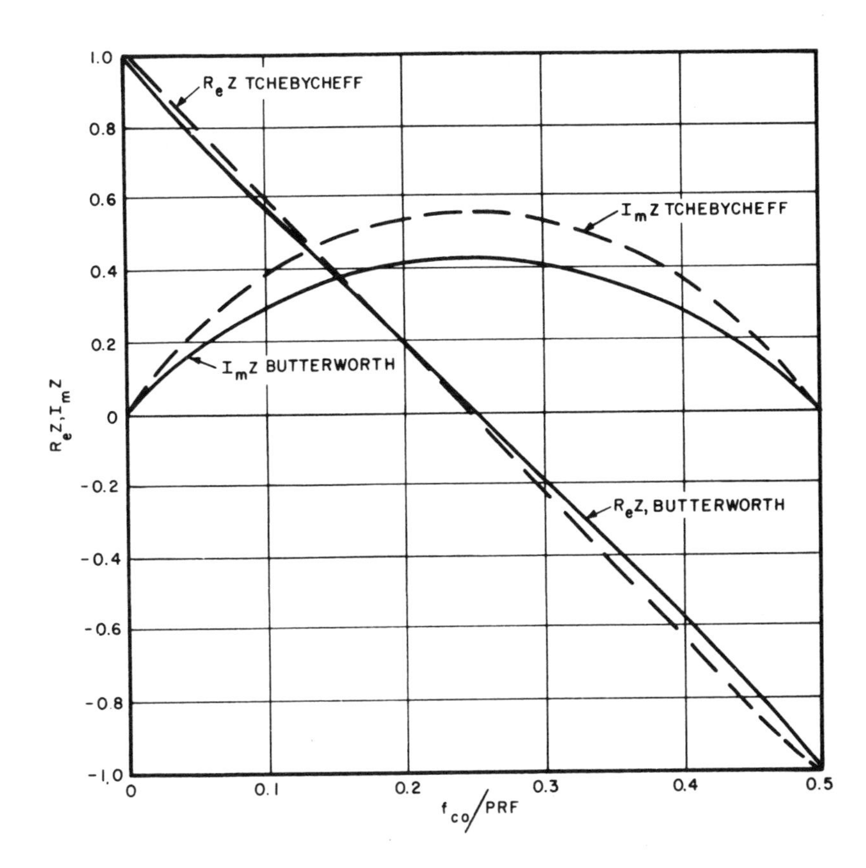

Figure 16. Pole Locations for Butterworth and Tchebycheff (1 dB) Filter Characteristics

The filter differs from the delay-line filters in that each RGF processes only one radar range resolution element. This allows the processing to be performed at audio frequencies where stable filters are easily built with good stop- and pass-band properties. Where MTI operation is desired over a large number of range cells microelectronic circuit techniques, as described in reference 14, can be employed to produce an MTI equipment that is actually smaller than a delay-line unit of lesser performance. Cancellation ratios of 45 dB are easily obtainable with RGF's. Adjustments and controls are eliminated, as is the need for precise PRF control.

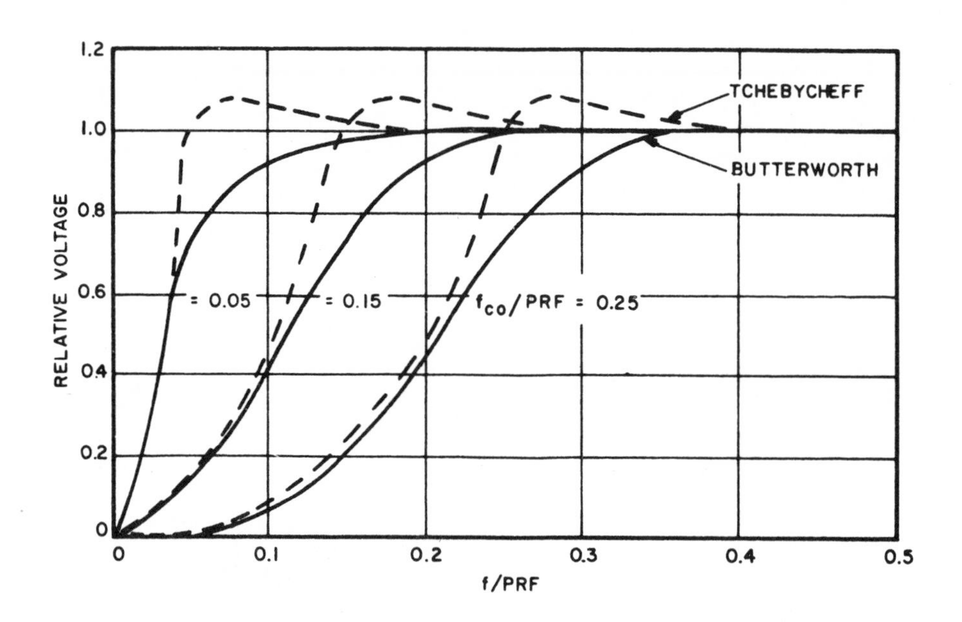

Figure 17. Shaped Delay-Line Filter Amplitude Response

A block diagram of an RGF is shown in Fig. 18. The box-car detector samples the radar video in a particular range cell once each radar PRF and holds this value during the interpulse interval. This has the effect of folding all the target and clutter information into the interval between the zero PRF line and a frequency of PRF/2. The clutter filter is composed of a high- and a low-pass filter. The high-pass filter is adjusted to provide the desired clutter attenuation. The low-pass filter cutoff frequency is placed at a frequency of PRF/2 in order to attenuate sampling transients from the boxcar and clutter components at the PRF and its harmonics. The output of the clutter filter is applied to a gain stabilized audio amplifier, detected, integrated and gated in time synchronism with the boxcar gate. This produces an output video pulse whose amplitude is related to the doppler signal amplitude from the clutter filter.

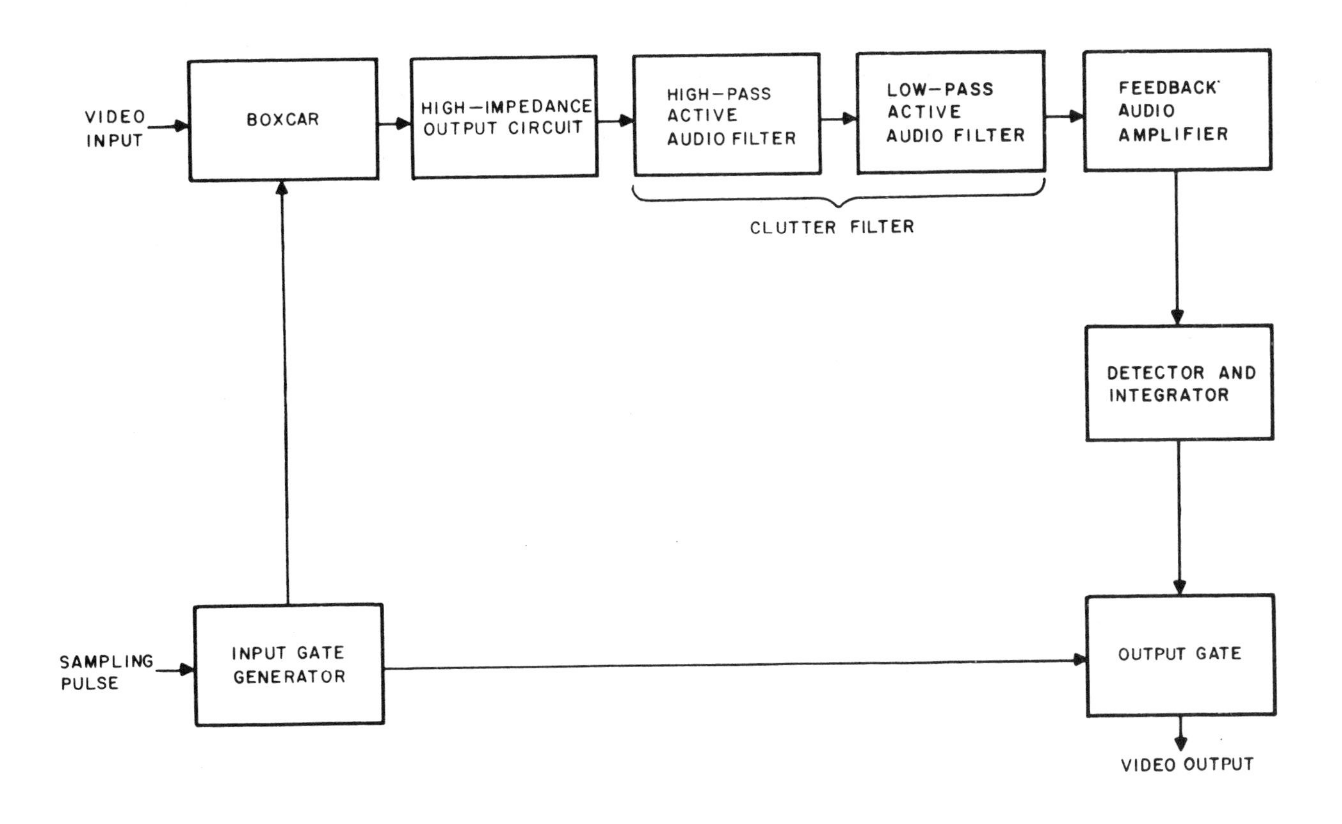

Figure 18. Range-Gated Filter Block Diagram

The normalized transfer function of the boxcar detector is given by:

$$|H(\omega)|^2 = \frac{4 \sin^2 \frac{\omega T}{2}}{\omega^2 T^2} \qquad (58)$$

This function processes the spectrum of Fig. 1, attenuating all clutter spectral components except those about dc. Optimum speed targets with dopplers in the vicinity of PRF/2 are also partially attenuated.

An active high-pass filter is employed to allow synthesis of approximations to an ideal high-pass filter. An active two-pole, high-pass configuration frequently employed is given in Fig. 19. If a unity gain amplifier is employed, its voltage transfer function is given by:

$$H(p) = \frac{p^2}{p^2 + \left(\alpha_1 \frac{R_1}{R_2} + \alpha_2\right) p + \alpha_1 \alpha_2} \tag{59}$$

If $C_1 = C_2$, the filter cutoff frequency is given by:

$$\omega_n = \sqrt{\alpha_1 \alpha_2} \tag{60}$$

while the damping factor equals:

$$\delta = \sqrt{\frac{R_1}{R_2}} \tag{61}$$

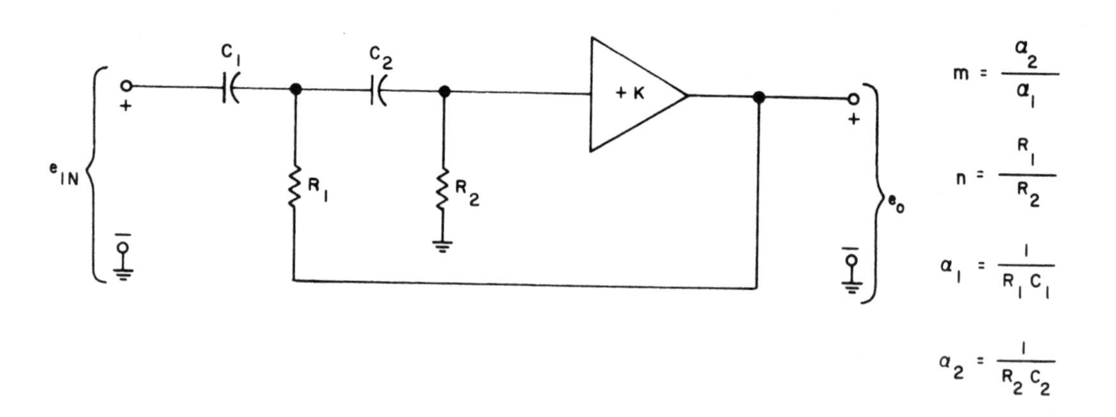

Figure 19. Two-Pole Active Filter

Table IV. High-Pass Butterworth and Tchebycheff Filter Parameters

Order B = Butterworth C = Tchebycheff (1 dB)		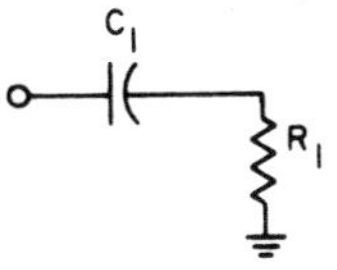	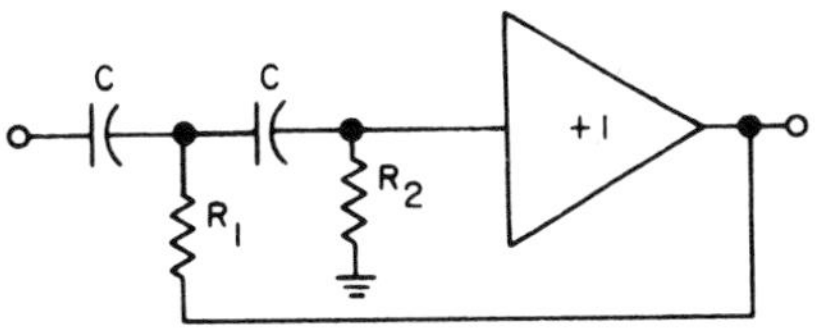			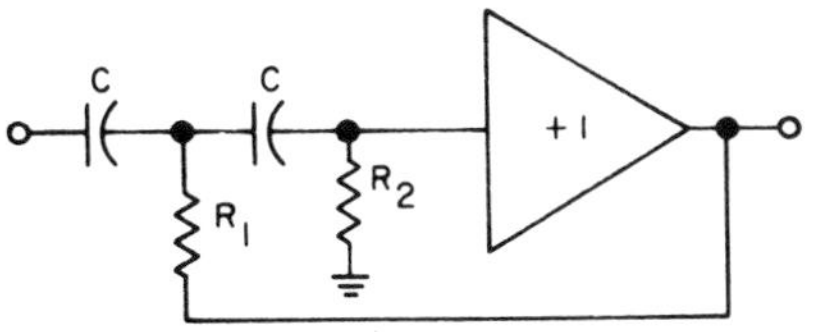		
		$\alpha_1 = \frac{1}{R_1 C_1}$	$\alpha_1 = \frac{1}{R_1 C_1}$	$\alpha_2 = \frac{1}{R_2 C_2}$	$n = \frac{R_1}{R_2}$	$\alpha_1 = \frac{1}{R_1 C_1}$	$\alpha_2 = \frac{1}{R_2 C_2}$	$n = \frac{R_1}{R_2}$
1	B	f_{co}						
	C	0.514 f_{co}						
2	B		1.41 f_{co}	0.707 f_{co}	0.5			
	C		1.83 f_{co}	0.5 f_{co}	0.275			
3	B	f_{co}	2.0 f_{co}	0.5 f_{co}	0.25			
	C	2.02 f_{co}	4.0 f_{co}	0.25 f_{co}	0.063			
4	B		1.08 f_{co}	0.924 f_{co}	0.86	2.6 f_{co}	0.38 f_{co}	0.15
	C		7.0 f_{co}	0.141 f_{co}	0.02	12.95 f_{co}	1.21 f_{co}	0.41

High-pass filters with Butterworth or Tchebycheff amplitude characteristics can be synthesized by employing active and simple RC filters. Table IV gives values for Butterworth and Tchebycheff filters with one to four poles.

Figures 20 and 21 give the amplitude characteristics of Butterworth and Tchebycheff (1-dB ripple) high-pass filters of order one to four.

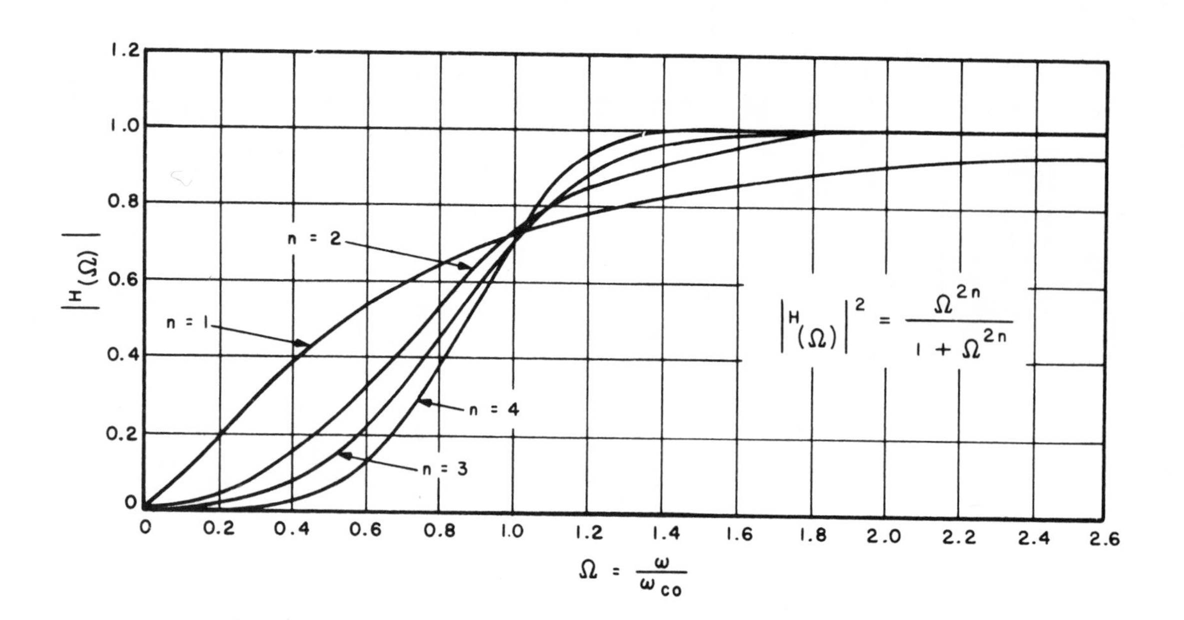

Figure 20. High-Pass Butterworth Filter

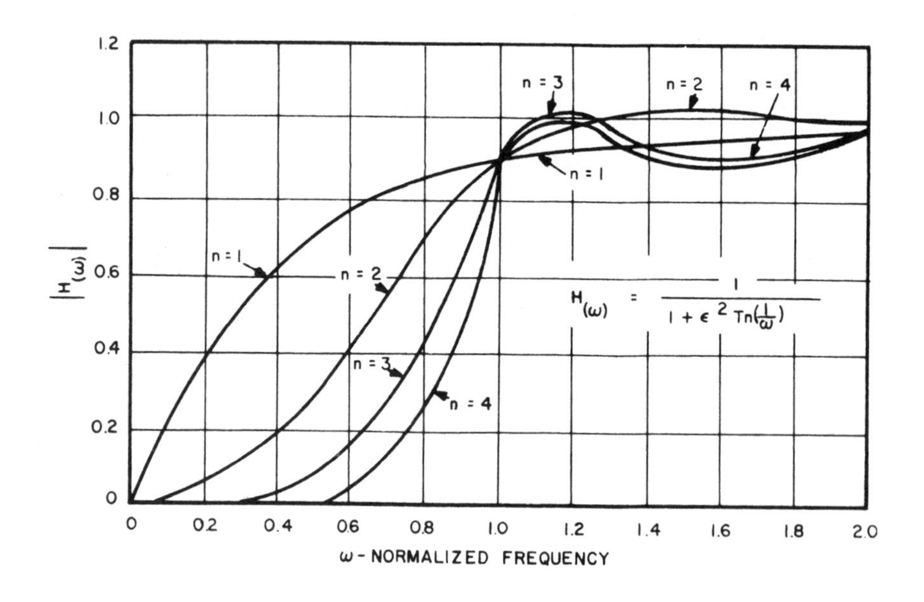

Figure 21. Tchebycheff High-Pass Filter

VI. PERFORMANCE CALCULATIONS

The overall performance of an MTI system can be found by combining the performance due to the doppler filter operating on the target-clutter spectrum with that due to system instabilities. For example, if the doppler filter provides a CA of 30 dB while system instabilities are 30 dB then the overall CA found from Fig. 14 is 27 dB.

It is difficult to predict system instabilities and the best procedure is to measure their effect by searchlighting on a hard target. However, in the initial design phase, some estimate of system stabilities is required and Table V gives a rough approximation of practical values for various systems. The following general rules were used in determining the values of Table V.

1. Coherent magnetron systems are limited to approximately 35 dB due to COHO lock, drift, and STALO stability considerations.
2. Single delay-line filters are limited to approximately 35 dB cancellation ratios; double and shaped delay-line filters to 40 dB cancellation ratios; and RGF's to 50 dB cancellation ratio.
3. Completely coherent MOPA transmitters are limited by their phase stability to approximately 50 dB.

The CA performance limitation due to the MTI filter acting on the clutter spectrum is given in the following paragraphs for the various filters discussed.

A. Single Delay-Line Filter

The CA performance for the single delay-line filter is given by:

$$CA_1 = \frac{0.5}{\frac{4}{\sqrt{2\pi\sigma_c^2}} \int_0^{\pi/T \to \infty} \epsilon^{-\omega^2/2\sigma_c^2} \cdot \sin^2 \frac{\omega T}{2}\, d\omega} \tag{62}$$

Table V. Approximate Performance Limitation Due to System Instabilities

Magnetron or Noncoherent Transmitter	Coherent (dB)	Noncoherent (dB)
Single delay filter	25 to 35	25 to 35
Double delay filter	30 to 35	30 to 40
Shaped delay filter	30 to 35	30 to 40
RGF	30 to 35	30 to 40
MOPA or Coherent Transmitter		
Single delay filter	30 to 35	
Double delay filter	35 to 40	
Shaped delay filter	35 to 40	
RGF	40 to 50	

B. Double Delay-Line Filter

The CA performance for the double delay-line filter is given by (equation 63 is plotted in Fig. 22):

$$CA_1 = \frac{0.5}{\dfrac{16}{\sqrt{2\pi\sigma_c^2}} \displaystyle\int_0^{\pi/T \to \infty} \epsilon^{-\omega^2/2\sigma_c^2} \cdot \sin^4 \frac{\omega T}{2}\, d\omega} \tag{63}$$

C. Scanning Limitation for Single and Double Delay-Line Filters

When σ_c is determined by the scanning limitation, then the performance of single and double delay-line filters can be related to the number of pulses per 3 dB beamwidth (n_a). Table VI gives an evaluation of equations 62 and 63 in terms of n_a for coherent and noncoherent MTI systems.

Table VI. Clutter Attenuation-Scanning Limitation

	Coherent	Noncoherent
Single delay line	$n_a^2/1.96$	$n_a^2/3.92$
Double delay line	$n_a^4/11.5$	$n_a^4/46$

Clutter attenuation as a function of n_a is given in Fig. 22. The values given in Table VI were obtained by matching a gaussian spectrum to the spectrum found for a uniformly illuminated scanning antenna and are relatively insensitive to antenna pattern shape (references 1 and 2).

D. Shaped Delay-Line Filters

Both shaped delay-line and range-gated filters can be designed to have a variety of transfer functions. Tchebycheff and Butterworth transfer functions generally provide good performance and can be readily synthesized using the procedures given in Sections V-3 and V-4. However, in general, the transfer function is determined by the application, and exact performance calculations are complicated.

A useful general procedure is to assume an asymptotic approximation given by:

$$\left|H_{A(\omega)}\right|^2 = \left(\frac{\omega}{\omega_{co}}\right)^{2n} \quad \omega \leq \omega_{co}$$
$$= 1 \quad \omega \geq \omega_{co} \tag{64}$$

where n is the number of filter poles and ω_{co} is the radian doppler frequency dividing the stop and pass band.

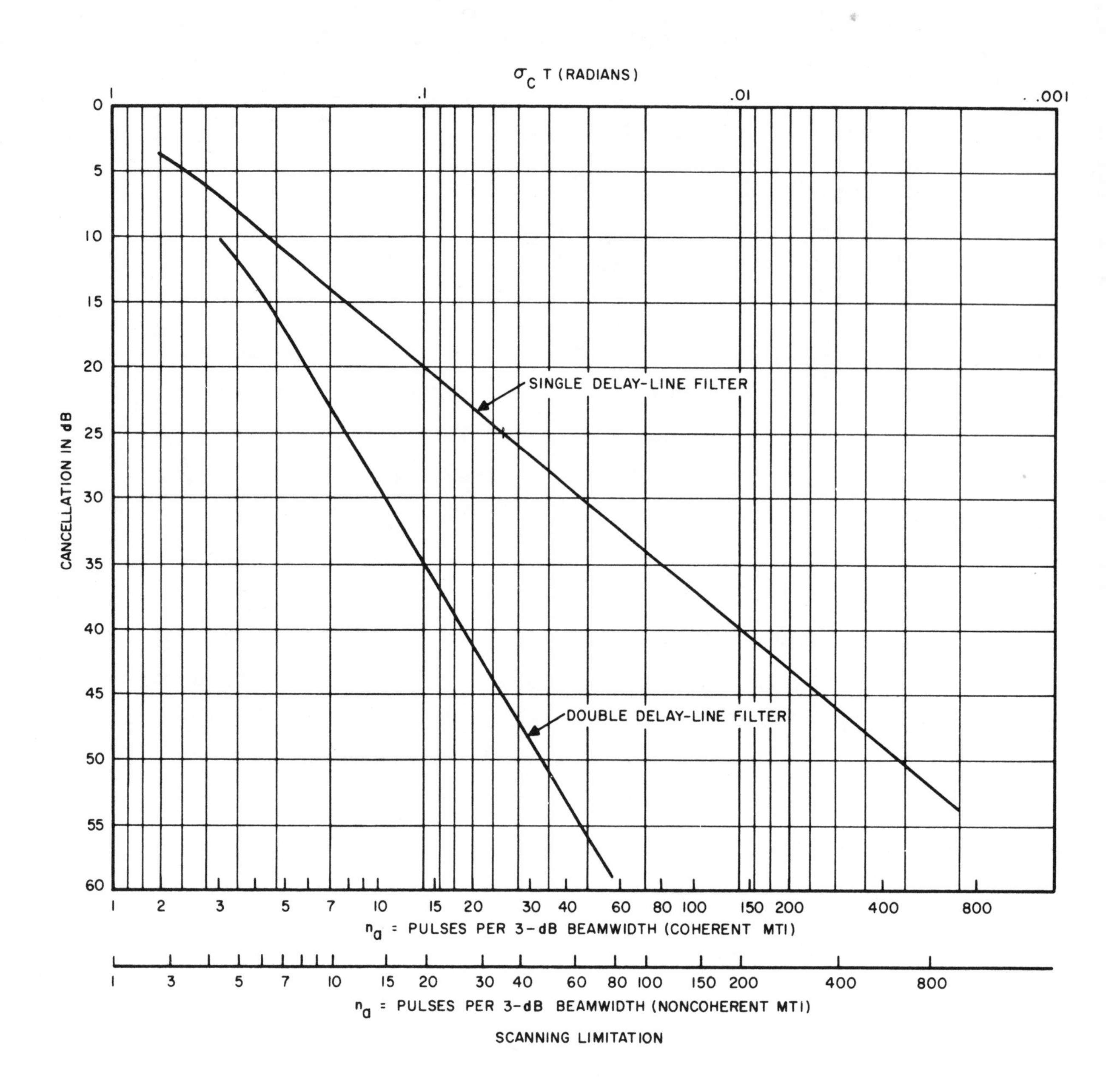

Figure 22. Clutter Attenuation for Single and Double Delay-Line Filters

The CA of the two pole shaped delay-line filter of Fig. 4 using the asymptotic approximation is given by:

$$CA_1 = \frac{0.5}{\frac{1}{\sqrt{2\pi\sigma_c^2}} \int_0^{\omega_{co}} \epsilon^{-\omega^2/2\sigma_c^2} \left(\frac{\omega}{\omega_{co}}\right)^4 d\omega + \frac{1}{\sqrt{2\pi\sigma_c^2}} \int_{\omega_{co}}^{\pi/T \to \infty} \epsilon^{-\omega^2/2\sigma_c^2} d\omega} \qquad (65)$$

This equation is plotted in Fig. 23 ($n = 2$) with ω_{co} normalized to the standard deviation (σ_c) of the clutter spectrum.

CA and SCV are equivalent criterion when the target spectrum is located in the filter's pass band.

E. Range-Gated Filters

The performance analysis of RGF's is similar to that used for shaped delay-line cancelers, with two exceptions. First, the effect of the boxcar response and the low-pass filter on the clutter component at the fundamental PRF must be considered. This is important because the basic high-pass filter does not attenuate this clutter component. Secondly, the effect of the boxcar response and low-pass filter on the target spectrum must be considered. It should be noted that it is possible to peak the low-pass filter so that the combined boxcar/low-pass filter response is flat over the expected range of target frequencies.

The CA due to the clutter component about dc is found from:

$$CA_1 = \frac{0.5}{\frac{1}{\sqrt{2\pi\sigma_c^2}} \int_0^{\omega_{co}} \epsilon^{-\omega^2/2\sigma_c^2} \left(\frac{\omega}{\omega_{co}}\right)^{2n} d\omega + \frac{1}{\sqrt{2\pi\sigma_c^2}} \int_{\omega_{co}}^{\pi/T \to \infty} \epsilon^{-\omega^2/2\sigma_c^2} d\omega} \qquad (66)$$

The equation is plotted in Fig. 23 for n equal to 1, 2, 3 and 4 poles.

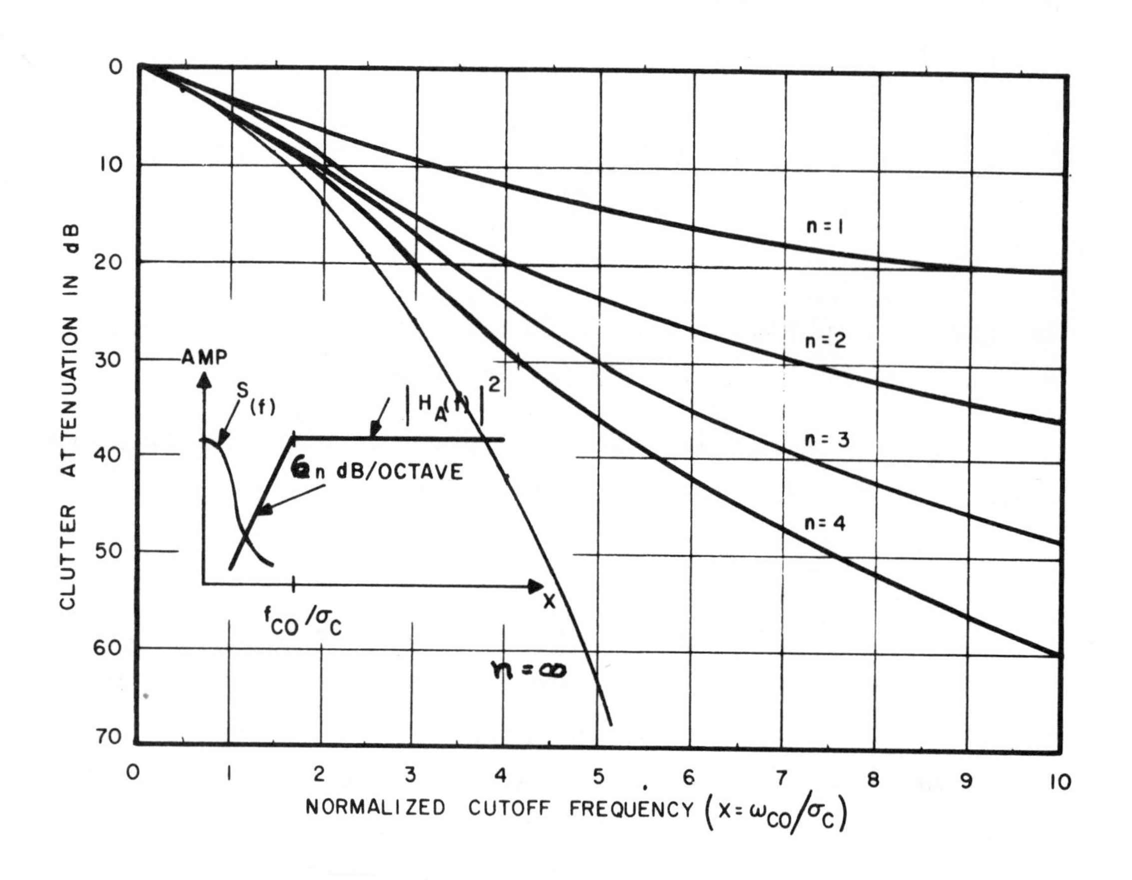

Figure 23. Clutter Attenuation for Various Filter Cutoff Rates

The CA due to the clutter component about the fundamental PRF line is given by:

$$CA_2 = \frac{1}{\frac{4}{\sqrt{2\pi\sigma_c^2}} \int_0^{\pi/T} \frac{\sin^2 \frac{\omega T}{2}}{\omega^2 T^2} \epsilon^{-(\omega - 2\pi/T)^2/2\sigma_c^2} d\omega + \frac{4}{\sqrt{2\pi\sigma_c^2}} \int_{\pi/T}^{\infty} \frac{\sin^2 \frac{\omega T}{2}}{\omega^2 T^2} \left(\frac{\pi}{\omega T}\right)^{2n} \epsilon^{-(\omega - 2\pi/T)^2/2\sigma_c^2} d\omega} \quad (67)$$

Equation 67 is plotted in Fig. 24 for a boxcar, and a boxcar in combination with a 1, 2, 3 and 4 pole asymptotic filter.

The net clutter attenuation of the range-gated filter acting on the clutter spectrum can be found by combining the results of equations 66 and 67 using Fig. 14.

F. Effect of Antenna Side Lobes

In airborne applications, the platform motion spectral density usually predominates, especially as the antenna points away from the aircrafts ground track. Under this condition, the antenna side-lobe structure is important in determining system performance. Fig. 25 shows the clutter attenuation of a two pole asymptotic filter acting on the spectrum of equation 35 with 13 and 25 dB side lobes. Fig. 25 also shows the clutter attenuation provided by the gaussian approximation given by equation 33.

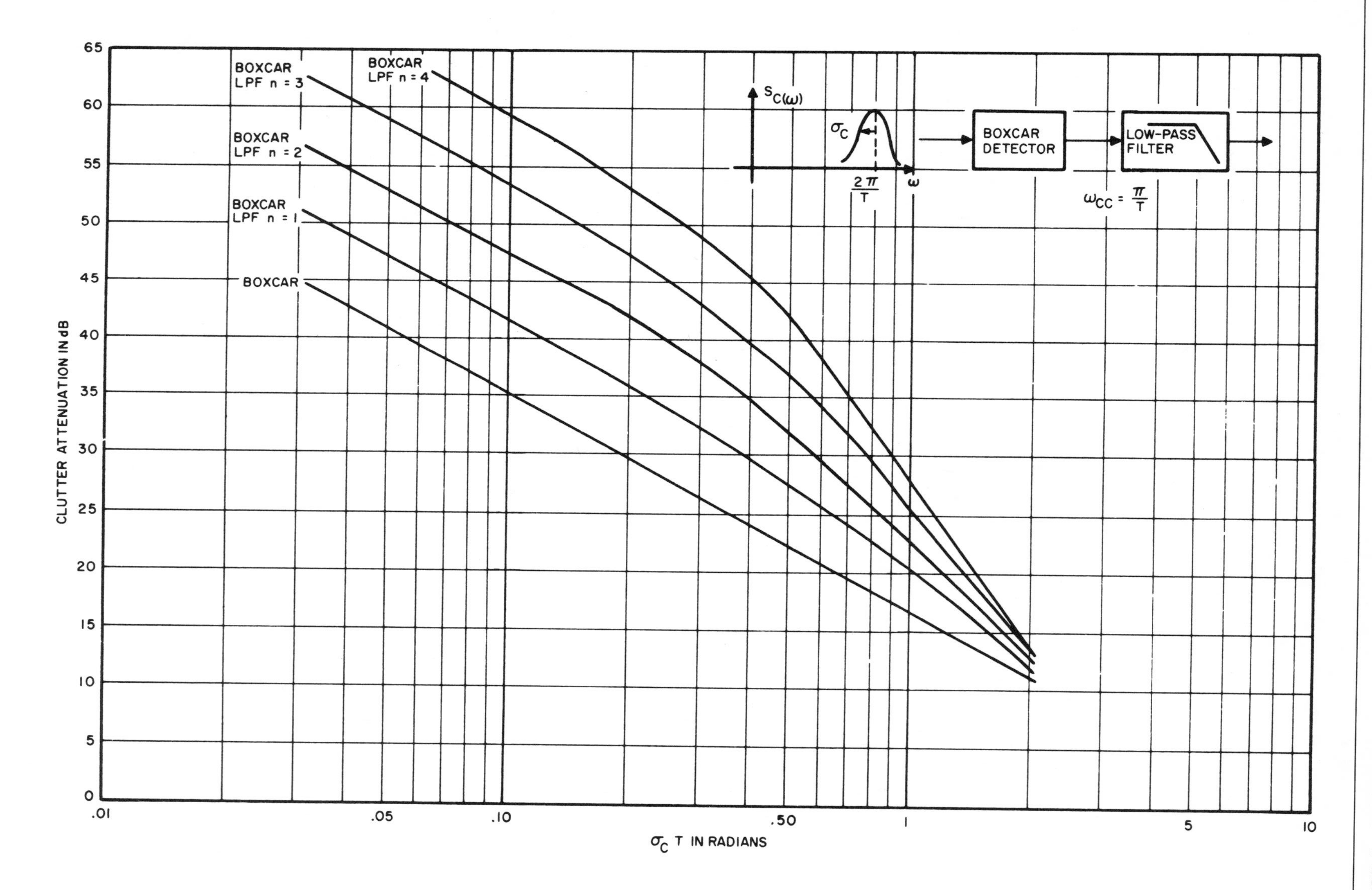

Figure 24. Clutter Attenuation of First PRF Spectrum

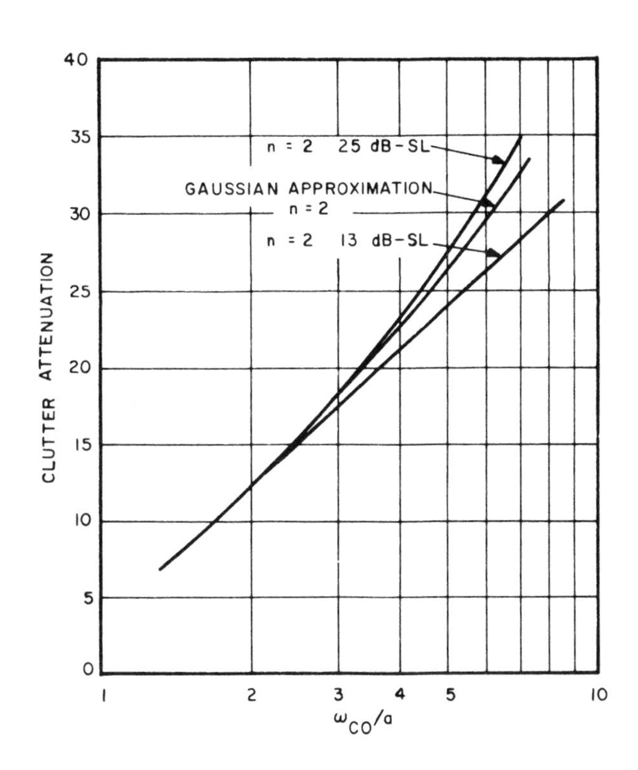

Figure 25. Clutter Attenuation for Platform Motion Spectrum

VII. EXAMPLES

The performance of three MTI radar systems is analyzed. These systems utilize both coherent and noncoherent MTI systems.

The first system is a ground combat surveillance radar whose function is to detect slowly moving personnel and vehicles in a heavy clutter environment.

The second system is an airborne radar system whose function is to detect slowly moving ground vehicles.

The third system is an airport surveillance radar whose function is to detect aircraft for traffic control purposes.

A. Combat Surveillance Radar

The characteristics of this radar pertinent to MTI operation are given in Table VII.

Table VII. Combat Surveillance Radar Characteristics

Item	Characteristics
Frequency	16.3 GHz
Wavelength	1.8 cm
Pulse width	0.25 μs
Pulse repetition frequency	4000 pps
Azimuth beamwidth	1.15 degrees
Scan speed	18 degrees/second
Sector width	120 degrees
MTI type	Noncoherent
System instability	35 dB
Transmitter instability	25 kHz pulse to pulse
Backscattering coefficient	-20 dB

Table VII (cont)

Item	Characteristics
Doppler filter	Range-gated filter
Doppler cutoff frequency	100 Hz
Doppler cutoff rate	12 dB/octave

The system requirements are to detect: (1) personnel (1 m^2) moving at radial velocities greater than 2 knots at ranges up to 5000 meters, and (2) vehicles (10 m^2) moving at radial velocities up to 30 knots. Performance is to be investigated in winds of 10 and 25 knots.

The equivalent clutter cross-sectional area is calculated from equation 22 as:

$$\begin{aligned} \sigma_e &= 38 \text{ m}^2 \text{ at 5000 meters} \\ &= 76 \text{ m}^2 \text{ at 10,000 meters} \end{aligned} \tag{68}$$

The signal-to-clutter ratios before doppler filtering are:

$$\left.\frac{S}{C}\right|_{1 \text{ m}^2} = \frac{1 \text{ m}^2}{38\text{m}^2} = -15.8 \text{ dB} \tag{69}$$

on a 1 m^2 target at 5000-meters range and:

$$\left.\frac{S}{C}\right|_{10 \text{ m}^2} = \frac{10 \text{ m}^2}{76 \text{ m}^2} = -9 \text{ dB} \tag{70}$$

on a 10 m^2 target at 10,000 meters range.

The clutter spectral variance, as found from Table III, is ($\sigma_c^2 = 2\,\sigma_\theta^2 + 2\,\sigma_f^2 + \sigma_w^2$; $\sigma_\theta = 25.6$ rad/s, $\sigma_f = 67$ rad/s, $\sigma_w = 27.8$ rad/s, 81 rad/s for 10 and 25 knot winds)

$$\begin{aligned} \sigma_c &= 108 \text{ rad/s (17.4 Hz) for 10 knot wind} \\ &= 152 \text{ rad/s (24.2 Hz) for 25 knot wind} \end{aligned} \tag{71}$$

Clutter attenuation is found from Fig. 23 to be 26 dB ($\omega_{co}/\sigma_c = 5.8$) for a 10 knot wind and 20 dB ($\omega_{co}/\sigma_c = 4.1$) for a 25 knot wind. The clutter attenuation of the first PRF line is found from Fig. 24 to be 55 dB ($\sigma_{cT} = 0.027$) for a 10 knot wind and 54 dB ($\sigma_{cT} = 0.0395$) for a 25 knot wind. The overall clutter attenuation is found from Fig. 14 to be 26 dB for the 10 knot wind and 20 dB for the 25 knot wind, since the clutter attenuations due to system instabilities (35 dB) and the first PRF line are insignificant compared to that of the doppler filter.

The target spectral width is found from Table III to be:

$$\begin{aligned} \sigma^2_{tar} &= 2\,\sigma_\theta^2 + \sigma^2_f + \sigma^2_w \\ &= 2\,(25.6^2) + 67^2 + 27.8^2 = 6602 \end{aligned} \tag{72}$$

$$\begin{aligned} \sigma_{tar} &= 81 \text{ rad/s for 10 knot wind} \\ &= 111 \text{ rad/s for 25 knot wind} \end{aligned}$$

For a 2 knot ($\omega_{tar} = 690$ rad/s) target, the loss of target power due to the doppler clutter filter ($\omega_{co} = 628$ rad/s) is less than 1 dB for a 10 knot wind and less than 1.5 dB for a 25 knot wind.

The signal-to-clutter ratio for a personnel target at 5000 meters fully in the doppler filters pass band is +10.2 dB for a 10 knot wind and +4.2 dB for a 25 knot wind. The signal-to-clutter ratio for a vehicle target at 10,000 meters is +17 dB for a 10 knot wind and +11 dB for a 25 knot wind.

The display used for this radar has an angular width of 120 degrees and a range of 5000 meters (0 to 5000; 5000 to 10,000). This results in 125-range elements and 104-angle elements for a total of 1.3×10^4 elements on the display. If it is assumed that 10 false alarms per scan are allowed, then the probability of a false alarm is approximately 10^{-3}. Reference 2 (Fig. 1.12) gives a probability of detection on a fluctuating target of 50 percent for a signal-to-noise of 10 dB and 18 percent for a signal-to-noise of 4 dB. It should be noted that even though the MTI output is integrated, there is very little integration gain because of the high correlation between successive doppler filtered clutter returns. Table VIII summarizes the performance of the radar.

Table VIII. Performance Summary of CS Radar

Range (km)	σ_e (m^2)	$\frac{1\ m^2}{\sigma_e}$	$\frac{10\ m^2}{\sigma_e}$ (db)	$\frac{S}{C}\Big\vert_{10\ k}$ (db)	Pd (percent)	$\frac{S}{C}\Big\vert_{25\ k}$ (db)	Pd (percent)
5	38	-15.8		10.2	50	4.2	18
10	76		-9.0	17.0	85	11.0	55

$P_{fa} = 10^{-3}$; fluctuating target

B. Airborne Surveillance Radar

The characteristics of this airborne radar are assumed to be the same as given in Table VII with the exception that: (1) the scan speed is increased to 60 degrees per second, (2) the scan sector is ± 45 degrees, (3) the pulse-to-pulse transmitter frequency jitter is 50 kHz, (4) the backscattering coefficient (σ_0) is -15 dB, and (5) the doppler filter characteristics are to be investigated. The radar is required to detect a small (4 m^2) moving ground vehicle at a range of 18 km. The aircraft has a velocity of 200 knots and a maximum ground wind velocity of 10 knots is expected.

It is desired to calculate the minimum target velocity that can be detected as a function of azimuth scan angle. A four-pole, range-gated filter with a variable cutoff frequency, a single delay-line filter, and a double delay-line filter are to be analyzed.

The equivalent clutter cross-sectional area at 18-km range is calculated from equation 22 as 430-square meters.

The signal-to-clutter ratio on a 4-square meter target at 18 km before doppler filtering is:

$$\left.\frac{S}{C}\right|_{4\ m^2} = \frac{4}{430} = -20\ \text{dB} \tag{73}$$

The display has an angular width of 90 degrees (78 elements), and a range of 18,000 meters (480 elements), resulting in 3.75×10^4 elements. If 10 false alarms per scan are allowed, the probability of a false alarm is approximately 3×10^{-4}. If a probability of detection on a fluctuating target of 40 percent is desired at maximum range, then a signal-to-clutter ratio of +9 dB is required.

The spectral variance of the clutter is found from Table IV; σ_θ = 71 rad/s, σ_f = 134 rad/s, σ_w = 28 rad/s, and σ_{pm} = 360 sin θ rad/s. The video clutter and target spectral standard deviations are given in Table IX as a function of antenna scan angle.

The cutoff frequency of the range-gated filter is varied as a function of antenna scan angle to provide the required clutter attenuation while providing minimum attenuation of the target. A good compromise is to allow the filter to attenuate the minimum velocity target by 1 dB. This allows the center of the target spectrum to be located approximately 1σ higher than ω_{co}. Making this assumption results in a required overall clutter attenuation of 30 dB. The required clutter attenuation of the range-gated filter is found from Fig. 14 to be 32 dB. From Fig. 24, it is found that ω_{co} must be located at 4.5 σ_c to provide 32-dB clutter attenuation. The values of ω_{co} for various azimuth scan angles is given in Table IX. Fig. 26 plots the minimum target velocity at 18-km range that can be detected with the range-gated filter as a function of azimuth scan angle.

Table IX. Video Clutter and Target Standard Deviations

θ (degrees)	$\sigma_c\|_{vid}$ (rad/s)	$\sigma_{tar}\|_{vid}$ (rad/s)	ω_{co} (rad/s)
0	217	170	980
5	222	172	995
10	232	180	1040
20	276	208	1240
30	335	247	1500
45	420	305	1890

The clutter attenuation provided by single and double delay-line filters, as a function of azimuth scan angle as found from Fig. 22, is given in Fig. 27. The overall clutter attenuation is found from Fig. 14. Minimum target velocities that provide the desired probability of detection for single and double delay-line filters are given in Fig. 25.

C. Airport Surveillance Radar

The pertinent characteristics of the airport surveillance radar are given in Table X.

The radar is required to detect low-flying aircraft approaching the radar at speeds up to 600 knots. Over the expected velocity range there are many blind zones that occur at multiples of approximately 80 knots ($v_b = \lambda \cdot PRF/102$; λ in cm, PRF in pps). PRF staggering is

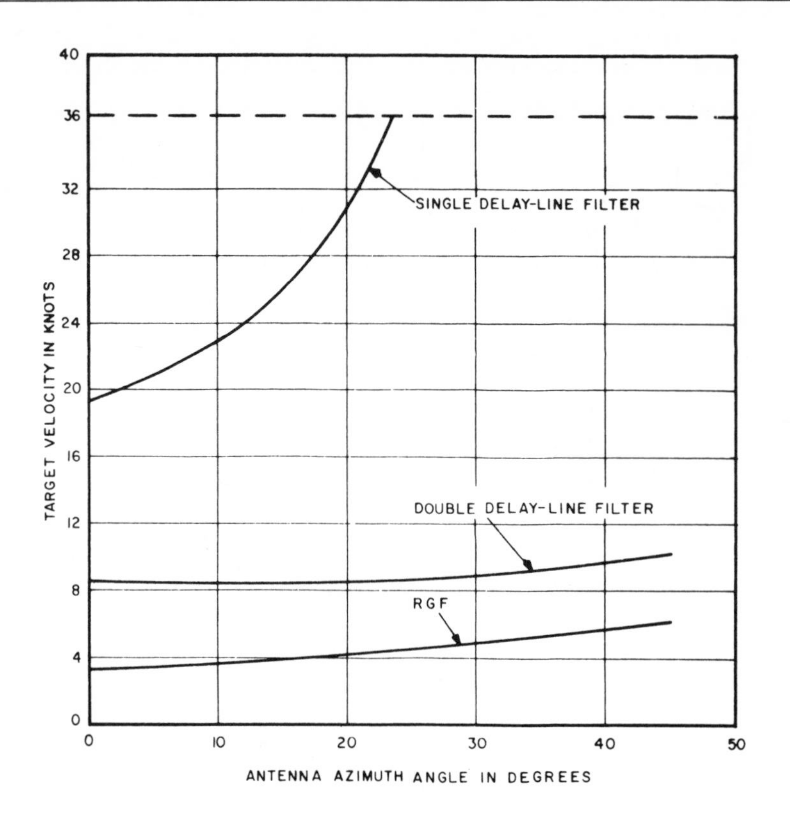

Figure 26. Minimum Target Velocities that Provide S/C = +9 dB at 18-km Range

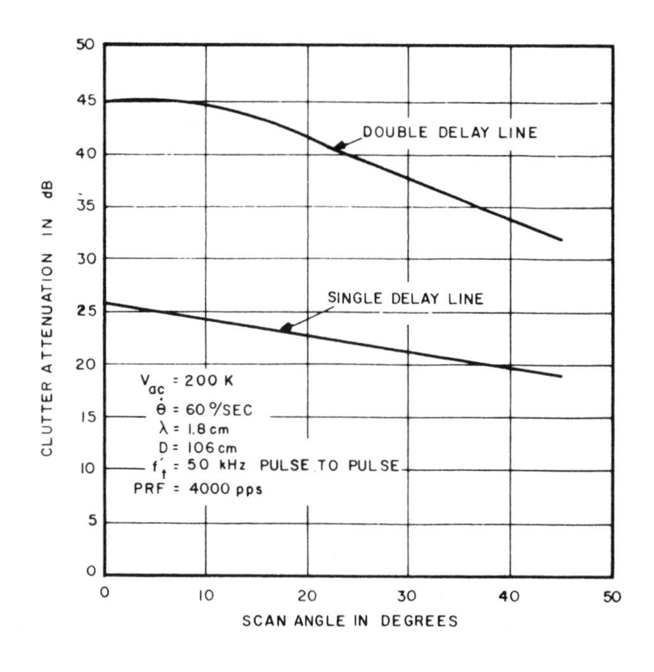

Figure 27. Clutter Attenuation of Single and Double Delay-Line Filters

used to smooth out the blind zones in the velocity response. The average velocity response is determined by the MTI gain of the system.

Ground clutter is assumed to extend out to 20 nmi. The system performance on a 10 m^2 target of single, double, and shaped delay-line filters is to be calculated.

The equivalent clutter cross-sectional area at 20 nmi is calculated from equation 22 as 7600 $feet^2$.

The signal-to-clutter ratio on a 10 m^2 target before doppler filtering is:

$$\left.\frac{S}{C}\right|_{10\ m^2} = \frac{10}{710} = -18.5\ \text{dB} \tag{74}$$

The clutter spectral variance, σ_c, for a 25-knot wind as found from Table III ($\sigma_\theta = 44$, $\sigma_f = 7.2$, $\sigma_w = 6.4$, $\sigma_c^2 = \sigma_e^2 + \sigma_f^2 + \sigma_w^2$) is 45 rad/s.

Table X. Characteristics of Airport Surveillance Radar

Item	Characteristic
Frequency	1280 to 1350 MHz
Wavelength	23 cm
Pulse width	2 μs
PRF	360 pps
Azimuth beamwidth	1.2 degrees
Scan speed	6 r/min
MTI type	Coherent
System instability	33 dB
Subclutter visibility	27 dB
Transmitter instability	10 kHz pulse to pulse (estimated)
Doppler filter	Shaped delay-line filter
Backscattering coefficient	-25 dB

The clutter attenuation provided by single and double delay-line filters as found from Fig. 22 (σ_{cT} = 0.125) is 18 and 31 dB. The overall clutter attenuation found from Fig. 14 is 18 dB for the single delay-line filter and 29 dB for the double delay-line filter. The specified SCV for the system is 27 dB. Fig. 23 gives a cutoff frequency (ω_{co}) of six times the clutter spectrums standard deviation to provide 27-dB clutter attenuation, or equivalently for a shaped delay-line filter 27 dB SCV.

The MTI gain of the three systems can be found from equation 21. The MTI gain of the single delay filter is 0.5, while the double delay filter provides a gain of 0.375. The MTI gain of a RGF or a shaped delay-line asymptotic filter is given by:

$$\text{MTI gain}\Big|_{RGF} = 1 - \frac{2n}{2n+1} \frac{\omega_{co}}{\frac{\omega_{PRF}}{2}} \tag{75}$$

where n is the number of filter poles, ω_{co} is the filter cutoff frequency and ω_{PRF} is the radar's pulse repetition frequency. The MTI gain of the shaped delay-line filter as found from equation 75 is 0.8.

The radar's MTI performance is summarized in Table XI.

Table XI. Summary of Airport Surveillance Radar Performance

Filter Type	Clutter Attenuation (dB)	MTI Gain
Single delay line	18	0.5
Double delay line	29	0.375
Shaped delay line	27	0.80

REFERENCES

1. Skolnik, Introduction to Radar Systems, McGraw-Hill, 1962.

2. Barton, Radar System Analysis, Prentice-Hall, 1965.

3. Berkowitz, Modern Radar, Wiley, 1965.

4. Waterman, Airborne Radar, Van Nostrand, 1961.

5. Schwartz, Bennett, and Stein, Communication Systems and Techniques, McGraw-Hill, 1966.

6. Kuo and Freeny, "Hilbert Transforms and Modulation Theory," National Electrics Conference, p 51-58, 1962.

7. Truxal, Automatic Feedback Control Systems Synthesis, McGraw-Hill, 1955.

8. Davenport and Root, Introduction to Random Signals and Noise, McGraw-Hill, 1958.

9. RTD-TR-65-AWACS, Clutter Data Appendixes, Airborne Instruments Laboratory, September 1965.

10. Crist, Scanning Modulation and MTI Filter Design, Unpublished AIL Memorandum, 1963.

11. Marshall and Hitschfeld, "Interpretation of the Fluctuation Echo From Randomly Distributed Scatterers," Canadian Journal of Physics, Vol 31, 1953.

12. White and Ruvin, "Recent Advances in the Synthesis at Comb Filters," IRE National Convention Record, 1957.

13. Weinberg, Network Analysis and Synthesis, McGraw-Hill, 1962.

14. Cappadona and Kenneally, "A Microminiature MTI System For Surveillance Radars," IEEE International Convention Record, 1966.

Section 2 Optimum MTI Filters

EDITOR'S COMMENTS

Theoretical criteria for the optimization of MTI filters are developed in this series of papers. Two situations are examined for detecting doppler shifted target signals in Gaussian clutter.

One criterion assumes that all target velocities are equally likely, and that the MTI filter functions to minimize the clutter output under this constraint. This criterion was initially developed by Emerson (Paper 1.1), and its properties are examined by Capon (Paper 2.1), Robinson (Paper 2.2), and Hsiao (Paper 2.4).

Emerson's procedure is modified by Schleher and Schulkind (Paper 2.5) to include additional constraints that suppress discrete fixed targets and allow a preferred MTI velocity characteristic.

The second criterion, examined by Brennen and Reed (Paper 2.3), assumes that the target's radial doppler frequency is known a priori. The resulting optimum likelihood receiver for detecting a target of known doppler frequency in Gaussian clutter corresponds to a bank of doppler filters which cover the expected range of target velocities.

Many useful properties of these optimum MTI filter designs are discussed in this series of papers.

Capon (Paper 2.1) shows that the optimum MTI filter weights for equally likely target dopplers can be found by solving for the eigenvalues of the clutter's covariance matrix. The improvement factor is the reciprocal of the minimum eigenvalue, while the weights are given by the elements of the eigenvector that corresponds to the minimum eigenvalue. Expressions for the improvement factor and optimum weights of two, three and four pulse MTI processors are given.

The asymptotic solution ($n \rightarrow \infty$) for the optimum weights approaches a very narrow band filter whose center frequency is tuned to the minimum value of the clutter spectrum.

An analytic solution is given for the case of a Markov random process (exponential covariance function). The improvement factor for this case shows a monotonically increasing value as a function of the number of pulses processed.

Robinson (Paper 2.2) shows that the z-plane zeros of the optimum MTI filters transfer function lie on the unit circle. This is an important analytic property of the optimum MTI filter.

Brennan and Reed (Paper 2.3), using a likelihood ratio statistical detection test, show that the optimum filter weights for detecting a known doppler shifted signal in Gaussian clutter are given by the vector product of the inverse of the clutter covariance matrix and the complex conjugate of the signal vector. They also derive an expression for the optimum weights using a maximum signal-to-clutter ratio criteria and show that the weights are identical to those obtained using the likelihood ratio test. This well-known result of estimation theory states that for Gaussian noise the maximum likelihood ratio filter, maximum signal-to-noise filter and minimum variance (Wiener or Kalman) filter are all identical linear filters.

Brennan and Reed (Paper 2.3) also derive the optimum weights for Emerson's procedure using a Lagrangian multiplier approach. They provide an expression for the improvement factor when unequally spaced pulses are used and the clutter's covariance matrix is exponential. This allows selection of optimum spacing codes for staggering the radar's PRF.

Hsiao (Paper 2.4), using a matrix algebra approach, examines in detail the properties of Emerson's optimal weight solution. Some of the interesting properties of the optimum solution are:

- The power transfer function increases monotonically from 0 to the ½ PRF frequency.
- The clutter power output increases as the width of the clutter spectrum increases.
- The clutter power output for a given clutter spectrum width decreases as the number of processed pulses increases.
- The improvement factor asymptotically approaches a limit as the number of processed pulses becomes large.
- For large clutter spectral widths, the output clutter power becomes independent of the number of processed pulses.

Hsiao provides a curve for the improvement factor as a function of the ratio of the clutter spectrum width to the PRF for n = 2 to 20 processed pulses. He also shows that the improvement factor for a staggered PRF MTI lies between the performance provided by the lowest and highest PRFs of the stagger.

Schleher and Schulkind (Paper 2.5) discuss several limitations of Emerson's procedure when applied to the design of a practical MTI. These limitations are:

1. Fixed discrete targets are eliminated only when an even number of pulses are processed.
2. The MTI's velocity response has significant variation in the passband. As the number of processed pulses becomes large this variation asymptotically approaches an impulse at the velocity where the clutter spectrum is a minimum.
3. The specification of a preferred velocity response causes a complete breakdown of Emerson's optimization procedure.

A quadratic programming approach is described that allows the essential features of Emerson's procedure to be retained while specifying a preferred velocity response and suppressing fixed discrete targets.

2.1 OPTIMUM WEIGHTING FUNCTIONS FOR THE DETECTION OF SAMPLED SIGNALS IN NOISE
J. Capon
IEEE Transactions on Information Theory, Vol. IT-10, No. 2, April 1964, pp. 152-159.

Optimum Weighting Functions for the Detection of Sampled Signals in Noise

JACK CAPON, MEMBER, IEEE

Summary—The problem of designing a linear predetection filter for the detection of a sampled random signal in additive noise is considered. The design of the filter is based on an optimality criterion which maximizes the signal-to-noise ratio enhancement. The optimum weighting function obtained in this manner has the advantage that it is independent of signal characteristics and depends only on the covariance function of the noise.

The optimum filter, for general covariance functions, is obtained for $N = 2$, 3 and 4 samples. The asymptotic solution for large N is also presented by employing results from the theory of Toeplitz forms. In addition, the complete solution for all N is given for several particular covariance matrices.

An application of the results is made to the problem of designing a linear predetection filter in a moving target indication (MTI) radar system. The optimum weighting function for $N = 2$ is a single-cancellation unit, while that for $N = 3$ is similar but not quite the same as a double-cancellation unit. It is shown that the signal-to-noise ratio enhancement provided by the double-cancellation scheme is 1.76 db worse than that of the optimum filter when the noise has a Gaussian covariance function.

Manuscript received July 31, 1963.
The author is with the Applied Research Laboratory, Sylvania Electronic Systems, Sylvania Electric Products, Inc., Waltham, Mass.

Introduction

A PROBLEM of considerable importance is that of detecting a random signal in additive noise or clutter. More precisely, we will consider the problem of detecting signals of the form $S(t, \theta)$, where $S(t, \theta)$ is a real-valued bounded and measurable function of t and θ, t is a continuous time variable and θ is a random variable which is independent of the noise. In addition, it will be assumed that the signal is available only in the form of samples taken at uniformly spaced intervals of time. That is, the only data which is available to the detector is in the form $S(kT, \theta)$, $k = 0, \pm 1, \pm 2, \cdots$.

As an example, let us consider the case of a pulsed-radar system, and in particular the situation in which it is desired to obtain clutter rejection by using moving target indication (MTI). A discussion of MTI radar systems has been given by Ridenour [1]. In this case the signal is given by $S(t, \theta) = A \cos (\omega_d t + \theta)$, where A is a constant, ω_d denotes the Doppler frequency and θ is a random phase angle.

It has been recognized that the problem of detecting signals in noise can be treated as one of testing statistical hypotheses. In fact, a vast literature exists on this subject and these results are summarized to a great extent in four recently published books [2]–[5]. In order to treat the present problem with these methods, we must have complete information about the *a priori* probability of occurrence of the signal, the probability distribution of the noise as well as that of θ, and a loss or cost function indicating the relative loss incurred by false alarms or false dismissals. If a Neyman-Pearson detector is used then only the probability distributions of the noise and of θ are required. In any event the optimum detector is based on a likelihood ratio, the construction of which usually necessitates the assumption that the noise be a Gaussian random process. The likelihood ratio must be averaged with respect to the probability measure of θ, an operation which usually requires the assumption that θ be uniformly distributed in some interval, *e.g.* Reich and Swerling [6].

In spite of all these detailed assumptions concerning the structure of the problem, there are very few results in particular problems. For example, in the MTI radar problem the structure of the optimum detector is known only in the two-pulse case, *i.e.*, in the situation where only two samples are taken from the unknown waveform at the detector input terminals. In such a case, assuming a uniform distribution of phase, the optimum detector consists of a pulse-to-pulse subtraction system, or single cancellation unit, and threshold logic circuitry.[1] The solution for higher-order pulse-sampling cases is difficult to obtain, and in addition the solution in these cases depends upon the signal, or target, characteristics. This is an undesirable feature, due to such effects as target scintillation or fading.

Since there are serious difficulties associated with the detection methods based on statistical decision-theoretic optimality criteria, it is advantageous to introduce an optimality criterion which is more suitable to the problem under consideration. In particular, we will be concerned only with the problem of designing a linear predetection filter or weighting function. The optimality criterion will be based on a certain signal-to-noise ratio and will allow the design of weighting functions in many important practical problems. In addition, the only information required in the problem is the covariance function of the noise and some knowledge of the effect of averaging with respect to the probability distribution of θ. Another advantage is that the weighting functions obtained are independent of signal characteristics. This is important for those applications where signal characteristics are not completely specified.

The Optimality Criterion

We will consider only linear filters or weighting functions. The input of the filter is denoted by $E_1(kT, \theta)$ and is equal to $N(kT) + S(kT, \theta)$, where $\{N(t)\}$ represents a real-valued stationary noise process. The output of the filter, $E_0(NT)$, is related to the input by

$$E_0(NT) = \sum_{k=1}^{N} u_k E_1(kT) \tag{1}$$

where $\bar{u}_N = \{u_1, u_2, \cdots, u_N\} = \{x_1, x_2, \cdots, x_N\} + i\{y_1, y_2, \cdots, y_N\}$, $i = \sqrt{-1}$, is the complex weighting function or vector, the u's are complex numbers and the x's and y's are real numbers. It is, of course, apparent that in the applications a real physical weighting function is desired. It will be obvious subsequently how the real and imaginary parts of $\bar{u}_N$ should be combined to obtain the physical weighting functions.

The average output power of the filter is

$$\begin{aligned}\overline{|E_0(NT)|^2} &= \overline{\left|\sum_{k=1}^{N} u_k[S(kT, \theta) + N(kT)]\right|^2} \\ &= \sum_{j=1}^{N}\sum_{k=1}^{N} u_j u_k^* \overline{S(jT, \theta)S(kT, \theta)} \\ &\quad + \sum_{j=1}^{N}\sum_{k=1}^{N} u_j u_k^* \overline{N(jT)N(kT)} \\ &\quad + \sum_{j=1}^{N}\sum_{k=1}^{N} (u_j u_k^* + u_j^* u_k)\, \overline{S(jT, \theta)}\ \overline{N(kT)}\end{aligned} \tag{2}$$

where the asterisk denotes the complex conjugate. The bars in (2) indicate averages taken with respect to either the probability measure of θ or $\{N(t)\}$. The appropriate measure to use should be obvious from the context. It will be assumed that $\{N(t)\}$ is a weakly stationary random process with mean zero and covariance function

$$\overline{N(t)N(t + \tau)} = C\rho(\tau) \tag{3}$$

where C is the noise power and $\rho(\tau)$ is the normalized covariance function, $\rho(0) = 1$. Thus, (2) can be written as

$$\overline{|E_0(NT)|^2} = \sum_{j=1}^{N}\sum_{k=1}^{N} u_j u_k^* \overline{S(jT, \theta)S(kT, \theta)} + C\sum_{j=1}^{N}\sum_{k=1}^{N} u_j u_k^* \rho_{j-k}, \tag{4}$$

so that the output signal-to-noise ratio is

$$(S/N)_{\text{OUT}} = \frac{\sum_{j=1}^{N}\sum_{k=1}^{N} u_j u_k^* \overline{S(jT, \theta)S(kT, \theta)}}{C\sum_{j=1}^{N}\sum_{k=1}^{N} u_j u_k^* \rho_{j-k}} \tag{5}$$

where

$$\rho((j - k)T) = \rho_{j-k}. \tag{6}$$

The input signal-to-noise ratio is

$$(S/N)_{\text{IN}} = \frac{1}{C}\overline{S^2(NT, \theta)}, \tag{7}$$

[1] See Wainstein and Zubakov [5], pp. 212–222.

so that the system gain is

$$G_N = \frac{(S/N)_{\text{OUT}}}{(S/N)_{\text{IN}}} = \bar{G}_N F_N(S) \tag{8}$$

where

$$\bar{G}_N = \frac{\sum_{k=1}^{N} |u_k|^2}{\sum_{j=1}^{N} \sum_{k=1}^{N} u_j u_k^* \, \rho_{j-k}} \tag{9}$$

$$F_N(S) = \frac{\sum_{j=1}^{N} \sum_{k=1}^{N} u_j u_k^* \frac{\overline{S(jT, \theta) S(kT, \theta)}}{\overline{S^2(NT, \theta)}}}{\sum_{k=1}^{N} |u_k|^2}. \tag{10}$$

It will be assumed hereafter that $\sum_{k=1}^{N} |u_k|^2$ is unity. The quantity $\bar{G}_N$ is the reference gain and is seen to be independent of the signal characteristics and, for fixed N, to depend only on $\rho(\tau)$ and $\bar{u}_N$. The expression $F_N(S)$ is the signal factor and, for fixed N, depends only on signal characteristics and $\bar{u}_N$. The over-all system gain G_N is the product of these two factors.

In formulating an optimality criterion it is desirable that the criterion be independent of signal characteristics, for reasons which have been mentioned previously. This leads us to the following definition for the optimum weighting function:

> The optimum weighting function is given by the vector $\bar{u}_N$ which maximizes the reference gain $\bar{G}_N$.

The main drawback associated with the above definition for optimality is that the optimum weighting function may produce a signal factor $F_N(S)$ which is poor, in the sense that it lowers, or deteriorates, the over-all system gain G_N. This, however, is unavoidable and is the penalty incurred for introducing an optimality criterion which is independent of the signal's characteristics.

It is seen from (9) that $\bar{G}_N$ is the reciprocal of a normalized quadratic form. Therefore its maximum value is equal to the reciprocal of the smallest eigenvalue, $\lambda_{N,1}$, of the covariance matrix

$$R_N = \begin{bmatrix} 1 & \rho_1 & \cdots & \rho_{N-1} \\ \rho_{-1} & 1 & \cdots & \rho_{N-2} \\ \cdots & \cdots & \cdots & \cdots \\ \rho_{-N+1} & \rho_{-N+2} & \cdots & 1 \end{bmatrix}, \tag{11}$$

and this value is assumed when $\bar{u}_N$ is any eigenvector corresponding to this eigenvalue [7].

Since R_N is a non-negative definite matrix its eigenvalues are real and non-negative. The minimum value of $(\bar{G}_N)^{-1}$ over an N-dimensional vector space can be viewed as the minimum value of $(\bar{G}_{N+1})^{-1}$ taken over an N-dimensional subspace of an $(N + 1)$-dimensional vector space. It therefore follows that the minimum value of $(\bar{G}_{N+1})^{-1}$ can never be greater than the minimum value of $(\bar{G}_N)^{-1}$, so that the sequence $\{\lambda_{N,1}\}$ is non-increasing. This implies that the optimum reference gain $\bar{G}_N$ increases, or possibly remains the same, as N increases. In many cases it can be said that $\{\lambda_{N,1}\}$ is a strictly decreasing sequence so that the optimum reference gain can only increase as the memory, N, of the filter increases. The particular value of N to be used will be dictated by a compromise between the amount of reference gain required and the allowable equipment complexity.

General Solution for $N = 2, 3, 4$

In this section the optimum weighting functions will be derived for general covariance matrices for the special cases $N = 2, 3, 4$. If $N = 2$ we have

$$R_2 = \begin{bmatrix} 1 & \rho_1 \\ \rho_1 & 1 \end{bmatrix}, \tag{12}$$

and the eigenvalues satisfy the determinantal equation

$$\begin{vmatrix} (1-\lambda) & \rho_1 \\ \rho_1 & (1-\lambda) \end{vmatrix} = 0 \tag{13}$$

so that

$$\begin{aligned} \lambda_{2,1} &= 1 - \rho_1, \\ \lambda_{2,2} &= 1 + \rho_1, \qquad \rho_1 > 0 \end{aligned} \tag{14}$$

$$\begin{aligned} \lambda_{2,1} &= 1 + \rho_1, \\ \lambda_{2,2} &= 1 - \rho_1, \qquad \rho_1 < 0. \end{aligned} \tag{15}$$

The components of the eigenvector $\bar{u}_2$ must satisfy the equation

$$(1 - \lambda_{2,1})u_1 + \rho_1 u_2 = 0, \tag{16}$$

so that

$$\bar{u}_2 = \left\{\frac{1}{\sqrt{2}}, -\frac{1}{\sqrt{2}}\right\}, \qquad \rho_1 > 0 \tag{17}$$

$$\bar{u}_2 = \left\{\frac{1}{\sqrt{2}}, \frac{1}{\sqrt{2}}\right\}, \qquad \rho_1 < 0. \tag{18}$$

Thus, when $N = 2$, the optimum filter performs pulse-to-pulse subtraction when $\rho_1 > 0$ and pulse-to-pulse addition when $\rho_1 < 0$.

If $N = 3$,

$$R_3 = \begin{bmatrix} 1 & \rho_1 & \rho_2 \\ \rho_1 & 1 & \rho_1 \\ \rho_2 & \rho_1 & 1 \end{bmatrix}, \tag{19}$$

and the eigenvalues satisfy the determinantal equation

$$\begin{vmatrix} (1-\lambda) & \rho_1 & \rho_2 \\ \rho_1 & (1-\lambda) & \rho_1 \\ \rho_2 & \rho_1 & (1-\lambda) \end{vmatrix} = 0$$

which can be written as

$$(1 - \lambda - \rho_2)[(1 - \lambda)^2 + \rho_2(1 - \lambda) - 2\rho_1^2] = 0$$

so that the eigenvalues are

$$\lambda_{3,1} = 1 + \frac{\rho_2}{2}\left(1 - \sqrt{1 + 8\left(\frac{\rho_1}{\rho_2}\right)^2}\right)$$

$$\lambda_{3,2} = 1 - \rho_2$$

$$\lambda_{3,3} = 1 + \frac{\rho_2}{2}\left(1 + \sqrt{1 + 8\left(\frac{\rho_1}{\rho_2}\right)^2}\right) \tag{20}$$

provided $0 \leq \rho_2 \leq \rho_1 < 1$, which is the case of most interest. The components of the eigenvector $\bar{u}_3$ satisfy the equations

$$-\frac{\rho_2}{2}\left(1 - \sqrt{1 + 8\left(\frac{\rho_1}{\rho_2}\right)^2}\right)u_1 + \rho_1 u_2 + \rho_2 u_3 = 0$$

$$\rho_1 u_1 - \frac{\rho_2}{2}\left(1 - \sqrt{1 + 8\left(\frac{\rho_1}{\rho_2}\right)^2}\right)u_2 + \rho_1 u_3 = 0$$

so that

$$u_1 : u_2 : u_3 = 1 : -\frac{\rho_2 + \sqrt{\rho_2^2 + 8\rho_1^2}}{2\rho_1} : 1. \tag{21}$$

If $N = 4$,

$$R_4 = \begin{bmatrix} 1 & \rho_1 & \rho_2 & \rho_3 \\ \rho_1 & 1 & \rho_1 & \rho_2 \\ \rho_2 & \rho_1 & 1 & \rho_1 \\ \rho_3 & \rho_2 & \rho_1 & 1 \end{bmatrix}, \tag{22}$$

and

$$\begin{vmatrix} (1-\lambda) & \rho_1 & \rho_2 & \rho_3 \\ \rho_1 & (1-\lambda) & \rho_1 & \rho_2 \\ \rho_2 & \rho_1 & (1-\lambda) & \rho_1 \\ \rho_3 & \rho_2 & \rho_1 & (1-\lambda) \end{vmatrix}$$

$$= [(1-\lambda)^2 + (\rho_1 + \rho_3)(1-\lambda) + \rho_1\rho_3 - (\rho_1 + \rho_2)^2]$$
$$\cdot [(1-\lambda)^2 - (\rho_1 + \rho_3)(1-\lambda) + \rho_1\rho_3 - (\rho_1 - \rho_2)^2] = 0$$

so that the eigenvalues are

$$\lambda_{4,1} = 1 - \frac{\rho_1 + \rho_3}{2} - \sqrt{\left(\frac{\rho_1 - \rho_3}{2}\right)^2 + (\rho_1 - \rho_2)^2}$$

$$\lambda_{4,2} = 1 + \frac{\rho_1 + \rho_3}{2} - \sqrt{\left(\frac{\rho_1 - \rho_3}{2}\right)^2 + (\rho_1 + \rho_2)^2}$$

$$\lambda_{4,3} = 1 - \frac{\rho_1 + \rho_3}{2} + \sqrt{\left(\frac{\rho_1 - \rho_3}{2}\right)^2 + (\rho_1 - \rho_2)^2}$$

$$\lambda_{4,4} = 1 + \frac{\rho_1 + \rho_3}{2} + \sqrt{\left(\frac{\rho_1 - \rho_3}{2}\right)^2 + (\rho_1 + \rho_2)^2} \tag{23}$$

provided $0 \leq \rho_3 \leq \rho_2 \leq \rho_1 < 1$. The components of the eigenvector $\bar{u}_4$ satisfy the equations

$$\beta u_1 + \rho_1 u_2 + \rho_2 u_3 + \rho_3 u_4 = 0$$

$$\rho_1 u_1 + \beta u_2 + \rho_1 u_3 + \rho_2 u_4 = 0$$

$$\rho_2 u_1 + \rho_1 u_2 + \beta u_3 + \rho_1 u_4 = 0$$

so that

$$u_1 : u_2 : u_3 : u_4 = (\beta^2\rho_3 + \rho_1^3 + \rho_1\rho_2^2 - 2\beta\rho_1\rho_2 - \rho_1^2\rho_2):$$
$$(\beta^2\rho_2 + \rho_1\rho_2\rho_3 + \rho_1^2\rho_2 - \rho_2^3 - \beta\rho_1^2 - \beta\rho_1\rho_3):$$
$$(\beta^2\rho_1 + \rho_1\rho_2^2 + \rho_1^2\rho_3 - \beta\rho_2\rho_3 - \beta\rho_1\rho_2 - \rho_1^3):$$
$$(\rho_2 - \beta)(\beta^2 + \beta\rho_2 - 2\rho_1^2) \tag{24}$$

where

$$\beta = 1 - \lambda_{4,1}. \tag{25}$$

It is not advantageous to pursue the general solution for $N > 4$. Further properties of the general solution will be discussed next for the asymptotic case, $N \to \infty$, by recognizing that $\bar{G}_N$ is the reciprocal of a Toeplitz form [8].

Asymptotic Properties of the Eigenvalues and Eigenvectors of Toeplitz Forms

Let $f(x)$ be a real-valued function of the class L, *i.e.*, $L_1(-\pi, \pi)$, and

$$f(x) = \sum_{k=-\infty}^{\infty} \rho_k \epsilon^{ikx} \tag{26}$$

be its Fourier series, where

$$\rho_k = \frac{1}{2\pi}\int_{-\pi}^{\pi} f(x)\epsilon^{-ikx}\,dx, \qquad k = 0, \pm 1, \pm 2, \cdots. \tag{27}$$

The function $f(x)$ is, of course, related to the sampled-power spectrum of the noise. Since $\rho_k = \rho_{-k}$ we have $f(x) = f(-x)$.

The Hermitian form

$$T_N(f) = \sum_{j,k=1}^{N} \rho_{j-k} u_j u_k^*$$
$$= \frac{1}{2\pi}\int_{-\pi}^{\pi} \left|\sum_{k=1}^{N} u_k \epsilon^{ikx}\right|^2 f(x)\,dx \tag{28}$$

is defined as the Toeplitz form associated with the function $f(x)$, and the matrix R_N is called a Toeplitz matrix.[2] It is known,[3] that $f(x)$ is non-negative, except for a set of Lebesgue measure zero, if and only if its Toeplitz forms are non-negative for all values of N. Since R_N is a covariance matrix, and thus non-negative definite, it follows that $f(x)$ is non-negative almost everywhere with respect to Lebesgue measure.

It is seen from (9) that $\bar{G}_N$ is the reciprocal of the Toeplitz form associated with $f(x)$. Hence, the asymptotic behavior of $\lambda_{N,1}$ and of the optimum weighting function can be obtained directly from some known results for the eigenvalues and eigenvectors of Toeplitz forms.

We will assume that $f(x)$ is continuous in $[-\pi, \pi]$, with $f(-\pi) = f(\pi)$. Let $\min f(x) = f(x_0)$ and let $x = x_0$ be the only value of x (mod 2π) for which this minimum is attained. Moreover, assume that $f(x)$ has a continuous second derivative in a certain neighborhood of x_0. Finally, let $f''(x_0) \neq 0$. Since x_0 is a minimum point of $f(x)$ we

[2] See Grenander and Szegö [8], p. 17.
[3] *Ibid.*, p. 19.

must have $f''(x_0) > 0$. Under these conditions, it is known that,[4]

$$\lim_{N\to\infty} N^2(\lambda_{N,1} - f(x_0)) = \frac{\pi^2}{2} f''(x_0). \tag{29}$$

Thus, the reference gain $\bar{G}_N$ can never be greater than $1/f(x_0)$, and approaches this value asymptotically, in a nondecreasing manner, at the rate indicated in (29). Therefore, the number $f(x_0)$ sets a natural limit on the performance of the predetection filter.

We will say that a vector $\bar{u}_N$ with Toeplitz form $T'_N(f)$ is asymptotically an optimum weighting function, if for each $\delta > 0$ there exists $N(\delta)$ such that

$$|T'_N(f) - \lambda_{N,1}| < \delta, \qquad N \geq N(\delta). \tag{30}$$

Physically speaking, the justification for this definition of an asymptotically optimum weighting function is that, if $f(x_0) \neq 0$, it leads to a reference gain which can be made as close to the limiting gain $1/f(x_0)$ as desired by simply increasing N. If $f(x_0) = 0$, then the asymptotically optimum weighting functions produce a sequence of reference gains which tend, in a nondecreasing manner, to infinity as $N \to \infty$.

If $f(x)$ satisfies the conditions stated previously, and if $f(x)$ is not identically equal to $f(x_0)$, then asymptotically optimum weighting functions are given by,[5]

$$\bar{u}_N = \{N^{-1/2}\epsilon^{i(x_0+\pi/N)}, N^{-1/2}\epsilon^{2i(x_0+\pi/N)}, \cdots, N^{-1/2}\epsilon^{Ni(x_0+\pi/N)}\} \tag{31}$$

or

$$\bar{u}_N = \{N^{-1/2}\epsilon^{i(x_0-\pi/N)}, N^{-1/2}\epsilon^{2i(x_0-\pi/N)}, \cdots, N^{-1/2}\epsilon^{Ni(x_0-\pi/N)}\}. \tag{32}$$

The physical asymptotically optimum weighting functions are

$$\bar{u}_N = \left\{\left(\frac{2}{N}\right)^{1/2} \cos\left(x_0 + \frac{\pi}{N}\right), \left(\frac{2}{N}\right)^{1/2} \cos 2\left(x_0 + \frac{\pi}{N}\right), \cdots, \left(\frac{2}{N}\right)^{1/2} \cos N\left(x_0 + \frac{\pi}{N}\right)\right\}, \tag{33}$$

$$\bar{u}_N = \left\{\left(\frac{2}{N}\right)^{1/2} \cos\left(x_0 - \frac{\pi}{N}\right), \left(\frac{2}{N}\right)^{1/2} \cos 2\left(x_0 - \frac{\pi}{N}\right), \cdots, \left(\frac{2}{N}\right)^{1/2} \cos N\left(x_0 - \frac{\pi}{N}\right)\right\}, \tag{34}$$

$$\bar{u}_N = \left\{\left(\frac{2}{N}\right)^{1/2} \sin\left(x_0 + \frac{\pi}{N}\right), \left(\frac{2}{N}\right)^{1/2} \sin 2\left(x_0 + \frac{\pi}{N}\right), \cdots, \left(\frac{2}{N}\right)^{1/2} \sin N\left(x_0 + \frac{\pi}{N}\right)\right\}, \tag{35}$$

$$\bar{u}_N = \left\{\left(\frac{2}{N}\right)^{1/2} \sin\left(x_0 - \frac{\pi}{N}\right), \left(\frac{2}{N}\right)^{1/2} \sin 2\left(x_0 - \frac{\pi}{N}\right), \cdots, \left(\frac{2}{N}\right)^{1/2} \sin N\left(x_0 - \frac{\pi}{N}\right)\right\}. \tag{36}$$

It should be noted that we have used $\bar{u}_N$ to denote an asymptotically optimum weighting function in (31)–(36) and that it does not necessarily represent the same vector from one equation to the next. It is interesting to note that the asymptotically optimum weighting functions have a sinusoidal or cosinusoidal variation and are completely determined by x_0. It is thus possible to give, in a certain sense, a rather complete asymptotic solution to the problem of finding optimum weighting functions.

It is interesting to observe that the sinusoidal or cosinusoidal variation of the asymptotically optimum weighting functions turns out to be at the frequency corresponding to the minimum value of the sampled-power spectrum of the noise. An immediate consequence of this is the intuitively appealing result that the optimum weighting functions approach very narrow-band filters centered around a minimum in the sampled-power spectrum of the noise. This interpretation also points out a difficulty in the present approach in that the asymptotically optimum weighting functions tend to ignore that part of the signal outside a narrow band centered around the minimum of the sampled-power spectrum of the noise.

Complete Solutions for Particular Covariance Matrices

A complete solution for all N will now be given for certain covariance matrices.

Case A—Markov Process

In this case we have $\rho(\tau) = \epsilon^{-\alpha|\tau|}$, $\alpha > 0$, and $R_N = \{\rho^{|j-k|}\}$, $j, k = 1, \cdots, N$, where $\rho = \epsilon^{-\alpha T}$. It can be shown that the eigenvalues of this matrix are given by[6]

$$\lambda_{N,k} = \frac{1-\rho^2}{1 - 2\rho\cos\phi_{N,N-k+1} + \rho^2}, \qquad k = 1, \cdots, N \tag{37}$$

where $0 < \phi_{N,1} < \phi_{N,2} < \cdots < \phi_{N,N} < \pi$ are the solutions of the equation

$$\frac{\sin(N+1)\phi}{\sin\phi} - \frac{2\rho\sin N\phi}{\sin\phi} + \rho^2\frac{\sin(N-1)\phi}{\sin\phi} = 0. \tag{38}$$

The smallest eigenvalue $\lambda_{N,1}$ is obtained from $\phi_{N,N}$ and it can be shown that[7]

$$\phi_{N,N} = \frac{N\pi}{N+1} + o(1). \tag{39}$$

The function $f(x)$ in this case is

$$f(x) = \frac{1-\rho^2}{1 - 2\rho\cos x + \rho^2} \tag{40}$$

so that $x_0 = \pi$ and

$$f(x_0) = \frac{1-\rho}{1+\rho} \tag{41}$$

$$f''(x_0) = \frac{2\rho(1-\rho)}{(1+\rho)^3}. \tag{42}$$

[4] *Ibid.*, p. 72.
[5] *Ibid.*, pp. 112–114.
[6] *Ibid.*, pp. 69–71.
[7] *Ibid.*, p. 71.

Hence, using (29),

$$\lim_{N\to\infty} N^2\left(\lambda_{N,1} - \frac{1-\rho}{1+\rho}\right) = \frac{\pi^2\rho(1-\rho)}{(1+\rho)^3}. \tag{43}$$

An explicit expression can also be found for the eigenvectors by recognizing that the eigenvectors of the inverse matrix are the same as those of the original matrix and[8]

$$R_N^{-1} = \frac{1}{1-\rho^2} \cdot \begin{bmatrix} 1 & -\rho & 0 & 0 & \cdots & 0 & 0 \\ -\rho & 1+\rho^2 & -\rho & 0 & \cdots & 0 & 0 \\ 0 & -\rho & 1+\rho^2 & -\rho & \cdots & 0 & 0 \\ \cdot & \cdot & \cdot & \cdot & \cdots & \cdot & \\ 0 & 0 & 0 & 0 & \cdots & 1+\rho^2 & -\rho \\ 0 & 0 & 0 & 0 & \cdots & -\rho & 1 \end{bmatrix}. \tag{44}$$

Thus, the components of an eigenvector satisfy the system of equations

$$\begin{aligned} (2\cos\phi - \rho)u_1 - u_2 &= 0 \\ -u_1 + (2\cos\phi)u_2 - u_3 &= 0 \\ \ddots & \\ -u_{N-3} + (2\cos\phi)u_{N-2} - u_{N-1} &= 0, \end{aligned} \tag{45}$$

and hence the difference equation

$$-u_{k+1} + 2\cos\phi\, u_k - u_{k-1} = 0 \tag{46}$$

with the boundary conditions

$$\begin{aligned} \frac{u_k}{u_1} &= 1, \quad k = 1 \\ &= 2\cos\phi - \rho, \quad k = 2, \end{aligned} \tag{47}$$

where we have used ϕ to denote any one of the N solutions of (38). It is easily seen that the solution of the difference equation yields

$$\frac{u_k}{u_1} = \frac{1}{\sin\phi}[\sin k\phi - \rho\sin(k-1)\phi], \quad k = 1, \cdots, N. \tag{48}$$

A straightforward calculation for the normalizing constant shows that

$$u_k = \left[\frac{N}{2}(1 - 2\rho\cos\phi + \rho^2) + \frac{1-\rho^2}{2}\right]^{-1/2} \cdot [\sin k\phi - \rho\sin(k-1)\phi], \quad k = 1, \cdots, N. \tag{49}$$

The components of the eigenvector corresponding to the smallest eigenvalue are obtained by putting $\phi_{N,N}$ for ϕ in (49). It is interesting to compare this eigenvector with the asymptotically optimum eigenvectors given in (35) and (36).

In this case it is easily seen that the eigenvalues $\{\lambda_{N,1}\}$ form a strictly decreasing sequence. Hence, the optimum reference gain can only increase as N increases. However, $\lambda_{2,1} = 1 - \rho$ and $\lambda_{N,1} \to (1-\rho)/(1+\rho)$ as $N \to \infty$, so that if ρ is slightly less than unity, the optimum reference gain for very many pulses is only 3 db greater than that for just two pulses.

[8] *Cf.* Reich and Swerling [6], p. 293.

Case B—Off-Diagonal Correlation

In this case we have

$$R_N = \begin{bmatrix} 1 & r & & & & 0 \\ r & 1 & r & & & \\ & r & 1 & r & & \\ & & \ddots & \ddots & \ddots & \\ & & & r & 1 & r \\ 0 & & & & r & 1 \end{bmatrix}, \tag{50}$$

and the eigenvalue equation is

$$\begin{vmatrix} (1-\lambda) & r & & & & 0 \\ r & (1-\lambda) & r & & & \\ & r & (1-\lambda) & r & & \\ & & \ddots & \ddots & \ddots & \\ & & & r & (1-\lambda) & r \\ 0 & & & & r & (1-\lambda) \end{vmatrix} = 0.$$

If D_N denotes the value of the above determinant, we then have the difference equation

$$-D_{k+1} + (1-\lambda)D_k - r^2 D_{k-1} = 0 \tag{51}$$

with boundary conditions

$$\begin{aligned} D_k &= 1 - \lambda, \quad k = 1 \\ &= (1-\lambda)^2 - r^2, \quad k = 2. \end{aligned} \tag{52}$$

Therefore

$$D_k = \frac{|r|^k \sin(k+1)\gamma}{\sin\gamma} \tag{53}$$

where

$$\cos\gamma = \frac{1-\lambda}{2|r|}, \tag{54}$$

so that the eigenvalues are

$$\lambda_{N,k} = 1 - 2|r|\cos\frac{k\pi}{N+1}, \quad k = 1, \cdots, N. \tag{55}$$

It should be noted that

$$|r| \le \frac{1}{2\cos\dfrac{\pi}{N+1}} \tag{56}$$

in order that R_N be non-negative definite. In addition we must have $|r| \le \frac{1}{2}$ if R_N is to be non-negative definite for all N.

The components of the eigenvector corresponding to the smallest eigenvalue satisfy the difference equation

$$ru_{k+1} + 2\,|r| \cos \frac{\pi}{N+1} u_k + ru_{k-1} = 0 \tag{57}$$

and the boundary conditions

$$\begin{aligned} \frac{u_k}{u_1} &= 1, \qquad k = 1 \\ &= -\frac{2\,|r|}{r} \cos \frac{\pi}{N+1}, \qquad k = 2. \end{aligned} \tag{58}$$

Therefore, we have

$$u_k = \sqrt{\frac{2}{N+1}} \sin \frac{k\pi}{N+1}, \qquad r < 0, k = 1, \cdots, N, \tag{59}$$

$$u_k = (-1)^k \sqrt{\frac{2}{N+1}} \sin \frac{k\pi}{N+1}, \qquad r > 0, k = 1, \cdots, N. \tag{60}$$

In this case

$$f(x) = 1 + 2r \cos x \tag{61}$$

and $x_0 = \pi$ if $r > 0$ and $x_0 = 0$, $r < 0$. We have $\lambda_{2,1} = 1 - |r|$ and $\lambda_{N,1} \to 1 - 2\,|r|$ as $N \to \infty$, so that the optimum gain is strictly increasing from $(1 - |r|)^{-1}$ to $(1 - 2\,|r|)^{-1}$. It is also interesting to compare the eigenvectors for this case with the asymptotically optimum eigenvectors given in (35) and (36).

Case C—Circular Matrices

In this case we will find a solution only for a particular value of N for which the covariance matrix has the circular form

$$R_N = \begin{bmatrix} a_1 & a_2 & a_3 & \cdots & a_N \\ a_N & a_1 & a_2 & \cdots & a_{N-1} \\ & & \cdots & & \\ a_2 & a_3 & a_4 & \cdots & a_1 \end{bmatrix} \tag{62}$$

with $a_1 = 1$. It is known that[9]

$$R_N = \sum_{k=1}^{N} a_k J^{k-1} \tag{63}$$

where

$$J = \begin{bmatrix} 0 & 1 & \cdots & 0 & 0 \\ 0 & 0 & \cdots & 0 & 0 \\ & & \cdots & & \\ 0 & 0 & \cdots & 0 & 1 \\ 1 & 0 & \cdots & 0 & 0 \end{bmatrix}. \tag{64}$$

The eigenvalues of J are

$$\beta_k = \epsilon^{2\pi i k/N}, \qquad k = 1, \cdots, N \tag{65}$$

[9] See Grenander and Rosenblatt [9], p. 98.

so that the eigenvalues of R_N are

$$\lambda_N^{(k)} = \sum_{j=1}^{N} a_j \beta_k^{j-1} = \sum_{j=1}^{N} a_j \epsilon^{2\pi i k(j-1)/N}. \tag{66}$$

The normalized eigenvectors of R_N are the same as those of J, which are easily obtained as

$$u_{j,k} = N^{-1/2} \epsilon^{2\pi i j k/N}, \qquad j = 1, \cdots, N, \tag{67}$$

where $u_{j,k}$ denotes the jth component of the kth eigenvector.

In general, we cannot say which β_k leads to the smallest eigenvalue of R_N, since that particular value of k will depend on the choice for the a_j's. Due to symmetry we have $a_2 = a_N$, $a_3 = a_{N-1}$, $\cdots$, so that

$$\lambda_N^{(k)} = 1 + 2a_2 \cos \frac{2\pi k}{N} + 2a_3 \cos \frac{4\pi k}{N} + \cdots \tag{68}$$

where the series has a finite number of terms and the last term depends on whether N is even or odd.

Application to MTI Radar System

A serious problem in radar reception occurs due to the background clutter introduced by rain clouds, foliage, etc. One method for reducing the effects of the clutter is to take advantage of the fact that the clutter is essentially stationary whereas the desired targets are not. An MTI radar system is one which uses coherent pulse-to-pulse processing to utilize the Doppler effect in this manner, *cf.* [1].

Our results are directly applicable to the problem of designing a linear predetection filter for an MTI radar system. In this case we have

$$S(t, \theta) = A \cos (\omega_d t + \theta) \tag{69}$$

where θ is a random variable which is uniformly distributed from 0 to 2π radians, ω_d is the Doppler radian frequency and is related to target velocity v by

$$\omega_d = \frac{4\pi v}{\lambda} \tag{70}$$

and λ is the transmitter wavelength. Thus,

$$\overline{S(jT, \theta) S(kT, \theta)} = \frac{A^2}{2} \cos \omega_d (j - k) T \tag{71}$$

and

$$F_N(S) = \sum_{j=1}^{N} \sum_{k=1}^{N} u_j u_k^* \cos \omega_d (j - k) T. \tag{72}$$

If $N = 2$ and $\rho_1 > 0$ the optimum weighting function is given by (17). This corresponds to the usual single-cancellation method used in MTI radar. The signal factor, which in this case is also known as the blind speed characteristic, is given by

$$F_N(S) = 1 - \cos \omega_d T. \tag{73}$$

Thus, there are blind speed frequencies occuring at integral multiples of the pulse repetition frequency, $1/T$.

If $N = 3$ and $0 \leq \rho_2 \leq \rho_1 < 1$, the optimum weighting function is given by (21). If ρ_1 and ρ_2 are close to unity, then the optimum system is very similar to the usual double-cancellation system, with weights 1: −2:1, whose blind speed characteristic is [*cf.* (72)]

$$F_N(S) = 1 - \tfrac{4}{3}\cos\omega_d T + \tfrac{1}{3}\cos 2\omega_d T. \tag{74}$$

It is instructive to compare the reference gain of the optimum three-pulse system with that for double cancellation. We obtain easily

$$M = \frac{\bar{G}_3(\text{Optimum})}{\bar{G}_3(1:-2:1)} = \frac{1 - \frac{4}{3}\rho_1 + \frac{1}{3}\rho_2}{1 + \frac{\rho_2}{2}\left(1 - \sqrt{1 + 8\left(\frac{\rho_1}{\rho_2}\right)^2}\right)}. \tag{75}$$

We assume that $\rho(\tau)$ has at least four continuous derivatives so that

$$\rho(\tau) = 1 + b_2\tau^2 + b_3\tau^3 + b_4\tau^4 + o(\tau^4) \tag{76}$$

where

$$b_2 = \frac{\rho''(\tau)}{2}\bigg|_{\tau=0} \qquad b_3 = \frac{\rho'''(\tau)}{6}\bigg|_{\tau=0} \qquad b_4 = \frac{\rho''''(\tau)}{24}\bigg|_{\tau=0}. \tag{77}$$

Therefore

$$M = \frac{2b_3T^3 + 6b_4T^4}{2b_3T^3 + (6b_4 - b_2^2)T^4} + 0(1). \tag{78}$$

If $b_3 = 0$, this simplifies to

$$M = \frac{1}{1 - \frac{1}{6}\frac{b_2^2}{b_4}} + o(1). \tag{79}$$

If $\rho(\tau) = \epsilon^{-\tau^2}$, then $b_2 = -1$, $b_3 = 0$, $b_4 = \frac{1}{2}$ and

$$M = \tfrac{3}{2} + o(1). \tag{80}$$

Thus, the reference gain of the double-cancellation system is, for small T, within 1.76 db of that for the optimum system. The optimum weighting function for $N = 4$ is given by (24), while the solution for $N \to \infty$ is given by (33)–(36).

Concluding Remarks

An optimality criterion has been presented leading to the design of linear predetection filters. This criterion was based on the maximization of the reference gain $\bar{G}_N$, independent of the signal characteristics. It was pointed out that this procedure could produce a signal factor which deteriorates, or lowers, the over-all gain G_N. For example, in the *MTI* radar case, when $N = 2$, the over-all gain is zero for those signals whose Doppler frequencies are integral multiples of the pulse repetition frequency $1/T$.

If the signal's characteristics are completely known, then it is appropriate to define the optimum weighting function as that vector which maximizes the over-all gain G_N. It is easily seen that the maximum value of G_N is equal to the reciprocal of the smallest eigenvalue of the matrix equation

$$R_N\bar{u}_N = \lambda S_N\bar{u}_N$$

where

$$S_N = \left\{\frac{\overline{S(jT,\theta)S(kT,\theta)}}{\overline{S^2(NT,\theta)}}\right\}$$

$j, k = 1, \cdots, N$, and this value is assumed for any eigenvector of this matrix equation corresponding to the smallest eigenvalue.

A straightforward calculation shows that for $N = 2$ the optimum weighting function is given by (17) when $\rho_1 > \overline{S(T,\theta)S(2T,\theta)}/\overline{S^2(2T,\theta)}$, and by (18) when $\rho_1 < \overline{S(T,\theta)S(2T,\theta)}/\overline{S^2(2T,\theta)}$, assuming $\overline{S^2(T,\theta)} = \overline{S^2(2T,\theta)}$. In the *MTI* case this means that the optimum filter performs pulse-to-pulse addition when ω_d is near a blind speed frequency and pulse-to-pulse subtraction otherwise, provided ρ_1 is close to unity.

In general, the optimum weighting function obtained will depend on signal characteristics. This is undesirable, since in many applications the signal is not completely specified. For example, in the MTI case the Doppler frequency is usually not known and a predetection filter which performs reasonably well for most Doppler frequencies is adequate. However, if signal characteristics are known, the weighting functions should be chosen to maximize G_N instead of $\bar{G}_N$.

References

[1] L. N. Ridenour, "Radar System Engineering," McGraw-Hill Book Co., Inc., New York, N. Y., ch. 16; 1947.
[2] W. B. Davenport, Jr. and W. L. Root, "An Introduction to the Theory of Random Signals and Noise," McGraw-Hill Book Co., Inc., New York, N. Y.; 1958.
[3] C. W. Helstrom, "Statistical Theory of Signal Detection," Pergamon Press, London, England; 1960.
[4] D. Middleton, "An Introduction to Statistical Communication Theory," McGraw-Hill Book Co., Inc., New York, N. Y.; 1960.
[5] L. A. Wainstein and V. D. Zubakov, "Extraction of Signals from Noise," Prentice-Hall, Inc., Englewood Cliffs, N. J.; 1962.
[6] E. Reich and P. Swerling, "Detection of a sine wave in Gaussian noise," *J. Appl. Phys.*, vol. 24, pp. 289–296; March, 1953.
[7] R. Courant and D. Hilbert, "Methods of Mathematical Physics," Interscience Press, New York, N. Y., vol. 1; 1953.
[8] U. Grenander and G. Szegö, "Toeplitz Forms and Their Applications," University of California, Berkeley; 1958.
[9] U. Grenander and M. Rosenblatt, "Statistical Analysis of Stationary Time Series, " John Wiley and Sons, Inc., New York, N. Y.; 1957.

2.2 ON "OPTIMIUM WEIGHTING FUNCTIONS FOR THE DETECTION OF SAMPLED SIGNALS IN NOISE"

E.A. Robinson

IEEE Transactions on Information Theory, Vol. IT-11, No. 3, July 1965, pp. 452-453.

Optimum Weighting Functions for the Detection of Sampled Signals in Noise

In his excellent paper,[1] Capon considers the class of optimum weighting functions $(u_1, u_2, \cdots, u_N)$ of unit norm which maximize the reference gain given by his (9). He shows that the maximum value of the reference gain is attained when the weighting function is any eigenvector corresponding to the smallest eigenvalue of the autocorrelation matrix of the noise. He also points out that the reference gain is equal to the reciprocal of the Toeplitz form given by his (28). We see that an optimum weighting function $(u_1, u_2, \cdots, u_N)$ enters into his (28) only as

$$\left| \sum_{k=1}^{N} u_k e^{ikx} \right|^2 .$$

If we let $z = e^{ix}$, then this expression becomes the quasi-polynomial in z given by

$$U(z) \equiv \left| \sum_{k=1}^{N} u_k z^{k-1} \right|^2 \equiv \sum_{n=-N+1}^{N-1} r_n z^n$$

where the autoproducts r_n are defined as

$$r_n \equiv \begin{cases} u_{1+n}u_1^* + u_{2+n}u_2^* + \cdots + u_N u_{N-n}^* \\ \qquad \text{for } n = 0, 1, \cdots, N-1 \\ r_{-n}^* \qquad \text{for } n = -1, -2, \cdots, -N+1. \end{cases}$$

(The asterisk denotes the complex-conjugate.) The point of our argument is this; if we can find another weighting function, say $(v_1, v_2, \cdots, v_N)$, whose quasi-polynomial $V(z)$ is equal to $U(z)$, then $(v_1, v_2, \cdots, v_N)$ would also yield the maximum value of the reference gain, and hence $(v_1, v_2, \cdots, v_N)$ would also be an optimum weighting function. Let us now explore this situation. If z_i is any root of the quasi-polynomial $U(z)$ then $1/z_i^*$ is also a root. This follows from the structure of the quasi-polynomial. Consequently the roots of $U(z)$ must occur in pairs, namely as z_i and $1/z_i^*$. Counting any root of multiplicity p as p roots of multiplicity unity, we see that $U(z)$ has $N-1$ such pairs of roots. Now let us arbitrarily select one root from each pair, where for convenience we denote our selection as $z_1, z_2, \cdots, z_{N-1}$. Then we form the polynomial

$$v_N(z - z_1)(z - z_2) \cdots (z - z_{N-1}) \equiv \sum_{k=1}^{N} v_k z^{k-1}$$

where the constant scale factor v_N is determined (within a scale factor of magnitude one) by requiring the vector $(v_1, v_2, \cdots, v_N)$ to have unit norm. By construction we see that the weighting function $(v_1, v_2, \cdots, v_N)$ satisfies the required condition, i.e., the quasi-polynomial

$$V(z) \equiv \left| \sum_{k=1}^{N} v_k z^{k-1} \right|^2$$

is equal to the given $U(z)$. At most there can be 2^{N-1} such vectors $(v_1, v_2, \cdots, v_N)$, including the original optimum vector $(u_1, u_2, \cdots, u_N)$. This is the well-known Fejér factorization.[2] Let us now return to the eigenvalue problem of Capon, and for simplicity let us assume that the smallest eigenvalue of the autocorrelation matrix of the noise has multiplicity unity. It then follows that the eigenvector associated with this eigenvalue is unique, except for a scale factor of magnitude unity. Hence $(u_1, u_2, \cdots, u_N)$ represents the unique eigenvector solution corresponding to this minimum eigenvalue, and consequently the 2^{N-1} possible vectors must indeed all be identical, except again for scale factors of unit magnitude. Returning now to the Fejér factorization, it follows that the two roots z_i and $1/z_i^*$ in each pair must be the same, i.e., $z_i = 1/z_i^*$. Hence all these roots must lie on the unit circle in the z plane, that is, $|z_i| = 1$, so each root is of the form $z_i = 1/z_i^* = e^{ix}$. For example, for the MTI radar system discussed by Capon in the case when $N = 2$ and $\rho_1 > 0$, the optimum weighting function is $(1/\sqrt{2}, -1/\sqrt{2})$ which has the associated polynomial $(1/\sqrt{2}) - (1/\sqrt{2})z$ with the root $z_1 = 1$ of unit magnitude. In the case when $N = 2$ and $\rho_1 < 0$, the optimum weighting function is then $(1/\sqrt{2}, 1/\sqrt{2})$ corresponding to the root $z = -1$ of unit magnitude. Finally, note that in all the examples given by Capon, the optimum weighting functions are all either symmetric or antisymmetric about their central point. For example, the first one we mentioned is antisymmetric and the second one symmetric. This result of symmetry or antisymmetry follows from the fact that all the roots of the quasi-polynomial $U(z)$ fall on the unit circle in the z plane. That is, the optimum weighting function is symmetric in case there are either no roots or an even number of roots equal to $+1$, and is antisymmetric in case there are an odd number of roots equal to $+1$. For example, in Capon's (59), there is no root equal to unity and consequently the optimum weighting function is symmetric; in his (60) when N is even, $z = 1$ is a root and consequently the optimum weighting function is antisymmetric. The same arguments apply in the more general case, i.e., when the optimum weighting function is defined as the vector that maximizes the overall gain. It follows then that the associated polynomial of the optimum weighting function in this general case has all its roots on the unit circle, and the optimum weighting function is either symmetrical (in case of an even number of $+1$ roots) or antisymmetrical (in case of an odd number of $+1$ roots).

ENDERS A. ROBINSON
Geoscience Inc.
Cambridge, Mass.

[1] Capon, J., Optimum weighting functions for the detection of sampled signals in noise, *IEEE Trans. on Information Theory*, vol IT-10, Apr 1964, pp 152–159.

[2] Fejér, L., Über trigonometrische Polynome, *Z. Reine Angew. Math.*, vol 146, 1915, pp 53–82.

Manuscript received December 22, 1964.

OPTIMUM PROCESSING OF UNEQUALLY SPACED RADAR PULSE TRAINS
FOR CLUTTER REJECTION

2.3

L.E. Brennan and I. Reed
IEEE Transactions on Aerospace and Electronic Systems, Vol. AES-4, No. 3, May 1968,

pp. 474-477.

Optimum Processing of Unequally Spaced Radar Pulse Trains for Clutter Rejection

LAWRENCE E. BRENNAN, Member, IEEE
IRVING S. REED
The RAND Corporation
Santa Monica, Calif.

Abstract

A train of radar pulses from one resolution cell can be processed coherently to reject echoes from external clutter and detect targets moving radially with respect to the clutter. Optimum methods of signal processing are defined for systems in which the interpulse spacings are multiply staggered to avoid target blind speeds. Likelihood ratio tests are developed for systems in which the target Doppler frequency is known a priori and for systems employing a bank of filters to cover the target Doppler band. To implement such tests, the *N* pulses in the train are added with complex weights and the amplitude of the sum compared with a detection threshold. The set of weights which maximizes the average signal-to-clutter ratio is also computed for a single-filter system with unknown target Doppler frequency. When the clutter autocorrelation function is exponential, the clutter covariance matrix can be inverted analytically. This latter result is useful for comparing different interpulse-spacing codes for a particular system application.

Manuscript received September 18, 1967.

I. Introduction

The performance of a radar is often limited by echoes from external clutter that are large compared with internal receiver noise. In many practical situations targets of interest have a radial velocity component with respect to the clutter scatterers, and Doppler discrimination can be used to enhance the signal-to-clutter ratio (SCR). A similar problem arises in sonar where the echoes from moving targets are masked by reverberation.

When the frequency spectrum of the clutter is narrow, conventional moving target indication (MTI) systems employing two-pulse or three-pulse cancellation can enhance SCR considerably. Broader clutter spectra are encountered in many cases (e.g., spectra due to wind shear of chaff or rain returns, or to platform motion in airborne MTI systems) and can be rejected by processing a coherent train of many radar pulses. In effect, a better reject filter can be synthesized with a longer pulse train.

When the pulses in a train are equally spaced, Doppler discrimination systems exhibit blind speeds—a target which moves an integral number of half-wavelengths between pulses cannot be distinguished from a fixed scatterer. These blind speeds can be avoided by varying the interpulse spacings in a coherent pulse train. Emerson[1] discussed the processing of coherent variably spaced pulse trains using a time-domain representation, i.e., describing the signal processor by a set of complex weights applied to the successive pulses rather than by a filter transfer function. The variations in clutter return during the pulse train are then expressed in terms of a covariance matrix rather than a frequency spectrum. Using this approach, methods of selecting the optimum set of complex weights for a pulse train are derived in Sections II and III. In Section II, a test based on the likelihood ratio is discussed and an optimum set of complex weights is obtained for detecting targets of known Doppler frequency. In Section III, a more tractable criterion, based on the maximum SCR, is developed for selection of the pulse weights.

The analogous problem for equally spaced pulse trains was discussed by Rummler.[2] With equally spaced trains, the design problem is one of selecting an optimum set of complex weights for the pulses and of specifying the interpulse spacing. With variable spacings there is the additional problem of selecting a good interpulse spacing code. No direct method for optimizing the spacing code was found. It is shown in Section IV, however, that for a particular form of the clutter autocorrelation function, the covariance matrix can be inverted and an analytic expression obtained for system performance as a function of target Doppler frequency. This should be useful in testing different spacing codes so that one can be selected for a particular system application.

II. Likelihood Ratio Detectors

Consider a coherent pulsed radar which transmits a train of N pulses in one direction at times $t_1, t_2, \cdots, t_N$. The interpulse spacings may all be different, but the minimum interpulse spacing is long enough that returns are not received simultaneously from different ranges within a region of extended clutter. The N returns from one range resolution cell can be represented by a set of vectors $\{z_n\}$, where z_n corresponds in amplitude and phase to the RF return on the nth pulse. When a signal reflected from a moving target is present, it adds a signal vector s_n to each z_n, which may be represented by

$$s_n = s_0 e^{i(\omega t_n + \phi)} \tag{1}$$

where ω is 2π times the target Doppler frequency.

When the clutter return in each resolution cell is due to many scatterers of comparable magnitude, it follows from the central limit theorem that each quadrature component of z_n has a Gaussian amplitude distribution. This model is representative of many types of clutter, such as rain, chaff, and sea returns, and is assumed to apply in the following analysis. The probability density function of the z_n is then[3],[4]

$$p(Z \mid s_0, \phi, \omega) = \frac{\det(Q)}{(2\pi)^N} \exp\left\{\frac{(Z-S)_T Q (Z-S)^*}{2}\right\} \tag{2}$$

where Z is the column vector of $\{z_n\}$, S is the column vector of $\{s_n\}$, and Q is the matrix inverse of H. The covariance matrix H of the complex clutter process has the elements

$$H_{kl} = \tfrac{1}{2}E\{z_k^* z_l\} \tag{3}$$

where the H_{kl} are evaluated in the absence of a signal. Receiver noise, independent from pulse to pulse, can be included in H, and appears as a real constant added to the diagonal elements of H.

The most efficient method[5] of choosing between the signal-plus-clutter and clutter-only hypotheses is that of comparing the likelihood ratio with a threshold:

$$L(Z \mid s_0, \phi, \omega) = \frac{p(Z \mid s_0, \phi, \omega)}{p(Z \mid 0, -, -)}$$

$$= \exp\left\{-\frac{S_T Q S^*}{2}\right\} \exp\{\mathrm{Re}[Z_T Q S^*]\}. \tag{4}$$

The signal phase ϕ is unknown a priori and uniformly distributed over 2π in most cases of interest. Averaging L over ϕ gives the optimum Bayes test,

$$L(Z \mid s_0, \omega) = \exp\left\{-\frac{S_T Q S^*}{2}\right\} I_0(\mid Z_T Q S^* \mid). \tag{5}$$

Only the I_0 term in (5) contains the observed signal Z. The optimum test, given the Doppler frequency ω, is to compare the amplitude of the weighted sum $W_T Z$ with a threshold. From (5), the optimum set of complex weights $W = (w_1, w_2, \cdots, w_N)_T$ is

$$W_{\mathrm{opt}} = kQS^* \tag{6}$$

where k is an arbitrary complex constant.

The weights of (6) can be used to design an optimum filter bank, i.e., a set of parallel processing channels each optimized for one target Doppler frequency. The problem is more difficult when a single filter is desired that will give the best overall preformance for some a priori distribution of target Doppler frequency, such as a uniform distribution over a frequency interval. The best test variable in this case is the average of (5) over the a priori distributions of ω and s_0.[1] The performance of the ω-averaged likelihood test has been evaluated for uniformly spaced pulse trains and a Rayleigh distributed s_0 of known mean-square value.[6] A similar analysis for the nonuniform pulse train, based on (5), does not appear tractable. An alternative, but less efficient, design of a signal processor to cover a band of target Doppler frequencies is discussed in the next section.

III. Linear Processors with Maximum Output Signal-to-Clutter Ratio

Let W again represent the set of complex weights applied to the set of radar returns Z before summation. The output is then

$$X = W_T Z = \sum_{n=1}^{N} w_n z_n. \tag{8}$$

[1] A simpler expression, which may be averaged over ω to approximate the average likelihood ratio test, is the maximum-likelihood ratio test with respect to s_0 and ϕ:

$$\max_{s_0,\phi} [L(Z \mid s_0, \phi, \omega)] = \exp\left\{\frac{\mid Z_T Q \mathcal{S}^* \mid^2}{2 \mathcal{S}_T Q \mathcal{S}^*}\right\} \tag{7}$$

where $\mathcal{S}$ is a column vector of $\{e^{i\omega t_n}\}$. This equation is proved by the fact that the maximum of L in (4) occurs when ln L is maximum, replacing S by $s_0 e^{i\phi}\mathcal{S}$, and completing the square:

$$-2 \ln L(Z \mid s_0, \phi, \omega) = \left| s_0 e^{i\phi} \sqrt{\mathcal{S}_T Q \mathcal{S}^*} - \frac{Z_T Q \mathcal{S}^*}{\sqrt{\mathcal{S}_T Q \mathcal{S}^*}} \right|^2 - \frac{\mid Z_T Q \mathcal{S}^* \mid^2}{\mathcal{S}_T Q \mathcal{S}^*}.$$

The maximum of L occurs when $s_0 e^{i\phi} = Z_T Q \mathcal{S}^* / (\mathcal{S}_T Q \mathcal{S}^*)$.

The average output clutter power, plus noise power if noise terms are included in the covariance matrix H, is

$$P_C = \tfrac{1}{2}E\{|X|^2\} = W_T^*HW. \quad (9)$$

For a target of Doppler frequency ω, the output signal power is

$$P_S = \tfrac{1}{2}|W_TS|^2. \quad (10)$$

Forming the sum $(P_C+\lambda P_S)$, where λ is a Lagrange multiplier, and differentiating this expression with respect to each w_k again gives (6) as the set of optimum weights. These weights minimize the output clutter power for a given output signal power. The MTI gain, or ratio of output to input SCR, is then

$$G = \frac{|W_TS|^2}{W_T^*HW}\frac{1}{s_0^2} = S_TQS^* \quad (11)$$

where the gain is expressed with reference to the input SCR for a single pulse. The z_N are normalized so that the diagonal terms of the covariance matrix H are unity, and S is a column vector of $\{e^{i\omega t_n}\}$.

The same method can be used to compute the optimum set of weights when the target Doppler frequency is unknown. Let ω have a uniform a priori distribution in the interval (ω_1, ω_2) where $(\omega_2-\omega_1)(t_{n+1}-t_n)\gg 1$ for all n. Averaging (10) over ω gives

$$\overline{P}_S \doteq \frac{s_0^2}{2}W_T^*W. \quad (12)$$

Again using the Lagrange multiplier λ, the output clutter power is minimized for a given $\overline{P}_S$ when

$$(P_C - \lambda\overline{P}_S) = W_T^*\left(H - \lambda\frac{s_0^2}{2}I\right)W$$
$$= W_T^*(H - \lambda' I)W \quad (13)$$

is minimum. The symbol I denotes the identity matrix. Setting the derivatives of this expression equal to zero with respect to w_k, $k=1, 2, \cdots, N$, gives

$$(H - \lambda' I)W = 0. \quad (14)$$

This is the eigenvalue problem with solutions for the set of λ'_k for which $\det(H-\lambda' I)=0$. The average MTI gain is then

$$\overline{G} = \frac{2}{s_0^2}\frac{\overline{P}_S}{P_C} = \frac{W_T^*W}{W_T^*HW} = \frac{1}{\lambda'}. \quad (15)$$

The smallest eigenvalue gives the largest average MTI gain, and the corresponding set of optimum weights is given by (14). As noted earlier, this set of weights does not necessarily yield the optimum performance for a single filter. While the weights of (14) maximize the Doppler average SCR, better performance in terms of detection probability for a given false-alarm probability may be obtained with the Doppler average likelihood-ratio test discussed in Section II.

IV. An Exact Analysis with Exponential

When the elements of the covariance matrix have the form

$$H_{kl} = \exp\{-K|t_k - t_l|\} \quad (16)$$

the elements of the inverse matrix $Q=H^{-1}$ can be obtained analytically. Let $\rho_k=\exp\{-K(t_{k+1}-t_k)\}$, so that elements below the diagonal in H are

$$H_{kl} = \prod_{n=l}^{k-1}\rho_n \quad (17)$$

and H is real and symmetric. To eliminate all elements below the diagonal in H, multiply the kth row successively by ρ_k and subtract it from the $(k+1)$th row. Beginning this elimination with the bottom row, it can be shown that

$$\det(H) = \prod_{n=1}^{N-1}(1 - \rho_n^2). \quad (18)$$

The same elimination procedure can be used to obtain the cofactors of H.

The elements of $Q=H^{-1}$ are

$$Q_{11} = \frac{1}{1-\rho_1^2}$$
$$Q_{NN} = \frac{1}{1-\rho^2_{N-1}}$$
$$Q_{kk} = \frac{(1-\rho_k^2\rho^2_{k-1})}{(1-\rho_k^2)(1-\rho^2_{k-1})} \qquad 1 < k < N$$
$$Q_{k,k+1} = Q_{k+1,k} = \frac{-\rho_k}{(1-\rho_k^2)}$$
$$Q_{kl} = 0 \qquad |k - l| \geq 2. \quad (19)$$

For an equally spaced pulse train, this reduces to the form given by Capon[7] and others.

Combining (11) and (19) gives an analytic expression for the composite performance of a bank of signal processors, with one channel optimized for each target Doppler frequency. The MTI gain is expressed as a function of the interpulse spacing code and target Doppler frequency. When the correlation between all adjacent pulse pairs is large, i.e., $K(t_{n+1}-t_n)$ is small for all n, the gain is easily seen to be

$$G \doteq \sum_{n=1}^{N-1}\frac{1-\cos\omega\tau_n}{K\tau_n} \quad (20)$$
$$\tau_n = t_{n+1} - t_n.$$

For a bank of processors, the average gain over a large interval in target Doppler frequency $[(\omega_2-\omega_1)\tau_n \gg 1$ for all $n]$ is a monotonically decreasing function of the τ_n, and when the $K\tau_n$ are small

$$\operatorname*{avg}_{\omega}[G] = \frac{1}{K}\sum_{n=1}^{N-1}\frac{1}{\tau_n}. \tag{21}$$

When the covariance matrix of a clutter process has the exponential form, (11) and (19) give an analytic expression for MTI gain. This should be very useful in evaluating unequally spaced pulse trains and in selecting good spacing codes. This undoubtedly will require the evaluation of many alternative codes. For example, if detection probability is expressed as a function of gain, it can be averaged over an a priori distribution of target Doppler frequencies to obtain a useful measure of performance for each spacing code.

REFERENCES

[1] R. C. Emerson, "Some pulsed Doppler, MTI, and AMTI techniques," The RAND Corp., Santa Monica, Calif. Rept. R-274, March 1954.

[2] W. D. Rummler, "Clutter suppression by complex weighting of coherent pulse trains," *IEEE Trans. Aerospace and Electronic Systems*, vol. AES-2, pp. 689–699, November 1966.

[3] E. J. Kelly and I. S. Reed, "Some properties of stationary Gaussian processes," M.I.T. Lincoln Lab., Lexington, Mass., Tech. Rept. 157, June 4, 1957.

[4] R. Arens, "Complex processes for envelopes of normal noise," *IEEE Trans. Information Theory*, vol. IT-3, pp. 204–207, September 1957.

[5] E. J. Kelly, I. S. Reed, and W. L. Root, "The detection of radar echoes in noise—I" *J. SIAM*, vol. 8, pp. 309–341, June 1960.

[6] L. E. Brennan, I. S. Reed, and W. Sollfrey, "A comparison of average-likelihood and maximum-likelihood ratio tests for detecting radar targets of unknown Doppler frequency," *IEEE Trans. Information Theory*, vol. IT-14, pp. 104–110, January 1968.

[7] J. Capon, "Optimum weighting functions for the detection of sampled signals in noise," *IEEE Trans. Information Theory*, vol. IT-10, pp. 152–159, April 1964.

Lawrence E. Brennan (S'47–A'51–M'57) was born in Oak Park, Ill., on January 29, 1927. He received the B.S. and Ph.D. degrees in electrical engineering from the University of Illinois, Urbana, in 1948 and 1951, respectively.

Most of his work since leaving the University of Illinois has been concerned with radar and infrared systems. He joined the Department of Electronics, The RAND Corporation, Santa Monica, Calif., in October, 1958. From April, 1962, to October, 1963, on leave of absence from RAND, he was employed by the SHAPE Technical Center, The Hague, Netherlands. He is presently a Senior Scientist with Technology Service Corporation, Santa Monica, Calif., and a Consultant to RAND. His current research interests include radar detection theory, phased array antennas, and MTI.

Irving S. Reed was born in Seattle, Wash., on November 12, 1923. He received the B.A. and Ph.D. degrees in mathematics from the California Institute of Technology, Pasadena, in 1944 and 1949, respectively.

He was associated with Lincoln Laboratory, Massachusetts Institute of Technology, Lexington, from 1951 to 1960. From 1960 to 1963, he was a Senior Staff Member of The RAND Corporation, Santa Monica, Calif. He has been a Professor of Electrical Engineering at the University of Southern California, Los Angeles, since 1963. He is also a Consultant to RAND and is associated with Technology Service Corporation, Santa Monica. His interests include mathematics, computer design, coding theory, stochastic processes, and information theory.

2.4
ON THE OPTIMIZATION OF MTI CLUTTER REJECTION
J.K. Hsiao
IEEE Transactions on Aerospace and Electronic Systems, Vol. AES-10, No. 5, September 1974, pp. 622-629.

On the Optimization of MTI Clutter Rejection

J.K. HSIAO
Naval Research Laboratory
Washington, D.C. 20375

Abstract

In this paper is formulated the problem of optimization of the improvement factor of a nonrecursive MTI by minimization of a quadratic form. The minimum normalized clutter output (a reciprocal of the average improvement factor) is the minimum eigenvalue of this quadratic form, and the corresponding eigenvector is the optimal weight to be used in this filter. By use of classic matrix theory, some properties of this MTI improvement factor are shown, namely, that it is bounded and is a monotonic function of the clutter spectrum variance. Also discussed is the limit of an MTI system having a large number of cancellers. Finally, the problem of a staggered-PRF MTI filter is examined, for which it is shown that its improvement factor is bounded by two equivalent constant-PRF MTI systems. One of these systems has a PRF equal to the lowest PRF of the staggered-PRF system, while the other has a PRF equal to the highest PRF of the staggered system.

I. Introduction

The performance of a radar is often limited by echoes from external clutter that are large compared with internal noise. In many practical situations, targets have a radial velocity component with respect to the clutter scatterer. In this case, Doppler discrimination can be used to enhance the signal-to-clutter ratio (SCR). Usually, this type of Doppler discriminator uses a train of identical pulses uniformly spaced. This type of processor has been discussed extensively in the literature [1]-[4]. When the frequency of the clutter is narrow, two or three pulses are enough to achieve the required improvement for detection. When the clutter spectrum broadens, it is generally believed that by processing a large number of pulses the same amount of improvement can be achieved. However, a quantitative analysis of the limitations to this approach is not available.

When the pulses in a train are equally spaced, the Doppler discrimination system described above exhibits blind speeds; a target which moves at an integral number of half-wavelengths between pulses can not be distinguished from a fixed scatterer. These blind speeds can be avoided by varying the interpulse spacing (staggered PRF). This has been discussed by Emerson [5], Brennan [6], and Jacomini [7]. One additional problem associated with the staggered-PRF system is the selection of good interpulse spacing. No direct method for optimizing the spacing code has been found.

This paper is intended to discuss some of these problems. For a uniformly spaced pulse-train system, the limitations on SCR improvement will be presented in terms of clutter spectral spread and the number of pulses in the train. As for the staggered-PRF system, the performance bounds of such a system will be examined in terms of various interpulse spacing.

II. AVERAGE SCR

Consider a coherent pulse radar which transmits a train of N pulses in one direction at times $t_1, t_2, \cdots, t_n$. The pulse-to-pulse spacing may be varied in the train, but the minimum interpulse spacing is long enough that returns are not received simultaneously from different ranges within a region of extended clutter. Let the signal reflected from a target be

$$S_n = S_0 \exp [j(2\pi f t_n + \phi] \tag{1}$$

where f is the Doppler frequency of the target. It is well know that, for best detection, the received signal train should be weighted with the following optimal weights [6]:

$$x_i = k \sum_j b_{ij} S_j^* \tag{2}$$

$$B = [b_{ij}] \tag{3}$$

where b_{ij} are the elements of matrix B and B is the inverse of the clutter covariance matrix. Receiver thermal noise can

Manuscript received October 22, 1973.

This work was supported by the U.S. Naval Ship Systems Command.

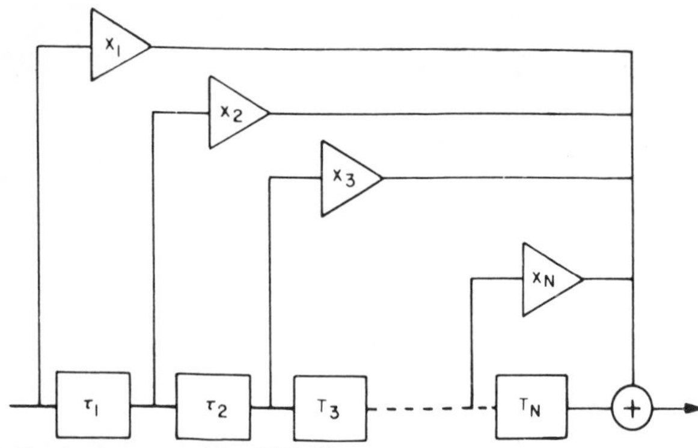

Fig. 1. A nonrecursive MTI filter.

be included as a real constant added to the diagonal element of this covariance matrix. The x_i are the set of complex weights to be used in the filter (see Fig. 1). Since these weights are a function of the target Doppler frequency, a filter bank of many channels is required. Each of these channels is designed for a specific Doppler frequency.

An alternative approach to this problem is to design a single filter channel such that the SCR is optimized to the average Doppler frequency band. This approach is less efficient, but it is more economical. The circuit described in the first case is usually referred to as a Doppler processor, while the latter is called MTI (moving target indication). In this paper we shall concentrate our attention on the MTI system.

The output signal power of this MIT system is

$$P_s = 1/2\,|\sum_i x_i S_i|^2. \tag{4}$$

By use of (1), this can be written as

$$P_s = (S_0^2/2)\sum_i\sum_j x_i x_j \cos\,[2\pi f(t_i - t_j)]. \tag{5}$$

This function is periodic with a period of

$$f_p = l_p/(t_i - t_j) \tag{6}$$

where the l_p are integers for all i and j[1]. This f_p is commonly referred to as the blind Doppler frequency. When the target Doppler frequency is unknown, one may assume that it has a uniform distribution. In this case, the expected signal output is then

$$\bar{P}_s = (S_0^2/2)\sum_i |x|_i^2. \tag{7}$$

The average output power, plus noise if noise terms are included in the covariance matrix A, is

$$P_c = \sum_i\sum_j x_i x_j^* a_{ij} \tag{8}$$

where the a_{ij} are the elements of the clutter covariance matrix A.

[1] T_i and T_j must be rational numbers. For a detailed discussion, see [8].

The MTI improvement factor (I.F.) or ratio of output to input SCR is then

$$I.F. = (\sum_i x_i x_j^*)/(\sum_i\sum_j x_i x_j^* a_{ij}) \tag{9}$$

where the elements of the covariance matrix are normalized, so that $a_{ij} = 1$. This same result has been derived by Brennan [6].

The quantities a_{ij}, elements of the covariance matrix A, are the Fourier transforms of the clutter power density function. One of the convenient and widely assumed models of this power density function is

$$G(f) = [1/(\sqrt{2\pi}\,\sigma_c)]\,\exp\,(-f^2/2\sigma_c^2) \tag{10}$$

where f is the Doppler frequency and σ_c is the standard deviation. Without losing generality, we have assumed that the mean Doppler frequency is zero. Accordingly, the elements of the covariance matrix A are real and have the form

$$a_{ij} = \exp\,[-2\pi^2\sigma_c^2(t_i - t_j)^2]. \tag{11}$$

For a constant-PRF MTI filter, this becomes

$$a_{ij} = \exp\,(-2\pi^2\sigma^2 k^2\tau^2) \tag{12}$$

where

$$k = i - j.$$

One may notice here that the only difference between a staggered-PRF MTI and a constant-PRF MTI is the difference between these matrix elements. One may also notice that this matrix is symmetrical; that is,

$$a_{ij} = a_{ji}.$$

The diagonal element has the following form:

$$a_{ii} = 1 + \eta_0 \tag{13}$$

where η_0 is the normalized receiver noise level. This diagonal element has the maximum value; hence, one may normalize this matrix such that

$$0 < a_{ij} < a_{ii} = 1; \quad i, j = 1, 2, \cdots, N. \tag{14}$$

In practice, the receiver noise component η_0 is usually very small. Furthermore, the effect of this noise is to reduce the value of the off-diagonal element, which is equivalent to increasing σ_c or the correlation time $t_i - t_j$. For convenience of discussion, we shall ignore this effect.

Since the covariance matrix is real and symmetrical, and, as will be shown later, the optimal filter weights x_i are components of the eigenvector of this matrix, these x_i are real.

The clutter output P_c is actually a quadratic form. Representing the filter weights $x_1, x_2, \cdots, x_N$ by an

N-dimensional vector x, this quadratic form can be compactly written as

$$Q(\mathbf{x}) = (\mathbf{x}, A\mathbf{x}) \quad (15)$$

where the parentheses represent the inner product of the vector x and the vector **A**x.

The improvement factor thus becomes

$$I.F. = (\mathbf{x}, \mathbf{x})/(\mathbf{x}, A\mathbf{x}).$$

that the reciprocal of this improvement factor is actually the clutter output, or

$$P_c = (\mathbf{x}, A\mathbf{x}) \quad (16)$$

when **x** is normalized.

In an MTI filter design, one of the optimization criteria is to maximize the improvement factor or, equivalently, to minimize the clutter output factor (x, Ax). The problem then becomes to find a vector x such that the quadratic form (x, Ax) is a minimum.

Before we can find such a vector, we will show that this quadratic form is positive definite. The elements of the matrix A can be represented by the inner products of a set of N-dimensional vectors, such that

$$a_{ij} = (\mathbf{a}^i, \mathbf{a}^j). \quad (17)$$

This can be done because the diagonal elements in matrix A are always larger than the off-diagonal elements, and the matrix is symmetrical. According to the theorem of Gram-Schmidt [9], the quadratic form of this kind of matrix is always positive definite. This can also be seen intuitively. Since $Q(\mathbf{x})$ represents the clutter output power, it should always be positive, unless all filter weights are zero.

III. Minimum Eigenvalue and its Eigenvector

In the previous section we discussed the fact that the optimal design of the MTI filter requires finding an N-dimensional vector x such that the quadratic form $Q(\mathbf{x})$ is minimum, under the constraint that this vector lies on a sphere,

$$(\mathbf{x}, \mathbf{x}) = 1.$$

This amounts to solving a classic eigenvalue problem. It is well known that these eigenvectors and eigenvalues possess the following properties [9], [10]:

1) The eigenvalues of a real, symmetrical matrix are real. Furthermore, if A is positive definite, these eigenvalues are positive.

2) The eigenvectors associated with distinct eigenvalues of a real symmetrical matrix A are real and orthogonal. Moreover, these vectors can be normalized so that they form an orthonormal set.

3) By virtue of property 2), the quadratic form $Q(\mathbf{x})$ can be represented as

$$Q(\mathbf{x}) = \sum_i d_i^2 \lambda_i \quad (18)$$

where

$$\mathbf{x} = \sum_i d_i \mathbf{x}^i \quad (19)$$

and λ_i are eigenvalues and $\mathbf{x}^i$ are the associated eigenvectors. It is evident that the minimum $Q(\mathbf{x})$ occurs when x takes the eigenvector which is associated with the minimum eigenvalue λ_{min}, since

$$\lambda_1 \geqslant \lambda_2 \geqslant \cdots \geqslant \lambda_{min}$$

due to property 1).

Then.

$$Q(\mathbf{x}) = \lambda_{min}.$$

Thus, the quadratic form has a minimum value which is equal to the smallest eigenvalue of matrix A, and the eigenvector associated with this eigenvalue is the required minimization vector. This same result has been shown by Emerson [5] and Brennan [6] by use of Lagrange's multiplier.

If one expands the determinant $|A - I|$, one finds the following valuable relations between the eigenvalues and the elements in matrix A:

$$\sum_{i=1}^{N} \lambda_i = \sum_{i=1}^{N} a_{ii} \quad (20)$$

$$\prod_{i=1}^{N} \lambda_i = \det A. \quad (21)$$

We pointed out earlier that the diagonal elements of matrix A are unity; hence,

$$\sum_{i=1}^{N} \lambda_i = N. \quad (22)$$

Since the quadratic form $Q(\mathbf{x})$ is positive definite, we have the following properties [9]:

$$\det A > 0 \quad (23)$$

$$\det A \leqslant \prod_{i=1}^{N} a_{ii}. \quad (24)$$

Hence,

$$0 < \prod_{i=1}^{N} \lambda_i \leqslant 1. \quad (25)$$

From (22) and (25), one can deduce that

$$0 \leqslant \lambda_{min} \leqslant 1. \quad (26)$$

Here, λ_{min} is the smallest eigenvalue, which then, in turn, represents the optimal clutter output of an MTI filter. The equalities in the above equations represent two degenerate cases. When either the clutter spectrum standard deviation $\sigma_c \to \infty$ or the correlation time $t_i - t_j$ becomes infinite, the off-diagonal elements in matrix element A approach zero [see (13)]. In this case,

$$\lambda_1 = \lambda_2 = \cdots = \lambda_N = 1$$

and any normalized vector can satisfy the eigen equations.

On the other hand, if σ_c or the correlation time $t_i - t_j$ approaches zero, then every element of matrix A approaches unity. In this case, the determinant A is zero. According to (21), the smallest eigenvalue is zero. Under this condition, the eigen equations are identical and have the following form:

$$x_1 + x_2 + \cdots + x_N = 0.$$

There are an infinite number of solutions. One solution is to set these x_i according to the binominal distribution. This is well known. One sees that the binominal weighting MTI minimizes the clutter output only when the standard deviation of the clutter spectrum is zero.

We thus conclude that the clutter output of an MTI system always lies in a range from 0 to 1. As the clutter spectrum standard deviation σ_c or the correlation time $t_i - t_j$ increases, the clutter output increases.

IV. Constant-PRF MTI Filter

If the PRF is constant, elements of matrix A, as pointed out earlier, can be represented as

$$a_{ij} = \exp\left[-2\pi^2\sigma_c{}^2(i-j)^2\tau^2\right] \tag{27}$$

One introduces a new variable

$$\sigma = \sigma_c\tau$$

and sets

$$\xi = \exp(-2\pi^2\sigma^2). \tag{28}$$

Then,

$$a_{ij} = \xi^{(i-j)^2} \tag{29}$$

With this normalized σ, the $\lambda_{\min}$ (or clutter output) can be represented as

$$\lambda_{\min} = [2/(\sqrt{2\pi}\sigma)]\int_0^\infty \exp - [f^2/(2\sigma^2)] \cdot \sum_i\sum_j x_i x_j \cos 2\pi f(i-j)\, df \tag{30}$$

where the x_i are the optimal filter weights for a given σ. Notice that in this equation the independent variable f is normalized, so that the filter power transfer function

$$|F(f)|^2 = \sum_i\sum_j x_i x_j \cos 2\pi f(i-j) \tag{31}$$

is a periodic function when f is an integer Furthermore, it can be shown that this power transfer function is symmetrical within each period [8]. Therefore, it is evident that for a working MTI system, σ must be limited to a very small value (much smaller than 0.5). Otherwise, the clutter spectrum spreads so far into the filter passband that no effective filtering can be achieved. Thus, the value of σ plays an important role in an MTI filter design.

With this assumption, we shall show several interesting properties of the minimum clutter output $\lambda_{\min}$ of a constant-PRF MTI system.

Monotonic Property

Assume that, for a given σ_0, the corresponding optimal filter weights x_i' are inserted into (30) and a new function is defined:

$$\lambda_{\min}(\sigma_0, \sigma) = [2/(\sqrt{2\pi}\sigma)]\int_0^\infty \exp - [2f^2/(2\sigma^2)] \cdot \sum_i\sum_j x_i'(\sigma_0)x_j'(\sigma_0) \cos 2\pi f(i-j). \tag{32}$$

The derivative of the filter power transfer function

$$(\delta/\delta f)\,|F(f)|^2 = 2\pi\sum_i\sum_j (i-j)x_i'(\sigma_0)x_j'(\sigma_0) \sin 2\pi f(i-j) \tag{33}$$

is zero when $f = 0, 1/2, 1, 3/2, \cdots$, no matter how the x_i are chosen. Since the clutter power density function is a bell-shaped function and heavily concentrated at the neighborhood of $f = 0$, it is evident that the optimal weights x_i must be so chosen that a minimum occurs at $f = 0$. Furthermore, the x_i are constrained such that the average frequency response of the filter is constant; that is, $\Sigma x_i'^2 = 1$. Therefore, the envelope of the power transfer function must increase monotonically from $f = 0$ to $f = 0.5$. The area under the curve of the function $(1/\sigma)\exp[-f^2/(2\sigma^2)]$ is a constant. However, as σ decreases, more area will be concentrated at the neighborhood $f = 0$. Therefore, it is not difficult to see that

$$\lambda(\sigma_0, \sigma) \leqslant \sigma(\sigma_0, \sigma_0) \qquad \text{when } \sigma < \sigma_0. \tag{34}$$

This may be interpreted as follows. The improvement factor for an MTI filter designed for a certain σ_0 will not be degraded if the actual spectrum standard deviation σ is less than σ_0.

Furthermore, it is evident that

$$\lambda_{\min}(\sigma, \sigma) \leqslant \lambda_{\min}(\sigma_0, \sigma) \leqslant \lambda(\sigma_0, \sigma_0) \qquad \text{if } \sigma < \sigma_0 \tag{35}$$

due to the nature of the optimization.

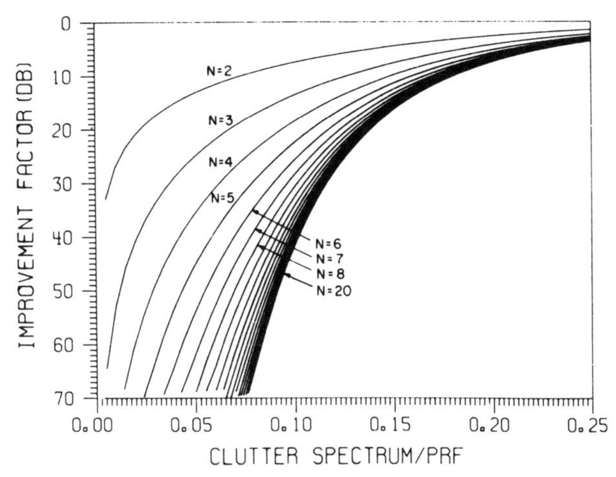

Fig. 2. Improvement factor versus normalized clutter-spectrum standard deviation for a number of cancelling pulses from 2 to 20.

Hence, one may also conclude that $\lambda_{\min}(\sigma)$ is a monotonic function and its value increases from 0 to 1 as σ increases from 0 to infinity.

Improving Property

Consider matrixes A_N as a sequence of matrixes; that is,

$$A_2 = \begin{bmatrix} 1 & \xi \\ \xi & 1 \end{bmatrix} \quad (36a)$$

$$A_3 = \begin{bmatrix} 1 & \xi & \xi^4 \\ \xi & 1 & \xi^{\circ} \\ \xi^4 & \xi & 1 \end{bmatrix} \quad (36b)$$

$$A_N = \begin{bmatrix} 1 & \xi & \xi^4 & \cdots & \xi^{N^2} \\ \xi & 1 & \xi & \cdots & \xi^{(N-1)^2} \\ & & \vdots & & \\ \xi & & \cdots & & 1 \end{bmatrix}. \quad (36c)$$

According to the Sturmian separation theory [11],

$$\lambda_{N+1}(A_{N+1}) \leqslant \lambda_N(A_N) \quad (37)$$

where $\lambda_N(A_N)$ represents the minimum eigenvalue of matrix A_N. This means that the minimum clutter output decreases as the number of cancelling pulses in the MTI system increases.

Limiting Property

It is well known that each eigenvalue of a non-negative matrix lies in the interior or on the boundary of at least one of the circles [12] - [15]

$$|\lambda - a_{ii}| \leqslant \sum_{j=1}^{N} a_{ij}, \quad j \neq i. \quad (38)$$

From (36), we see that as N, the number of MTI canceling pulses, increases, the off-diagonal element a_{iN} approaches zero. Thus, the quantities $|\lambda - a_{ii}|$ approach a limit as N increases. Since $a_{ii} = 1$, the a_{ij} are positive, and $\lambda_{\min}$ is always less than unity, one may write

$$\lambda_{\min} \geqslant 1 - \sum_j a_{ij}, \quad i \neq j.$$

From (37), one may conclude that although the improvement factor $(1/\lambda_{\min})$ improves as N increases, it reaches a limit as N becomes sufficiently large.

Closely related to this result is the clutter spectrum deviation σ. One may see that as σ increases, all high-order terms become essentially zero. We may thus write

$$A \approx \begin{bmatrix} 1 & \xi & 0 & \cdots & 0 \\ \xi & 1 & \xi & \cdots & 0 \\ & & \cdots & & \\ 0 & 0 & 0 & \cdots & 1 \end{bmatrix}. \quad (39)$$

Under this condition, the limiting circle in (38) remains about the same no matter what N is. Thus, we can establish another limit. This limit says that as the σ increases to a point, the clutter output of an MTI system will be independent of the number of stages of the MTI system.

V. Numerical Results of a Constant-PRF MTI System

The optimal improvement factors are computed versus the normalized clutter-spectrum standard-deviation factor σ. This is shown in Fig. 2. The number of MTI canceling pulses N is used as the parameter, and a family of curves is obtained. For each N at each σ point, the minimum eigenvalue is computed. The reciprocal of this value, which is the improvement factor, is plotted on a decibel scale. At

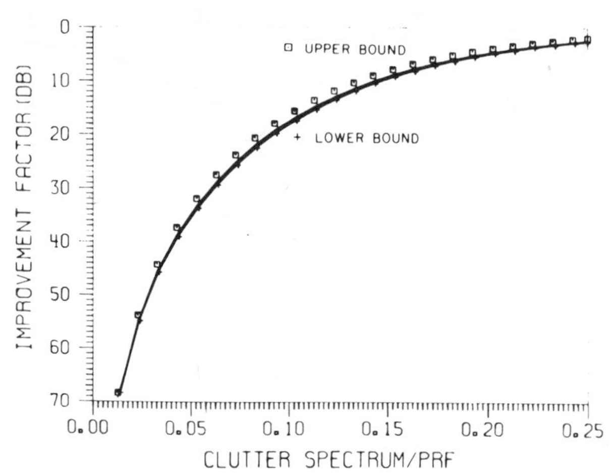

Fig. 3. Staggered-PRF MTI filter with a 3-pulse canceller and 10 percent interpulse variation.

each point for a given σ, a different eigenvector or filter weight is used. Thus, each of these curves represents the maximum improvement factor one can expect. It clearly demonstrates the properties discussed in Section IV. The curves are monotonic. For a given σ, the improvement factor approaches a limit as the number of MTI cancelling pulses increases. Furthermore, as σ increases, all these curves converge closely to a single curve, and the improvement factor approaches 0 dB.

These curves have significant value in the MTI design. For example, if the normalized-clutter spectrum standard deviation has a frequency of about 1/10 of the PRF for a 10-pulse MTI, one would expect an improvement factor of 34 dB. A further increase in the number of MTI pulses does not appreciably improve this situation. The improvement factors shown on these curves represent an upper bound. Better improvement can not be expected. In the next section we will show that the use of a staggered-PRF system does not improve this situation.

VI. Staggered-PRF MTI System

To avoid blind velocities, staggered-PRF MTI sytems are generally used. In these systems, the interpulse durations vary from pulse-to-pulse. The choices of this variation are numerous. It thus makes the analysis of such a problem extremely difficult. However, in a practical radar system this variation is somewhat bounded by the need for efficient usage of radar time. One may therefore assume that the interpulse durations are varying between a lower and upper bound:

$$T_1 < T_s < T_2$$

$$T_2 = T_1(1 + \alpha) \tag{40}$$

where α represents the fractional variation of the interpulse duration.

In a previous section it was shown that the optimal improvement factor of a constant-PRF MTI filter is a monotonic function of σ ($\sigma = \sigma_c T$). Increasing the interpulse correlation time, in general, increases the clutter output λ_{min}. Since the covariance matrix of a staggered-PRF MTI system has properties similar to those of a constant-PRF MTI system, we would anticipate that this similar monotonic property would be maintained. Therefore, we conjecture that the minimum clutter output of a staggered-PRF MTI system is bounded by the performance of two equivalent constant PRF MTI systems. One of them has a PRF equal to the lowest PRF of the staggered PRF system, while the other has a PRF equal to the highest PRF of the staggered system. That is,

$$\lambda_{min}(T_1) \leqslant \lambda_{min}(T_s) \leqslant \lambda_{min}(T_2). \tag{41}$$

To investigate this effect, many numerical examples were computed. In each of these examples, a lower interpulse time T_1 and an upper interpulse time T_2 are set. The interpulse times of the staggered-PRF system are then generated randomly (with a uniform distribution). However, they are kept within these bounds. The optimal improvement factors are then computed as a function of σ. These improvement factors were then compared with constant-PRF MTI systems having interpulse times T_1 and T_2. Many samples have been computed with the number of cancelling pulses ranging from 2 to 20 pulses and with a variation of interpulse time from 10 to 40 percent. So far we have not found a single case which violates this conjecture. Fig. 3 shows a typical example. The lower bound of these curves, which is represented by crosses, is the improvement factor of a constant-PRF MTI filter having an interpulse time T_1, while the upper bound, represented by squares, is the corresponding improvement factor with an interpulse time T_2. In this particular example, $T_2 = 1.1T_1$. Curves (almost a solid line in Fig. 3) which lie within these two bounds are those of staggered-PRF MTI systems.

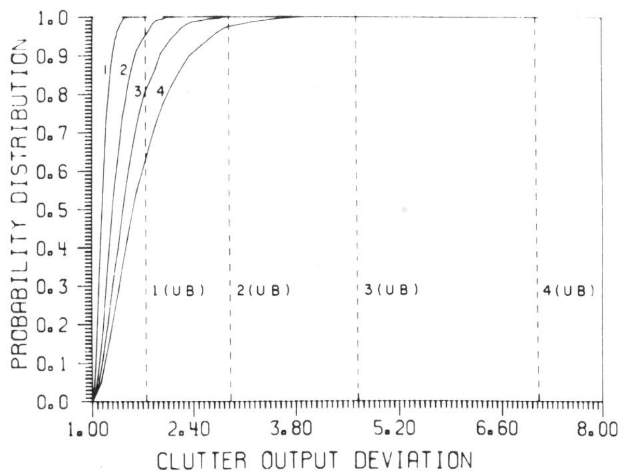

Fig. 4. Clutter output statistic distribution of a 4-pulse staggered-PRF MTI filter. $\sigma = 0.05$; interpulse variation: curve 1 = 10 percent, curve 2 = 20 percent, curve 3 = 30 percent, curve 4 = 40 percent; U.B. denotes upper bound.

In Fig. 4 we show the statistical property of the improvement factor of a staggered-PRF MTI system. The interpulse duration is again generated randomly according to a constraint shown in (40), which is repeated as follows:

$$T_1 \leqslant T_s \leqslant T_2$$

$$T_2 = T_1(1+\alpha)$$

where T_1 and T_2 are, respectively, the lower and upper bound of the interpulse duration, while T_s is the interpulse duration of the staggered-PRF MTI filter. The clutter power density function is assumed to be Gaussian. In this example, its normalized values of standard deviation for the lower and upper bounds are, respectively,

$$\sigma_{\mathrm{LB}} = \sigma_c T_1 = 0.05$$

$$\sigma_{\mathrm{UB}} = \sigma_c T_2 = 0.05(1+\alpha).$$

There are four curves plotted in this figure. The curves correspond to maximum interpulse variations of 10, 20, 30, and 40 percent. In each of these cases the optimal clutter output of a constant-PRF MTI having an interpulse duration of the lower bound is computed. Subsequently, a set of interpulse durations of the staggered-PRF MTI is generated randomly within the specified bounds, and its minimum clutter output is computed and then normalized with respect to the clutter output of the lower bound. The clutter output of the upper bound (constant-PRF with interpulse time T_2) is also computed and normalized. This is shown by dotted lines in Fig. 4. It is also assumed that at least one of the interpulse durations of the staggered-PRF system is equal to the lower bound T_1.

A large number of samples are computed. The cumulative probability of these samples is then plotted against the normalized clutter output. These curves rise sharply. This means that the majority of the mass of the probability density function lies within a narrrow region of the clutter output. For example, curve 1 shows the samples of a 10 percent variation of interpulse duration. It shows that a randomly chosen interpulse sample has a probability of 0.99 that its clutter output lies in the middle of the upper and lower bound, while the probability of a sample having a clutter output about 10 percent higher than the lower bound is no more than 0.1.

VII. Conclusion

In this paper, we have formulated the problem of optimization of the improvement factor of a nonrecursive MTI filter by minimization of a quadratic form. It was shown that the minimum normalized clutter output (a reciprocal of the improvement factor) is the minimum eigenvalue of this quadratic form, and that the corresponding eigenvector is the optimal weight to be used in this filter. It was also shown that the improvement factor of a nonrecursive MTI filter has the following properties:

1) The normalized clutter output is always greater than zero and less than unity.

2) The improvement factor is a monotonic function of the normalized standard deviation of the clutter spectrum. As the clutter spectrum increases or the PRF decreases, the improvement factor is reduced.

3) An increase in the number of cancellers in an MTI system improves its performance; however, as this number increases to a point (say 8 to 10 cancellers), this improvement reaches a limit. Furthermore, the normalized clutter spectrum standard deviation also limits the performance of an MTI system. At a value of approximately 0.25, the improvement factor that an MTI can achieve is independent of the number of cancelling pulses used.

4) The improvement factor achieved by a staggered-PRF MTI system lies within two bounds. The lower bound is the improvement factor which can be achieved by an equivalent constant-PRF MTI having a PRF equal to the lowest PRF of the staggered MTI system, while the upper bound is equal to the equivalent constant-PRF MTI system, which has a PRF equal to the highest PRF in the staggered system.

In deriving the above conclusions, it is assumed that the clutter power density function is Gaussian. We suspect, however, that other even and bell-shaped clutter power density functions may have these same properties.

References

[1] W.D. Rummler, "Clutter suppression by complex weighting of coherent pulse trains," *IEEE Trans. Aerospace and Electronic Systems,* vol. AES-2, pp. 689-699, November 1966.

[2] R.L. Mitchell and A.W. Rihaczek, "Clutter suppression properties of weighted pulse trains," *IEEE Trans. Aerospace and Electronic Systems,* vol. AES-4, November 1968.

[3] L.J. Spafford, "Optimal radar signal processing in clutter, "*IEEE Trans. Information Theory,* vol. IT-14, pp. 734-743, September 1968.

[4] D.F. Delong and E.M. Hofstetter, "On the design of optimum radar waveforms for clutter rejection," *IEEE Trans. Information Theory,* vol. IT-13, pp. 454-463, July 1967.

[5] R.C. Emerson, "Some pulse Doppler and MTI techniques," Rand Rept. R-274, March 1954.

[6] L.E. Brennan and I.S. Reed, "Optimum processing of unequally spaced radar pulse trains for clutter rejection," *IEEE Trans. Aerospace and Electronic Systems,* vol. AES-4, pp. 474-477, May 1968.

[7] O.J. Jacomini, "Weighting factor and transmission time optimization in video MTI systems," *IEEE Trans. Aerospace and Electronic Systems,* vol. AES-8, pp. 517-527, July 1972.

[8] J.K. Hsiao and F.F. Kretschmer, Jr., "Design of a staggered-PRF moving target indication filter," *Radio and Electronic Eng.*, vol. 43, November 1973.

[9] R. Bellman, *Introduction to Matrix Analysis,* 2nd ed. New York: McGraw-Hill, 1970, ch. 4.

[10] C.R. Wylie, Jr., *Advanced Engineering Mathematics.*

[11] W.S. Burnside and A.W. Panton, *Theory of Equations,* vol. 2. New York: Logmans, Green, 1928, ch. 7.

[12] H. Rohbach, "Bemenkungen zu einem Determinantensatz von Minkowski," *Jahr. Deut. Math. Ver.* vol. 40, p. 49, 1931.

[13] E.W. Barankin, "Bounds for the characteristic roots of a matrix," *Bull. Am. Math. Soc.*, vol. 51, pp. 767-770, 1945.

[14] E. Bodewig, *Matrix Calculus,* 2nd ed. Amsterdam: 1959.

[15] A. Brauer, "On the characteristic roots of non-negative matrices," in *Recent Advances in Matrix Theory*, H. Schneider, Ed. Madison, Wisc.: University of Wisconsin Press, 1964.

James K. Hsiao received the B.E.E. degree from Hunan University, China, the M.S. degree in electrical engineering in 1957 from Montana State College, and the Ph.D. degree in 1962 from Iowa State University.

He was with the Chinese Nationalist Air Force for 10 years, and he taught electrical engineering at Taipli Institute, China. From 1962 to 1967 he was with Bell Telephone Laboratories in the Digital System Department, working on research and application of super-conductor devices. Since 1967 he has been with the Radar Division of the Naval Research Laboratory, Washington, D.C., specializing in array antennas. His current interests are in the areas of phased arrays and radar signal processing.

2.5 OPTIMIZATION OF NONRECURSIVE MTI

D.C. Schleher and D. Schulkind

IEE International Radar Conference, London, October 1977, pp. 182-185.

ABSTRACT

Procedures for optimization of nonrecursive and recursive digital MTI systems are reviewed and compared. A limitation in the current method used to synthesize nonrecursive MTI filters is identified and discussed. It is shown how the general nonrecursive MTI optimization problem can be put in the form of a Quadratic Programming problem. A general solution method using Dantzig's modified simplex Linear Programming Algorithm is presented and discussed. A numerical example is provided to illustrate the method, and the results are compared against those obtained using current methods. Several design examples are presented.

I. INTRODUCTION

Digital MTI filters can be designed using either recursive or nonrecursive digital filtering techniques. Nonrecursive digital MTI filters have become popular due to their ease of implementation and superior transient response. The response due to external interference or clutter transients, lasts only for a duration consistent with the finite impulse response (FIR) of this type of MTI. The FIR also controls the shape of the MTI filter passband, thereby limiting the velocity response which may present a serious problem in a number of practical MTI applications.

Optimization procedures for recursive MTI filters are well developed. A straightforward transformation given in [1] allows standard network theory to be applied to the problem of designing a recursive MTI. However, comparable optimization procedures for nonrecursive MTI's are not available.

The best currently available design procedure for nonrecursive MTI filters is due to Emerson [2], and was further developed by Capon [3]. This procedure maximizes the signal-to-clutter ratio at the output of the filter when the target's Doppler shift is randomly distributed over all possible values and the clutter spectrum is known. An important property of this procedure is: by appropriate normalization of the filter weights, the zeros of the filter's response in the Z-plane can be made to lie on the unit circle [4]. The results of this property is that fixed discrete targets are eliminated only when an even number of pulses (odd number of delay sections) are processed.

There are several limitations of Emerson's procedure. First, all practical MTI's are designed to eliminate fixed discrete targets (the sum of the filter weights is zero). This is not an explicit constraint of Emerson's procedure. Secondly, in a practical MTI, there is usually a velocity acceptance band where uniform response is desired and a rejection band where targets are attenuated. Emerson's procedure provides a velocity response that has significant variations in the acceptance band, and furthermore, this variation is emphasized as the number of processed pulses is increased [3] asymptotically approaching an impulse at the minimum value of the clutter spectrum. Thirdly, the optimization is predicated on a complete ignorance of the velocity characteristic of the target, and hence, precludes any shaping of the MTI response to provide a desired target velocity response. This limitation is overly restrictive, but its removal causes a complete breakdown of Emerson's procedure.

Quadratic programming [5] provides a method for optimizing a nonrecursive MTI that alleviates many of the limitations of Emerson's procedure. The general structure of this approach is similar to Emerson's procedure, in that a quadratic form (the output clutter power) is minimized, subject to constraints. However, rather than requiring a complete ignorance of target velocity, a desired MTI velocity response is specified. In addition, a further constraint is imposed that requires the sum of the filter weights be zero thereby causing the MTI to reject discrete targets.

The quadratic programming algorithm [6] represents a systematic procedure for the minimization of a quadratic form (objective function) subject to linear constraints. It is guaranteed to converge [7] in a finite number of steps. Furthermore, if overly restrictive constraints are imposed, thereby preventing minimization of the quadratic objective function, the algorithm will identify this problem and allow the constraints to be relaxed for the purpose of obtaining a solution. In the design of MTI filters, this property can be used to determine the best velocity characteristic a given number of delay sections can achieve, or alternatively, the number of delay sections required to achieve a specified velocity response.

II. MTI OPTIMIZATION BY QUADRATIC PROGRAMMING

The nonrecursive MTI filter depicted in Figure 1 consists of a tapped delay line whose outputs are weighted and summed. The optimization problem is to determine appropriate weights that provide the best performance in accordance with the optimization criteria.

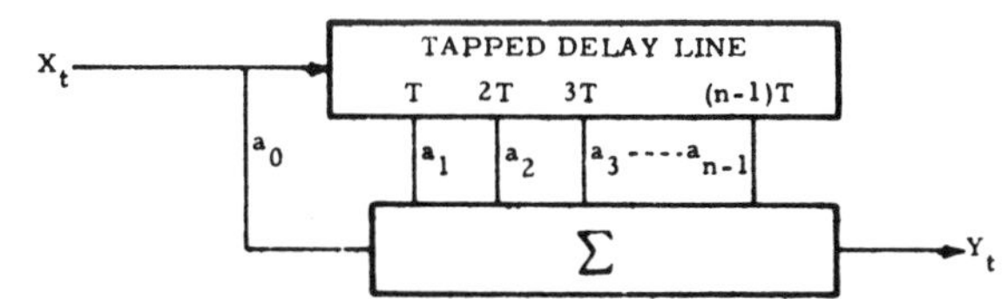

Figure 1 Nonrecursive MTI filter

To apply quadratic programming, the problem must be put in the form of a minimization of a quadratic objective function (whose matrix is positive semi-definite) subject to linear constraints. The linear constraints can be in the form of equalities or inequalities. When inequalities are used, slack variables must be added to put the problem in canonical form

Consider the nonrecursive MTI filter of Figure 1. The clutter output power is given by:

$$P_c = \sum_{j=0}^{n-1} \sum_{k=0}^{n-1} a_j a_k \rho_c [(j-k)T] \quad \cdots\cdots\cdots (1)$$

where $a_{j,k}$ are the filter weights, $\rho_c(\cdot)$ are the elements of the covariance matrix of the distributed clutter, and n is the number of pulses processed. As in the Emerson procedure [2], (1) is the function that is selected to be minimized. Furthermore, it meets the requirement of a quadratic programming objective function since the covariance matrix $[\rho_c(\cdot)]$ is always positive semi-definite [8].

To eliminate clutter discretes, a zero response is required at zero frequency. This is accomplished by causing the filter weights to sum to zero

$$\sum_{i=0}^{n-1} a_i = 0. \quad \cdots\cdots (2)$$

Since this is a linear constraint, it satisfies the requirements of the quadratic programming problem.

In order to specify the MTI's velocity characteristic in the form of a linear constraint, it is necessary to restrict the class of MTI filters to those with linear phase. This results in the filter weights being symmetric about their central point when an odd number of pulses (even number of delays) are processed, and anti-symmetric about their central point when an even number of pulses (odd number of delays) are processed. Therefore, the filter transfer function for n even ($n \geq 2$) is given by:

$$H(e^{j\omega t}) = e^{-j\left(\frac{n-1}{2}\omega T + \frac{\pi}{2}\right)} \left[\sum_{i=0}^{\frac{n}{2}-1} (-1)^{i+1}\, 2\,|a_i|\, \mathrm{SIN}\left(\frac{n-1}{2} - i\right)\omega T\right] \quad \cdots\cdots (3)$$

and for n odd ($n \geq 3$):

$$H(e^{j\omega t}) = e^{-j\frac{n-1}{2}\omega T}\left[(-1)^{(n-1)/2}\left|a_{\frac{n-1}{2}}\right| + \sum_{i=0}^{\frac{n-3}{2}} (-1)^i\, 2\,|a_i|\, \mathrm{COS}\left(\frac{n-1}{2} - i\right)\omega T\right] \quad \cdots\cdots (4)$$

The first term of the response represents a linear phase term, while the second term is the magnitude of the transfer function. The magnitude function is linear in the filter weights and hence, satisfies the conditions for a linear constraint of a quadratic programming problem. Note that the class of linear phase MTI's includes those with binominal weights obtained by cascading single cancelers, and also includes those obtained by Emerson's procedure where solutions for the eigenvector results in symmetric and anti-symmetric weights.

The quadratic programming problem for the optimization of a nonrecursive MTI can be summarized as follows:

$$\text{Min } P_c = \sum_{j=0}^{n-1}\sum_{k=0}^{n-1} a_j a_k \rho_c|(j-k)T| \quad \cdots\cdots (1)$$

subject to

$$\sum_{i=0}^{n-1} a_i = 0, \quad \cdots\cdots (2)$$

and m contraints evaluated throughout the MTI's velocity response of the form

$$|H(\omega)| = \sum_{i=0}^{\frac{n}{2}-1} (-1)^{i+1}\, 2\,|a_i|\, \mathrm{SIN}\left(\frac{n-1}{2} - i\right)\omega T;\ n \text{ even} \quad \cdots (5)$$

or,

$$|H(\omega)| = (-1)^{\frac{n-1}{2}}\left|a_{\frac{n-1}{2}}\right| + \sum_{i=0}^{\frac{n-3}{2}} (-1)^i\, 2\,|a_i|\, \mathrm{COS}\left(\frac{n-1}{2} - i\right)\omega T;\ n \text{ odd}. \quad \cdots\cdots (6)$$

In addition, the quadratic programming algorithm requires that the variables (weights) be equal to or greater than zero ($a_i \geq 0$). This is easily handled for the MTI case by using the absolute value of the weights in the algorithm and including the signs in the constraints:

$$a_i = |a_i| = x_i;\ a_i \text{ positive}$$
$$a_i = -|a_i| = -x_i;\ a_i \text{ negative} \quad \cdots\cdots (7)$$

The quadratic programming problem specified by equations (1), (2), (5) and (6) can be solved by application of an algorithm originally developed by Dantzig [5]. This algorithm is a straightforward modification of the Simplex method for linear programming. Reference [6] describes a streamlined version of Dantzig's quadratic programming algorithm developed by van de Panne and Whinston; they proved that the algorithm converges in a finite number of steps [7].

III. QUADRATIC PROGRAMMING EXAMPLE

In this section, a simple numerical example is provided to illustrate the procedure. Extension to more complex problems can be readily accomplished in a straightforward manner.

As an example, assume a requirement exists for a nonrecursive MTI filter design with a clutter attenuation of at least 30 dB; the clutter spectrum is Gaussian shaped with $\sigma_f/PRF = 0.025$, and the velocity characteristics are to have a 3 dB point $f_{co}/\sigma_f = 10$. A four pulse processor ($n = 4$) is to be investigated.

Applying equation (5) at the PRF/2 point ($\omega = \pi/T$) and at f_{co} ($\omega = \pi/2T$) results in:

$$H\left(\omega = \frac{\pi}{T}\right) = 2a_0 + 2a_1 = 1$$
$$H\left(\omega = \frac{\pi}{2T}\right) = \sqrt{2}(a_1 - a_0) \geq b. \quad \cdots\cdots (8)$$

Applying equation (1) using

$$\rho_c = e^{-i^2\Omega^2/2} \quad \cdots\cdots (9)$$

where $\Omega = 2\pi\sigma_f/PRF$ results in

$$P_c = 0.2108\, a_0^2 + 0.02452\, a_1^2 - 0.14356\, a_0 a_1 \quad \cdots\cdots (10)$$

The sum of the weights is zero because n is even.

The quadratic programming problem [9] is then:

$$\max.\ [0.05255 - P_c] = 0.28196\, x_1 - 0.37826\, x_1^2$$

Subject to

$$x_1 - x_2 + y = 0.5 \text{ or } 0.4240$$
$$x_1, x_2 \geq 0, \qquad \cdots\cdots (11)$$

Where x_2 is a slack variable and y is an artificial variable unrestricted in sign. The use of an artificial variable is required to find an initial basic feasible solution [6]. The parameters of the quadratic problem are:

$$C = \begin{bmatrix} 0.75652 & 0 \\ 0 & 0 \end{bmatrix} \quad p = \begin{bmatrix} 0.28196 \\ 0 \end{bmatrix} \cdots\cdots (12)$$
$$b = 0.5 \text{ or } 0.4240 \quad A = \begin{bmatrix} 1 & -1 \end{bmatrix}$$

where the alternate values for b permit either the velocity constraint (b = 0.5) or the CA constraint (b = 0.4240) to be achieved.

The application of the algorithm is given in Table 1 for b = 0.424 and illustrates the initialization procedure given in Reference [6]. This procedure requires an identification of the values of x that are in the solution set (x_s). This is obviously x_1 in the problem of (11). The x_s variables are moved into the Basis replacing the artificial (y) variables. In addition, the v - variables are introduced into the Basis replacing the dual - u - variables (u_s) of the solution set. The procedure terminates with a standard tableau when the v - variables in the Basis are unrestricted in sign while the u - variables are non-negative.

The optimum value of 2F from Table 1 is 0.10310 obtained with weights (a_0 = 0.076, a_1 = 0.424) resulting in P_c = 0.001 thereby obtaining a CA of 30 dB. The weights for Emerson's procedure can be found by maximizing (11) without constraints (a_0 = 0.1273, a_1 = 0.3727). The velocity characteristic for both Emerson's procedure and the quadratic programming solution is depicted in Figure 2. Note that the desired velocity characteristic and CA cannot be obtained with n = 4 requiring the processing of additional pulses.

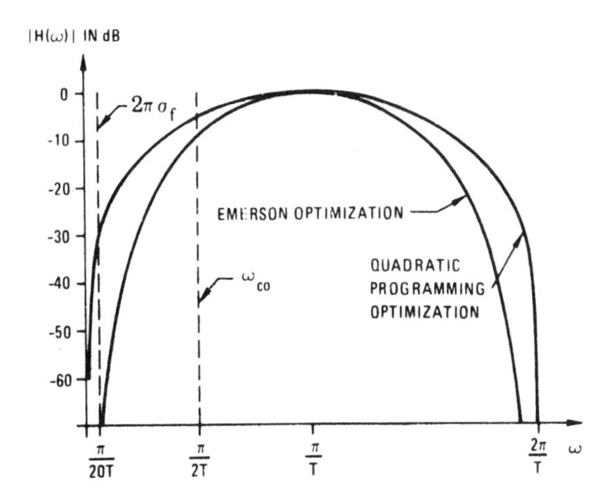

Figure 2 MTI velocity response

TABLE 1 Quadratic Programming Algorithm

Basis	Value	x_1	x_2	v	y	u_1	u_2
u_1	-0.28196	-0.75652	0	[-1]	0	1	0
u_2	0	0	0	1	0	0	1
y	0.4240	1	-1	0	1	0	0
2F	0	-0.28196	0	-0.4240	0	0	0
v	0.28196	0.75652	0	1	0	-1	0
u_2	-0.28196	-0.75652	0	0	0	1	1
y	0.4240	[1]	-1	0	1	0	0
2F	0.11955	0.03880	0	0	0	-0.424	0
v	-0.03880	0	0.75652	1	-0.75652	-1	0
u_2	0.03880	0	-0.75652	0	0.75652	1	1
x_1	0.4240	1	-1	0	1	0	0
2F	0.10310	0	0.3880	0	-0.03380	-0.4240	0

IV. MTI DESIGN EXAMPLE EXTENSION

The problem of the previous section is extended in Figure 3 where the MTI velocity characteristic for a five pulse processor is depicted for various velocity constraints. Note that for this case ($\alpha = .95$) the maximum MTI bandwidth consistent with a CA = 30 dB results in considerable variation in the MTI passband.

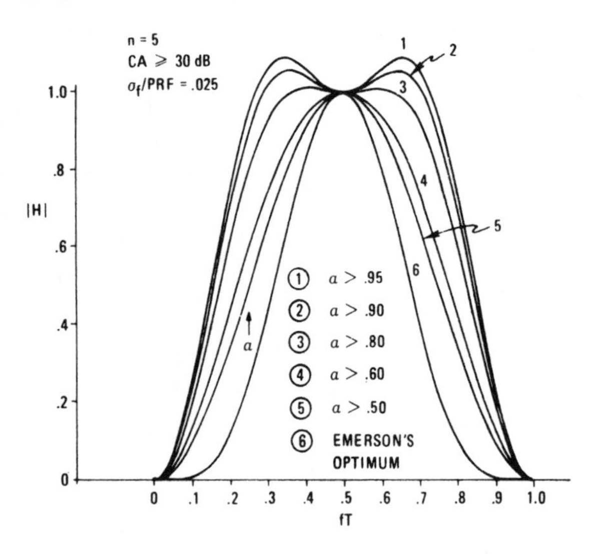

Figure 3 MTI velocity response - quadratic programming

In Figure 4 the design is extended to a six pulse processor with constraints in both the acceptance and rejection bands.

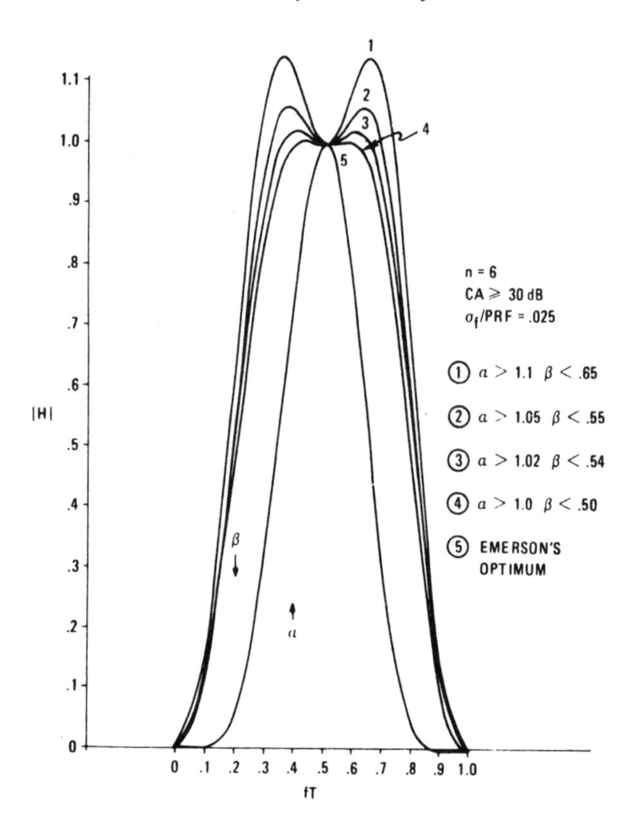

Figure 4 MTI velocity response - quadratic programming

Figure 5 depicts the MTI velocity characteristic for a ten pulse processor with multiple constraints in both the acceptance and rejection bands.

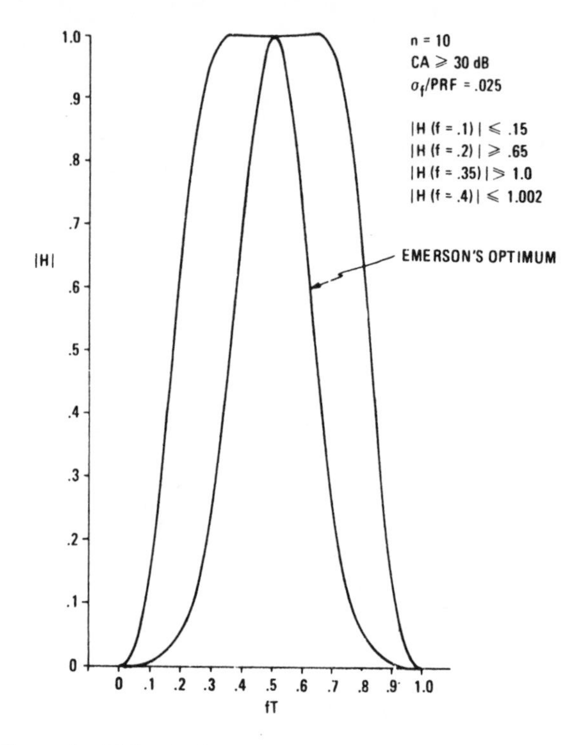

Figure 5 MTI velocity response - quadratic programming

REFERENCES

1. White, W., Ruvin, A., "Recent Advances in the Synthesis of Comb Filters," IRE Natl Conv Record, Vol 5, Part 2, 1957.

2. Emerson, R., "Some Pulse Doppler MTI and AMTI Techniques," Project Rand R-274, March 1954.

3. Capon, J., "Optimum Weighting Functions for the Detection of Sampled Signals in Noise," IEEE Trans, Vol IT-10, April 1964.

4. Robinson, E., "Optimum Weighting Functions for the Detection of Sampled Signals in Noise," IEEE Trans, Vol IT-11, (Corresp), July 1965.

5. Dantzig, G., "Linear Programming and Extensions," Princeton University Press, Princeton 1963.

6. van de Panne, C., Whinston, A., "The Simplex and the Dual Method for Quadratic Programming," Operational Research Quarterly 15, 355-388, 1964.

7. van de Panne, C., Whinston, A., "Simplicial Methods for Quadratic Programming," Naval Research Logistics Quarterly, 11, 273-302, 1964.

8. Bellman, R., "Introduction to Matrix Analysis," McGraw-Hill, N.Y., 1970.

9. Schleher, D.C., Schulkind, D., "Optimization of Digital MTI Using Quadratic Programming," IEEE Int. Conv. on Acoustics, Speech and Sig. Proc., May 1977.

Section 3 Pulse-Burst MTI Radars

EDITOR'S COMMENTS

This series of papers extends the theoretical treatment of the previous set of papers on optimum MTI filter design to include the effects of waveform design. The papers illustrate that, in general, both transmitter waveform and receiver processing designs are necessary to achieve an optimum MTI system design.

The particular transmitter waveform examined in these papers consists of a burst of relatively high PRF pulses which are repeated at a low PRF rate determined by the desired unambiguous target range. The pulse burst waveform provides many of the features associated with both low and high-PRF MTI systems. The use of the pulse burst waveform is advantageous in situations where the doppler clutter spread is large (e.g., airborne radar), making the use of low-PRF designs impractical. In general, both amplitude and phase modulation is employed across the burst to optimize the design. This results in a complex transmitter design that may present a severe hardware problem.

Spafford (Paper 3.1) considers the general optimization problem where signal and filter design are jointly optimized. The following design criteria are determined:

- **The use of a constant-amplitude, uniformly spaced, pulse-burst can achieve close to the maximum signal-to-interference ratio (SIR), provided the range-time extent of the clutter is less than the total equivalent signal extent and an optimum mismatched filter is used. An expression for the optimum filter weights is given.**
- **Only limited performance can be obtained with a uniform burst in range extended clutter. Receiver weighting is not effective in this case since it leads to an improvement of only 3 dB maximum over a matched filter.**
- **In range extended clutter, amplitude and phase weighting (complex weighting) of the transmitted burst combined with optimum processing provides considerably improved performance over that provided by a uniform burst. The maximum SIR improvement is bounded by the clutter-to-noise ratio (λ).**
- **Performance primarily depends on the number of pulses in the transmitted burst and the weighting applied. Optimization of the receiver plays a secondary role, and excellent performance is achieved by a filter matched to the transmitted signal.**
- **When the target-to-clutter doppler characteristics are unknown, adaptive signal processing is required to provide high performance.**

Ares (Paper 3.2) presents a physical picture of the considerations involved in optimum waveform and filter design by using the cross-ambiguity function $[\chi(\tau, \Delta)]$. Using the cross-ambiguity diagram, the author shows that a burst waveform with a constant interpulse period is desirable in range extended clutter and that the pulse spacing within the burst is determined by the doppler extent of the clutter. A short description of the range ambiguities with pulse burst waveforms is given.

A particularly lucid explanation is given for the maximum 3 dB improvement provided by the optimum filter over a matched filter when a uniform burst waveform in range extended clutter is employed.

An example of a six-pulse burst is given that illustrates the dramatic improvement that results from waveform optimization. When waveform weighting is employed, the optimum filter is found (using a nonlinear optimization procedure) to correspond to a filter matched to the transmitted waveform. This illustrates that the maximum performance that can be achieved with a pulse-burst radar in range extended clutter is determined primarily by the transmitter waveform characteristics, while the role of the receiver is that of the usual matched filter.

Mitchell and Rihaczek (Paper 3.3) discuss pulse-burst waveforms designed to reduce the response to interference uniformly within the entire interval between doppler ambiguities. This reflects the practical consideration that, in general, the clutter distribution (in both range and doppler) is not known. The effects are illustrated by high quality computer generated cross-ambiguity response functions for weighted pulse trains.

The interesting results presented by Mitchell and Rihaczek are:

- Complex weighting (phase and amplitude) is effective in changing the detail sidelobe structure between the doppler ambiguities, and hence is effectively utilized only when the clutter space is of a detailed form. In general, real weights (amplitude) are as effective in reducing the overall clutter level as complex weights.
- The doppler sidelobe level depends on the degree of weighting while the doppler width of the ambiguity spikes depends on the length of the transmitted burst. Adding low energy pulses at either end of the burst significantly reduces the sidelobe level between the burst without significantly broadening the width of the doppler ambiguity spikes.

3.1 OPTIMUM RADAR SIGNAL PROCESSING IN CLUTTER
L. Spafford
IEEE Transactions on Information Theory, Vol. IT-14, No. 5, September 1968, pp. 734-743.

Optimum Radar Signal Processing in Clutter

LLOYD J. SPAFFORD, MEMBER, IEEE

***Abstract*—This paper considers the joint optimization of a class of radar signals and filters in a number of clutter-plus-noise environments. The radar signal processor in this case will be optimum in the sense that its output at the time of target detection yields the maximum ratio of peak signal power to total interference power. If the interference at the input to this signal processor is a Gaussian random process, this processor also yields the maximum probability of detection for a given value of false-alarm probability.**

The signals used are pulse trains and the filters are tapped delay lines. The purpose of signal design is to determine the optimum complex weighting for each pulse of the pulse train. Filter design yields the optimum complex weighting for the output taps of the delay line. Filter design for a specified signal is considered first. This is followed by combined signal and filter design and matched filter design. Constrained signal and filter design is investigated last. It should be emphasized that the optimizations require a knowledge of the clutter time-frequency distribution. For practical situations, when the clutter distribution is unknown, an adaptive filter is proposed that automatically provides the optimum filter weights for a given transmitted signal.

When the clutter has a range-time extent less than the equivalent range-time extent of the signal, filter design alone yields nearly optimum performance. As the clutter becomes extended in range-time, it is necessary to consider jointly the design of signal and filter to obtain an optimum radar signal processor. In this report it is suggested that the signal be designed under the assumption of the clutter being extended over a broad range of Dopplers and that the signal processor consist of a bank of adaptive filters. Then each filter output yields the maximum ratio of peak signal to total interference power for this signal design.

Manuscript received September 25, 1967; revised March 21, 1968. This paper is taken from a dissertation submitted to the Faculty of the Polytechnic Institute of Brooklyn, in partial fulfillment of the requirements for the Ph.D. degree (Systems), 1967. Funded in part by the Rome Air Development Center, Rome, N. Y., under Contract AF30(602)-4105.
The author is with the Heavy Military Electronics Department, General Electric Company, Syracuse, N. Y.

I. Introduction

The RADAR statistical detection problem is one of choosing between signal and clutter-plus-noise or clutter-plus-noise alone at the receiver input. If the receiver input is given by $r(t)$, one is to choose between H_0 (no target) or H_1 (target) as follows.

$$H_0: r(t) = c(t) + n(t) \tag{1}$$

or

$$H_1: r(t) = s(t) + c(t) + n(t) \tag{2}$$

where $c(t)$ represents the clutter voltage and $n(t)$ the white thermal noise voltage contributed by the radar receiver itself. The performance of the matched filter [1] in the presence of white Gaussian noise is well known. Not so well known is the radar signal processing scheme desired for detection of targets when the interference is clutter-plus-noise. It is apparent from the differing characteristics of clutter and noise discussed in [8] that a different processing scheme is required for clutter-plus-noise than for noise alone. Most significant of these characteristics is that clutter appears as a signal-like return, whereas the noise is independent of the signal transmitted. The importance of signal design is then readily apparent if the response due to clutter is to be reduced. In fact, combined signal and filter design should be considered.

The first to investigate optimum filter design in a clutter environment was Urkowitz [2], but, neglecting the noise, he assumed clutter was the only form of interference present. Noise, however, must be present to limit performance in those cases where the response due to clutter can be reduced to a point below the anticipated level of the noise.

Manasse [3] first investigated radar signal design for both clutter and noise where the clutter possessed the same Doppler value as the target. When the signal energy and bandwidth are constrained, Manasse found the best signal to be a wide-band signal or a very short pulse. One signal that meets this requirement is a high time–bandwidth signal such as a linear frequency-modulated (FM) pulse. This result is no longer applicable for target and clutter at different Doppler values. In fact, a very wide bandwidth signal is suboptimum when target and clutter can be resolved in Doppler. This paper will consider the joint optimization of signal and filter when target and clutter have different Dopplers.

After this paper was written, it was brought to the attention of the author that the signal and filter design problem considered in this report had also been investigated by Rummler [12] and by Delong and Hofstetter [13]. This paper provides additional results for the signal and filter design problem and introduces an adaptive implementation for the optimum filter.

The radar signal to be considered will have the following real form.

$$s(t) = a(t) \cos [w_0 t + \phi_s(t)]. \tag{3}$$

Complex envelope [1] notation will be used throughout this paper; it defines the amplitude and phase modulation portions of the real transmitted signal as follows.

$$u(t) = a(t) \exp [i\phi_s(t)]. \tag{4}$$

A typical signal processing scheme employed in radar and used for this paper is shown in Fig. 1. The complex envelope of one received signal is shown with target range-time τ_T, target Doppler-frequency shift Δ_T and unknown amplitude and carrier phase A and θ.

For a filter matched [1] to white noise, the impulse response of $h(t)$ is the conjugate mirror image of the expected received signal. If the filter output is to peak at $t = 0$ when the expected arrival time of the signal is $t = \tau_0$, the filter impulse response must have the form

$$h(t) = u^* (-t - \tau_0). \tag{5}$$

In the case of clutter however, performance may be improved by using a filter that is not perfectly matched to the expected received signal, but rather is matched to some altered time function, $v(t)$. (In this report "matched" is synonymous with "conjugate mirror image.") In this case, the filter is said to be mismatched to the expected received signal. Its impulse response is

$$h(t) = v^* (-t - \tau_0). \tag{6}$$

In terms of the mismatched filter impulse response, the filtered output becomes[1]

$$z(t) = A \exp (i\phi) \int u(s) v^*(s - t + \tau) \cdot \exp (-2\pi i \, \Delta s) \, ds \tag{7a}$$

[1] All integrals without limits are meant to have infinite limits.

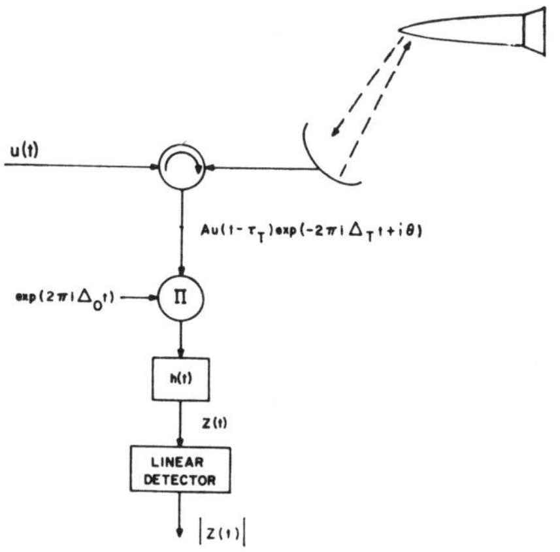

Fig. 1. Radar signal processor.

where

$$\Delta = \Delta_T - \Delta_0 \tag{7b}$$

$$\tau = \tau_T - \tau_0 \tag{7c}$$

and

$$\phi = \theta - 2\pi(\Delta_T - \Delta_0)\tau_T. \tag{7d}$$

Note that τ and Δ are, respectively, the target range-time and Doppler relative to the expected or reference values.

The output of the processing filter at $t = 0$, designated by $z(0)$, is related by the complex constant $A \exp (i\phi)$ to the time-frequency $(t - f)$ cross-correlation function [4] defined as

$$\chi_{uv}(\tau, \Delta) = \int u(s) v^*(s + \tau) \exp (-2\pi i \, \Delta s) \, ds. \tag{8}$$

The expression to be used most throughout this paper is the ambiguity function, which will be defined as the normalized power output of the signal processor.

$$\psi_{uv} (\tau, \Delta) = |\chi_{uv} (\tau, \Delta)|^2 = |z(0)|^2/A^2. \tag{9}$$

This definition will be applied to point targets with the target of interest placed at the origin, and the value of the ambiguity function at any other range-time or Doppler then may be used to represent the possible interference or clutter power present.

Whenever the ambiguity function is encountered in this paper, it will be assumed that the signal $u(t)$ and filter $v(t)$ are normalized to unit energy as follows.

$$\int |u(t)|^2 \, dt = 1 \tag{10a}$$

and

$$\int |v(t)|^2 \, dt = 1. \tag{10b}$$

The volume under the ambiguity function is then the same for all signals and filters, as shown by Stutt [4].

$$\iint \psi_u(\tau, \Delta)\, d\tau\, d\Delta = \iint \psi_{uv}(\tau, \Delta)\, d\tau\, d\Delta = 1. \tag{11}$$

Rearrangement of this volume will be important in the reduction of clutter and is the subject of this paper.

Many of the important signal processing parameters can now be defined in terms of the ambiguity function. First, the peak detected signal power is defined as

$$P_s = A^2\, \psi_{uv}\, (0, 0) \leq A^2. \tag{12}$$

One sees that by mismatched filtering the peak signal power can be reduced. Only when $u(t)$ and $v(t)$ are equal, as with the matched filter, is the equality achieved. The ratio of peak signal power to mean noise power, abbreviated SNR, is defined as

$$\text{SNR} = A^2\psi_{uv}(0, 0)\cdot\left[2N_0 \int |v(t)|^2\, dt\right]^{-1} \tag{13}$$

where $2N_0$ represents the power density level in the complex envelope of the noise. Clutter is characterized as a signal-like radar return and, as such, can be defined in terms of the ambiguity function. If $R_0(s, t)$ represents the correlation function of clutter at the output of Fig. 1, the total clutter power at point of target detection is

$$J = R_0(0, 0) = 2P_c \iint p(\tau, \Delta)\psi_{uv}(\tau, \Delta)\, d\tau\, d\Delta \tag{14}$$

where $2P_c$ represents the returned power in the complex envelope of the clutter and $p(\tau, \Delta)\, d\tau\, d\Delta$ is the probability of finding a clutter scatterer in the elemental area $d\tau\, d\Delta$ at τ and Δ. This definition is consistent with Fowle *et al.* [5], as well as Van Trees [6]. Hence, the ratio of peak signal power to average clutter power (SCR) is defined as

$$\text{SCR} = \tfrac{1}{2}A^2\psi_{uv}(0, 0) \cdot\left[P_c \iint p(\tau, \Delta)\psi_{uv}(\tau, \Delta)\, d\tau\, d\Delta\right]^{-1}. \tag{15}$$

Likewise, the ratio of peak signal power to total interference power (SIR), where the interference represents both clutter and noise, is defined as follows.

$$\text{SIR} = \tfrac{1}{2}A^2\psi_{uv}(0, 0) \cdot\left[N_0 \int |v(t)|^2\, dt + P_c \iint p(\tau, \Delta)\psi_{uv}(\tau, \Delta)\, d\tau\, d\Delta\right]^{-1}. \tag{16}$$

It is this SIR that will be maximized to find the optimum signal and filter combination in a given clutter environment. Maximizing this SIR is shown in reference [7] to be equivalent to maximizing the probability of detection for a given probability of false alarm in the Neyman–Pearson criterion, when the input to the radar signal processor shown in Fig. 1 is a Gaussian random process.

II. Optimum Filter Design

In this section, the optimum filter $v(t)$ obtained from maximizing the SIR of (16) will be found for a given signal $u(t)$ and clutter environment $p(\tau, \Delta)$. This is the first step in the combined signal and filter optimization problem. Some interesting conclusions result from filter design alone. Dividing numerator and denominator of (16) by P_c yields the following equivalent and more convenient expression to be maximized.

$$\text{SIR} = (A^2/2P_c)\psi_{uv}(0, 0) \cdot\left[N_0/P_c \int |v(t)|^2\, dt + \iint p(\tau, \Delta)\psi_{uv}(\tau, \Delta)\, d\tau\, d\Delta\right]^{-1}. \tag{17}$$

This form of the SIR refers the return clutter power to the normalized cross-ambiguity function of the signal and filter.

The optimum filter problem has been solved by Stutt and Spafford [8] and the purpose of this section will be to supplement these results. The previous paper considers the trajectories of all possible solutions for $v(t)$; in this section only the point of maximum SIR will be considered. The optimum filter as derived by Stutt and Spafford [8], for the point of maximum SIR, is the function $v(s)$ satisfying the following integral equation.

$$\mu k^* u(t) = \int [H(s, t) + \lambda\, \delta(s - t)]v(s)\, ds \tag{18a}$$

where the clutter correlation function is given in terms of the signal by

$$H(s, t) = \int M(\tau, s - t)u(t - \tau)u^*(s - \tau)\, d\tau \tag{18b}$$

with the characteristic function of the clutter defined as

$$M(\tau, s - t) = \int p(\tau, \Delta) \exp\,[i2\pi\, \Delta(s - t)]\, d\Delta \tag{18c}$$

and also where the following definitions have been used.

$$\lambda \triangleq N_0/P_c \tag{18d}$$

$$\mu \triangleq A^2/2P_c(\text{SIR}) \tag{18e}$$

$$k^* \triangleq \int u^*(s)v(s)\, ds. \tag{18f}$$

The parameter λ represents the noise-to-clutter ratio and is inversely proportional to the maximum SIR improvement possible when one is able to cancel the clutter completely. This solution will now be specialized to pulse trains with nonuniform amplitude and phase [10].

The most important property of pulse trains when used in a clutter environment is its simultaneously good range and velocity resolution. Another advantage of pulse trains is the ease with which a filter function may be implemented and changed through the use of a tapped delay line.

The radar signals to be considered in this report can adequately be described by Fig. 2. Here δ represents the pulse width, T the interpulse period, D the duration of the signal, and a_n a complex number describing the amplitude and phase of each pulse $p(t)$. The entire signal can be expressed mathematically in the following form when $\delta < T/2$.

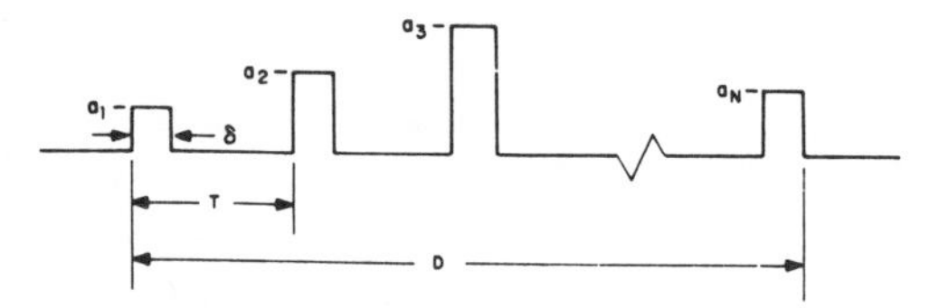

Fig. 2. Equally spaced envelope recurrent pulse train.

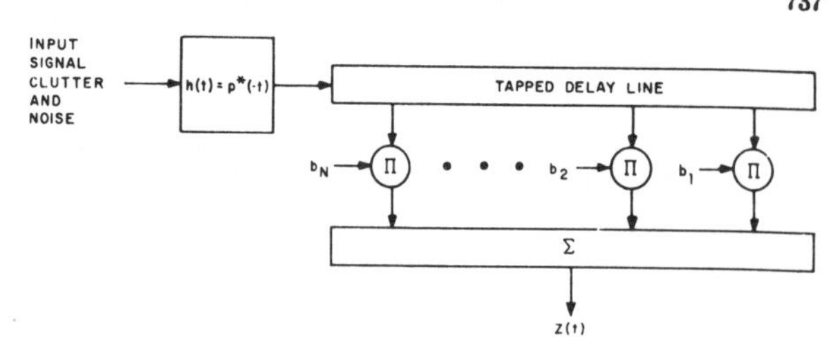

Fig. 3. Tap delay line representation of receive filter.

$$u(t) = \sum_{n=1}^{N} a_n p(t - nT). \tag{19}$$

Using a signal such as a pulse train requires the filter to have the same discrete spacing, which can be realized by the weighting tapped delay line shown in Fig. 3. The weights b_n, when properly combined with the outputs of the tapped delay line, represent the received filter function and can be written mathematically as

$$v(t) = \sum_{n=1}^{N} b_n p(t - nT). \tag{20}$$

It has the same form as the signal and is matched only when $a_n = b_n$ for all n. By using the pulse train representations of (19) and (20), it is shown in reference [7] that the linear integral (18a) has the following equivalent matrix counterpart [8].

$$[\mathbf{C} + \lambda\mathbf{I}]\,\mathbf{B} = \mu k^* \mathbf{A} \tag{21}$$

where **A** represents the transmitted vector, **B** the received processing vector, and **C** the matrix corresponding to the clutter correlation function $H(s, t)$. Hence the optimum filter is simply

$$\mathbf{B} = \mu k^* [\mathbf{C} + \lambda\mathbf{I}]^{-1} \mathbf{A}. \tag{22}$$

Because the matrix $[\mathbf{C} + \lambda\mathbf{I}]$ is Hermitian and

$$|\mathbf{A} * \mathbf{B}|^2 = |k|^2 \tag{23}$$

the SIR can be written as

$$\mathrm{SIR} = \mathbf{A}^* [\mathbf{C} + \lambda\mathbf{I}]^{-1} \mathbf{A} \tag{24}$$

where the superscript * indicates conjugate transpose when applied to a matrix.

The relative Doppler differences between target and clutter, in the case of ground and weather clutter, can be quite large. In addition, the Doppler spreads of these two types of clutter will usually be small compared with the ambiguous Doppler. In this light, the clutter density function may have the following normalized form.

$$p(\tau, \Delta) = p(\tau/T)\, \delta[(\Delta - \Delta_0)\, T] \tag{25}$$

where Δ_0 represents the relative target-to-clutter Doppler difference. Furthermore, if there is no reason to favor one range-time over another, rain or ground clutter may have the following special form.

$$p(\tau, \Delta) = \delta[(\Delta - \Delta_0)\, T]. \tag{26}$$

Filter design is extremely important when the signal, for some reason, is constrained to a constant value. First, consider the clutter to be stationary and distributed as given by (26). The SIR obtainable for a uniform signal and the filter optimized for each $\Delta_0 T$ value is shown in Fig. 4, where the SIR is referred to that of a single pulse. The ordinate of Fig. 4, defined as SIR gain, can be interpreted as the SIR improvement over a single pulse. For small target-to-clutter Doppler separations, the results show that the pulse train reduces to a single pulse, as obtained by Manasse [3]. The SNR loss occurs when the filter becomes mismatched to the signal, and it indicates the tradeoff between the peak signal power and clutter cancellation. Throughout this paper, a noise-to-clutter ratio λ of -50 dB will be assumed unless otherwise indicated. Also a pulse train of $N = 11$ pulses is used unless otherwise indicated. Each figure will have an insert showing the clutter distribution, as in Fig. 4. The SIR gain by filter design for this clutter distribution is significant in that an overall gain of approximately 3 dB is obtained by filter design for range-extended or stationary clutter.

Chaff and ground clutter are extended in range, but not usually infinitely extended or stationary. In this light, let the clutter be represented by the following density function, where x is a variable proportional to the range extent of the clutter.

$$p(\tau, \Delta) = \mathrm{rect}\,(\tau/Tx)\; \delta\,(\Delta T - 0.5) \tag{27a}$$

where for $N = 11$ pulses

$$0 < x \leq 20.0. \tag{27b}$$

In this case, one obtains the SIR and SNR curves shown in Fig. 5. This shows that filter design is very effective to approximately the point where the clutter has range-time extent equal to the signal duration. Up to this point, very close to maximum SIR improvement is obtained (that is, $1/\lambda$). The SIR improvement does not change noticeably if the center of the clutter is moved to some other point in τ, but the design of the filter changes drastically.

Besides the uniform transmitted signal, another signal that one might transmit is the Chebyshev [11] amplitude-weighted pulse train with zero phase. Fig. 6 shows the performance of the matched Chebyshev signal and filter combination together with the Chebyshev signal and its optimum filter for the clutter distribution of (26). For large target-to-clutter Doppler separations, a SIR gain of nearly three orders of magnitude was obtained over the uniform signal shown in Fig. 4. This result thus

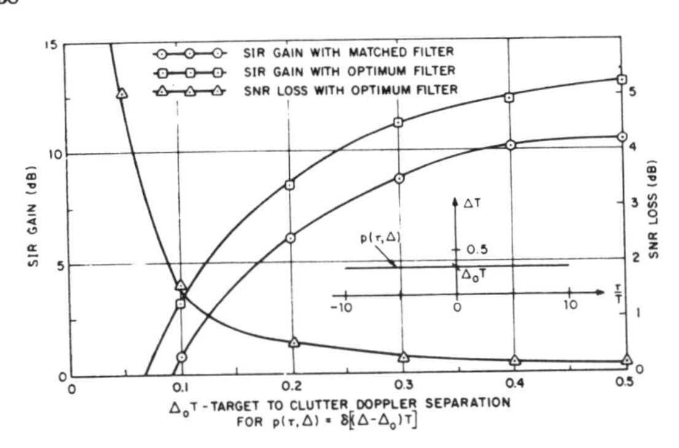

Fig. 4. Signal is a uniform pulse train of 11 pulses and clutter is 50 dB above the noise level.

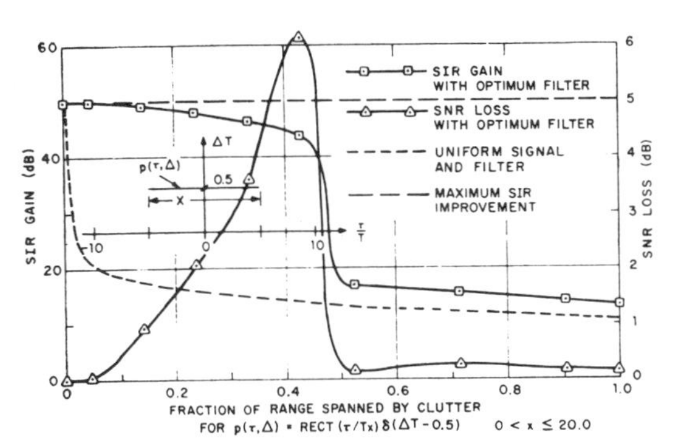

Fig. 5. Signal is a uniform pulse train of 11 pulses and clutter is 50 dB above the noise level.

indicates the importance of signal design as well as filter design when the clutter is fully extended in range.

III. Optimum Signal–Filter Design

The optimum signal–filter combination can now be found by maximizing the SIR given in (17) first with respect to the filter function $v(t)$ and then separately with respect to the signal function $u(t)$. This will yield two linear integral equations, one of which yields $v(t)$ in terms of $u(t)$, and the other of which yields $u(t)$ in terms of $v(t)$. Combined, these two equations would represent a nonlinear integral equation, but separately they are linear. The first step in this optimization procedure of finding the best filter in terms of a known signal is given by (18a). The second step in the optimum signal–filter procedure is symmetrical to the first step, the details of which can be found in reference [7], with the result

$$\int [G(s, t) + \lambda\, \delta(s - t)]u(s)\, ds = \alpha K^* v(t) \tag{28a}$$

where the clutter correlation function is now defined in terms of the filter as

$$G(s, t) = \int M(\tau, t - s) v(t + \tau) v^*(s + \tau)\, d\tau \tag{28b}$$

with the characteristic function of the clutter

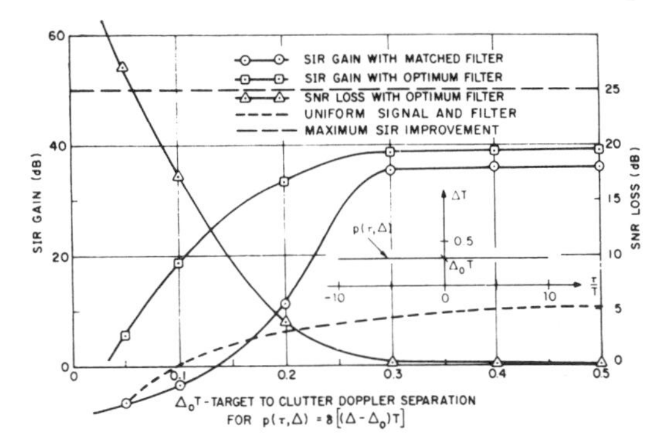

Fig. 6. Signal is a Chebyshev weighted pulse train of 11 pulses with design sidelobes of 45 dB; clutter is 50 dB above the noise level.

$$M(\tau, t - s) = \int p(\tau, \Delta) \exp [2\pi i\, \Delta(t - s)]\, d\Delta \tag{28c}$$

and also where

$$\alpha \triangleq A^2/2P_c \text{ (SIR)}. \tag{28d}$$

For pulse trains, the equivalent matrix representation becomes

$$\mathbf{A} = \alpha k^*[\mathbf{D} - \lambda \mathbf{I}]^{-1}\mathbf{B} \tag{29}$$

and the maximum SIR is given by

$$\text{SIR} = \mathbf{B}^*[\mathbf{D} - \lambda \mathbf{I}]^{-1}\mathbf{B}. \tag{30}$$

Finding the optimum signal–filter combination requires an iterative procedure, starting with a known signal. First, the optimum filter is found in terms of this given signal, and next, the optimum signal is found for this new filter. The SIR in each case is tabulated, and the procedure is repeated over and over again until no change in the SIR is observed. The proof that this iteration procedure converges is given in reference [7] and assumes that noise is present. The upper bound on performance is

$$\text{SIR} \le 1/\lambda. \tag{31}$$

This proof does not show that the maximum SIR is a global maximum, but physical considerations such as clutter size and location, as well as the dimensionality of the signal, will usually indicate whether a global maximum SIR has been reached.

Next, combined signal–filter design is applied to the clutter distribution given by (26). Figs. 7 ($N = 11$) and 8 ($N = 21$) indicate the SIR performance obtainable when both signal and filter are optimized as a function of Doppler for this range-extended clutter environment. The SIR improvement for higher target-to-clutter Doppler differences is limited only by the noise since the upper limit of performance given by (31) is reached. The situation described in Figs. 7 and 8 corresponds to the radar tracking problem where the Doppler of the target is known with respect to the clutter. For each new or anticipated target-to-clutter Doppler difference, the transmitter simply selects the optimum signal to be transmitted and its corresponding optimum processing filter. Tables I and II list

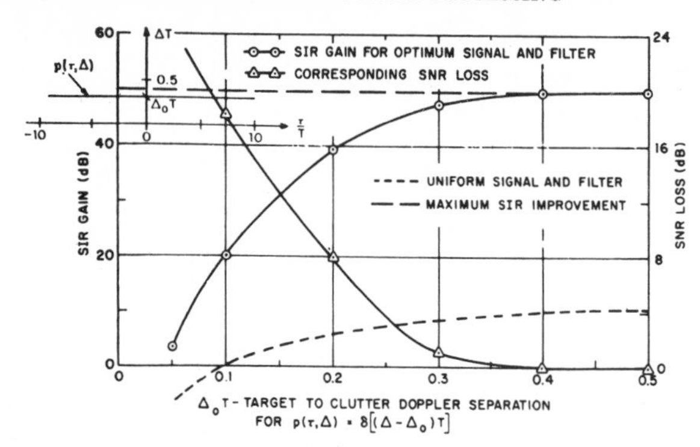

Fig. 7. Optimum signal-filter performance for pulse train of 11 pulses and clutter is 50 dB above the noise level.

Fig. 8. Optimum signal-filter performance for pulse train of 21 pulses and clutter is 50 dB above noise level.

TABLE I
SIGNAL AMPLITUDE AND PHASE (DEGREES) WEIGHTS

	$\Delta_0 T$									
	0.1		0.2		0.3		0.4		0.5	
n	a_n	ϕ_n	a_n	ϕ_n	a_n	ϕ_n	a_n	ϕ_n	a_n	ϕ_n
1	0.080	117.9	0.032	137.1	0.031	79.5	0.017	27.4	0.008	0.0
2	0.135	−17.9	0.114	97.3	0.121	46.9	0.083	12.4	0.051	0.0
3	0.239	134.5	0.235	66.9	0.240	26.2	0.202	3.5	0.157	0.0
4	0.403	−9.9	0.355	41.3	0.355	14.2	0.342	0.0	0.319	0.0
5	0.327	161.7	0.440	20.0	0.438	6.2	0.457	−0.6	0.475	0.0
6	0.547	0.0	0.472	0.0	0.468	0.0	0.502	0.0	0.538	0.0
7	0.327	−161.7	0.440	−20.0	0.438	−6.2	0.457	0.6	0.475	0.0
8	0.403	9.9	0.355	−41.3	0.355	−14.2	0.342	0.0	0.319	0.0
9	0.239	−134.5	0.235	−66.9	0.240	−26.2	0.202	−3.5	0.157	0.0
10	0.135	17.9	0.114	−97.3	0.121	−46.9	0.083	−12.4	0.051	0.0
11	0.080	−117.9	0.032	−137.1	0.031	−79.4	0.017	−27.4	0.008	0.0

TABLE II
FILTER AMPLITUDE AND PHASE (DEGREES) WEIGHTS

	$\Delta_0 T$									
	0.1		0.2		0.3		0.4		0.5	
n	b_n	ϕ_n	b_n	ϕ_n	b_n	ϕ_n	b_n	ϕ_n	b_n	ϕ_n
1	0.080	−139.3	0.051	−170.1	0.031	−85.5	0.017	−31.6	0.008	0.0
2	0.135	−143.3	0.183	−109.8	0.121	−51.7	0.083	−15.7	0.051	0.0
3	0.239	−75.3	0.296	−63.3	0.240	−29.9	0.202	−6.0	0.157	0.0
4	0.403	−70.6	0.340	−37.0	0.355	−16.6	0.342	−1.7	0.319	0.0
5	0.327	−21.9	0.400	−21.6	0.438	−7.4	0.457	−0.3	0.475	0.0
6	0.547	0.0	0.448	0.0	0.468	0.0	0.502	0.0	0.538	0.0
7	0.327	21.9	0.400	21.6	0.438	7.4	0.457	0.3	0.475	0.0
8	0.403	70.6	0.340	37.0	0.355	16.6	0.342	1.7	0.319	0.0
9	0.239	75.3	0.296	63.3	0.240	29.9	0.202	6.0	0.157	0.0
10	0.135	143.3	0.183	109.8	0.121	51.7	0.083	15.7	0.051	0.0
11	0.080	139.3	0.051	170.1	0.031	85.5	0.017	31.6	0.008	0.0

the optimum signal and corresponding optimum filter weights for five target-to-clutter Doppler values when one considers a signal of $N = 11$ pulses.

Optimization of the radar signal and filter for all target-to-clutter range-times and Doppler values of interest may present a difficult problem. First, the target range-time and Doppler may not be known. Second, the signal should be optimized for all target-to-clutter range-times and Doppler values of interest since it cannot be changed after transmission. Therefore, one should consider a clutter distribution with an appreciable Doppler extent for optimization of the signal, but should adopt a narrow clutter region such as (26) for the filter optimization. Such a wide clutter region of interest may be given by the following clutter distribution, whereby no one Doppler value in this region is weighted any more than another.

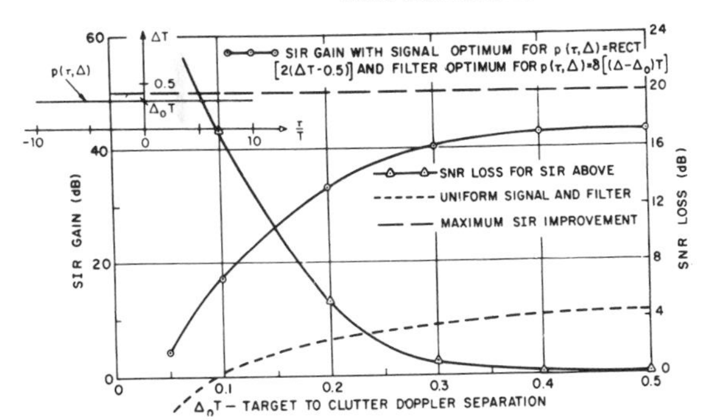

Fig. 9. Signal is a pulse train of 11 pulses optimized for $p(\tau, \Delta) = \text{rect}\,[2(\Delta T - 0.5)]$ and clutter is 50 dB above the noise level.

TABLE III

FILTER AMPLITUDE AND PHASE (DEGREES) WEIGHTS OPTIMIZED FOR $p(\tau, \Delta) = \delta[(\Delta - \Delta_0)T]$ WHEN SIGNAL IS OPTIMUM FOR $p(\tau, \Delta) = \text{RECT}\ [2(\Delta T - 0.5)]$

	$\Delta_0 T$									
	0.1		0.2		0.3		0.4		0.5	
n	b_n	ϕ_n	b_n	ϕ_n	b_n	ϕ_n	b_n	ϕ_n	b_n	ϕ_n
1	0.143	−7.3	0.084	120.9	0.045	−128.7	0.026	−48.8	0.021	0.0
2	0.392	116.3	0.264	−153.8	0.169	−76.9	0.114	−26.7	0.100	0.0
3	0.354	−140.9	0.326	−87.8	0.279	−39.6	0.240	−11.8	0.230	0.0
4	0.274	−100.0	0.321	−58.8	0.343	−23.8	0.352	−5.3	0.359	0.0
5	0.296	−43.4	0.382	−27.8	0.420	−12.0	0.438	−3.0	0.443	0.0
6	0.275	0.0	0.368	0.0	0.441	0.0	0.476	0.0	0.472	0.0
7	0.296	43.4	0.382	27.8	0.420	12.0	0.438	3.0	0.443	0.0
8	0.274	100.0	0.321	58.8	0.343	23.8	0.352	5.2	0.359	0.0
9	0.354	140.9	0.326	87.8	0.279	39.6	0.240	11.8	0.230	0.0
10	0.392	−116.3	0.264	153.8	0.169	76.9	0.114	26.7	0.100	0.0
11	0.143	7.3	0.084	−120.9	0.045	128.7	0.026	48.8	0.021	0.0

$$p(\tau, \Delta) = \text{rect}\ [2(\Delta T - 0.5)]. \tag{32}$$

Such a situation would correspond to the search radar problem where the target-to-clutter Doppler is unknown and one transmits a signal optimized for all target-to-clutter Doppler values of interest indicated by (32). Since the phase weights are zero, the optimum signal in this case is given by eleven symmetric amplitude weights.

$$\mathbf{A}^* = [0.036, 0.114, 0.228, 0.350, 0.444, 0.479, \cdots]. \tag{33}$$

The SIR improvement with this signal when the filter is optimized as a function of target-to-clutter Doppler for the clutter distribution of (26) is given in Fig. 9. If five Doppler channels are used, the corresponding Doppler filter weights are those shown in Table III.

IV. Optimum Matched Filter Design

It may be of interest in some clutter environments to know the limits of matched filter processing. Matched filter processing yields maximum SNR at the signal processor output and for that reason is an appealing processing approach. The optimum matched condition is achieved by finding the function $u(t)$ that maximizes the SIR of (17) when $v(t) = u(t)$. Since all functions $u(t)$ yield equally good performance against noise, finding the filter which maximizes the SIR is equivalent to determining the filter which maximizes the SCR of (15) for $v(t) = u(t)$. The result of maximizing this SCR is the following eigenvalue problem [7].

$$\sigma u(t) = \int H_m(s, t) u(s)\, ds \tag{34a}$$

where

$$H_m(s, t) = \frac{1}{2} \int [M(\tau, s - t) + M(-\tau, t - s)] \cdot u(t - \tau) u^*(s - \tau)\, d\tau \tag{34b}$$

and

$$\sigma = A^2/2P_c\ (\text{SCR}). \tag{34c}$$

The minimum eigenvalue σ, or the point of maximum SCR, is sought. The solution for $u(t)$ is the eigenfunction satisfying the integral equation above, whose kernel $H_m(s, t)$ is also a function of $u(t)$. The matrix equivalent of this integral equation, where the matrix $\mathbf{C}$ corresponds to the symmetric clutter function $H_m(s, t)$, can be written as

$$\mathbf{CA} = \sigma \mathbf{A}. \tag{35}$$

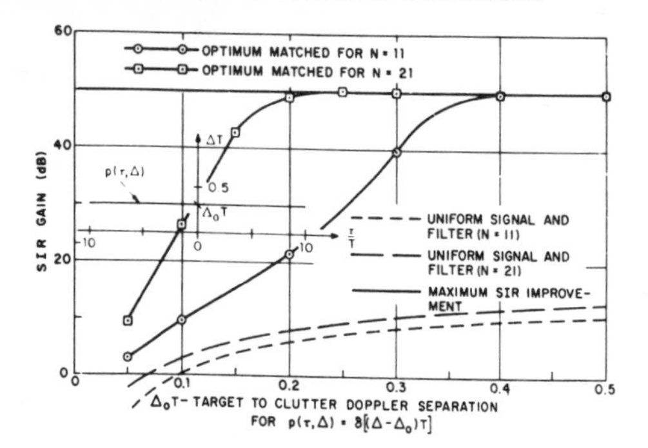

Fig. 10. Optimum matched filter performance for a pulse train of N pulses and clutter is 50 dB above the noise level.

Comparing the matched and mismatched filter processors, one sees that because the mismatched processor has twice the number of degrees of freedom it will perform as well as and probably better than the matched filter processor in most clutter environments, hence

$$(\mathrm{SCR})_{\mathrm{mismached}} \geq (\mathrm{SCR})_{\mathrm{matched}}. \tag{36}$$

The equality sign would hold only in those cases where the clutter regions are symmetric;

$$M(\tau, s - t) = M(-\tau, t - s) \tag{37}$$

because this condition yields identical clutter correlation functions for the matched and mismatched filter formulations. That is,

$$H_M(s, t) = H(s, t). \tag{38}$$

To obtain a matched filter processor in clutter, $p(\tau, \Delta)$ must be symmetric. When a nonsymmetric clutter distribution $p(\tau, \Delta)$ is encountered, the optimum matched filter processor results if one also includes $p(-\tau, -\Delta)$. Thus, the iteration scheme developed previously for the optimum signal–filter combination can also be used for this completely symmetric clutter region to obtain the optimum matched filter processor. This iteration scheme converges, however, only when λ is included. Use of the nonsymmetric stationary clutter distribution of (26) results in a matched filter SIR improvement over a single pulse, shown in Fig. 10. These results compare quite favorably with the optimum signal–filter results of Figs. 7 and 8 for high target-to-clutter Dopplers.

V. Optimum Constrained Signal–Filter Design

Optimum signal–filter combinations may also be obtained when, for some reason or other, part of the signal or filter elements are constrained to a given amplitude and phase value. Such constraints may be limitations on the radar transmitter tubes, or they may be in the interest of reducing the receiver complexity. Constraints may also be required in order to maintain a prescribed SNR. To reemphasize, one may constrain a portion of the signal, the filter, or both, and obtain the constrained optimum signal–filter combination in any case.

The problem of finding the best filter $\mathbf{B}$ in a clutter-plus-noise environment was solved in Section III. This procedure will be modified only slightly to obtain the best constrained signal. In the unconstrained case, the result of maximizing the SIR of (17) is equivalent to the following matrix equation (where δ denotes the first variation).

$$0 = \delta\mathbf{B}^*[-\mu_0 k^*\mathbf{A} + \mathbf{CB} + \lambda\mathbf{B}]. \tag{39}$$

However, in the constrained case, one must partition these matrix relations into their constrained and unconstrained parts. In the case of $\mathbf{B}$, the unconstrained portion is $\mathbf{b}$, and the constrained portion is $\mathbf{B}_0$, as follows.

$$\mathbf{B} = \begin{bmatrix} \mathbf{b} \\ \hline \mathbf{B}_0 \end{bmatrix}. \tag{40a}$$

Likewise, $\mathbf{A}$ and $\mathbf{C}$ can be partitioned in a similar manner.

$$\mathbf{A} = \begin{bmatrix} \mathbf{a} \\ \hline \mathbf{A}_0 \end{bmatrix}, \qquad \mathbf{C} = \left[\begin{array}{c|c} \mathbf{C}_c & \mathbf{C}_0 \\ \hline \mathbf{C}_0^* & \mathbf{X} \end{array}\right] \tag{40b}$$

where $\mathbf{a}$ and $\mathbf{C}_c$ are the unconstrained parts. In the constrained case the result equivalent to (39) is

$$0 = [\delta\mathbf{b}^* \mid 0][-\mu_0 k^*\mathbf{A} + \mathbf{CB} + \lambda\mathbf{B}] \tag{41}$$

from which the unconstrained portion of the filter function becomes

$$\mathbf{b} = [\mathbf{C}_c + \lambda\mathbf{I}]^{-1}[\mu_0 k^*\mathbf{a} - \mathbf{C}_0\mathbf{B}_0] \tag{42a}$$

where

$$\mu_0 k^* = \frac{\mathbf{B}_0^*[\mathbf{X} + \lambda\mathbf{I} - \mathbf{C}_0^*(\mathbf{C}_c + \lambda\mathbf{I})^{-1}\mathbf{C}_0]\mathbf{B}_0}{\mathbf{B}_0^*\mathbf{A}_0 - \mathbf{B}_0^*\mathbf{C}_0^*[\mathbf{C}_c + \lambda\mathbf{I}]^{-1}\mathbf{a}}. \tag{42b}$$

Likewise, if one desires the constrained optimum transmitted waveform for a given filter, it can be derived in a fashion completely analogous to that obtained above, with the result

$$\mathbf{a} = [\mathbf{Q}_c + \lambda\mathbf{I}]^{-1}[\alpha_0 k\mathbf{b} - \mathbf{Q}_0\mathbf{A}_0] \tag{43a}$$

where

$$\alpha_0 k = \frac{\mathbf{A}_0^*[\mathbf{Y} + \lambda\mathbf{I} - \mathbf{Q}_0^*(\mathbf{Q}_c + \lambda\mathbf{I})^{-1}\mathbf{Q}_0]\mathbf{A}_0}{\mathbf{B}_0^*\mathbf{A}_0 - \mathbf{A}_0^*\mathbf{Q}_0^*[\mathbf{Q}_c + \lambda\mathbf{I}]^{-1}\mathbf{b}}. \tag{43b}$$

The $\mathbf{Q}$ matrix has been partitioned as follows, where $\mathbf{Q}_c$ corresponds to the unconstrained part.

$$\mathbf{Q} = \left[\begin{array}{c|c} \mathbf{Q}_c & \mathbf{Q}_0 \\ \hline \mathbf{Q}_0^* & \mathbf{Y} \end{array}\right]. \tag{43c}$$

With these two solutions, one can use the iteration procedure of Section IV to find the constrained optimum signal–filter combination. An example will illustrate one of the many variations of this constrained optimum formulation. Fig. 11 shows the optimization as a function of Doppler when all signal elements except the outer two elements are constrained to a constant value and the filter is unconstrained for the clutter distribution of (26). Compared with Fig. 4, almost double the SIR improvement is obtained by allowing the signal an additional four degrees of freedom.

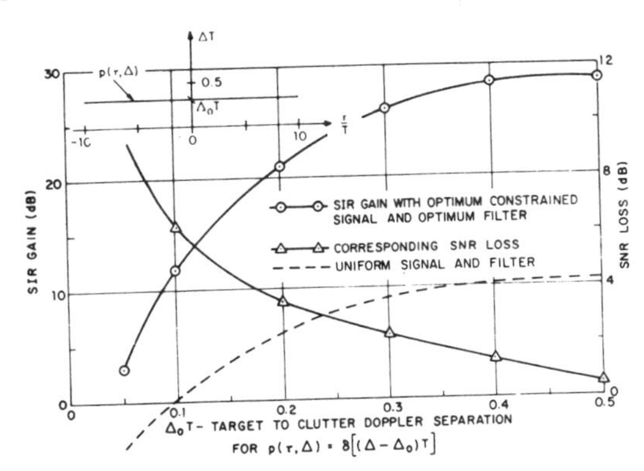

Fig. 11. Signal is uniform pulse train of 11 pulses except edge pulses are unconstrained; clutter is 50 dB above the noise level.

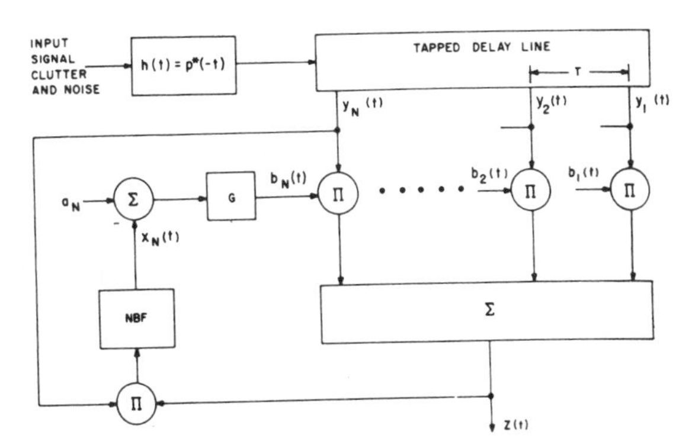

Fig. 12. Adaptive-implementation of the optimum filter.

VI. Adaptive Signal Processing

It is appropriate at the outset of this discussion to define what is meant by adaptive signal processing and how the term "adaptive" will be used. An adaptive signal processor will be defined as a system that continually adjusts its own parameters in the course of time to meet a certain performance criterion. The performance criterion here is the maximization of the signal-to-interference ratio. Since the clutter is normally time-varying, some adaptive mechanism is required to continually update the filter weights to meet this changing interference. A signal processor or filter is proposed that senses the interference and automatically adjusts the filter weights to their proper values.

Perhaps the strongest argument for adaptability comes from interpretation of the definition of clutter power employed in this paper and given by (14). In theory it is assumed that $p(\tau, \Delta)$ is known in order for the signal and filter weights to be computed. But the very nature of $p(\tau, \Delta)$ influences the signal and filter weights through the C and D matrices in (22) and (29). The proposed adaptive filter automatically computes $p(\tau, \Delta)$ and determines the best set of filter weights for processing.

Since any signal that is transmitted must be a good signal for all range-times of interest, its selection will no doubt be a function of a number of target-to-clutter Doppler differences. In this light, one might choose a transmitted signal that performs equally well against clutter at all target-to-clutter Dopplers of interest, and then use the adaptive filter to maximize the SIR for all range-times. This would correspond to the search radar case discussed in Section III.

The generation of the optimum filter weights is automatically accomplished by a set of feedback loops like the one shown in Fig. 12. The basic feedback loop was developed by P. W. Howells, formerly of the General Electric Company, under U. S. Patent 3 202 990. The design shown in Fig. 12 is a modification of this basic device with the addition of the signal weights a_m. This modification was suggested by J. J. Gostin of General Electric's Heavy Military Electronics Department. The N feedback circuits are all alike and are attached to the N outputs of the tapped delay line. The narrow-band filter (NBF) in each circuit operates in such a fashion as to take an approximate time average of its input $z^*(t)y_m(t)$. The optimum filter weights $b_m(t)$ then become the differences between the transmitted signal weights a_m and the NBF output quantities $x_m(t)$.

It will now be shown that the feedback scheme of Fig. 12 yields the optimum filter weights. By assuming that the NBF is a single pole device, the adaptive filter can be described by the following differential equation (where the $\cdot$ indicates a time derivative).

$$T_f \dot{x}_m(t) + x_m(t) = G \sum_{n=1}^{N} y_m(t) y_n^*(t)[a_n - x_n(t)] \tag{44}$$

for $m = 1, 2, \cdots N$

where T_f is the NBF time constant and G is the forward gain of the feedback loop. If one assumes the statistics of the clutter are slowly varying functions of time compared with the signal processing interval, then taking an ensemble average of the terms in (44) yields the following matrix equation.

$$T_f \dot{\mathbf{X}}_t + \mathbf{X}_t = G\mathbf{C}[\mathbf{A} - \mathbf{X}_t] \tag{45}$$

where the elements of the $\mathbf{C}$ matrix are given by

$$c_{mn} = \overline{y_m(t) y_n^*(t)} \tag{46}$$

representing the elements of the covariance matrix, which for zero mean Gaussian statistics completely describes the random process. (Here the overbar indicates expected value.)

The steady-state solution of matrix equation (46) is of primary interest, that is

$$\mathbf{X}_t = (\mathbf{I} + G\mathbf{C})^{-1} G\mathbf{C}\mathbf{A}. \tag{47}$$

Upon substituting for $\mathbf{X}_t$ into $\mathbf{B}_t = \mathbf{A} - \mathbf{X}_t$, one has

$$\mathbf{B}_t = G(\mathbf{I} + G\mathbf{C})^{-1}\mathbf{A}. \tag{48}$$

The time subscript may be dropped since **A** is a constant vector. If the interference is strong, the norm of $G\mathbf{C}$ will be considerably greater than the norm of the identity matrix. Then

$$I + G\mathbf{C}^{-1} \cong \mathbf{C}^{-1}/G \tag{49}$$

yielding

$$\mathbf{B} = \mathbf{C}^{-1}\mathbf{A} \tag{50}$$

which are the optimum filter weights providing this **C** matrix corresponds to the **C** matrix obtained in the theory. The two **C** matrices are equivalent for zero mean Gaussian statistics because the covariance matrix in this case completely describes the random process. Hence the SIR performance curves shown in Figs. 4, 5, and 6 are also the steady-state performance curves of the adaptive signal processor for Gaussian input statistics.

The assumptions made in obtaining the optimum weights are not limited assumptions. It will be assumed that the signal processing interval will be chosen so that any time variations in the statistics of the clutter will be slowly varying or approximately constant over the processing interval. Also, the assumption of strong interference, an order of magnitude greater than the signal, is not limiting since the interference must be strong to be troublesome.

VII. Conclusions

It has often been pointed out in the literature that signal design is paramount to good performance in clutter. Signal design is extremely important for range-extended or stationary clutter situations. However, if the clutter range-time extent is less than the equivalent extent of the signal, then filter design alone may yield close to the maximum SIR improvement without regard to signal design. These results are shown in Fig. 5. The SIR performance is referred to a single pulse of the pulse train. For zero target-to-clutter Doppler differences, the SIR curves approach the single pulse result of Manasse [3]. Since a single pulse is a high bandwidth signal, the SIR results may also be referred to a linear FM coded pulse, or for that matter to any signal with a high bandwidth.

Optimum signal–filter design is very important in a tracking radar where the target-to-clutter Doppler is known. In this situation, a catalog of optimum signal and filters should be available for each anticipated target-to-clutter Doppler, from which the radar would simply select the required combination. The search radar situation is quite different in that the target-to-clutter Doppler value is unknown. For this situation it is proposed the signal be optimized for all target-to-clutter Doppler values of interest and that the adaptive filter provide the required optimization as a function of range-time. That is, the adaptive filter would supply the optimum filter for each Doppler processed.

The optimum matched filter results shown in Fig. 10 closely approach the optimum signal–filter design shown in Figs. 7 and 8. The results differ appreciably only where a high SNR loss is encountered in obtaining the optimum signal–filter combination. Therefore, the optimum matched filter results are significant in that optimum performance against noise alone is also obtained. The constrained optimum signal–filter curves shown in Fig. 11 are interesting in that double the SIR performance of Fig. 4 is obtained with four additional degrees of freedom for the signal.

The pulse train representation of (19) is not limited to pulse trains alone. It can also be thought of as a complex sample function representation with a_n being the sample value at time $t = nT$ and $p(t)$ representing the sampling aperture. Hence the theory as developed in this paper is quite general, applying to sampled representations of general signals and not limited only to pulse trains.

Acknowledgment

The author acknowledges the guidance of Prof. D. L. Schilling, his thesis advisor at Polytechnic Institute of Brooklyn, and of Dr. C. A. Stutt, of the General Electric Research and Development Center, who first introduced him to the subject of this paper. Also, it has been his privilege over the past three years to work closely with J. J. Gostin, whose knowledge of radar signal detection has contributed much to this paper. Thanks are also due to the General Electric Company for their support of the Graduate Studies Program.

References

[1] C. W. Helstrom, *Statistical Theory of Signal Detection.* New York: Pergamon, 1960.

[2] H. Urkowitz, "Filters for detection of small signals in clutter," *J. Appl. Phys.*, vol. 24, p. 1024, August 1953.

[3] R. Manasse, "The use of pulse coding to discriminate against clutter," M.I.T. Lincoln Lab., Rept. 312-12, June 1961, AD 260-230.

[4] C. A. Stutt, "A note on invariant relations for ambiguity and distance functions," *IRE Trans. Information Theory*, vol. IT-5, pp. 164–167, December 1959.

[5] E. N. Fowle, E. J. Kelly, and J. A. Sheehan, "Radar system performance in a dense-target environment," *1961 IRE Internat'l Conv. Rec.*, pt. 4, pp. 136–145.

[6] H. L. Van Trees, "Optimum signal design and processing for reverberation-limited environments," *IEEE Trans. Military Electronics*, vol. MIL-9, pp. 212–229, July/October 1965.

[7] L. J. Spafford, "Optimum radar signal processing in clutter," Ph.D. dissertation, Polytechnic Institute of Brooklyn, June 1967 (obtainable from University Microfilms, Ann Arbor, Mich.).

[8] C. A. Stutt and L. J. Spafford, "A 'best' mismatched filter response for radar clutter discrimination," *IEEE Trans. Information Theory*, vol. IT-14, pp. 280–287, March 1968.

[9] L. J. Spafford, "Optimum radar receive waveforms in the presence of clutter," M.S. thesis, Polytechnic Institute of Brooklyn, June 1965.

[10] A. W. Rihaczek, "Radar resolution properties of pulse trains," *Proc. IEEE*, vol. 52, pp. 153–164, February 1964.

[11] G. J. Van der Mass, "A simplified calculation for Dolph–Tchebycheff arrays," *J. Appl. Phys.*, vol. 25, pp. 121–124, January 1954.

[12] W. D. Rummler, "A technique for improving the clutter performance of coherent pulse train signals," *IEEE Trans. Aerospace and Electronic Systems*, vol. AES-3, pp. 898–906, November 1967.

[13] D. F. Delong, Jr., and E. M. Hofstetter, "On the design of optimum radar waveforms for clutter rejection," *IEEE Trans. Information Theory*, vol. IT-13, pp. 454–463, July 1967.

3.2
OPTIMUM BURST WAVEFORMS FOR DETECTION OF TARGETS IN UNIFORM RANGE EXTENDED CLUTTER
M. Ares
IEEE Transactions on Aerospace and Electronic Systems, Vol. AES-3, No. 1, January 1967, pp. 138-141.

Optimum Burst Waveforms for Detection of Targets in Uniform Range-Extended Clutter

The problem of selecting a waveform and a filter to achieve acceptable performance in a clutter environment can usually be separated into two parts. The first consists of choosing a suitable waveform and the second consists of designing a filter which optimizes the response to a Doppler shifted delayed replica of this waveform in the presence of clutter interference. Both the waveform selection and filter design have received considerable emphasis in recent years. The problem of designing a filter which is optimum for a specified waveform has been formulated in terms of the cross ambiguity function and recently solved for arbitrary clutter distributions by Stutt and Spafford.[1]

The full problem solution is an optimization process which specifies both the waveform structure to be transmitted as well as the filtering process to be used upon reception. As yet, that problem, which is nonlinear except in a few special cases, has not been solved rigorously. In this paper, the problem of waveform design for the case of uniform range extended clutter is considered. Specifically, the pulse amplitudes of an envelope recurrent pulse train and the weights on the tapped delay line receiver filter are determined.

The study of waveforms proceeds most naturally in terms of the signal cross ambiguity function which is based on the work of Woodward,[2] Stutt,[3] and others. In the discussion that follows, the emphasis will be on the development of a physical picture rather than on mathematical rigor.

The cross-ambiguity function is defined as

$$\chi_{12}(\tau, \Delta) = \int_{-\infty}^{\infty} u_1(t)u_2^*(t+\tau)e^{-j2\pi\Delta t}dt \quad (1)$$

where $u_1(t)$ is the complex envelope of the transmitted waveform and $u_2(t)$ is the complex envelope of the receiver impulse response.

With the energy in the waveforms normalized to unity,

$$\int_{-\infty}^{\infty} |u_1(t)|^2 dt = \int_{-\infty}^{\infty} |u_2(t)|^2 dt = 1$$

the volume under $|\chi_{12}|^2$ becomes

$$\int_{-\infty}^{\infty}\int_{\infty}^{\infty} |\chi_{12}(\tau, \Delta)|^2 d\tau d\Delta = 1. \quad (2)$$

For the envelope recurrent N-pulse burst with uniform interpulse spacing T, the ambiguity function is[4]

$$|\chi_{12}(\tau, \Delta)|^2 = \sum_{k=-(N-1)}^{N-1} |\chi_p(\tau - kT, \Delta)^2| \cdot \left| \sum_{n=0}^{N-1-|k|} a_n b_{n+|k|} e^{-i2\pi n\Delta T} \right|^2 \quad (3)$$

where the a_n and b_n are the complex weights of the transmit and receive waveforms, respectively, and χ_p is the self-ambiguity function of each of the pulses in the burst. Thus, the ambiguity function for the uniformly spaced train of N pulses consists of $2N-1$ subsurfaces separated in τ by the interpulse spacing T. These subsurfaces will be called the "distal" ambiguities. For low duty factor waveforms; that is, when the time resolution of each pulse is much smaller than the interpulse period, $\chi_p(\tau, \Delta)$ is a slowly varying function of Δ and can be considered to be the "envelope" of the periodic function within the bars in (3). As an example, the self-ambiguity function for a uniformly space uniform amplitude burst is shown in Fig. 1.

For the situations of interest in this correspondence, the clutter is extended in range and is confined in Doppler so that it appears as a narrow strip parallel to the τ axis on

Manuscript received July 7, 1966.

[1] C. A. Stutt and L. J. Spafford, "A 'best' mismatched filter response for radar clutter discrimination," presented at the 1966 IEEE Internat'l Communications Conf., Philadelphia, Pa.

[2] P. M. Woodward, *Probability and Information Theory with Applications to Radar.* New York: Pergamon, 1963.

[3] C. A. Stutt, "A note on invariant relations for ambiguity and distance functions," *IRE Trans. on Information Theory,* vol. IT-5, pp. 164–167, December 1959.

[4] A. W. Rihaczek, "Radar resolution properties of pulse trains," *Proc. IEEE,* vol. 52, pp. 153–164, February 1964.

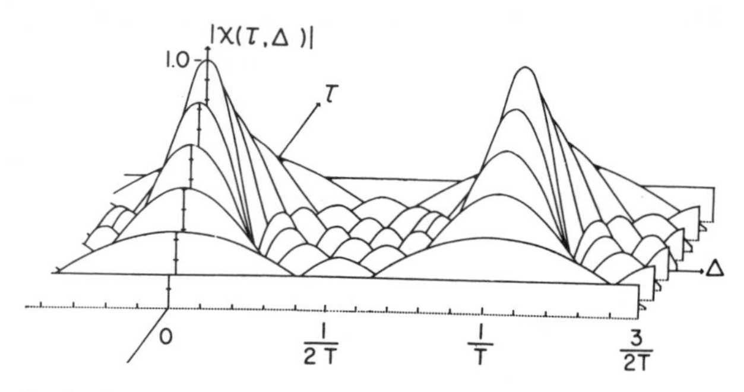

Fig. 1. Self-ambiguity function of uniform amplitude burst.

Fig. 2. Cross-ambiguity function of uniform amplitude burst and best m.s-matched filter.

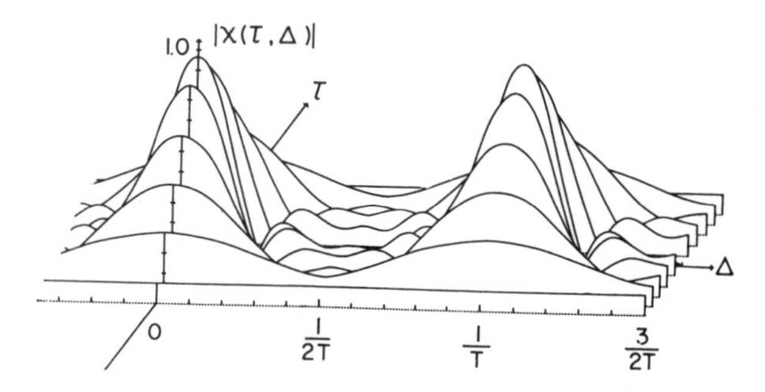

the τ, Δ plane. Under these circumstances, burst waveforms having constant interpulse periods are desirable. The reason for this is that the volume under the ambiguity function must remain constant—the only effect of different waveform choices being that of rearranging the volume. The uniformly spaced burst waveform has $2N-1$ peaks on the τ axis, thereby reducing the volume off the axis where the clutter is presumed to exist. The price paid for the performance in extended clutter is the introduction of ambiguous target range.

If we now consider incoherent scattering, that is, if returns from the elemental scatterers add on a power basis, then the problem can be simply stated to be one of choosing the phase and amplitude of the individual pulses in the burst, and the receiver filter (the a_n and b_n) so that each distal of the ambiguity function has a low value at the Doppler shift corresponding to the clutter. When this problem is expressed mathematically, a nonlinear matrix equation results. A general solution to this problem has not been found.

The problem of designing a *filter* which is optimum for a specified waveform has been recently solved for arbitrary clutter or interference distributions. The problem has been formulated in terms of the cross-ambiguity function by Stutt of the General Electric Research Laboratory. Application of Stutt's work by Spafford has led to computer programs which minimize the clutter-to-signal ratio with a minimum signal-to-noise ratio degradation. The optimum filter theory has been applied to the problem of determining the optimum weights on a tap delay-line filter for the case of a uniform amplitude, envelope recurrent waveform with uniform spacing.

In Fig. 2 the cross-ambiguity function for a uniform amplitude six pulse burst and a filter optimized to reject range-extended clutter at $\Delta = 1/2T$ is shown. The signal-to-clutter improvement relative to a single pulse was found to be 10 dB. The matched filter for this case yields an improvement of 7.8 dB. The improvement of the best mismatched filter is 2.2 dB in this case.

A heuristic explanation of the reason for the relatively small improvement obtained with the mismatched filter can be made with the aid of the following integral:

$$\int_{-\infty}^{\infty} |\chi_{12}(\tau, \Delta)|^2 d\tau = \int_{-\infty}^{\infty} \chi_{11}(\tau, 0)\chi_{22}^*(\tau, 0) \cdot e^{-i2\pi\tau\Delta} d\tau. \tag{4}$$

Equation (4) is a generalization of the results first given by Westerfield, Prager, and Stewart.[5] For the matched filter case, (4) reduces to Westerfield's result,

$$\int_{-\infty}^{\infty} |\chi(\tau, \Delta)|^2 d\tau = \int_{-\infty}^{\infty} |\chi(\tau, 0)|^2 e^{-i2\pi\tau\Delta} d\tau. \tag{5}$$

Note that for the case being considered, the clutter power is proportional to the integral on the left in (4); but, the right-hand side is the Fourier transform of the product of two autocorrelation functions $\chi_{11}(\tau, 0)$ and $\chi_{22}^*(\tau, 0)$. The latter is equivalent to the convolution of their power spectra so that the clutter power P_c can be written as

$$P_c = k \int_{-\infty}^{\infty} |U_1(f)|^2 \cdot |U_2(f + \Delta)|^2 df \tag{6}$$

where k is a constant.

Thus, the clutter power is proportional to the convolution of the power spectra of the waveform and the receiver impulse response. Since power spectra are positive everywhere, it is obvious that their convolution is everywhere positive. Thus, the clutter power can never go to zero even

[5] E. C. Westerfield, R. H. Prager, and J. L. Stewart, "Processing gains against reverberation (clutter) using matched filters," *IRE Trans. on Information Theory*, vol. IT-6, pp. 342–348, June 1960.

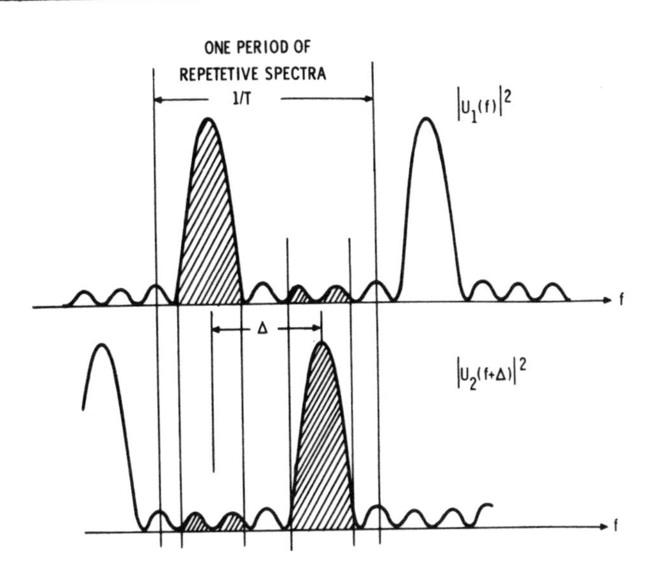

Fig. 3. Spectra for matched filter case.

F g. 4. Spectra for mismatched filter case.

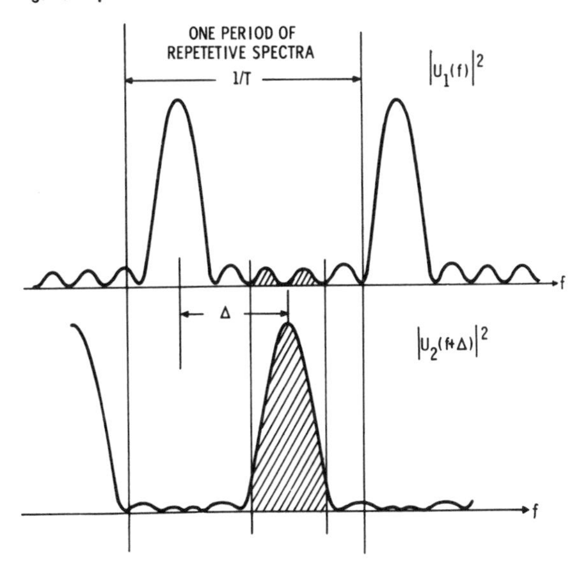

for stationary clutter. It is instructive to pursue the interpretation of (6) a little further. Consider again the matched filter case in which the bandwidth of the individual pulses in a burst is much greater than the pulse repetition frequency. The power spectrum $|U_1(f)|^2$ of such a waveform is depicted in Fig. 3. Also depicted is the frequency response of the processing filter $|U_2(f)|^2$ (the receiver waveform) shifted by Δ relative to $|U_1(f)|^2$.

Considering just one period of the repetitive spectra shown, the principal contribution to the convolution integral will be from integration of the product over the two shaded regions—that is, the product of the mainlobe of U_1 with the sidelobe of U_2, plus the product of the mainlobe of U_2 with the sidelobe of U_1.

If, instead of a matched filter receiver, an optimum mismatched filter is considered, then the sketch of Fig. 4 applies. For this case, the response of U_2 in the region around $f=\Delta$ has been reduced by choice of the receive filter. The principal contribution to the convolution integral is now the product of mainlobe of U_2 with the sidelobe of U_1.

It now becomes apparent that the most that can be gained by using a mismatched filter is about 3 dB relative to the matched filter. It must be emphasized, however, that this statement is applicable only to uniform stationary clutter extended in time over the entire ambiguity function. For other distributions of clutter, the improvement obtained with mismatched filter can be much greater than 3 dB.

It is also obvious from this discussion that appreciable reductions in the response to clutter can be made only if both the waveform and the filter are optimized.

A case of particular interest is the one for which $\Delta=1/2T$. Here, the clutter is located midway between the two ambiguous peaks of the function $\chi_{12}(\tau, \Delta)$. In reality, clutter is considered to be stationary and the target is presumed to be at the optimum Doppler. If the clutter power is minimized, the signal-to-clutter ratio will be maximized because the signal is at the peak of the response. Such a case was programmed for a digital computer. Minimization of the response to clutter was achieved by using nonlinear optimization routines. Solutions for N from 3 to 8 have been obtained.

The solutions were later verified by using the waveform obtained by the computer optimization as the input to Spafford's optimum mismatched filter program. As expected, the best mismatched filter turned out to be matched to the transmitted waveform.

As an example, the results for a six pulse burst will be presented. The waveform is shown in Fig. 5. Its ambiguity function is shown in Fig. 6. The ambiguity function is seen to have a valley at the location of the clutter and the "lobes" of each distal have all been widened as a result of rearranging the volume under the ambiguity function.

In regards to the signal-to-clutter ratio, the optimized waveform and its matched filter resulted in an improvement of 35.3 dB relative to a single pulse. This is impressive when compared to the uniform burst with its matched filter which yields only 7.8 dB or when compared with the uniform amplitude burst with the best mismatched filter which led to an improvement of 10 dB.

The preceding example confirms the notion that waveform design can lead to improvements in performance which overshadow receiver optimizations.

MANUEL ARES
Heavy Military Electronics Dept.
General Electric Company
Syracuse, N. Y.

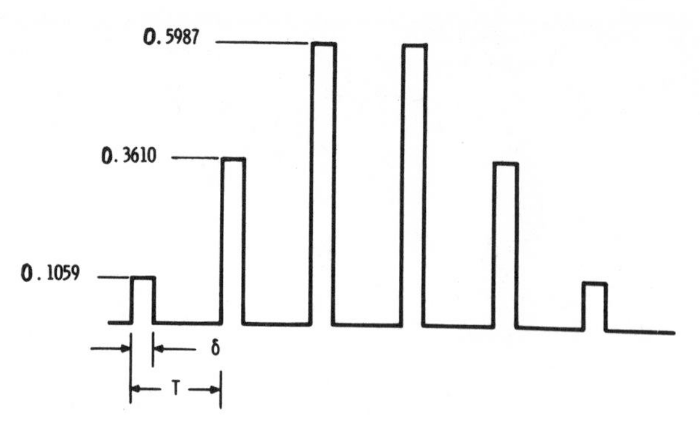

Fig. 5. Envelope of six pulse burst optimized for range-extended clutter at $\Delta = 1/2T$.

Fig. 6. Self-ambiguity function of optimum burst.

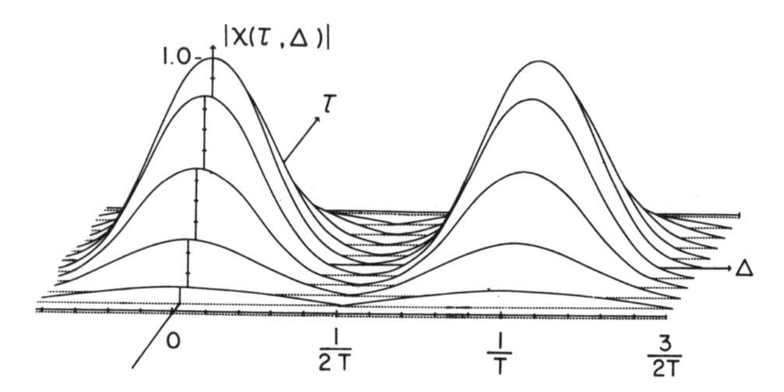

3.3 CLUTTER SUPPRESSION PROPERTIES OF WEIGHTED PULSE TRAINS
R.L. Mitchell and A.W. Rihaczek
IEEE Transactions on Aerospace and Electronic Systems, Vol. AES-4, No. 6, November 1968, pp. 822-828.

Clutter Suppression Properties of Weighted Pulse Trains

R. L. MITCHELL, Member, IEEE
A. W. RIHACZEK, Member, IEEE
Technology Service Corporation
Santa Monica, Calif.

Abstract

A common but troublesome requirement on radar sensors is the detection of a target in the interference from undesired scatterers, or clutter. Systems with coherent processing of pulse trains are uniquely suited for the purpose because, with pulse trains, it is possible to concentrate the receiver output for particular values of Doppler and thus suppress the clutter by Doppler filtering. This paper discusses to what degree the effectiveness of the method can be enhanced by tapering, or weighting, of the pulse amplitudes. The general results are illustrated by computer-plotted response functions for weighted pulse trains. The clutter suppression efficiency of weighting is calculated both for unilateral weighting in the receiver and for bilateral weighting in both receiver and transmitter. The significance of additional phase weighting is discussed and the results for pure amplitude weighting are compared with published work on phase and amplitude weighting.

I. Introduction

The problem of target detection in clutter has received considerable attention, both in the past and relatively recently. The severity of clutter interference depends on the type of clutter, whether it has a Doppler spread, is shifted in Doppler with respect to the target return, and similar operational characteristics. In any situation, however, the radar waveform particularly well suited to reduce the masking effects of the clutter is the coherent pulse train. These signals have good clutter suppression properties because their signal energy is confined to narrow intervals, both in time and in frequency.

If the pulse train is of finite duration, as it must be when we mean the coherently processed signal, the confinement of the signal energy to narrow intervals in both domains can be only imperfect. The remaining side lobes of the receiver response often will contribute sufficient interference to mask the desired return signal. It then becomes necessary to modify the pulse train by changing the amplitude, and possibly also the phase, from one pulse to the next. Such weighting can improve the clutter suppression performance of coherent pulse trains to a remarkable degree.

In a recent publication, Rummler [1] considers the problem of finding optimum weighting functions for reducing the response side lobes within a specified interval in range and Doppler. The weighting function turns out to be complex, that is, both the amplitudes and the phases of the pulses must be modified. A somewhat more general approach to the problem is described by DeLong and Hofstetter [2], who avoid specification of the exact shape of the clutter space. They investigate briefly the approach to an optimum waveform for the task, but then also limit the study to weighted pulse trains. In both papers, the objective is to find the type of weighting that maximizes the ratio of signal to the sum of clutter and thermal noise.

In addition, clutter suppression by means of pulse train weighting is discussed by Ares [3], Stutt [4], and Spafford [5].

The investigation in this paper is guided by a different philosophy. In practice, clutter distributions usually are not accurately specified and, moreover, often change with time. It then is impractical to fix the number of pulses and tailor the weighting function to a given type of clutter space. Instead of suppressing the clutter within a given Doppler interval, for example, it will generally be more desirable to reduce the interference uniformly within the entire interval between Doppler ambiguities, so as to accommodate uncertainties and changes in the Doppler distribution of the clutter. For this reason, we shall study the general relation between pulse train weighting and the behavior of the Doppler side lobes, and determine the clutter suppression potential of pulse trains in the more general case, where the clutter space is not perfectly known.

Manuscript received February 13, 1968.

This work was performed at Aerospace Corporation, El Segundo, Calif., under U. S. Air Force Contract F04695-67-C-0158.

II. Receiver Response for the Weighted Pulse Train

Let $\mu_1(t)$ be the complex envelope of the received signal and $\mu_2(t)$ be the waveform to which the receiver is matched. The receiver response in delay and Doppler, or the cross-ambiguity function, then is

$$\chi_{12}(\tau, \nu) = \int_{-\infty}^{\infty} \mu_1(t)\mu_2^*(t-\tau)e^{j2\pi\nu t}dt, \quad (1)$$

where τ is the differential delay and ν is the differential Doppler. For a pulse train with uniform repetition period T_r and a total of N pulses, the complex envelope has the form

$$\mu_1(t) = \sum_{n=0}^{N-1} a_n\mu_c(t - nT_r). \quad (2)$$

The coefficient a_n is the (generally complex) weight of the nth pulse and $\mu_c(t)$ is the complex envelope of the repeated pulse. In (2), we have assumed negligible Doppler distortions of the complex envelope, which is a good approximation for range rates $\dot{R} \lesssim 0.1c/TB$, where c is the velocity of light, and TB the time–bandwidth product of the pulse train. As will be seen later, we are interested in this study in Doppler shifts up to the first ambiguity $\nu_{\max} = 1/T_r$. With a conversion of Doppler into range rate and use of the above inequality, the condition for the Doppler distortions to be negligible then becomes

$$f_0/B \gtrsim 5N. \quad (3)$$

The waveform $\mu_2(t)$ to which the receiver is matched is to have the same form as $\mu_1(t)$, except that the pulse weights are b_n. From (1), the receiver response is

$$\chi_{12}(\tau, \nu) = \sum_{n=0}^{N-1}\sum_{m=0}^{N-1} a_n b_m^* e^{j2\pi\nu nT_r} \cdot \chi_c[\tau - (n-m)T_r, \nu], \quad (4)$$

where

$$\chi_c(\tau, \nu) = \int_{-\infty}^{\infty} \mu_c(t)\mu_c^*(t-\tau)e^{j2\pi\nu\tau}dt \quad (5)$$

is the ambiguity function of the pulse. We collect the terms centered at the same position $\tau = pT_r$ by rewriting the double sum as

$$\sum_{n=0}^{N-1}\sum_{m=0}^{N-1} = \sum_{p=0}^{N-1}\sum_{m=0}^{N-1-p}\Bigg|_{n=m+p} + \sum_{p=-(N-1)}^{-1}\sum_{n=0}^{N-1-|p|}\Bigg|_{m=n-p}. \quad (6)$$

Use of (6) in (4) expresses $\chi_{12}(\tau, \nu)$ in terms of components with fixed p. If the duty ratio T_p/T_r (T_p = pulse length) of the pulse train is less than 50 percent, none of these components overlap (and the overlap is insignificant for larger duty ratios if the pulses have large TB products). The envelope of the cross-ambiguity function then is the sum of the envelopes of the p-components, or

$$|\chi_{12}(\tau, \nu)| = \sum_{p=0}^{N-1} |\chi_c(\tau - pT_r, \nu)| \cdot \left|\sum_{m=0}^{N-1-p} a_{m+p}b_m^* e^{j2\pi\nu mT_r}\right| + \sum_{p=-(N-1)}^{-1} |\chi_c(\tau - pT_r, \nu)| \cdot \left|\sum_{m=0}^{N-1-|p|} a_m b^*_{m+|p|} e^{j2\pi\nu m T_r}\right|. \quad (7)$$

With the preceding result, the cross-ambiguity surface in the τ, ν plane is represented by p-surfaces centered at $\tau = pT_r$ on the delay axis.

Since we shall discuss the detailed properties of the receiver response to pulse trains, it is best to illustrate the behavior of $|\chi_{12}(\tau, \nu)|$ by actual plots of the function as a surface above the τ, ν plane. Fig. 1 shows such a plot for $a_n = b_n = 1$, that is, the ambiguity function of the uniform pulse train without weighting. The train length is $N = 5$ and the duty ratio is 30 percent. The p-surfaces are separated by empty strips. The breakup in Doppler is most pronounced for the central p-surface, diminishing with increasing $|p|$ until there is no breakup at all for $p = N - 1 = 4$. A detailed discussion of this ambiguity function is given in [6]. As the number of pulses increases, the central part of the ambiguity surface assumes the shape of a regular array of sharp spikes, with Doppler side lobes between the spikes. This is indicated in Fig. 2, which shows such a plot of $|\chi(\tau, \nu)|$ for $N = 8$ and a duty ratio of 10 percent.

The problem we shall consider is that of detecting a target within the range interval but outside the Doppler interval occupied by clutter, as discussed in [1]. An idealized clutter space thus might have the shape shown in Fig. 3, where the target is assumed located at the origin, $\tau = \nu = 0$. Interference from the clutter due to the concentrated ambiguous spikes in $|\chi(\tau, \nu)|$ can be avoided only if the Doppler spread of the clutter does not exceed $1/T_r$, the Doppler spacing of the ambiguities. In other words, to exclude such interference, we must choose the PRF high enough to meet this condition, even though this may mean that the range spread of the clutter becomes large compared with the unambiguous range interval. Clutter still will cause interference through the Doppler side lobes between the ambiguous spikes. The question is to what degree these side lobes can be suppressed by weighting of the pulse train.

From (7), a given p-surface is defined by

$$|\chi_p(\tau, \nu)| = |\chi_c(\tau - pT_r, \nu)| \left|\sum_{m=0}^{N-1-p} a_{m+p}b_m^* e^{j2\pi\nu m T_r}\right|. \quad (8)$$

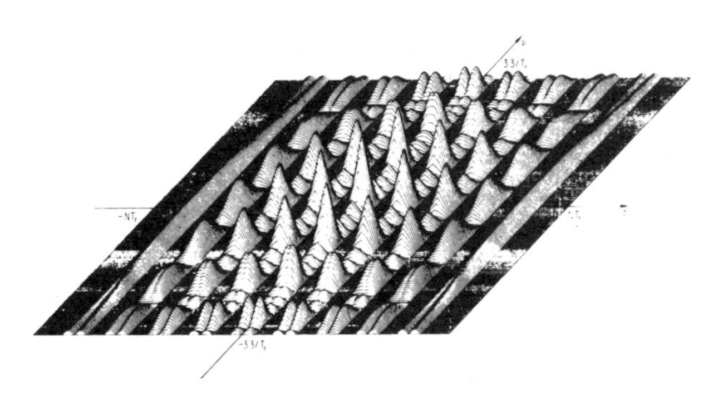

Fig. 1. Envelope of the matched filter response of a uniform pulse train, with $N=5$ and 30 percent duty ratio.

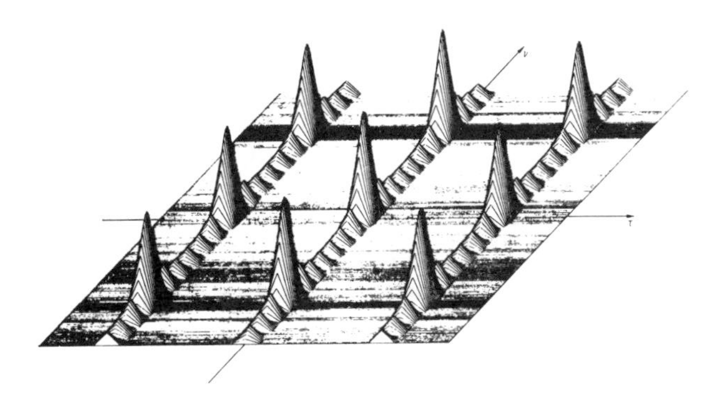

Fig. 2. Central part of uniform pulse train response for $N=8$ and 10 percent duty ratio.

Fig. 3. Location of the clutter space in the τ, ν plane.

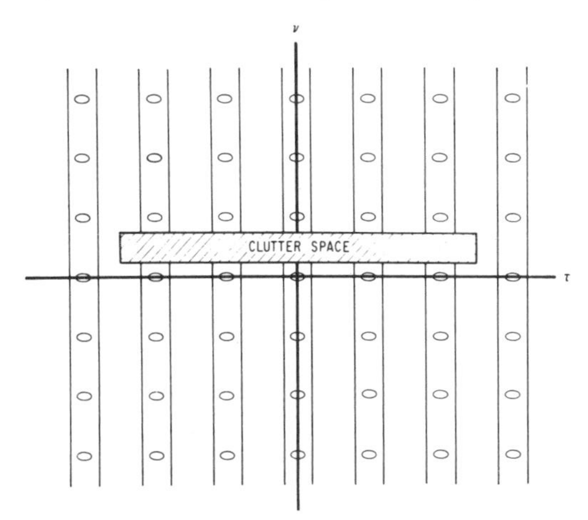

Although selection of $p \geq 0$ is arbitrary, it is seen from (7) that the same results will apply for $p<0$. In order to understand the effect of the factor $a_{m+p}b_m^*$, we take the Fourier transform of the sum in (8),

$$\mathfrak{F}\left\{\sum_{m=0}^{N-1-p} a_{m+p}b_m^* e^{j2\pi\nu m T_r}\right\} = \sum_{m=0}^{N-1-p} a_{m+p}b_m^*\delta(t - mT_r), \quad (9)$$

where $\delta(t)$ is the Dirac delta function. Now, let a continuous weighting function $a(t)$ be defined such that sampling of $a(t)$ at intervals mT_r yields the coefficients a_m. Outside the time interval occupied by the pulse train, we assume $a(t)=0$. A weighting function $b(t)$ for generating the coefficients b_m is defined similarly. Since $\delta(t-mT_r)=0$ for $t \neq mT_r$, the quantity $a_{m+p}b_m^*$ in (9) can be replaced by $a(t+pT_r)b^*(t)$. Thus

$$\sum_{m=0}^{N-1-p} a_{m+p}b_m^*\delta(t - mT_r) = \sum_{m=-\infty}^{\infty} a(t + pT_r)b^*(t)\delta(t - mT_r), \quad (10)$$

where the extension of the limits is permissible, because the weighting functions are zero outside the pulse train interval. Substitution of (10) in (9) and inversion of the Fourier transform gives

$$\sum_{m=0}^{N-1-p} a_{m+p}b_m^* e^{j2\pi\nu m T_r} = \mathfrak{F}^{-1}\{ a(t + pT_r)b^*(t)\} \star (1/T_r) \sum_{m=-\infty}^{\infty} \delta(\nu - m/T_r), \quad (11)$$

the star denoting convolution. If this result is applied to (8), we find the following. The ambiguity function of the pulse, $\chi_c(\tau-pT_r, \nu)$, is multiplied in the Doppler domain by a function generated by repetition of the transform of $a(t+pT_r)b^*(t)$ at intervals $1/T_r$. Hence, if this transform has low side lobes, the ambiguity function of the pulse train will have low Doppler side lobes between the spikes. The problem of clutter suppression thus is one of finding weighting functions that minimize the Doppler side lobes within all p-strips crossing the clutter space.

III. Real and Complex Weighting Functions

To see the difference between real and complex weighting functions, let us assume that the clutter space is centered on the Doppler axis, and consider the effect of the real weighting function $a(t)$. Since one desires the same behavior for positive and negative p for the symmetrical clutter space, the weighting function should also be symmetrical. In each p-strip, the main lobes and the Doppler side lobes are determined by the Fourier transform of the product $a(t+pT_r)a(t)$. Now, in general, the relative translation pT_r of one part of this product with respect to the other causes a narrowing of the function $a(t+pT_r)a(t)$ and a sharpening of the edges; therefore, from Fourier theory, the main lobe of the transform is widened and the side-lobe level is increased. Since the translation pT_r depends on p, the side-lobe level of the ambiguity function will change from one p-strip to the other.

If we now assume that $a(t)$ is chosen to give low side lobes of the transform, the effects of side lobes due to "aliases" centered at Doppler values removed from the Doppler interval of interest can be neglected. With increasing p, the widening of the response peak and the increase in the side-lobe level thus will apply to the ambiguity function and not just to the individual repeated transforms. Of course, if the clutter space is centered at a point off the Doppler axis, corresponding to $p \neq 0$, the same arguments apply with respect to the p-strips adjacent to the particular strip at which the clutter space is centered.

Both effects, the broadening of the main spikes and the increase in the side-lobe level, are equally detrimental to clutter performance. A Doppler broadening of the main spikes forces us to increase the repetition frequency, in order to accommodate a clutter space of given Doppler width. However, this results in a closer spacing of the p-strips and inclusion within the clutter space of the side lobes of the higher order p-strips. A general increase in the side-lobe level, of course, leads to a corresponding increase in the interference from the clutter. It follows that clutter performance depends approximately on the area under the squared envelope of the transform of $a(t+pT_r)a(t)$, where the squared function is taken because power superposition applies for the clutter.

For our purposes then, the question of real versus complex weighting depends on whether or not the area under the squared transform of $a(t+pT_r)a^*(t)$ can be reduced by complex weighting functions. If the combined weighting function is written in the form $u(t)e^{j\phi(t)}$, we see that this area is simply the energy of the "signal" $u(t)e^{j\phi(t)}$. However, the signal energy is not changed when phase modulation is applied. This means that the general clutter level cannot be reduced by complex weighting. Such weighting merely allows us to modify the lobe structure in each p-strip and adapt it better to a clutter space of specified form.

The preceding reasoning could be repeated for the more general case of two different but symmetrical weighting functions, $a(t)$ and $b(t)$. Quite generally, complex weighting is useful when the ambiguity function is to be matched to a clutter space with more complicated properties than considered here. This assumes that the clutter space is known to the required accuracy and does not change too rapidly, and that the approach is not limited by unavoidable system instabilities and nonlinearities. As already remarked in the Introduction, this approach is not treated here. In studying side-lobe suppression, we thus will restrict our investigation to real weighting functions.

We shall analyze the two cases of particular practical interest. In the first case, where the weighting functions in transmitter and receiver are the same, $a(t)=b(t)$. The radar then transmits the weighted pulse train and the receiver is matched to it. In the second case, the transmitted pulses all have the same amplitude, so that $a(t)=1$ within the pulse interval and $a(t)=0$ outside. The receiver weights the pulses in accordance with $b(t)$, which means that it is mismatched. This type of operation is important because it simplifies the system design and permits full use of the peak power capability of the transmitter. The two cases will be referred to as bilateral and unilateral weighting, respectively.

IV. Receiver Responses for Weighted Pulse Trains

If the clutter space were confined to a single ambiguous range interval, only the side lobes for the central p-strip would be of interest. These side lobes depend on the exact shape of the weighting function (and the tolerances with which it can be implemented). For extended clutter spaces, on the other hand, the primary effect is the increase of the interference through side lobes and main-lobe broadening from one p-strip to the next. For our purposes, we can thus gain little by considering differences among various specific weighting functions. In illustrating the general behavior of ambiguities and side lobes, we may choose any convenient weighting function. We select a Gaussian function truncated to the pulse train duration.

Fig. 4 gives the central part of $|\chi(\tau, \nu)|$ for the uniform pulse train with unilateral weighting in the receiver, with a degree of weighting such that the amplitudes of the pulses on the fringes are 8 percent of that of the central pulse. The pulse train length is $N=11$ and the ambiguous spikes are truncated to 20 percent of maximum height for better visibility of the side lobes. The plot shows a substantial increase of the side-lobe level over the p-strips. The broadening of the main spikes with increasing $|p|$ also is substantial.

The effect should be less pronounced for bilateral weighting, because both the narrowing of $a(t)b(t+pT_r)$ and the sharpening of the edges of this function are less severe if $a(t)$ is smoothly tapered, rather than being rectangular. The expectation is confirmed by the plot of Fig. 5, which applies for the same conditions as Fig. 4, except that the former weighting function $b(t)$ for unilateral weighting now is split into two parts $\sqrt{b(t)}$, one for the transmitter and the other for the receiver. The side lobes on the Doppler axis for both figures thus must be identical. The broadening of the main spikes and the side-lobe increase for larger $|p|$ is seen to be considerably slower.

We now investigate how the degree of weighting affects the receiver response. Fig. 6 shows the plot of $|\chi(\tau, \nu)|$ for the same case as Fig. 4, but with the same weighting function used also on transmission. The weighting function thus is the square of that for Fig. 5. At first glance, the heavier weighting seems to solve the side-lobe problem; however, note the strong broadening of the bases of the ambiguous spikes. The corresponding decrease in the width of the "clear" Doppler zone may easily offset the gain from the reduction of the side lobes. The heavy weighting has reduced the effective length of the pulse train.

The behavior of the receiver responses indicates that for suppression of the side lobes without main-lobe broadening, one should not simply increase the degree of weighting of a given pulse train. Rather, the heavier weighting should be accomplished by addition of low-amplitude pulses on the fringes. The effectiveness of the method is illustrated in Fig. 7. In Fig. 7(A), we show an amplified version of the part for $|\nu| \leq 1/2T_r$ of Fig. 5, truncated to

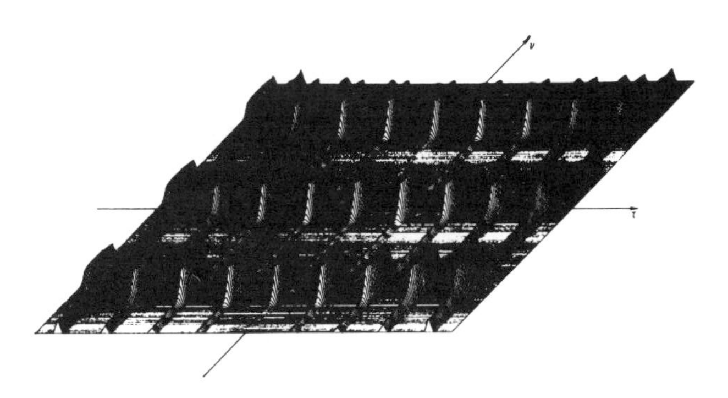

Fig. 4. Central part of the response for unilateral Gaussian weighting, N = 11, duty ratio of 10 percent, peaks truncated to 20 percent of maximum height.

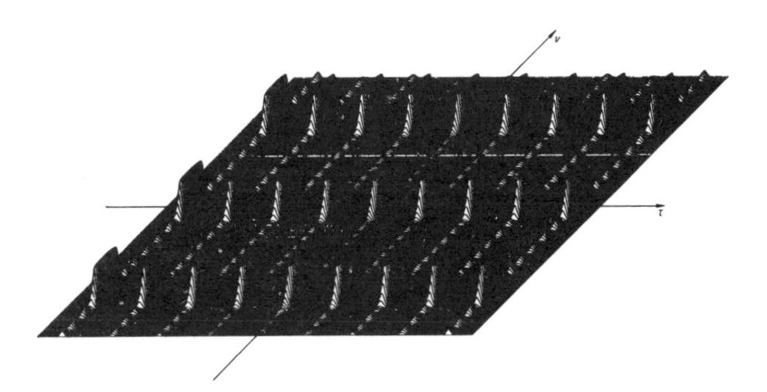

Fig. 5. Response of Fig. 4, except bilateral weighting.

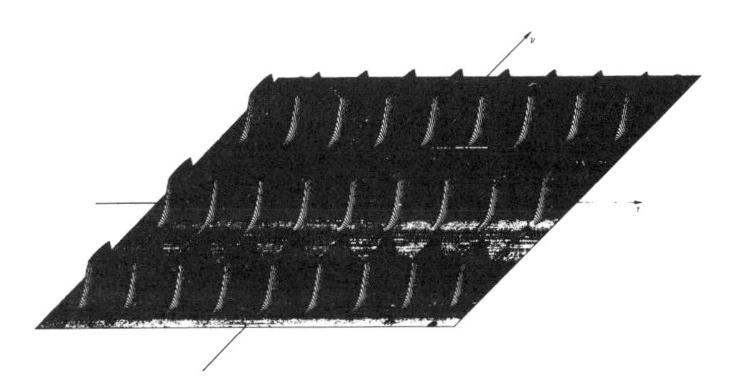

Fig. 6. Response for bilateral weighting with the function used in Fig. 4.

Fig. 7. Side-lobe reduction by addition of low-energy pulses peaks truncated to 2 percent of maximum height; (A) N = 11 (B) N = 15, and (C) N = 19.

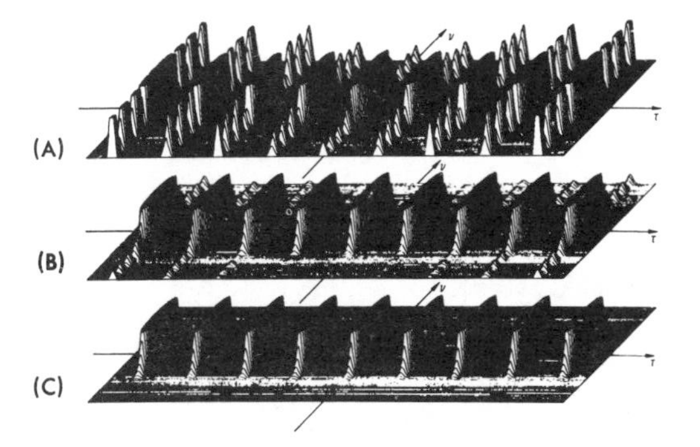

2 percent of peak amplitude. The Doppler side lobes in the outermost p-strips exceed the truncation level. In Fig. 7(B), two pulses were added at each end, their amplitudes corresponding to the Gaussian weighting function. The degree of weighting roughly equals that of Fig. 6, except that we achieved it by lengthening the pulse train from $N=11$ to $N=15$. A rather pronounced reduction in the side-lobe level without significant broadening of the Doppler spikes can be observed. In Fig. 7(C), two more pulses were added at each end, so that $N=19$. The last pulse has an amplitude of 1.7 percent of that of the central pulse. Addition of these low-energy pulses suppresses the side lobes until they cannot be seen on the scale used in the figure, again without broadening of the base of the main ambiguities.

In summary, the side-lobe level, in general, depends on the degree of weighting. The Doppler width of the ambiguous spikes is determined by the length of the pulse train between, say, the half-power points. In other words, it is not affected by low-amplitude pulses at the fringes. The side-lobe level increases with $|p|$ or, more precisely, with $|p|/N$. For a given p-strip, the side-lobe enhancement is less severe for larger N.

V. Clutter Suppression Effectiveness of Weighting

The preceding discussion and the plots provide an understanding for weighting and side-lobe suppression. To obtain quantitative results, however, we must sum all side-lobe contributions over the entire clutter space. As an example, we perform these calculations for a representative clutter space, which we choose to be the one considered by Rummler [1], for better comparison with complex weighting. The detection problem to be considered is that of Fig. 3. The target is to be outside the Doppler interval of the clutter (at the origin in the figure). The clutter space does not extend to the first Doppler ambiguity, which can always be ensured by proper choice of the pulse repetition frequency. The penalty is the inclusion of several p-strips within the clutter space.

Real weighting was shown to reduce the side lobes of the receiver response uniformly between the Doppler spikes. Hence, for the clutter space of Fig. 3, the side lobes also would be reduced in the region outside the clutter space, if the latter does not extend over the entire ambiguous Doppler interval. If indeed the clutter space is specified so accurately that side-lobe suppression only within a particular Doppler interval is adequate, then complex weighting will be useful. It essentially allows reshuffling of the side-lobe energy such that it is removed from the clutter space to the free Doppler region (within the same ambiguous Doppler interval). On the other hand, in the absence of precise, known, and stable boundaries on the clutter space, we must try to reduce the side lobes everywhere between the ambiguous spikes.

As is seen from the plot of ambiguity functions, sidelobe suppression improves with increasing degree of weighting. At the same time, the signal-to-noise ratio is degraded, because weighting reduces the signal energy when the number of pulses is fixed. However, while the loss in signal energy is undesirable, it is not of primary concern in a typical clutter situation. Signal detectability in the thermal noise can be improved by raising the transmitter power, but signal detectability in the clutter cannot. An optimum compromise between noise and clutter performance thus does not appear to be a suitable criterion for actual system design. Rather, one will choose the type of weighting that gives adequate clutter performance and then raise the total signal energy until noise performance also is acceptable.

As a first case, we calculate the clutter suppression efficiency of real weighting for one of the clutter spaces considered by Rummler [1], in order to permit compàrison of real and complex weighting. The clutter space is assumed to extend in Doppler $0.5/T_r$ to $0.83/T_r$ and in range over 5 p-strips to either side of the Doppler axis. Thus we must integrate over the side lobes of the ambiguity surface falling within this region, both with and without weighting. As a measure for the degree of weighting, we retain Rummler's "detection efficiency," which simply is the reduction of the receiver output peak due to the weighting. The results are shown in Fig. 8, where the curve for the 21-element pulse train studied by Rummler is supplemented by corresponding curves for 23 and 25 pulses.

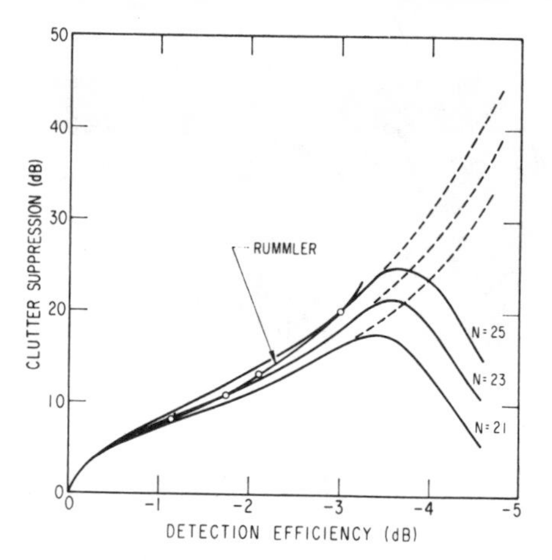

Fig. 8. Clutter suppression ratio for unilateral weighting.

For suppression ratios of less than about 20 dB, complex weighting is seen to be slightly more effective, but the advantage can be offset by addition of a few pulses. When the degree of weighting is increased, the Doppler spikes broaden in the case of pure amplitude weighting and this causes the curves to bend downward. This effect is avoided to some degree with complex weighting, which is better tailored to the particular clutter space. The dashed lines indicate clutter suppression in the limit as the clutter becomes more concentrated halfway between the ambiguous spikes, in which case the broadening of the spikes has no effect. This is not at all an artificial situation. The effects of the bending of the curves can be avoided by increase of the number of pulses, which decreases the Doppler width of the ambiguous spikes.

Fig. 9. Comparison of clutter suppression for unilateral and bilateral weighting.

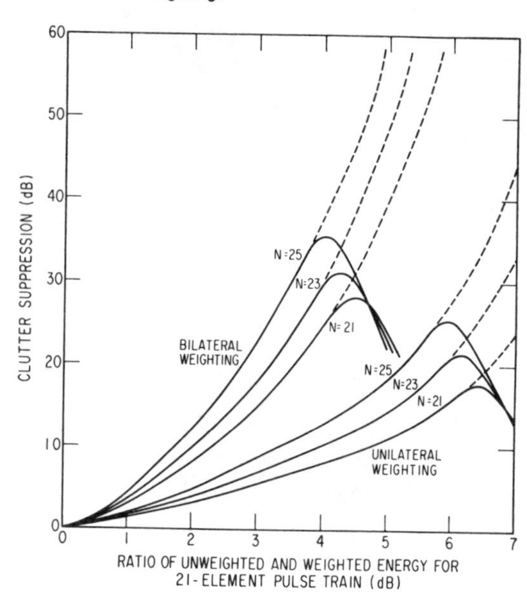

In Fig. 9, we compare clutter suppression for unilateral and bilateral weighting. The abscissa is the ratio of the energies in the unweighted and weighted 21-member pulse train, which is merely a measure of the degree of weighting for this particular pulse train. With bilateral weighting, the same type of weighting also is used for the transmitted signal. The curves illustrate the gain in clutter suppression due to bilateral weighting. However, the system is more difficult to implement and does not fully use the peak power capability of the transmitter.

The other curves of Fig. 9 demonstrate how the clutter can be further suppressed by addition of low-energy pulses on the fringes, so that the overall pulse train (now longer) is more heavily tapered. The effect is shown qualitatively in Fig. 7. The curves show, for example, that close to the turning point for $N=21$ we gain an additional 10 dB by adding 2 pulses at each end. Increasing the number of pulses shifts the turning point to higher values of the suppression ratio, both for unilateral and for bilateral weighting. We can shift it to the point where it is beyond practical system capabilities, considering the problem of approximating the theoretical performance in practice.

VI. Conclusions

The results presented here demonstrate the effectiveness of amplitude weighting in suppressing the clutter caused by the Doppler side lobes of the receiver response. Complex weighting is of interest if the clutter space is exactly specified and, primarily, if it is located unsymmetrically between ambiguous spikes. However, both assumptions are somewhat artificial, the first because the characteristics of the clutter space will not be known to such precision and the second because the PRF can be chosen such that the clutter space will be centered halfway between the spikes. Most noteworthy, the results for amplitude weighting were derived without any type of waveform optimization, which means simplicity in practice and good clutter suppression for a broad range of clutter spaces. As a general conclusion, we find that adding a few pulses can be a highly effective substitute for complicated tailoring of weighting functions.

REFERENCES

[1] W. D. Rummler, "Clutter suppression by complex weighting of coherent pulse trains," *IEEE Trans. Aerospace and Electronic Systems*, vol. AES-2, pp. 689–699, November 1966.
[2] D. F. DeLong and E. M. Hofstetter, "On the design of optimum radar waveforms for clutter rejection," *IEEE Trans. Information Theory*, vol. IT-13, pp. 454–463, July 1967.
[3] M. Ares, "Optimum burst waveforms for detection of targets in uniform range-extended clutter," *IEEE Trans. Aerospace and Electronic Systems*, vol. AES-3, pp. 131–141, January 1967.
[4] C. A. Stutt, "A 'best' mismatched filter response for radar clutter discrimination," G. E. Research and Development Center, Schenectady, N. Y., Rept. 66-C-020, March 1966.
[5] L. J. Spafford, "Optimum radar receiver waveforms in the presence of clutter," M.S. thesis, Polytechnic Institute of Brooklyn, Brooklyn, N. Y., June 1965.
[6] A. W. Rihaczek, "Radar resolution properties of pulse trains," *Proc. IEEE*, vol. 52, pp. 153–164, February 1964.

Richard L. Mitchell (S'60–M'64), for a photograph and biography, please see page 432 of the May, 1968, issue of this TRANSACTIONS.

August W. Rihaczek (M'59), for a photograph and biography, please see page 432 of the May, 1968, issue of this TRANSACTIONS.

Section 4 Special MTI Effects

EDITOR'S COMMENTS

This series of papers describes effects that cause MTI performance to degrade from the ideal performance detailed in the previous papers. The topics discussed in this series are MTI signal limiting, pulse repetition rate (PRF) staggering, and the power spectral density of wind driven clutter.

Grasso and Guarguaglini (Paper 4.1) discuss the effect of limiting the IF signal before MTI processing. IF limiting is a necessary process in all but the highest performance MTI systems. The limiting functions to compress the clutter's wide dynamic range so that the MTI processor can reduce the clutter residue to the receiver noise level. Unfortunately, the limiting spreads the clutter spectrum thereby reducing the improvement factor of the MTI.

The effect of a smooth limiter is analyzed as a function of the clutter-to-limit (C/L) level. In general, the harder the limiting and the greater the inherent (no limiting) performance of the MTI, the greater the performance degradation. Performance degradations as large as 25 dB are determined for a double canceler processing 30 to 40 pulses.

Prinsen (Papers 4.2 and 4.3) discusses PRF staggering to eliminate MTI "blind speeds." In uniform PRF MTI systems "blind speeds" occur in the velocity response at velocity nulls that coincide with multiples of the radar's PRF. The problem is most severe for long range radars that operate in the higher radar frequency bands.

Prinsen's first paper (4.2) considers a nonrecursive MTI filter and develops a nonlinear gradient search procedure for finding the optimum weights and stagger interval in accordance with the performance criteria. The performance criteria he uses attempts to minimize the depth of valleys in the signal-to-clutter ratio (SCR) gain while maximizing the average SCR. Several general results determined by Prinsen are:

- **Both interpulse spacings and MTI filter weightings must be optimized in a staggered PRF MTI.**
- **The weightings optimized for a uniform PRF MTI are no longer optimum for a staggered PRF MTI.**
- **The performance of MTI filters using binomial weightings can degrade severely when PRF staggering is used.**
- **A class of MTI filters using Vandermonde weighting, although suboptimal is recommended for staggered PRF MTI systems.**

The class of MTI filters using Vandermonde weightings is described in Paper 4.3. This set of MTI weightings has the property that it provides a maximally flat stopband transfer function about zero frequency. The weights are easily determined for an arbitrary PRF stagger and coincide with binomial weights for a uniform PRF system.

Fishbein, Graveline and Rittenbach (Paper 4.4) examine MTI systems whose performance is determined by internal clutter motion caused by wind effects. This situation occurs in Combat Surveillance radars of the type discussed in this paper (non-scanning) and also in other types of radars where the spectrum spreading caused by internal clutter motion is significant with respect to the spectrum spreading caused by other effects (See Paper 1.6).

The paper determines, on the basis of experimental measurements, that the spectral spreading induced by wind driven clutter has a cubic dependence. The cubic dependence has significantly greater magnitude at high frequencies than the normally assumed Gaussian shaped spectrum. This results in poorer MTI performance than would be predicted using the normal Gaussian shaped clutter spectrum. An example of a range-gated filter MTI in a 20 knot wind shows a 7.5 dB degradation in performance with respect to the performance predicted using a Gaussian clutter shape assumption.

4.1 CLUTTER RESIDUES OF A COHERENT MTI RADAR RECEIVER
G. Grasso and P.F. Guarguaglini
IEEE Transactions on Aerospace and Electronic Systems, Vol. AES-5, No. 2, March 1969, pp. 195-204.

Clutter Residues of a Coherent MTI Radar Receiver

GIANCARLO GRASSO
PIER FRANCESCO GUARGUAGLINI, Member, IEEE
Selenia S. p. A.
Rome, Italy

Abstract

In moving target indicator radar receivers a saturation is always present in the IF stages in order to compress the dynamic range of large clutter inputs. This nonlinearity greatly affects the performances of the cancelling circuits, because of the spreading of the spectra. A quantitative analysis of this effect has been conducted, with an analytical expression chosen for the limitation characteristic.

A high degradation of both single- and double-canceller improvement factors has been obtained. Losses of 20 dB for the double canceller are not uncommon. Clutter residue levels at the receiver output have also been computed in order to see how the performance compares with ideal constant false alarm rate behavior.

Introduction

The level of clutter residues at the output of a moving target indicator (MTI) receiver, as compared to the noise level, is established by the filtering capacity of the cancelling circuits. This capacity of the canceller is based on the different frequency distribution of echoes coming from moving targets. In the literature the behavior of the canceller has been evaluated by comparing the clutter power of the output with that at the input of the canceller [1], [2]. The analyses have been conducted assuming a clutter spectrum at the input of the cancelling circuits equal to that at the input of the receiver. In other words, all the evaluations are conducted for a linear receiver. This is not the case for real MTI receivers, since a saturation always occurs so that the large dynamic range of the input signals can be handled. The introduction of nonlinearities causes a distortion and a spreading of the spectra. This effect has been analyzed for concentrated clutter, and a large degradation in cancellation occurs [3], [4]. In the present paper the case of distributed clutter is examined, with the level of clutter residues being evaluated for the single and double canceller when a limitation occurs. In order to carry out calculations a smooth limitation shape which is expressed in an easy analytical form has been assumed, according to [5].

Manuscript received August 4, 1967.

A General Expression for the Improvement Factor For a Smoothly Limited Receiver

The calculation of clutter residues implies the evaluation of the improvement factor, which is the parameter that describes the behavior of the cancelling circuits. This parameter, defined in the literature [6] as the average improvement in the signal-to-clutter ratio introduced by the canceller, describes the cancellation including the canceller average gain:

$$I = \frac{S_0/C_0}{S_i/C_i} \tag{1}$$

where the signal is intended to be averaged over echoes from moving targets, considering all possible radial speeds equiprobable.

The improvement factor is also related to the more widely used clutter attenuation [1] by the noise gain G between the output and the input of the canceller [2]:

$$I = G \cdot CA. \tag{2}$$

Assume the analytical expression of the frequency transfer functions given by [1]:

$$\begin{aligned} |H_1(f)| &= 2\left|\sin\left(\frac{\pi f}{f_r}\right)\right| \\ |H_2(f)| &= 4\sin^2\left(\frac{\pi f}{f_r}\right) \end{aligned} \tag{3}$$

for the single and the double canceller, respectively. In such a case the improvement factor is 3 dB larger than the clutter attenuation for the single canceller and is 7.8 dB larger than the clutter attenuation for the double canceller.

In the time domain the canceller effect may be easily investigated through the autocorrelation function at the input to the canceller, $R_0(t)$. The single and double canceller perform, respectively, a comparison between two and three successive pulses, spaced by pulse repetition period T.

Therefore, the improvement factor in the two cases is expressed by

$$I_1 = \frac{R_0(0)}{R_0(0) - R_0(T)} \tag{4}$$

$$I_2 = \frac{3R_0(0)}{3R_0(0) - 4R_0(T) + R_0(2T)} \, . \tag{5}$$

From the above expressions, it is obvious that the presence of nonlinearities before the canceller, which alter the autocorrelation, also has an influence on the improvement factor. It is therefore necessary to evaluate how the $R_0(t)$ at the input of the canceller is bound to the input autocorrelation $R_i(t)$.

With reference to Fig. 1, let v_i be the input and v_l the output of the IF stages. The smooth limiting characteristic is expressed by [5]

$$v_l = \int_0^{v_i} \exp\left[- \frac{z^2}{1.28 l^2} \right] dz \tag{6}$$

where l is the saturation level. In this situation the output autocorrelation function is related to the input autocorrelation function by [5]

$$R_l(t) = \frac{2}{\pi} \, l^2 \sin^{-1}\left[\frac{\rho_i(t)}{1 + \alpha} \right] \tag{7}$$

where $\rho_i(t) = R_i(t)/C^2$ and the parameter α is defined as

$$\alpha = 0.64 \frac{l^2}{C^2} \tag{8}$$

with C^2 the clutter input power.

The arc sine of (7) may be developed in a series of the odd powers of its argument:

$$R_l(t) = \frac{2}{\pi} \, l^2 \sum_{p=0}^{\infty} a_p \left[\frac{\rho_i(t)}{1 + \alpha} \right]^{2p+1} . \tag{9}$$

Consider now that $\rho_i(t)$ is the normalized autocorrelation function of a signal at an IF frequency $\omega_0/2\pi$. It may therefore be expressed as

$$\rho_i(t) = \phi(t) \cos \omega_0 t. \tag{10}$$

The odd powers of $\rho_i(t)$ are given by

$$\rho_i(t)^{2p+1} = [\phi(t)]^{2p+1} b_p \cos^{2p+1} \omega_0 t. \tag{11}$$

Considering that only the components around the frequency $\omega_0/2\pi$ are of interest because of the filtering around the IF frequency, the autocorrelation function before the canceller is expressed by

$$R_0(t) = \frac{2}{\pi} \, l^2 \sum_{p=0}^{\infty} a_p b_p \frac{[\phi(t)]^{2p+1}}{(1 + \alpha)^{2p+1}} \tag{12}$$

where

$$a_0 = 1$$
$$a_p = \frac{1 \cdot 3 \cdot 5 \cdots (2p - 1)}{2 \cdot 4 \cdot 6 \cdots 2p(2p + 1)} \quad \text{for } p = 1, 2, \cdots \tag{13}$$

and

$$b_p = \frac{1}{2^{2p}} \binom{2p + 1}{p} . \tag{14}$$

The use of (12) with (13) and (14) permits one to obtain, given the input autocorrelation function of the video signal $\phi(t)$, the improvement factor for the single and double cancellers from (4) and (5).

Losses on the Improvement Factor for a Gaussian Clutter Spectrum

The expression of the autocorrelation function at the input to the canceller, obtained in the preceding section, is completely general and allows the evaluation of the improvement factor for any form of input autocorrelation. In order to carry out numerical computations of the influence of a limitation on MTI performance, the case of a Gaussian input autocorrelation function has been examined.

This choice is justified by the fact that the clutter spectrum due to internal fluctuations is always approximated with a Gaussian shape [1]. Furthermore, one of the main causes of clutter spreading, antenna motion, may be considered to give rise to a Gaussian spectrum, since the Gaussian form may closely approximate the usual antenna beam shape.

It has been shown [3] that for concentrated clutter the approximation of the antenna pattern by a Gaussian shape cannot be accepted with a strong limitation, since the side lobes acquire an increasing importance. Considering instead a distributed clutter, the approximation may always be accepted, since the whole antenna diagram is involved at any time. Therefore, the contribution from all the clutter elements will be reduced, more or less proportionally, by the limiter.

Also taking into consideration radar internal instabilities, the best choice of clutter spectrum shape is still the Gaussian. Another advantage offered by a Gaussian spectrum is that the corresponding autocorrelation function is still Gaussian and its powers are still Gaussian. It is therefore particularly easy to use (12), and it is possible to obtain immediately, by a Laplace transform, the terms of the corresponding spectrum.

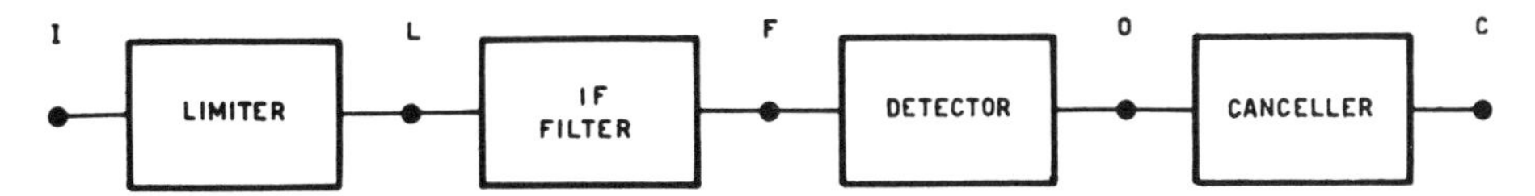

Fig. 1. Simplified block diagram of a limited coherent MTI.

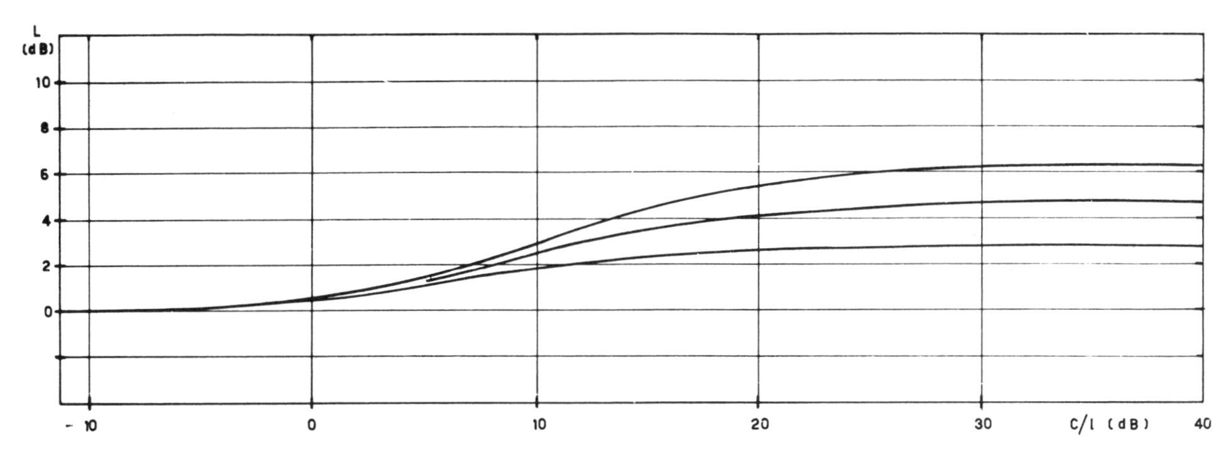

Fig. 2. Loss in improvement factor of a single canceller versus clutter input relative to limit level.

Fig. 3. Loss in improvement factor of a double canceller versus clutter input relative to limit level.

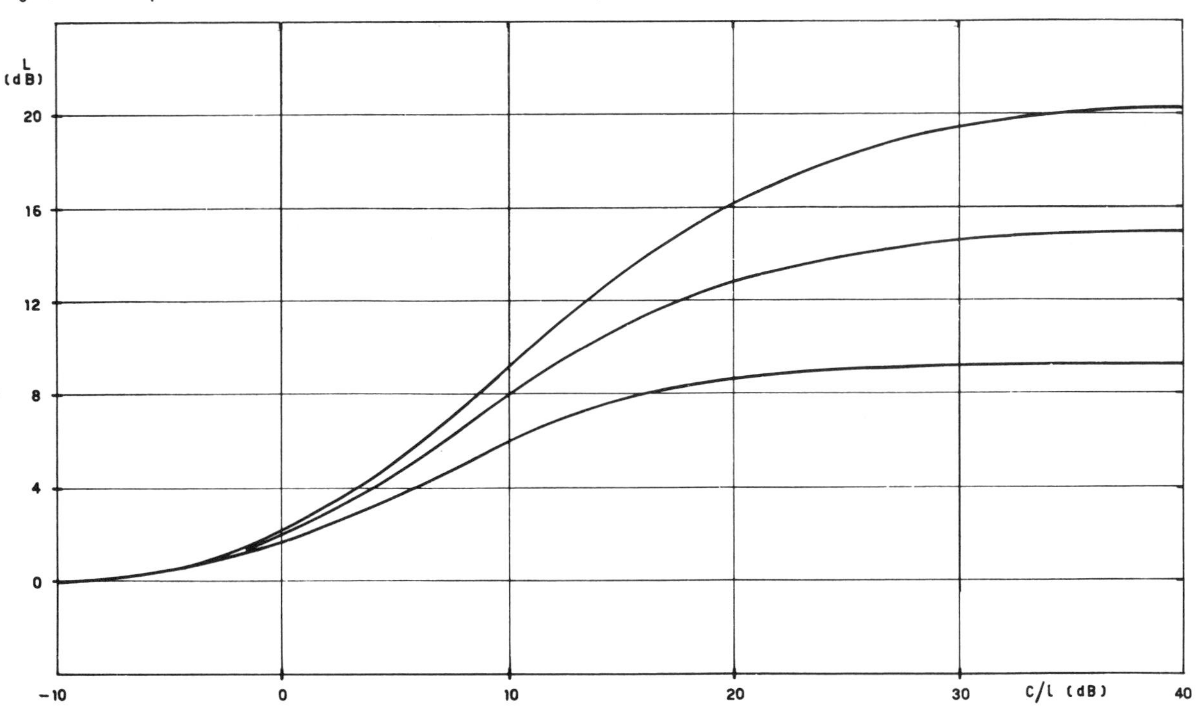

Assume, then, a Gaussian input autocorrelation function, expressed by

$$R_i(t) = C^2 \exp\left[-2\pi^2\sigma_c^2 t^2\right] \cos \omega_0 t \tag{15}$$

where σ_c^2 is the clutter spectrum variance. In this case the improvement factor in the presence of IF limiting, after some easy computations, results [7] in

$$I_1^{(l)} = \frac{\sum_{p=0}^{\infty} \frac{a_p b_p}{(1+\alpha)^{2p+1}}}{\sum_{p=0}^{\infty} \frac{a_p b_p}{(1+\alpha)^{2p+1}}\left(1 - \exp\left[-\frac{2\pi^2(2p+1)\sigma_c^2}{f_r^2}\right]\right)} \tag{16}$$

for the single canceller, and

$$I_2^{(l)} = 3\frac{\sum_{p=0}^{\infty} \frac{a_p b_p}{(1+\alpha)^{2p+1}}}{\sum_{p=0}^{\infty} \frac{a_p b_p}{(1+\alpha)^{2p+1}}\left\{3 - 4\exp\left[-\frac{2\pi^2(2p+1)\sigma_c^2}{f_r^2}\right] + \exp\left[-\frac{8\pi^2(2p+1)\sigma_c^2}{f_r^2}\right]\right\}} \tag{17}$$

for the double canceller.

As stated above, antenna motion is the prime cause of clutter spectrum spreading. In this case the spectrum may be expressed as a function of the number of pulses on target, N, considering that

$$\sigma_c = 0.264\frac{f_r}{N} \tag{18}$$

where f_r is the pulse repetition frequency. The results obtained have consequently been expressed as a function of N. From the ratio between the clutter spectrum rms value and the radar pulse repetition frequency, an equivalent number of pulses on target may be easily evaluated. Therefore, the results obtained in this section may be generalized to include any cause of clutter spectrum spreading. Equations (16) and (17) have been numerically evaluated for several numbers of pulses on target, and a comparison has been made with the improvement factors in linear conditions, given, for the single and the double canceller, by

$$I_1 = \frac{N^2}{1.89} \tag{19}$$

$$I_2 = \frac{N^4}{3.85}\,. \tag{20}$$

Figs. 2 and 3 show the losses in improvement factor due to limitation versus the ratio between input clutter and limitation level l. The computations, carried out for $N = 5$, $N=10$, and $N=20$, show a loss which increases as does the input clutter power. An asymptotic level is almost reached for $C/l=40$ dB. Losses in improvement factor, for single and double cancellers, have also been plotted in Fig. 4 as a function of the number of pulses on target for several values of C/l.

The asymptotic values of the losses are given by the curve $C/l=\infty$. It is evident that the losses reach higher values as the clutter amplitude and the value of the improvement factor increases.

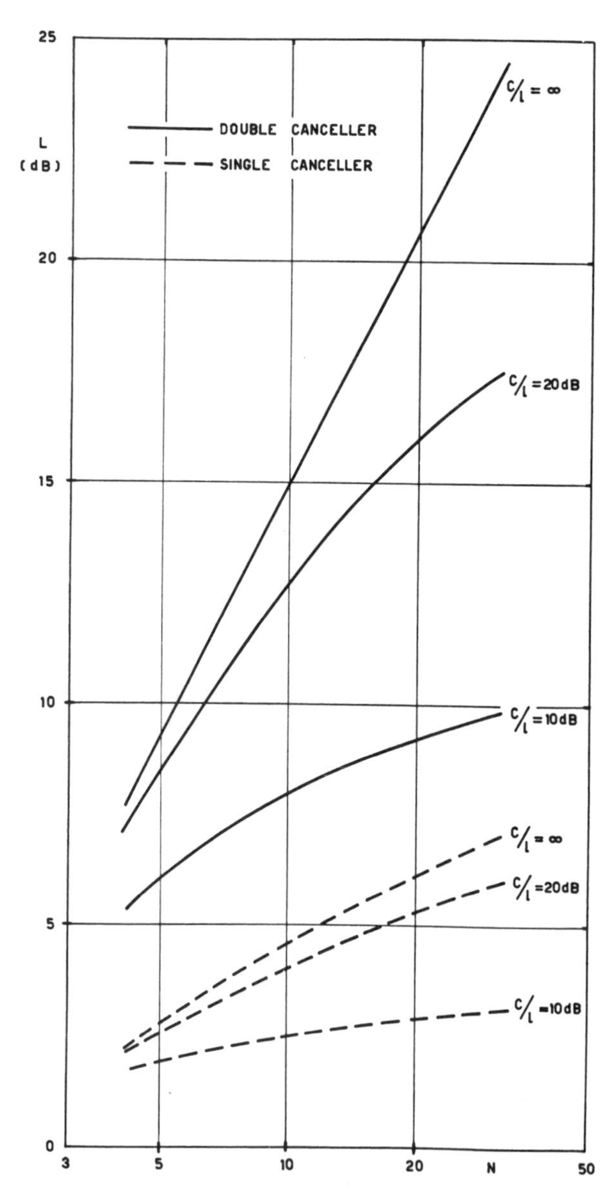

Fig. 4. Loss in improvement factor versus number of pulses on target, for several limiting conditions.

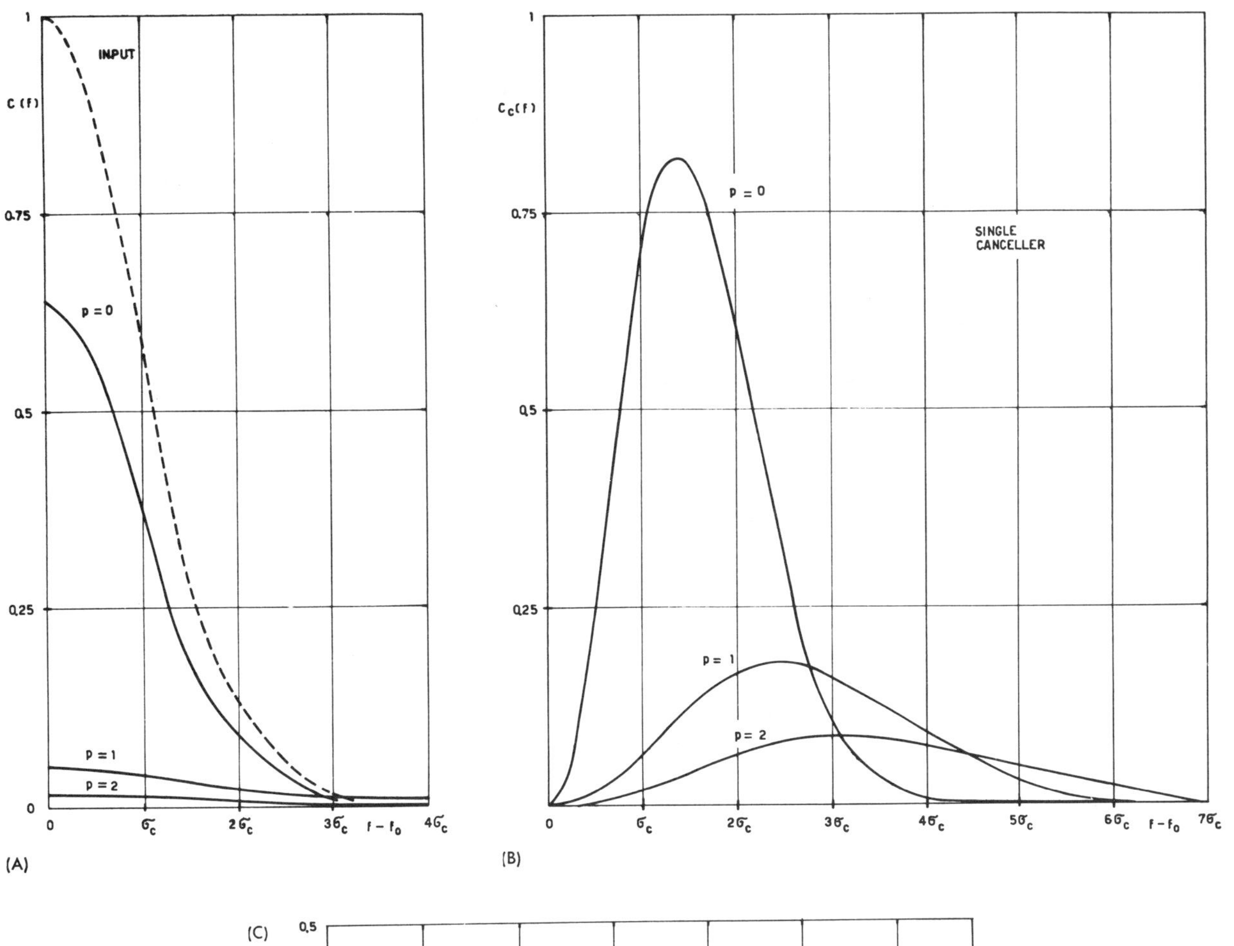

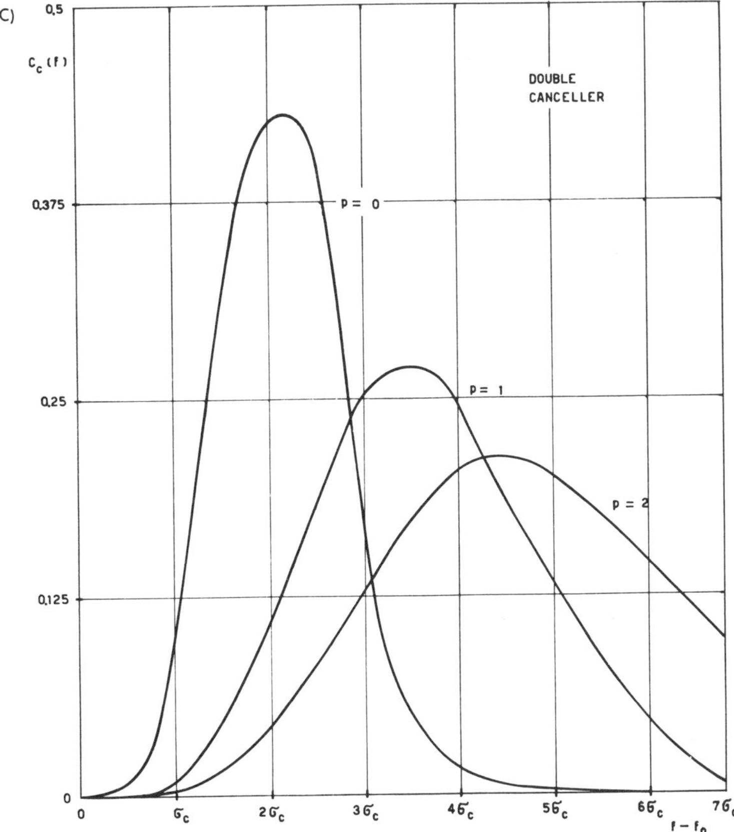

Fig. 5. First components of clutter spectrum (A) at the output of IF limiting stages, (B) at the output of a single canceller, and (C) at the output of a double canceller.

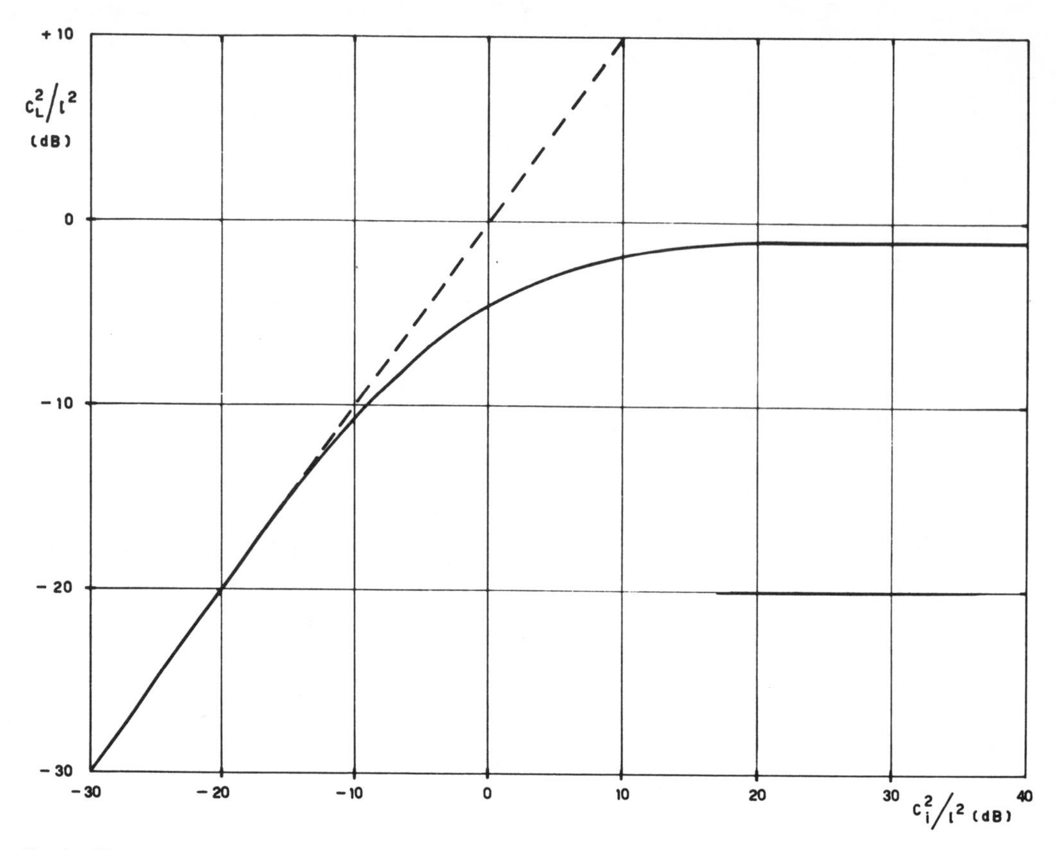

Fig. 6. Clutter power at the output of a smooth limiter as a function of input clutter.

The total of the losses due to limitation is by no means negligible. With 20 pulses on target, for example, losses of 5 dB are reached for the single canceller and 20 dB for the double canceller. The great variation introduced by a limitation on the efficiency of the cancellers, in spite of the fact that the input spectrum is little affected by the limitation [2], is due to the high-pass characteristics of the cancellers. In fact, for a Gaussian input spectrum, the spectrum at the output of a limiter is made up of an infinite number of Gaussian terms whose amplitudes decrease rapidly according to the coefficients of (13). But from (12) it should be noted that the variances of the successive terms increase according to the odd numbers [Fig. 5(A)]. There is, in other words, a progressive translation of power toward higher frequencies, and therefore the high filtering effect of the cancelling circuits gives more importance to terms of larger spreads.

This phenomenon is naturally more evident as the high-pass characteristic of the filter is accentuated. This explains why the losses for the double canceller are so much higher than for the single canceller. In Fig. 5(B) and (C) the first three components of the clutter $C_c(f)$, analogous to Fig. 5(A), have been drawn at the output of both the single and the double canceller. It is evident that terms of negligible importance before the canceller contribute greatly to the power of the clutter residues.

Clutter Residue Variation Due to a Limitation

The improvement factor, while describing completely the behavior of the canceller, does not describe the performance of the whole receiver. In fact, it is necessary to take into account not only the shape at the output of the limiting stages, but also the power level. As the clutter input level increases approaching the saturation level, the output power is compressed in comparison to the input. The output clutter power C_l is shown in Fig. 6 versus the input power C_i, where both powers are normalized to the limit level. For an input clutter power which is equal to the crossover level of the ideal sharp-limiting conditions, the output is 6 dB below the corresponding linear level.

As the input clutter increases, a level is reached about 1 dB lower than the level corresponding to ideal saturation. In order to get a single parameter which takes care of the effect of all parts of the receiver, a computation of the clutter residues at the output ot the receiver as compared to the clutter input level has been carried out.

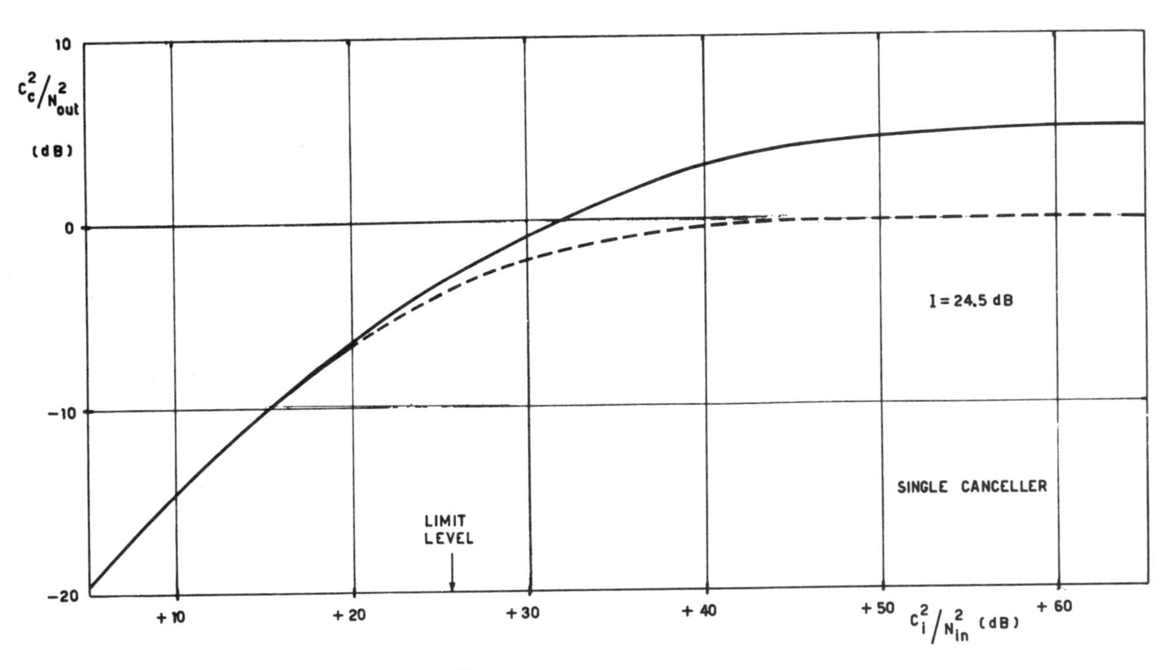

Fig. 7. Clutter residues at the output of a single canceller.

Fig. 8. Clutter residues at the output of a double canceller.

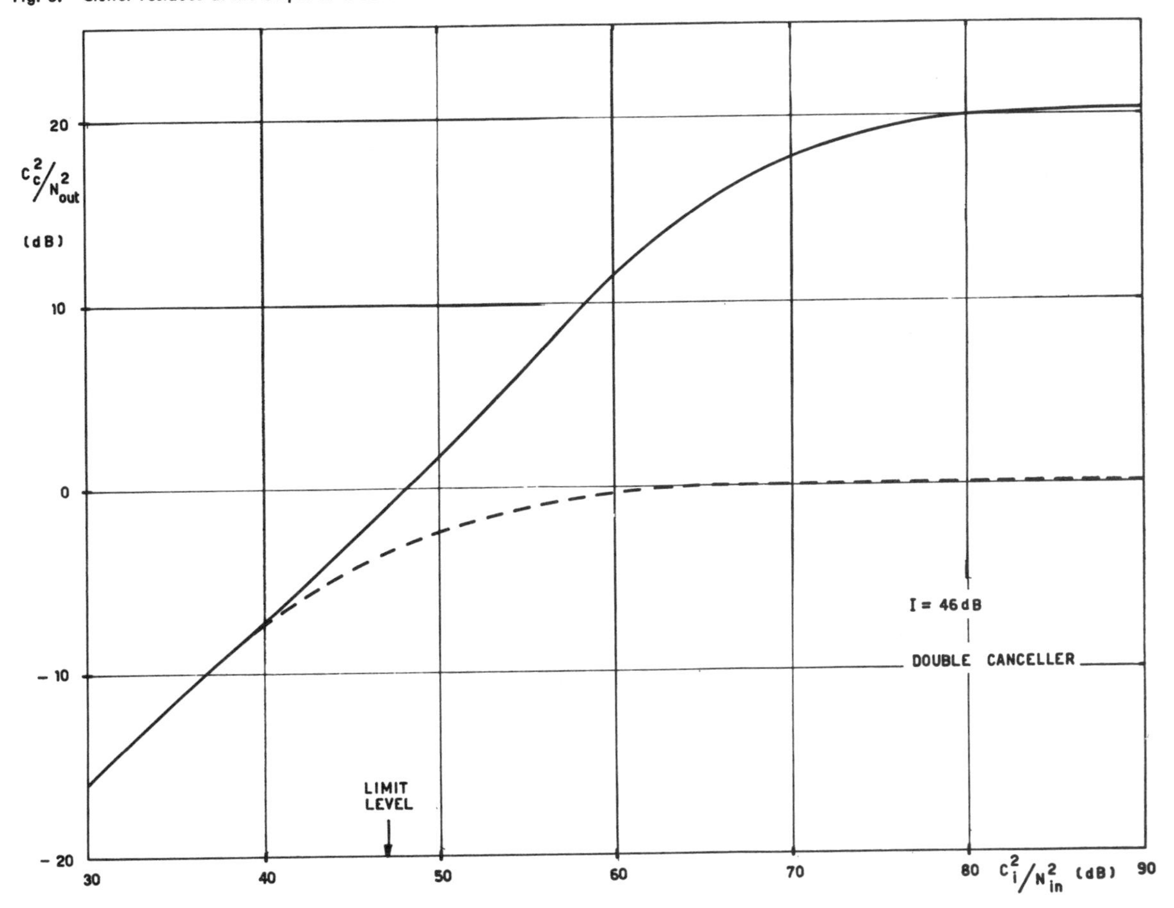

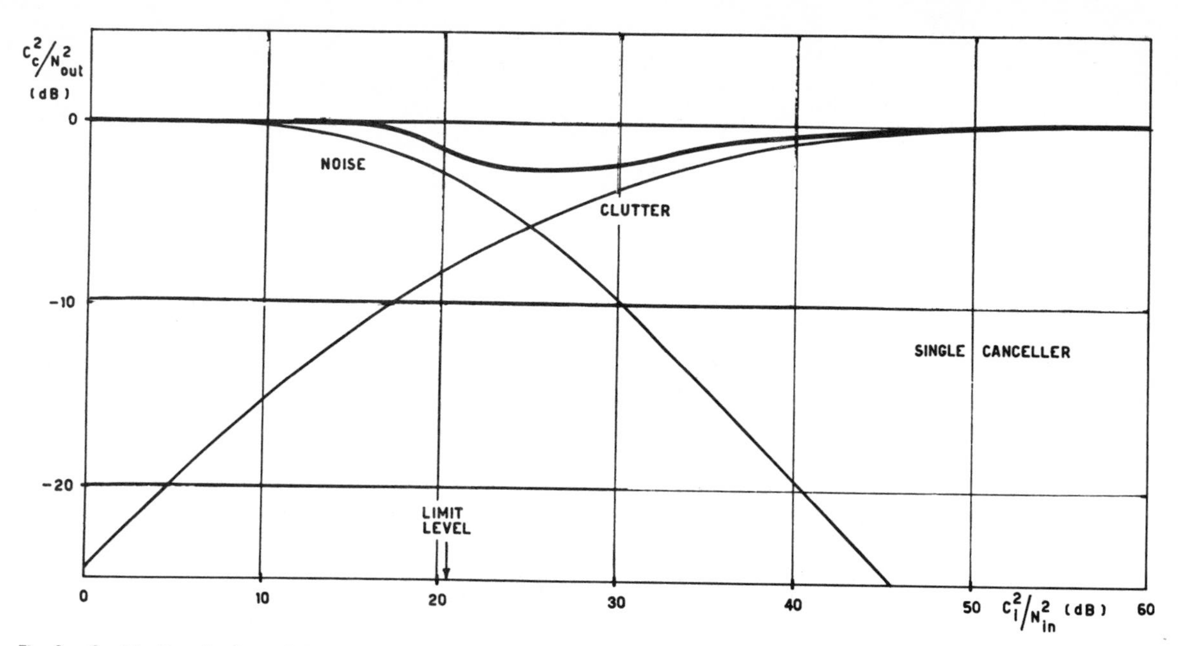

Fig. 9. Combination of noise and clutter residues at the output of a single canceller.

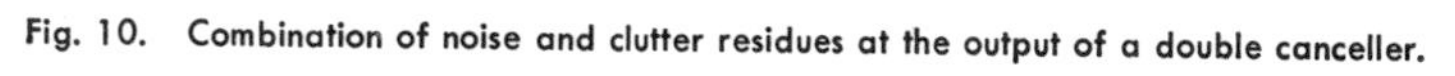

Fig. 10. Combination of noise and clutter residues at the output of a double canceller.

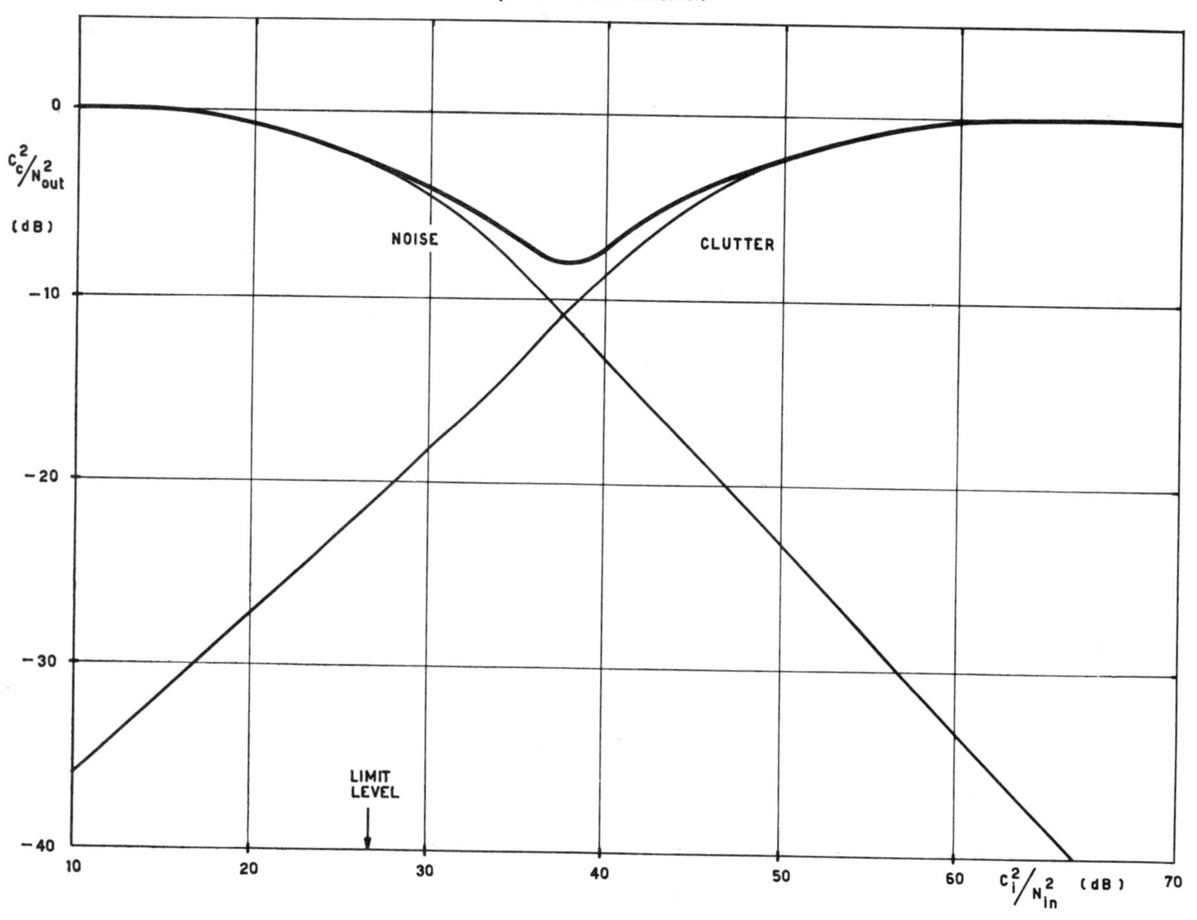

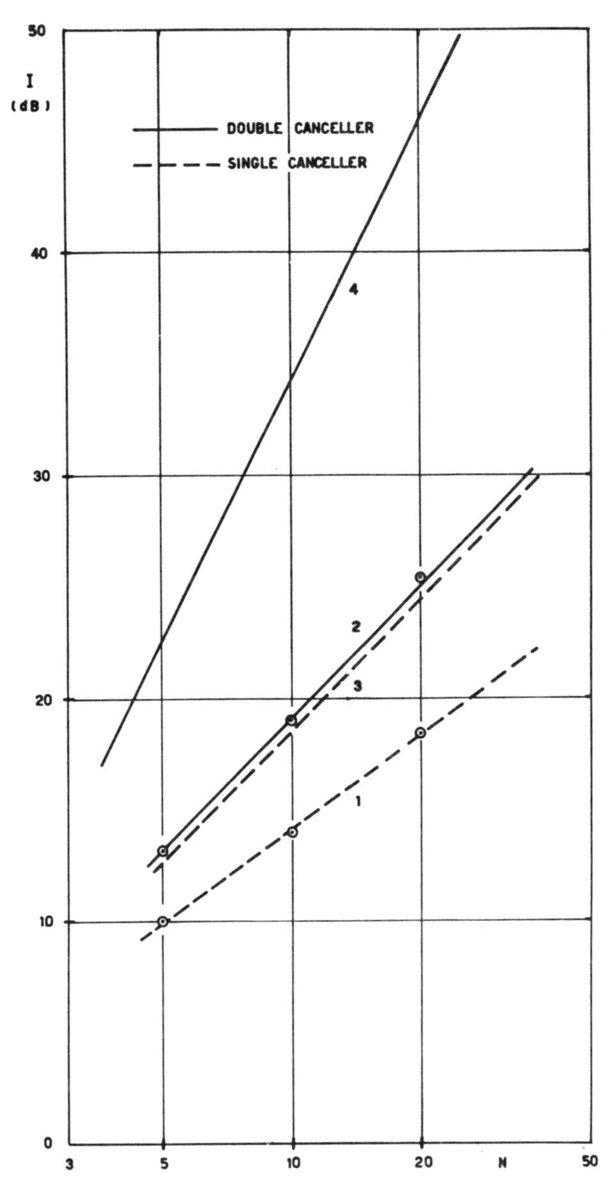

Fig. 11. Improvement factor versus number of pulses on target with linear and asymptotic conditions.

The dynamics of the output residue are greatly influenced by a limitation. In fact, besides the power saturation shown in Fig. 6, the output level is influenced by the large variation in the canceller's improvement factor.

However, the residue reaches an asymptotic level for input clutter high over limitation level. In Figs. 7 and 8 the residue power relative to a radar employing 20 pulses on target is shown versus input clutter power for the single and the double canceller, respectively. Also, dashed lines in these figures show the residue power evaluated, not taking into consideration the variation in improvement factor due to clutter spreading caused by a limitation. The limit levels chosen for both Figs. 7 and 8 are the improvement factor plus 1 dB above noise level. One decibel is added to take care of the power saturation of Fig. 6. The improvement factor is 24.5 dB for the single canceller and 46 dB for the double canceller [2]. It is obvious that the real situation is very different from that so far assumed by radar project engineers. In order to really obtain a clutter residue at the level of the output noise level the limit level must be chosen considering the great reduction of improvement factor caused by the nonlinearity. The asymptotic value of I is 19.5 dB for the single canceller and 25.8 dB for the double canceller. Therefore, a better choice of the limit levels is one decibel higher than this value. The corresponding situation of the residues, as compared to output noise, is that shown in Figs. 9 and 10. In these figures the sum of noise and clutter residue is also shown. This curve is to be compared with the practically constant false alarm rate situation which should be obtained if no deterioration of the improvement factor has occurred.

Conclusion

From the preceding analysis it is obvious that the traditional analysis of MTI receivers with linear conditions is greatly in error when a limitation occurs. The presence of a nonlinearity may degrade the canceller improvement factor by a large amount. A deterioration of 20 dB is not unusual in characteristic situations. Fig. 11 shows the asymptotic values of the improvement factor (curves 1 and 2) valid for infinite input clutter, for the single and double canceller, respectively. The same figure shows typical values of the improvement factor evaluated for single (curve 3) and double (curve 4) cancellers with linear conditions. The results are expressed as a function of the number of pulses on target, but may be readily extended to any cause of clutter spectrum spreading.

Finally, because of the deterioration of the improvement factor, the clutter residue level is very sensitive to clutter input power, and the system is very far from a constant false alarm rate situation.

REFERENCES

[1] M. I. Skolnik, *Introduction to Radar Systems.* New York: McGraw-Hill, 1962.
[2] D. K. Barton, *Radar System Analysis.* Englewood Cliffs, N. J.: Prentice-Hall, 1964.
[3] G. Grasso, "Influenza della rotazione d'antenna su MTI non lineare in presenza di clutter concentrato," *Alta Frequenza,* vol. 27, March 1968.
[4] W. W. Shrader, "Recent results in MTI theory and practice" (to be published).
[5] R. F. Baum, "The correlation function of smoothly limited Gaussian noise," *IRE Trans. Information Theory,* vol. IT-3, pp. 193–197, September 1957.
[6] R. S. Berkowitz, *Modern Radar.* New York: Wiley, 1965, pt. 6, ch. 2.
[7] G. Grasso and P. F. Guarguaglini, "Performances for distributed clutter with IF limiting," Selenia S. p. A., Rome, Italy, Internal Rept. RT-67/465, April 12, 1967.

Giancarlo Grasso was born in Taranto, Italy, on August 18, 1940. He received the Doctor's degree in electrical engineering from the University of Rome, Rome, Italy, in 1963. In 1965 he attended postgraduate courses on radar and telecommunications at the Consiglio Nazionale delle Ricerche.

In 1963 he joined the Advanced System Group of Selenia S.p.A., Rome, where he has been chiefly engaged in the study of radar detection devices. He has been mainly concerned with signal processing, and in particular he has been working on MTI and ECCM design and analysis. From 1964 to 1965 he served in the Italian Air Force Technical Service. Also, since 1963 he has been at the Institute of Electronics of the University of Rome as a Teaching Assistant.

Pier Francesco Guarguaglini (M'63) was born in Donoratico, Italy, on February 25, 1937. He received the Doctor's degree in electrical engineering from the University of Pisa, Pisa, Italy, in 1961.

In 1961, he held a combined teaching and research assistantship in electronics at the University of Pisa. His research dealt with dc amplifiers. From 1961 to 1963 he served as a Technical Air Force Officer. In 1963 he joined the Advanced Systems Group of Selenia S.p.A., Rome, Italy, where he has worked on electronic countermeasures and electronic counter-countermeasures, signal processing, sequential detection, and radar system analysis. From 1965 to 1966 he was on leave of absence at the University of Pennsylvania, Philadelphia, where he studied the effects of interference on an interferometer tracking system. Also, since 1962 he has been a Teaching Assistant in radar systems at the University of Rome.

4.2 ELIMINATION OF BLIND VELOCITIES OF MTI RADAR BY MODULATING THE INTERPULSE PERIOD
P. Prinsen
IEEE Transactions on Aerospace and Electronic Systems, Vol. AES-9, No. 5, September 1973, pp. 714-724.

Elimination of Blind Velocities of MTI Radar by Modulating the Interpulse Period

PETER J.A. PRINSEN
Physics Laboratory TNO
The Hague, The Netherlands

Abstract

The performance of a coherent MTI is determined by two sets of parameters: the interpulse periods and the weighting coefficients. The latter are expressed in terms of the former so as to maximize clutter attenuation. Interpulse periods are optimized so as to remove blind velocities. The many local optima of the interpulse periods have improvement factors that show differences of up to 6 dB for a 5-pulse canceler. A computer program has been developed that combines random search and the gradient method to produce points that combine a high value of the improvement factor with good blind speeds performance.

I. Introduction

A. General

In a pulsed-radar moving target indicator (MTI), echoes with zero Doppler shift (clutter) are suppressed, while other Doppler shifted echoes, reflected by a moving target, are passed. In practical situations, the Doppler shift is small with respect to the inverse of the radar pulse width. Therefore, for each range cell a train of echoes (if a target is present) from two or more subsequent coherent radar pulses is processed in the MTI filter to discriminate between stationary and moving targets. This processing is known as high-pass filtering, and the processor acts as a sampled data system for which the radar transmitter is the sampler with sample period T, where T is the radar interpulse period.

Using digital techniques, the MTI processor can be implemented as a digital filter with many degrees of freedom.

As sampled data systems with uniform interpulse periods have a periodic frequency characteristic, the MTI filter not only rejects clutter echoes, but also echo signals with a Doppler shift equal to multiples of the pulse repetition frequency (PRF). Thus, "blind speeds" are introduced by this type of MTI system. These blind speeds or nulls, in the frequency characteristic, can be avoided by varying the interpulse periods. As an analytical description is lacking, a numerical optimization method is used to achieve optimum interpulse periods. A method is developed which combines an analytical optimization of the filter coefficients with the numerical optimization of the interpulse periods.

B. MTI Processor

This study is confined to radar systems with step scan antenna (phased array antenna). During a number of periods, say N, the beam can be fixed in a particular direction. At the beginning of each period a radar pulse is transmitted. During one period the received signal, as it appears at the output of the last mixer, is sampled, converted from analog to digital form (ADC), and stored in a digital register (Fig. 1). The sampling rate of the ADC is related to the range resolution. The length of the register is related to the observed range. The duration of the shortest period between two radar pulses determines the maximum unambiguous range. After the Nth period,[1] a filter operation is performed for each range cell on the information that is gathered from that range cell. Thus, for each range cell a weighted sum y_k is calculated of the N consecutive radar echoes reflected from that cell:

$$y_k = \sum_{n=0}^{N-1} w_n x_{nk} \qquad (1)$$

[1] As far as implementation is concerned, it should be noted that the adder in Fig. 1 does not necessarily operate on N x-registers in parallel. The x-registers essentially are redundant, apart from possible buffer requirements due to computing time. Moreover, only one multiplier is required.

Manuscript received July 6, 1972.

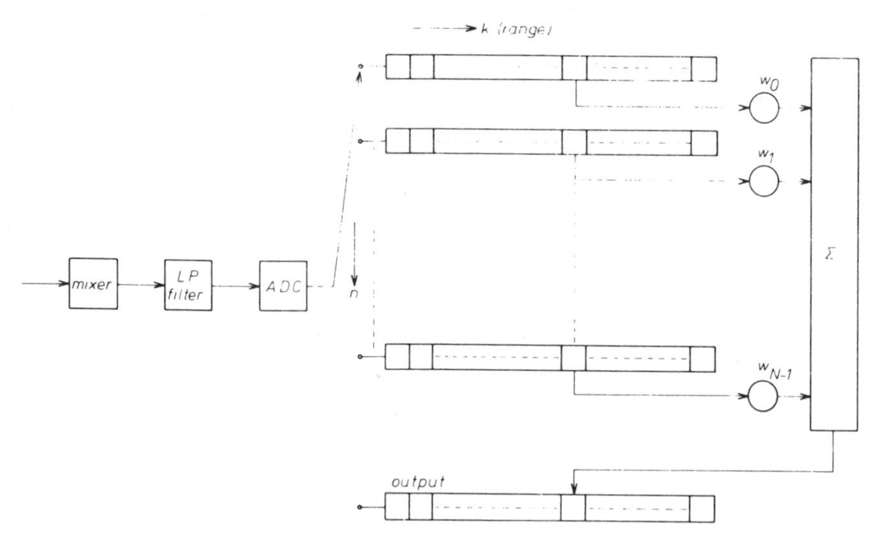

Fig. 1. MTI processor with nonrecursive high-pass digital filter (single channel).

with

x_{nk} = echo signal of the nth radar pulse, reflected from the kth range cell

w_n = nth weighting coefficient

y_k = weighted sum of all echoes from the kth range cell.

The adder in Fig. 1, together with the weighting coefficients, constitutes a nonrecursive digital filter. Obviously, the filter output is not affected by transient responses as the output signal is observed only after reception of the entire sequence of N radar echo pulses. A quadrature channel is assumed to complete the MTI processor (Fig. 2) (c.f. [2, p. 474]), although this has no effect on the system parameters considered in this paper (e.g., improvement factor).

As we are not interested in an asymmetric frequency characteristic, all coefficients w_n are real. When adjusting the interpulse periods $T_1, T_2, \cdots, T_{N-1}$ between the N radar pulses, the following requirements should be satisfied:

1) $T_{\min} \geqslant T_r$. Here T_r is the time corresponding to the specified maximum unambiguous range of the radar system. $T_{\min}$ is the smallest time in the sequence $T_1, \cdots, T_{N-1}$. In the rest of this paper, $T_{\min}$ will be denoted by T.

2) A detection decision is made after N pulses for one direction. The time on target should be bounded. The time on target equals

$$\sum_{i=1}^{N-1} T_i .$$

Another argument for this requirement is discussed in Section II-C.

3) The frequency characteristic should have a rejection band around zero frequency, while the passband should be as flat as possible, according to a criterion which will be specified in one of the following sections. The passband (velocity region) extends to some multiple of $1/T$ (e.g., 10 or 20 times $1/T$).

Fig. 2. MTI processor with phase quadrature channel.

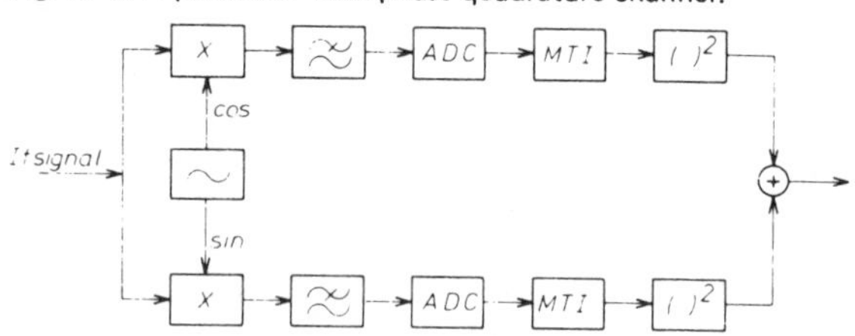

C. Frequency Characteristic

The response $y(f)$ of the MTI system on an echo signal from a target with Doppler shift f and unity amplitude can be represented by substituting x_{nk} in (1) by $e^{-j2\pi f t_n}$:

$$y(f) = \sum_{n=0}^{N-1} w_n e^{-j2\pi f t_n} \qquad (2)$$

with[2]

$$t_n = \sum_{m=1}^{n} T_m .$$

The ideal MTI filter has a high-pass frequency characteristic. This ideal characteristic has a rejection band between zero frequency and a frequency f_c which is specified by the clutter model. This frequency band will be referred to as the clutter region. The frequency region, corresponding to the Doppler frequencies of moving targets, that must be visible for the radar will be called the velocity region. The transition region between clutter region and velocity region is of no interest.

In an actual uniformly sampled data system, the high-pass requirements cannot be satisfied because the

[2] If the upper bound of the running index of a series representation is smaller than the lower bound (e.g., n=0), the sum by definition equals zero.

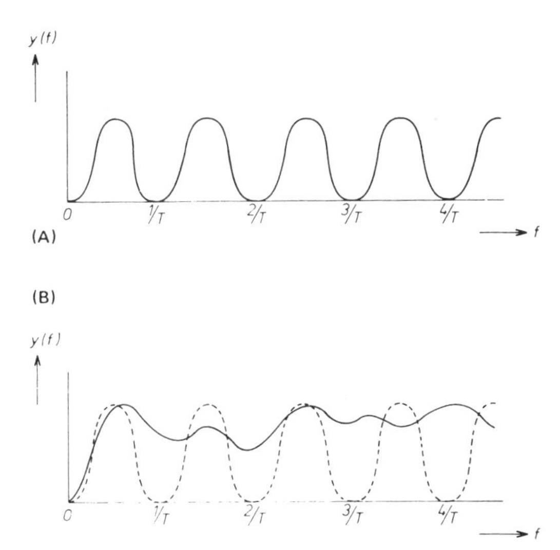

Fig. 3. Frequency characteristic of MTI filter. (A) Uniform PRF. Blind velocities occur at multiples of $1/T$. (B) Staggered PRF.

sampling rate should be higher than the maximum expected Doppler frequency. This requirement will be in conflict with the required maximum unambiguous range.

A PRF which is sufficiently low for the required maximum unambiguous range will produce blind velocities in the MTI filter characteristics; i.e., apart from $f = 0$ there is a rejection notch at each frequency which is a multiple of the PRF [Fig. 3(A)]. By staggering the PRF, i.e., by using unequally spaced radar pulses, the undesired rejection notches can be removed while keeping the attenuation in the stopband around $f = 0$ sufficiently high [Fig. 3(B)].

We define a unitary frequency band B_1 as

$$B_1 = \frac{1}{T} . \tag{3}$$

In a nonuniformly sampled system, the minimum interpulse period is T, so B_1 is related to the maximum unambiguous range. Let the upper bound of the velocity region be denoted by B_u and the lower bound by B_l. We assume that

$$B_l > f_c . \tag{4}$$

If a Gaussian clutter power spectrum with variance σ^2 [1] is assumed, the frequency f_c is closely related to σ.

Example (C.f., Fig. 4)

An X-band radar having an unambiguous range of 50 km has a first blind speed at about 200 km/h if equispaced pulses are used. If targets having a speed of twice the velocity of sound should be detectable (say 2000 km/h), then $B_u = 10B_1$. A reasonable value of σ is 4 km/h [1], which is 2 percent of B_1 [3].

Let the filter coefficients be determined so as to produce maximum attenuation of the clutter power for equispaced radar pulses. These coefficients can be said to be optimum in the equidistant case according to some criterion. The optimum coefficients depend on the clutter spectrum.

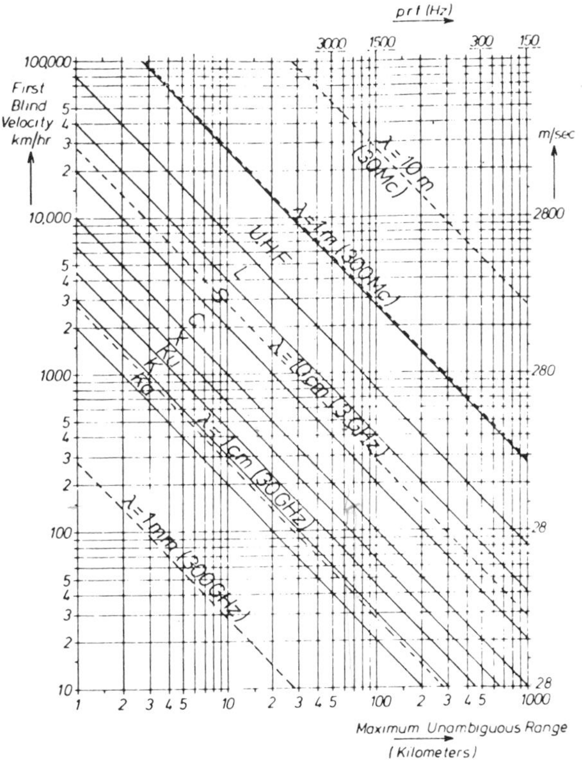

Fig. 4. First blind velocity as a function of the maximum unambiguous range and the radar frequency. The diagonal lines represent the boundaries of the different radar frequency bands.

Now let the interpulse periods be staggered, while the coefficients are kept fixed. This causes a deterioration of the clutter attenuation, mainly because the filter coefficients are no longer optimum in the new situation. The optimum coefficients now depend on both clutter spectrum and interpulse periods and can be found as the solution to an eigenvalue problem.

First the solution for optimum filter coefficients in the case of nonuniform periods will be derived in the next sections. The criterion according to which the coefficients are optimum concerns only a part of the frequency domain, viz., the clutter region. Next a performance index is defined, concerning the velocity region. This performance index is optimized by means of a numerical algorithm (gradient method) in such a way that at each stage of the procedure the optimality conditions for maximum clutter attenuation are kept satisfied.

II. Optimization Of The Weighting Coefficients

A. The Improvement Factor and the SCR gain

A computer program has been developed by which the interpulse periods are optimized by a numerical method. The optimization procedure repeatedly calculates a performance index and determines what corrections should be applied to the variables (the interpulse periods) to improve the performance index.

As a starting point for the definition of an appropriate performance index, as well as for the definition of the optimum filter, the gain in the signal-to-clutter power ratio as a function of frequency can serve. This gain will be referred to as SCR gain or S^2. It is closely related to the improvement factor. The improvement factor I is defined as the target-to-clutter power ratio at the output of the filter, divided by the target-to-clutter power ratio at the input, averaged over all target speeds.

If

P_{ti} = target power at the input of an MTI filter
P_{to} = target power at the output of an MTI filter
P_{ci} = clutter power at the input of an MTI filter
P_{co} = clutter power at the output of an MTI filter,

then

$$I = \overline{\left(\frac{P_{to}/P_{co}}{P_{ti}/P_{ci}}\right)} \tag{5}$$

$$= \frac{\overline{(P_{to}/P_{ti})}}{(P_{co}/P_{ci})} \quad \text{(averaged over all targets speeds.)}$$

$$\doteq \text{mean SCR gain.} \tag{6}$$

In this expression P_{co} and P_{ci} refer to integrals of the clutter power spectrum over all clutter frequencies at the output and input, respectively, of the filter. P_{to} and P_{ti} refer to values of the respective power of a target signal at a particular frequency. The ratio of these values must be averaged over all target speeds, as indicated in the formula.

A clutter signal can be characterized by its power spectrum. The total power is obtained by integrating its power spectrum. This can be approximated by summing over a finite number of discrete frequencies.

If, in general, a (clutter) signal with power spectrum $c^2(f)$ is applied to a system with transfer function $y(f)$ (see Fig. 5), its output power spectrum is

$$c^2(f)y(f)y^*(f) = c^2(f)|y(f)|^2 \tag{7}$$

where * denotes a complex conjugate.

The output clutter power is approximated by summing over a number of discrete frequencies $f_{-K}, f_{-K+1}, \cdots, f_K$:

$$P_{co} = \Delta f \sum_{k=-K}^{K} c^2(f_k)|y(f_k)|^2 \tag{8}$$

$$= 2\Delta f \sum_{k=0}^{K} c_k^{\,2}|y_k|^2 \tag{9}$$

Where $c^2(f) = 0$ for $|f| > f_K$, $c_0^{\,2} = \frac{1}{2}c^2(0)$, $c_k^{\,2} = c^2(f_k), k = 1, \cdots, K$, $y_k = y(f_k)$ for $k = 0, \cdots, K$, and Δf = frequency increment. f_K is the same frequency as f_c, the upper bound of the clutter region (Section I-C).

The input power is

$$P_{ci} = 2\Delta f \sum_{k=0}^{K} c_k^{\,2}. \tag{10}$$

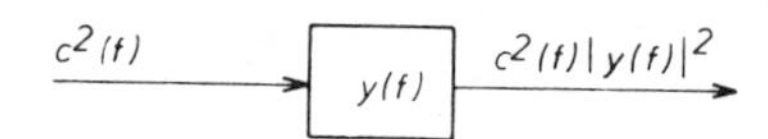

Fig. 5. Power spectrum of input and output signal.

Thus, the denominator of (6) becomes

$$\frac{P_{co}}{P_{ci}} = \frac{\sum_{k=0}^{K} c_k^{\,2}|y_k|^2}{\sum_{k=0}^{K} c_k^{\,2}}. \tag{11}$$

This ratio represents the reciprocal clutter attenuation. The filter coefficients are chosen so as to minimize this ratio (see next section).

The ratio in the numerator of I is easily obtained as the power transfer function of the filter. The mean of this ratio may again be approximated by a finite sum:

$$\overline{\left(\frac{P_{to}}{P_{ti}}\right)} = \frac{1}{L}\sum_{i=0}^{L} |y(f_i)|^2 \tag{12}$$

where L is defined by the velocity region. The improvement factor becomes

$$I = \frac{\frac{1}{L}\sum_{i=1}^{L} |y(f_i)|^2}{\left(\sum_{k=0}^{K} c_k^{\,2}|y_k|^2 \Big/ \sum_{k=0}^{K} c_k^{\,2}\right)} \tag{13}$$

The SCR gain S^2 can be represented by

$$S^2(f) = \frac{(P_{to}/P_{ti})}{(P_{co}/P_{ci})}$$

$$= \frac{|y(f)|^2}{\left(\sum_{k=0}^{K} c_k^{\,2}|y_k|^2 \Big/ \sum_{k=0}^{K} c_k^{\,2}\right)} \tag{14}$$

B. Optimum Filter

Let the transfer function of the MTI filter be represented by $y(f)$ and the power spectrum of the clutter by $c^2(f)$ $[c(f) \geq 0$ for all $f]$. The clutter spectrum at the output of the filter is (see Fig. 5)

$$c^2(f)y(f)y^*(f) = z(f)z^*(f) \tag{15}$$

with

$$z(f) \triangleq c(f)y(f). \tag{16}$$

Let these functions be sampled as in Section II-A at the frequencies f_k, $k = 0, \cdots, K$ $[c(f) = 0$ for $f > f_K]$. The samples are represented by the vectors $\mathbf{z}' = (z_0, \cdots, z_K)$ and

$y' = (y_0, \cdots, y_K)$. A prime denotes a transpose of a vector or matrix. Then

$$z = Cy \tag{17}$$

with

$$C \triangleq \begin{pmatrix} c_0 & \cdots & 0 \\ \vdots & \ddots & \vdots \\ 0 & \cdots & c_K \end{pmatrix} \tag{18}$$

where $c_0 = (1/\sqrt{2})c(0)$ and $c_k = c(f_k)$ for $k = 1, \cdots, K$. Note that $C' = C^* = C'^* = C$. y is a vector function of $w_0, \cdots, w_{N-1}$ [c.f., (2)]:

$$y_k = \sum_{n=0}^{N-1} w_n e^{-j2\pi f_k t_n}, \quad k = 0, \cdots, K \tag{20}$$

or

$$y = Ew \tag{21}$$

with

$$w' = (w_0, \cdots, w_{N-1}) \tag{22}$$

and

$$\{E\}_{ij} = e^{-j2\pi f_i t_j}. \tag{23}$$

Applying the least-squares criterion, the expression to be minimized by selecting $w_0, \cdots, w_{N-1}$ is

$$\begin{aligned} F(w) &= z'^* z \\ &= y'^* C C y \\ &= w'^* E'^* C C E w. \end{aligned} \tag{24}$$

However, in order to obtain a nontrivial solution, it is required as a side constraint that $w'^* w > 0$. This can be formulated as

$$G(w) \triangleq w'^* w = w^2. \tag{25}$$

The optimization problem can now be stated as follows:

$$\min_w F(w) = w'^* E'^* CCEw \tag{26}$$

subject to

$$G(w) = w'^* w = w^2. \tag{27}$$

This problem will be solved using the Lagrange-multiplier technique.

As all components of w are real (Section I-B), we have

$$\begin{aligned} \nabla F &= E'^*CCEw + (E'^*CCE)'w \\ &= \{E'^*CCE + (E'^*CCE)'\}w \\ &= 2\mathrm{Re}\{E'^*CCE\}w \\ &\triangleq 2E_r w. \end{aligned} \tag{28}$$

Consequently, as ∇F is independent of Im $\{f\}$, we can, instead of $F(w)$, minimize a function $H(w)$ that is defined as

$$H(w) \triangleq w' E_r w \tag{29}$$

so that we have

$$\min_w H(w) = w' E_r w \tag{30}$$

subject to

$$G(w) = w'w = w^2. \tag{31}$$

Now, requiring

$$\nabla H - \lambda \nabla G = 0 \tag{32}$$

yields

$$E_r w - \lambda w = 0, \tag{33}$$

which is an eigenvalue problem.

The solution yields N distinct real eigenvectors w_i, each of which is associated with an eigenvalue λ_i. As E_r is symmetric, the λ_i are real but not necessarily distinct. Combining (29), (31), and (33) yields

$$H(w) = \lambda w'w = \lambda w^2. \tag{34}$$

From this expression we see that we are looking for an eigenvector that is associated with the smallest eigenvalue (c.f., [4], [5]).

Conclusion

1) The elements of an eigenvector associated with the smallest eigenvalue of the matrix E_r are the weighting coefficients of the filter that maximizes the clutter attenuation.

2) The smallest eigenvalue is a measure for the maximum clutter power attenuation that can be obtained for a given matrix E_r, which in turn depends on clutter model and interpulse periods. A small eigenvalue corresponds to a high clutter attenuation.

Some Numerical Aspects of the Eigenvalue Problem

1) To calculate the eigenvector that is associated with the smallest eigenvalue, a modified power method is used. In this method an iteration consists of solving a system of linear inhomogeneous equations instead of a matrix-vector multiplication.

2) If the smallest eigenvalue λ_1 has multiplicity k, there are k orthogonal eigenvectors $u_1, \cdots, u_k$ associated with λ_1. In this case any vector w that is a linear combination of u_1 through u_k is an eigenvector and satisfies if (31) is satisfied.

3) The (modified) power method will fail if there are at least two eigenvalues that have maximum (minimum) magnitude and different sign. However, this case is never met. This can be seen as follows.

As z'^*z is real and positive, we can write

$$z'^*z = w'E'^*CCEw > 0 \tag{35}$$

and

$$\mathrm{Im}\{w'E'^*CCEw\} = 0 \tag{36}$$

for any w. So,

$$\begin{aligned} w'E'^*CCEw &= w'\mathrm{Re}\{E'^*CCE\}w \\ &= w'E_r w > 0 \end{aligned} \tag{37}$$

for any w.

From this expression we see that the matrix E_r is positive definite. Hence, all eigenvalues are positive.

C. The Relation Between σ, $\sum_i T_i$, f_c, and I

If σ increases with respect to B_1, the improvement factor decreases, because in this case more and more clutter spectral components enter into a region with less relative attenuation.[3] Therefore, it should be noticed which parameters govern the mapping of the velocity axis on the Doppler frequency axis.

Radar Carrier Frequency

Let the carrier frequency of the radar signal be denoted by f_c. Then the Doppler shift f for a target with velocity v is

$$f = 2\frac{v}{c} f_c. \tag{38}$$

where c = speed of light. So, if the standard deviation of the clutter velocity is denoted by σ_v, then the clutter spectrum standard deviation σ is

$$\sigma = 2\frac{\sigma_v}{c} f_c. \tag{39}$$

Thus it is seen that, with respect to I, a small value of f_c is preferred.

Pulse Spacings

The improvement factor also depends on the interpulse periods, because these periods provide a scaling of the frequency axis of Fig. 3 with respect to the function $y(f)$, while the clutter spectrum is fixed. So decreasing T or $\sum_{i=1}^{N} T_i$ is equivalent to decreasing σ.[3] Thus, requiring a large unambiguous range is seen to be in conflict with a large improvement factor requirement.

[3] It can be seen that the rows of the matrix E'^*CCE of (24) tend more and more to linear dependence if σ and T decrease. In this case the condition of the matrix is said to degrade. The determinant D tends to zero. Then we have

$$D = \prod_{i=1}^{N} \lambda_i \to 0.$$

On the other hand, we have

$$\text{trace}\left\{E'^*CCE\right\} = N \sum_{k=1}^{K} c_k^2 = \text{constant}$$

$$= \sum_{i=1}^{N} \lambda_i.$$

From this it follows that the smallest eigenvalue tends to zero, corresponding with an increasing improvement factor. For the same reason, it follows that if T is kept constant, the stagger rate should be as small as possible. Thus, for a given unambiguous range, the best improvement factor is attained by uniform pulse spacings.

Conclusion

Increasing T or $\sum_i T_i$, as well as increasing f_c, is equivalent to increasing σ. Large values of σ cause small values of the improvement factor.

III. Optimization of the Interpulse Periods

A. The Performance Index

Using the definitions of Section II-A, there are several possibilities to define a performance index. Two extreme alternatives are:

1) the improvement factor I(= mean SCR gain) [c.f., (13)]
2) the minimum value of S^2 in the velocity region [c.f., (14)].

The first alternative has the disadvantage that narrow rejection notches have little effect on the performance index. The disadvantage of the second alternative is that the performance index is determined by one point of the frequency characteristic. Neither of these alternatives exactly represents the intuitive criterion for judging an MTI system.

To meet these objections the performance index P could be defined as

$$P = \int_{B_l}^{B_u} \frac{1}{S^{2M}}\, df \tag{40}$$

where B_u and B_l are upper and lower bounds, respectively, of the velocity region (Section I-C).

If a high value is chosen for M (e.g., 8), only points of the frequency characteristic which lie in a narrow strip, the height of the actual minimum, contribute appreciably to the integral. When the integral (40) is approximated by a finite sum of samples in the frequency domain, a drawback of this definition becomes apparent. Due to the high value of the exponent of the integrand, the associated harmonics give rise to the necessity of a high sampling rate. A sampling rate which is too small will cause the numerical optimization scheme to force the actual minima of S^2 between the grid points of the frequency axis, so that the gradient points into the direction of the best fit into the frequency grid. Increasing the sampling rate is an expensive remedy. The eventual definition of the performance index is, therefore, the sum of the Mth power of the reciprocal minima of the SCR gain:

$$P = \sum \frac{1}{S_{\min}^{2M}}. \tag{41}$$

The sum is to be taken over all minima in the velocity region. Starting with an SCR gain characteristic which is sampled at twice the Nyquist rate, the minima of S^2 can be found quite efficiently using a one-dimensional minimum-seeking scheme.

It should be noted that the performance index, defined above, and the improvement factor have an important factor in common, viz., the clutter attenuation. A high clutter attenuation corresponds to a good performance index, as well as a high improvement factor.

Computer results show differences in the improvement factor of up to 6 dB for different local optima in the interpulse period space, due to 6-dB differences of the corresponding minimum eigenvalue. From this it should be clear what the aim of the optimization procedure must be. It must investigate the interpulse period space and produce points that combine high minimum values of the SCR gain characteristic with high values of the improvement factor. This is accomplished by the optimization procedure that will be introduced in the next subsection. It combines a random search and a gradient algorithm to locate the minima of (41) in the interpulse period space.

B. Optimization Procedure

In the above the performance index is defined. The performance index is a function of the $N-1$ interpulse periods. A computer program has been developed that minimizes this function using a gradient algorithm. Since the function has many local minima, the gradient algorithm is combined with a random search. At random, a starting point is generated for the gradient algorithm to start. As soon as an optimum has been found, another starting point is generated at random. Each time the new optimum is compared with the best optimum thus far. This process is stopped as soon as the computing time exceeds a preset value.

On the randomly chosen starting points the following restriction is imposed:

$$T \leq T_i \leq 1.5T, \quad i = 1, \cdots, N-1. \tag{42}$$

By (42), boundaries are defined for maximum unambiguous range and time on target. One of the variables T_i is always set to a constant T, which has the same value in all optimization searches. The relevant variable is kept constant during one trial. This makes all optima that have been found comparable with respect to maximum unambiguous range. An attendant advantage is the fact that the dimension of the space, defined by the state variables T_i, is decreased by 1 to N-2. Actually, the constant T is normalized to 1 in the program.

According to the gradient algorithm, the gradient must be evaluated to find a direction in which a one-dimensional search is done. The gradient is approximated by differences. In the point that results from the one-dimensional search the gradient is computed again and the procedure is repeated. This process is continued until the improvement obtained by the one-dimensional search is less than a present value. Many variations on this scheme are possible.

In the actual version, the following method is used for the one-dimensional search. At some distance from a given point, in the direction determined by the gradient, the function is evaluated. If this appears to be an improvement, the step size is increased and, at a distance equal to the increased step size, the function is evaluated again, and so on. As soon as a new point appears to be a deterioration, an optimum obviously is enclosed on a segment equal to the last two step sizes. This segment is the starting interval of an optimization procedure according to the golden section method, which can be summarized as follows.

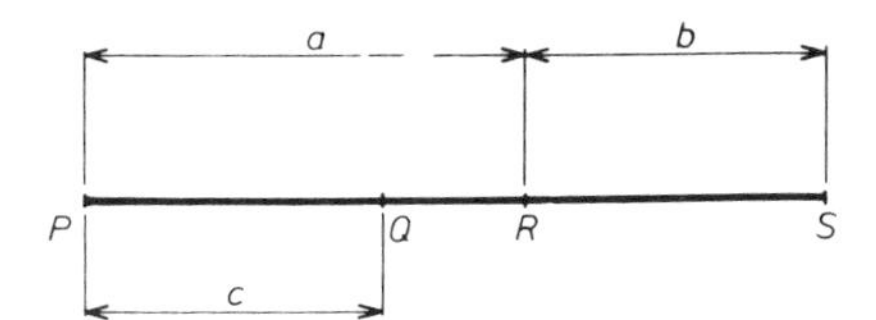

Fig. 6. Golden section ratio.

The interval is divided into three parts (see Fig. 6). If the function values at P, Q, R, and S are known, it can be reasoned whether PQ or RS should be cut off so as to enclose the optimum on a smaller segment. The ratio in which the segment is divided is optimum and represents the golden section ratio. The golden section ratio requires that

$$\frac{a+b}{a} = \frac{a}{b} \equiv \tau. \tag{43}$$

This yields a quadratic equation in a and b which yields

$$\tau = \frac{1}{2}(1+\sqrt{5}) \approx 1.61. \tag{44}$$

c is chosen equal to b.

If either c or b is cut off, the remaining parts satisfy the golden section condition (43), and so does the ratio of all two subsequent segments. The procedure stops as soon as the segment is small enough. It will be clear that only in unimodal cases (only one optimum on a closed segment) this method is unambiguous.

An important quantity is the improvement obtained by one one-dimensional search. If the function which must be optimized is unimodal, and if the procedure has monotone convergence, there can be sense in requiring that the improvement decreases to some value that is comparable with round-off noise. If there are many relative optima and the gradient method is combined with random search, much time would be wasted by accurately determining the position of optima, all but one of which will be exceeded. From a large number of observations, it appeared that the improvement decreases very rapidly after two or three one-dimensional iterations. Seldom is a substantial improvement gained by subsequent searches. For this reason, the gradient algorithm is stopped after two one-dimensional searches. The result is compared with the best optimum thus far. If it is worse, a new starting point is generated. If it is better, its value is defined to be "the best optimum thus far," and next the optimum is approximated more accurately by another procedure, comprising two one-dimensional searches. Then a new random starting point is generated for the next trial.

The program has the following parameters, which can be chosen arbitrarily:

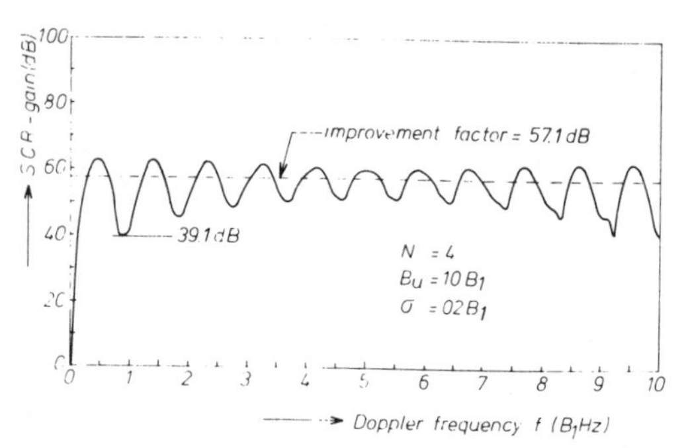

Fig. 7. SCR gain as a function of Doppler frequency; $N = 4$, $B_u = 10B_1$, $\sigma = 0.02B_1$. Optimum intervals: 1.100, 1.094, 1.000. Optimum coefficients: 1.000, −3.003, 3.155, −1.152.

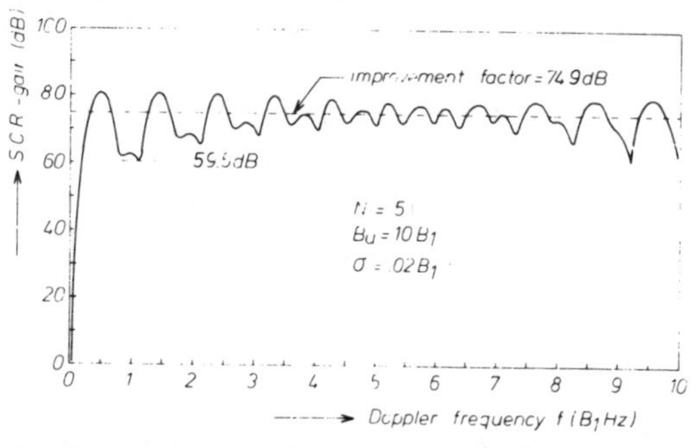

Fig. 8. SCR gain as a function of Doppler frequency; $N = 5$, $B_u = 10B_1$, $\sigma = 0.02B_1$. Optimum intervals: 1.111, 1.000, 1.091, 1.058. Optimum coefficients: 1.000, −4.258, 6.242, −4.031, 1.048.

1) upper bound of velocity region B_u
2) clutter spectrum variance σ^2 (program uses a Gaussian clutter spectrum model)
3) number of radar pulses N.

The program produces the following output:

1) optimum interpulse periods (normalized to $T_{min} = 1$)
2) optimum weighting coefficients (normalized to $w_0 = 1$)
3) a plot of the SCR gain characteristic
4) improvement factor
5) smallest eigenvalue.

Fig. 9. SCR gain as a function of Doppler frequency; $N = 5$, $B_u = 40B_1$, $\sigma = 0.02B_1$. Optimum intervals: 1.007, 1.021, 1.000, 1.299. Optimum coefficients: 1.000, −3.784, 5.432, −3.273, 0.626.

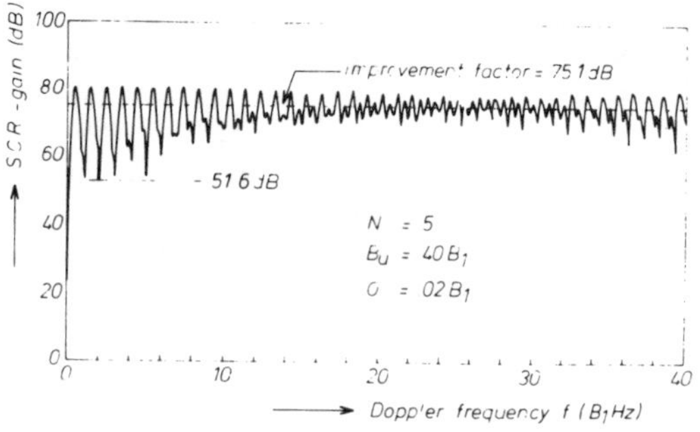

IV. Numerical Results

In this section some numerical examples will be given which show the influence of the parameters B_u, N, and σ, as well as of the stagger rate. Optimum interpulse periods are found for different values of N, B_u, and σ, using optimum filter coefficients. The effect of varying σ while the filter coefficients, that are optimum only for a specific value of σ, are fixed is also considered. The initial values of the interpulse periods with which the gradient algorithm starts are random numbers between 1 and 1.5 and differ from start to start. (It is recalled that the optimization scheme for the interpulse periods is a combination of gradient algorithm and random search.)

In all computations the minimum interpulse period is normalized to 1 ($T \equiv T_{min} = 1$), so that all results refer to the same maximum unambiguous range. Unit frequency range is $B_1 = 1/T$. (With uniform pulse spacing this would correspond to the first blind velocity.) The standard deviation of the clutter spectrum is expressed relative to B_1. The number of pulses is denoted by N. (Number of interpulse periods is $N - 1$.) In all examples the constant M of (40) is 8. (This constant governs the trade-off between improvement factor and minimum SCR gain.) For the lower bound B_l of the velocity region a frequency of $(1/2)B_1$ is chosen.

A. SCR Gain Characteristics

An optimum SCR gain characteristic for $N = 4$, $B_u = 10B_1$, and $\sigma = 0.02B_1$ is shown in Fig. 7. The minimum value in the velocity region is 39.1 dB. The improvement factor is 57.1 dB. The interpulse periods and the filter coefficients that produce these values are shown in the caption.

The same program produced another set of interpulse periods with a minimum which was slightly lower (38.3 dB), but with a better improvement factor (58.2 dB). During the search, values for the improvement factor of 51.0 to 58.3 dB were encountered.

Fig. 8 shows an optimum characteristic for $N = 5$, $B_u = 10B_1$, and $\sigma = 0.02B_1$. Here the minimum value in the velocity region is 59.5 dB, while the improvement factor is 74.9 dB. During the search the improvement factor varies from 68.6 dB to 75.0 dB.

Fig. 9 shows the effect of increasing the upper bound of the velocity region to $40B_1$. The optimum characteristic for this case has a minimum of 51.6 dB.

Fig. 10 shows an optimum characteristic for $N = 6$, $B_u = 10B_1$, and $\sigma = 0.02B_1$, with a minimum of 73.9 dB and an improvement factor of 88.7 dB. The improvement factor varies during optimization from 82.0 dB to 92.4 dB.

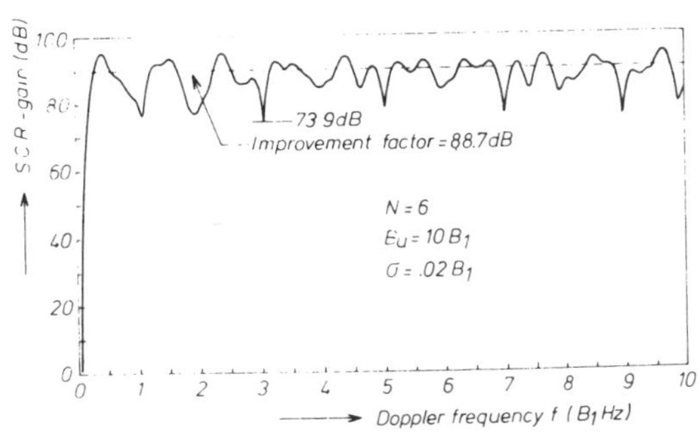

Fig. 10. SCR gain as a function of Doppler frequency; $N = 6$, $B_u = 10B_1$, $\sigma = 0.02B_1$. Optimum intervals: 1.006, 1.489, 1.000, 1.084, 1.152. Optimum coefficients: 1.000, –3.533, 8.670, –10.327, 5.213, –1.022.

Again, the program produces some good alternatives, e.g., a curve with minimum at 69.4 dB and an improvement factor of 91.7 dB.

B. Clutter Region Characteristic

Many effects and properties can be illustrated by studying the stopband characteristic of different types of filters. Fig. 11 is a plot of five filter characteristics (power transfer functions) and a Gaussian clutter power spectrum. All transfer functions are normalized so as to have approximately equal mean values of the passband characteristics (about 10^8). In all cases, $N = 5$, $\sigma = 0.02B_1$.

1) Curve 1 (solid line) represents the same filter as is represented in Fig. 8. Improvement factor $I = 74.9$ dB.

2) Curve 2 (broken line) refers to a filter with uniform pulse spacings and optimum coefficients (adapted to uniform PRF) in the ratio 1.00000, –3.90747, 5.81567, –3.90751, 1.00003. This filter has an improvement factor $I = 76.8$ dB.

3) In MTI literature, binomial weighting coefficients, which originate implicity from cascaded delay-line cancelers, are frequently discussed. Curve 3 shows that this type of filter initially provides large attenuation, but it soon degrades. The improvement factor is 70.5 dB. The weighting coefficients are 1, –4, 6, –4, 1.

4) Curve 4 illustrates the degradation of the binomial filter if the PRF is staggered. Here the same pulse spacings as in curve 1 were used. The improvement factor degrades to 46.3 dB.

5) It can be shown [6] that the binomial filter with uniformly spaced pulses belongs to a class of filters defined by the property that a maximum number of derivatives of the transfer function at $f = 0$ vanishes for a given set of pulse spacings. Solving for the filter coefficients requires an inversion of a Vandermonde matrix. In [6] it is shown that this leads to an extremely simple algorithm,

$$w_k = -w_0 \prod_{\substack{j=1 \\ j \neq k}}^{N=1} \frac{t_j}{t_j - t_k}, \quad k = 1, \cdots, N-1. \tag{45}$$

This algorithm produces filter coefficients that are adapted to the given set of pulse spacings. Curve 5 shows the stopband characteristic of this filter for the same set of pulse spacings as was used for curve 1. The improvement factor of this filter is 68.6 dB. The minimum SCR gain value is 53.4 dB. The weighting coefficients are 1.0000, –4.1846, 6.1953, –3.9766, 0.9659. For arbitrary pulse spacings, the improvement factor of this type of filter is about 6.2-dB less than the improvement factor of the optimum filter of Section II-B. A comparison of minimum values of the SCR gain give roughly the same degradation. Some ad hoc filters of this type are discussed in the literature, e.g., [1, p.45] and [2, p.349].

C. Varying σ

In all optimization runs (Figs. 7 through 10), the filter coefficients were adapted to the set of actual interpulse periods and σ, throughout the optimization procedure, according to Section II-B. As was discussed in Section II-C, increasing σ is equivalent to increasing T or $\sum_i T_i$, or to increasing the radar frequency f_c. This is summarized in an experiment in which σ was varied to study the effect on the improvement factor I. The result is shown in Fig. 12. In this case $N = 5$ and $B_u = 10B_1$.

The solid curve in Fig. 12 represents the improvement factor as a function of σ when optimum filter coefficients that are matched to σ are used, while the interpulse periods, which are optimum for $\sigma = 0.02$, are fixed. (It was found that the value of σ does not appreciably influence the optimum interpulse periods.) The dotted line represents the improvement factor as a function of σ when those coefficients are used that are optimum for $\sigma = 0.02$. This effect arises when the characteristics of the clutter model for which the MTI system is optimized differ from the actual (local) clutter characteristics. If the standard deviation of the clutter spectrum is assumed smaller than the actual clutter standard deviation, a loss of 4 dB in the improvement factor occurs.

D. Varying the Stagger Rate

If the maximum unambiguous range is fixed (or, equivalently, if T is fixed), varying the stagger rate has a

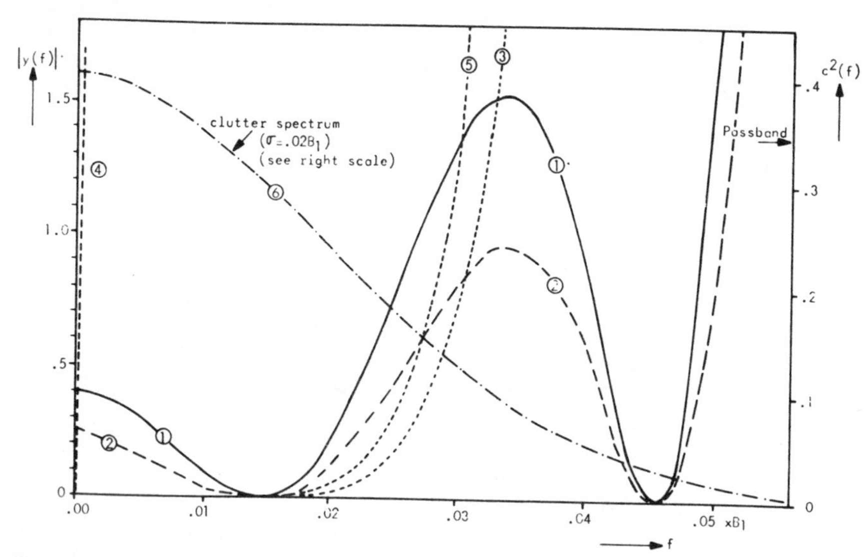

Fig. 11. Clutter region characteristic and clutter spectrum; $N = 5$, $\sigma = 0.02$. Curve 1: Optimum filter, staggered PRF, as in Fig. 8, with $I = 74.9$ dB. Curve 2: Optimum filter, uniform PRF, $I = 76.8$ dB. Curve 3: Binomial filter, uniform PRF, $I = 70.5$ dB. Curve 4: Binomial filter, staggered PRF, as in Fig. 8, with $I = 46.3$ dB. Curve 5: Vandermonde filter, staggered PRF, as in Fig. 8, with $I = 68.6$ dB. Curve 6: Clutter power spectrum.

Fig. 12. Improvement factor as a function of clutter spectral width σ; $N = 5$, $B_u = 10B_1$. Coefficients for dotted curve (optimum for $\sigma = 0.02B_1$): 1.000, −4.258, 6.242, −4.031, 1.048. Intervals for both solid and dotted curves (optimum for $\sigma = 0.02B_1$): 1.111, 1.000, 1.091, 1.058 (c.f., Fig. 8).

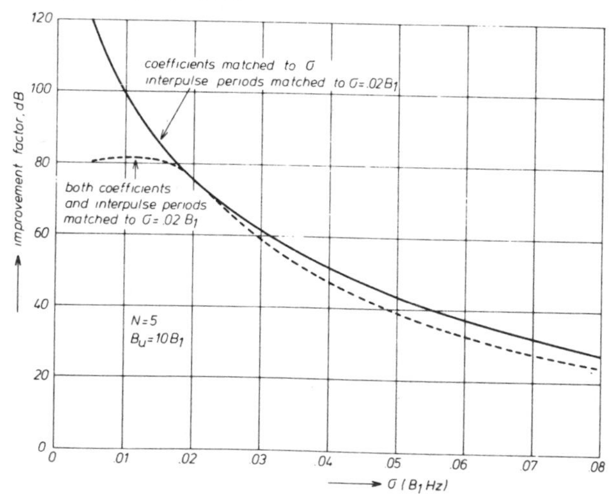

Fig. 13. Improvement factor as a function of stagger rate.

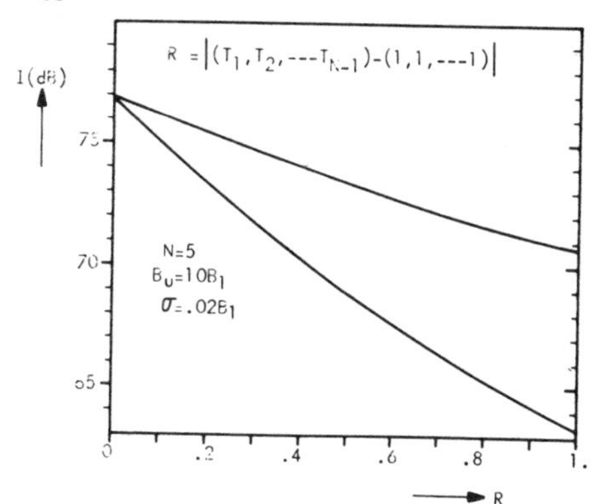

similar effect as varying σ. This is shown in Fig. 13. Let each point in an $(N - 2)$-dimensional space (interpulse period space) correspond to a particular combination of pulse spacings. (Remember that the smallest pulse spacing T is normalized to 1 throughout this experiment.) The point $(1, 1, \cdots, 1)$ represents uniform pulse spacing. Now define the stagger rate R of a given set of spacings as the distance from the actual point in the interpulse period space to the point $(1, 1, \cdots, 1)$. In a computer run, random direction vectors in the interpulse period space were generated. In each direction, starting in $(1, 1, \cdots, 1)$, the improvement factor was calculated as a function of R, while the filter coefficients were adapted to the spacings. Two extreme results are shown in Fig. 13 ($N = 5$, $\sigma = 0.02$, $B_u = 10 \times B_1$). In all directions, the improvement factor (in decibels) appeared to be a monotonic smooth function of R.

V. Conclusion

Blind velocities of MTI radar can be removed to a great extent by optimizing the PRF stagger. For a given clutter spectral width σ, a given number of pulses involved, and a given upper bound of the velocity region, there are many relative optima which show great differences with respect to minimum value of SCR gain (residue of a blind velocity) and improvement factor. Among the local optima, some good alternatives can be found that combine high improvement factor and high minimum SCR gain. With respect to the improvement factor, small values of the time on target and of the radar frequency are preferred.

Acknowledgment

The author acknowledges the valuable suggestions and comments of C. van Schooneveld and G.A. van der Spek.

References

[1] W. Shrader, "MTI radar" in M. Skolnik, Ed. *Radar Handbook*, New York: McGraw-Hill, 1970.
[2] F. Nathanson, *Radar Design Principles*. New York: McGraw-Hill, 1969.
[3] A. Zverev, "Digital MTI radar filters," *IEEE Trans. Audio and Electroacoustics*, vol. AU-16, September 1968.
[4] J. Capon, "Optimum weighting functions for the detection of sampled signals in noise," *IEEE Trans. Information Theory*, vol. IT-10, April 1964.
[5] T. Murakami and R.S. Johnson, "Clutter suppression by use of weighted pulse trains," *RCA Rev.*, vol. 32, September 1971.
[6] P.J.A. Prinsen, "A class of high-pass digital MTI filters with non-uniform p.r.f." *Proc. IEEE*, to be published.

Peter J.A. Prinsen was born in Eindhoven, The Netherlands, on March 4, 1941. He received the M.S. degree in electrical engineering from the Eindhoven University of Technology in 1969.

During his military service he worked on measuring problems at the Physics Laboratory of the Royal Dutch Powder Factory. In 1970 he joined the Physics Laboratory TNO of the National Defence Research Organization, The Hague, The Netherlands, where he has been working on MTI Radar.

Mr. Prinsen is a member of the Royal Institution of Engineers in the Netherlands (KIvI).

4.3 A CLASS OF HIGH-PASS DIGITAL MTI FILTERS WITH NONUNIFORM PRF
P. Prinsen
Proceedings of the IEEE, Vol. 61, No. 8, August 1973. pp. 1147-1148.

A Class of High-Pass Digital MTI Filters with Nonuniform PRF

PETER J. A. PRINSEN

***Abstract*—A class of digital feed-forward filters is developed, satisfying the requirement of a maximally flat stopband characteristic at zero frequency if the pulse-repetition frequency is modulated. A simple algorithm is given to obtain the filter coefficients. Some properties are summarized.**

Introduction

High-pass digital feed-forward filters with alternating binomial weighting coefficients are frequently discussed in literature. A property of these filters is that they cancel input signals which are polynomials of degree $N-2$ (N equals the number of weighting coefficients). They provide good relative attenuation in the stopband and are therefore often proposed as clutter suppression filters in moving target indicator (MTI) radar (e.g., [3]). Besides, many authors refer to binomial filters for comparing the performance of other types of filters. However, if the pulse-repetition frequency (PRF) is staggered to eliminate unwanted stopbands at multiples of the PRF ("blind velocities" in MTI radar), the performance of the binomial filter degrades heavily [5] because the typical properties are no longer valid. In this letter a generalization of the binomial filter for staggered PRF is developed from the requirement of a maximally flat stopband at zero frequency.

Theory

Let the weighting coefficients of a digital feed-forward filter with N elements be denoted by $\{w_n\} = w_0 \cdots w_{N-1}$. Then the transfer function $y(f)$ of this filter for a nonuniformly sampled signal is (see Fig. 1)

$$y(f) = \sum_{n=0}^{N-1} w_n e^{j2\pi f t_n} \tag{1}$$

where $t_0 \cdots t_{N-1}$ are the sample moments with $t_0 = 0$. The Taylor series expansion of $y(f)$ around $f=0$ is

$$y(f) = y(0) + \frac{y'(0)}{1!} f + \frac{y''(0)}{2!} f^2 + \cdots \tag{2}$$

with

$$y^{(k)}(0) = (j2\pi)^k \sum_{n=0}^{N-1} t_n^k w_n, \qquad k = 0, 1, 2, \cdots. \tag{3}$$

Now we develop a class of filters with maximally flat stopband by requiring that the first M coefficients $y^{(k)}(0)$, $k = 0, \cdots, M-1$ in (2) be zero.

Then the remaining series has a leading term which is proportional to f^M. If M is maximum, then $y(f)$ is maximally flat around $f=0$. Each coefficient being set to zero yields a homogeneous linear equation in $\{w_n\}$. As there are N variables w_n, we can form a system of maximally $N-1$ linearly independent equations with a nontrivial solution, so $M = N-1$. This yields [using (3)]

Manuscript received March 30, 1973.
The author is with the Physics Laboratory TNO, The Hague, The Netherlands.

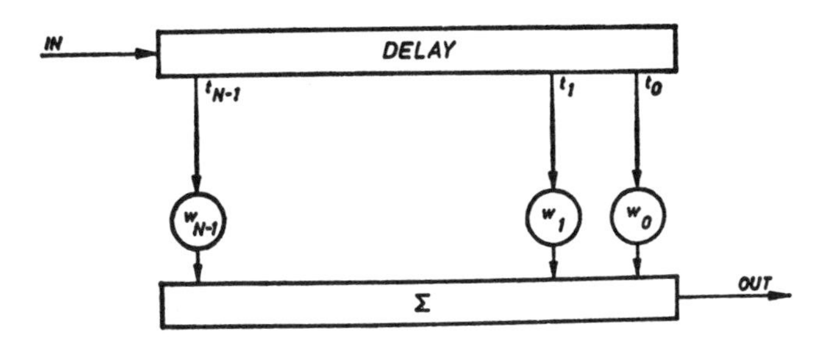

Fig. 1.

$$\sum_{n=0}^{N-1} t_n^k w_n = 0, \qquad k = 0, 1, \cdots, N-2. \tag{4}$$

Since $t_0 = 0$, only the first equation of the system in (4) has a term containing w_0. Then we have

$$\left.\begin{aligned} w_1 + w_2 + \cdots + \quad w_{N-1} &= -w_0 \\ t_1 w_1 + t_2 w_2 + \cdots + t_{N-1} w_{N-1} &= 0 \\ \cdots\cdots\cdots\cdots\cdots\cdots\cdots \\ t_1^{N-2} w_1 + t_2^{N-2} w_2 + \cdots + t_{N-1}^{N-2} w_{N-1} &= 0 \end{aligned}\right\} \tag{5}$$

or

$$Tw = -w_0 u_1 \tag{6}$$

with

$$w' = (w_1, w_2, \cdots, w_{N-1})$$

$$u_1' = (1, 0, \cdots, 0)$$

and

$$T = \begin{pmatrix} 1, & 1, & \cdots, & 1 \\ t_1, & t_2, & \cdots, & t_{N-1} \\ \cdots & \cdots & \cdots & \cdots \\ t_1^{N-2}, & t_2^{N-2}, & \cdots, & t_{N-1}^{N-2} \end{pmatrix}. \tag{7}$$

The system (6) can be solved by inverting the matrix T if T^{-1} exists:

$$w = -w_0 T^{-1} u_1. \tag{8}$$

The matrix T is a Vandermonde matrix [6] and T^{-1} exists since $t_i \neq t_j$ for $i \neq j$.

Let the inverse of T be denoted by R with elements r_{ij}. Then in the Appendix it is shown that

$$r_{ij} = (-1)^{N-1-j} \prod_{\substack{n=1 \\ n \neq i}}^{N-1} (t_i - t_n)^{-1} \sum_{C_{N-2}^{N-1-j}} t_{l_1} \cdot t_{l_2} \cdots t_{l_{N-1-j}}, \qquad i, j = 1, \cdots, N-1. \tag{9}$$

Here C_{N-2}^{N-1-j} denotes that the sum has to be taken over all $\binom{N-2}{N-1-j}$ products having a combination of $N-1-j$ indices

$$(l_1, l_2, \cdots, l_{N-1-j})$$

out of a set of $N-2$ values $1, 2, \cdots, i-1, i+1, \cdots, N-1$. Each product comprises $N-1-j$ factors (a product comprising 0 factors being 1).

Since u_1 in (8) has only one nonzero component, only the first column of R needs to be evaluated:

$$w_i = -w_0 \cdot r_{i,1}, \qquad i = 1, \cdots, N-1. \tag{10}$$

This is the column with $j=1$ [see (9)].

Since

$$C_{N-2}^{N-2} = \binom{N-2}{N-2} = 1$$

(10) simplifies to

$$w_i = -w_0 \prod_{\substack{n=1 \\ n \neq i}}^{N-1} \frac{t_n}{t_n - t_i}, \qquad i = 1, \cdots, N-1 \tag{11}$$

(cf. [1, p. 44] and [2, p. 349]).

Application of this type of filter in MTI radar for suppression of echoes from stationary targets (clutter) is discussed in [5].

Properties

A. Frequency Domain

According to the definition (4), a filter defined by (11) has a maximum number of derivates of the transfer function in $f=0$ being zero. (The first $N-1$ terms of the Taylor series expansion of the frequency characteristic are zero.)

B. Time Domain

Suppose that the filter input $\{x_n\}$ consists of N samples of a function $x = t^k$ (k integer):

$$x_n = t_n^k, \qquad n = 0, \cdots, N-1. \tag{12}$$

Let the output at t_{N-1} be denoted by z. Then

$$z = \sum_{n=0}^{N-1} t_n^k w_n. \tag{13}$$

Comparing (13) with (4), one learns that a filter defined by (11) rejects signals that are proportional to t^k if $k \leq N-2$.

So the filter has the property of rejecting the first $N-2$ terms of the Taylor series expansion of the input signal (cf. [1, p. 43]).

C. Uniform PRF

It is easily verified that if $t_j = j, j = 0, \cdots, N-1$, then (11) can be written as

$$w_k = (-1)^k w_0 \binom{N-1}{k}. \tag{14}$$

This expression can be recognized as the binomial filter which can be obtained by cascading $N-1$ single delay line cancelers.

Appendix

The Inverse of the Vandermonde Matrix (cf. [6, p. 125])

Let T be a Vandermonde matrix of dimension $N-1$.

$$T = \{t_j^i\}, \qquad i = 0, \cdots, N-2, \quad j = 1, \cdots, N-1 \tag{15}$$

and let the inverse of T be denoted by $R = \{r_{ij}\}$. Then

$$RT = I. \tag{16}$$

Observe the fundamental polynomials

$$\pi_i(t_k) = \prod_{\substack{j=1 \\ j \neq i}}^{N-1} (t_k - t_j), \qquad i, k = 1, \cdots, N-1$$

$$\triangleq \sum_{j=0}^{N-2} c_{ij} t_k^j \begin{cases} \neq 0, & \text{if } i = k \\ = 0, & \text{if } i \neq k \end{cases} \tag{17}$$

or

$$\sum_{j=1}^{N-1} c_{i,j-1} t_k^{j-1} = \delta_{ik} \cdot \pi_i(t_i), \qquad i, k = 1, \cdots, N-1 \tag{18}$$

where δ_{ik} is the Kronecker delta and [from (17)]

$$c_{i,j} = (-1)^{N-2-j} \sum_{C_{N-2}^{N-2-j}} t_{l_1} \cdot t_{l_2} \cdots t_{l_{N-2-j}}. \tag{19}$$

In (19) C_{N-2}^{N-2-j} denotes that the sum has to be taken over all $\binom{N-2}{N-2-j}$ products having a combination of $N-2-j$ indices

$$(l_1, l_2, \cdots, l_{N-2-j})$$

out of a set of $N-2$ values $1, 2, \cdots, i-1, i+1, \cdots, N-1$. Each product comprises $N-2-j$ factors (a product comprising 0 factors being 1).

On the other hand, since the i, jth element of RT must equal δ_{ij} in order to satisfy (17) we can write

$$\sum_{j=1}^{N-1} r_{ij} t_k^{j-1} = \delta_{ik}, \qquad i, k = 1, \cdots, N-1. \tag{20}$$

Comparing (20) with (18) and using (19) we find that

$$r_{ij} = \frac{c_{i,j-1}}{\pi_i(t_i)} = (-1)^{N-1-j} \{\pi_i(t_i)\}^{-1} \sum_{C_{N-2}^{N-1-j}} t_{l_1} \cdot t_{l_2} \cdots t_{l_{N-1-j}},$$

$$i, j = 1, \cdots, N-1. \tag{21}$$

References

[1] W. Shrader, "MTI radar," in *Radar Handbook*, M. Skolnik, Ed. New York: McGraw-Hill, 1970.

[2] F. Nathanson, *Radar Design Principles*. New York: McGraw-Hill, 1969.

[3] C. Benning and D. Hunt, "Coefficients for feed-forward MTI radar filters," *Proc. IEEE* (Lett.), vol. 57, pp. 1788–1789, Oct. 1969.

[4] W. W. Shrader and V. G. Hansen, "Comments on 'Coefficients for feed-forward MTI radar filters,'" *Proc. IEEE* (Lett.), vol. 59, p. 101, Jan. 1971.

[5] P. J. A. Prinsen, "Elimination of blind velocities of MTI radar by modulating the interpulse period," *IEEE Trans. Aerosp. Electron. Syst.*, vol. AES-9, July 1973.

[6] R. W. Hamming, *Numerical Methods for Scientists and Engineers*. New York: McGraw-Hill, 1962, p. 125.

4.4

CLUTTER ATTENUATION ANALYSIS
W. Fishbein, S. Graveline and O.R. Rittenbach
Technical Report ECOM-2808, U.S. Army Electronics Command, Fort Monmouth, N.J., March 1967, pp. 1-30

Reprinted by Permission.

TECHNICAL REPORT ECOM-2808

CLUTTER ATTENUATION ANALYSIS

by

William Fishbein

Stanley W. Graveline

Otto E. Rittenbach

Radar Technical Area
Combat Surveillance and Target Acquisition Laboratory

March 1967

Subtask Nr. 1P6-20901-A-188-03-05

UNITED STATES ARMY ELECTRONICS COMMAND . FORT MONMOUTH, NEW JERSEY

ABSTRACT

The performance of moving target indication (MTI) systems for combat surveillance radars depends to a large extent on the clutter spectrum. The clutter spectrum is especially important when the radar attempts to detect slowly moving ground targets. This spectrum has long been assumed to be Gaussian shaped. However, MTI system performance predicted by this assumption was not achieved in practice. This report describes the results of an investigation conducted to determine the performance to be expected from an MTI system.

The approach was to measure the clutter rejection ratios afforded by various high-pass filters. The signal was taken from the boxcar demodulator of an X-band radar observing different clutter targets under varying wind conditions. Clutter rejection ratios of 10 to 40 db were measured. These results were then used to obtain a theoretical expression for the clutter power spectrum. This expression differs from the usual Gaussian assumption. Some credence is given to the results by a direct spectral analysis performed on a clutter signal.

Two methods of filtering clutter signals which will result in acceptable MTI performance are suggested in this report.

The results of this investigation are significant in that they have led to establishing criteria for a better MTI system design.

CONTENTS

FIGURES

TABLES

APPENDICES

INTRODUCTION

The purpose of MTI systems for combat surveillance radars is to process the radar video output to eliminate all undesired signals and have an output only for moving target input signals. The primary source of undesired signals which limit the performance of an MTI system is the spectrum resulting from wind blown natural clutter targets. A problem arises in that this spectrum contains doppler frequencies which overlap those resulting from slowly moving targets of interest, e.g., a walking man.

The clutter spectrum was long assumed to be Gaussian shaped. However, MTI system performance predicted by this assumption was not achieved in practice. An investigation was conducted to determine the system performance to be expected. This investigation led to the establishment of criteria for a better MTI system design.

The approach used was to measure the clutter rejection ratios afforded by various high-pass filters. The clutter signal was taken from the boxcar demodulator of an X-band noncoherent radar observing different clutter targets under varying wind conditions. A theoretical expression for the clutter spectrum was obtained from these data. The results of a direct spectral analysis performed on a clutter signal gave some credence to the theoretical expression obtained from the measurements. Comparisons with a Gaussian clutter spectrum were made and the error obtained with this assumption was calculated.

This report discusses the results of the investigation and the significance of the results.

BACKGROUND

An important function of certain combat surveillance radars is to provide information which will enable one to distinguish between fixed and moving ground-based targets. These radars are designed to make use of the doppler effect, i.e., to have as an output a signal which contains the frequencies resulting from target motion.

The most prevalent class of radar in use for combat surveillance is a pulse doppler noncoherent system. Noncoherent means that a fixed target reference must exist within the same range resolution cell as a moving target if moving target information is to be recovered. The output signal of such a radar is a video pulse train amplitude modulated according to the doppler frequency. The doppler modulation can be recovered by means of a circuit called a boxcar demodulator. The boxcar is a sample-and-hold device, usually gated at the radar pulse repetition frequency. The boxcar output contains the doppler frequency. This audio signal is processed and fed to an indicator. The indicator in many MTI combat surveillance radars is a simple aural display. An operator notes the presence of a moving target by the sounds he hears.

Video Processing

While the operator does an excellent job of detecting moving targets, he can observe only a small area at a time. To increase the information rate of a radar and also to reduce operator fatigue, modern radar systems employ video processors.* A video processor provides an output only for moving target input signals in a form suitable for a visual display. This is accomplished by recovering the doppler frequencies via a boxcar demodulator, filtering to remove all undesired frequencies, rectifying, integrating, and gating the remaining signal. The final output is a video pulse occurring at the correct range.

Filtering Need

The basic sources of undesired frequencies that must be eliminated (or at least reduced) in the video processor are the returns from wind blown natural clutter targets. Conventional processors employ high-pass filters for this task. These filters must be critically designed. One which gives maximum MTI sensitivity on calm days would be ineffective on windy days. On the other hand, a filter that eliminated all clutter signals on a windy day would result in a needless loss of information on clamer days. A compromise is needed if this type of filtering is used. The nature of the clutter target suggests that a different type of filtering might be employed. The clutter analysis described in this report yields the information required to arrive at some filter design suggestions.

*See Appendix I

Nature of a Clutter Target

Foliage is the primary cause of doppler signals that compete with those from slowly moving targets in an MTI receiver. The detection cell size of a radar at a typical operating range might be 75 meters in both range and azimuth. If an area fairly dense in trees were under observation, the clutter return would consist of a complex addition of the returns from several trees. The total return will consist of a large d-c component plus a fluctuating component. The d-c return is due to the trunks of the trees and the large branches (plus any other fixed targets that fall within the detection cell). The a-c component is due to many small individual scatterers. These are the leaves and small branches of the trees. The a-c component varies according to the wind conditions.

CLUTTER ATTENUATION MEASUREMENTS

The original effort in this work was to experimentally determine the performance of various high-pass filters in attenuating clutter signals. The procedure used was to measure the clutter rejection-ratios afforded by five filters when attenuating the signal provided by a radar observing several different clutter targets.

Clutter Rejection Ratio

A method of describing the performance of a filter in attenuating clutter signals is by its clutter rejection ratio. This is defined as:

$$R = \text{Clutter Rejection Ratio} = \frac{\int_0^\infty P(f)\,df}{\int_0^\infty P(f)\,|H(p)|^2\,df} \tag{1}$$

where $P(f)$ = clutter power spectrum

$H(p)$ = filter voltage transfer function.

R is thus the ratio of the total clutter power into the filter to the total power out.

Test Procedure

The clutter rejection ratios of five high-pass filters were obtained by means of an extensive series of measurements. An AN/TPS-25 radar system* was used to collect data for the measurements. This radar is of the noncoherent pulse doppler type. It operates at a 3-cm wave length (X-band). Target information was taken from the boxcar demodulator output of the radar. This output contains the clutter spectrum when the radar is observing a natural clutter site.

Figure 1 is a block diagram of the experimental equipment used. The radar boxcar output was fed into five parallel filter channels and a reference channel. The filter output was amplified, rectified, and integrated via a thermocouple, chopped, and fed into a pen recorder. The thermocouple output was chopped to eliminate any errors due to bias drifts and to provide a more sensitive input to the pen recorder. The output of each filter channel was compared with the output of the reference (unfiltered) channel on the recorder. The ratio of these two values is the clutter rejection ratio.

*Characteristics of Radar Set AN/TPS-25 are given in Appendix II.

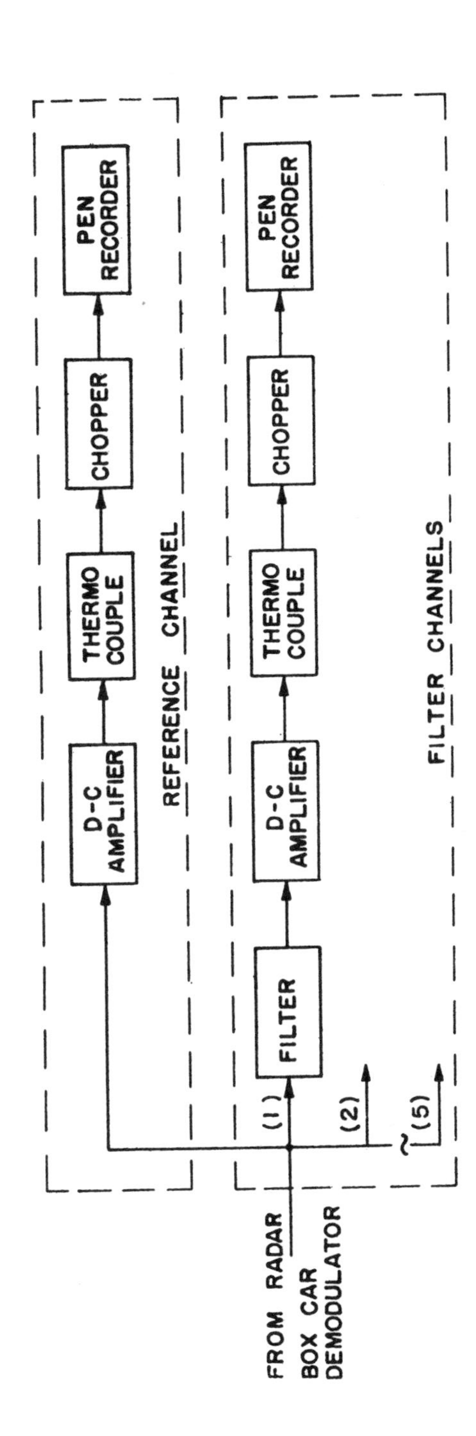

FIG. 1 SETUP FOR MEASURING CLUTTER REJECTION RATIOS

The high-pass filters** for which clutter rejection ratios were measured are described in Table I below.

TABLE I

CLUTTER FILTER CHARACTERISTICS

ATTENUATION SLOPE	CUTOFF FREQUENCY	TYPE
db/octave	Hz	
6	50	passive
6	100	passive
12	100	active
40	50	passive
40	100	passive

Six different clutter targets were observed by the radar. The range to the targets was always such that many trees fell within the radar detection cell. The procedure used was to measure the clutter rejection ratio (R) obtained with each filter when operating on the spectrum provided by a particular clutter site. Observations on different days provided a value of R as a function of the wind conditions to which each target was subjected. Finally, the rejection ratios obtained with the six targets were averaged. The results are summarized in Table II below.

TABLE II

MEASURED CLUTTER REJECTION RATIOS

	50	100	100	50	100	← f_p
V	6	6	12	40	40	← α
3	24	26	37	36	40	
6	21	26	36	33	38	
9.5	17	17	30	27	34	
13.5	15	18	26	—	29	R
18	11	16	21	—	23	
20	9	13	18	—	19	

f_p = Filter low-frequency cutoff (Hz)

α = Filter attenuation per octave below f_p (db)

V = Average wind velocity (knots)

R = Clutter rejection ratio (db)

Clutter Spectrum Expression

An attempt was made to find an expression for the clutter power spectrum which would yield calculated values for the clutter rejection ratio in agreement with Table II. An expression that gives good agreement is

$$P(f) = \frac{1}{1 + \left(\frac{f}{f_c}\right)^3} \qquad (2)$$

**Frequency response plots of the filters are given in Appendix III.

where f_c = clutter spectrum characteristic frequency

$$= k \exp(\beta V) \quad (\text{Hz}) \tag{3}$$

$k = 1.33$ (Hz)

$\beta = 0.1356$ (knots)$^{-1}$

V = wind velocity (knots).

The expression for f_c is plotted in Figure 2. Calculated values* of R using (2) and (3) are compared with the measured values in Table III below.

TABLE III

COMPARISON OF MEASURED AND CALCULATED CLUTTER REJECTION RATIOS

		50		100		100		50		50		← f_p
		6		6		12		40		40		← α
V	f_c	M	C	M	C	M	C	M	C	M	C	
3	2	24	24	26	29**	37	38	36	35	40	41	R
6	3	21	21	26	26	36	35	33	31**	38	38	R
9.5	5	17	17	17	22**	36	30	27	27	34	33	R
13.5	8	15	14	18	19	26	26			29	29	R
18	15	11	10	16	15	21	21			23	24	R
20	20	9	9	13	13	18	19			19	21**	R

f_p = filter low-frequency cutoff (Hz)

α = filter attenuation per octave below f_p (db)

V = average wind velocity (knots)

f_c = clutter spectrum characteristic frequency (Hz)

R = clutter rejection ratio (db)

M = measured value

C = calculated value

** denotes a difference of more than one db between the measured and calculated values.

It can be seen in the table that, of 27 values compared, 23 differ by one db or less. Two of the remaining four differ by 2 db. The measured and calculated values are thus seen to be in very good agreement.

The results of the experimental program are summarized by Equations (2) and (3). The performance of any filter in attenuating the clutter spectrum provided by an X-band radar can be predicted from these expressions.

*See Appendix III

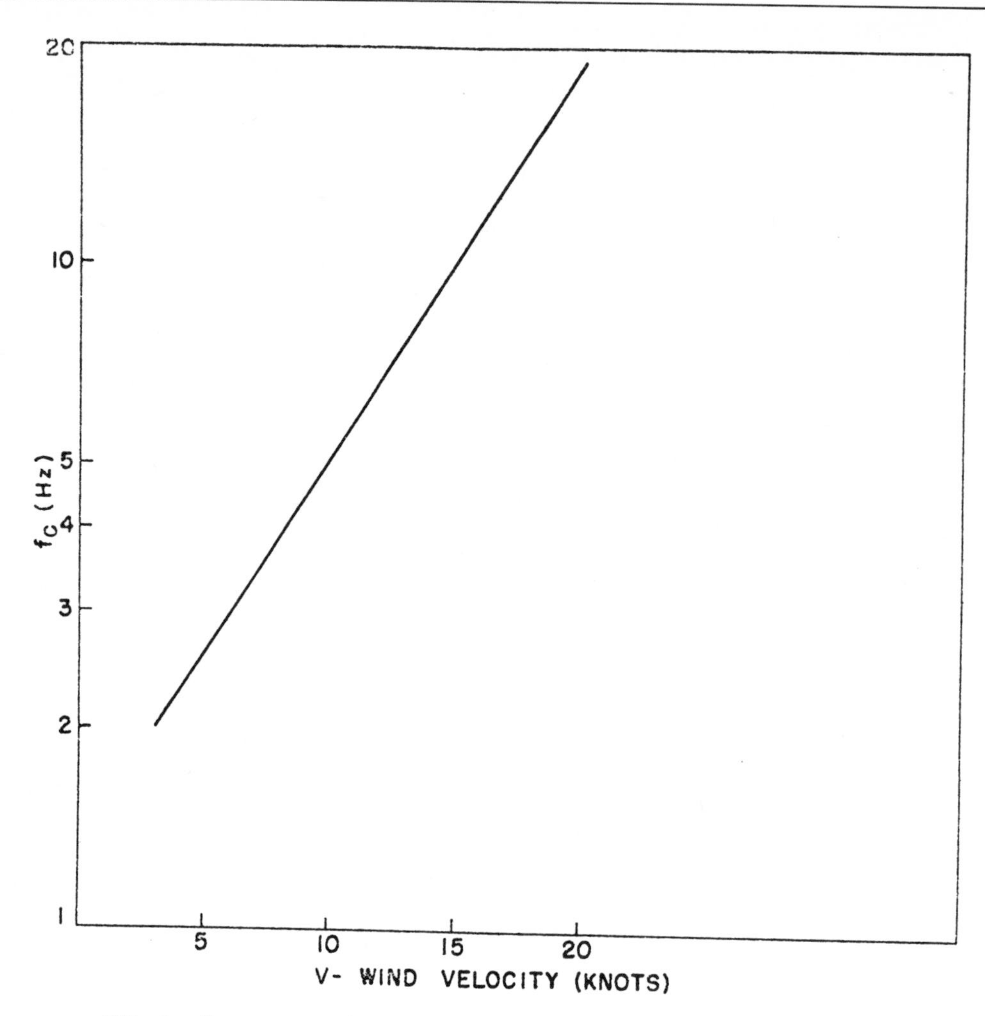

FIG. 2 f_C vs V, f_C = CHARACTERISTIC FREQUENCY OF THE CLUTTER SPECTRUM OBTAINED WITH AN X-BAND RADAR

GAUSSIAN COMPARISON

The clutter spectrum has long been assumed to be Gaussian shaped. However, MTI system performance predicted by this assumption was not achieved in practice. A comparison will be made here between the cubic power spectrum described by Equation (2) and the long assumed Gaussian spectrum.

Gaussian Spectrum

Kerr[1] and Barlow[2], in early work, stated that the power spectrum of fluctuating clutter targets can be approximated by

$$P(f) = \exp\left[-af^2\right] \tag{4}$$

where a = a constant determined by the type of target, radar frequency, and wind conditions (sec^2).

Skolnik[3] quotes a value of $a = 2.3 \times 10^{-3}$ for the spectrum obtained with a radar operating at a frequency of 10 GHz observing a heavily wooded area subjected to a wind velocity of 20 mph.

*See Appendix IV

Gaussian Comparison

One method of comparing the spectra described by Equations (2) and (4) is to assume that the two spectra have equal total powers. Then*

$$\int_0^{\infty} \frac{df}{1+\left(\frac{f}{f_c}\right)^3} = \int_0^{\infty} \exp\left[-af^2\right] df$$

and $$f_c = \frac{3}{4}\left[\frac{3}{\pi a}\right]^{1/2} \qquad (5)$$

Substituting the value of a given above into Equation (5) results in an f_c of 15.3 Hz.

It is to be noted that a value for f_c of 14 Hz can be found from Equation (3) for a 20 m.p.h. (17.4 knot) wind. This is in good agreement with the 15.3 Hz value when one considers the two were obtained through completely independent investigations. For this comparison, f_c is taken equal to 15.3 Hz. The two spectra are plotted in Figure 3. The curves are seen to be similar. The main difference is that the Gaussian spectrum decays more quickly beyond 30 Hz. This difference is emphasized by comparing the clutter rejection ratios obtained when the spectra are attenuated by 12-db-per-octave active 100-Hz high-pass filters. R for the Gaussian case is calculated (Appendix V) to be 28.5 db. From Table II, R for the cubic spectrum is seen to be 21 db. The 7.5 db difference is attributed to the difference in energy available in the two spectra at the higher frequencies. It is seen then that assumption of a Gaussian spectrum gives a more optimistic prediction of clutter attenuation and thus MTI performance than is actually achieved. The difference in the rejection ratios increases with increasing wind conditions.

SPECTRAL ANALYSIS

The process described up to this point has been to measure the clutter attenuation afforded by various filters and then find a theoretical expression for the clutter spectrum based on these measurements. One might ask why a different approach was not used, i.e., why not perform a direct analysis on a clutter signal? This would yield the clutter power spectrum from which the performance of the selected filters could be determined. The answer to the question of methods is that filter performance can be determined with significantly less effort by making attenuation measurements. A spectral analysis would result in meaningful data only after many clutter samples were observed for lengthy intervals. An extensive analysis on these samples would then have to be made; finally, the attenuation afforded by the various filters would have to be calculated.

A spectral analysis for the clutter return from a single target and single wind condition was performed in an attempt to verify the cubic clutter power spectrum expression derived previously.

Procedure and Results

The boxcar demodulator output of the radar was recorded while the radar was observing a clutter target subjected to a wind velocity of 12 knots. A one second length of the recording was sampled 200 times ($\Delta t = 5$ milliseconds). A computer was programmed to determine the power spectrum via a direct Fourier analysis.[4]

The power spectra for six one-second samples of the recorded clutter data were obtained. A mean was taken of the six values for each frequency. The data points are plotted in Figure 4. A plot of the spectrum for a 12-knot wind given by Equations (3) and (4) is also shown in Figure 4. The question is, how well does the curve fit the data points? The curve is seen to be a good approximation above 10 Hz. In fact, above 20 Hz, the curve is probably the best possible approximation to the data points.

The good agreement between the data points and the curve at the higher frequencies gives strong credence to the accuracy of assuming a cubic clutter spectrum. Also plotted in Figure 4 is a Gaussian-shaped spectrum having the same total power as the cubic spectrum. It can be seen that this curve is a poorer approximation for the data points at the higher frequencies. A linear plot of the two curves is seen in Figure 5. It again shows the similarity between the two curves, the main difference being that the Gaussian spectrum decays more quickly.

FILTER SUGGESTIONS

The results obtained up to this point can now be used to suggest filters which will give adequate performance in an MTI video processor. What is sought is a filter which will sufficiently attenuate clutter signals with a minimum loss of signals from slowly moving targets of interest. The clutter filter may be repeated up to a few hundred times in some video processors. This fact dictates that the filter be of simple design in order that resulting sizes and weights be reasonable.

Two types of filtering are discussed in the sections that follow.

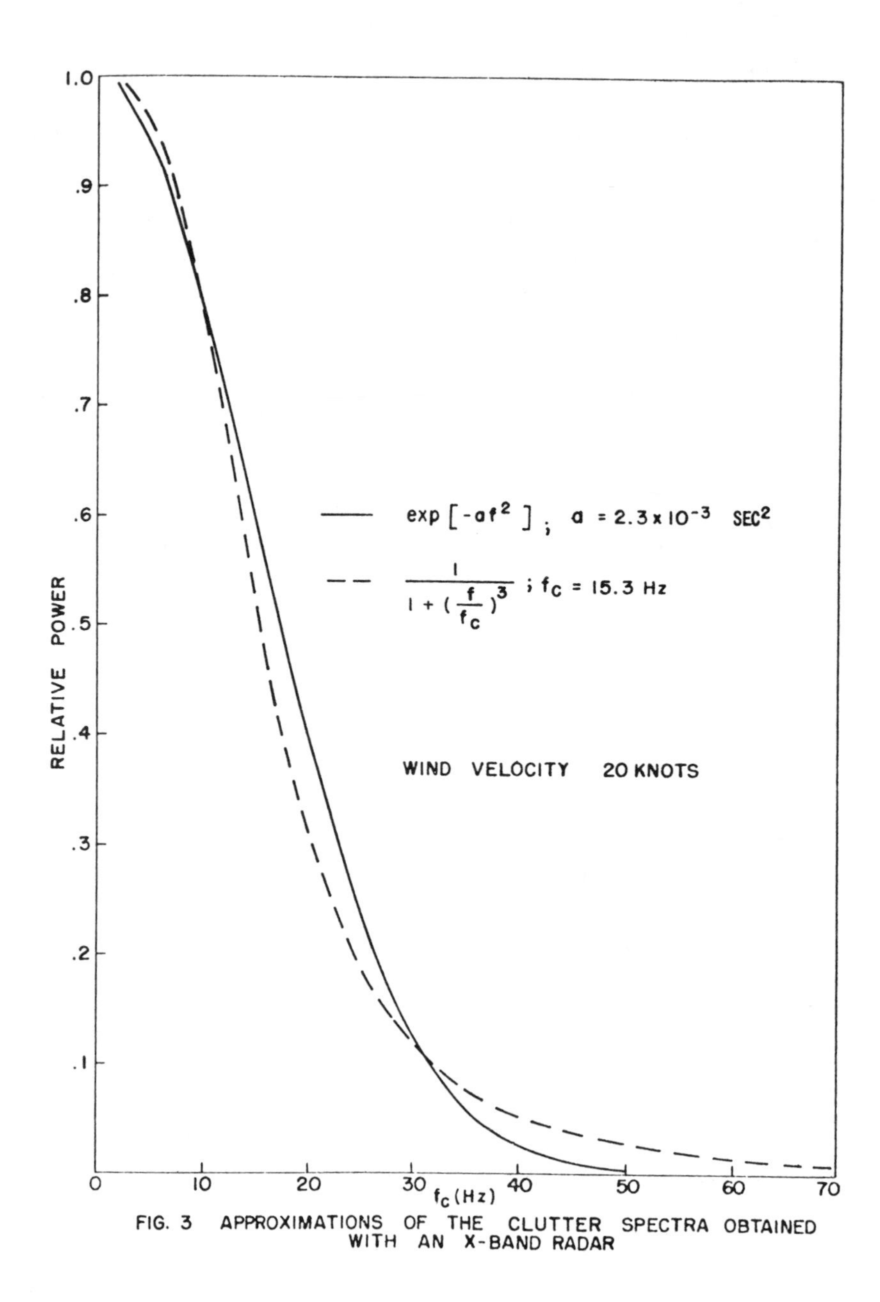

FIG. 3 APPROXIMATIONS OF THE CLUTTER SPECTRA OBTAINED WITH AN X-BAND RADAR

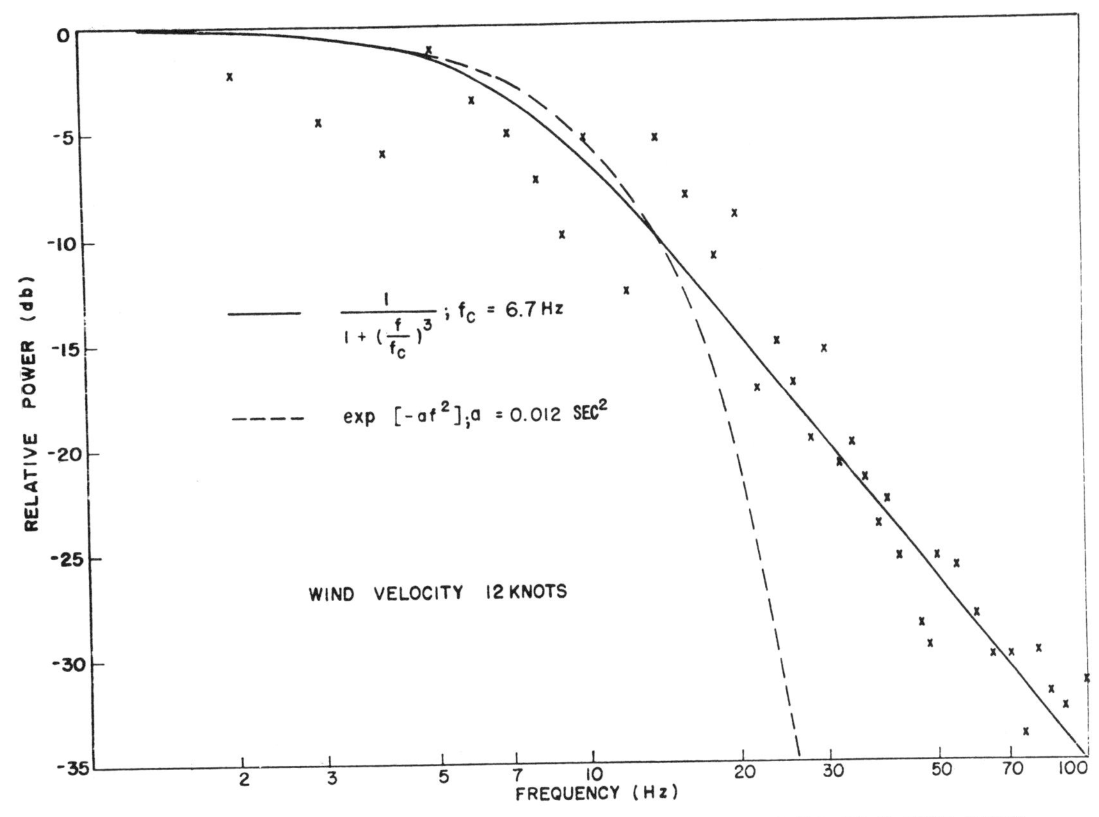

FIG. 4 APPROXIMATIONS OF THE CLUTTER SPECTRA OBTAINED WITH AN X-BAND RADAR

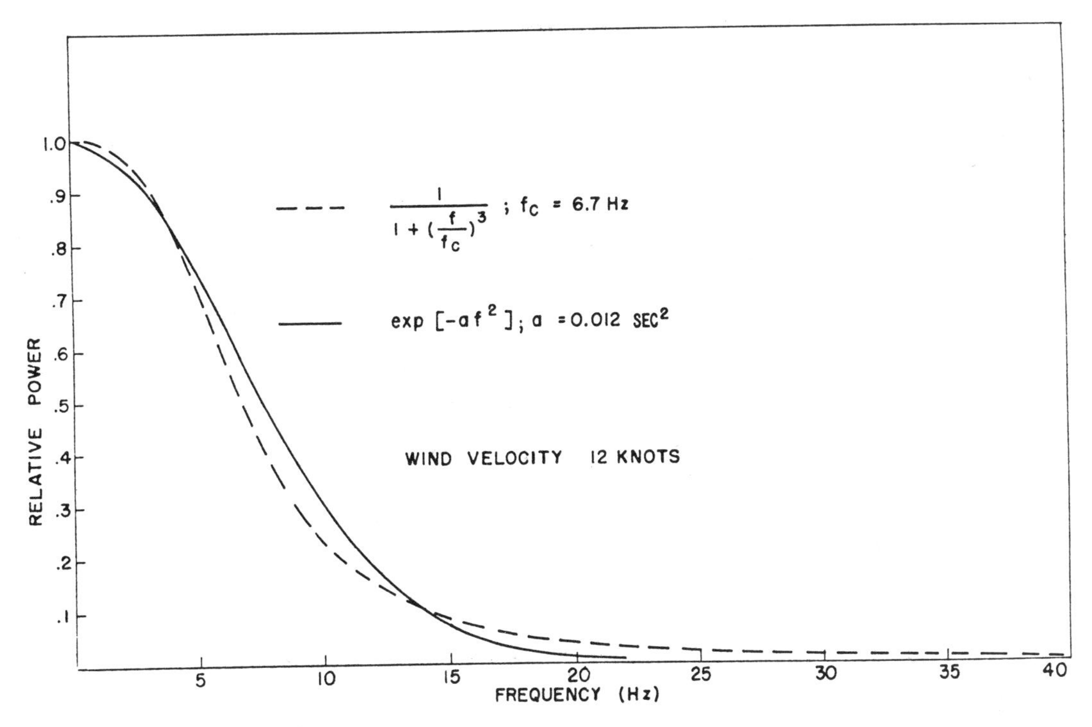

FIG. 5 APPROXIMATIONS OF THE CLUTTER SPECTRA OBTAINED WITH AN X-BAND RADAR

Conventional Filtering

Typical video processors (Appendix I) employ high-pass filtering to reduce signals resulting from clutter targets. A filter having a clutter rejection ratio of at least 20 db will usually give satisfactory MTI performance. An active filter with an attenuation slope of 12 db per octave meets this requirement. The clutter rejection ratios afforded by these types of filters having cutoff frequencies of 50 and 100 Hz are plotted in Figure 6. The rejection ratios for the 50-Hz filter were calculated; those of the 100-Hz filter are repeated from Table II. The rejection ratio provided by the 50-Hz filter is greater than 20 db for wind velocities up to 10 knots. The 100-Hz filter provides an attenuation of at least 20 db for nearly all wind conditions.

A circuit diagram of the active filter is shown in Figure 7. It contains two active and six passive elements. The two transistors may be replaced by a single high input impedance device such as a field-effect transistor. The filter can be packaged so that its contribution in size and weight compared to that of an overall MTI system is small.

The active elements in the filter sharpen the break point of the frequency response of the filter, also shown in Figure 7. The filter attenuates an input signal at a frequency of $f/f_p = 1$ by less than 1 db; a passive dual RC filter, while having the same attenuation slope as the active filter for $f/f_p \ll 1$, would attenuate a signal by 6 db at the break point. The sharp break point allows a good approximation of the frequency response of the active filter, e.g.

$$|H(P)| = \left(\frac{f}{f_p}\right)^2, \quad f < f_p$$

$$= 1, \quad f > f_p.$$

The filter has no insertion loss in the pass band.

Because of its simplicity, small size and weight, and performance, a 12-db-per-octave active filter has been selected for use in several MTI systems. A filter having a cutoff frequency of 50 Hz is used in most applications; one system offers a choice between a 50-Hz and a 100-Hz cutoff. The 50-Hz cutoff is preferred, since it allows a greater range of target frequencies to pass.

A 6-db-per-octave passive filter having a cutoff frequency of 100 Hz also meets the requirements of performance and simplicity. However, the 50-Hz active filter discussed above is preferred because of its lower cutoff frequency. A 6-db-per-octave passive filter does not provide sufficient clutter attenuation.

Although the 40-db-per-octave filter meets attenuation requirements, the complexity of the filter circuit results in excessive size and weight.

Narrow Band Filtering

The preceding section indicates that a 12-db-per-octave active filter with a cutoff frequency of 50 Hz will give adequate performance in an MTI video processor. This filter still has the basic limitation that its frequency response is fixed while the signal it is designed to attenuate, i.e., the clutter spectrum, varies significantly according to the wind conditions to which the clutter target is subjected. On a calm day, the suggested filter would do a good job in eliminating clutter signals. For a clam condition, it would be possible (and desirable) to lower the cutoff frequency of the filter to allow a greater range of target frequencies to pass. On windy days, i.e., with winds above 10 knots, the filter will not provide sufficient clutter attenuation. It would be desirable under these conditions to raise the cutoff frequency of the filter.

A method of overcoming the limitations imposed by a single high-pass filter is to divide the entire audio bandwidth over which frequencies resulting from any target motion occur into several narrow band filters. This scheme allows the gain of the lower frequency bands to be adjusted according to the clutter conditions. A further improvement would be a system in which the gain of the low frequency bands would vary simultaneously with changing clutter conditions. An example of how this might be accomplished is shown in Figure 8. The outside boxes in the block diagram illustrate the method by which doppler information is processed. The boxcar output contains clutter and target signals.

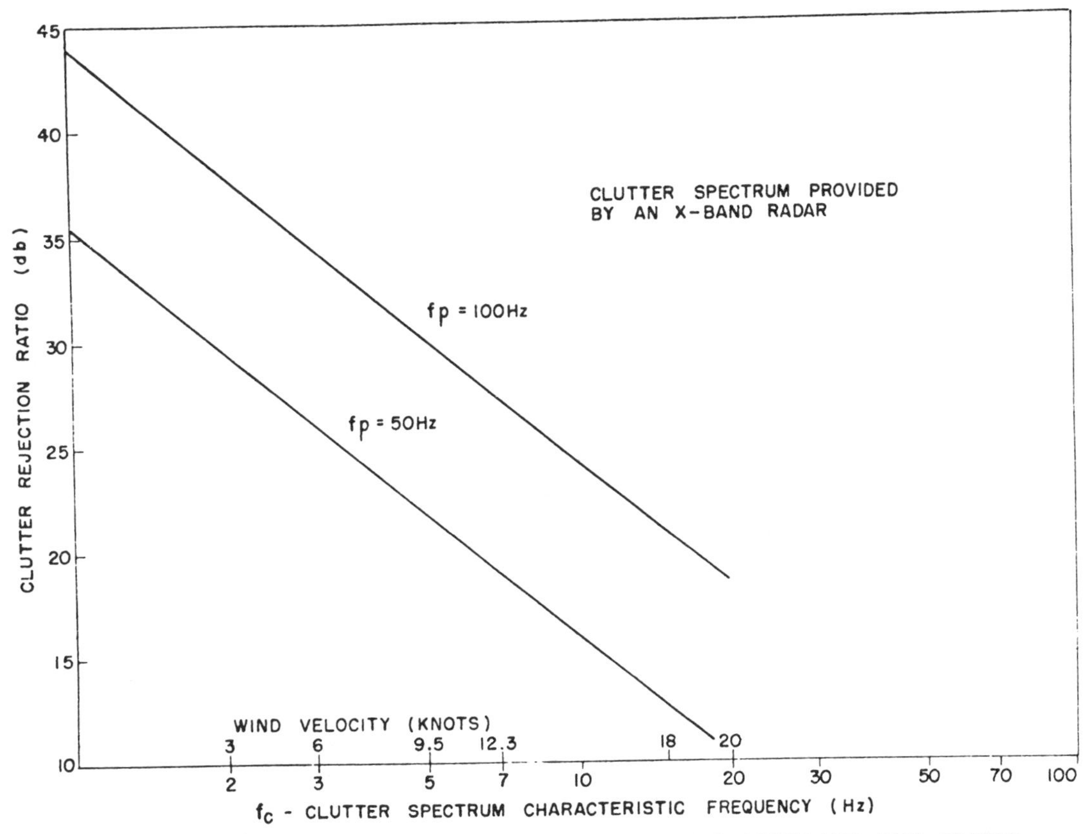

FIG. 6 CLUTTER REJECTION RATIOS OBTAINED WITH 12db/OCTAVE HIGH-PASS FILTERS

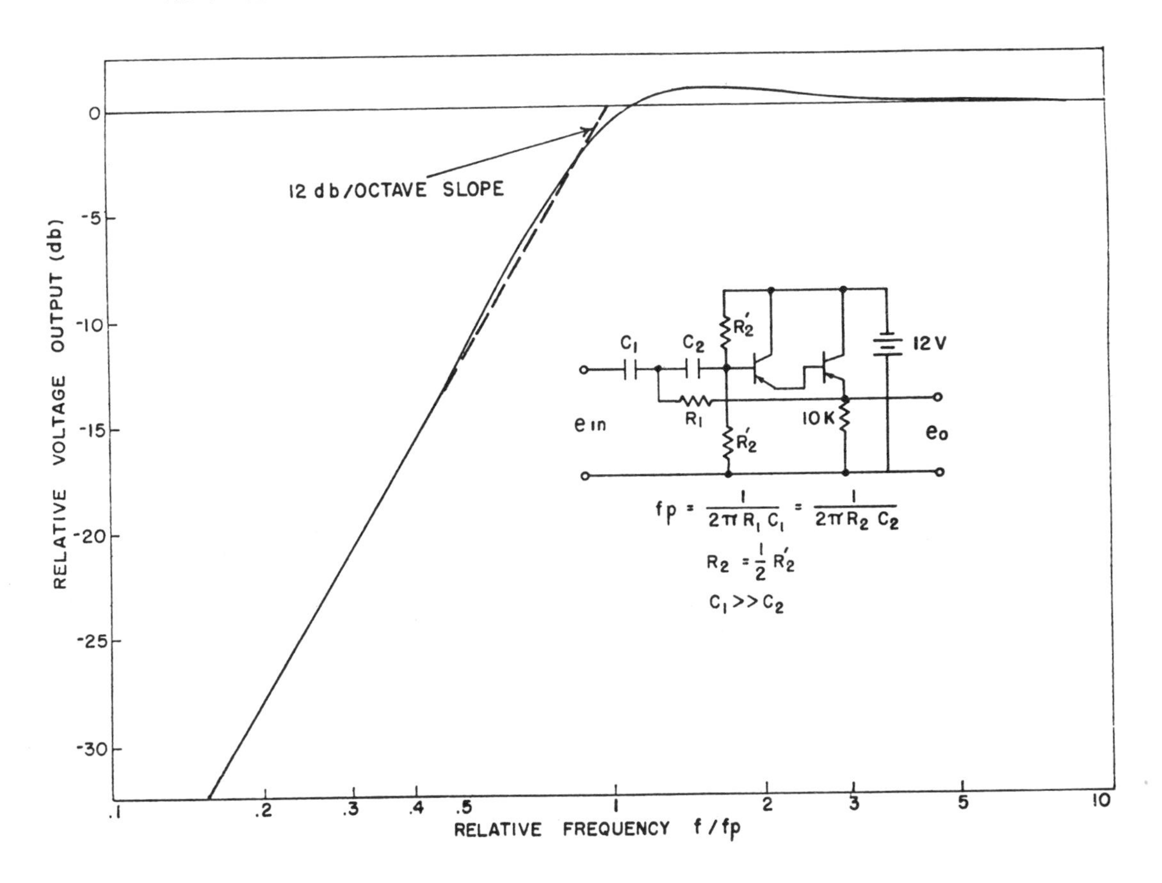

FIG. 7 12db/OCTAVE ACTIVE HIGH-PASS FILTER

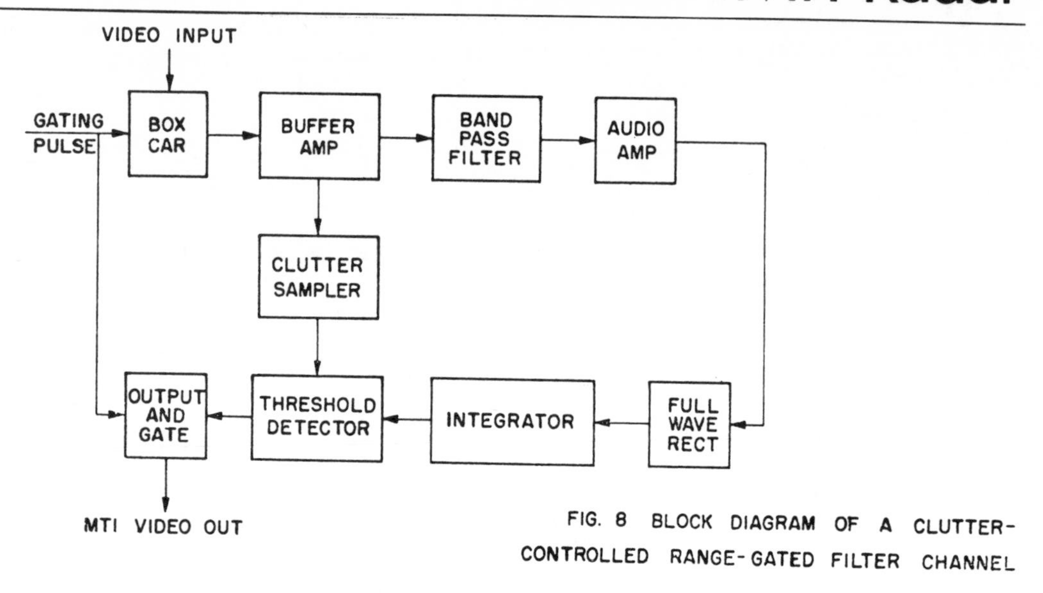

FIG. 8 BLOCK DIAGRAM OF A CLUTTER-CONTROLLED RANGE-GATED FILTER CHANNEL

A narrow band filter follows an amplification stage; the purpose of this filter is to pass only target signals. After further processing, the signal is threshold detected. The threshold level might be set by the signal level out of a clutter sampler. The sampler could be a narrow band filter centered at a frequency below that of the target filter, i.e., the filter designed to pass target frequencies. Under calm conditions, little energy would enter the clutter sampler and the threshold level would be at a minimum level. As the wind increases, clutter signals will be passed by the target filter. However, the energy in the clutter sampler will also increase, thus raising the threshold level which a signal passed by the target filter must exceed. No output signal due to clutter signals will exist at the threshold detector output.

Use of this scheme will yield the maximum signal-to-clutter level allowable under all conditions, thus obtaining a considerable enhancement in the performance of MTI systems.

CONCLUSIONS

The clutter spectrum received by a noncoherent X-band radar observing a group of natural clutter targets can be approximated by

$$P(f) = \frac{1}{1 + \left(\frac{f}{f_c}\right)^3}$$

where $f_c = k \exp \quad \beta v \quad \text{Hz}$

$k = 1.33$ Hz

$\beta = .1356$ knots^{-1}

v = wind velocity (knots).

This equation is significant in that the performance of any filter in attenuating clutter signals can now be predicted. The cubic expression gives a more accurate prediction of filter performance than the long assumed Gaussian spectrum.

A 12-db-per-octave active high-pass filter having a cutoff frequency of 50 Hz will provide a clutter attenuation of at least 20 db for winds up to 10 knots, which will afford satisfactory MTI performance under most conditions.

MTI systems performance can be improved by covering the audio spectrum with several narrow band filters. Applying adaptive filtering techniques to the lower frequency bands will result in the maximum MTI sensitivity possible under all wind conditions.

RECOMMENDATIONS

Future effort should be expended in three areas. Firstly, the variations in the clutter spectrum with changes in the operating frequency of the radar should be determined. The return from coherent radars should also be analyzed.

Secondly, work should be done in developing a suitable adaptive filtering technique to yield maximum MTI sensitivity.

Finally, a study should be made to see if moving target signals of interest can be separated from clutter signals by using the statistical properties of clutter.

APPENDIX I

A VIDEO PROCESSOR

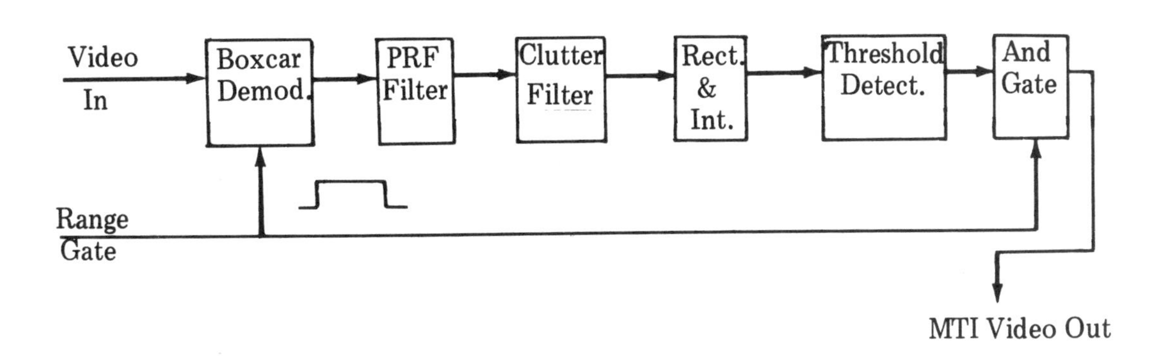

FIGURE I-1

RANGE-GATED FILTER CHANNEL

The basic building block of a video processor for pulsed radar systems is a range-gated filter channel. Figure I-1 shows a block diagram of a range channel. A boxcar demodulator is used to recover the doppler modulation information present in the video output of a radar system. The boxcar is a sample-and-hold device. It is usually gated at the radar PRF, with a pulse width equal to that of the transmitted pulse.

The delay of the gating pulse in relation to the transmitted pulse determines the range of the target being sampled. The boxcar output is a complex audio signal. Ambiguous signals, high frequency noise, and components of the PRF are removed by a low-pass filter. This filter has a high frequency cutoff of one-half the radar pulse repetition frequency. D-c signals resulting from fixed targets and other undesired low-frequency signals are eliminated by a high-pass filter. These latter signals result primarily from windblown natural clutter targets. The remaining signal is rectified, integrated, and threshold detected. The detector output is regated with the same pulse that gated the boxcar. The final output is a video pulse occurring at the correct radar range. A pulse will appear only for moving target inputs.

It is possible to use several range channels, each corresponding to a specific radar range interval. In this manner, moving targets appearing anywhere within the antenna beamwidth can be displayed simultaneously on a radar indicator.

APPENDIX II

RADAR SET AN/TPS-25 CHARACTERISTICS

The Radar Set AN/TPS-25 is a long-range noncoherent pulse doppler combat surveillance radar. The radar characteristics are as follows:

Frequency	9375 megahertz
Pulse Width	0.5 microseconds
PRF	1850 pulses per second
Peak Power	65 kilowatts
Azimuth Beamwidth	2 degrees
Elevation Beamwidth	4 degrees
Sector Scan	30 degrees in 22 seconds
Maximum Range	20,000 yards

APPENDIX III

CLUTTER REJECTION RATIO COMPUTATIONS

A method of calculating the clutter rejection ratios provided by various filters when attenuating the clutter spectrum is discussed here. The clutter rejection ratio is defined by

R = Clutter Rejection Ratio

$$= \frac{\int_0^\infty P(f)\,df}{\int_0^\infty P(f)\,|H(P)|^2\,df} \qquad \text{(III-1)}$$

where $P(f)$ = clutter power spectrum

and $H(P)$ = filter voltage transfer function.

The clutter spectrum can be approximated by

$$P(f) = \frac{1}{1+\left(\frac{f}{f_c}\right)^3} \qquad \text{(III-2)}$$

where f_c = clutter spectrum characteristic frequency (Hz).

The numerator of (III-1) can then be expressed as

$$\int_0^\infty \frac{df}{1+\left(\frac{f}{f_c}\right)^3}.$$

This integral can be evaluated directly.

Let $X = \frac{f}{f_c}$.

Then

$$\int_0^{\infty} \frac{df}{1+\left(\frac{f}{f_c}\right)^3} = f_c \int_0^{\infty} \frac{dx}{1+x^3} .$$

The substitution $S = x^3$ results in

$$f_c \int_0^{\infty} \frac{dx}{1+x^3} = \frac{f_c}{3} \int_0^{\infty} \frac{S^{1/3-1}}{1+S} \; dS.$$

Evaluation of the right-hand integral requires the use of Beta and Gamma functions.[6] The result is given in many handbooks of definite integrals. From Burington, [7]

$$\int_0^{\infty} \frac{y^{n-1}}{1+y} \; dy = \frac{\pi}{\sin n\pi}, \quad 0 < n < 1.$$

Using this,

$$\int_0^{\infty} \frac{df}{1+\left(\frac{f}{f_c}\right)^3} = \frac{2\sqrt{3}\,\pi\, f_c}{9}. \qquad \text{(III-3)}$$

The denominator of (III-1) cannot usually be evaluated by direct integration. Evaluation requires use of a form of the fundamental theorem of integral calculus,[8] i.e.

$$\int_0^{\infty} P(f) \mid H(P) \mid^2 df = \lim_{\substack{N \to \infty \\ \Delta f \to 0}} \sum_{n=1}^{N} P(f = n\Delta f) \mid H(p = n\Delta f) \mid^2 \Delta f . \qquad \text{(III-4)}$$

For all clutter spectra and filters used to attenuate clutter signals, the values $\Delta f = 1$ Hz and $N = 200$ are sufficient for the summation to be a good approximation to the integral.

The clutter rejection ratios of six filters acting on the clutter spectrum were obtained by programming a computer to perform the computations shown in Equations (III-3) and (III-4). Values of f_c, the clutter spectrum characteristic frequency, of from one to twenty Hz were taken in one Hz increment. Magnitudes of the filter transfer functions H(p) were taken directly from frequency response plots of the filters. The plots are shown in Figure (III-1).

The caluculated clutter rejection ratios are given in Table III-1.

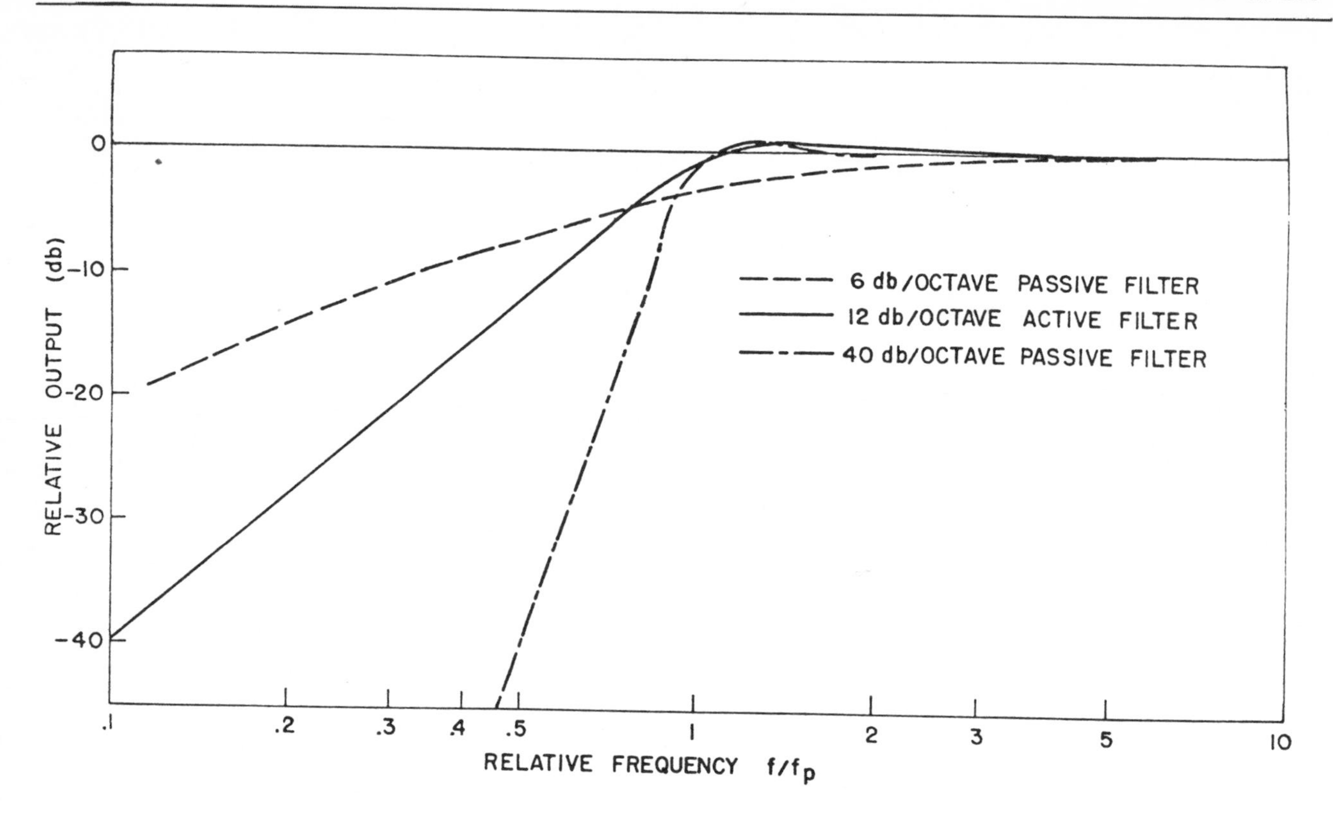

FIG. III-1 FREQUENCY RESPONSE CURVES OF CLUTTER FILTERS

TABLE III-1

CALCULATED CLUTTER REJECTION RATIOS

	50	100	50	100	50	100	f_p(Hz)
f_c(Hz)	6	6	12	12	40	40	α(db)
1	29	35	36	44	41	47	
2	24	29	30	38	35	41	
3	21	26	26	35	31	38	
4	19	24	24	32	29	35	
5	17	22	22	30	27	33	
6	16	21	20	29	25	32	
7	15	20	19	27	24	30	
8	14	19	18	26	23	29	
9	13	18	17	25	22	28	
10	13	17	16	24	21	27	R (db)
11	12	17	15	24	20	26	
12	12	16	14	23	19	26	
13	11	16	14	22	19	25	
14	11	15	13	22	18	24	
15	10	15	13	21	17	24	
16	10	14	12	20	17	23	
17	10	14	12	20	16	23	
18	9	13	11	19	16	22	
19	9	13	11	19	15	22	
20	9	13	10	19	15	21	

f_p = filter low-frequency cutoff
α = filter attenuation slope
f_c = clutter spectrum characteristic frequency
R = clutter rejection ratio

APPENDIX IV

CUBIC AND GAUSSIAN SPECTRA COMPARISON

Two approximations to the clutter power spectrum are

$$P(f) = \exp\left[-af^2\right]$$

and

$$P(f) = \frac{1}{1+\left(\frac{f}{f_c}\right)^3}$$

where a and f_c are constants determined by the frequency of the radar observing the clutter target and wind velocity to which the clutter target is subjected.

The constants a and f_c can be related by assuming that the two spectra have equal total powers. Then,

$$\int_0^\infty \frac{1}{1+\left(\frac{f}{f_c}\right)^3}\, df = \int_0^\infty \exp\left[-af^2\right] df . \qquad \text{(IV-1)}$$

The right-hand integral can be written as

$$\frac{\sqrt{2\pi}}{\sqrt{2a}} \int_0^\infty \frac{1}{\sqrt{2\pi}} \exp\left[-\frac{1}{2}(\sqrt{2a}\,f)^2\right] \sqrt{2a}\, df.$$

This integral is the normalized Gaussian function and has a value of 1/2.

$$\therefore \int_0^\infty \exp\left[-af^2\right] df = \frac{\sqrt{\pi}}{2\sqrt{a}} .$$

The left-hand integral in IV-1 can be evaluated from tables of definite integrals.* The result is

$$\int_0^\infty \frac{df}{1+\left(\frac{f}{f_c}\right)^3} = \frac{2\sqrt{3}\,\pi\, f_c}{9} .$$

Equating the results of these two integrals,

$$\frac{2\sqrt{3}\,\pi\, f_c}{9} = \frac{\sqrt{\pi}}{2\sqrt{a}}$$

and $f_c = \dfrac{9}{4\sqrt{3\pi a}}$

*See Appendix III

APPENDIX V

CLUTTER REJECTION RATIO CALCULATION FOR A GAUSSIAN SPECTRUM

The clutter rejection ratio obtained when a 12-db-per-octave active highpass filter attenuates a Gaussian-shaped clutter power spectrum is calculated here. The rejection ratio is defined by

$$R = \frac{\int_0^\infty P(f)\, df}{\int_0^\infty P(f)\,|\,H(p)\,|^2\, df}$$

where $P(f)$ = clutter power spectrum

and $H(p)$ = filter voltage transfer function.

A Gaussian spectrum may be expressed by

$$P(f) = \exp\left[-af^2\right]$$

where a is a constant determined by the frequency of the radar when observing the clutter target and wind velocity to which the clutter target is subjected.

A good approximation of the transfer function of a 12-db-per-octave highpass filter is

$$H(p) = \left(\frac{f}{f_p}\right)^2 \qquad f \leqslant f_p$$

$$= 1 \qquad f > f_p$$

where f_p is the filter cutoff frequency.

Then

$$R = \frac{\int_0^\infty \exp\left[-af^2\right] df}{\int_0^{f_p}\left(\frac{f}{f_p}\right)^4 \exp\left[-af^2\right] df + \int_{f_p}^\infty \exp\left[-af^2\right] df} \, . \tag{V-1}$$

The numerator of this expression can be written as follows:

$$\int_0^\infty \exp\left[-af^2\right] df = \frac{\sqrt{2\pi}}{\sqrt{2a}} \int_0^\infty \frac{1}{\sqrt{2\pi}} \exp - \left[\frac{(\sqrt{2a}\, f)^2}{2}\right] \sqrt{2a}\, df \, .$$

This integral is the normal function and has a value of 1/2.

$$\therefore \int_0^\infty \exp\left[-af^2\right] df = \frac{\sqrt{\pi}}{2\sqrt{a}} \, . \tag{V-2}$$

The first integral in the denominator of (V-1) can be expanded through successive integration by parts into

$$- \left[f_p^{\,2} + \frac{3}{2a} \right] \left[\frac{1}{2af_p^{\,3}} \right] \exp \left[-af_p^{\,2} \right] + \frac{3}{[2af_p^{\,2}]^2} \int_0^{f_p} \exp \left[-af^2 \right] df \; .$$

The second integral in the denominator of (V-1) can be written as

$$\int_0^{\infty} \exp \left[-af^2 \right] df - \int_0^{f_p} \exp \left[-af^2 \right] df.$$

The left-hand integral has been evaluated above. The denominator of (V-1) can now be written as

$$\frac{\sqrt{\pi}}{2\sqrt{a}} - \left[f_p^{\,2} + \frac{3}{2a} \right] \left[\frac{1}{2af_p^{\,3}} \right] \exp \left[-af_p^{\,2} \right] + m \left[\frac{3}{(2af_p^{\,2})^2} - 1 \right] \qquad \text{(V-3)}$$

where $m = \int_0^{f_p} \exp \left[-af^2 \right] df$

m can be put in the form of the normal function by making the substitution $x = \sqrt{2a}\; f$ into the above integral.

Then $$\int_0^{fp} \exp \left[-af^2 \right] df = \frac{\sqrt{\pi}}{\sqrt{a}} \int_0^{\sqrt{2a}f_p} \frac{1}{\sqrt{2\pi}} \exp \left[-\frac{x^2}{2} \right] dx$$

$$= \frac{\sqrt{\pi}}{\sqrt{a}} \, n$$

where $n = \int_0^{\sqrt{2a}f_p} \exp \left[-\frac{x^2}{2} \right]$ dx is the normal function.

Values for n as a function of the limits of the integral are given in many mathematical handbooks.*

(V-3) is now written as

$$\frac{\sqrt{\pi}}{2\sqrt{a}} \cdot \left[1 + 2\,n \left(\frac{3b^2}{f_p^{\,2}} - 1 \right) \right] - b \left[1 + \frac{3b}{f_p} \right] \exp \left[-af_p^{\,2} \right] \qquad \text{(V-4)}$$

where $b = \dfrac{1}{2af_p}$.

Substituting (V-2) and (V-4) into (V-1),

$$R = \left\{ 1 - 2n + 6n \frac{b^2}{f_p^{\,2}} - \frac{2\sqrt{a}\,b}{\sqrt{\pi}} \left[1 + \frac{3b}{f_p} \right] \exp \left[-af_p^{\,2} \right] \right\}^{-1} \; . \qquad \text{(V-5)}$$

*For example, Burington, R.S., op. cit. See Reference 7.

Evaluation of this expression gives the attenuation of a Gaussian approximation to a clutter spectrum, when a 12-db-per-octave active high-pass filter is used. Consider the case when a filter having a cutoff frequency of 100 Hz attenuates the spectrum provided by a radar operating at 10 GHz observing a clutter target subjected to a wind having a velocity of 20 m.p.h.

Skolnik* gives a value of a for the conditions cited of 2.3×10^{-3}. The constants in (V-5) are now as follows:

$$f_p = 100 \text{ Hz}$$

$$a = 2.3 \times 10^{-3} \text{ sec}^2$$

$$n = \int_0^{\sqrt{2af_p}} \exp\left[-\frac{x^2}{2}\right] dx = 0.5$$

$$b = \frac{1}{2af_p} = 2.17 \;.$$

Substituting these values into the last term of (V-5) reveals this term to have a magnitude of the order of 10^{-11}. This value is negligible compared to that of the $6\,n\,b^2/f_p^{\,2}$ term. Hence, if this term is disregarded, (V-5) can be written simply as

$$\left[6\,n\,\frac{b^2}{f_p^{\,2}}\right]^{-1} \;.$$

$$\text{Finally, } R = \frac{4a^2 f_p^{\,4}}{3} \;.$$

$$= 705$$

$$= 28.5 \text{ db.}$$

*Skolnik, M.I., op. cit. See Reference 3.

REFERENCES

1. Kerr, D.E., Propagation of Short Radio Waves, pp 553-588, McGraw-Hill Book Company, Inc., New York (1951).
2. Barlow, E.J., "Doppler Radar," Proceedings of the IRE, Vol. 37, pp 340-355 (April 1949).
3. Skolnik, M.I., Introduction to Radar Systems, p 146, McGraw-Hill Book Company, Inc. New York (1962).
4. Davenport. W.B. and Root, W.R., Random Signals and Noise, Chapter 6, McGraw-Hill Book Company, Inc., New York (1958).
5. Lawson, J.L. and Uhlenbeck, G.E., Threshold Signals, pp 26-29, McGraw-Hill Book Company, Inc., New York (1950).
6. Hildebrande, F.B., Advanced Calculus for Applications, pp 80-91 Prentice-Hall, Inc. Englewood Cliffs (1964).
7 Burington, R.S., Handbook of Mathematical Tables and Formulas, Third Edition, p 88, Handbook Publishers, Inc., Sandusky (1954).
8. Gay, H.J., Analytic Geometry and Calculus, pp 156-159, McGraw-Hill Book Company, Inc., New York.

Section 5 Digital MTI (DMTI)

EDITOR'S COMMENTS

The six papers presented in this section describe the rapidly evolving field of digital signal processing as applied to MTI radars.

The current trend is to perform all radar signal processing functions in a general purpose programmable digital processor. The rapid development of LSI digital circuits and microprocessors support this trend.

The desire for computational efficiency, which allows an increasing number of functions to be performed in the digital signal processor at an increasing processing speed, has given the FFT algorithm a special place in the digital processing world. Recent research in this area has resulted in algorithms (e.g., the Winograd Super FFT algorithm) that provide greater processing speed than the original Cooly-Tukey FFT algorithm.

The "Achilles Heel" of the digital signal processor is the associated analog-to-digital (A-D) converter. The A-D converter samples the analog input signal at the Nyquist rate consistent with the signal bandwidth, and provides a digital message that represents the signal's amplitude at the sample time. The A-D output is necessarily a quantized version of the signal, and this process generates quantization noise that must be considered in addition to the usual noise encountered in analog signal processors.

The first two papers (5.1 and 5.2) cover various theoretical and practical aspects of A-D converters. The present state-of-the-art provides much greater processing rates (200 to 400 MHz) and message bits (10 to 12 bits) than the A-D converters described in Paper 5.2. (The rates and bits quoted above are not currently achieved in the same A-D converter).

Paper 5.2 describes a digital implementation of a conventional analog type MTI. Of significance is the advanced capability provided by this digital MTI (DMTI) over the analog MTIs available at the time this paper was first published in 1967.

Papers 5.3 through 5.6 describe an intensive effort at Lincoln Laboratories to develop digital processors for application to Air-Traffic Control radars. The digital processors described in these papers are representative of the current state-of-the-art in digital processing.

Paper 5.3 is a summary of the theoretical considerations involved in the design of modern digital signal processors. The significant result of this paper is that a three pulse canceler followed by a 16 point DFT approaches the performance provided by the optimum complex weight MTI processor detailed in Paper 2.3.

Papers 5.4 and 5.5 describe the system philosophy and hardware implementation of a modern pipeline digital processor (called the MTD) applied to a modified coherent linear FPS-18 radar.

Paper 5.6 describes a general purpose programmable digital processor as applied to MTI radars and gives some of the theoretical considerations involved in the use of the FFT.

The intersting results presented in this series of papers are:

Paper 5.1 *(Brennan and Reed)*

- **Presents an in-depth analysis of quantization noise as applied to digital MTI.**
- **Identifies a quieting effect due to correlation between quantization errors on successive pulses.**

Paper 5.2 *(Linder and Kutz)*

- **Describes a 5-pulse nonrecursive digital MTI (DMTI) optimized using a least-mean-square fit to an ideal MTI frequency response.**
- **Uses both inphase-and-quadrature (I and Q) channels to eliminate blind phases.**
- **Uses variable interpulse intervals (VIP) to eliminate blind speeds and provides post MTI digitally implemented feedback video integration.**
- **Describes a 9 bit, 10 MHz sampling rate A-D converter.**

Paper 5.3 *(McAulay)*

- **Using a prewhitening filter approach (classic Bode-Shannon optimal linear filtering approach) the author determines that an optimum MTI filter can be approximated by a clutter rejection filter (MTI canceler) followed by a doppler filter bank (DFT).**
- **Compares the signal-to-interference ratio (SIR) for an optimum complex weight MTI,**

a clutter-filter-16 point DFT processor, a clutter-filter-32 point DFT processor and a 2 and 3 pulse canceler. Both the clutter-filter - DFT processors approach the optimum performance, while the MTI cancelers provide substantially less performance than the optimum. The performance of the MTI cancelers do not include post detection integration so that their performance is understated.

Paper 5.4 *(Muehe, Cartledge, Drury, Hofstetter, Labitt, McCorison and Sferrino)*

- In-depth discussion of limitations of current Air Traffic Control radars. Problems include noncoherent operation, receiver limiting, second-time-around clutter, weather and angel clutter, inadequate MTI performance, rejection of tangential targets, and clutter residues saturating the data processing system.
- Solution embodied in MTD digital processor includes: coherent operation, wide dynamic range linear receiver, quasi-optimum clutter filter — DFT MTI processor, multiple PRFs, adaptive thresholding through digital ground clutter map, and weather clutter suppression through desensitization of doppler cells using mean-level threshold.

Paper 5.5 *(O'Donnell, Muehe, Labitt, Drury, Cartledge)*

- Describes hardware details of the MTD processor. Uses a 10 bit A-D converter sampling I and Q channels at a 2.6 MHz sampling rate, a 8192 bit memory, a 3 pulse MTI canceler followed by a cosine-on-pedestal weighted 8 point FFT, a disc stored clutter map, multiple PRF, and mean-leveled adaptive thresholding in doppler cells 2 through 6.
- Modified FPS-18 transmitter allows 52 dB MTI improvement.

Paper 5.6 *(Gold and Muehe)*

- Describes a general purpose signal processor (GPSP) that performs data weighting, DFT, recursive and nonrecursive digital filtering, matched filtering via FFT, post detection integration, doppler centroid estimation using FFT, and adaptive thresholding.
- Determines that the basic figure of merit for a signal processor equals the time it takes to perform a basic FFT "butterfly" operation. In 1970, a rate of 31,000 FFTs per second was possible thus allowing the doppler processing of thousands of range gates.

QUANTIZATION NOISE IN DIGITAL MOVING TARGET INDICATION SYSTEMS
5.1 L.E. Brennan and I. Reed
IEEE Transactions on Aerospace and Electronics Systems, Vol. AES-2, No. 6, November, 1966, pp. 655-658.

Quantization Noise in Digital Moving Target Indication Systems

LAWRENCE E. BRENNAN, Member, IEEE
The RAND Corporation
Santa Monica, Calif.

IRVING S. REED
The RAND Corporation and
Department of Electrical Engineering
University of Southern California
Los Angeles, Calif.

Abstract

This paper deals precisely with the effect of quantum or step size in an analog-to-digital converter on the quantization noise in the output of MTI (moving target indication) radar processors. The correlation between quantization errors on successive pulses is considered in the analysis and is shown to have a small quieting effect on the output quantization noise.

Key Words—Digital, indication, moving target, noise, quantization, systems.

Manuscript received April 22, 1966.

Introduction

In the past, most MTI (moving target indication) radars have used analog signal processing. With the present capability in analog-to-digital converters, digital memories, and arithmetic units, it is possible to perform this processing digitally in some systems. A trend toward digital MTI processors can be expected with future improvements in digital technology. This paper deals with the effect of quantum size, or number of bits in the analog-to-digital conversion, on quantization noise in the output of an MTI processor. The correlation between quantization errors on successive pulses is considered in the analysis. This correlation has a quieting effect[1] in some cases, reducing the output quantization noise.

An MTI processor forms the coherent sum of two or more returns from one radar resolution cell, employing a set of weights a_n, and can be represented by

$$X = \sum a_n x_n. \tag{1}$$

In the most general case, both the weights a_n and the train of radar returns x_n can be complex quantities. If only one quadrature component of the RF or IF signal is retained, the x_n can be represented as real quantities. If both quadrature components are processed, the x_n are complex; but it will suffice to consider only real x_n. When both quadrature components are present, the following analysis is applicable to each of the two components, and the output powers in the two channels are additive.

Two common types of MTI are the single-delay canceller, which combines two returns with the weights $a_2 = -a_1$, and the double-delay canceller, which combines three returns with $a_1 = a_3 = -a_2/2$. In range-gated filter-bank systems, more returns are added coherently, and the a_n represent time samples of the impulse response function of a filter.

Formulation of Problem

Let $[x_n]$ represent the digital value of x_n, and $e(x_n)$ the error in analog-to-digital conversion:

$$[x_n] = x_n + e(x_n). \tag{2}$$

The error is a sawtooth function as shown in Fig. 1. It is convenient to use the Fourier series representation of $e(x)$ in the following analysis:

$$e(x) = \frac{\Delta}{\pi} \sum_{k=1}^{\infty} \frac{(-1)^k}{k} \sin\left(\frac{2\pi k x}{\Delta}\right). \tag{3}$$

[1] The possibility of quieting due to correlation was suggested to the authors by Glenn W. Preston, a Consultant to The RAND Corporation.

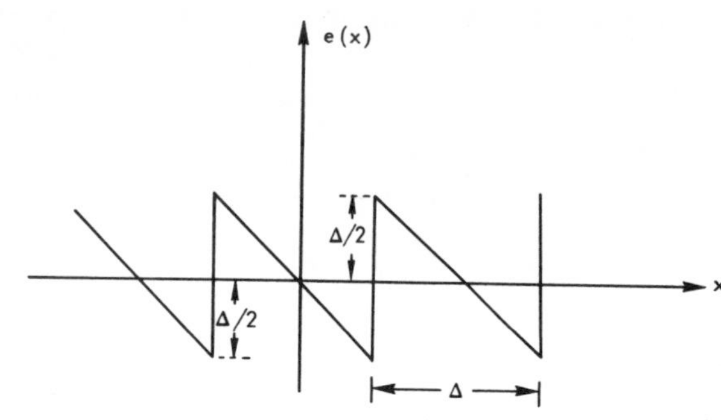

Fig. 1. The error e(x).

If the clutter return in a resolution cell is due to reflections from many randomly phased scatterers, a model commonly assumed and representative of sea clutter and backscatter from many types of terrain, the variables x_n are normally distributed. The joint probability density of x_m, x_n is then

$$p(x_m, x_n) = \frac{1}{2\pi\sigma^2\sqrt{1-\rho_{mn}^2}} \exp\left\{-\frac{(x_m - A_m)^2 + (x_n - A_n)^2 - 2\rho_{mn}(x_m - A_m)(x_n - A_n)}{2\sigma^2(1-\rho_{mn}^2)}\right\} \quad (4)$$

where

σ^2 = variance of x_n
A_n = mean value of x_n due to a signal
ρ_{mn} = normalized correlation coefficient.

The output of a digital MTI processor, neglecting errors in the computation and in the weights a_n, is

$$[X] = \sum a_n [x_n]. \quad (5)$$

A measure of the performance of a MTI system, based on the deflection criterion,[2] is the output ratio

$$R_0 = \frac{|E\{[X] \mid A_n \neq 0\} - E\{[X] \mid A_n = 0\}|^2}{E\{|[X] - E\{[X]\}|^2 \mid A_n = 0\}}. \quad (6)$$

The numerator is the square of the incremental output voltage due to a signal, and the denominator is the variance of output voltage due to clutter plus quantization noise. Since the expectation of $[X]$ is zero in the absence of signal, as shown in the next section, the numerator of (6) reduces to $|E\{[X] \mid A_n \neq 0\}|^2$, and the denominator becomes

$$E\{|[X]|^2 \mid A_n = 0\} = \sum\sum a_m a_n^* E\{[x_m][x_n]\}. \quad (7)$$

The problem then reduces to computing the first and second moments of $[x_n]$.

When a signal is present, the mean value of x_n is generally nonzero. The set of mean values $\{A_n\}$ is a function of the input signal-to-clutter power ratio R_1 and the signal phases with respect to the reference phase $\{\psi_n\}$.

$$R_1 = \frac{A_{\max}^2}{2\sigma^2}$$

$$A_n = A_{\max} \cos \psi_n. \quad (8)$$

Expression for First Moment of $[x_n]$

The first moment of $[x_n]$ is

$$E\{[x_n]\} = E\{x_n\} + E\{e(x_n)\}$$
$$= A_n + \frac{1}{\sqrt{2\pi}\sigma}\int_{-\infty}^{\infty} e^{-(x_n - A_n)^2/2\sigma^2} e(x_n) dx_n. \quad (9)$$

Replacing $e(x_n)$ by its Fourier series and inverting the order of summation and integration gives

$$E\{[x_n]\} = A_n + \frac{\Delta}{\pi}\sum_{k=1}^{\infty}\frac{(-1)^k}{k} \cdot e^{-2\pi^2 k^2(\sigma/\Delta)^2} \sin\left(\frac{2\pi k A_n}{\Delta}\right); \quad (10)$$

when $\sigma > \Delta$, the case of most interest is

$$E\{[x_n]\} = A_n + 0(e^{-2\pi^2(\sigma/\Delta)^2}). \quad (11)$$

Expression for Second Moments of $[x_n]$

From (2), the second moments can be expanded in the form

$$E\{[x_m][x_n]\} = E\{x_m x_n\} + E\{e(x_m)e(x_n)\}$$
$$+ E\{x_m e(x_n) + x_n e(x_m)\}. \quad (12)$$

The first term is independent of quantum size and is given by

$$E\{x_m x_n\} = \sigma^2\rho_{mn} + A_m A_n. \quad (13)$$

In the second term

$$E\{e(x_m)e(x_n)\} = \iint_{-\infty}^{\infty} p(x_m, x_n) e(x_m) e(x_n) dx_m dx_n; \quad (14)$$

replacing $e(x_m)$ and $e(x_n)$ by Fourier series expansions, substituting $u = (x_m - A_m)$ and $v = (x_n - A_n)$, and interchanging the order of summation and integration gives

$$E\{e(x_m)e(x_n)\} = \left(\frac{\Delta}{\pi}\right)^2 \sum_{k,l=1}^{\infty} \frac{(-1)^{k+l}}{kl} I_{kl}$$

$$I_{kl} = \iint_{-\infty}^{\infty} e^{-(u^2+v^2-2\rho_{mn}uv)/2\sigma^2(1-\rho_{mn}^2)} \sin\left[\frac{2\pi k}{\Delta}(u + A_n)\right] \cdot \sin\left[\frac{2\pi l}{\Delta}(v + A_n)\right] \frac{du dv}{2\pi\sigma^2\sqrt{1-\rho_{mn}^2}}. \quad (15)$$

[2] See, for example, J. L. Lawson and G. E. Uhlenbeck (Eds.), *Threshold Signals*. New York: McGraw-Hill, 1950, p. 161.

This integral can be evaluated by expanding the sine terms to obtain integrals of the form[3]

$$\int_{-\infty}^{\infty} e^{-q^2(x^2-2\lambda x)} \begin{pmatrix}\sin\\ \cos\end{pmatrix} px \, dx = \frac{\sqrt{\pi}}{q} e^{-(p^2/4q^2)+q^2\lambda^2} \begin{pmatrix}\sin\\ \cos\end{pmatrix} p\lambda. \quad (16)$$

The resulting expression for the second moments of $e(x_n)$ is[4]

$$E\{e(x_m)e(x_n)\} = \frac{\Delta^2}{2\pi^2} \sum_{k,l=1}^{\infty} \frac{(-1)^{k+l}}{kl} \cdot \Big\{ e^{-2\pi^2(\sigma/\Delta)^2(k^2+l^2-2kl\rho_{mn})} \cos\left[\frac{2\pi}{\Delta}(kA_m - lA_n)\right] - e^{-2\pi^2(\sigma/\Delta)^2(k^2+l^2+2kl\rho_{mn})} \cdot \cos\left[\frac{2\pi}{\Delta}(kA_m + lA_n)\right]\Big\}. \quad (17)$$

In cases of interest, the quantization interval Δ will be smaller than σ, and (14) reduces to

$$E\{e(x_m)e(x_n)\} = \frac{\Delta^2}{2\pi^2} \sum_{k=1}^{\infty} \frac{1}{k^2} e^{-4\pi^2(\sigma/\Delta)^2 k^2(1-\rho_{mn})} \cdot \cos\left[\frac{2\pi k}{\Delta}(A_m - A_n)\right] + 0(e^{-2\pi^2(\sigma/\Delta)^2}). \quad (18)$$

Further, the expectation of $e^2(x_n)$ is

$$E\{e^2(x_n)\} = \frac{\Delta^2}{2\pi^2} \sum_{k=1}^{\infty} \frac{1}{k^2} + 0(e^{-2\pi^2(\sigma/\Delta)^2}) = \frac{\Delta^2}{12} + 0(e^{-2\pi^2(\sigma/\Delta)^2}). \quad (19)$$

The remaining two terms in (12) have the form

$$E\{x_m e(x_n)\} = \iint_{-\infty}^{\infty} x_m e(x_n) p(x_m, x_n) dx_m dx_n. \quad (20)$$

Again replace $e(x_n)$ by its Fourier series representation and substitute $u=(x_m-A_m)$ and $v=(x_n-A_n)$:

$$E\{x_m e(x_n)\} = \frac{\Delta}{\pi} \sum_{k=1}^{\infty} \frac{(-1)^k}{k} \int_{-\infty}^{\infty} \cdot \sin\left[\frac{2\pi k}{\Delta}(v + A_n)\right] \frac{dv}{\sqrt{2\pi}\sigma} \cdot I \quad (21)$$

where

$$I = \int_{-\infty}^{\infty} (u + A_m) \exp\left\{-\frac{u^2 + v^2 - 2\rho_{mn}uv}{2\sigma^2(1-\rho_{mn}^2)}\right\} \cdot \frac{du}{\sqrt{2\pi}\sigma\sqrt{1-\rho_{mn}^2}} = (A_m + \rho_{mn}v)e^{-v^2/2\sigma^2}. \quad (22)$$

Substituting (22) into (21) and integrating with respect to v gives

$$E\{x_m e(x_n)\} = \frac{\Delta}{\pi} \sum_{k=1}^{\infty} \frac{(-1)^k}{k} e^{-2\pi^2 k^2(\sigma/\Delta)^2} \cdot \left[A_m \sin\left(\frac{2\pi k A_n}{\Delta}\right) + \frac{\pi k \rho_{mn}\sigma^2}{\Delta} \cos\left(\frac{2\pi k A_n}{\Delta}\right)\right]. \quad (23)$$

As a check on (23), note that when $\sigma \to 0$, the expectation of $x_m e(x_n)$ reduces to $A_m e(A_n)$. Of more practical interest is the case when $\sigma > \Delta$ and

$$E\{x_m e(x_n)\} = 0 + 0(e^{-2\pi^2(\sigma/\Delta)}). \quad (24)$$

Combining (12), (13), (18), and (24) gives

$$E\{[x_m][x_n]\} = \sigma^2\rho_{mn} + A_m A_n + \frac{\Delta^2}{2\pi^2} \sum_{k=1}^{\infty} \frac{1}{k^2} \cdot e^{-4\pi^2 k^2(\sigma/\Delta)^2(1-\rho_{mn})} \cos\left[\frac{2\pi k}{\Delta}(A_m - A_n)\right] + 0(e^{-2\pi^2(\sigma/\Delta)^2}). \quad (25)$$

Example—Two-Pulse Canceller

In a conventional single-delay canceller with $a_1 = -1$, $a_2 = 1$,

$$X = x_2 - x_1. \quad (26)$$

Substituting (25) into (6), the output power ratio R_0 is

$$R_0 = \frac{(A_1 - A_2)^2}{2\sigma^2(1-\rho_{12}) + \frac{\Delta^2}{6}(1 - Q)} \quad (27)$$

[3] Bierens de Haan, *Nouvelles tables d'Intégrales Définies*. New York: Hafner, table 269.

[4] It was recently brought to the authors' attention that the method used above for computing $E\{e(x_m)e(x_n)\}$ was suggested originally by S. O. Rice; see W. R. Bennett "Spectra of quantized signals," *Bell System Technical Journal*, vol. 27, July 1948, pp. 446–472. For the special case of $A_1 = A_2 = 0$, Bennett obtains a more complex expression for this quantity by a different method. It can be shown by an appropriate change of indices that the two expressions are identical in this special case.

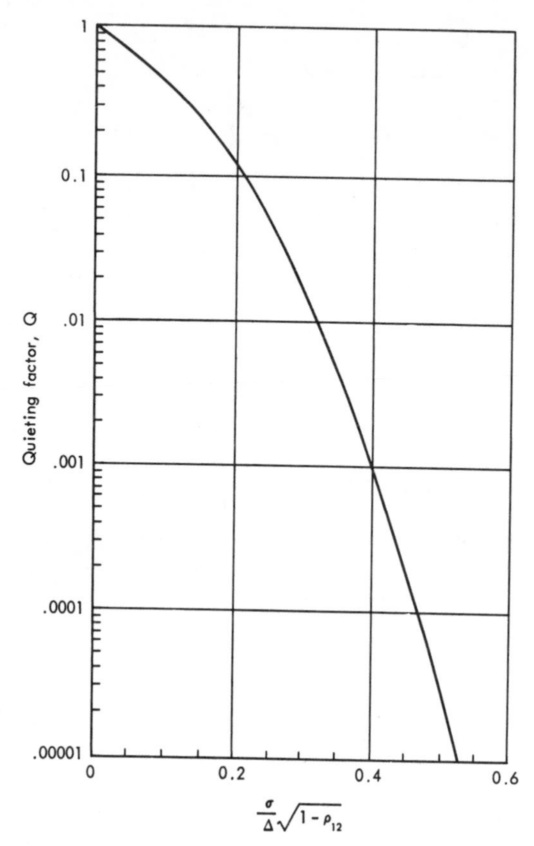

Fig. 2. Quieting factor vs. $\sigma\sqrt{1-\rho_{12}}/\Delta$.

neglecting terms of order $e^{-2\pi^2(\sigma/\Delta)^2}$. The factor Q is given by

$$Q = \frac{6}{\pi^2}\sum_{1}^{\infty}\frac{1}{k^2}e^{-4\pi^2k^2(\sigma/\Delta)^2(1-\rho_{12})} \tag{28}$$

and is a measure of the quieting due to correlation between quantization errors on successive pulses. The quieting factor Q is shown in Fig. 2 as a function of $\sigma/\Delta\sqrt{1-\rho_{12}}$.

The denominator of (27) contains two terms, the first of which is the output clutter power in the absence of quantization noise, and the second is the noise power due to quantization. In the absence of quieting, the second term becomes $\Delta^2/6$, which can be derived directly by assuming that the quantization errors in x_1 and x_2 are independent and uniformly distributed between $-\Delta/2$ and $\Delta/2$.

Conclusions

Expressions have been obtained for the first and second moments of $[x]$, the output of an analog-to-digital converter, where the input x is normally distributed. The results were given as a function of the variance of x, σ^2, the mean values of successive samples of x, A_n, and the step size of the A-D converter (Δ). The effects of a finite range for the A-D converter and possible overflow of this range in x were not considered. The results were applied to a MTI canceller, where it was shown that the correlation between quantization errors on successive pulses has a small quieting effect on the output quantization noise.

Lawrence E. Brennan (S'47–A'51–M'57) was born in Oak Park, Ill., on January 29, 1927. He received the B.S. and Ph.D. degrees in electrical engineering from the University of Illinois, Urbana, in 1948 and 1951, respectively.

Most of his work since leaving the University of Illinois has been concerned with radar and infrared systems. He joined the Electronics Department of The RAND Corporation, Santa Monica, Calif., in October, 1958. From April, 1962, to October, 1963, on leave of absence from RAND, he was employed by the SHAPE Air Defense Technical Center, The Hague, Netherlands. Since returning to RAND, he has been engaged primarily in research on radar detection theory and MTI.

Irving S. Reed was born in Seattle, Wash., on November 12, 1923. He received the B.A. and Ph.D. degrees in mathematics from the California Institute of Technology, Pasadena, in 1944 and 1949, respectively.

He was associated with Lincoln Laboratory, Massachusetts Institute of Technology, Lexington, from 1951 to 1960. From 1960 to 1963, he was a Senior Staff Member of The RAND Corporation, Santa Monica, Calif. He is presently both a Consultant to The RAND Corporation and a Professor of Electrical Engineering at the University of Southern California, Los Angeles. He is concerned primarily with mathematics, computer design, coding theory, stochastic processes, and information theory.

5.2

DIGITAL MOVING TARGET INDICATORS

R.A. Linder and G.H. Kutz

Supplement to IEEE Transactions on Aerospace and Electronic Systems, Vol. AES-3, No. 6, November 1967, pp. 374-385.

ABSTRACT

The chief limitations to moving target indicator (MTI) radars in the past have been the hardware complexity and its difficulty of maintenance. Rigorous requirements are imposed on analog implementation of even the simplest concept of fixed target signal cancellation. Technological advances in digital devices now make it feasible to implement the concept of video storage and cancellation digitally. This not only eliminates the analog maintenance and reliability problems but allows complete flexibility in the choice of period between radar pulses. The velocity response is readily shaped as desired by the use of multiple pulse comparison, while blind velocities are eliminated by the use of variable interpulse period lengths. Radar systems using the digital MTI approach have been built and demonstrated in a practical environment using practical hardware. The performance of these systems has been outstanding.

INTRODUCTION

One of the basic requirements for present-day tactical radars is acceptable performance in a heavy clutter environment. This requirement often leads to a radar system approach which utilizes a moving target indicator (MTI) to extract moving targets from the clutter background.

In the past, most MTI systems have been implemented by using analog techniques. The most common form uses an ultrasonic delay line to store received phase information for one repetition period. The rapid technological advances in the area of integrated digital devices, however, has made it practical to use digital techniques to implement the delay element and associated processing. In fact, digital techniques are not only practical but they are essential to meet the present day tactical requirements of light weight, small volume, high mobility, high availability, and optimum system performance at a reasonable cost.

Implementation of an analog MTI system usually requires modulating an IF carrier frequency by the output of a receiver phase detector. This modulated carrier is then applied to an ultrasonic detector line where the phase information is delayed one interpulse period. The output of the delay line is detected and subtracted from the undelayed output of the receiver phase detector during the next interpulse period. Thus, under ideal conditions, fixed targets exhibit no phase difference between successive radar returns and hence cancel in the video subtraction process, while moving targets do not cancel.

Matching the gain and phase characteristics of the delayed and undelayed analog channels and controlling the absolute delay time places severe limitations on the maximum cancellation ratios obtainable (approximately 30 db) in a practical environment. Automatic gain control circuitry and oven-controlled temperature for the delay line unit are necessary, at the very least, if any reasonable performance is to be realized.

The velocity response of the simple two-pulse canceller just described has a sinusoidal response as shown in Figure 1C. This system is "blind" at multiples of target radial velocities which

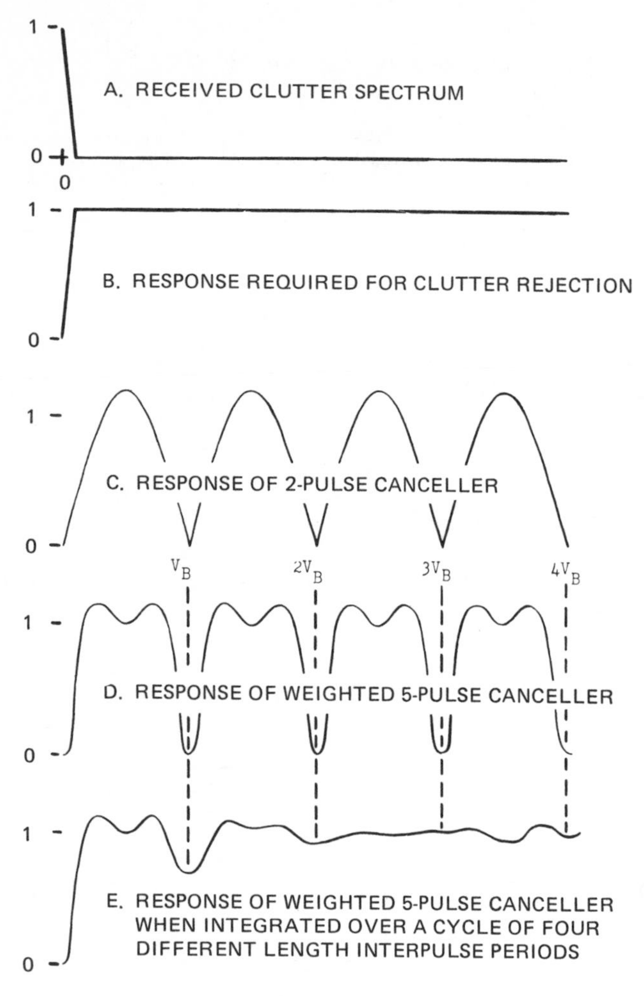

Evolution of ideal velocity response

Figure 1.

cause a target phase shift of 360° between successive radar transmission. Thus, the "blind velocities" are given by

$$v_B = \frac{n\lambda}{1.03T}$$

where v_B = blind velocity in knots
λ = wavelength in meters
T = interpulse period in seconds
n = 1, 2, 3 . . .

For a response without "blind velocities" such as shown in Figure 1E, it is desirable to use multiple interpulse periods. For an analog canceller, either a separate delay line or a trimmer delay line is required for each specific interpulse period, thereby adding considerable complication for a variable interpulse period (VIP).

The sinusoidal velocity response is often inadequate, from a radar performance standpoint, in the rejection notch around zero velocity and in the velocity passband region. The clutter spectrum to be rejected will have some finite width, as shown in Figure 1A, due to environmental conditions and radar scanning modulation. The ideal canceller rejection notch would be as shown in Figure 1B, where there is zero response to the clutter spectrum and max-

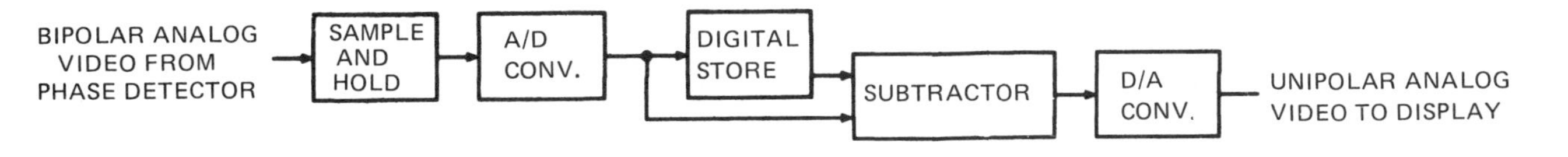

Figure 2 Basic Digital MTI Canceller

imum response throughout the target velocities of interest. This overall velocity filter characteristic can be approximated by using combinations of multi-pulse cancellers, various weighting coefficients, and feedback. Figure 1D illustrates the effect of using a 5-pulse canceller and optimizing the weighting of each pulse. The rejection notch is widened and the passband is flattened. Once again, in an analog canceller a separate delay line is required for each additional pulse (2-pulse canceller being the reference) to be remembered, and critical gains have to be controlled for the proper weighting coefficients.

Analog MTI systems have been used for a number of years with limited success. They have suffered from the inherent difficulty of controlling the delay time and system gains. Their on-line availability has been generally poor, operating well when tuned-up, but experiencing rapid deterioration in performance with time. Their effective contribution to the radar system performance has been limited (Usually 2- to 3-pulse cancellers, one to four inter-pulse periods, and relatively small amounts of feedback to maintain stability) due to the inflexibility of the hardware.

The digital MTI (DMTI), however, eliminates all the disadvantages of the analog MTI, as will be shown in the remainder of this paper. In addition, the DMTI has the inherent flexibility and stability required to obtain greatly improved performance in a practical radar system.

BASIC DIGITAL MTI

In the simple 2-pulse canceller shown in Figure 2, the radar receiver's phase detector output is sampled by the A/D converter. The voltage output of the phase detector at any instant of time is represented by $A\sin\phi$ (or $A\cos\phi$) where A is the amplitude of the echo being received and ϕ is its phase relative to the signal transmitted. This must be sampled at discrete range intervals and converted to a series of digital numbers.

Ideally, this sampling would be done at least twice per radar pulse width to prevent a loss in detectability due to sampling rate and for quadrature phase detectors to prevent a loss in detectability due to "blind" phases. A system performance versus hardware tradeoff has to be made for each individual system to determine the proper sampling rate and whether to use quadrature channels. Historically, analog MTI's have not used quadrature channels because the performance versus hardware tradeoff has not been able to justify two channels.

The sampling rate should be at least once per radar effective pulse width. This results in a theoretical average loss of detectability of about 1.5 db as compared to analog MTI's having single phase detection. The loss figure is established by assuming the signal pulse is of the form sinx/x and that the sample time is equally likely to occur anywhere within the 4 db points. The loss due to processing only one coherent video channel is equal in both the analog and digital systems.

The most important system timing relationship in the DMTI system is that between the transmitter pulse and the sample control. After the information has been digitized, the timing stability is relatively unimportant. There are no time delay or stability problems, the information being handled in strictly a binary form.

A typical sample width is normally in the 10 to 100 nano-second range. Typical sample rates are in the 0.5 to 6μs range.

Each A/D converter output is a digital word representing the receiver phase and amplitude in terms of its video amplitude and polarity for one range interval. From a hardware standpoint, the A/D converter is the most difficult circuit in the DMTI for a tactical application. Figure 3 shows a mil-spec A/D converter which weighs 0.85 lbs (0.39 Kg), converts 9 bits of information, and operates at a 10 MHz bit rate. Westinghouse also has a microminiature version which weights 0.5 oz. (14 g) and has dimensions of 0.5 x 0.5 x 1.2 inches (1.27 x 1.27 x 3 cm). These devices are presently being used in several DMTI systems.

The A/D output for each range interval is sent to a digital store. After one interpulse period the stored digital words are read out in time sequence and digitally subtracted from the current A/D converter output for each range interval. Thus, the digital store has introduced a delayed digital word for each range increment analogous to the delay line of the analog canceller system.

The actual device used for the digital store is determined by a tradeoff study once the system performance has been specified. Such devices as magnetic core memories, integrated shift registers, integrated scratch pad memories and other similar devices should be considered for each application. In a sophisticated system requiring a storage greater than 5000 bits, the flexibility of a core memory is usually preferred.

The number of bits of storage required is determined in the following manner. The number of digital words (W_n) per radar "look" may be given by

$$W_n = \frac{12.3 \times R}{S_r}$$

where R = instrumented range in nautical miles

S_r = sample rate in microseconds

The total number of digital bits (B_t) of storage required is given by:

$$B_t = W_n \times B_w \times (N - 1)$$

where B_w = bits/word

N = total number of pulses in cancellers

For example, a system requiring an instrumented range of 80 nautical miles, a four-pulse canceller, 8 bits per word, and a sampling rate of 2 μs would require:

W_n = 492 words/radar "look"

B_t = 11,808 total bits of storage

A magnetic core memory of size 512 words x 24 bits would probably be the choice for the digital store. The memory cycle time would probably be 2 μs. This is a modest "scratch pad" memory.

After the subtraction process, the magnitude of the resulting digital word is D/A converted to a unipolar analog voltage for use in the radar displays.

CANCELLATION RATIO

The degree of fixed target rejection possible in the digital canceller depends on the number of bits of resolution in the A/D conversion. A theoretical cancellation capability of approximately 6 db per bit results since each bit represents a factor of two in amplitude resolution of threshold levels. When a fixed target signal is at a threshold level of the least significant bit, a possible error of

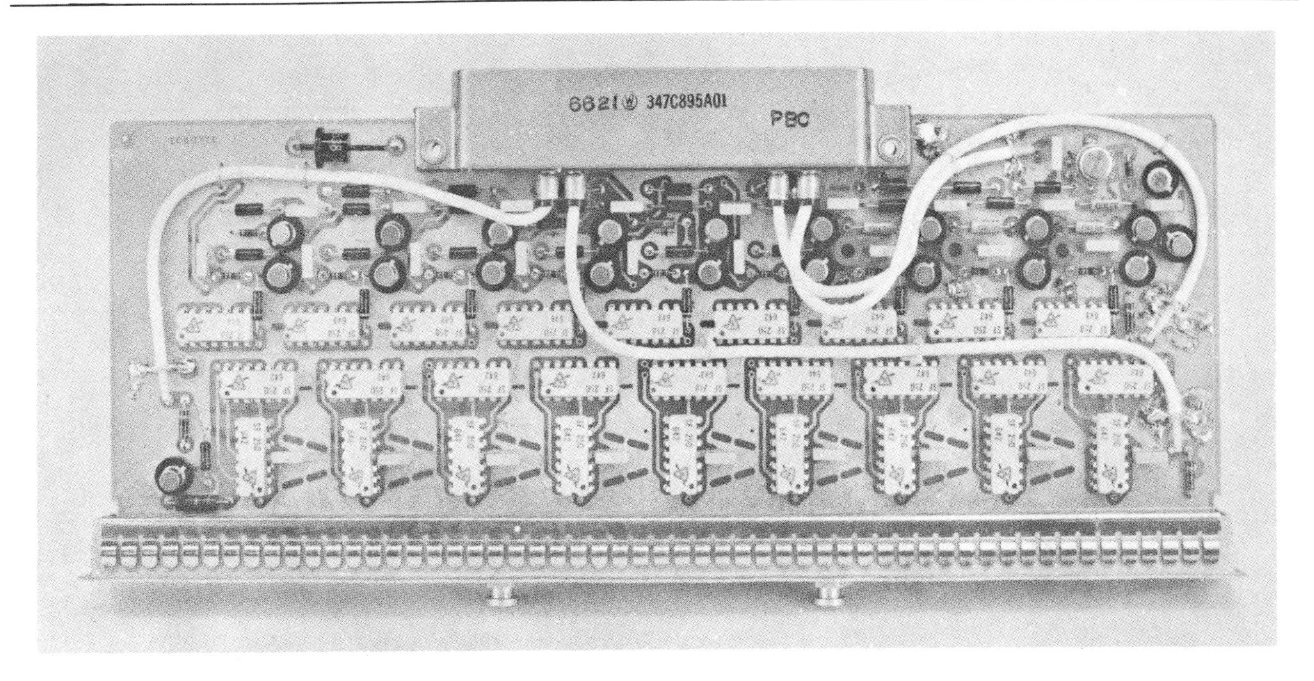

Figure 3. A/D Converter, 10 MHz Bit Rate

one amplitude increment could occur. Thus, the maximum cancellation resolvable by, say, a 9 bit A/D converter is 1 out of 511 levels (0 to ± 255), or approximately 54 db.

The number of bits is only a cancellation ratio limitation and does not imply that a digital system's subclutter visibility is limited to this value. Other coherent and noncoherent signal processing techniques provide additional target detectability. More specifically, the least significant bit residue is noise-like in character, and subsequent video integration yields the same improvement in target detectability as would be obtained in the presence of thermal noise.

In practice, the number of bits implemented is chosen to provide somewhat more cancellation ratio than the limitations resulting from other radar system parameters, particularly antenna scanning modulation. This modulation becomes significant with high-performance radar specifications which require high data rates and high azimuth resolution.

MULTI-PULSE CANCELLER

Since the digital MTI system is insensitive to the length of time it can hold video information, it would be easy to pass the output of the canceller of Figure 2 through a second, similar canceller. This would result in a $\sin^2$ response and provide a wider clutter rejection notch than the single canceller.

In practice, the digital double canceller would be implemented as an equivalent 3-pulse comparison canceller. Only unprocessed video amplitudes are stored. These and the incoming value are used in one calculation rather than two separate calculations, the second of which would use a result of the first. This procedure keeps the computation time at a minimum. This digital MTI would be similar to Figure 2, except that the subtractor would become a processor performing the computation A-2B+C where A, B, and C are successive video amplitudes in the same range cell. A is the current incoming value while B and C are stored values from the preceding periods.

In a similar manner, a four-pulse canceller could be implemented by storing three previous radar periods (B, C, D) and performing the computation A-3B+3C-D. This will give a $\sin^3$ velocity response. Higher order cancellers can be implemented by simply increasing the digital storage and performing the necessary computations.

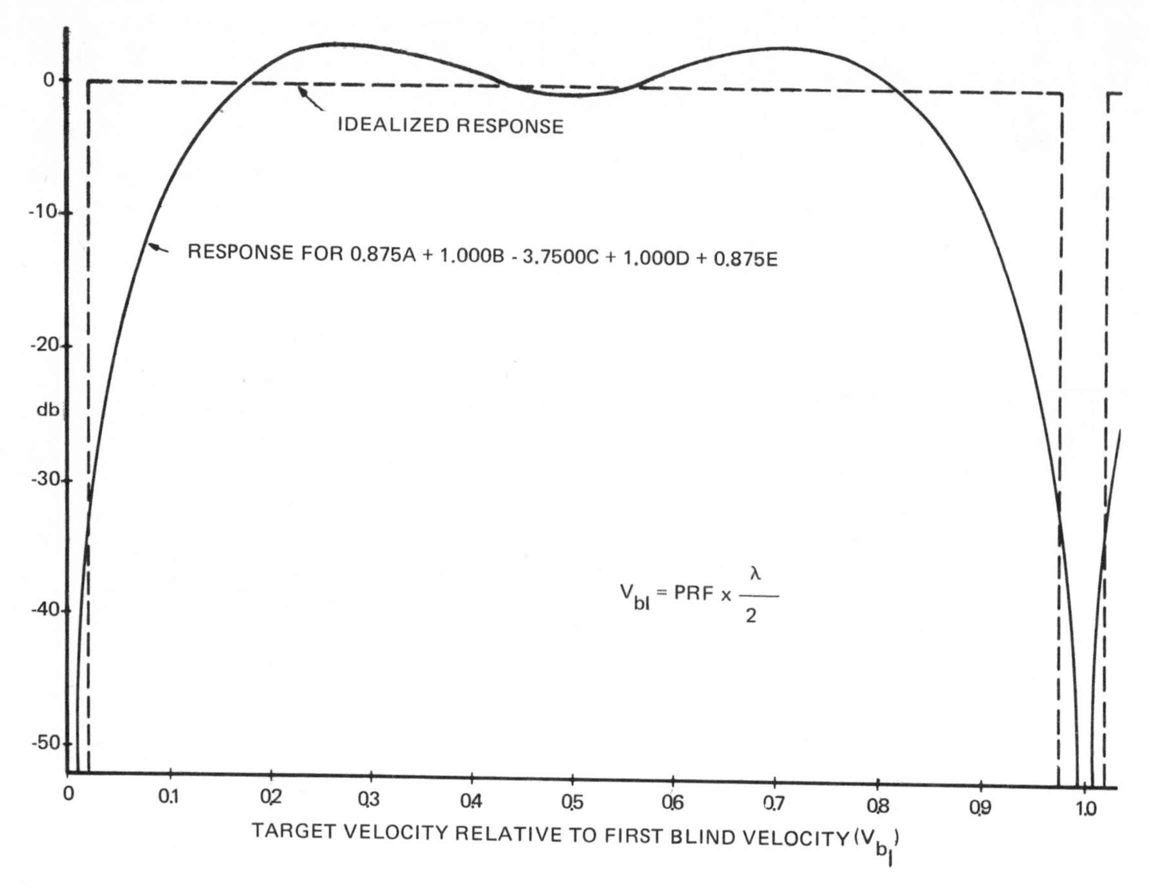

Figure 4. Response of a Weighted 5-Pulse Canceller, Fixed PRF

WEIGHTING CONSTANTS

A multi-pulse canceller is equivalent to (N-1) cascaded subtractors if successive pulses are weighted with binomial coefficients. However, it is possible to modify the velocity response from the sin (N-1)x shape by weighting successive pulses in other ways.

As an example, one weighting technique is to select coefficients that provide the best mean square fit to a specified ideal response curve such as shown in Figure 1B to reject the clutter spectrum of Figure 1A. These wieghts can be determined by representing the required response as a Fourier series expansion of terms. In a 5-pulse canceller for instance, the canceller output could be written as follows:

$$a_2 e^{-j2\omega t} + a_1 e^{-j\omega t} + a_0 + a_1 e^{j\omega t} + a_2 e^{2j\omega t}$$

time being referred to zero on the center pulse of a symmetrical set of five.

This series simplies to $a_0 + 2a_1 \cos\omega t + 2a_2 \cos 2\omega t$, with the values for a_0, a_1, a_2 dependent on the width of the passband with respect to the PRF. In a typical high-performance radar, the received clutter spectrum might contain frequencies up to about 2% of the PRF around zero frequncy and each PRF line. Calculation of the coefficients for this example does not provide zero response at zero velocity. However, by modifying the value for a_0, this can be corrected. The resulting Fourier coefficients obtained are:

$$a_0 = \frac{1}{\pi}\left(1 + \frac{\sqrt{3}}{2}\right)$$

$$a_1 = -\frac{1}{2\pi}$$

$$a_2 = -\frac{1}{4\pi}\sqrt{3}$$

or ratios of 3.732 to -1 to -.866 respectively.

One further, but important, simplification is now made to the weighting coefficients. Since digital methods are used throughout the canceller it is highly desirable to use simple digitally derived numbers for weighting purposes, i.e. powers of 2 and combinations thereof. By choosing

$$a_0 = 3.750 \quad = 2^2 - \frac{1}{2^2}$$

$$a_1 = -1.000$$

$$a_2 = -0.875 \quad = -1 + \frac{1}{2^3}$$

this end is obtained.

A ditial system to implement this canceller would evaluate the expression

$$-7/8A - B + 3\ 3/4C - D - 7/8E$$

where A, B, C, D, and E are successive radar "looks" in the same range cell. A is the current incoming value while B, C, D, and E are stored values from the preceding periods.

The resulting velocity response for this canceller is shown in Figure 4 along with the idealized goal. The response is down 3 db at 0.145 V_{bl} and 40 db down at 0.014 V_{bl}. The 0 db reference level is the normalized gain of the canceller to noise or the average of all velocities. This allows direct comparison of all cancellers without confusion by such terms as optimum velocity. The abscissa is labelled in terms of the first blind velocity (V_{bl}).

In the example chosen, symmetrical weighting was used which resulted in only three different values for the five weighting constants. If desired, five different values could be used. The important characteristic is the flexibility the digital implementation provides in controlling the weighting constants for velocity shaping.

VARIABLE INTERPULSE PERIOD

To minimize the loss of detectability of targets near the "blind speeds", variable interpulse periods are necessary. Indiscriminate choice of these steps however, can seriously degrade the clutter notch characteristics while providing only marginal improvement in the blind velocity regions.

To provide a consistent velocity response at least one complete cycle of VIP must be completed during the time to sweep through the radar antenna two-way beam pattern. The interpulse period must change in relatively small increments to prevent degradation of the clutter notch, and it must cover a wide enough range of values to smooth out the lowest "blind" velocity as required.

For the 5-pulse canceller described in preceding paragraphs a

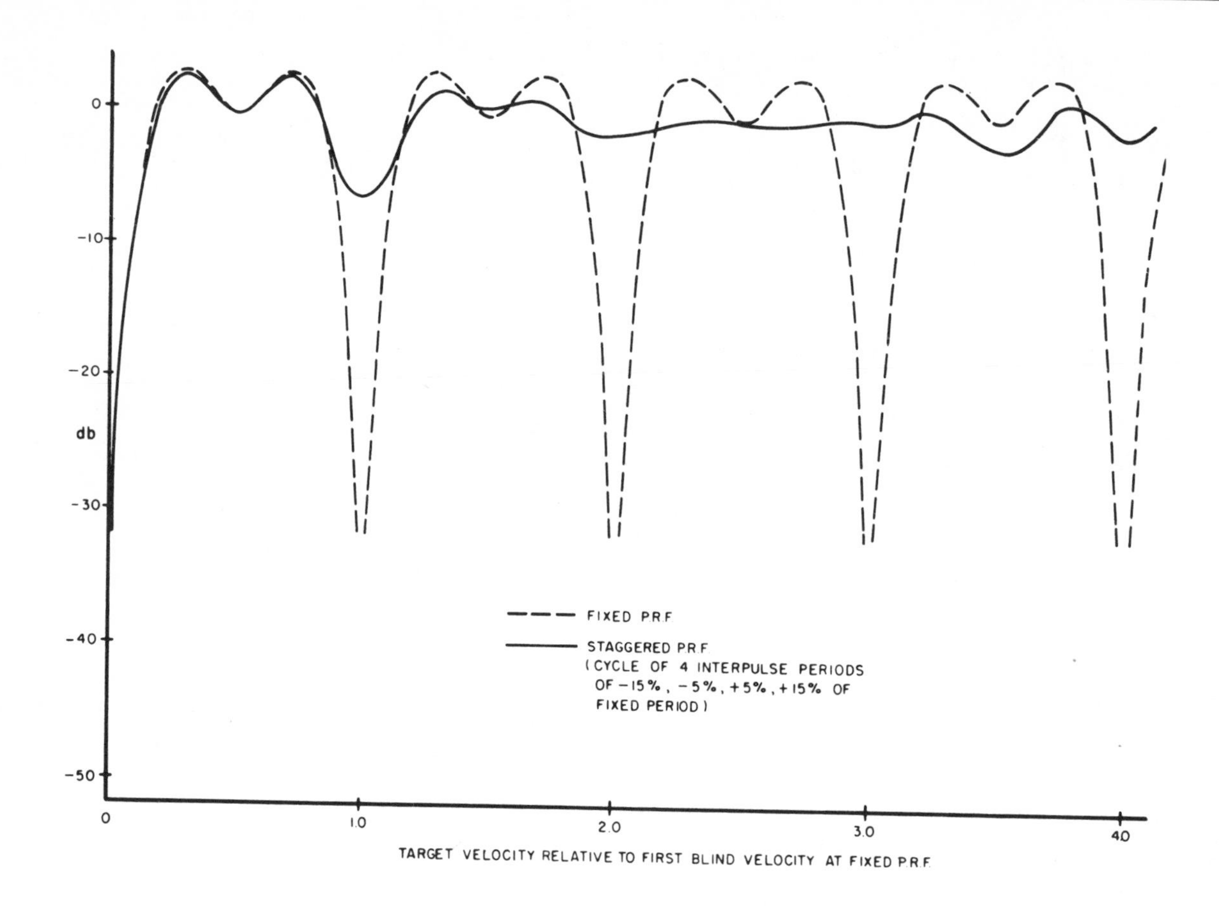

Figure 5. Response of a Weighted 5-Pulse Canceller, Staggered PRF

simple cycle through four interpulse period lengths of equally increasing amounts results in the response shown in Figure 5. The ratio of maximum to minimum periods is 23 to 17 and the velocity ordinate is in terms of the first blind velocity that would occur with operation at the average interpulse period value. The response at the first "blind" velocity is down only 6.6 db. This improvement is obtained with only 1.6 db increase in response at .02 times the blind velocity.

More sophisticated VIP programs have been investigated and implemented with excellent results. As many as 70 discrete steps have been implemented in a sinusoidal VIP program. The sine wave function was chosen to minimize the discontinuities and hence cause a minimum degradation in the rejection notch. Other VIP functions have been and are being investigated in an attempt to optimize the velocity passband ripple with minimum degradation to the rejection notch. Again, the important point is that the DMTI approach provides the flexibility to program the VIP in ways that, from a practical standpoint, were impossible in the analog MTI's. Thus, the DMTI approach opens the way for improved system performance.

DMTI FLEXIBILITY

The inherent basic flexibility of the DMTI has already been pointed out in the discussion of multi-pulse cancellers, weighting constants and VIP. Further general comments, however, are appropriate before leaving this feature.

Feedback can be used in DMTI cancellers for velocity shaping.

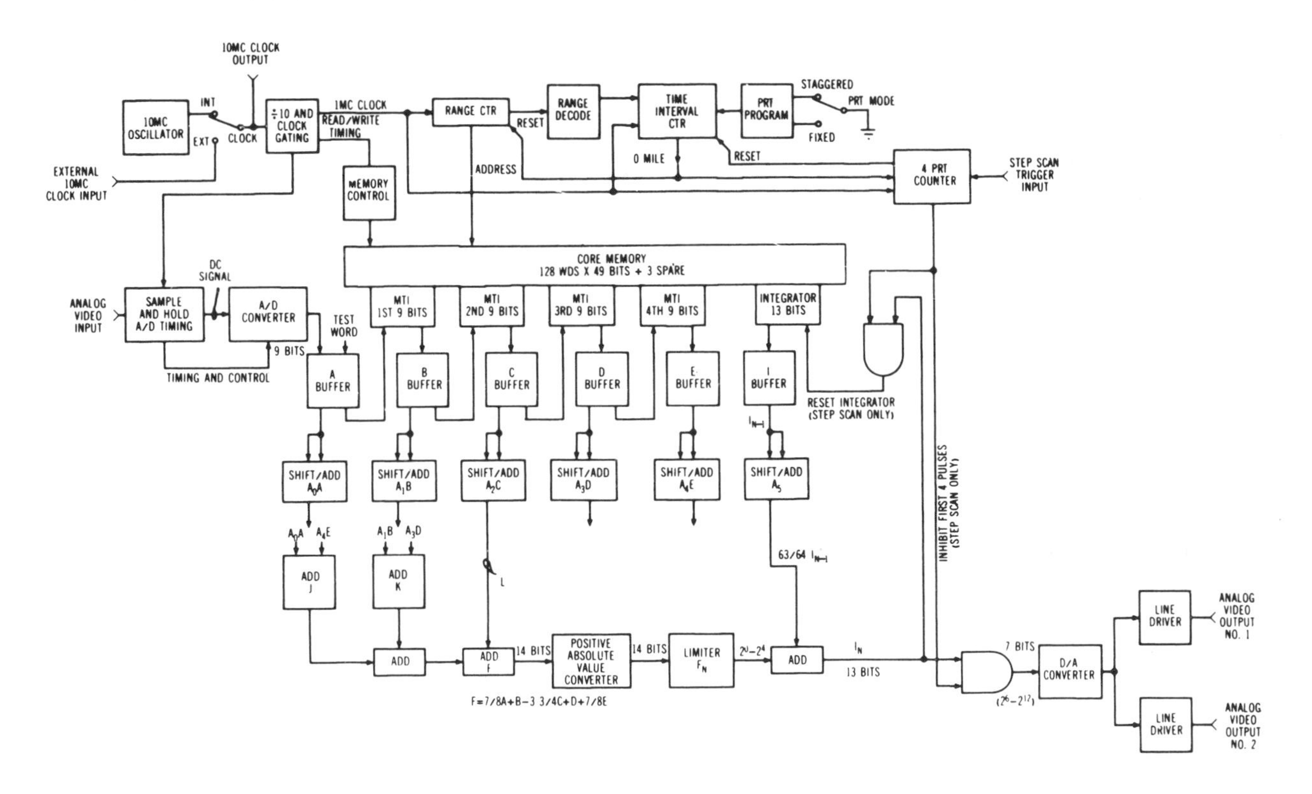

Figure 6. A 5-Pulse DMTI System

Feedback ratios near unity can be used without fear of oscillations. Although the velocity response can be made more rectangular by this process, the resulting response to pulse interference from other radars or jammers is seriously degraded. This is a system tradeoff for each application. Digital feedback is being investigated, primarily toward implementation of a given performance with fewer store location requirements.

It is practical, in a DMTI, to implement quadrature channels to avoid the loss in detectability due to "blind" phases. This requires doubling the amount of digital storage but would not double the cost. The computation circuitry can be multiplexed if time is available or doubled if it is not. Hardware has been built using quadrature channels.

Digital video integration is a natural process to use following a DMTI (see Figure 6). The video information is already in binary form and it is simple to make a feedback video integrator by using a small portion of the canceller storage device for the integrator store. The feedback constant can be optimized for the desired number of pulses by picking an appropriate binary term. For a step scan application, the integrator can be cleared by normal digital switching. Digital video integration is being implemented as a part of many DMTI systems.

Before digital integration, it may be desirable to use some degree of limiting to decrease the effect of random pulse interference and to decrease the output dynamic range. A digital limiter can easily be implemented by monitoring an overflow bit. The limit level can be varied as a function of radar mode or real time.

The DMTI approach is compatible with frequency diversity, which may be desirable for additional clutter decorrelation or for ECCM purposes. Separate store locations are necessary for each frequency. Frequency diversity in combination with different VIP programs per frequency and multi-pulse cancellers leads to a wide variety of very interesting radar system concepts.

Some forms of radar pulse coding, such as Barker phase coders, are especially compatible with DMTI. The DMTI can be applied to the received information first and then followed by a digital pulse compression network. This appraoch keeps the system

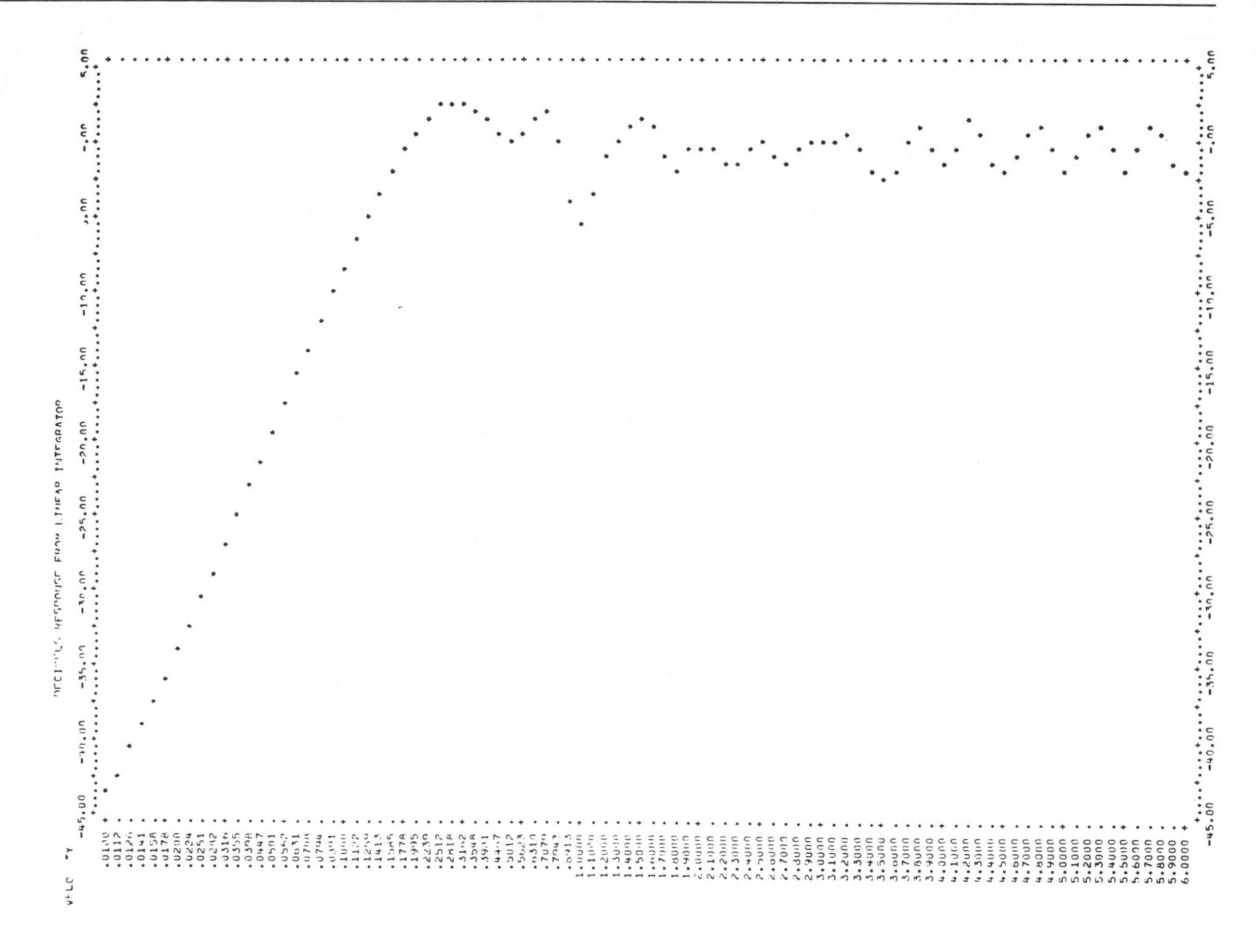

Figure 7. Computer printout of a velocity response

dynamic range at a minimum, and the digital pulse compression uses the information which is already in binary form.

For radar systems which are mounted on moving platforms it is often necessary to provide motion compensation to keep the rejection notch centered on the clutter spectrum as a function of antenna azimuth. In addition, to handle weather clutter it is sometimes desirable to change the velocity position of the rejection notch as a function of range. By proper sampling and weighting of the phase information within the DMTI processing, it is feasible to generate the necessary corrections with digital techniques.

Figure 6 is a block diagram of a DMTI unit that includes many of the features discussed. It is a 5-pulse canceller having individually set weighting constants, a VIP program, a limiter, an integrator, a step scan control, and a synchronizer.

The digital implementation of signal processing functions, such as the DMTI, opens the door to a wide variety of system options that were not practical in analog form. It is truly a step to the adaptive radar systems of the future.

COMPUTER PROGRAM

The data for the curves shown in Figures 4 and 5 were obtained from a computerized program for canceller velocity response. This technique provides a powerful tool for fast evaluation and optimization of DMTI parameters. Responses for various tradeoffs can be obtained in minutes. Because of the wide number of system options available with a DMTI, an automated technique for evaluation such as this is a necessity.

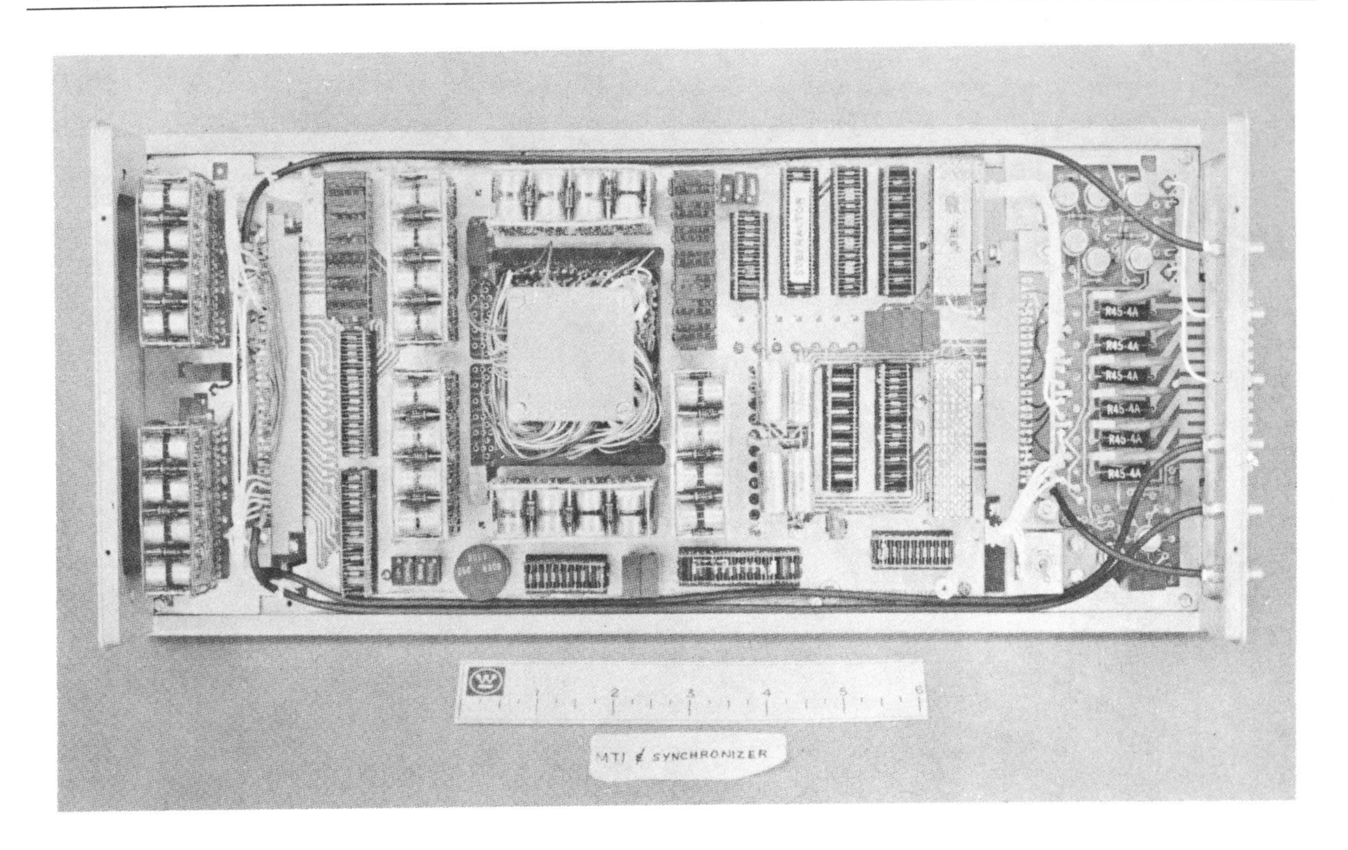

Figure 8. A Microminiaturized DMTI Unit

The program simulates the operation of digital cancellers and is implemented to tabulate the data and to plot the velocity response for a wide variety of multipulse cancellers, VIP programs, and pulse weightings. Figure 7 shows a typical curve printed out by the digital computer. Note that by choosing successive velocities for computation that are logarithmically related, and by computing the output level in db, the line-by-line printout is equivalent to a log-by-log graph and the slope of the velocity response toward zero velocity may be directly viewed on the printout as a straight line.

The abscissa is normalized with respect to the first "blind" velocity (V_{bl}) corresponding to the average interpulse period. After one cycle of V_{bl}, computations are made for linear velocity steps. Thus, the response characteristics for both the zero velocity region and the higher "blind" velocities may be studied on the same graph without loss of detail on either as would occur with a purely linear or purely logarithmic abscissa.

The curve shown in Figure 7 is for a 5-pulse canceller with weighting

$$-7/8, -1, +3\ 3/4, -1, -7/8$$

over a cycle of eight interpulse periods of

-15, -5, +5, +15, +15, +5, -5, -15%

of the average interpulse period.

The computer program is presently implemented to handle up to seven pulse cancellers with each pulse individually weighted from 0 to $\pm 1.0000 \times 10^3$. Nine hundred and ninety-nine different interpulse periods may be programmed and computations can be made for up to 20 bits of video resolution.

Figure 9. DMTI Radar Signal Processor

DMTI HARDWARE

Several DMTI radar systems have been and are being built and tested. Figure 8 is a photograph of a microminiature DMTI. This chassis contains the radar synchronizer as well as the complete canceller unit. The signal input to this chassis is bipolar analog video to the A/D converter and the output is unipolar analog video from the D/A converter.

This is a 2-pulse canceller with two interpulse periods. The cancellation ratio is 42 db (7 bits). There are 1000 range bins available to be used with sampling rates of either 1.4 or 4.2 μs. A 1024 words by 7 bits magnetic core store is used for the digital storage element.

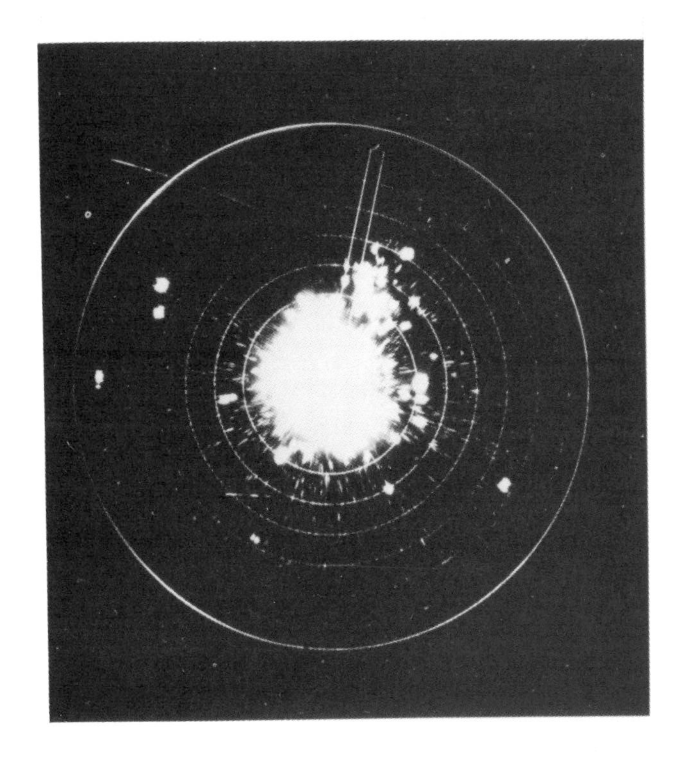

A. LOG VIDEO

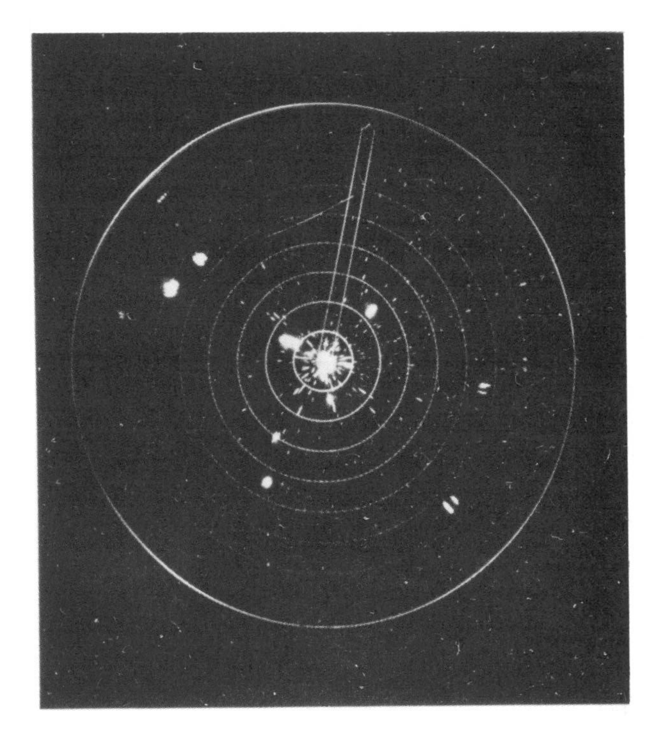

B. MTI VIDEO

Figure 10. PPI Display of 4-Pulse DMTI Performance

The unit weight is 3.5 lbs (1.6 Kg) with a volume of 210 in^3 (3442 cm^3). The linear circuitry is implemented with thin film techniques. The timing and control circuitry uses integrated "flat packs" and the core drivers use discrete components in welded module construction. This equipment completed in-plant testing and is presently undergoing field evaluation.

Another DMTI unit, which uses more conventional packaging, is shown in Figure 9. The radar synchronizer and DMTI unit is contained on the printed circuit card rack (four rows) in the center of the cabinet. The rest of the cabinet contains other radar subsystems. This DMTI uses a core memory which is mounted on the back of the card rack.

The basic canceller design is a sophisticated multi-pulse canceller with quadrature channels and numerous interpulse periods in a VIP program. It utilizes all plug-in subassemblies and built-in test facilities. It contains the system flexibility which is inherent in this digital technique. An analog MTI, with the same system performance, would not be practical to implement. This DMTI has been tested in a radar system, both in-plant and in the field, with outstanding results in both performance and reliability.

In addition to several other DMTI radar systems, modification kits are being added to existing radars. Some improvement in performance can normally be obtained from the kit, depending upon the existing radar parameters. Usually the velocity coverage can be improved because of the VIP flexibility and some improvement in clutter rejection can be obtained by shaping of the rejection notch. There is certainly a large improvement in system availability.

Typical results of a multi-pulse DMTI radar are shown in Figure 10 on a standard PPI. Figure 10A is the system log video display of the Baltimore Friendship International Airport area.

The range markers are 2 miles, showing approximately 16 miles of total coverage. Clutter contour measurements revealed the highest clutter peaks to be about 45 db stronger than a minimum detectable signal.

Figure 10B shows the same radar with the DMTI processed video. Figures 10A and 10B were taken with the same time exposure (one complete azimuth. scan) and within two minutes of real time. The results speak for themselves. All targets shown are targets of opportunity. The results are even more impressive in more rugged terrain having a heavier clutter environment.

CONCLUSION

The problem of extracting radar target information from a heavy clutter environment has been, and still is, a very difficult one. It is still one of the major problems of every radar systems engineer. The digital MTI technique described in this paper opens the door to many system options that were not previously available in analog MTI's. These new options, or system flexibility, give the potential to greatly improved system performance. This potential has been demonstrated in practical radar systems — it is here today.

In addition to the performance advantages, the DMTI technique gives high system availability through high reliability and ease of maintenance. It is also compatible with the tactical concept of light weight, small volume, and high mobility. Further, the DMTI technique is compatible with most other signal processing techniques which are or will be performed in digital form. This technique is a very important weapon in the battle of radar versus clutter.

ACKNOWLEDGMENTS

The authors wish to acknowledge the contributions by Dr. E.C. Watters, Mr. J. W. Taylor, and Mr. T.M. Moore of Westinghouse for many of the digital signal processing concepts described in this paper and for the computer programming and resulting analyses which led to the optimization of the parameters chosen for the implementation of these techniques.

Although no specific references have been cited, the basic terminology and principles of MTI may be reveiwed in the following books:

L. N. Ridenour, "Radar System Engineering," New York, New York, McGraw-Hill, 1947, Chapter 16.

M. I. Skolnik, "Introduction to Radar Systems," New York, New York, McGraw-Hill, 1962, Chapter 4.

D. K. Barton, "Radar System Analysis," Englewood Cliffs, New Jersey, Prentice-Hall, 1964, Chapter 7.

5.3 A THEORY FOR OPTIMAL MOVING TARGET INDICATOR (MTI) DIGITAL SIGNAL PROCESSING (SUPPLEMENT 1)
R. McAulay
Technical Note 1972-14, MIT Lincoln Laboratory, October 1972, pp. 1-16.
Reprinted by Permission.

A Theory for Optimal MTI Digital Signal Processing

Supplement I

ABSTRACT

In the report, "A Theory for Optimal MTI Digital Signal Processing, Part I: Receiver Synthesis," [1], the problem of eliminating scanning ground clutter from an aircraft surveillance radar was examined from a statistical decision theoretical point of view. An optimum processor was derived which could be approximated by a clutter filter followed by a discrete Fourier transform (DFT). In this report, additional numerical work is documented that compares the performance of the pulse cancellers, pulse cancellers with feedback and the DFT with that of the optimum processor. The issue of coherent vs incoherent integration gain is considered by comparing the filters only on their ability to reject clutter. A clutter rejection improvement factor is defined and used to compare the various filters. It is shown that the pulse cancellers can be quite effective in rejecting clutter provided the input clutter power is not too large and that additional gains are possible using the DFT.

Accepted for the Air Force
Joseph J. Whelan, USAF
Acting Chief, Lincoln Laboratory Liaison Office

I. INTRODUCTION

In the report, "A Theory for Optimal MTI Digital Signal Processing, Part I: Receiver Synthesis," [1], the problem of eliminating scanning ground clutter from an aircraft surveillance radar was examined from a statistical decision theoretic point of view. In this way, an optimum MTI processor was derived whose performance could be used as a benchmark to compare practical receivers that have been in use for the last two decades. Furthermore, it was of interest to determine whether or not digital processing techniques would be of any use in improving the ability of a radar to reject clutter. It was found that the optimum filter could be interpreted as a clutter filter followed by a bank of doppler filters matched to the two-way antenna scanning modulation. It was suggested that a good approximation to the optimum processor might be a classical clutter filter followed by a discrete Fourier transform (DFT). This would then provide the link between digital signal processing techniques and improved clutter rejection.

It was originally intended that Part I be principally a theoretical document to demonstrate the thought process linking the digital processing of data to MTI clutter rejection and to show the derivation of the tools needed to effect a comparison of the old schemes with the new. In our haste to get the ideas in print, a figure was drawn which compared the performance of the pulse canceller MTI filters with the optimum performance possible. It was intended to show how the signal-to-interference ratio (SIR) performance criterion could be used to evaluate filter performance. Unfortunately, a conclusion was drawn from the curves which has become quite controversial. In fact, the comparison was somewhat unfair because the optimum processor was permitted full use of coherent integration gain, while the pulse cancellers were evaluated allowing for no incoherent averaging. Of course, if the clutter is of such a level that the canceller leaves little residual clutter, then there will be little loss in using incoherent, rather than coherent integration since the number of pulses available for integration is small. On the other hand, if the clutter saturates the cancellers, such that significant residual clutter is produced, then incoherent integration ought to result in little improvement in the overall performance.

To clarify these issues we have performed more numerical work to compare the performance of the pulse cancellers, pulse cancellers with feedback and the DFT with that of the optimum processor. This is done in Sections II and III. Then, in Section IV, we address the issue of coherent vs incoherent integration gain, by comparing the filters only on their ability to reject clutter. We define a clutter rejection improvement factor and compare the various filters once again. It is shown that the pulse cancellers can be quite effective in rejecting clutter provided the input clutter power is not too large and that additional gains are possible using the DFT.

II. PRELIMINARY DEFINITIONS

In this section, we plan to perform a more detailed comparison of the performance of many of the MTI filters that are found in practice. The criterion on which this comparison is based is the signal-to-interference ratio (SIR) derived in Part I, [1]. For the optimum linear processor it was shown, in Eq. (89), that the SIR was given by

$$\rho_{opt}(\nu_o) = \frac{|\gamma_o|^2}{2N_o T_p} \int_{-1/2T_p}^{1/2T_p} \frac{|F_g(f - \nu_o)|^2}{\left[\frac{\sigma_c^2}{2N_o} \frac{1}{T_E T_p} |F_g(f)|^2 + 1\right]} \, df \quad , \tag{1}$$

where $F_g(f)$ is the Fourier Transform of the two-way antenna pattern and

$\frac{|\gamma_o|^2}{N_o}$ = predetection signal-to-noise ratio (SNR),

$\frac{\sigma_c^2}{N_o}$ = predetection clutter-to-noise ratio (CNR),

ν_o = target Doppler,

T_p = interpulse period,

T_E = effective time on target,

$= \theta_B/\omega_s$,

θ_B = one-way antenna 3 dB beamwidth,

ω_s = rate of antenna scan.

It was shown that the optimum filter could be realized as a clutter filter followed by a Doppler filter bank. For any other linear filter the SIR performance was shown to be given by

$$\rho_{sub}(\nu_o) = \frac{|\gamma_o|^2}{2N_oT_p} \cdot \frac{\left| \int_{-1/2T_p}^{1/2T_p} H(f)F_g(f - \nu_o)\, df \right|^2}{\frac{\sigma_c^2}{2N_o} \cdot \frac{1}{T_pT_E} \cdot \int_{-1/2T_p}^{1/2T_p} |H(f)|^2 |F_g(f)|^2\, df + \int_{-1/2T_p}^{1/2T_p} |H(f)|^2\, df} , \tag{2}$$

where H(f) is the transfer **function** of the filter of interest.

All of the results that follow are based on a Gaussian antenna pattern. In this case, the one-way antenna voltage pattern is

$$G(\theta) = e^{-\left(\frac{\theta}{\Delta\theta}\right)^2} , \tag{3}$$

where $\Delta\theta$ is chosen to make the 3 dB beamwidth θ_B. From this we compute the two-way pattern as

$$g(t) = G(\omega_s t) , \tag{4}$$

and taking its Fourier Transform we obtain

$$F_g(f) = \sqrt{\frac{2}{\pi}}\ e^{-2\alpha^2 f^2} , \tag{5}$$

where $\alpha = \frac{\pi}{2}\frac{\Delta\theta}{\omega_s}$. The system parameters used in all of the comparisons are those used in the FAA Airport Surveillance Radar. They are:

$$T_p = 1/1200 \text{ sec},$$
$$\theta_B = 1.5 \text{ deg}, \tag{6}$$
$$\omega_s = 15 \text{ rpm}.$$

The SNR parameter is chosen such that in the absence of clutter the SIR of the optimum processor is 0 dB. For the above parameter values this requires that the SNR be -8.75 dB.

In the next section, we will specify several MTI filters of current interest and compare their performance with the optimum as a function of target Doppler and CNR.

III. MTI FILTER SPECIFICATION

In this section, we shall briefly review the MTI filters that will be used in the comparison. Then in Section IV, their performance will be compared in a variety of operating environments.

A. The Optimum Filter

In Part I it was shown that the best detection performance was achieved by the filter having the transfer function

$$H(f) = \frac{F_g^*(f - \nu_o)}{\frac{\sigma_c^2}{T_E} |F_g(f)|^2 + 2N_oT_p} , \tag{7}$$

provided the true target Doppler is ν_o. Using a bank of these filters then gives an upper bound on the SIR that can be achieved by the class of linear processors. This bound is given by (1). In addition to the clutter rejection properties of this filter, the overall performance is enhanced by the target matched filter which provides the maximum coherent integration gain for the target in receiver noise.

B. The Pulse Cancellers

In Part I, it was shown that the denominator in (7) could be interpreted as a clutter filter as it produced a null about DC. Although optimum, this would be hard to realize in practice because it requires precise knowledge of the antenna pattern and the average clutter power. Based on classical theory it seems reasonable to approximate this clutter filter by the pulse canceller filters that have the transfer function

$$H(f) = \left(1 - e^{-j2\pi fT_p}\right)^{n_c} , \tag{8}$$

where $n_c + 1$ is the number of pulses involved in the cancellation. In other words for the simplest two-pulse canceller $n_c = 1$. Since

$$|H(f)| = |\sin \pi fT_p|^{n_c} , \tag{9}$$

the pulse cancellers locate a zero at DC and in addition, as n_c increases, the width of the null increases.

C. Feedback Cancellers

Although the above clutter filters can effectively eliminate clutter, the price paid is a loss in signal detectability because of the overall poor shaping of the velocity response curve. In order to regain some of this loss in detectability, feedback is introduced to shape the overall response curve. It is obvious that the best clutter filter would provide a wide notch about DC to null out the clutter and then a flat response elsewhere. This type of response curve can be achieved using feedback. A common realization is the dual delay-line canceller with feedback. This has the transfer function

$$H(f) = \frac{(z - 1)^2}{z^2 - (\alpha_1 + \alpha_2) z + \alpha_1} , \tag{10}$$

where $z = e^{j2\pi fT_p}$. It is expected that as the response is shaped to give better target detectability the clutter rejection capabilities will degrade because the depth of the notch about DC must move as the bulk of the response moves upward.

D. The DFT Processor

The optimum processor was shown to be a clutter filter in cascade with a Doppler filter bank. In addition to the difficulty is realizing the optimum clutter filters the velocity filters would be very difficult to construct using analog hardware especially if many range gates are to be considered. Using digital hardware, however, the problem becomes tractable since the Doppler filter bank is well approximated by a Discrete Fourier Transform (DFT). If the data is first passed through a standard pulse canceller before the DFT is taken, we should have a fairly good approximation to the optimum filter. In this case, if $r(nT_p)$ represents samples of the incoming data, and $r_c(nT_p)$ the output of the clutter filter, then the N-point DFT of this latter sequence yields the frequency samples

$$\xi(kT_p;m\Delta\nu) = \sum_{n=k-N+1}^{k} r_c(nT_p)g\left[\left(n + \frac{N}{2} - k\right) T_p\right]e^{-j2\pi\frac{nm}{N}}, \tag{11}$$

where $\Delta\nu = 1/NT_p$. This can be expressed as the output of a filter whose impulse is

$$h(nT_p;m\Delta\nu) = w(nT_p)\ g\left[\left(- n + \frac{N}{2}\right) T_p\right] e^{j2\pi\frac{nm}{M}}, \tag{12}$$

where

$$w(nT_p) = \begin{cases} 1 & 0 \leq n \leq N-1 \\ 0 & \text{otherwise} \end{cases} \quad . \tag{13}$$

Furthermore, the output of the clutter filter is

$$r_c(nT_p) = \sum_{k=-\infty}^{\infty} r(kT_p)\, h_c[(n-k)\, T_p] \tag{14}$$

where $h_c(kT_p)$ is the sampled-data impulse response of any one of the previously described filters. Then the overall DFT-clutter filter processor has the transfer function

$$H(f;m\Delta\nu) = \frac{1}{T_p} H_c(f)\, F_{wg}(f - m\Delta\nu) \ , \tag{15}$$

where $F_{wg}(f)$ is the Fourier Transform of the waveform $w(t)\, g(-t + \frac{N}{2})$. It is worth noting that the pulse canceler frequency response changes slowly relative to that of $F_{wg}(f - m\Delta\nu)$. Therefore, the detection performance of the processor can be improved with no loss in clutter rejection by normalizing each of the DFT coefficients by $H(m\Delta\nu)$. Therefore, the approximation to the optimum MTI processor is taken to be

$$H(f;m\Delta\nu) = \frac{1}{T_p} \cdot \frac{H_c(f)}{H_c(m\Delta\nu)} \cdot F_{wg}(f - m\Delta\nu) \quad . \tag{16}$$

This expression is used in (2) to generate its SIR performance. In the results to follow we shall take $H_c(f)$ to be the three pulse canceler. (i.e., n_c=2).

IV. COMPARISON OF PERFORMANCE

In the last section, several MTI filters of theoretical and practical interest were proposed. In this section, curves showing their SIR performance vs target Doppler for various CNR's will be discussed for the ASR system parameters. We begin with Figure 1 which shows the optimum, two and three pulse cancellers and the DFT processor for a CNR of 48 dB. The curves show that the DFT-3 pulse canceller is a good approximation to the optimum. It appears that the classical pulse cancellers are performing significantly poorer than the DFT processor. However, part of this performance loss is due to the fact that the DFT implicitly utilizes coherent integration gain since each DFT coefficient represents the output of a perfectly matched filter. Since the pulse cancellers will undoubtedly be followed by some incoherent integration of pulses or at least by an operator at a cathode ray tube, the SIR performance measure is an unfair criterion for comparing the clutter rejection capabilities of the various filters. It is useful in evaluating various DFT processors (i.e., using fewer data samples) as the degradation from the overall optimum SIR performance can then be determined directly. However, to fairly compare the pulse canceller with the DFT processor, we adopt another performance measure, the output peak signal to average clutter ratio (SCR). This is obtained from (1) and (2) by neglecting the effect of filtering the receiver noise. In this case, the optimum performance is given by

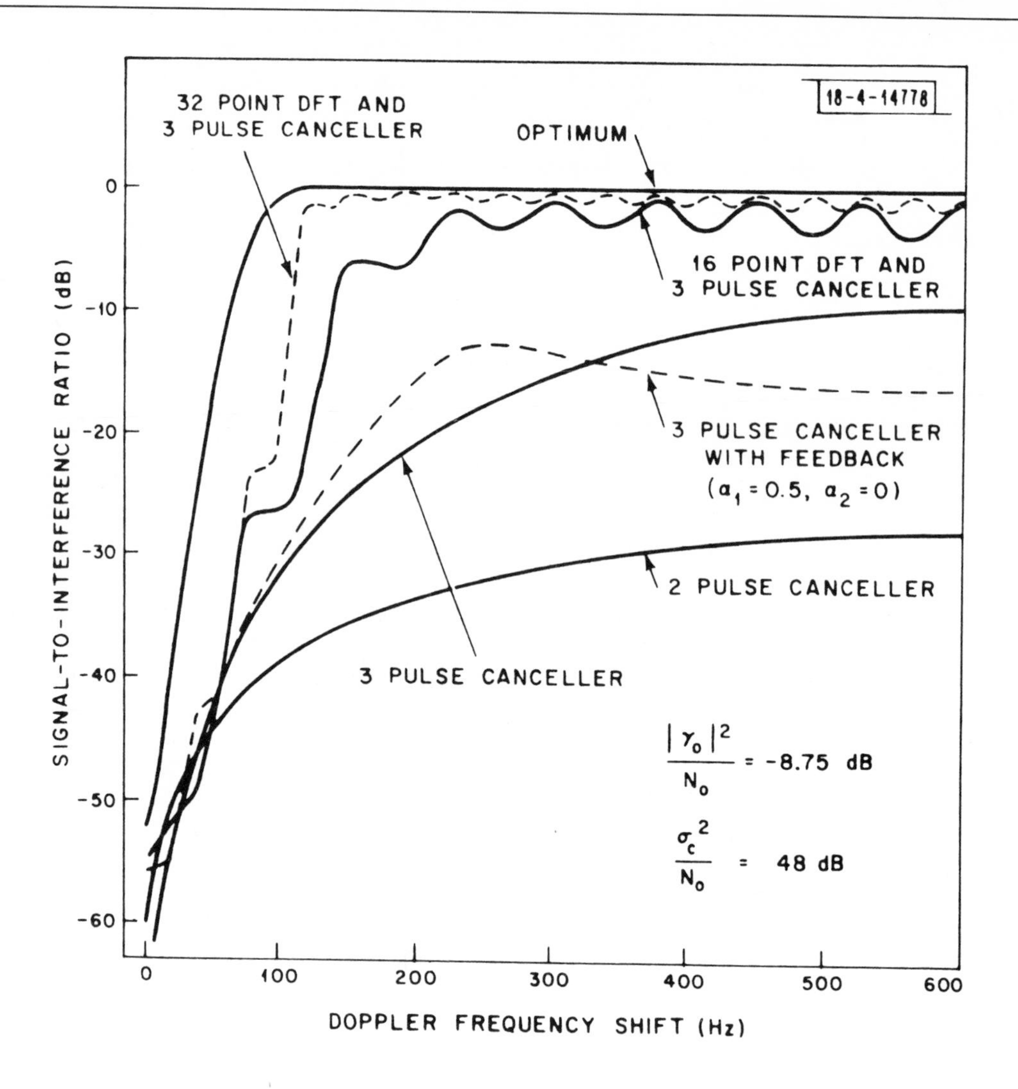

Figure 1. Signal-to-interference ratio for several practical MTI processors.

$$\beta^o_{opt}(\nu_o) = T_E \cdot \frac{|\gamma_o|^2}{\sigma_c^2} \cdot \int_{-1/2T_p}^{1/2T_p} \frac{|F_g(f - \nu_o)|^2}{|F_g(f)|^2} \, df \quad , \tag{17}$$

while that of the suboptimal processors is given by

$$\beta^o_{sub}(\nu_o) = T_E \cdot \frac{|\gamma_o|^2}{\sigma_c^2} \cdot \frac{\left| \int_{-1/2T_p}^{1/2T_p} H(f) \, F_g(f - \nu_o) \, df \right|^2}{\int_{-1/2T_p}^{1/2T_p} |H(f)|^2 |F_g(f)|^2 \, df} \quad . \tag{18}$$

It was shown in Part I that the average clutter power per sample was given by

$$\overline{|C(nT_p)|^2} = \frac{\sigma_c^2}{T_E} \int_{-1/2T_p}^{1/2T_p} |F_g(f)|^2 \, df \quad . \tag{19}$$

Therefore, the input peak signal-to-clutter ratio is

$$\beta^i = \frac{|\gamma_o|^2}{\overline{C(nT_p)}^{\,2}} \quad . \tag{20}$$

Then, we define the improvement factor to be

$$I(\nu_o) = \frac{\beta^o(\nu_o)}{\beta^i} \quad . \tag{21}$$

For the optimum processor this becomes

$$I_{opt}(\nu_o) = \int_{-1/2T_p}^{1/2T_p} |F_g(f)|^2 \, df \int_{-1/2T_p}^{1/2T_p} \frac{|F_g(f - \nu_o)|^2}{|F_g(f)|^2} \, df \ , \tag{22}$$

while the suboptimal processors result in

$$I_{sub}(\nu_o) = \int_{-1/2T_p}^{1/2T_p} |F_g(f)|^2 \, df \cdot \frac{\left| \int_{-1/2T_p}^{1/2T_p} H(f) \, F_g(f - \nu_o) \, df \right|^2}{\int_{-1/2T_p}^{1/2T_p} |H(f)|^2 |F_g(f)|^2 \, df} \ . \tag{23}$$

The improvement factors were computed for the optimum, DFT and pulse canceller processors and the results are shown in Figure 2.

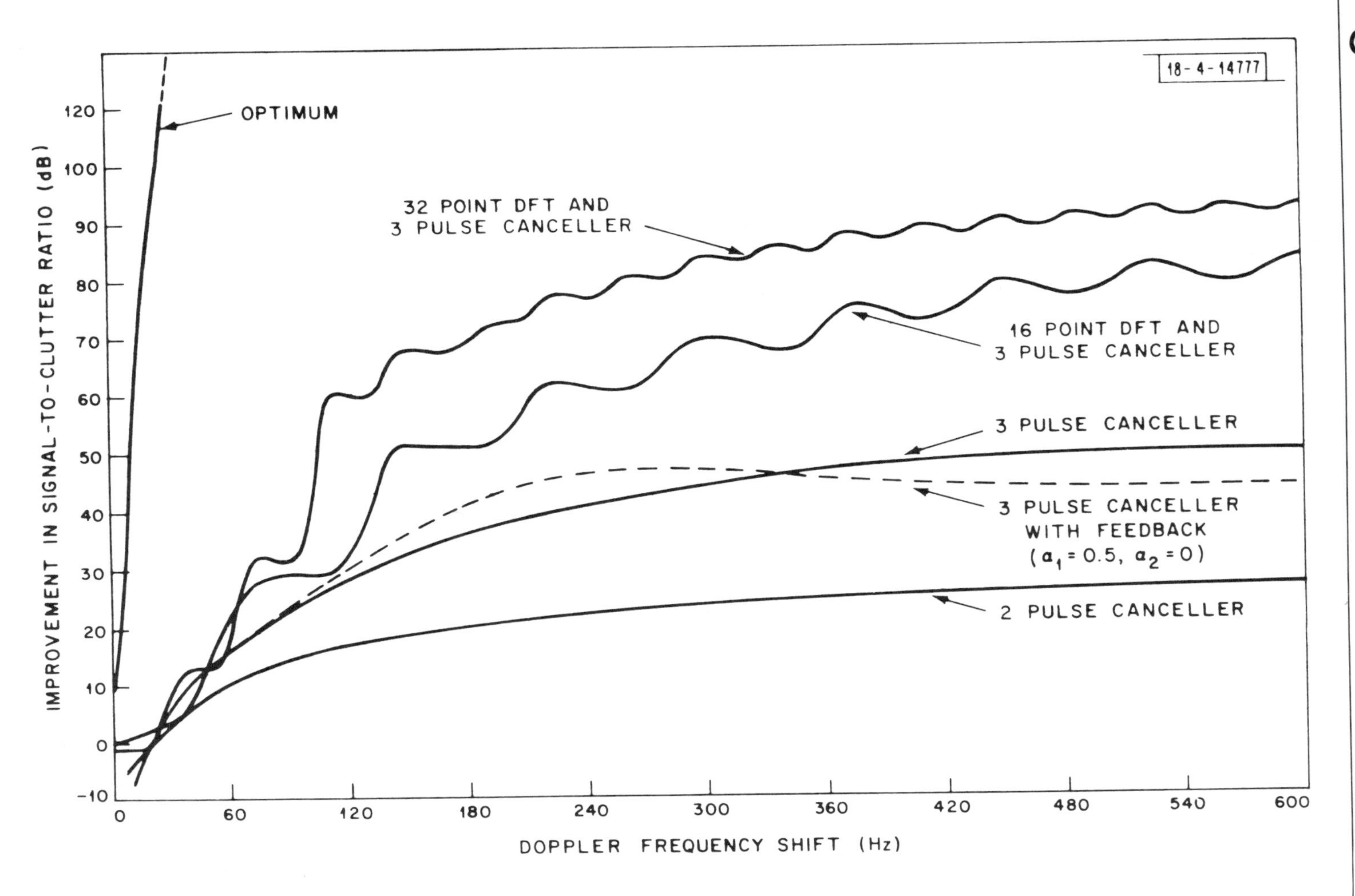

Figure 2. Improvement in clutter reflection by filtering.

V. CONCLUSIONS

In Figure 1, it is clearly demonstrated that shaping the velocity response of the clutter filter can improve the low-frequency performance of the filter at the expense of a greater loss in the high frequency region. Furthermore, the loss in performance is of the order of 15 dB and is due principally to the presence of residual clutter which will not be eliminated by incoherent integration.

Figure 2 shows that the pulse canceller and DFT can be very effective in eliminating scanning ground clutter. This curve shows that much of the improvement shown in the SIR performance curves is due to the ability of the DFT to further reject the residual clutter. By making the data window longer (16 T_p to 32 T_p), the frequency sidelobes of the matched filters are reduced, resulting in less interaction with the residual clutter. This is the principal reason the DFT can lead to significant improvements in the rejection of clutter.

Finally, it can be concluded that if the clutter background is not too severe, then the pulse cancellers can eliminate it effectively. For example, Figure 2 shows that the improvement factor for the three-pulse canceller is more than 30 dB over 75% of the total frequency range. Hence, if the input SCR is at least -15 dB then the output SCR will be +15 dB and the clutter will become a fractional part of the noise background.

REFERENCES

[1] R.J. McAulay, "A Theory for Optimal MTI Digital Signal Processing. Part I. Receiver Synthesis," Technical Note 1972-14, Lincoln Laboratory, M.I.T. (22 Feb. 1972).

[2] R.J. McAulay, "A Theory for Optimal MTI Digital Signal Processing. Part II. Signal Design," Technical Note 1972-14, Lincoln Laboratory, M.I.T. (4 October 1972).

[3] W.W. Shrader, "MTI Radar," Chap 17 in Radar Handbook, M.I. Skolnik, ed. (McGraw Hill, New York, 1970).

5.4
NEW TECHNIQUES APPLIED TO AIR-TRAFFIC CONTROL RADARS
C. Muehe, L. Cartledge, W. Drury, E. Hofstetter, M. Labitt, P. McCorison and V. Sferrino
Proceedings of the IEEE, Vol. 62, No. 6, June 1974, pp. 716-723.

New Techniques Applied to Air-Traffic Control Radars

CHARLES E. MUEHE, LINCOLN CARTLEDGE, WILLIAM H. DRURY, EDWARD M. HOFSTETTER, MEMBER, IEEE, MELVIN LABITT, MEMBER, IEEE, PETER B. McCORISON, AND VINCENT J. SFERRINO

***Abstract*—During the past two years a program has been carried out to show how new techniques can greatly improve the performance of radars used for air-traffic control. A survey of problems associated with presently used radars was undertaken. This survey indicates that primary radar in an automated air-traffic control system can be made significantly more effective by the use of new techniques. The radar's handling of extraneous reflections (clutter) is critical to its performance.**

Three types of interfering clutter were found to predominate: ground clutter, weather clutter, and angels. Angels are generally accepted to be radar returns from flocks of birds. In addition, second-time-around clutter is often troublesome. For each type of clutter, all known remedies for improving the signal-to-clutter ratio were studied and radar systems were configured using appropriate sets of remedies.

Some specific solutions incorporated in the resulting radar systems are: a) the use of linear large dynamic range, near-optimum digital signal processors to filter signals from clutter, b) the use of electronically step-scanned antennas to improve the correlation of aircraft and clutter returns from pulse to pulse, c) the use of multiple PRF's instead of staggered PRF's together with coherent transmitters to keep second-time-around clutter returns well correlated while still overcoming blind speeds, d) the use of a fine grained ground clutter map to give superclutter visibility on tangential targets, and e) the use of lower operating frequencies to greatly reduce weather and angel returns.

Two demonstration radar systems have been implemented, an *S*-band radar using a mechanically rotating antenna and a UHF radar using an electronically step-scanned cylindrical antenna. Experimental results are described.

I. Introduction

PERHAPS the most important application of radar in this country, especially to the frequent air traveler, is as a sensor in the nation's air-traffic control system. The system is served by 84 high-power long-range *L*-band enroute radars serving the Air Route Traffic Control Centers and approximately 125 medium-power *S*-band Airport Surveillance Radars (ASR) serving the terminal areas.

These radars, both terminal and enroute, while generally giving good service in a manual control environment, have certain deficiencies which make their use difficult in the fully automated systems being implemented by the Federal Aviation Administration (FAA). The principal difficulties are either an excessive number of false alarms which will overload the automation computers or else aircraft detection so spotty that the computer cannot make long continuous tracks of all aircraft.

Over the last two years, a concerted effort was made to gain a good understanding of the root causes of these deficiencies, to study all the ways the radars can be improved, and, finally, to design and demonstrate new signal processing techniques which should, together with certain antenna techniques, eliminate the problems.

We will first, briefly, describe the basic radar problems then discuss improvements to solve them. Finally, we describe signal processing techniques being built and tested.

Manuscript received September 4, 1973; revised January 16, 1974. This work was sponsored by the Federal Aviation Administration under Interagency Agreement DOT-FA72-WAI-242 and the Department of the Air Force.
The authors are with M.I.T. Lincoln Laboratory, Lexington, Mass. 02173.

II. Radar Problems

The poor performance previously mentioned is due to the competition of aircraft radar returns with the so-called "clutter" returns and the response of circuits used in the radar to overcome these clutter returns. It is convenient then to classify the problems according to the type of clutter return. These are fixed ground clutter, second-time-around effect, precipitation clutter, angels, and surface vehicles. We will describe these in terms of the ASR radars. The extension to the enroute radars is fairly obvious.

A. Fixed Ground Clutter

By far the largest undesired radar reflections come from fixed objects on the ground. Ground clutter generally extends out to about 20 nmi except in very hilly or mountainous areas where it may extend out to the maximum radar range (~60

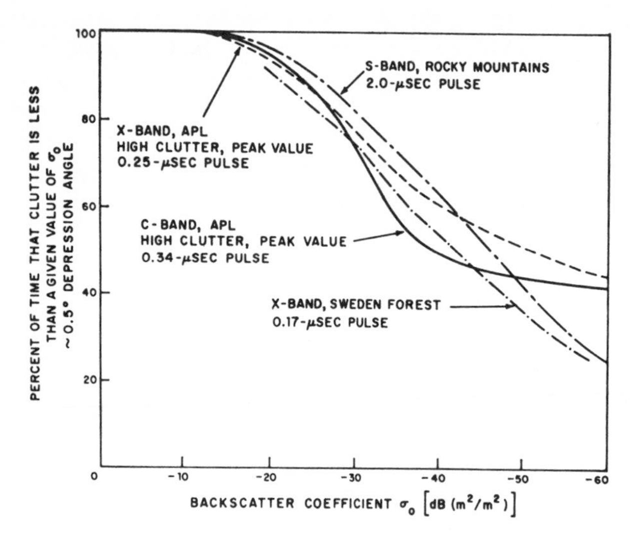

Fig. 1. Land clutter backscatter distribution from surface radars (from [1]).

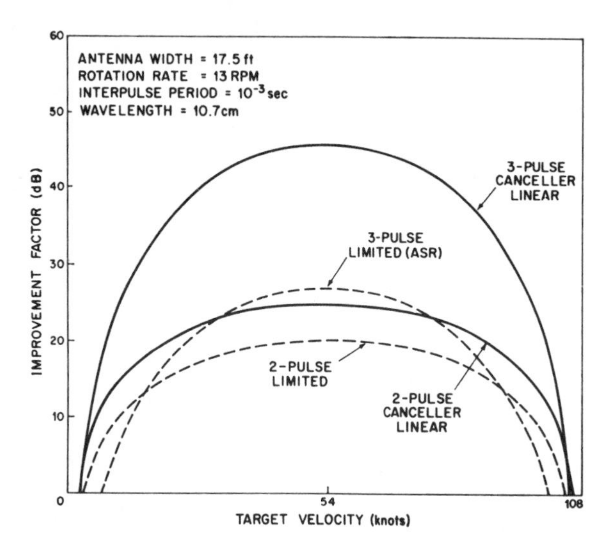

Fig. 2. Performance of S-band cancelers. The improvement factor is the ratio of signal-to-clutter out of the canceler to signal-to-clutter in. The latter ASR radars use three-pulse cancelers following a limiter shown above by the curve "3-Pulse Limited (ASR)."

nmi). Its natural or intrinsic spectrum is very narrow compared to the spectral spread caused by antenna scanning motion.

Ground clutter varies appreciably from spot to spot in the area of coverage. Typical distributions of the mean values σ_0 are shown in Fig. 1. It tends to be highest from cities.

In the present ASR radars, ground clutter is reduced by three mechanisms: MTI, antenna tilt, and by mounting the antenna close to the ground to take advantage of the shielding effect of nearby objects. Fig. 2 shows the MTI filtering performance achievable using one and two delay line cancelers with and without limiting. Previous ASR radars have all employed limiting in the IF followed by a phase detector. The purpose of the limiting is to normalize the video output so that clutter residue from the MTI filter is reduced to the average noise level. This allows the video gain to be adjusted so the clutter will not show up on the controller's scope. Unfortunately, this limiting action spreads the clutter spectrum

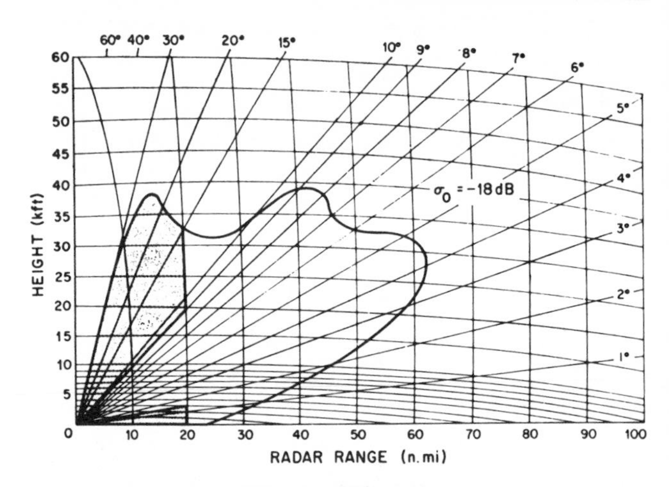

Fig. 3. Solid curve is coverage of ASR-7 radar against a 2-m² target in receiver noise [2]. Cross hatched region shows region of poor performance against ground clutter.

so that considerably poorer subclutter visibility (SCV) is achieved than if the normalization had been done by some other mechanism not involving nonlinearities.

If we consider the parameters of a typical terminal radar (ASR-7, see Section III-A 1) at 15 nmi, and σ_0 from Fig. 1 that is exceeded only 5 percent of the time, we find that for a 1-m² target (typical small aircraft) the input signal-to-clutter ratio is -31 dB. Since an output signal-to-clutter ratio of about 15 dB is needed for adequate target visibility, an improvement factor of 46 dB is required. We see from Fig. 2 that this is not achievable with the present configuration. It is, thus, common practice to achieve greater signal-to-clutter advantage by tilting the antenna upward (see Fig. 3) by 2° to 5° depending on the local clutter situation. If tilted, as shown in Fig. 3, there is a 17-dB advantage (maximum range divided by zero elevation range to the fourth power) in input signal to clutter for an aircraft flying in the peak of the antenna pattern. This advantage is degraded as the aircraft gets out of the peak of the antenna pattern so that, typically, detection gets spotty due to competition with ground clutter for small aircraft below about 1.5° or above about 9°. These angles change depending on the antenna tilt and ground clutter intensity. It is estimated that a 20-dB increase in improvement factor would be required for really adequate detection of small aircraft at all altitudes.

Another undesirable feature of the improvement curves in Fig. 2 is the very wide notch around zero and the first blind speed. The notch around the first blind speed is usually removed by using staggered PRF. The notch around zero means that targets will be lost for a considerable distance on the scope when the aircraft flies tangential to the radar. It will be observed that the three-pulse canceler with limiting is worse in this respect than the two-pulse canceler with limiting. Below, we shall describe how more advanced signal processing techniques and antenna techniques can both provide a large degree of improvement in SCV, and much better performance near zero velocity.

A further limitation in performance of existing ASR's is the presence at many sites of buildings or hills which limit the minimum elevation visible to the radar. Increasing the height of the antenna to overcome this limitation causes an undesirable increase in ground clutter level which could be overcome by improvement in SCV.

B. Second-Time-Around Clutter

In the so-called "second-time-around" clutter effect, returns are being received due to illumination of clutter beyond the nonambiguous range by the next-to-last pulse transmitted. These returns are prevalent where conditions for anomalous propagation exist such that the radar waves are bent back downward with range and intercept the ground at great distances (greater than that corresponding to the interpulse period). This effect is also prevalent in regions where mountains exist beyond the nonambiguous range.

Present ASR's use magnetron transmitters that transmit pulses with random phase from pulse to pulse. Thus it is impossible to maintain the phase relation between the first- and second-time-around clutter returns and the two cannot be filtered out simultaneously.

Further, the present ASR's use pulse trains with staggered interpulse periods so as to avoid Doppler blind speeds. But this causes the second-time-around clutter return to be from a different range cell on succeeding returns so there is no hope of filtering it out. To effectively filter out second-time-around clutter, a fully coherent transmitter (one coherent from pulse to pulse) and a constant PRF must be used. The PRF need not be constant forever, but only over an interval sufficient to collect a group of pulses for processing. If the PRF is changed from group to group of pulses, the radar is said to use "multiple PRF."

C. Precipitation Clutter

The backscatter from precipitation has been studied extensively. Fig. 4 shows the mean volume reflectivity from rain at 15 mm/h. This is considered a heavy rain found only 0.04 percent of the time at New Orleans [1]. This heavy rainfall is usually found only in relatively small-size cells in the center of storms. The radar should be designed to reject at least this level and as much higher a level as possible.

Also marked on Fig. 4 is the point where the volume reflectivity is such as to cause a 1-m² return at 30 mi in an ASR radar (rain return from a typical cell with precipitation extending from the surface to 10 000 ft). Rain at 15 mm/h is about 13 dB above this value. Remembering that these are average reflectivities and that ~15-dB signal-to-noise ratio is required for automatic detection, we need about 30-dB rain rejection for good performance.

The rain clutter spectrum is spread around some mean value determined by the wind velocity. The spectral spread observed by the radar is fixed by wind shear conditions [1]. The standard deviation of the rain velocity spectrum typically reaches values of 4 m/s at 30 nmi and increases with range.

Circular polarization is normally used to reduce rain clutter by about 15 dB while reducing the signal level to some extent. The use of MTI helps reduce rain clutter except when the antenna is looking toward or away from the wind direction. In these directions, the rain clutter spectrum is such that all of the rain clutter signals may pass through the MTI filters.

Log-FTC-antilog circuits [2], [3] reduce the receiver gain in proportion to the average level of rain clutter for about a mile in range surrounding the cell of interest. It, thus, normalizes the rain clutter level just as limiting is used to normalize ground clutter at the output of the MTI circuit. Its purpose is to suppress the rain clutter on the scope. At the same time, of course, it suppresses the signal. For adequate detection, the signal amplitude must be appreciably above the clutter residue from the MTI filters.

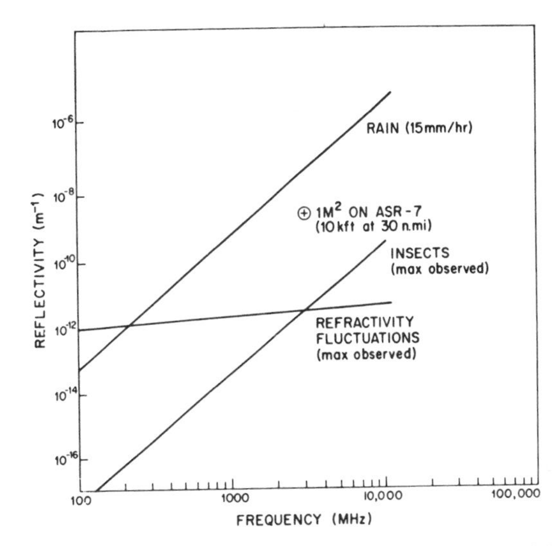

Fig. 4. Reflectivity of various moving clutter sources [1], [4].

D. Angel Clutter

The so-called "angel clutter" refers to all returns which cannot be explained as being ground or precipitation clutter or targets. Much effort has been spent in studying angels. It is now believed that nearly all, if not all, angels are caused by bird flocks. Two other sources were considered possible.

Swarms of insects have been observed by meteorologists using powerful radars [4]. The swarms may cover large areas and, in general, drift with the winds. Well organized layers of turbulent refractivity in the atmosphere associated with changes in the refractive index have been observed. The maximum volume reflectivity associated with these types of returns is plotted in Fig. 4. It will be observed that both are much lower than a 1-m² target so should cause little difficulty when trying to detect aircraft.

Returns from single birds [5] at *S* band range in size between 10^{-4} and 10^{-2} m². The return is principally from the body with very little from the wings. For large birds, the body is resonant near *L* band (1300 MHz) and is in the Rayleigh region at UHF. Typically, there may be anywhere from one to several hundred birds in a resolution cell. Although the mean return from a typical flock of birds may be low ($\sim 10^{-2}$ m²), the tail of the distribution has been observed to return up to 10 m². Although birds have been seen as high as 12 000-ft altitude, they usually fly less than 7000 ft. The usual appearance on the scope is as so called "dot angels." "Ring angels" are also caused by birds as a large group leave their nesting place at sunrise.

Of particular interest are the bird migrations in spring and fall. These have been described as "night effect," "falling leaves," "seasonal AP angel clutter," and have been reported by many terminals in the eastern part of the United States. The appearance on the scope when the radar is using MTI is that of two well-defined lobes. In Fig. 5, there is a strong migration in an easterly direction so MTI notches appear north to south. The lobes appear to be made up of a multitude of spots which move like falling leaves.

These migrations occur at night when there is a favorable wind. Migration will be very heavy on favorable nights so that most of the migration occurs on relatively few nights (5 to 15) each spring and fall. The number of birds associated with these migrations may be very large. One author esti-

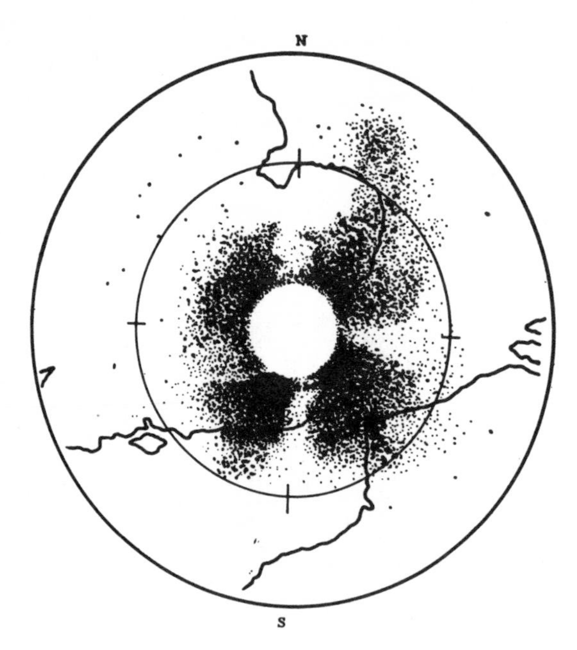

Fig. 5. Migrating birds as seen using MTI radar (from [5]).

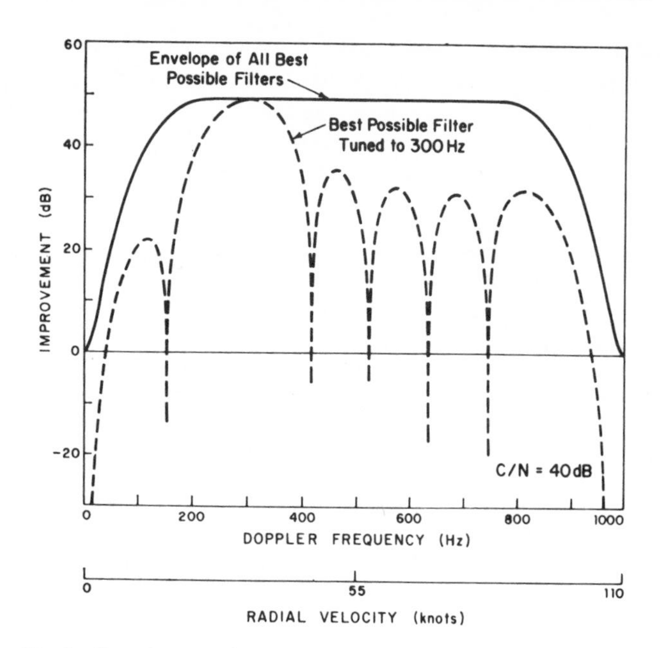

Fig. 6. Improvement of target-to-interference ratio, scanning antenna.

mated that a few million birds crossed a 100-mi front during one of the busy nights of the autumn migration in the Cape Cod region [5].

Birds fly between 15 and 45 knots true air speed. Taking into account winds, radial velocities over the range ± 80 knots or so may be observed.

A fairly effective radar improvement used against bird clutter is a carefully tailored sensitivity time control (STC) [6]. The STC varies the radar gain with range and is adjusted so that the minimum detectable target is a specific value, say, 1 m^2. This calls for an R^{-4} attenuation law.

E. Surface Vehicles

The cross section of ground vehicles is in the same range as aircraft; namely, from 1 to 100 m^2. Radial velocities range over ± 60 knots.

Some reduction in ground vehicle returns is achieved by tilting the antenna upward. The only other solution found so far, as is practiced at the Atlanta airport, is to blank out areas on the scope known to contain visible roads carrying cars with radial velocities outside the notch at zero velocity. This has proven effective and causes only small holes in the coverage.

III. Radar Improvements

From the previous discussion, it is obvious that there is no single solution to the problem of providing high probability of detection with low false alarm rate in an automated system under all conditions. Rather, several changes must be made and the resulting system examined to see how well it corrects all the various problems.

We will now discuss a series of improvements which could be made to the radar and tell how each would help solve the above problems.

A. Signal Processing Techniques

1) Optimum Signal Processing: We first concentrate on fixed ground clutter since this presents the biggest clutter problem. To give complete flexibility in siting and tilting the antenna while still rejecting ground clutter, an approximate 20-dB improvement in performance should be provided in the signal processor. It is obvious that this requires processing linear clutter and target signals and that one must avoid all nonlinearities in the receiver. To narrow the blind speed region, one should also process more pulses.

In order to assess quantitatively what could be considered a "good" MTI processor for improving the performance of ASR radars against fixed ground clutter, calculations were made of the performance of the so-called "optimum processor." Given the initial conditions, the optimum processor has the highest target-to-interference (interference is defined as clutter plus front-end noise) ratio improvement of any processor. By knowing the performance of such a processor, one can judge how well a conventional easily implemented or any other processor (i.e., suboptimum) can approach the theoretical limit. The processor considered here can be defined as a device that takes M complex signal returns V_i, multiplies these returns by a complex filter weight $\overline{W}_i$, adds them, and then takes the square of the amplitude

$$R = \left| \sum_{i=1}^{M} \overline{W}_i V_i \right|^2 .$$

V_i is composed of target, noise, and clutter. The theory of optimization will not be shown here but follows that of DeLong and Hofstetter [7]. The clutter spectrum, which in this case is essentially all caused by the antenna scanning motion, is modeled by an antenna having a Gaussian beam shape as in Emerson [8].

Two general cases have been studied: the mechanically rotating antenna as in the ASR radars and the step-scan antenna. In both of these cases the transmitter pulses are assumed uniformly spaced. Fig. 6 shows the target-to-interference improvement in decibels that is possible (optimum) for the mechanically rotating antenna. The results in this section assume the use of a sufficiently stable coherent transmitter. Poorer, as yet undetermined, results will be obtained using a magnetron transmitter. The parameters are (similar in most

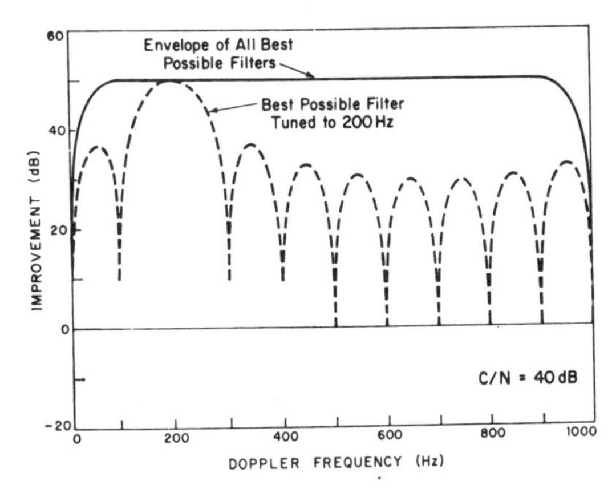

Fig. 7. Improvement of target-to-interference ratio, step-scan antenna.

respects to the ASR-7):

antenna width	5.25 m
antenna rotational speed	1.36 rad/s
wavelength	0.107 m
PRF	1000 pulses/s
number of pulses processed/look	10
clutter-to-noise ratio	40 dB.

The maximum clutter-to-noise ratio which can be handled will be set by the dynamic range of available analog-to-digital (A/D) converters. Mean clutter-to-noise ratios of 40 to 50 dB can be handled in available A/D converters with adequate sampling rates.

The upper curve in Fig. 6 is the improvement obtained when the optimum filter is tuned to the Doppler frequency of the target as the target Doppler is varied. The lower curve is the improvement when the optimum filter is tuned to a fixed Doppler (300 Hz) as the target Doppler is varied. The lower curve represents the frequency response of the particular optimum filter tuned to 300 Hz.

The following general characteristics of the optimum processor should be noted:

a) The upper curve (Fig. 6) levels out at about $M \times C/N = 10^5$ where M is the number of pulses processed and C/N is the clutter-to-noise ratio, unless M is small. This points up the need for wide dynamic range A/D converters as explained above.

b) At the so-called "blind speeds" (0 and 1000 Hz), there is no improvement, but there is no deterioration either, thus a target whose cross section is sufficiently above clutter can be seen.

c) For filters that are not tuned on or close to blind speeds, there are very deep nulls at the blind speeds.

d) The width of the notch about the blind speeds increases with antenna rotational speed when all other parameters are held constant.

e) The filter cannot, in general, be approximated by a discrete Fourier transform (DFT) except in certain special cases.

f) Because the optimum weights W_i are a function of clutter-to-noise ratio, the optimum processor requires some *a priori* knowledge. However, this ratio can be determined in principle by the application of a proper algorithm in the receiver, together with ground clutter memory from scan to scan.

The step-scan case is shown in Fig. 7. Because the antenna is not scanned and ground clutter can be considered to be time-stationary (constant voltage), the clutter spectrum is just an impulse at zero frequency. The intrinsic ground clutter spectrum (motion of trees) is ignored because it is too narrow to have any effect upon the results. It should be pointed out that the only input parameters needed for the step-scan case are the PRF, the number of pulses per look (M), and the clutter-to-noise ratio. Thus these curves are directly applicable to other radar frequencies as well as S band. The parameters used in Fig. 7 are the same as in the scanning case.

The following properties of the step-scan curves (Fig. 7) should be noted:

g) The notches at the blind speeds are now very narrow. There would be much less chance of losing a target in clutter with near tangential velocity.

h) In most instances a DFT can replace the optimum filter and thus improve computation efficiency.

i) Although it cannot be seen on this figure, the improvement at the "blind speeds" is 0 dB as in the scanning case of Fig. 6.

By comparing these results with those of Fig. 2, we see the amount of clutter rejection achieved in the present radars as well as other conventional MTI systems is far less than the best that can be done, whether scanning or not.

2) Near-Optimum Signal Processing: In the scanning antenna case, the implementation of the optimum processor for every range–azimuth cell calls for M complex multiplications for each target velocity examined. Usually, if M pulses are being processed, a filter bank with M filters will give adequate coverage for all target velocities. Thus M^2 complex multiplication must be performed for every range cell. For a typical ASR, 800 range cells per sweep must be sampled on 10 sweeps and processed every 10 ms. If optimum filters were used, 8 000 000 complex multiplications per second would be required or 32 million simple multiplications.

A simpler processor can be built. The optimum processor can be broken into two parts, a clutter filter followed by a target filter. The filter used to reduce clutter multiplies the signal vector by the antenna weighting and by the inverse of the interference covariance matrix. The target filter used to enhance the target is a DFT. The near-optimum processor could consist of a digital filter which approximates as closely as possible the frequency response of the clutter filter followed by a DFT for the target filter. This combination will give improvement factors within a few decibels of the optimum shown in Fig. 6 and require fewer multiplications per second than previously indicated. It will also provide Doppler information on the target.

Through direct comparison it has been found that a simple three-pulse canceler without feedback forms the clutter filter portion of a near-optimum processor for the scanning antenna.

For the step-scan case, Fig. 7, the clutter filter is nothing more than a dc removal filter and as such is very easy to implement.

3) Ground Clutter Map: The near-optimum ground clutter processor would not be complete without adequate thresholding. For a typical ASR, ground clutter will appear only in the zero Doppler filter and the filters immediately adjacent on each side. Ground clutter is very spotty in character. It varies greatly in size from one resolution cell to the next. Thus averaging nearby cells will not give a good estimate for thresholding purposes.

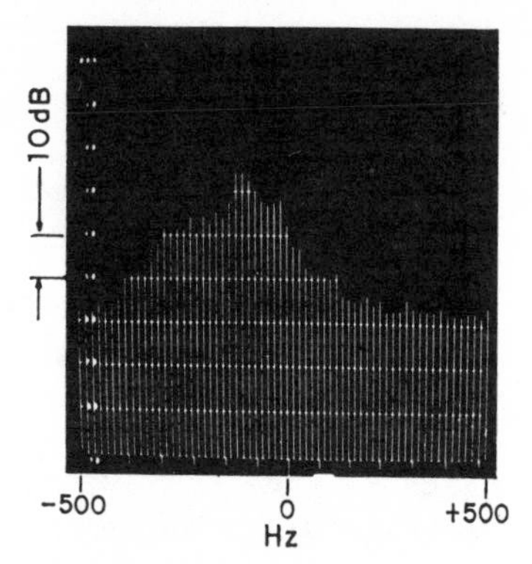

Fig. 8. Typical storm spectrum taken using an S-band radar. Receiver noise level is −57 dB.

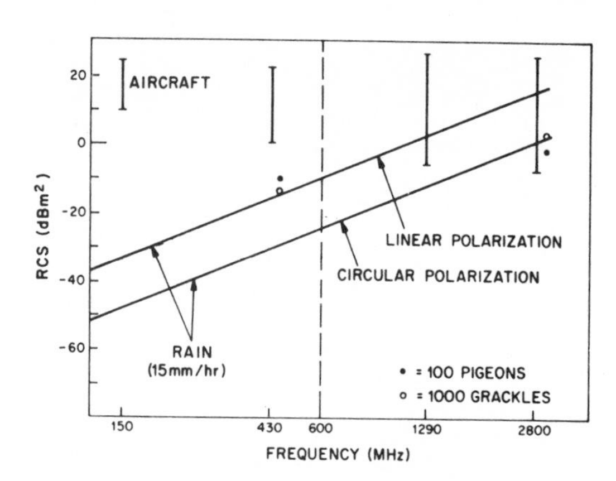

Fig. 9. Radar cross section of aircraft and moving clutter (rain and birds) as a function of frequency [1].

A practical way to accurately set ground clutter thresholds is to use a digital ground clutter map which remembers the ground clutter in every range–azimuth resolution cell averaged over a sufficient time period (number of scans).

Such a clutter map provides superclutter visibility. Whenever an aircraft's cross section is larger than the clutter over which it is flying it will be seen even if it has zero radial velocity (tangential target). The operator makes no decision concerning the selection of MTI or normal video to be displayed on the scope. The best detection is provided automatically in each range–azimuth cell because detection in each velocity is optimized separately.

4) Filter Bank for Precipitation Filtering: It is fortunate that the optimum or near-optimum filtering against ground clutter utilizes a filter bank since this is a good approach to eliminating weather clutter. We need about 30-dB weather clutter rejection of which 15 dB is provided by circular polarization. Filtering of some sort is a viable solution to obtaining the remaining 15 dB. The filtering could be near-optimum as in the case of ground clutter except for the fact that the weather clutter spectrum (see Fig. 8) changes with time. This change could be measured and the filter adapted to the spectrum, but this would result in an intolerable amount of hardware.

A good alternative is to use the filter bank produced by the near-optimum ground clutter filter. It is only necessary to set the threshold on each filter adaptively. A so called "mean-level" threshold is employed. Since storms are rarely less than about 1 mi in extent, the moving clutter is averaged over a half mile on either side of the cell being examined for a target. Each velocity is averaged separately so that filters containing only noise and not weather clutter will not be penalized.

If further, a multiple PRF system is used so that high-speed aircraft typically will fall in different filters in the filter bank on successive PRF's, there is a very high likelihood that the target return will be competing with noise only on one of two PRF's. Only for aircraft whose true (not aliased) radial velocity coincides with that of the rain will there be a degradation in detection performance.

In summary, a modern radar for air-traffic control use employing a scanning antenna would have a fully coherent transmitter; a linear large dynamic range receiver; a signal processor containing a near-optimum ground clutter filter bank; a fine grained ground clutter map to set ground clutter thresholds; mean-level thresholding on weather; and would employ multiple PRF's for elimination of blind speeds.

B. Lower Frequency Radar

Another approach to the clutter problem is to choose radar parameters such that the target-to-clutter ratio at the input to the radar is much more favorable. For instance, the clutter cell size could be reduced by using very narrow beams or the range resolution reduced by using pulse compression. These are not very attractive solutions because they imply large antennas or very wide bandwidths in already overcrowded radar bands. A viable solution would be to go to lower frequencies.

Fig. 9 shows a summary of aircraft and moving clutter cross-section data as a function of frequency. For rain we have assumed a fan beam with a 1.5° azimuth beamwidth at 30-nmi range. It is quite clear that the aircraft return is maintained at the same size or larger as the frequency decreases, whereas the moving clutter, precipitation and birds, decreases in size. The clutter sizes shown are mean values so that an approximate 15-dB ratio between target and clutter is required for automatic detection. The dashed line at 600 MHz is the highest frequency where rain and birds are not considered a problem. Below 600 MHz, the radar need not even use circular polarization to combat rain.

C. Antenna Improvements

1) Dual Beam Antennas: The present ASR antenna suffers because of two facts. First, it is tilted up several degrees (see Fig. 3) to reduce the ground clutter signals, thus degrading its low altitude performance at long ranges. Secondly, it has a cosecant-squared pattern with inadequate gain at the high elevation angles. Because of the latter fact, an aircraft at, say, 9 mi and 15 000 ft suffers a 20-dB disadvantage because of its position in the antenna pattern compared to some moving clutter at the peak of the beam (9 mi and 3000-ft altitude). A cosecant-squared pattern is a poor pattern to use when the elimination of birds and weather clutter is desired.

The ideal pattern is one that is uniform with elevation angle. Then an R^{-4} STC curve together with thresholding against noise will produce a constant cross-section discrimination against birds and weather. This ideal pattern would however have about 9 dB less peak gain than the present ASR antennas so it is out of the question. An antenna whose vertical pattern changes with range on receive is needed. It could maintain long-range low-altitude coverage and correct the high-elevation short-range problem as well. The ASR-8 radar

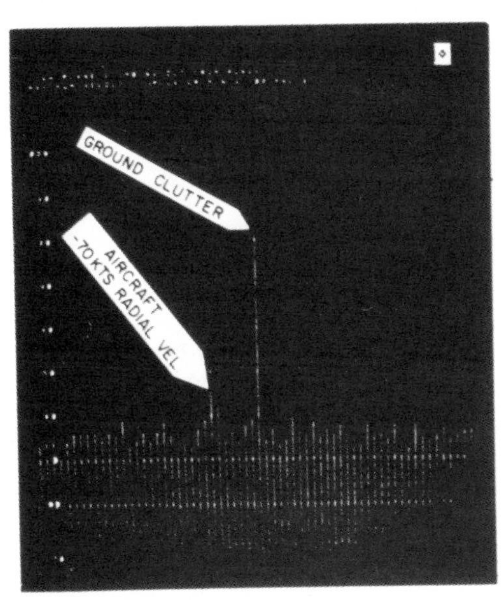

Fig. 10. Doppler spectrum of ground clutter and light single engine aircraft as observed by a UHF radar employing an electronically step-scanned antenna.

presently being developed will provide a dual beam on receive which will improve the long-range low-altitude coverage but will do little for the short-range high-angle coverage. Improvements of the SCV by 20 dB or more using improved processing techniques should remove enough restrictions on the antenna designer to solve the second problem.

2) Step-Scanned Antennas: An electronically step-scanned antenna is highly desirable for incorporation in an ASR because it reduces the spectral width of the ground clutter return to a narrow band. It is so narrow in fact that it is possible to completely separate the aircraft with slow velocities from the ground clutter. This is illustrated in Fig. 10 which is the spectral output of a 435-MHz radar using an electronically step-scanned cylindrical array antenna (Fig. 11). Notice that all of the ground clutter appears in one of the filter outputs and that the aircraft competes only with receiver noise. The SCV of the radar is limited only by the size of clutter the radar can handle and this, in turn, is limited only by the dynamic range of available A/D converters.

Although not entirely ruled out at *S* band, these cylindrical arrays are easier to build at lower frequencies and the large aperture they provide substantially reduces the required transmitter power, which in turn permits all the antenna switching to be accomplished with solid-state components. Cylindrical arrays have been used in a few radars at UHF and one is being designed for ATCRBS use at *L* band by Hazeltine.

Step-scanning a mechanically rotating antenna would produce a meaningful improvement although probably not as dramatic as that shown in Fig. 10.

IV. Some Results

Two demonstration radars utilizing the radar signal processing techniques described in Section III are being tested at Lincoln Laboratory, one at UHF and the other at *S* band.

A. UHF Radar with Step-Scan Antenna

This radar utilizes a fully coherent transmitter, an electronically step-scanned cylindrical array (see Fig. 11), and a wide dynamic-range linear signal processor. The antenna scans over a 45° sector and $\frac{1}{4}$-nmi range gates are processed out to 30 nmi every 3 s. A picture of typical synthetic video output for a period of 60 s is shown in Fig. 12. The processor was implemented using a minicomputer. It simply removes the dc from the quadrature video and noncoherently integrates two groups of 16 pulses on each azimuth. Detections on both groups of pulses are required before display. No ground clutter was seen confirming the results shown in Fig. 10. False alarms occurred in two regions and are believed due to automobile traffic. These could easily be removed by a tracking computer. Absolutely no rain was seen even on a rainy day. However, rain measuring equipment was not available to determine the rate of rainfall.

Fig. 11. Semicircular electronically step-scanned antenna used with UHF radar.

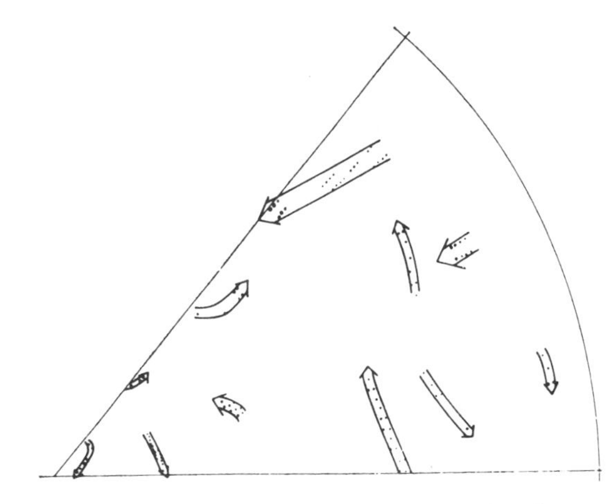

Fig. 12. Scan history display showing the detection of many aircraft. Arrows were added for emphasis.

A few words should be said concerning the obvious breaks in the data. The antenna has a 15° elevation beamwidth centered at 0° elevation. It is, thus, very susceptible to ground lobing effects. The breaks in the tracks occur at about the right interval for ground lobing to be the cause. A properly designed antenna would have a sharp cutoff in its vertical pattern at zero elevation to reduce the strong antenna gain variations with elevation. Typically, we could see single-engine aircraft over the whole field of view and almost until they actually landed on a nearby runway.

B. S-Band Radar with Rotating Antenna

A greatly modified FPS-18 radar has been put into operation to test some of the signal processing concepts outlined in this paper. The stability of its klystron transmitter was improved so as not to be a limit in system performance. The antenna was servo driven so that the ground clutter map would repeat itself every scan. A new, wide-bandwidth, linear receiver was provided. A processor employing the techniques described in Section III was simulated in real time on a general-purpose digital signal processor (the Lincoln Laboratory Fast Digital Processor, FDP) [9]. Signals were processed over 35° of arc by 8 mi in range. Ten complex video samples were processed as a batch using $\frac{1}{16}$-nmi range gates.

As of the present date, only a few results have been obtained using the *S*-band system. These results have been in consonance with the predictions of Section III.

V. Conclusions

The problems associated with air-traffic control surveillance radars to be used in a highly automated environment have been studied in detail. The general problem is one of aircraft returns competing with various kinds of clutter returns; ground clutter, weather clutter, birds, second-time-around clutter, and ground traffic. Radar solutions applicable to the elimination of each type of clutter have been studied. Sets of these solutions were brought together to define radars which would be suitable to use in automated systems. Two such radars, a UHF radar employing an electronically step-scanned array and an *S*-band radar using a scanning antenna, have been built for demonstration purposes.

Acknowledgment

The work described above depended for its success on all the members of Group 43 at Lincoln Laboratory. R. G. LeBlanc, H. P. McCabe, and M. A. Nader are to be particularly commended for their contribution. T. M. Hendricks helped work out the suboptimum ground clutter processor. The authors also wish to thank H. G. Weiss and P. R. Drouilhet for their constructive direction and the Federal Aviation Administration for their sponsorship of this work and, in particular, K. Coonley and D. Hopson for their help and encouragement.

References

[1] F. E. Nathanson, *Radar Design Principles.* New York: McGraw-Hill, 1969.
[2] "A brief description of the solid-state ASR-7 airport surveillance radar," Texas Instruments Inc. Rep., May 1970.
[3] J. Croney, "Clutter on radar displays, reduction by use of logarithmic receivers," *Wireless Eng.*, pp. 83–96, Apr. 1956.
[4] K. R. Hardy, "CPS-9 radar investigation of clear-air convection," in *13th Radar Metecrology Conf. Proc.*, pp. 236–240, Aug. 1968.
[5] E. Eastwood, *Radar Ornithology.* London, England: Methuen & Co., Ltd., 1967.
[6] R. E. Richardson, J. M. Stacey, H. M. Kohler, and F. R. Naka, "Elimination of a type of natural clutter in *L*-band radars," M.I.T. Lincoln Lab., Tech. Rep. 178, April 15, 1958, ASTIA Doc. AD 133851.
[7] D. F. DeLong, Jr., and E. M. Hofstetter, "On the design of optimum radar waveforms for clutter rejection," *IEEE Trans. Informat. Theory*, vol. IT-13, pp. 454–463, July 1967.
[8] R. C. Emerson, "Some pulse Doppler MTI and AMTI techniques," Rand Corp. Rep. 274, p. 104, Mar. 1, 1954.
[9] B. Gold, I. L. Lebow, P. G. McHugh, and C. M. Radar, "The FDP, a fast programable signal processor," *IEEE Trans. Comput.*, vol. C-20, pp. 33–38, Jan. 1971.

5.5 ADVANCED SIGNAL PROCESSING FOR AIRPORT SURVEILLANCE RADARS

R. O'Donnell, C. Muehe, M. Labitt and L. Cartledge

EASCON-74 Record, 1974, pp. 71-71F.

M. I. T. Lincoln Laboratory
Lexington, Massachusetts 02173

ABSTRACT

The inclusion of airport surveillance radars (ASR) in an automated air traffic control system, such as the ARTS-III, has been limited by the present radar's capability to automatically reject ground clutter, weather clutter and angels while still maintaining good detectability on all aircraft within their coverage patterns. Analytical and experimental studies have been performed which indicate that new techniques[1] can significantly enhance the automated capability of these radars. A special-purpose, hard-wired, digital signal processor has been designed, built and tested which provides near-optimum target detection over the entire ASR coverage out to 48 nmi. The processor which coherently integrates eight pulses has both a fine grained clutter map for optimal thresholding in high ground clutter environments and a mean-level thresholding scheme for filtering those Doppler cells which contain heavy precipitation. Because of the processor's ability to detect targets in a high ground clutter environment, the ASR's will be able to operate their antennas at lower elevation angles and, thus, have better coverage of low flying aircraft near the terminal. The processor is initially being tested on a highly modified, coherent S-band, FPS-18 radar. The stability of the klystron transmitter was improved so that it would not limit system performance and a new, wide dynamic range, linear receiver was provided.

I. PROBLEM DESCRIPTION

Automated airport surveillance radars must exhibit simultaneously low false alarm rates and high probabilities of detection of aircraft in the presence of extraneous clutter reflections. Three types of interfering clutter have been found to predominate: ground clutter, weather clutter and angels.

The ground clutter radar backscatter coefficient varies appreciably from spot to spot in the area of coverage. The present ASR radars suppress ground clutter by three mechanisms: MTI, antenna tilt and by mounting the antenna close to the ground to take advantage of the shielding effect of nearby objects. The MTI processors in these radars employ limiting in the IF, followed by a phase detector. The purpose of the limiting is to normalize the video output so that clutter residue from the MTI filter is reduced to the average noise level. This limiting action spreads the clutter spectrum so that considerably poorer subclutter visibility (SCV) is achieved that if the normalization had been done by some other mechanism not involving nonlinearities. The performance of S-band, three-pulse cancellers with and without limiting is presented in Figure 1. In order to obtain reasonable signal-to-clutter ratios, it is commonplace for the present ASR's to tilt the antenna upward by 2 to 5 degrees depending on the local clutter situation. This advantage is offset by the degraded detectability of aircraft flying at low elevation angles.

In the so-called "second-time-around" clutter effect, returns are being received due to illumination of clutter beyond the nonambiguous range by the next-to-last pulse transmitted. These returns are prevalent where conditions for anomalous propagation exist or in regions where mountains exist beyond the nonambiguous range. Present ASR's use magnetron transmitters that transmit pulses with random phase from pulse to pulse. Thus, it is impossible to maintain the phase relation between the first and second-time-around clutter returns and the two cannot be filtered out simultaneously.

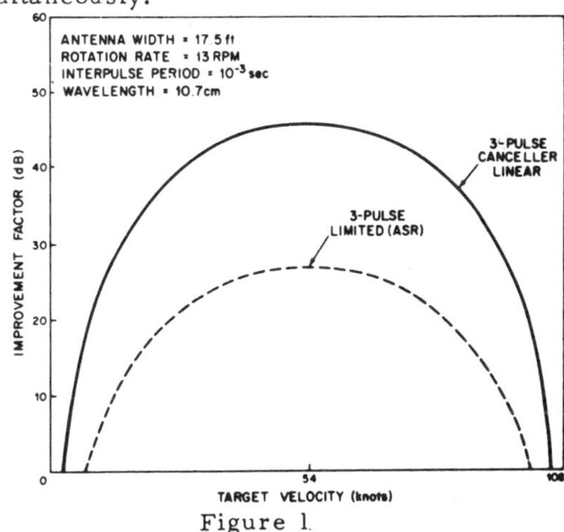

Figure 1

Performance of S-Band 3-Pulse Cancellers. The improvement factor is the ratio of signal-to-clutter out of the canceller to signal-to-clutter in. The latter ASR radars use 3-pulse cancellers following a limiter shown above by the curve "3-pulse limited (ASR)"

The backscatter from precipitation has been studied extensively[2, 3, 4]. Circular polarization is normally used to reduce rain clutter by about 15 dB while reducing the signal level to some extent. The use of MTI helps reduce rain clutter except when the antenna is looking toward or away from the wind direction. In these directions the rain clutter spectrum is such that all of the rain clutter

* The work reported was prepared for the Federal Aviation Administration under Interagency Agreement DOT-FA-72-WAI-242 by Lincoln Laboratory, a center for research operated by Massachusetts Institute of Technology under Air Force Contract F19628-73-C-0002.

signals may pass through the MTI filters. Log-FTC-antilog circuits[3,4] reduce the receiver gain in proportion to the average level of rain clutter for about a mile in range surrounding the cell of interest. It, thus, normalizes the rain clutter level just as limiting is used to normalize ground clutter at the output of the MTI circuit. Its purpose is to suppress the rain clutter on the scope. At the same time, of course, it suppresses the signal. For adequate detection, the signal amplitude must be appreciably above the clutter residue from the MTI filters.

So-called "angel clutter" refers to all returns which cannot be explained as being ground or precipitation clutter or targets. Much effort has been spent studying angels. It is now believed that most, if not all, angels are caused by bird flocks[6,7]. Swarms of insects[5] and returns from well organized layers of turbulent refractivity in the atmosphere[6] are two other sources which have been considered possible causes of some angels. Birds fly between 15 and 45 knots true air speed. Taking into account winds, radial velocities over the range ± 80 knots or so may be observed. A fairly effective radar improvement used against bird clutter is a carefully tailored sensitivity time control (STC)[7]. The STC varies the radar gain with range and is adjusted so that the minimum detectable target is a specific value, say, one m^2. This calls for an R^{-4} attenuation law. Some modification of this law may be required to handle ground clutter.

II. SOLUTION

A. Optimum Processing

In order to assess quantitatively what could be considered a "good" MTI processor for improving the performance of ASR radars against fixed ground clutter, calculations were made of the performance of the so-called "optimum processor"[1]. Given the initial conditions, the optimum processor has the highest target-to-interference (interference is defined as clutter plus front-end noise) ratio improvement of any processor. By knowing the performance of such a processor, one can judge how well a conventional, easily implemented or any other processor (i.e., suboptimum) can approach the theoretical limit[8]. The clutter spectrum which in this case is essentially all caused by the antenna scanning motion, is modeled by an antenna having a Gaussian beam shape as in Emerson[9].

Figure 2 shows the target-to-interference improvement in decibels that is possible (optimum) for the mechanically rotating antenna. The results in this section assume the use of a sufficiently stable, coherent transmitter. Poorer results will be obtained using a magnetron transmitter. The parameters are (similar in most respects to the ASR-7):

Antenna Width	5.25 meters
Antenna Rotational Speed	1.36 radians/sec
Wavelength	0.107 meter
PRF	1000 pulses/sec
No. of Pulses Processed/Look	10
Clutter-to-noise Ratio	40 dB

The maximum improvement factor shown in Figure 2 equals the clutter-to-noise ratio assumed (40 dB) times the number of pulses coherently integrated[8].

The maximum clutter-to-noise ratio which can be handled will be set by the dynamic range of available analog-to-digital (A/D) converters. Mean clutter-to-noise ratios of 40 to 50 dB can be handled in available A/D converters with adequate sampling rates.

The upper curve in Figure 2 is the improvement obtained when the optimum filter is tuned to the Doppler frequency of the target as the target Doppler is varied. The lower curve represents the frequency response of the particular optimum filter tuned to 300 Hz.

By comparing these results with those of Figure 1, we see that the amount of clutter rejection achieved in the present radars as well as other conventional MTI systems is far less than the best that can be done.

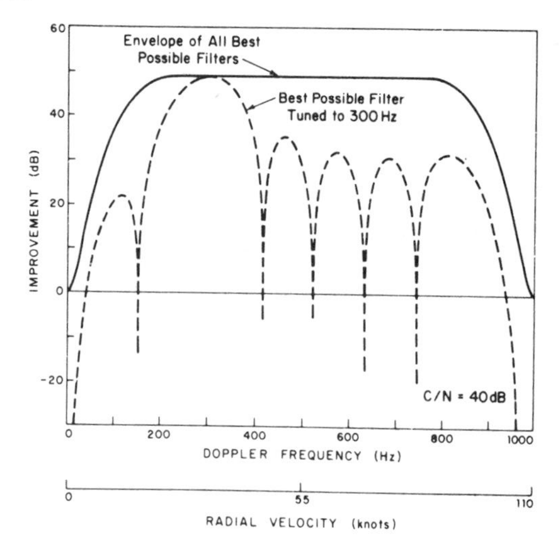

Figure 2

Improvement of Target-to-Interference Ratio, Scanning Antenna

B. Implementation of a Near-Optimum Processor

The optimum processor can be broken into two parts; a clutter filter followed by a target filter. The filter used to reduce clutter multiplies the signal vector by the antenna weighting and by the inverse of the interference covariance matrix. The optimum target filter used to enhance the target is a Discrete Fourier Transform (digital filter bank). The near-optimum processor consists of a three-pulse canceller without feedback which closely approximates the frequency response of the clutter filter followed by a Discrete Fourier Transform for the target filter. This combination gives improvement factors within a few dB of the optimum shown in Figure 2 and requires many fewer multiplications per second.

C. Thresholding

The near-optimum ground clutter processor would not be complete without adequate thresholding. For a typical ASR utilizing a near-optimum processor, ground clutter will appear only in the zero Doppler filter and the filters immediately adjacent on each side. Ground clutter is very spotty in character. It varies greatly in size from one resolution cell to the next. Thus, averaging nearby cells will not give a good estimate for thresholding purposes.

A practical way to accurately set ground clutter thresholds is to use a digital ground clutter map which remembers the ground clutter in every range-azimuth resolution cell averaged over a sufficient time period (number of scans). Whenever an aircraft's cross section is sufficiently larger than the clutter over which it is flying, it will be seen even if it has zero velocity (tangential target).

It is fortunate that the optimum or near-optimum filtering against ground clutter utilizes a filter bank since this is a good approach to eliminating weather clutter. About 30 dB weather clutter rejection is needed of which 15 dB is provided by circular polarization. Filtering of some sort is a viable solution to obtaining the remaining 15 dB. The filtering could be near-optimum as in the case of ground clutter except for the fact that the weather clutter spectrum (see Figure 3) changes with time. This change could be measured and the filter adapted to the spectrum, but this would result in an intolerable amount of hardware.

A good alternative is to use the filter bank produced by the near-optimum ground clutter filter. It is only necessary to set the threshold on each filter adaptively. A so-called "mean level" threshold is employed. Since storms are rarely less than about one mile in extent, the moving clutter is averaged over a half mile on either side of the cell being examined for a target. Each velocity is averaged separately so that filters containing only noise and not weather clutter will not be penalized.

Further, a multiple PRF system, rather than a staggered PRF system[1], is used so that high speed aircraft typically fall in different filters in the filter bank on successive PRF's. There is a very high likelihood that the target return will be competing with noise only and not weather on one of two PRF's. Only for aircraft whose true (not aliased) radial velocity coincides with that of the rain will there be a degradation in detection performance.

In summary, a modern radar for air traffic control use employing a scanning antenna would have a fully coherent transmitter; a linear, large dynamic range receiver; a signal processor containing a near-optimum ground clutter filter bank; a fine grained ground clutter map to set ground clutter thresholds; mean-level thresholding on weather and would employ multiple PRF's for elimination of blind speeds.

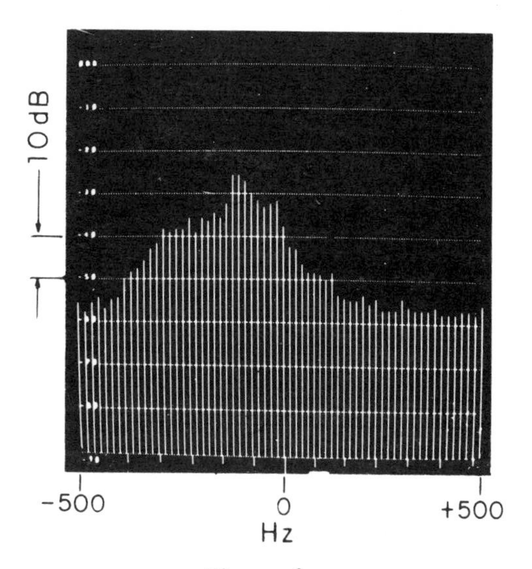

Figure 3

Typical Storm Spectrum Taken Using an S-Band Radar. Receiver Noise Level is -57 dB.

III. SOME EXPERIMENTAL RESULTS

A. Radar Used

A highly modified klystron-type, S-band, FPS-18 SAGE gap filler radar was used to test the signal processing concepts outlined in this paper.

An antenna servo control has been designed, constructed and installed which directs the FPS-18 antenna from digital instructions originating in the radar system timing. In addition to its normal scanning mode, the antenna is also capable of being pointed to a particular azimuth (searchlight mode) by setting the appropriate azimuth coordinate on a set of thumbwheels. Precise speed and position control are desired to allow variation in the speed as a parameter and so that the ground clutter level at each spot can be accurately correlated from scan to scan. Five antenna speeds are available from 5 to 15 rpm.

In the FPS-18 receiver, the original FPS-18 mixer and preamplifier have been replaced by an assembly which includes a microwave sensitivity-time control (STC) attenuator, a balanced hot carrier diode mixer and a semiconductor preamplifier. This change was made to enhance the dynamic range and stability of the system. The original system used an analog STC system in the IF preamplifier. The new STC is controlled by the digital control system through a D/A converter. It generates an R^{-4} curve which will alleviate the angel problem and which incorporates a gain control at the digital control panel.

Instabilities in the transmitter, local oscillators, coherent oscillator and range gate timing give rise to fundamental limitations on the clutter-rejecting capability of MTI radars. Amplitude or angle modulations of the signals in the radar generate sidebands which may fall in the Doppler passband and look to the processor like moving target signals. Similarly, jitter in the timing of the transmitted pulses or the range gates can cause amplitude modulation of the received signals and the consequent sidebands in the Doppler passband.

The FPS-18 was originally required to produce a "cancellation ratio" of 40 dB while operating at a fixed pulse repetition rate. This implies that the total spurious sideband power within the Doppler passband of the radar had to be at least 40 dB below the return from a fixed echo at the same range. The radar as designed met this requirement with some margin.

As modified for the FAA, the FPS-18 meets more stringent requirements. The total spurious power in any of the eight Doppler filters is at least 52 dB below the return from fixed clutter at the same range. In other words, the MTI improvement should be 52 dB. This level has been achieved with the FPS-18 operating in a variable PRF mode. Quantitative measurement of these spurious responses is quite difficult. It turns out that the radar receiver and processor make up the best spectrum analyzer available to us. A diagnostic routine known as SGP (Single Gate Processor) is used with the radar antenna stopped in the direction of a known fixed echo (e.g., a large smokestack or radio tower). When the range gate corresponding to the range of the fixed echo is selected, the SGP program performs a Digital Fourier Transform on the received signal and presents the Fourier coefficient (Figure 4). The stability levels shown in Figure 4 were obtained after extensive modification of the original FPS-18 radar.

B. Two Phases of Processor Development

1. General

The signal processing concepts outlined thus far have been tested in a two-phase approach. First, the signal processor was implemented in software on the Lincoln Laboratory Fast Digital Processor (FDP)[10], a very fast, special-purpose computer whose architecture was designed to optimally perform signal processing tasks such as fast Fourier transforms. Second, the signal processing concepts and algorithms tested and proven with the FDP tests were embodied in a hard-wired version, the MTD or Moving Target Detector. The MTD has been designed, built and tested at Lincoln Laboratory and was delivered in June to the FAA at NAFEC where it is undergoing further tests.

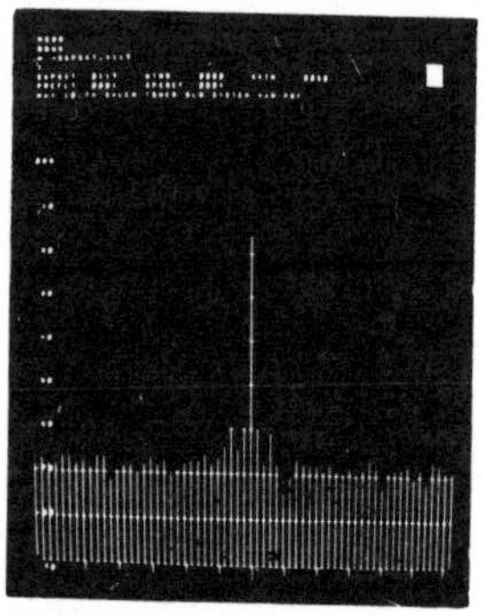

KNOWN FIXED ECHO FROM
500-ft STACK AT SALEM MASS.
(range 18 nmi)

VARIABLE PRF — 50 dB gain
HORIZONTAL SCALE — ± 60 knots
VERTICAL SCALE — 10 dB per line

Figure 4

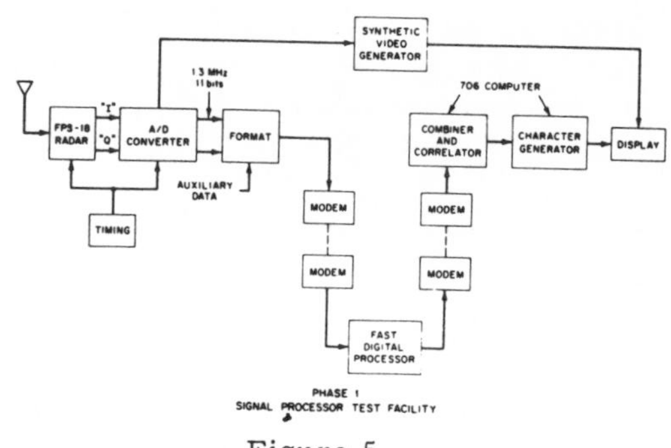

Figure 5

2. Demonstration of Concept Feasibility Using the Fast Digital Processor (FDP)

A block diagram of this phase of the processor development is shown in Figure 5. The FPS-18 receiver output is quadrature-detected and the in-phase and quadrature video signals are digitized. These signals were assembled into a message and sent over coaxial cable from the Lexington Field Station to the main Laboratory where the FDP is located. Since the FDP is a programmable machine, it is easy to adjust processing parameters and alter the processing algorithms. The FDP has the capability of processing an area eight nautical miles in range by 40 degrees in azimuth.

After the radar data was processed by the FDP, the target hit reports were sent over coaxial cable back to the Field Station. A Raytheon 706 minicomputer was used to combine hits from the same target and to interpolate between hits prior to displaying a target character on the ARTS-III display console.

A second signal path shown at the top of the diagram is used to superimpose analog video on the display. The A/D converter outputs are combined to form a synthetic reconstruction of normal video.

Figures 6 and 7 are results obtained from the radar data correlation/interpolation system. This system incorporates software routines to correlate all the reports from a single aircraft and to interpolate between them for increased position accuracy. Figure 6 shows the unprocessed primitive targets. The cluster of threshold crossings at 19 nmi and about 110° is typical of the multiple returns per target. Figure 7 shows 40 scans of radar data that have been correlated and interpolated on a scan-by-scan basis. Only those targets which were derived from more than two contiguous threshold crossings are displayed. The photograph shows the tracks of three moving targets which entered the area of coverage at separate times during the run. Notice that the software routine discriminates against single point returns so that a much higher false alarm rate for primitive returns can be tolerated. Therefore, it is necessary to take into account the correlation/interpolation feature when calculating false alarm rates and detection probability.

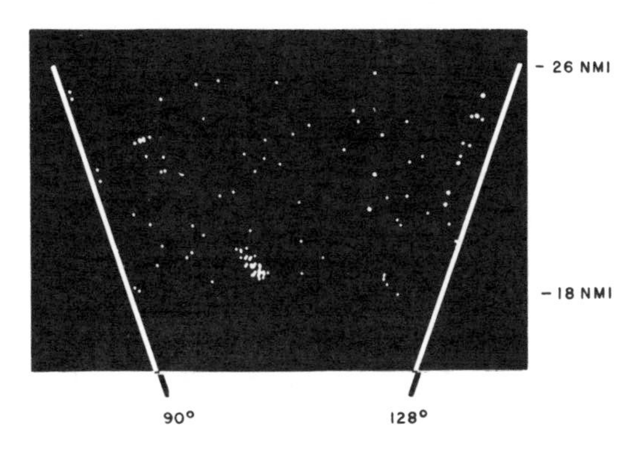

Figure 6

Ten Scans Uncorrelated

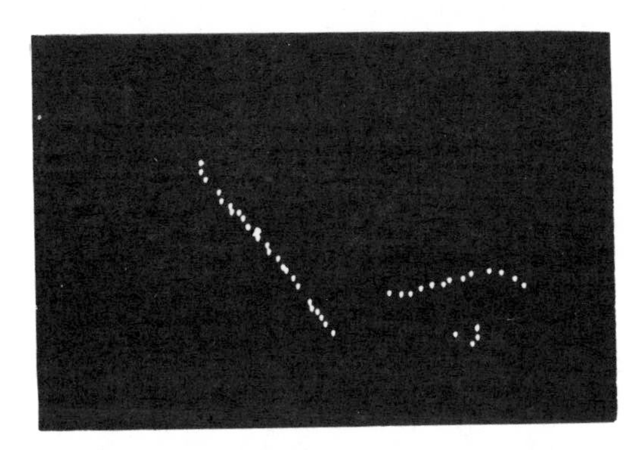

Figure 7

Forty Scans with Correlation and Interpolation

The Phase I tests on the FDP system indicated that the processor concepts performed as theoretical studies has indicated. The processor performed with low false alarm rates and high probability of detection in both heavy ground clutter and rain environments.

3. Moving Target Detection (MTD) - Design, Construction and Testing

a. General

The exact design of the Moving Target Detector (MTD) was proven on a conceptual basis by the tests of the algorithms on the FDP during Phase I. The exact implementation of these algorithms in the MTD was also tested by an extensive computer simulation which uses integer arithmetic and performs truncations and round-offs exactly as they are performed in the hard-wired MTD. The simulation indicated that some of the algorithms used in the thresholding generation were faulty, and that the clutter values stored on disc memory required greater accuracy than was possible with the 10 bits available. As a result, it was necessary to redesign the entire thresholding and disc interface portions of the MTD hardware. Since the new design requires that the clutter values be stored on the disc in floating-point format, floating-to-fixed-point and fixed-to-floating-point converters were included in the disc interface.

b. Moving Target Detection (MTD)

The MTD processor is a special purpose, hard-wired, signal processor which is capable of processing the full 360° coverage of the FPS-18 radar out to a nominal range of 48 nmi. A block diagram of the processor is presented in Figure 8. The I and Q (In-Phase and Quadrature) signals are sampled at a 2.6 MHz rate by a 10-bit A/D converters. The I and Q channels are then added coherently, two at a time, to produce 11-bit I and Q channel words at a 1.3-MHz rate. Ten consecutive samples of both the I and Q channels for each of 760 range gates (47.5 miles of range) are stored in a 8192-word, 36-bit memory. The I and Q channel samples are stored in two 11-bit sections of each 36-bit word; 14 bits are not used. These 7600 words of the 8192-word memory are than processed sequentially (ten time samples for each range cell) by a three-pulse MTI canceller. The I and Q channels are processed by separate hardware in the three-pulse canceller section of the processor. Note that the ten pulses of 11-bit I and Q channel samples exist after the three-pulse canceller as eight pulses of 13-bit words. The output of the three-pulse canceller for both the I and Q channels (real and imaginary parts of the signal) is fed into an 8-point Discrete Fourier Transform (DFT) which produces eight Doppler cells.

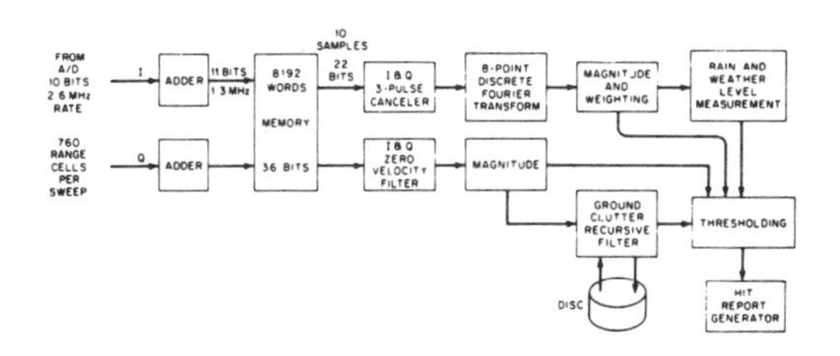

Figure 8

Moving Target Detector (MTD) Processor

Weighting of the I and Q channel signals to reduce the sidelobe level is done after the DFT. Subtracting 1/4 of the signal from the two Doppler cells adjacent to the one of interest is equivalent to a cosine on a pedestal weighting performed in the time-domain.

Since the three-pulse canceller has poor low Doppler velocity response, a zero velocity filter (ZVF) is employed to see low radial velocity targets. This low-pass filter is implemented by coherently adding the first five samples of each of the I and Q channels, respectively, taking their magnitude and adding to this the magnitude of the sum of the last five samples. This gives a somewhat broader frequency response than simply adding coherently all ten samples and then taking the magnitude. The magnitudes of the signals which come out of the three-pulse canceller, DFT and weighting chain are then taken.

After magnitudes are taken, adaptive background levels and thresholds are set and threshold crossings (detections) are noted and output. The adaptive background levels and thresholds are set depending on the clutter phenomenon which is present. The Doppler domain is divided into three domains; Doppler cell 0, Doppler cell 2 through 6, and Doppler cells 1 and 7.

In this Doppler cell 0, the clutter is generally due to ground backscatter. The average ground backscatter cross section varies from range-azimuth cell to range-azimuth cell. The average backscatter signal level for each cell is measured and stored on the disc (see Figure 8). A recursive filter is used to update on a scan-to-scan basis the average signal level stored on the disc. On each scan, 1/nth of the stored clutter level is subtracted from the stored level. One nth of the signal level output from the ZVF is added to the value remaining after subtraction. This new level is then stored on the disc for thresholding in the next scan. The threshold for the 0 Doppler cell is a fixed value between 4 and 8 times the level stored on the disc. This fixed value may be altered by changing a plug in the hardware.

In the Doppler cells 2 through 6, the clutter is due chiefly to rain. For each Doppler and signal cell, the average signal level is measured by averaging the received signal over 16 range cells (one mile) centered on the range cell of interest. The threshold for these cells is a fixed value set at 4 to 8 times the measured average signal level.

Doppler cells 1 and 7 can contain clutter due to rain or spillover from the ground backscatter in cell 0. The threshold set in these cells is the greater of two thresholds: (a) the threshold set as in Doppler cells 2 through 6, or (b) a fixed binary fraction [(1/2 n) n = integer] of the threshold set in cell 0, n is set by use of a wire jumper on the hardware.

Finally, it must be noted that if any I and Q channel sample is noted to have all the bits on (i.e., be in saturation), then any target detections for that range cell are deleted.

c. Moving Target Detector (MTD) Experimental Results

The MTD has been tested at Lincoln Laboratory before its shipment to NAFEC. A block diagram of the experimental set used in these tests is presented in Figure 9. The NOVA 1220 minicomputer is used to format the data from the MTD and to refresh the display. A time exposure photograph of the display for 40 scans of the radar is shown in Figure 10. The STC followed an R^{-4} law and the attenuation dropped to 0 dB at 8 miles. The weather threshold was set to be 6 times the mean signal level, and the ground clutter threshold was set to be 7 1/2 times the level stored on the disc. The range rings appear on the display at 10-mile intervals. There are several small airports in the area as evidenced by the presence of many small aircraft tracks. Logan Airport, Boston, is southeast at 13 nmi.

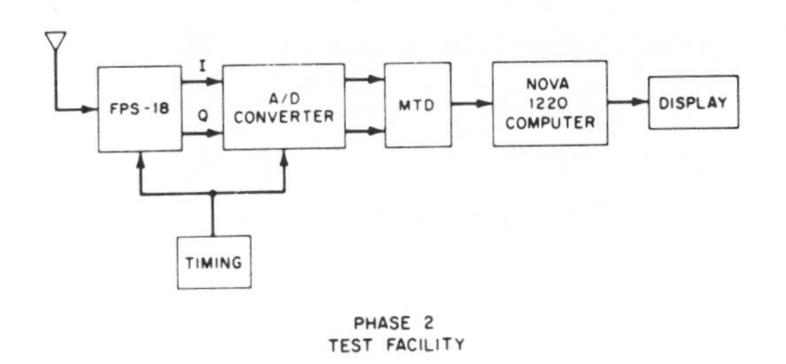

Figure 9

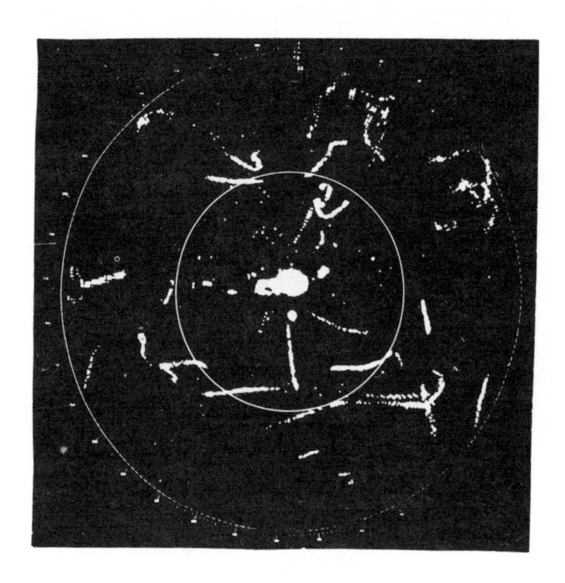

Figure 10

Time Exposure Photograph of PPI Display

ACKNOWLEDGEMENTS

The authors wish to thank all of their colleagues in Group 43 at Lincoln Laboratory without whose efforts this work would not have been completed. Among them are R. G. LeBlanc, M. A. Nader, H. P. HcCabe and P. B. McCorison. We thank H. G. Weiss and P. R. Drouilhet for the constructive direction. Also, we wish to thank the FAA for their sponsorship of this work and, in particular, K. Coonley and D. Hopson for their encouragement and patient criticism.

REFERENCES

[1] C. E. Muehe, L. Cartledge, W. H. Drury, E. M. Hofstetter, M. Labitt, P. B. McCorison and V. J. Sferrino, "New Techniques Applied to Air Traffic Control Radars," Proceedings of IEEE, Special Issue on Radar Techniques to be published in May 1974.

[2] F. W. Nathanson, Radar Design Principles, (McGraw-Hill, New York 1969).

[3] "A Brief Description of the Solid-State ASR-7 Airport Surveillance Radar," Texas Instruments Incorporated, (May 1970).

[4] J. Croney, "Clutter on Radar Displays, Reduction by Use of Logarithmic Receivers," Wireless Engineer, pp. 83-96 (April 1956).

[5] K. R. Hardy, "CPS-9 Radar Investigation of Clear-Air Convection." 13th Radar Meteorology Conference,pp. 236-240, (August 1968).

[6] E. Eastwood, Radar Ornithology, (Methuen & Company, Ltd., London, England, 1967).

[7] R. E. Richardson, J. M. Stacey, H. M. Kohler, F. R. Naka, "Elimination of a Type of Natural Clutter in L-Band Radars," Lincoln Laboratory M. I. T., Technical Report 178, (15 April 1958). ASTIA 133851.

[8] D. F. DeLong, E. M. Hofstetter, "On the Design of Optimum Radar Waveforms for Clutter Rejection," Trans. on Information Theory, Vol. It-13, No. 3 (July 1967).

[9] R. C. Emerson, "Some Pulse Doppler MTI and AMTI Techniques, "Rand Corporation, Report 274, p. 104 (1 March 1954).

[10] B. Gold, I. L. Lebow, P. G. McHugh, C. M. Radar, "The FDP, A Fast Programable Signal Processor," IEEE Trans. on Computers, Vol. C-20, No. 1, pp. 33-38 (January 1971).

5.6 DIGITAL SIGNAL PROCESSING FOR RANGE-GATED PULSE DOPPLER RADARS
B. Gold and C. Muehe
AGARD Conference Proceedings, No. 66, Advanced Radar Systems, May 1970, pp. 31-1—31-9.

DIGITAL SIGNAL PROCESSING FOR RANGE-GATED PULSE DOPPLER RADARS

B. Gold and C. E. Muehe
M. I. T. LINCOLN LABORATORY*
Lexington, Massachusetts, U. S. A.

SUMMARY

For many applications, digital signal processing can increase the effectiveness of range-gated pulse doppler radars. When many doppler filters are desired for each range gate, the fast Fourier transform algorithm is more efficient than digital filtering. In this paper, digital processing methods are briefly reviewed and the signal processing requirements are established in terms of the radar parameters. Equivalence relations are established between the fast Fourier transforms and a bank of velocity filters. Finally, the structure of a general purpose radar digital signal processor is described which should be applicable to a variety of range-gated pulse doppler systems.

I. INTRODUCTION.

In most cases the effectiveness of a range-gated pulse doppler radar is limited by its signal processing capabilities. The gradual introduction of digital signal processing techniques over the last few years[1,2] is a result of the general recognition of the superiority of digital techniques over analog. Digital processing offers filtering free of the usual analog errors due to zero drifts, non-linearities, stray capacities, and component tolerances. It offers exact reproducibility from gate-to-gate with no adjustments to make. The only noise introduced is quantization noise. This noise can be reduced and the dynamic range can be increased in most cases by increasing the number of bits used. The output is in digital form for easy transmission.

Finally, where a large number of range gates are to be processed, digital processing has a cost advantage over analog. The cost of analog filters varies linearly with the number used, costing about $10 a pole. Because a single arithmetic unit of a digital processor can provide filtering for a large number of gates, although the cost for processing a single gate is large, added gates require only the addition of more memory and therefore come cheaply. The crossover point for simple filters (say 5 poles) is several hundred gates. For more complicated filters it is usually more economical to use digital filters when more than about 100 gates are to be processed.

II. DIGITAL PROCESSING METHODS.

Two major processing schemes have been or are being implemented using digital techniques. The first scheme synthesizes high-pass filters by using the digital equivalent of delay lines. To produce high-pass filters these digital filters incorporate feed-forward connections to produce zeros and feed-back to produce poles. The second scheme makes use of the discrete Fourier transform (DFT) to digitally synthesize the equivalent of a bank of identical equally spaced filters.

The DFT provides one more dimension to radar target filtering. The others are range, azimuth, and elevation. This extra dimension allows one to eliminate (by blanking) large fixed targets (discretes) of known location. It allows for the reduction or elimination of moving clutter such as rain which is at a different velocity than ground clutter so is not eliminated by the digital high-pass filter approach. Finally, this separate filtering dimension allows a thresholding scheme to be employed wherein the clutter spectrum can be estimated from the DFT outputs. This estimate would be based on returns from the range cell being examined for a target and would not rely on the magnitude of the clutter from other range cells.

The DFT coherently integrates the signal over the time it takes to collect all the samples. The digital high-pass filter must be satisfied with non-coherent integration over this same period resulting in some signal-to-noise or signal-to-clutter advantage for the DFT when the target energy falls principally in one output of the DFT.

---

*This work was sponsored by the Department of the Air Force.

In the case of airborne radar, the use of the DFT obviates the need for changing the cut-off frequency of the high-pass filter as the antenna azimuth is changed, or of changing the local oscillator frequency to center the main-beam clutter return in the center of the high-pass filter. Instead, the DFT outputs containing main-beam clutter are either ignored or desensitized as far as thresholding is concerned. This technique is particularly applicable to tactical radars which are interested in close-in targets as well as distant ones. Because of the differences in depression angle (and consequently main-beam doppler frequency) the local oscillator cannot be adjusted simply when one wants to examine all target ranges.

Because the DFT gives a radial velocity sorting using two prf's, a partial sorting of ground targets from airborne targets can be effected. Of course a better method would be to observe motion on a scan-to-scan basis to sort according to speed rather than velocity, but this may require a larger computer capacity to handle all the targets.

The digital high-pass filter is a natural substitute for the analog delay lines previously used. A digital shift register, usually using MOS memory elements, replaces the analog delay lines. The feed-forward and feed-back coefficients are usually chosen to be a positive or negative integral power of two or else the sum of two such coefficients. These are then implemented with simple shifting and adding operations which are hard wired.

The DFT, on the other hand, uses a large core memory which, at the present time, is cheaper per bit than the MOS memory. It generally uses one or several array multipliers to sequentially process batches of data from various range cells using a fast Fourier transform (FFT) algorithm. Since the designs of the two types of processors are entirely different, one cannot make a general statement as to which is cheaper. In many cases, considering the advantages listed above, it is worth some added expense to employ the DFT approach. Where a large amount of memory is involved the DFT approach may be less expensive. Both approaches involve about the same complexity.

III. SIGNAL PROCESSING REQUIREMENTS.

Let M be the number of required range gates, f_d the highest doppler frequency to be measured and Δf the doppler resolution. Then, the sampling rate f_s must be at least twice f_d for unambiguous determination of velocity. If it is not possible to achieve this sufficiently high value of f_s, other techniques (not discussed in this paper) must be used to resolve doppler ambiguity. In performing spectrum analysis, the total number of frequency points is then $2f_d/\Delta f$. The total memory requirements are $2Mf_d/\Delta f$ registers (or twice that number if quadrature detection is assumed). There will be $M(\Delta f)$ spectrum analyses performed every second.

Figure 1 shows a range-gated pulse doppler block diagram. All blocks right of the dashed line represent digital hardware. Typically, the operation of this hardware would be equivalent to many banks of doppler filters, each bank receiving information from a specific region of space, as indicated in Figure 2. The plan of this paper is, first, to relate the signal processing requirements to the range and doppler specifications. Next, some theoretical questions are treated which arise when the fast Fourier transform is used to perform the doppler filtering. Finally, a block diagram of general purpose signal processing hardware will be proposed for performing range-doppler filtering for a variety of differing requirements. If such a proposal were to prove feasible, it would ultimately be possible to buy (rather than spend several years building) the requisite digital processing hardware for a given range-doppler radar.

Digital processing techniques need not be restricted to spectrum analysis to perform the doppler filtering. Instead, digital filters may be used. The trade-off depends primarily on the number of doppler filters desired per range gate; a large number implies that spectrum analysis is more efficient. In this paper, we restrict ourselves to the spectrum analysis approach, although the general purpose hardware described in the latter part of this paper could be programmed to perform either recursive or non-recursive doppler filtering.

Referring to Figure 1, we assume that the M range gates chosen by the operator are one of many possible sets that he could have chosen. Thus, the data rate from the A/D converter is higher than the data rate into the main memory. The preprocessor and high speed buffer is needed to scale down the data rate and, if necessary, to perform predetection integration and/or waveform decoding.

The type of main memory used depends mainly on the required storage size. If this size should be less than .5 to 1 megabits of memory, it would be most economical to use a commercial general purpose computer which could then be used as well for overall systems control, decision making and display. If the memory size is 1-10 megabits, a so called "window screen" memory (2 wires per core, rather than 3 or 4) may be used but beyond this a drum or a disc memory may prove most economical. Such a memory (the circulating type) is feasible because of the repetitive nature of the algorithms.

At this point, it seems logical to estimate speed and memory requirements of the signal processing hardware. Since such results depend greatly on the properties of the fast Fourier transform algorithm, we will divert our path to describe this algorithm, explain the theoretical filtering effects of the discrete Fourier transform, and then return to the hardware considerations.

IV. THE FAST FOURIER TRANSFORM AS A BANK OF SIGNAL DETECTORS.

The discrete Fourier transform (DFT) of a sequence of numbers x(n) is

$$X(k) = \sum_{n=0}^{N-1} x(n) W^{-nk}$$

where W exp $(-j2\pi/N)$. In general x(n) is a complex number. To compute N values of X(k) requires N^2 complex multiplications and additions.

The fast Fourier transform (FFT) refers to a set of algorithms for more rapid computation of the DFT. The most well known is the so called radix 2 algorithm which computes the FFT by repetitive computation of a "butterfly", defined as:

$$u(m+1) = u(m) + W^{\ell} v(m)$$

$$v(m+1) = u(m) - W^{\ell} v(m)$$

where u (m) and v (m) are complex numbers and ℓ is an integer which varies with m. It is necessary to perform $N/2 \log_2 N$ of these butterflies to compute the FFT, obtaining all N values of X(k).

In this section we will discuss the relations between the DFT, the filter banks of Figure 2 and signal detection. The detection problem is illustrated in Figure 3, which shows the spectrum of a moving target in a large clutter background. The "correct" filtering procedure is pre-whitening of the clutter spectrum followed by a filter matched to the resultant pre-whitened target spectrum, followed by detection (see Figure 4) and integration for as long a time as the target remains in the range gate.

It does not appear to be possible, with a DFT, to obtain the result indicated in Figure 4 if the input samples are contiguous. It is possible, however, to imagine the input appearing in bursts, as shown in Figure 5a. With this model we can now show how to perform matched filtering via the DFT. For simplicity we have set the on and off times and the impulse response duration of the filter equal. (If the filter is recursive, with an infinitely long impulse response, it is immediately apparent that the filter bank cannot be realized via DFT.) Now, the product of the DFT's of the augmented signals shown in Figure 6 is indeed the DFT of y (n) of Figures 4 and 5. Thus, the set of Y (k) of Figure 6 is the DFT of y (n) and Parseval's theorm yields,

$$\sum_{n=0}^{N-1} |y(n)|^2 \qquad \sum_{k=0}^{N-1} |Y(k)|^2$$

In practice, a given velocity filter covers a range of frequencies narrow compared to the total DFT spectral range. Thus, H (k) would be zero for most values of k. For a given doppler filter with H (k) the (augmented) DFT of h (n), the computation corresponding to q (L) in Figure 4 is:

$$q(L) = \sum_{K} |H(k)\ X(k)|^2$$

where K refers to the range of non-zero H (k)'s.

It is usually true that the response to a burst, as shown in Figure 5, is more useful after transients caused by switching the signal on and off have died down. For this reason, it may be desirable to weight the data. In effect, this weighting modifies the matched filter of Figure 4 so that the transient portion of q (n) (beginning and end) have the least effect on the subsequent measurements. Two ways of weighting are; (a) the augmented x (n) is multiplied by the weighting function w (n) prior to DFT or (b) the set of X (k) is convolved with W (k) (the DFT of w (n)).

V. GENERAL PURPOSE DIGITAL RADAR SIGNAL PROCESSOR.

The parameters for an air traffic control radar, a weather measurement radar and a ground mapping radar are very different and each system will require unique operations in addition to the common function of range-gated pulse doppler. For this reason, it is useful to develop a computer architecture which can be applied to a variety of radar signal processing systems and which is of sufficient generality to be adapted to the peculiarities of individual radars. The remainder of this paper, represents a preliminary attempt to develop such an architecture. Figure 1 shows the overall radar system for which this architecture is to be developed. In this paper we will not discuss signal design problems (such as chirp waveforms, Barker codes, etc.) but will assume that the preprocessor receives in-phase and quadrature signals which are sampled for inclusion into the range bins chosen by the processor program. Thus, samples for successive range bins can be made to enter the main memory at a uniform rate and, indeed, can be sent to the processor via the memory at the same uniform rate. In what follows, we restrict our discussion to the flow of data into and out of the main memory and into and out of the processor. A timing picture which fulfills the real time requirements is shown in Figure 7.

During the "Enter" phase,

(a) The pre-processor feeds samples from contiguous range gates into the main memory.
(b) The main memory feeds successive samples from a given range gate in the main memory to set up the processor data base.
(c) The main memory also sends results of past processing on the same range gate to the processor to be used for post-detection integration and thresholding.

During the "Do" phase the processor performs filtering and detection on a single range gate. These operations may include,

(a) Weighting the data.
(b) Discrete Fourier transform.
(c) Recursive or non-recursive digital filtering.
(d) Matched filtering via the FFT.
(e) Post-detection integration.
(f) Centroid measurements on the FFT to estimate frequency.
(g) Thresholding for yes/no decisions on targets.

The return phase consists of the return of all necessary post-detection, threshold and display information for storage by the main memory until the next turn of the just processed range gate information.

In order to perform these tasks in real time, the 3 phases must be overlapped as shown in Figure 7. In the present concept, control resides exclusively in the processor and it is only the processor which physically consists of high speed circuits capable of both processing and controlling the addressing of the main memory. Before describing the processor-main memory complex in more detail, let us describe an addressing algorithm which allows for the unimpeded flow of contiguous range data from the pre-processor to the main memory and (simultaneously) the flow of data from a single range gate in the main memory to the processor for the DFT. This can be accomplished by an addressing scheme illustrated in Figure 8 for the case of a 5 point DFT and 7 range gates. After the memory is filled by storing incoming samples sequentially (column 1) the address is then incremented by 7 (modulo 34) in column 2. For each new address a main memory word is first transferred to the processor and then that vacant location is filled by the incoming datum from the pre-processor. If M is the number of range gates and N the length of a DFT, then the addressing sequence obeys the rule,

$$A_n (k) = A_n (k - 1) + M^n$$

Where both the addition and the term M^n are understood to be taken modulo (MN-1) and n refers to the nth iteration through the entire memory (a single column in Figure 8).

VI. MORE DETAILS OF THE PROCESSOR AND MAIN MEMORY.

Figure 9 shows a structural diagram of the processor and intermediate memory. Since, in general, this memory will be large and does not have unduly high speed requirements, we can imagine it to be a core memory. The memory and arithmetic associated with the processor we imagine to consist of the fastest commercially available circuits. Since processing speed is a prime consideration, the computer has certain structural features for greater speed. These are:

(a) Separate program and data memories so that instruction fetch and data fetch take place in parallel.
(b) Overlap of data memory and arithmetic operations.
(c) An arithmetic element especially designed for high speed signal processing, in particular the FFT.
(d) Double length words so that two operations can be performed in parallel.
(e) Double instructions performed simultaneously.

The details of the large FFT array are shown in Figure 10. Maximum speed and minimum control of the butterfly algorithm is obtained at the cost of added hardware by making the array box purely combinational. Using emitter coupled logic gates and similar full adder integrated circuit packages results in an array speed of about 200 nsec. Processor memory cycle time is about 100 nsec. Thus, the timing diagram of Figure 11 where the array propagation time is taken to be twice the memory time, shows how a butterfly is accomplished in the time to perform two reads and two writes. For such a memory and for a 4 multiplier array, the structure does an FFT butterfly in minimum time (about 400 nsecs.)

When the array is not used to do a butterfly, it can be controlled to perform two simultaneous multiplications. This means that the weighting operation is very efficient, as is the matched filtering. The preliminary order code we have devised (Figure 12) allows completely general purpose programming. The F register, which communicates between program and data memory is also used to address the intermediate memory and the input-output instruction initiates the intermediate memory read or write pulse.

VII. PERFORMANCE OF SIGNAL PROCESSOR.

The time to do a butterfly is the basic figure of merit of the radar signal processor. This time can be translated into the maximum allowable number of range gates which can be processed in real time, given the sampling rate and the size of the FFT. For simplicity, let us assume that if the processing time of a butterfly is 400 nsec., then the expanded time due to other operations for a typical radar is 1 microsecond. Using this number, Table I shows how many FFT's (in round numbers) per second can be processed for various FFT sizes.

FFT Length	FFT's Per Second
16	31,000
32	12,000
64	5,200
128	2,200
256	1,000

We have thus demonstrated that it is possible to build digital signal processing equipment for range-gated pulse doppler radars with capability of measuring velocity or detecting moving targets in hundreds or thousands of gates. Such a capability is not economically feasible using analog signal processing. Furthermore, the structure we have defined, since it is general purpose, could be used with slight changes for different radar systems.

REFERENCES

1. Linder, R. A., Kutz, G. H., "Digital Moving Target Indicators" Aerospace and Electronics Systems Technical Convention 1967 Record, pp. 374-385.

2. Hall, E. L., Lynch, D. D., Young, R. E., "A Digital Modified Discrete Fourier Transform Doppler Radar Processor", Electronic and Aerospace Convention Record (1968), pp. 150-159.

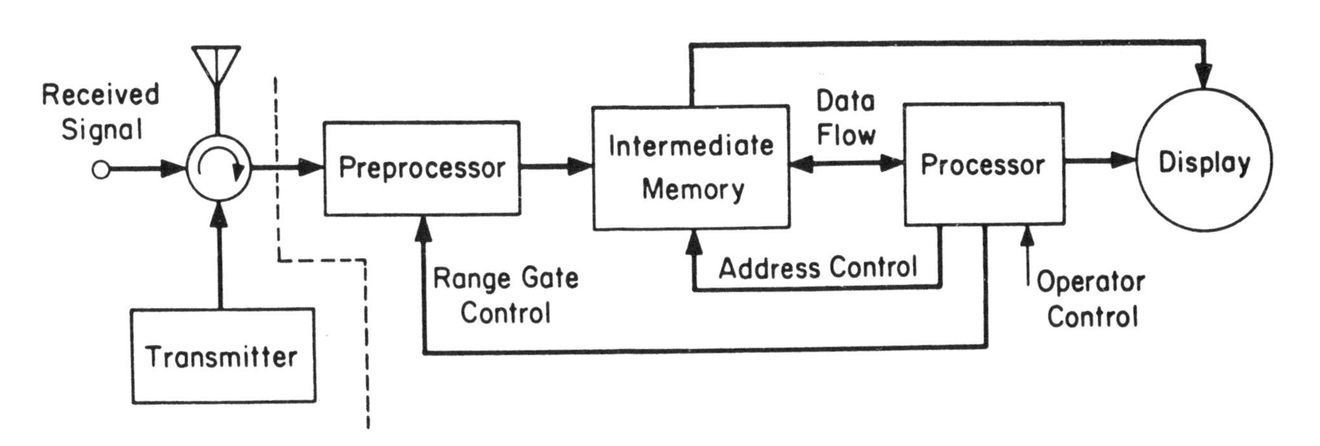

Figure 1. Block Diagram of a Range-Gated Pulse Doppler Radar

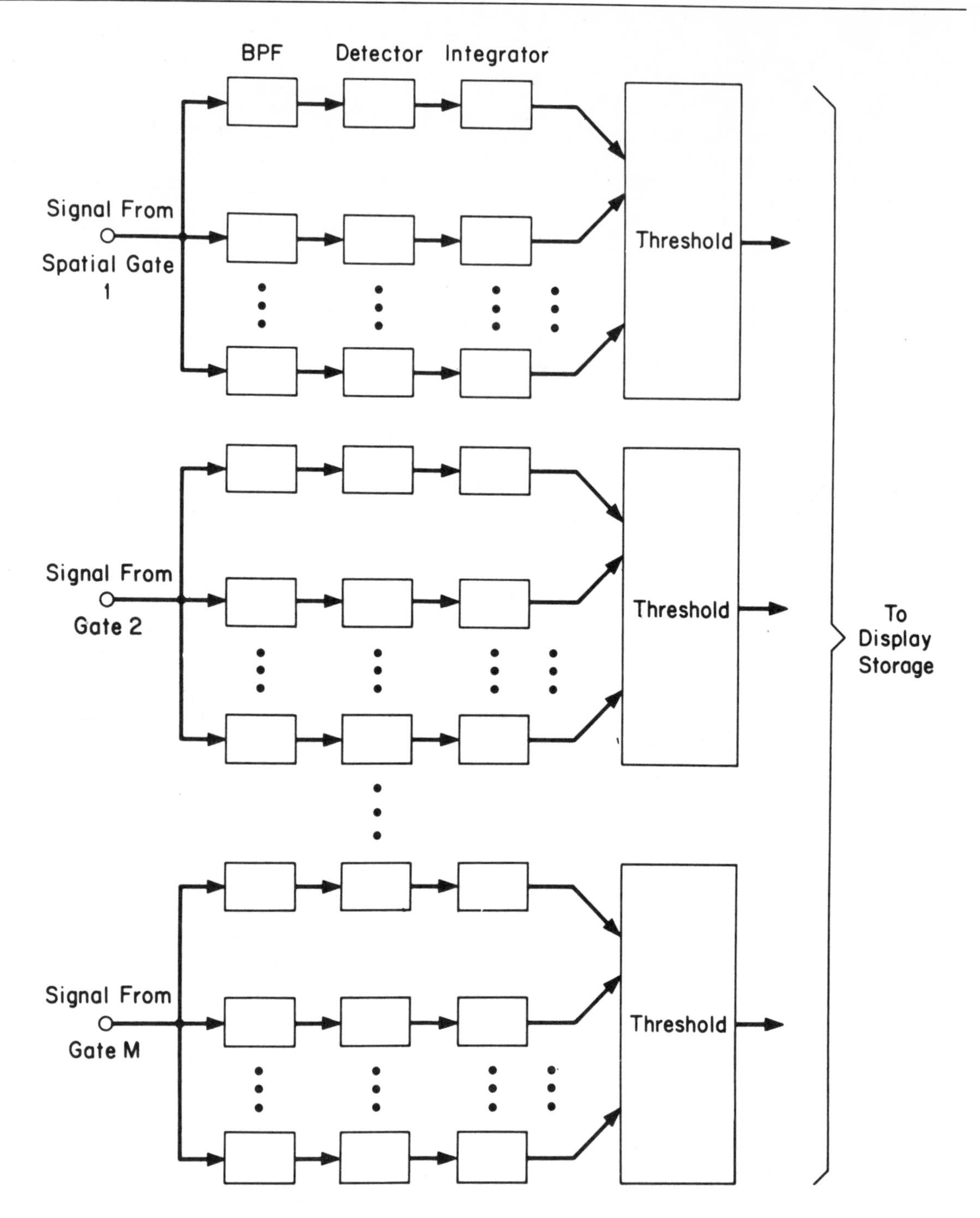

Figure 2. n Banks of Doppler Filters, Each Bank Processing the Gated Signal from a Specific Region of Space as Determined by the Radar Timing

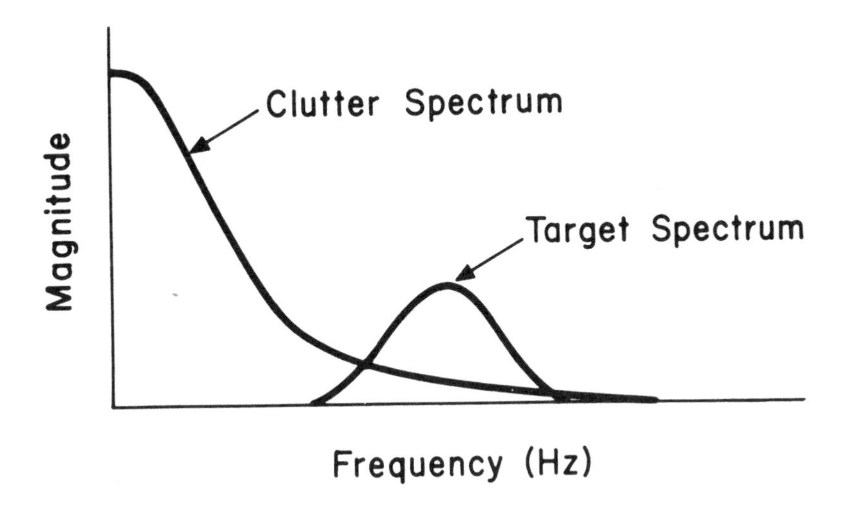

Figure 3. Spectrum of a Moving Target in a Large Clutter Background

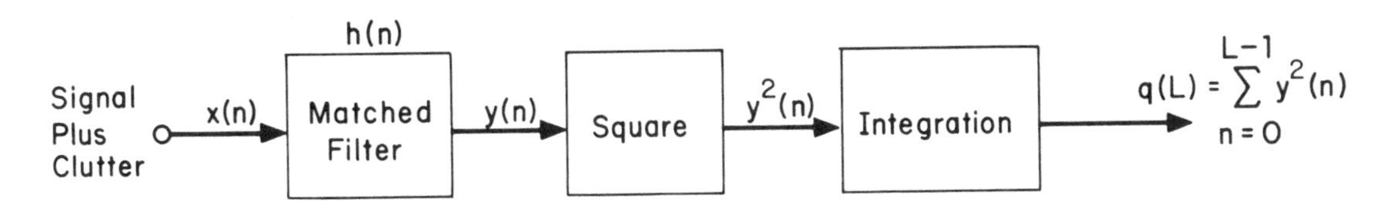

Figure 4. Matched Filter Detection of a Target in Clutter

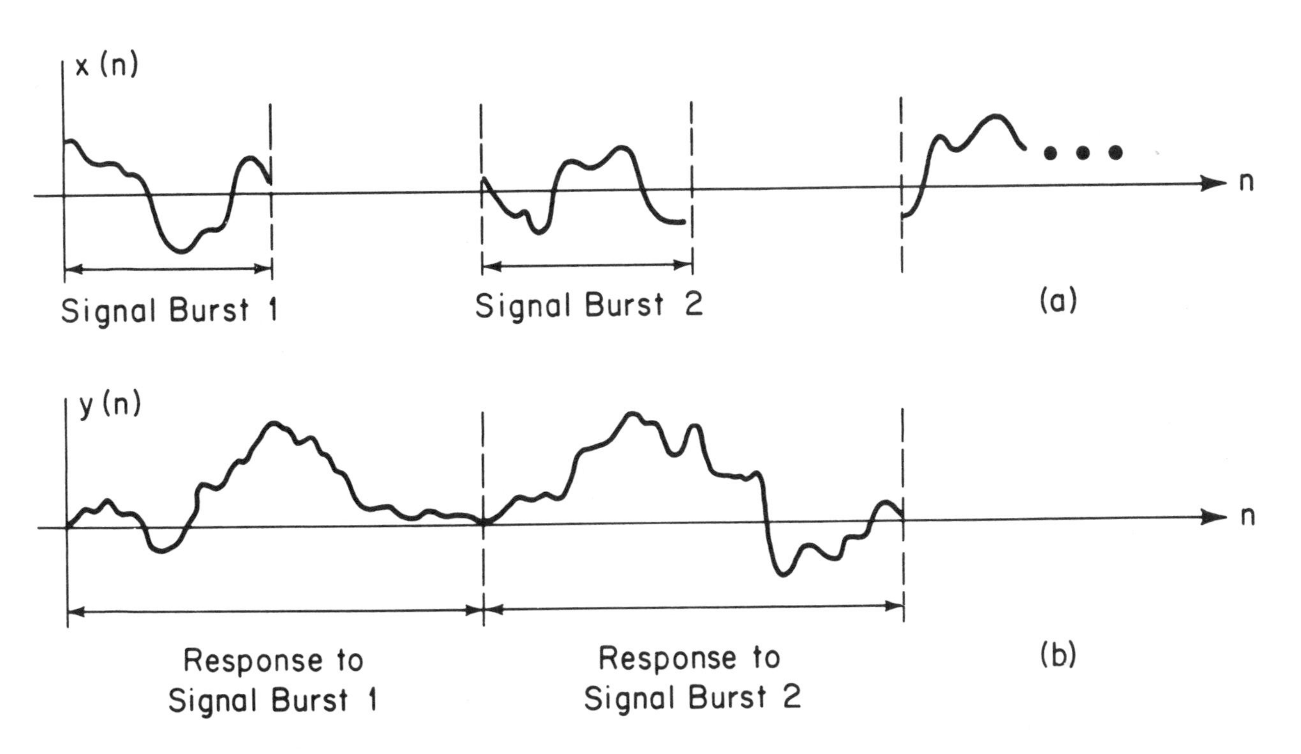

Figure 5. Input and Output Signals of Matched Filters

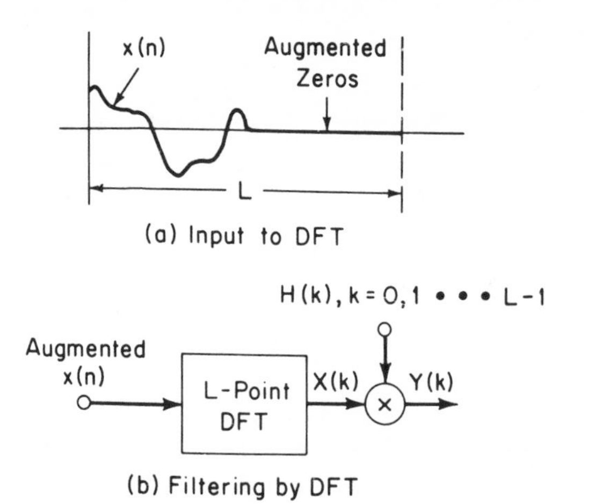

Figure 6. Fourier Transform which is Equivalent to Matched Filter of Fig. 4 when Excited by Signal Bursts

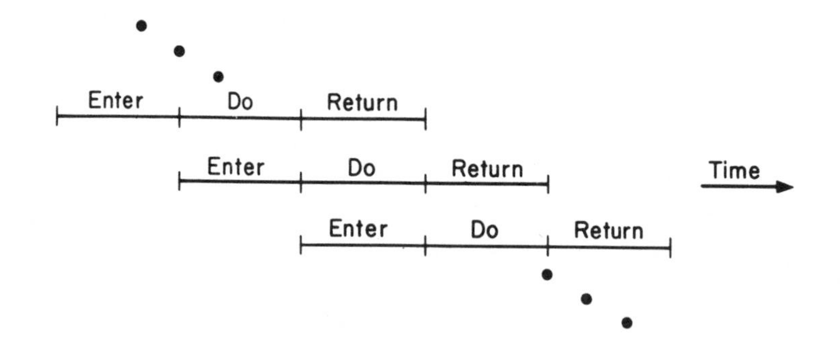

Figure 7. Overall Timing Diagram of the Radar Signal Processing

n	1	2	3	4	5	6
	0	0	0	0	0	0
	1	7	15	3	21	11
	2	14	30	6	8	22
	3	21	11	9	29	33
	4	28	26	12	16	10
	5	1	7	15	3	21
	6	8	22	18	24	32
	7	15	3	21	11	9
	8	22	18	24	32	20
	9	29	33	27	19	31
	10	2	14	30	6	8
	11	9	29	33	27	19
	12	16	10	2	14	30
	13	23	25	5	1	7
	14	30	6	8	22	18
	15	3	21	11	9	29
	16	10	2	14	30	6
	17	17	17	17	17	17
	18	24	32	20	4	28
	19	31	13	23	25	5
	20	4	28	26	12	16
	21	11	9	29	33	27
	22	18	24	32	20	4
	23	25	5	1	7	15
	24	32	20	4	28	26
	25	5	1	7	15	3
	26	12	16	10	2	14
	27	19	31	13	23	25
	28	26	12	16	10	2
	29	33	27	19	31	13
	30	6	8	22	18	24
	31	13	23	25	5	1
	32	20	4	28	26	12
	33	27	19	31	13	23
	34	34	34	34	34	34

Figure 8. Example Showing Address Sequence for Intermediate Memory for a 5-Point FFT and 7 Range Gates

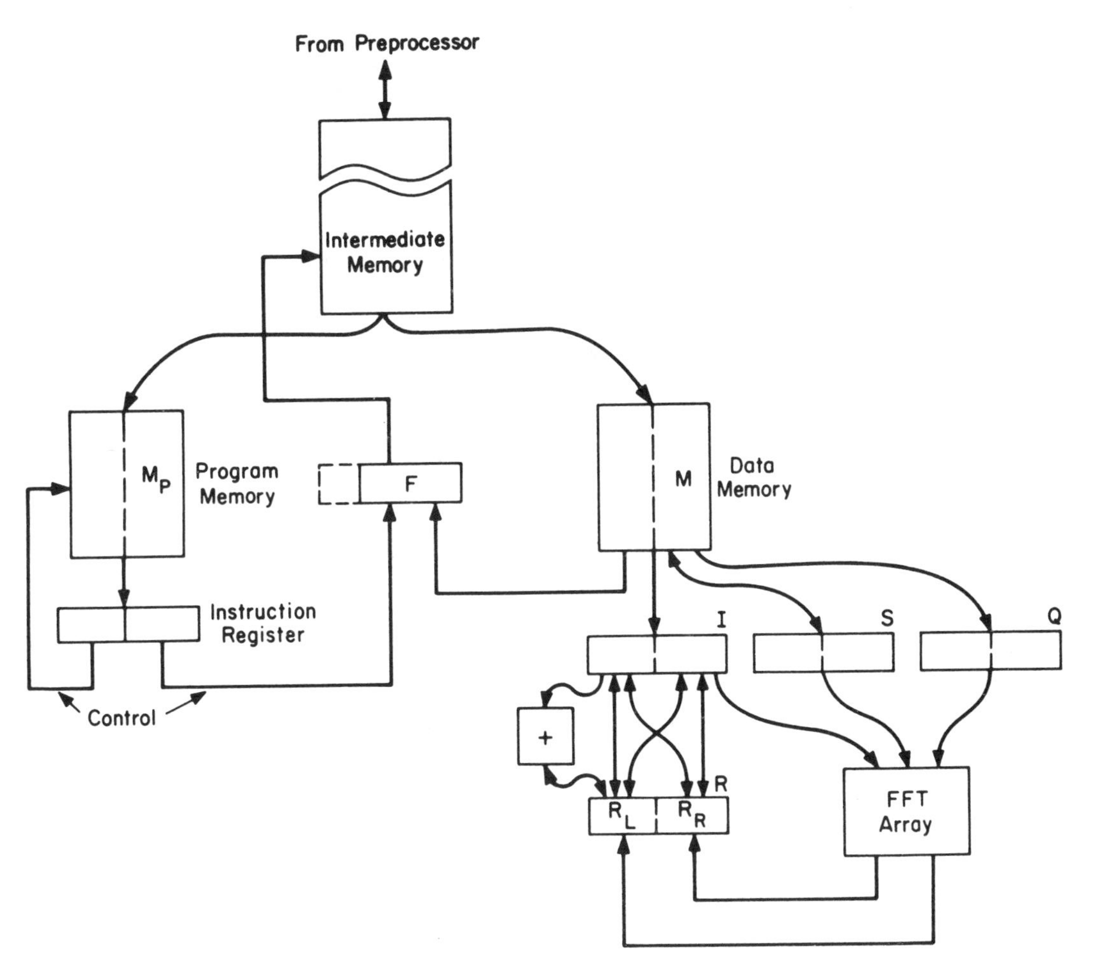

Figure 9. Structure of Processor and Intermediate Memory

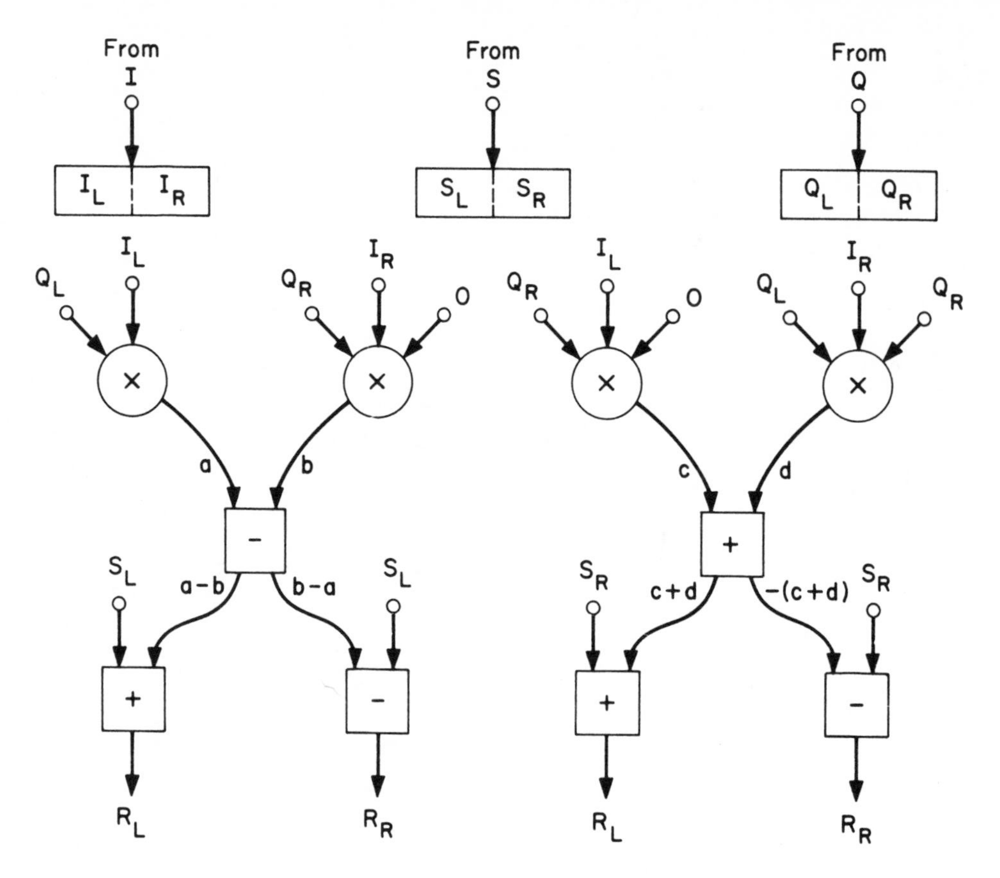

Figure 10. FFT Array for Performing Butterfly

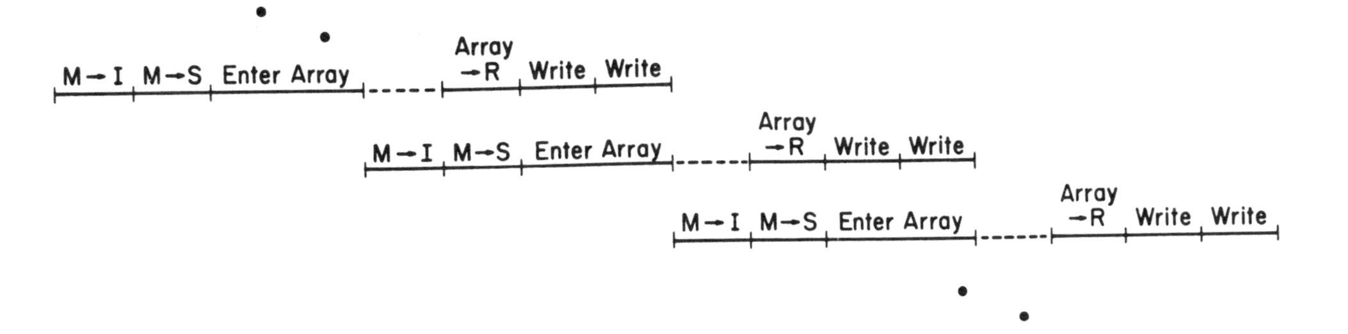

Figure 11. Timing of Executed Instructions Needed to Perform an Efficient Butterfly

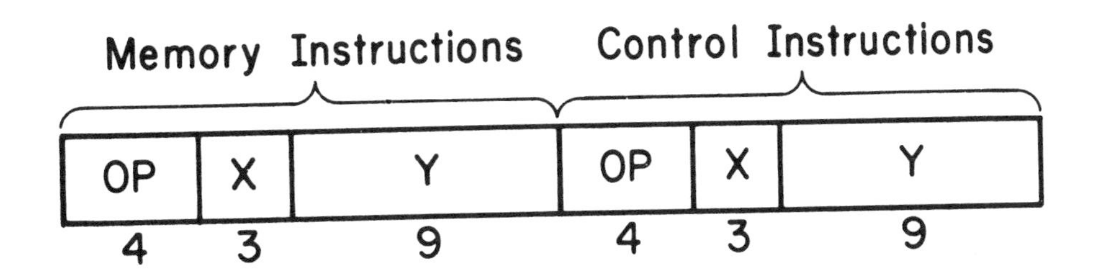

Memory Instructions

(1) M → F
(2) F → M
(3) M → I
(4) M → Q
(5) M → S
(6) R → M
(7) E → M
(8) M → E
(9) Arithmetic Group

Control Instructions

(1) Y + X → X
(2) X - Y → X
(3) Y - X → X
(4) Y → X
(5) Jumps
(6) Input-Output Control
(7) F-Group Instructions

Figure 12. Tentative Partial Instruction List for Structure of Fig. 10

Section 6 MTI Hardware

EDITOR'S COMMENTS

These two papers describe recent advances in MTI hardware at the system and sub-system levels.

Sugie, Tomita and Hagisawa (Paper 6.1) describe an Advanced Air Route Surveillance Radar (ARSR) with a number of advanced clutter suppression features. The system approach is emphasized in the design of this radar.

A dual beam antenna that transmits on the low elevation coverage beam and receives on both low and high elevation beams is utilized. At short ranges the high elevation beam is utilized thereby reducing close-in surface ground and sea clutter, while at longer ranges the low elevation beam is used on receive to provide high sensitivity on long range targets.

A digital MTI using a dual delay shaped feedback canceler is used. An MTI improvement of 40 dB is the limit set by the coherent Klystron transmitter.

A four PRF stagger eliminates blind speeds for targets up to 2100 knots, while dual I and Q channels are used to suppress blind phases.

Weather clutter is suppressed by a combination of circular polarization (CP) and a cell-averaging Log-CFAR adaptive thresholding.

Linear receiver operation is accomplished using an R-F sensitivity-time-control (STC).

Test results show that clutter is suppressed to a degree that the radar data is suitable for use in an automated data processing system.

Lobenstein and Ludington (Paper 6.2) describe a charge coupled device (CCD) MTI designed for use in a pulse-burst type radar. The CCDs used in this type of MTI represent a rapidly evolving technology with capabilities far beyond that described in this paper. Potentially high performance and economy coupled with the elimination of the complex expensive A-D converter used by digital MTIs are advantages of CCD MTIs.

The authors suggest that the CCD MTI can be used as a clutter filter in front of a digital processor. The considerably reduced dynamic range at the input to the digital processor results in a simplified A-D converter of the low bit — high speed class. The CCD MTI described provides a 50 dB cancellation ratio. The performance is apparently limited by a random pattern noise caused by utilization of the partial transfer mode of charge transfer from element-to-element.

6.1

ADVANCED AIR ROUTE SURVEILLANCE RADAR WITH VARIOUS UNIQUE CLUTTER ELIMINATION CAPABILITIES

M. Sugie, Y. Tomita and T. Hagisawa

NEC Research and Development, No. 39, October 1975, pp. 71-83.

ABSTRACT This paper describes the state-of-the-art design advances for modern L-band long range (200 nautical miles) Air Route Surveillance Radar (ARSR) which is implemented for up-to-date en-route air traffic control.

The new ARSR system design is aimed at improving target detection, assuring a constant low false-alarm output and reducing system downtime, which are essential factors in an en-route automatic air traffic control system.

The design concepts and performances, achieved for the major new technologies such as digital moving target indicator, dual beam antenna, digitial CFAR receiver, RF STC and range azimuth gating, are discussed.

1. INTRODUCTION

The volume of air traffic has made an unprecedented increase during the last ten years; existing air corridors the world over are approaching their maximum capacity, and both civil and military airports are ever expanding. The safety of flight operations depends on all-weather, fail-proof air-traffic control systems.

In particular, en-route air-traffic control by radar has assumed an increasingly important role, and various plans are underway for automatic radar control. For supporting such plans, an air route surveillance radar which proves to be an important improvement to serve as the sensor in an automatic air-traffic control system has been developed.

In the design of this radar system, the need of its adaptation into an automatic control system has led to the development of various anti-clutter techniques to provide unusually high and reliable detectability even in heavy clutter conditions.

The new ARSR system (shown in Fig. 1), incorporating the following newly-developed techniques, is described in this paper.

i) A klystron amplifier transmitter, resulting in greatly improved spectrum purity, MTI performance, and reliability.
ii) Various unique clutter elimination capabilities, resulting in the improvements of target detectability and a constant low false-alarm output data.
iii) An automatic performance monitoring and channel change-over system, realizing greatly reduced system downtime.

2. TARGET VISIBILITIES AGAINST VARIOUS CLUTTERS [1],[2]

Fundamentally, the improvement of target detection involves such problems as the S/N ratio regarding the receiver noise and probability deterioration due to a target fluctuation.

However, the target visibility against the environmental conditions or various clutters that requires to introduce advanced new technologies, is discussed here.

Clutters envisaged for ATC radar include ground clutter, sea clutter, weather clutter, angel echoes caused by groups of birds and insects and atmospheric density fluctuations, and artificial disturbances such as chaff.

These various types of clutter have different distributions in strength, area and spectrum characteristic respectively. A typical spectrum distribution of clutter is shown in Fig. 2, and an example of reflection power strength in Fig. 3.

In general, such clutters as caused by the sea, ground and angel echoes are distributed within relatively short range, while weather clutter and chaff are distributed over the entire range. Ground and sea clutters have smaller doppler shift and spread, while these are larger in weather clutter and chaff.

The conventional MTI well known as a means of clutter elimination, though usually adequate for eliminating ground clutter, is not equal to eliminating weather clutter, chaff and angel echoes. For this reason, various types of anti-clutter technology described in Section 3 are required.

(a) Dual beam antenna.

(b) Transmitter-receiver.

Fig. 1 New ARSR system.

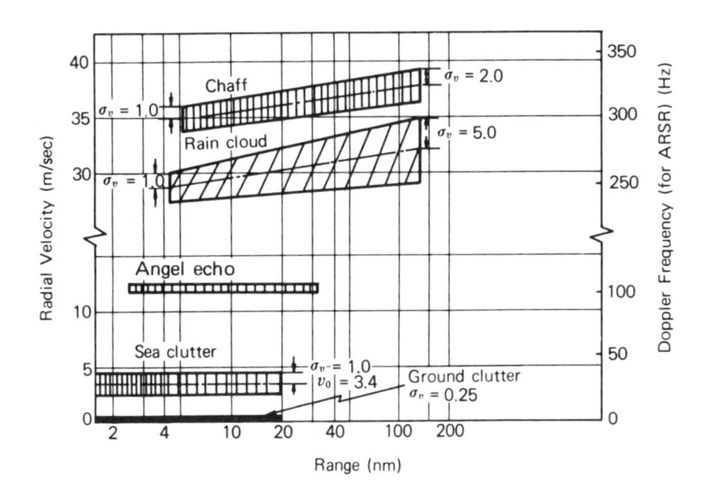

Antenna height : 20 m
Ground clutter : open land with 30 knot wind
Sea clutter : rough sea state

Fig. 2 Typical spectrum distribution of clutter.

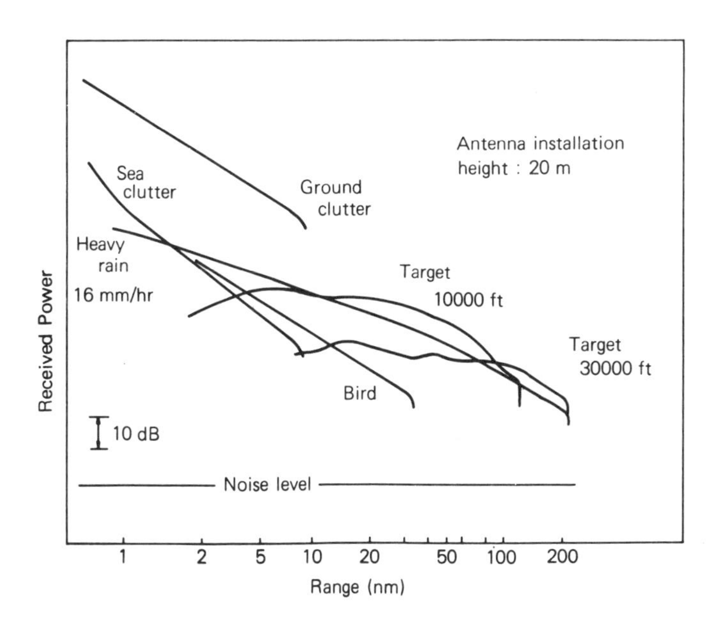

Fig. 3 Typical received power from clutter and targets.

3. NEW TYPES OF ANTI-CLUTTER TECHNOLOGY

For eliminating various types of clutter, the conventional ARSR has utilized the anti-clutter technologies such as the analog MTI, STC, FTC and IAGC. These, however, had such insufficiency as described below, which has kept the radar's clutter elimination and target detection capabilities inadequate for its use in an automatic air control system.

The analog MTI canceller had the virtual limitations in realizing the optimum MTI filter characteristic. It could realize only two or three staggered PRF and it could not select a suitable feedback factor because of the hardware restrictions. And especially, realization of an inphase and quadrature MTI canceller was limited by the complexity of the hardware. And other techniques such as STC, IAGC, FTC had the limitations caused by their analog circuitry and IF stage processing.

In order to solve the limitations of these types of anti-clutter technology, the new ARSR uses the following advanced anti-clutter technologies.

(1) Digital Moving Target Indicator (DMTI) [3]

(a) Improved MTI Filter Response

The MTI filter characteristic is required to have a sufficiently deep notch in the low frequency region in which clutter is widely distributed, and a flat and sufficient sensitivity in the frequency region where the target is distributed. A multi-stagger PRF has been used to increase flatness of the passband, and the use of the DMTI enables to control a complex multi-PRF and thus virtually eliminates blind speed completely.

Figure 4 shows a typical example of MTI velocity response applied to ARSR. The filter characteristic regarding the depth of notch can be improved by providing a stable feedback circuit in the canceller. The similar technique used in the conventional analog MTI has limitations in obtaining optimum response because the critical adjustment is required on the feedback factor. The DMTI system enables to realize a filter having a sharp clutter elimination characteristic and a high target detection probability. Figure 5 shows the filter improvement factor applied in the ARSR.

(b) Elimination of Blind Phase by I/Q Channel

Blind phase, like blind speed, was another factor

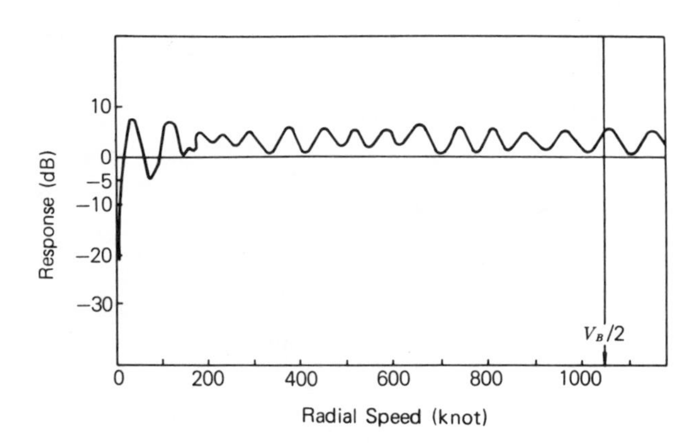

Stagger ratio : approx. 25:30:27:31
First blind speed (V_B) : approx. 2100 knots
(approx. 80 knots in unstaggered operation)

Fig. 4 MTI velocity response.

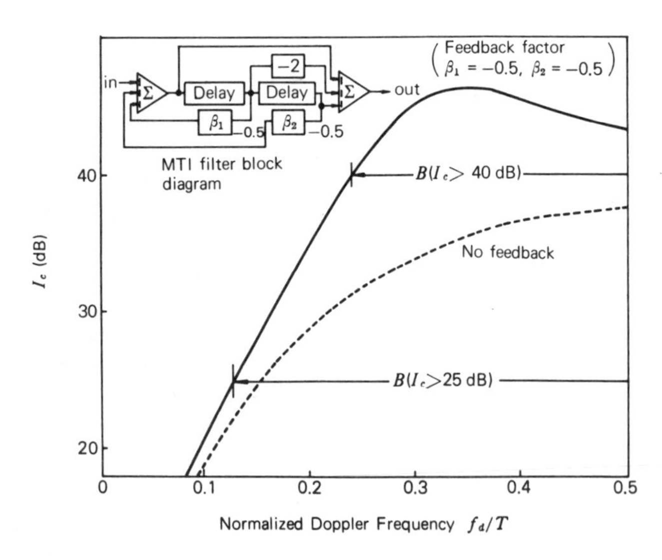

Note; $I_c = \dfrac{(\text{Target power/Clutter power})_{out}}{(\text{Target power/Clutter power})_{in}}$

The frequency spread of clutter power spectrum is due to the antenna scanning modulation.

Fig. 5 Characteristics of improvement factor.

in the MTI system. DMTI filter providing an I/Q channels by practically simple hardware, virtually eliminates the blind phase.

Figure 6 shows an example of target detectability improvement by the I/Q channel synthesis in the ARSR. The target visibility has been improved about 2 dB on the average.

Although it is desirable to obtain $\sqrt{I^2+Q^2}$ at the output of the canceller as the means of output synthesis of the I/Q channels, it is realized by simplified approximation of this calculation in the actual circuit design.

As described in the foregoing, the DMTI system has resolved the problems of blind speed and blind phase, and enables to obtain target visibility and stability as well as those of normal video signals in clear environment. None the less, the present MTI still has the following inherent limitations.

The MTI filter currently in use is incapable of sufficiently removing weather clutter, angel echoes, and even sea clutter under strong wind, because the notch is at zero speed.

If the notch width is enlarged to eliminate the clutter at low speed, it will degrade the visibility of aircraft of which the radial speed is low to the radar.

(2) Dual Beam Antenna [4]

For improving the MTI visibility in a shorter range against ground clutter and at the same time for eliminating sea clutter and angel echoes that cannot be eliminated by the MTI, a dual beam antenna system, as shown in Fig. 7, can be effectively used. These types of clutter usually exist at a low angle and within a short range, so that the receiver input level of these clutters can be reduced by lowering the low angle antenna gain at a short range through an addition of a high beam horn used only for receiving. For longer range reception, the usual horn is used by switching over from the high beam horn to keep a target at a low angle in a longer range in sight.

Figure 8 shows the coverage when the angle difference in the nose direction of the dual beam antenna is about 4 degrees. As can be seen from this figure, the degree of clutter suppression and target visibility is determined by the beam angles and the switching distance. For example, in this figure, taking the switching at point B will make target detection impossible in the hatched part. Consequently, it will be an important factor to

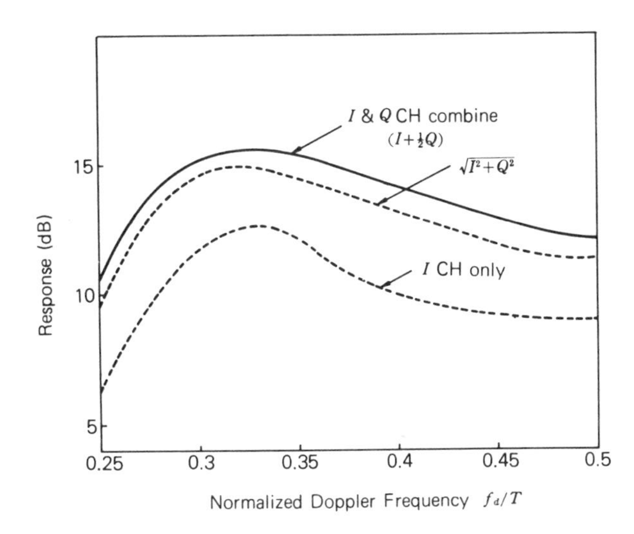

Fig. 6 Target output characteristics of I/Q MTI.

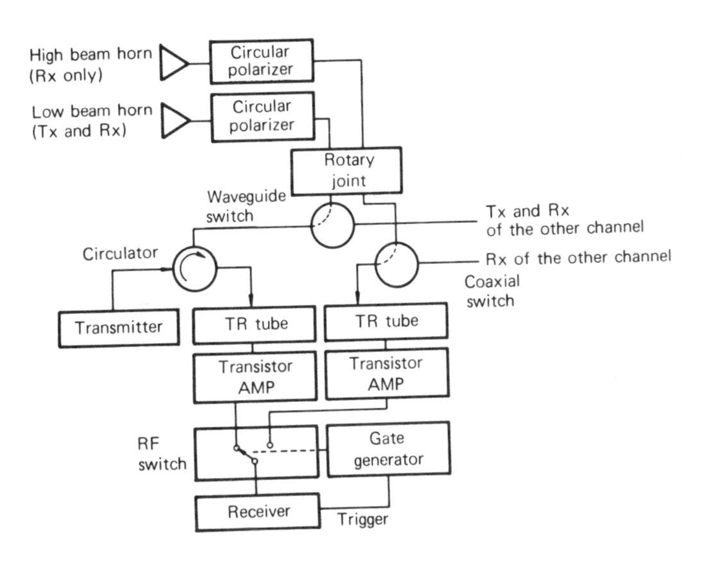

Fig. 7 Dual beam antenna system block diagram.

determine the beam angle settings and switching point, taking the clutter distribution and strength at the site location into account.

The dual beam antenna used in the new ARSR is shown in Fig. 1 and its measured patterns in Fig. 9.

The suppression effects on various types of clutter by means of the dual beam antenna as are determined as the results of simulation calculation of the target/clutter ratio are shown in Figs. 10 (a) to (c).

Compared with the calculation results of the suppression effects by the single beam antenna, shown in Figs. 10 (d) to (f), these results show that the dual beam antenna has the following distinguished advantages regarding suppression of the following types of clutter;

- o Ground clutter: An improvement of 15 to 20 dB is obtained. Taiget detection is possible in almost all area except the field within a short range of less than 5 nm (Note that an MTI having an SCV > 30 dB is taken into account).
- o Sea clutter: An improvement of 15 to 20 dB is obtained, making it possible to detect a target in almost all area except the field within a short range of less than 5 nm.
- o Angel echo: An improvement of about 15 dB is obtained. The T/C ratio can be made at least 15 dB, except in the area within a short range of less than 5 nm. Therefore, angel echo can be, virtually, completely eliminated without deteriorating the target detectability by means of STC.

The above results indicate that the dual beam antenna system is quite effective for suppressing various types of clutter. The ARSR system using the dual beam antenna has obtained the remarkable clutter suppression effects as shown in the PPI presentations in Fig. 11.

(3) LOG-CFAR [5]

Weather clutter cannot be eliminated by neither MTI nor the dual beam antenna system, because it has considerable radial velocity and is distributed in the high angle area.

A circular polarizer has been widely used to eliminate weather clutter, obtaining a suppression effect of about 15 to 20 dB.

Through suppression of weather clutter by the circular polarizer, and of sea clutter by the MTI and the dual beam antenna system, a nearly adequate improvement has been obtained with regard to the signal-to-clutter ratio.

None the less, sometimes, these types of clutter received at a level of a few tens dB higher than the receiver noise level saturate the receiver output, rendering target detection impossible. Thus the suppression is sometimes not sufficient for practical operation.

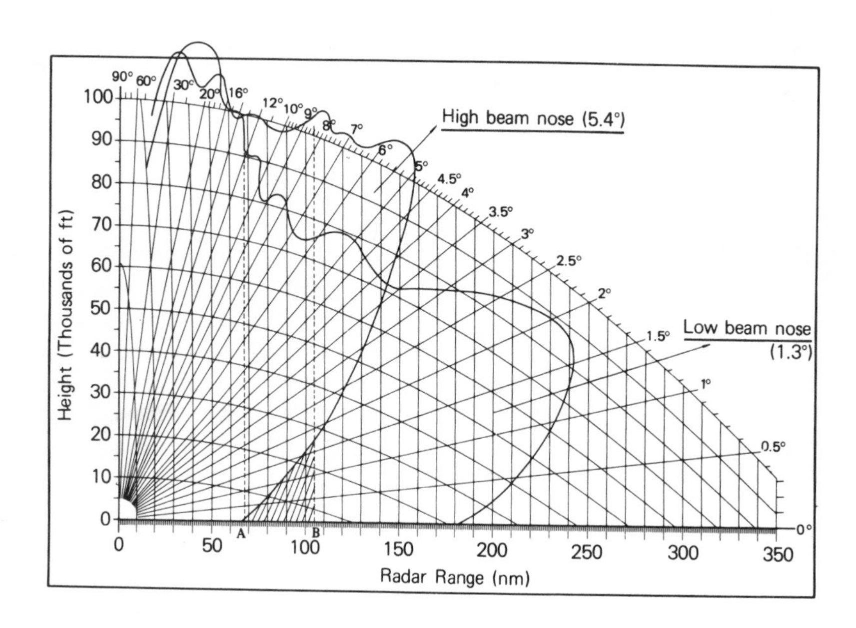

Fig. 8 Coverage of dual beam radar system.

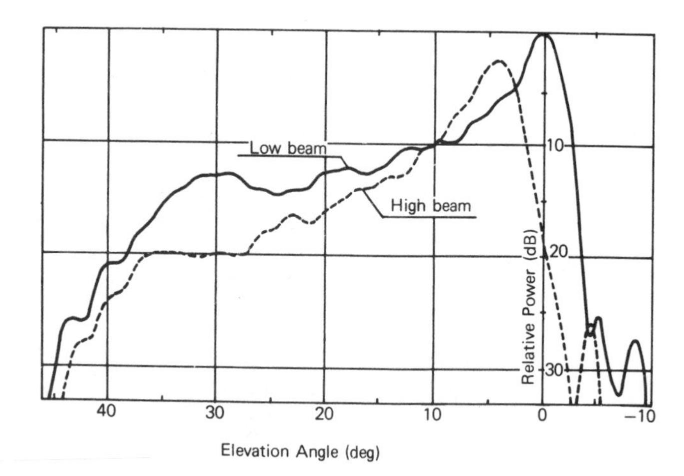

Frequency: 1300 MHz
Polarization: horizontal
3 dB beamwidth:
Low beam; 3.3°
High beam; 3.8°

Fig. 9 Elevation patterns of dual beam antenna.

The CFAR (Constant False Alarm Rate) technique which keeps the receiver output level constant against these types of clutter and suppresses them to the receiver noise level will be required.

This new ARSR system utilizes a cell averaging LOG-CFAR system which obtains a marked clutter suppression effect.

The LOG-CFAR system shown in Fig. 12 makes use of the fact that the amplitude variation of weather and sea clutters has a Rayleigh distribution, and is capable of reducing the clutter output to about the same level as the receiver noise level by means of a logarithmic amplifier and a CFAR circuit.

Figure 12 shows the principle of statical operation of the LOG-CFAR circuit, indicating that the output signal level of the anti-log converter is kept constant independently of the level of signal input to the LOG-CFAR

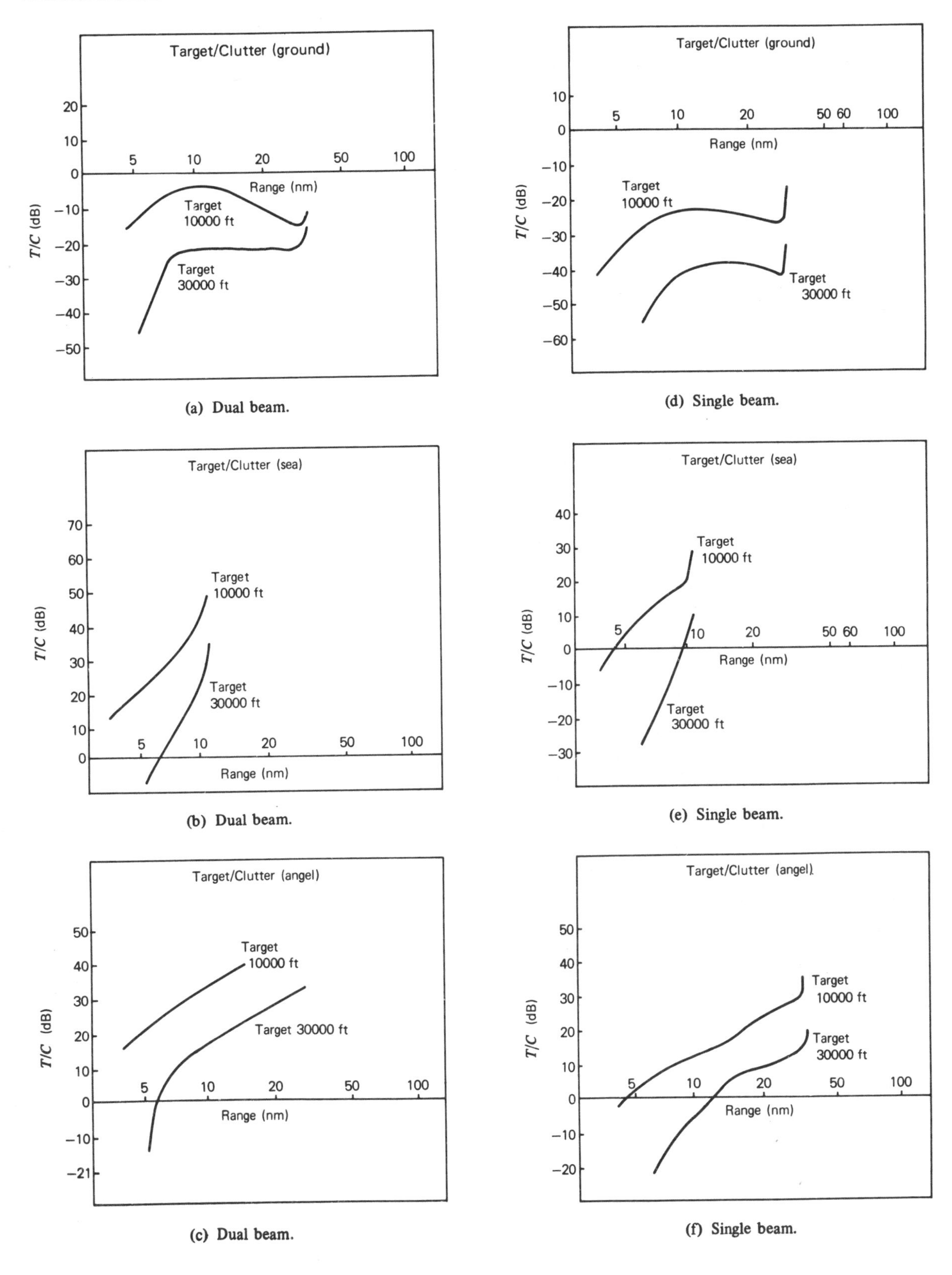

Fig. 10 Typical received target-to-clutter ratio for dual beam antenna system.

(a) Single beam.

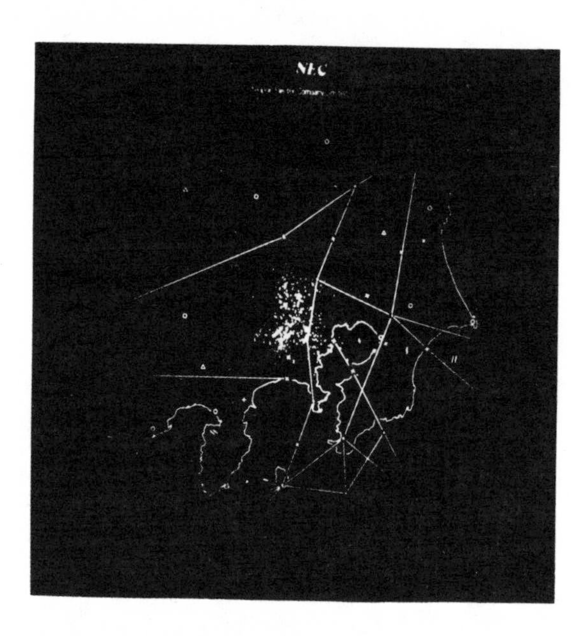

(b) Dual beam.

Fig. 11 Suppression effects for angel echoes.

circuit, and that the S/N ratio at the output of receiver is assured as the same as that at the input.

The sample number N for the average value detection shown in Fig. 12 is confined to a limited number in hardware. The determination of the value of N is an important factor in the LOG-CFAR design. It is determined by taking into account the CFAR processing loss and the clutter suppression characteristic of the LOG-CFAR receiver. The LOG-CFAR loss depending on the sample number N becomes as shown in Fig. 13 [6]. As in the new-ARSR, if $N=8$, when the number of hits from the target is about 13 and the LOG-CFAR receiver output is integrated, the CFAR loss theoretically becomes less than 2 dB.

This value has been confirmed by a number of flight checks and a LOG-CFAR signal detection test of this ARSR.

The frequency characteristic of the CFAR circuit with relation to the sample number N will become as shown in Fig. 14, and the cutoff frequency of the rejection band is given by

$$1/N\tau \qquad (\tau\text{: pulse width}).$$

When the spectrum of the mean value component of clutter is spread in a frequency region of less than $1/N\tau$, the mean value component of the clutter can be eliminated, and the clutter output becomes a constant amplitude with the similar characteristics of receiver noise.

If the LOG-CFAR loss should be kept between 1 and 2 dB, it needs a sample number N of about 8, so sufficient elimination of the mean value component can be attained for the clutter having an extension of more than $8\,\tau$, and the clutter can be suppressed to the receiver noise level. Thus, the Cell Averaging LOG-CFAR system has a marked CFAR characteristic even for weather and sea clutters

Figure 15 shows the effect of CFAR on the variation of the receiver noise level by the LOG-CFAR Processor. It indicates that the output noise level variation is suppressed within ±1 dB for an input noise level variation of more than 60 dB. Figure 16 shows the effect of CFAR on weather clutter, revealing that the weather clutter covering over 50 nm is almost completely suppressed down to the receiver noise level.

(4) RF-STC and Range Azimuth Gating

It has been observed that sufficient cancellation is obtained at the measurement of subclutter visibility, but the same degree of cancellation is not attained in the actual operation. One of the reasons of this is the spread of the clutter spectrum due to the saturation within the receiver.

The distribution of frequency spectrum of a fixed target essentially shows a Gaussian distribution, but this distribution spreads over a wide frequency range

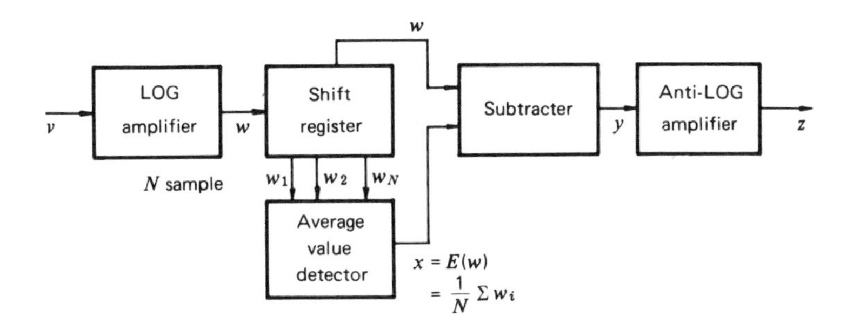

Input/Output Signal	v; Rayleigh signal $P(v)=\frac{2v}{\sigma^2}\exp\left[-\frac{v^2}{\sigma^2}\right]$	$w=a\ln[bv]$ a, b; constant	$y=w-x$ $=a\ln\left[\frac{v}{\sigma}\right]+\frac{a\gamma}{2}$	$z=c\exp[dy]$ $=\frac{cv}{\sigma}\exp\left[\frac{\gamma}{2}\right]$ c, d; constant $ad=1$
Average	$E(v)=\frac{\sqrt{\pi}}{2}\sigma$	$E(w)=a\ln[b\sigma]-\frac{a\gamma}{2}$ γ; Euler constant	$E(y)=0$	$E(z)=\frac{c\sqrt{\pi}}{2}\exp\left[\frac{\gamma}{2}\right]$
Power	$E(v^2)=\sigma^2$			$E(z^2)=c^2\exp[\gamma]$
Variance	$V(v)=\frac{4-\pi}{4}\sigma^2$	$V(w)=\frac{a^2\pi^2}{24}$ $=$constant	$V(y)=\frac{a^2\pi^2}{24}$ $=$constant	$V(z)=\frac{4-\pi}{4}c^2\exp[\gamma]$
Signal-to-Noise Ratio	$S/N=\frac{v^2}{E(v^2)}$ $=\frac{v^2}{\sigma^2}$			$S/N=\frac{z^2}{E(z^2)}$ $=\frac{v^2}{\sigma^2}$

Fig. 12 Operation analysis of cell averaging LOG/CFAR.

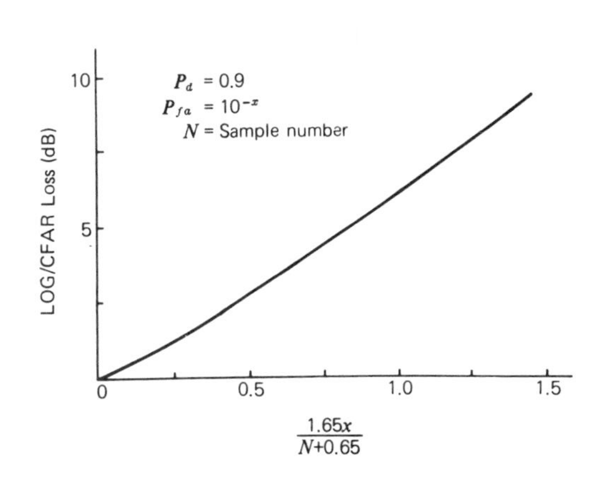

(a) LOG/CFAR loss for single pulse.

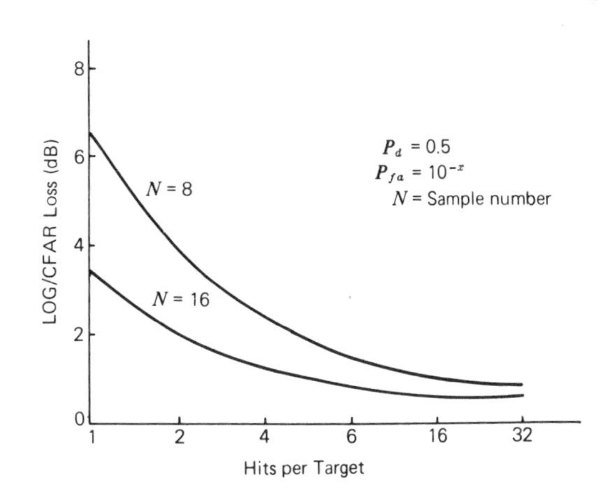

(b) LOG/CFAR loss with integration.

Fig. 13 LOG/CFAR loss.

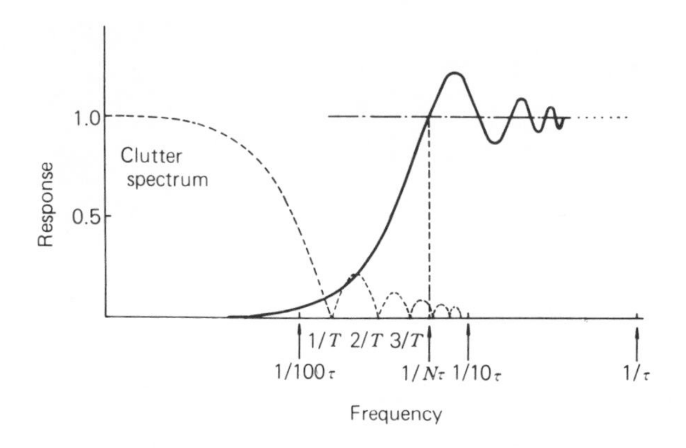

Fig. 14 Frequency characteristics of CFAR circuit.

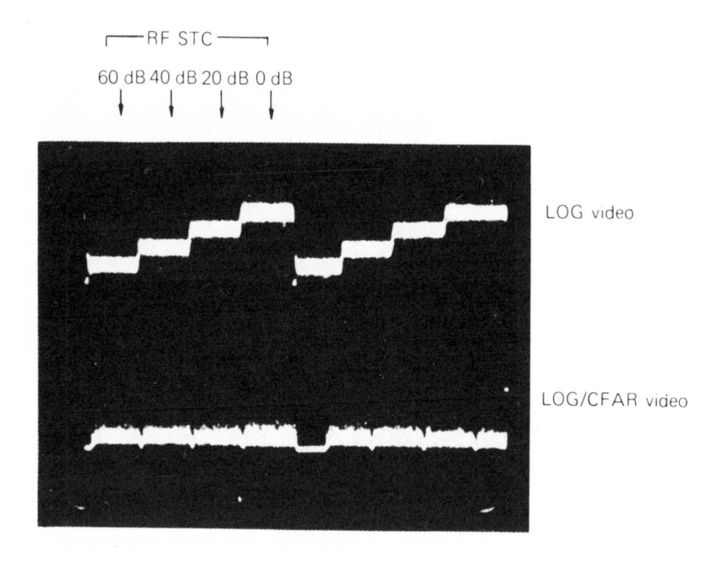

Fig. 15 Noise suppression effect of LOG/CFAR.

due to the limiting effect within the receiver.

In the case of the number of hits of 10 to 20, for example, the cancellation ratio will be deteriorated theoretically as much as about 20 dB during a hard limiting case [7].

This deterioration can be improved by the use of an RF STC having a range-azimuth gating function, by which the strength of clutter received from a short range can be attenuated to an appropriate level at the input stage of the receiver.

The recent progress of the MIC technology has brought it into practice to provide an RF STC which is capable of controlling an RF attenuation up to 80 dB with about 1 dB of insertion loss [8].

Moreover, it became much more practical to improve the S/C ratio by using the modified cosecant square antenna pattern in which the receive power of the target, approaching the radar site at a constant altitude, varies inversely as the fourth power of the distance, through fixing the gain of high angle of the antenna constant, and by giving an attenuation, which varies inversely as the fourth power of the distance, to the target and clutter by means of an RF STC.

Another feature of the RF STC is the Range-Azimuth Gating. It has the function of determining the start point and the end point at any area in the range as well as in the azimuth direction, and is capable of setting a window of any size at any area on the PPI scope. In this new ARSR, an optimum STC function to suit the distribution and nature of the clutter can be given through generation of an STC drive signal which gives an STC attenuation to meet a programmed function, thus minimizing the deterioration of target detectability due to the use of STC.

The utilization of the advanced anti-clutter technologies described previously enables this new ARSR to make adequate elimination of any clutters and to obtain a target output signal of high quality. Figure 17 shows the overall clutter suppression effect of the anti-clutter technologies applied in this system.

4. OTHER NEW TECHNOLOGIES

(1) Klystron Transmitter

The transmitter in the new ARSR is a power amplifier system with the klystron VA-963A, and has a capacity of 3 MW peak power output and 3 μs pulse width.

Compared with the magnetron transmitter in the conventional ARSR, the new transmitter is improved by more than 10 dB in spectrum purity, approx. 6 dB in MTI performance and more than 4 times in reliability.

Especially, the spectrum improvement is achieved by controlling the driver waveform of the solid-state driver, and this function can be realized only by a combination of the solid-state exciter and the klystron amplifier chain. And, for the equipment instability of MTI performance, the timing jitter and pulse width variation of RF pulse can be held to a minimum by the transmitter timing in "all coherent" relations with the crystal controlled STALO and COHO in the receiver. In addition, hit-to-hit phase shifts of RF

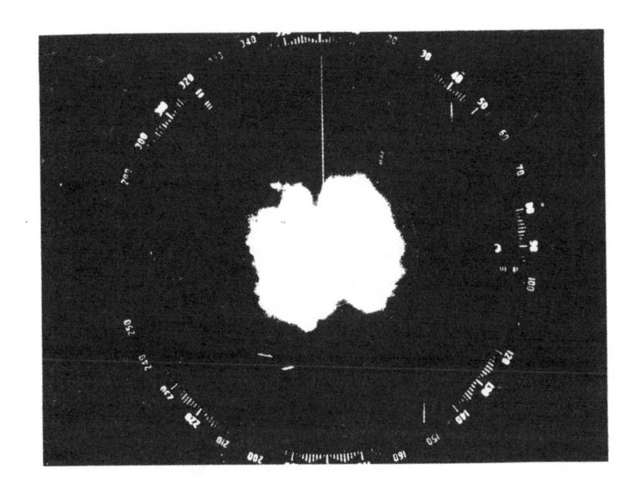

(a) MTI video without LOG/CFAR.

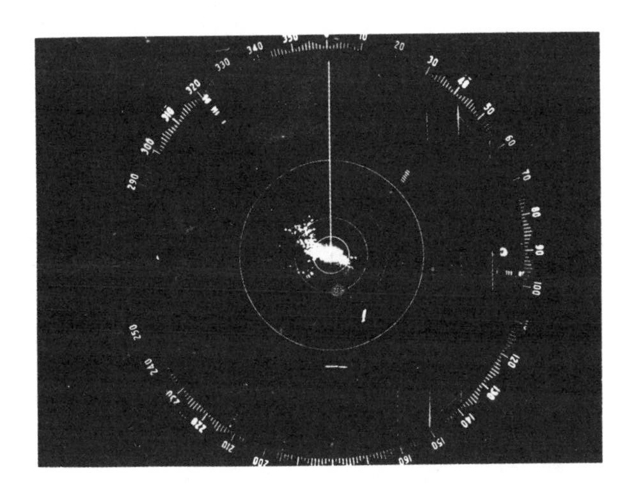

(b) MTI video with LOG/CFAR.

Fig. 16 Suppression effects for weather clutter (rain).

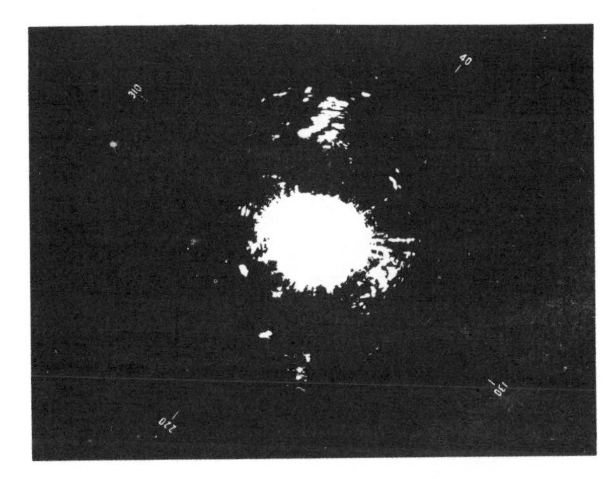

(a) Normal video.

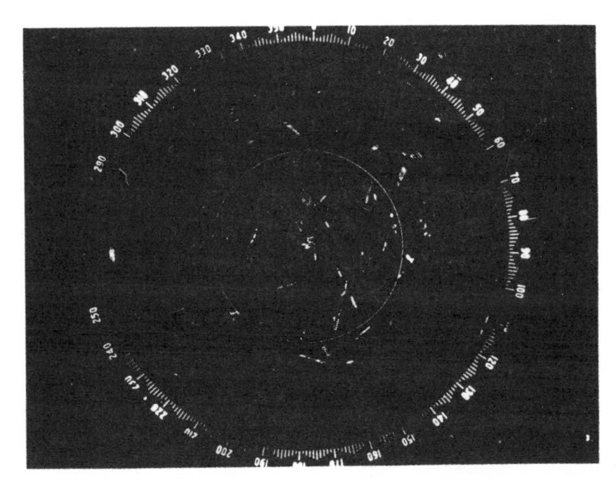

(b) MTI video.
Clutter is eliminated by DMTI, dual beam antenna and RF STC.

Fig. 17 Overall clutter suppression effects of anti-clutter technologies.

pulses due to the high-voltage stagger ripple is compensated for by shifting the phase of the COHO signal, so that MTI improvement factor of better than 40 dB can be obtained in the overall transmitter system.

(2) Automatic Performance Monitoring and Channel Change-over System

In order to minimize system downtime and make an unattended operation of the radar site possible, the new ARSR incorporates a radar control system for automatic monitoring, change-over and remote transmission of the performance information to the remote site. The following items of the transmitter, receiver and angle data system are monitored automatically.

Transmitter... POWER, FREQUENCY, TEMPERATURE, AIR FLOW, COOLANT FLOW and several kinds of VOLTAGE and CURRENT of the transmitter.
Receiver........ SENSITIVITY, PROCESSOR OPERATION and TR TUBE OPERATION.
Others........... ANGLE DATA.

By these automatical monitoring, when the fail condition is detected, the operation channel is changed over to the stand-by channel automatically, and greatly shortened system downtime is realized.

5. CONCLUSION

These new types of anti-clutter technologies have markedly improved the target visibility of the Air Route Surveillance Radar, and we believe that the ARSR is now capable of providing adequate radar information for Air Traffic Control Automation System. None the less, these new types of advanced technology still have the following inherent limitations.

- The inherent limitation of the MTI cancellation characteristic due to rotation of the antenna.
- Deterioration of the MTI detection probability of a target flying tangentially toward the radar.
- The inherent limitations of the MTI cancellation characteristic for the clutter distributed log-normally in a wide frequency range.

For resolving these problems, the signal processing technology basically different from the present MTI system has to be developed. An outlook for the new technology that would provide solutions of the limitations of the state of the art is summarized below. Targets and clutters can be analyzed to show their own peculiar characteristics such as amplitude variation, phase variation, and variations in the range and azimuth distribution.

Consequently, all received signals can be discriminated between target and clutter by analyzing the pulse-to-pulse, scan-to-scan, and range-bin-to-range-bin correlations of these received signals.

The target discrimination by the scan-to-scan correlations is effective for detecting a target whose radial speed is zero, while the range-bin-to-range-bin correlations help to detect a target in weather and sea clutters. Moreover, an on-line doppler filter bank based on the digital technology, which is further advanced than the MTI technology, enables to find the pulse-to-pulse correlations perfectly [9]. Through utilization of these new types of technology and pattern recognition methods, the anti-clutter technology will be further improved.

ACKNOWLEDGMENT

The authors wish to express their sincere appreciation for generous consultations and encouragements offered during the course of the system development by the personnel of Radio Engineering Division, Japan Civil Aviation Bureau, and of Development Division, Japan Electronic Navigation Research Institute.

They are also grateful for the valuable guidances and advices given by the staff of the 1st Radar Engineering Department, Radio Application Division.

REFERENCES

F. E. Nathanson, "Radar Design Principles," McGraw-Hill Book Company, New York, 1969.

M. Sugie and Y. Tomita, "Anti-clutter Technology," *Radar Symposium ; Present Radar Technologies, Four Institutes of Electrical Engineers of Japan*, October 1973.

I. Matsukasa, R. Inagaki and T. Hagisawa, "Digital MTI," *Inst. Electro. Com. Eng. Japan, SANE*72-4, May 1972.

Y. Tomita, H. Yokoyama and A. Mizutani, "Dual Beam Antenna System in ATC Radar," *Inst. Electro. Com. Eng. Japan, SANE*73-11, July 1973.

[5] Y. Tomita, T. Hagisawa and T. Irabu, "Rayleigh Clutter Suppression and Target Detection of Cell Averaging LOG/CFAR Receiver," *Inst. Electro. Com. Eng. Japan, SANE*75-13, July 1975.

[6] V. G. Hansen and H. R. Ward, "Detection Performance of the Cell Averaging LOG/CFAR Receiver," *IEEE Trans. of AES*, **AES-8**, p. 648, 1972.

[7] M. I. Skolnik, "Radar Handbook," McGraw-Hill Book Company, New York, 1970.

[8] H. Sato, M. Tan and E. Isobe, "Microwave Pin Diode Attenuator," *Inst. Electro. Com. Eng. Japan, MW*73-129, February 1973.

[9] A. Sawai and T. Irabu, "Moving Clutter Rejection Using Doppler Information and Its Range-to-Range Correlation Statistics," *IEEE 1975 International Radar Conf.*, Session II, 1975.

Received October 7, 1975

* * * * * * * * * * * * * * * *

MITSURU SUGIE was born on July 18, 1934. He received the B.E. degree in electrical engineering from Tokyo Metropolitan University in 1958.

He joined the Nippon Electric Co., Ltd. in 1958 and is now Engineering Manager of the First Radar Engineering Department, Radio Application Division. He has been engaged in the development of the various kinds of radar antennas, the development of phased array radar systems and design of ATC radar.

Mr. Sugie is a member of the Institute of Electronics and Communication Engineers of Japan.

* * *

YUICHI TOMITA was born on December 14, 1940. He received the B.E. degree in electronics from Tokyo Institute of Technology in 1964.

He joined the Nippon Electric Co., Ltd. in 1964 and is now Design Supervisor of the First Radar Engineering Department, Radio Application Division. He has been engaged in the development of MTI (moving target indicator) receivers and signal processors, and design of ATC radar.

Mr. Tomita is a member of Aerospace Electronic System and Acoustic and Signal Processing of IEEE and IECE of Japan.

TOSHIHIKO HAGISAWA was born on May 16, 1942. He received the B.E. degree in electronics from Hokkaido University in 1966.

He joined the Nippon Electric Co., Ltd. in 1966 and is now Design Engineer of the First Radar Engineering Department, Radio Application Division. He has been engaged in the development of digital MTI (moving target indicator) processors, radar signal processors and design of ATC radar.

Mr. Hagisawa is a member of the Institute of Electronics and Communication Engineers of Japan.

* * * * * * * * * * * * * * *

6.2 A CHARGE TRANSFER DEVICE MTI IMPLEMENTATION

H. Lobenstein and D. Ludington

IEEE International Radar Conference Proceedings, Washington, D.C., 1975, pp. 107-110.

ABSTRACT

This paper describes the successful application of Charge Transfer Device (CTD) technology to the design of a three-pulse, 50 dB clutter suppression, burst radar Moving Target Indicator (MTI). Design considerations, bench and system test results, plus the potential impact of advances in the CTD and CCD state of the art on future equipments, are discussed.

INTRODUCTION

Digital signal processing, with its inherent stability, has with the maturing of MSI and LSI technology quickly become an integral part of new radar systems. The one major difficulty, in high performance systems, has been the analog-to-digital (A/D) interface. Whenever dynamic range requirements approach eight or more bits, and bandwidths 5 MHz, the A/D conversion becomes expensive and complex.

An ideal solution would be preprocessing with a low cost analog device that combines the stability of digital integrated circuit operation with analog signal handling. This would leave A/D conversion within the more economical realm of low bit - high speed technology.

Bucket brigade techniques neatly fit into this category. They have not been utilized in the past for the same reason that digital processing was slow in being applied; i.e., the discrete part implementation problems, and cost. Integrated circuit versions of the bucket brigade are now emerging and, depending on device physics, are called Charge Transfer Devices (CTDs) or Charge Coupled Devices (CCDs).

With these devices, the analog signal is Nyquist sampled as in a digital system; however, the samples are stored and processed as discrete packets of charge. Hence, the CTDs and CCDs utilize a continuous range of amplitude, and for obvious reasons the MTI has been dubbed a Discrete Signal MTI (DSMTI).

What will be described is the use of linear CTDs to construct a feasibility model DSMTI to directly replace an acoustic delay line canceller in a burst radar system. The technical performance parameters are as follows:

Type of MTI	Three-pulse vector canceller (in-phase and quadrature)
Cancellation	50 dB
Dynamic Range	$\geq$ 50 dB
System Waveform	Multiple burst 250-microsecond intraburst PRI
Pulsewidth	1 microsecond
Input Format	In-phase (I) and quadrature (Q) baseband video
Output Format	I and Q video Reconstituted IF Detected IF
Synchronization	Coherent burst control

THEORY OF CHARGE TRANSFER DEVICES

The charge transfer device (CTD) is best characterized as an analog sampled-data delay line. The CTD equivalent circuit is shown in Figure 1. The CTD consists of capacitive storage elements separated by switches. When switch S1 is closed, capacitor C1 is charged to the input signal level. Then switch S1 is opened and switch S2 is closed, which allows the input sample to be transferred to C2. In this manner, the switches alternately open and close, sending the analog samples down the device. Isolation between samples is provided since all odd-numbered switches are operated together and all even-numbered switches are operated together. Complementary square wave clocks are used to drive the switches, thus preventing the odd-numbered switches and even-numbered switches from being closed at the same time.

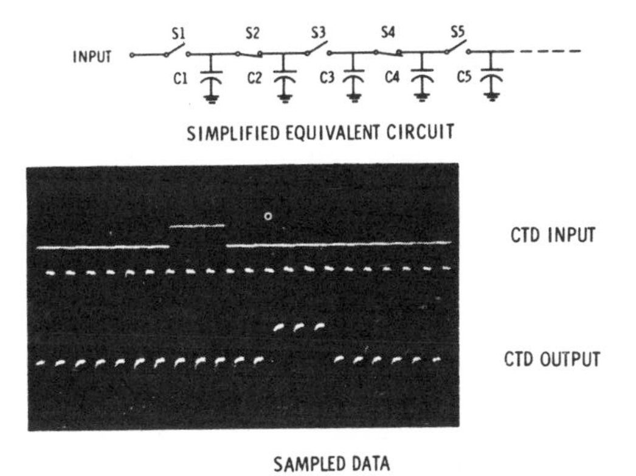

Figure 1. Charge Transfer Device (CTD)

The rate at which the analog information is transferred through the CTD is determined by the rate at which the switches are operated. For a given number of stages (a switch and associated capacitor are considered as one stage), the total delay is determined by the digital clock period, as given by equation (1):

$$\tau = \frac{N}{2f_c} \qquad (1)$$

where

τ = delay

N = number of stages in the CTD

f_c = CTD clock frequency

Thus the delay provided by the CTD can be varied either by changing the number of stages in the device or by changing the clock frequency.

CCDs provide the same function except that instead of capacitors being filled, charge packets are transferred to adjacent potential wells by clocked voltage gradients in a manner akin to raising and lowering the water level in a canal lock to transfer barges to the next stretch.

In practice, switches S_i can be bipolar, MOSFET, or JFET semiconductors which are implemented in the form of monolithic integrated circuits. This allows many stages to be fabricated on one chip. Capacitors C_i are an integral part of the monolithic device and are on the order of a few picofarads.

The actual transfer of information is accomplished by a charge-deficit transfer. The reader is referred to Refs. 1, 2, and 3 for further information.

There is never a complete transfer of charge from one storage element to the next; however, the amount of charge left behind at each transfer can be almost independent of the amplitude of the initial charge. Thus, it is not necessary to operate in a 100-percent transfer mode (which would severely restrict the frequency at which the device could be operated). Rather, the circuit can be operated in the so-called "partial transfer mode," in which the effect of charge left behind during one transfer is compensated by the charge which is picked up from the previous transfer. It follows that a bandwidth/dynamic range tradeoff can be expected from these devices; i.e., for a given length of delay line, the smaller the dynamic range required, the larger the bandwidth of the circuit.

The ultimate effect of incomplete charge transfer will be to band-limit the circuit. At low clocking frequencies, the bandwidth will be determined by the Nyquist sampling requirement. At high clocking or sampling rates, the band-limiting will result from the transfer characteristics of the circuit itself.

An elementary analysis of the effect of cascading a large number of delay sections was performed. It was assumed for this analysis that the transfer of charge from one storage capacitance to another can be roughly modeled by a simple RC low pass filter (Ref. 4).

The transfer function is given by:

$$T(s) = \frac{1}{1 + sCR} \tag{2}$$

where s is the complex frequency variable, and the 3-dB bandwidth of the circuit is given by

$$BW_1 = \frac{1}{2\pi CR} \tag{3}$$

It is further assumed that the time sequential transfer of charge through n sections may be modeled by cascading n isolated RC sections. Then the overall transfer function is given by

$$T(s) = \left(\frac{1}{1 + sCR}\right)^n \tag{4}$$

and the 3-dB overall bandwidth, BW_C, is given by

$$BW_C = \sqrt{2^{1/n} - 1}\, BW_1 \tag{5}$$

which, for large n, becomes

$$BW_C = \frac{0.83}{\sqrt{n}} BW_1 \tag{6}$$

Figure 2 shows the band-limiting effects of a MOSFET charge transfer device. In Figure 2(a), when the signal is picked off after the 12th stage, the pulse reproduction is very good. When the signal is picked off after the 108th stage, the band-limiting effect of the CTD is clearly evident.

The samples at the leading edge of the pulse are decreased in amplitude and the charge which is left behind builds up as samples at the trailing edge. These trailing voltage levels are not actual samples from the input but come from charge left behind by incomplete charge transer. This can be verified by counting the number of samples for the 12 taps; there are more samples for the 108th tap.

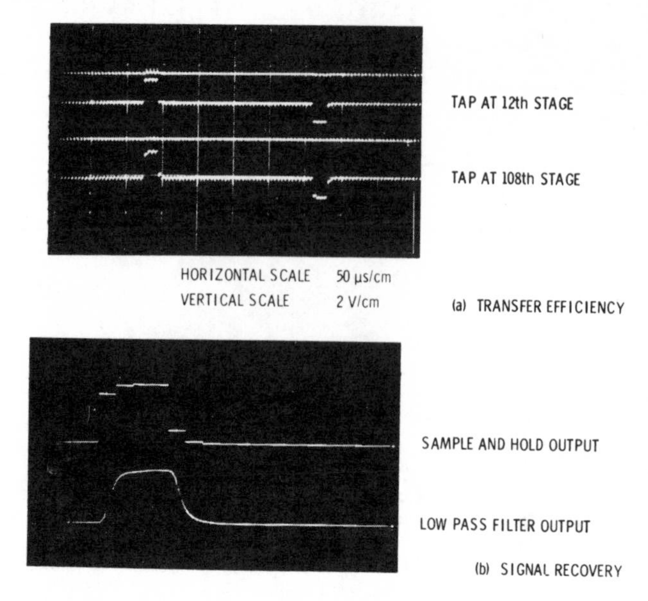

Figure 2. Charge Transfer Device Operation

MTI DESIGN

A moving target indicator is used to reduce the effects of sea or ground clutter while at the same time enhancing signals which represent moving targets. The method of implementation is to subtract a radar return which has been delayed one PRI from the incoming radar return. This is shown in Figure 3 where the CTD provides the delay. The advantage of the CTD as a delay line is that since the delay is controlled by a digital clock, the delay can be made very stable, while the signal path is still analog, and does not require A/D or D/A conversion as in a digital system.

The required IF bandwidth is 1 MHz. In-phase (I) and quadrature (Q) processing are used to preserve the amplitude and phase of the signal. The bandwidth of the I and Q signals is 500 kHz.

The Nyquist theorem states that the sampling frequency must be at least twice the bandwidth of the signal. Of course, this theorem assumes an ideal low-pass characteristic for the bandwidth. In the real world, the frequency response of the signal rolls off more slowly, and the bandwidth is usually taken as the 3-dB bandwidth. Because of this gradual rolloff

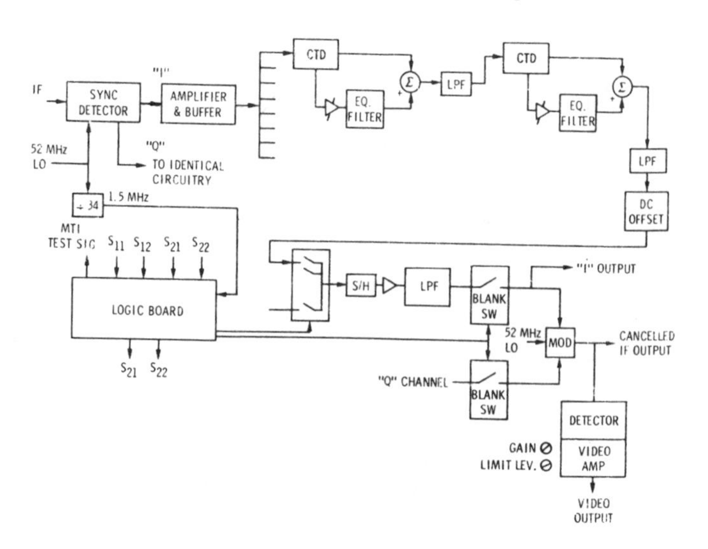

Figure 3. Discrete Signal MTI Simplified Block Diagram

in frequency, it was decided to sample at three times the bandwidth, or 1.5 MHz. By substituting the sampling frequency and PRI into equation (1), the required number of stages for the CTD can be found.

$$N = (2f_c)(\tau) = (2)(1.5 \times 10^6)(250 \times 10^{-6})$$

$$N = 750 \text{ stages} \qquad (7)$$

With the devices available at the time the design was initiated, it was not possible to cascade the required number of stages and still achieve the 500 kHz bandwidth. Therefore, a parallel scheme with multiplexing was required to achieve the required number of total stages while at the same time limiting the number of cascaded stages in each device.

Two methods are available for clocking the charge transfer devices which will preserve the sampling rate and delay. First, the CTDs can be clocked at a 1.5 MHz rate, but only for part of the PRI. After enough samples have been taken to fill a CTD, its clock is turned off, and the clock to the next CTD is turned on. This process is continued with each of the CTDs being clocked for a period equal to the PRI divided by N, where N is the number of parallel CTDs. At the next PRI, the sequence will be repeated (Ref. 5).

The second method of parallel operation uses lower frequency overlapping clocks. In this method, the clocks to each CTD are run continuously, but between adjacent CTDs the clocks are skewed so that only one CTD is sampling the input at a time.

It is evident that if N is greater than 3, the CTD clock frequency will be less than the 500-kHz analog bandwidth. This is permissible since, with the staggered clock arrangement, the time between samples is equal to the clock skew and is still equivalent to a 1.5-MHz sampling rate. In this case, the output is sampled only during the skew time.

Each method has advantages and disadvantages. The advantages of the overlapping clock mode are that the clock frequencies are lower and the clocks are running continuously throughout the PRI. A lower clock frequency meansthat more time is available for charge transfer between stages (thus reducing the charge loss in the CTD), while the continuous clock eliminates nonuniform leakage from the storage capacitors within the CTD.

The advantage of the gated clock mode is that the clock rate in each CTD is fast enough to satisfy the Nyquist rate, and the signal samples are stored in consecutive stages in the same device. A disadvantage is the nonuniform leakage of sites which evidence themselves as pattern noise on readout. The distribution of the pattern noise is random; some capacitors have large leakage because of processing problems and material imperfections.

Two problems are inherent in the overlapping clock configuration. Both of these are caused by incomplete charge transfer. Figure 2(a) shows that the first output sample is reduced in amplitude from the input. Since only one sample is initially taken by a CTD, the maximum signal output is reduced in amplitude. The input amplitude can be increased somewhat to compensate for this effect as long as the first stages of the CTD do not saturate. Essentially, this reduced output decreases the dynamic range.

The effects of charge transfer inefficiency also appear as additional samples that trail the delayed pulse. When the samples are in the same CTD (at a fast sampling rate), these extra samples cause the pulse to be extended in time similar to the effect of a RC low-pass filter. However, in the overlapping clock mode, each of several parallel CTDs takes one sample of the input. These samples are propagated down the CTDs with low clock frequencies. When the outputs of the CTDs are recombined, the input will be reconstructed and then, because of the lower clock frequency, the samples from the incomplete charge transfer will be reconstructed but displaced in time from the input. In essence, extraneous pulses have been created. The magnitude of the extraneous pulse depends on the charge transfer inefficiency while the relative delay time between the pulses is determined by the number of parallel CTDs (for a given clock frequency).

At the start of this endeavor when a decision was being made as to which multiplexing method to use, the Nyquist bandwidth of available devices was very limited, with a 200 kHz clock producing a linear relationship on a large sample of devices, and some approaching 800 kHz. Therefore, the overlapping clock approach was selected.

Taking the conservative approach, eight overlapping channels operating at 200 kHz were used (see Figure 3). This produced an effective 1.6 MHz sampling rate.

The transfer inefficiency of these devices was on the order of 1×10^{-3}, producing an extraneous pulse, in the uncancelled mode, approximately 20 dB below the main pulse. This does not appear to be of significance on a PPI presentation except for very large magnitude moving targets. (This is similar to pulse compression systems with 20 dB sidelobes.) In the case of clutter returns, the extraneous signal bears a precise relationship to the clutter patch and cancels as an entity.

Since this unit was designed, the state of the art has advanced rapidly, and CTDs having a transfer inefficiency of 5×10^{-5} at 1 MHz have been demonstrated. This should reduce the extraneous pulse level to below 40 dB. The dynamic range of the devices was excellent with a peak signal to peak spurious plus noise of 58 dB having been achieved.

Since an MTI needs to compare a direct to a delayed path, and subtract them, the signals should have very similar characteristics. Therefore, the direct signal was passed through a short length of the sampled data delay line; this is the significance of the 12th stage tap on Figure 2(a).

Referring to Figure 3, an equalization network is seen; this is a simple RC low-pass filter. Without the filter, an average of 26 dB cancellation was achieved for a single delay MTI channel; with the filter, this increased to 40 dB.

Signal recovery is by means of a sample and hold operation plus low-pass filtering following the demultiplexing (see Figures 2(b) and 3).

TEST RESULTS

The individual double delay channels achieved 58 dB of cancellation. The multiplexed system performance level decreased to approximately 51 dB due to multiplexing noise and residual unbalances between channels (see Figure 4). Figures 5 and 6 demonstrate actual system performance. Note that the hill in Figure 5, the PPI display without MTI, is approximately 60 dB above noise. This hill barely comes through in Figure 6, as it should. Although a little time elapsed between the changeover to the DSMTI mode, several targets (at 60°, 120°, etc.) previously inundated in clutter can be easily correlated. Note that several discrete returns due to the site location, plus a rain storm, have been eliminated.

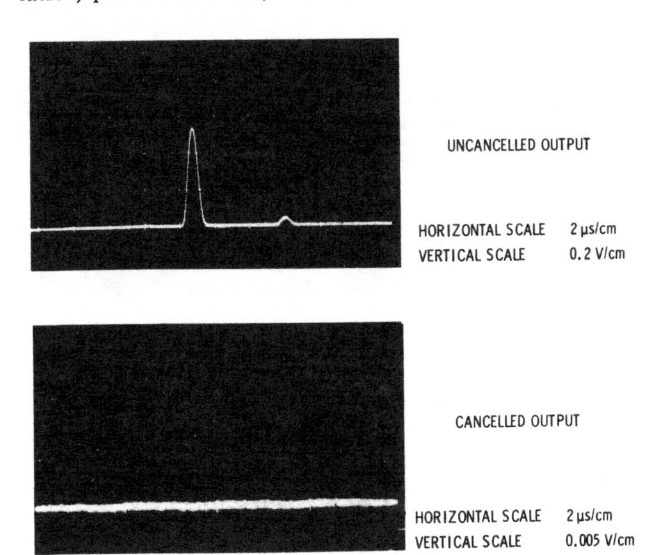

Figure 4. MTI Performance - Multiplexed System

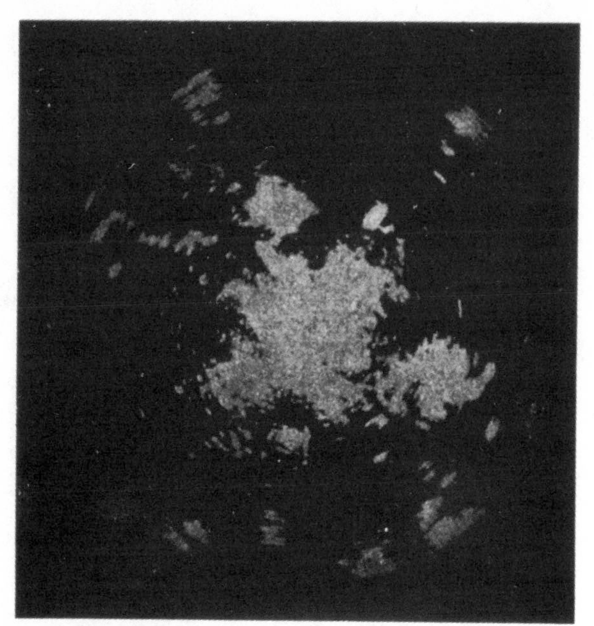

Range: 14 miles
Quaker Hill at 335° to 355° and 5 to 8 miles approx. 60 dB above noise. Rain storm from 310° to 335° at 10 to 12 miles.

Figure 5. Clutter Map with DSMTI Bypassed

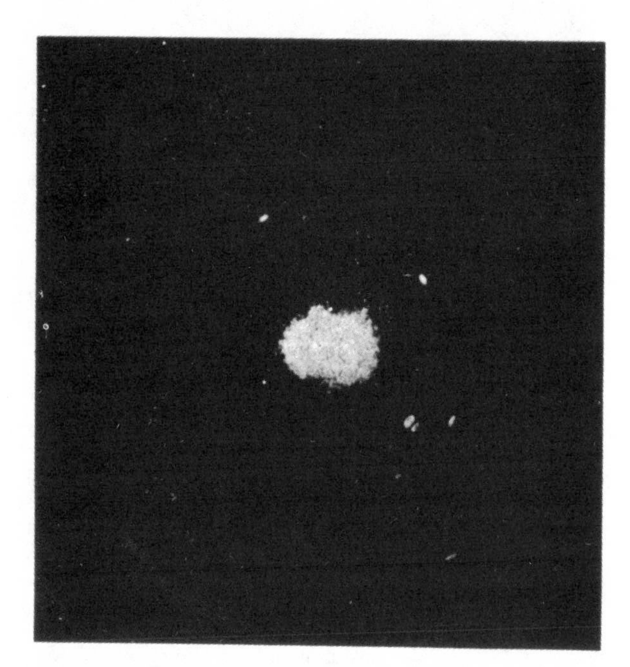

Note that Quaker Hill is barely visible. Double return at 132° and 6 miles due to combination of large target and transfer inefficiency.

Figure 6. Clutter Map with DSMTI

IMPACT OF CTD AND CCD ADVANCES

The state of the art in charge transfer, and charge coupled, devices remains very dynamic. A 136-stage device operating at 6 MHz with a 3×10^{-4} transfer inefficiency has recently been reported by the General Electric Corporate Research and Development Center (Ref. 6). Dr. Esser of Phillips has demonstrated a Peristaltic Charge Coupled Device (PCCD) operating at 135 MHz (Ref. 7). Various other companies are actively pursuing large scale CCDs for use as computer main frame memories as well as part of the readout devices in solid state imagers.

It should also be pointed out that functions other than MTIs can be easily implemented. For instance, a structure has been designed for providing 32 high-speed multiply and summing functions on a chip (Ref. 8); in addition, transversal filters can be easily generated (Ref. 9), so that in essence the system designer has another tool available to him.

The existing DSMTI model using RMOS devices consisted of eight thinly populated Cambion boards. A productionized version should bring this down to two multilayer boards. Utilizing newer devices, the complete DSMTI can be packaged on no more than a single 5-inch by 9-inch board.

CONCLUSIONS

It is evident from the test results that charge transfer devices can be used to implement an MTI with excellent clutter cancellation capability. Advances in the device state of the art should make it less necessary to multiplex next generation systems with similar parameters. However, it has been demonstrated that this technique is an important tool in extending the time-bandwidth product for a particular application.

ACKNOWLEDGEMENTS

This effort was supported by the Air Force Systems Command, Rome Air Development Center, Griffiss Air Force Base, New York, and monitored by W. Simkins, D. Budzynski, and C. Silfer.

The authors wish to thank J. Whitten and Drs. W. Butler and C. Puckette, General Electric Corporate Research and Development Center, for many enlightening discussions. Dr. Butler also provided the charge transfer devices used in this study.

REFERENCES

1. F.L.J. Sangster and K. Teer, "Bucket-brigade electronic - new possibilities for delay, time-axis conversion, and scanning," IEEE J. Solid-State Circuits, Vol. SC-4, pp 131-136, June 1969.

2. F.L.J. Sangster, "The bucket-brigade delay line," Phillips Tech. Rev., Vol. 31, pp 97-110, 1970.

3. W.J. Butler, et al., "Practical considerations for analog operation of bucket-brigade circuits," IEEE J. Solid-State Circuits, Vol. SC-8, pp 157-168, April 1973.

4. W.J. Butler, et al., "Bucket-brigade bandwidth characteristics," General Electric Company, TIS Report 72CRD002, Dec. 1971.

5. W.J. Butler, et al., "Implementation of an MTI by bucket-brigade circuits," Electronic Letters, No. 1972, Vol. 8.

6. W.E. Engeler, R.D. Baertsch, and H.S. Goldberg, General Electric Corporate Research and Development Center, Schenectady, N.Y., private communication.

7. L.J.M. Esser, et al., "The peristaltic charge coupled device," IEDM Technical Digest, pp 17-20, Dec. 1973.

8. J.J. Tieman, et al., "A surface charge correlator," General Electric Company, TIS Report 74CRD119, July 1974.

9. R.D. Baertsch, et al., "Two classes of charge transfer devices for signal processing," presented at International Conference on the Technology and Application of CCDs, Edinburgh, Scotland, Sept. 1974.

Section 7 Airborne MTI (AMTI)

EDITOR'S COMMENTS

Airborne MTI (AMTI) systems that function to detect aircraft or moving surface targets from a moving platform are described in this series of papers. The motion effects and increased clutter backscatter that occurs when looking down at the ground generally restrict performance to less than that achieved by comparable ground based MTI systems.

Dickey (Paper 7.1) in a paper written in 1953 analyzes the performance of a single canceler AMTI. This type of system was typical of the hardware limited systems operational through the early 1960's. The performance of this type of system is severely limited by clutter spectral spreading (induced by aircraft motion) as the antenna points away from the aircraft's ground track. A polar clutter attenuation contour plot illustrates this "butterfly effect" whereby residual clutter continually provides stronger returns as the antenna points away from the aircraft's ground track. As a note of interest this effect has been used to measure the aircraft's ground track that is symmetrically located with respect to the butterfly pattern observed on the radar's PPI.

The next paper by ap Rhys and Andrews (Paper 7.2) describes the performance of an Airborne Early Warning (AEW) coherent AMTI that utilizes motion compensation techniques.

Time Average Clutter Coherent Radar (TACCAR) is a technqiue developed by MIT Lincoln Laboratories whereby the average frequency of the clutter return is translated to a notch in the comb filter response of the AMTI. This effectively removes the mean velocity of the clutter in a direction normal to the antenna aperture. However, the angle of arrival to the clutter is a function of the clutter's range and complete correction would require a separate compensation in each range cell. In practice, usually one correction is sensed and applied to all range cells resulting in an error that is described in the paper.

Displaced Phase Center Antenna (DPCA) corrects for the component of aircraft motion parallel to the antenna aperture. This correction is accomplished by electronically causing the phase center of the antenna to remain motionless over the period corresponding to the number of radar returns processed by the MTI; i.e., two pulses for a single canceler. As currently applied, DPCA effectively compensates for spectral spreading induced by aircraft motion when used with a single canceler type MTI filter. However, if a multiple-stage MTI filter is employed, then DPCA corrects only the first stage thus resulting in imperfect performance as described in the paper.

The clutter spectrum due to platform motion has a shape that is approximately proportional to the shape of the two-way antenna pattern. Thus antenna sidelobes result in doppler components that fall within the passband of the MTI filter. This effect puts a premium on low antenna sidelobes, and the MTI performance is proportional to the integrated antenna sidelobe structure.

High AMTI performance for a low-PRF AEW application requires:

- Integrated two-way mainlobe-to-sidelobe ratio of -60 dB
- Pulse-to-pulse stability factor of 10^{-9}
- Transmitter pulse-to-pulse stability factor of 10^{-9}
- Triple canceler MTI
- 16 pulse coherent integration
- TACCAR correction at 3 points in range
- DPCA correction of all 3 cancelers

The last two papers (Paper 7.3 and 7.4), by Brennan, Reed and Mallett of Technology Service Corporation, describe an advanced type AMTI that adapts in both the spatial and temporal domains to maximize the radar's probability of detection. Both the antenna receiver phased array illumination function and the MTI filter parameters are controlled adaptively to reject both mainlobe and sidelobe clutter.

The algorithms presented in the paper simultaneously adjust the phase array illumination function to provide motion compensation while minimizing the sidelobe response in directions containing large clutter scatterers. In addition, the algorithms adjust the MTI filter response to minimize mainlobe clutter. Simulations are presented to show the excellent performance of this type of system. In the simulations, clutter is represented as 30 random scatterers uniformly distributed throughout the radar's sidelobe region.

Although this type of adaptive radar shows great theoretical promise, it still awaits an operational test. One of the remaining theoretical problems is concerned with the convergence of the algorithm in a dynamic situation.

The convergence problem is discussed in the last paper (Paper 7.4) where it is shown that the sample matrix inversion (SMI) algorithm provides rapid convergence. However, this rapid convergence is achieved at the expense of a formidable computational load (approximately N^3 complex multiplications, where N is the total number of adaptive weights). The Applebaum maximum-signal-to-noise (MSN) algorithm and the Widrow least-mean-square algorithm are recursive implementations that require modest computations but may converge slowly in particular dynamic situations.

7.1 THEORETICAL PERFORMANCE OF AIRBORNE MOVING TARGET INDICATORS

F. Dickey

IRE Transactions on Aeronautical and Navigational Electronics, Vol. PGAE-8, June 1953, pp. 12-23.

In the case of a radar moving target indicating system (MTI) in which video signals received on two successive cycles are subtracted, it is desired that signals reflected from fixed objects be completely cancelled. However, because of fluctuations in the signal strength, there is in general a residual ground clutter signal remaining after cancellation. The fluctuations may be due to slight motions of the supposedly stationary objects, to instabilities in the radar system, or to the motion of the antenna. In this paper only the effects of antenna motion are considered. That is, it is assumed that the ground is truly stationary, and that the radar system is perfectly stable. Under these conditions there is an effect due to rotation of the antenna, and, in the case of an airborne system, there will be effects due to the forward motion of the aircraft which may severely limit the performance.

The order of magnitude of most of the effects encountered has been given by Ridenour[1]. More recently, a rigorous method of calculation has been given by T. S. George[2], but his results are not in a form which is convenient for numerical calculations. The present paper describes a different method of setting up the problem which leads to simple formulas for ground clutter attenuation.

Figure 1 is an example of the type of results which can be obtained. The contours show, for a particular situation, the mean square value of residual ground clutter in decibels below the uncancelled clutter as a function of ground range and azimuth. The clutter is least, and hence the MTI performance is best along the ground track and deteriorates severely on either side.

The mean square value of the residual clutter is found to be the sum of four components, one due to rotation of the antenna, and three others due to aircraft motion. The equations for these components involve the parameters listed in Figure 1 plus the azimuth and elevation angles corresponding to positions on the radar scope. The equations make evident which ones of the system parameters are important when one wishes to design for a given performance in any particular region of the scope face.

GENERAL METHOD OF SOLUTION

In order to make calculations, a particular point on the scope, i.e., a particular range and azimuth, is selected. It is assumed that each time the antenna scans through this position an observation is made of (1) the signal voltage from the ground, and (2) the voltage remaining after subtraction of two successive signals.

[1]Ridenour, "Radar System Engineering" Rad. Lab. Series, Vol. I, McGraw Hill, 1947, pp. 657-658

[2]T. S. George, "Fluctuations of Ground Return in Airborne Radar Equipment" I.E.E. Proc., Pt. IV, Apr., 1952, v. 99, pp. 92-99

A quantity, ϵ is then defined after a large number of such observations as follows:

$$\epsilon = \frac{\text{Mean Square Pulse-to-Pulse Voltage Change}}{\text{Mean Square Voltage}} \qquad (1)$$

The received voltage is assumed to result from a large number of small component signals arriving from various angles and randomly phased with respect to each other. It is also assumed that, on a time average, the angular distribution of energy is uniform except as influenced by certain weighting functions.

The amplitudes of the individual components are weighted, as a function of azimuth angle of arrival, according to the azimuth antenna pattern. They also are weighted as a function of elevation angle, and in the usual case where the pulse duration is short, the elevation function depends on pulse width and shape. The changes in signal strength which prevent complete cancellation are caused by (1) changes in the relative phase of various components due to displacement of the antenna platform during the interpulse period, and (2) changes in the amplitude of various components due to rotation of the antenna during the interpulse period. Although the relative phases of the various components are random, the changes in relative phase during the interpulse period can be calculated.

The differences which are observed are the differences between successive voltage envelopes since the signals are rectified by an approximately linear detector, but the problem may be simplified by considering vector differences instead of envelope differences. The vector differences are those which would result from subtraction of successive signals before rectification (and with a first order compensation for Doppler frequency shift). This mathematical artifice may be employed with little error because, if the differences are small, the mean square values of the vector differences and the envelope differences are nearly the same except for a constant factor of 1/2. Figure 2 shows the vector difference, $\mathcal{V}$, and the envelope difference, ΔV, at a given time. The vector difference may be divided into a component in phase and in quadrature with the sum of the two signals and, upon averaging, these components each contain half of the mean square voltage. Referring to Figure 2, the in phase component, $\mathcal{V}'$, is very nearly equal to ΔV and, since $\mathcal{V}'$ has a mean square value exactly half that of $\mathcal{V}$, one can say that ΔV has a mean square value approximately half that of $\mathcal{V}$.

Assume that one has available the following functions:

$A(\theta)$ = round-trip voltage antenna pattern in azimuth,

$B(\phi)$ = elevation weighting function usually determined by pulse shape,

$F(\theta, \phi)$ = change in relative phase produced by a given displacement of the antenna,

θ, ϕ = azimuth and elevation angles, respectively, of individual signal components, measured with respect to the center of the effective beam.

The individual voltage components comprising a signal are weighted according to $A(\theta) B(\phi)$. The vector difference between two successive signals is composed of the same voltage components weighted according to the following function:

$$A(\theta + \tfrac{1}{2}\omega_a T) B(\phi) e^{+\frac{1}{2}jF(\theta,\phi)} - A(\theta - \tfrac{1}{2}\omega_a T) B(\phi) e^{-\frac{1}{2}jF(\theta,\phi)} \quad (2)$$

where

ω_a = angular rotation rate of antenna (about a vertical axis),

T = time interval between pulses.

With the assumptions regarding the statistical nature of the signal components, the ratio of mean square voltages is:

$$\epsilon = \frac{\tfrac{1}{2}\int_{-\infty}^{+\infty}\int_{-\infty}^{+\infty} \left| A(\theta + \tfrac{1}{2}\omega T) B(\phi) e^{+\frac{1}{2}jF(\theta,\phi)} - A(\theta - \tfrac{1}{2}\omega_a T) B(\phi) e^{-\frac{1}{2}jF(\theta,\phi)} \right|^2 d\theta d\phi}{\int_{-\infty}^{+\infty}\int_{-\infty}^{+\infty} |A(\theta) B(\phi)|^2 d\theta d\phi} \quad (3)$$

In order to put this expression into a more useful form, a rectangular coordinate system is oriented in such a way that the z axis is along the center of the effective beam, and the x axis is horizontal. The y axis in general is not vertical, but, as shown in Figure 3, it falls in the same vertical plane as the z axis. The displacement of the antenna during the interpulse period is expressed as a component along each axis. These components are:

$$X = vT \sin(\Theta - \Theta_0) \quad (4)$$

$$Y = vT \cos(\Theta - \Theta_0) \sin(\Phi - \Phi_0) \quad (5)$$

$$Z = vT \cos(\Theta - \Theta_0) \cos(\Phi - \Phi_0) \quad (6)$$

where v = velocity of the aircraft,

T = time interval between pulses

Θ, Φ = azimuth and elevation angles respectively of the z axis or center of the effective beam,

Θ_0, Φ_0 = azimuth and elevation angles respectively of the direction of motion.

The angles Θ and θ_0 are measured from some arbitrary azimuth angle, for example, the heading of the aircraft. The angles Φ and Φ_0 are measured downward from the horizon. For horizontal motion Φ_0 is zero. The value of Φ is:

$$\phi = \tan^{-1} \frac{h}{G} \tag{7}$$

where h = altitude

G = ground range corresponding to the chosen point on the radar scope.

The function $F(\theta, \phi)$ which expresses the phase shift produced by a given antenna displacement is:

$$F(\theta,\phi) = \frac{4\pi}{\lambda} [X \sin\theta + Y \cos\theta \sin\phi + Z (\cos\theta \cos\phi - 1)] \tag{8}$$

where λ is the wavelength, θ and ϕ are the angles of arrival with respect to the Z axis as previously defined and X, Y, and Z are the displacements as described above. It is assumed that all reflecting objects are at a great distance from the antenna.

If it is assumed that the beam widths are small, Equation 8 may be replaced by the following approximation:

$$F(\theta,\phi) = \frac{4\pi}{\lambda} [X \theta + Y \phi - Z (\theta^2 + \phi^2)] \tag{9}$$

The Z term is usually negligibly small, but is retained in order to take care of cases in which the coordinate system is oriented in such a way that X and Y are zero. This occurs at long range along the ground track.

When Equation 9 is substituted into Equation 3, with the provision that the beam widths are narrow, Equation 3 may be separated approximately into four terms as follows:

$$\epsilon = \epsilon_r + \epsilon_x + \epsilon_y + \epsilon_z \tag{10}$$

where

$$\epsilon_r = \frac{\frac{1}{2}\int_{-\infty}^{+\infty} | A(\theta+\frac{1}{2}\omega_a T)-A(\theta-\frac{1}{2}\omega_a T) |^2 \, d\theta}{\int_{-\infty}^{+\infty} |A(\theta)|^2 \, d\theta} \tag{11}$$

$$\epsilon_x = \frac{2\int_{-\infty}^{+\infty} |A(\theta)|^2 \sin^2\left(\frac{2\pi X\theta}{\lambda}\right) d\theta}{\int_{-\infty}^{+\infty} |A(\theta)|^2 \, d\theta} \tag{12}$$

$$\epsilon_y = \frac{2\int_{-\infty}^{+\infty} |B(\phi)|^2 \sin^2\left(\frac{2\pi Y\phi}{\lambda}\right) d\phi}{\int_{-\infty}^{+\infty} |B(\phi)|^2 \, d\phi} \tag{13}$$

$$\epsilon_z = \frac{2\int_{-\infty}^{-\infty}\int_{-\infty}^{+\infty} |A(\theta)\, B(\phi)|^2 \sin^2\left(\frac{2\pi Z}{\lambda}(\theta^2+\phi^2)\right) d\theta d\phi}{\int_{-\infty}^{+\infty}\int_{-\infty}^{+\infty} |A(\theta)\, B(\phi)|^2 \, d\theta d\phi} \tag{14}$$

These are the same expressions which one obtains by specializing Equation 3, first to rotation only, then to X displacement only, etc. Thus, to a first approximation, the various components of motion contribute independently to the mean square value of uncancelled ground clutter.

By making use of Parseval's theorem[3], the expressions for ϵ_r, ϵ_x, and ϵ_y, may be written in alternative forms. Alternative expressions for ϵ_r and ϵ_x, are:

$$\epsilon_r = \frac{2\int_{-\infty}^{+\infty} |F(\beta)|^2 \sin^2(\pi\,\omega_a T\beta)\, d\beta}{\int_{-\infty}^{+\infty} |F(\beta)|^2 \, d\beta} \tag{15}$$

$$\epsilon_x = \frac{\frac{1}{2}\int_{-\infty}^{+\infty} |F(\beta-\frac{X}{\lambda}) - F(\beta+\frac{X}{\lambda})|^2 \, d\beta}{\int_{-\infty}^{+\infty} |F(\beta)|^2 \, d\beta} \tag{16}$$

[3]The required identity may be found in Campbell and Foster, "Fourier Integrals for Practical Applications", D. Van Nostrand Company, 1948, Footnote page 39.

where $F(\beta)$ is the Fourier transform of $A(\theta)$.

Instead of starting with the antenna patterns, one can equally well start with an assumed aperture illumination function. $F(\beta)$ may be obtained by convolution of the aperture function as follows:

$$F(\beta) = \int_{-\infty}^{+\infty} G(\gamma)\, G(\beta - \gamma)\, d\gamma \tag{17}$$

where G = function describing aperture field strength,

γ = horizontal distance from center of the aperture in terms of wavelengths.

The ϵ_r term describes the effect of scanning while the ϵ_x term describes the effect of motion perpendicular to the center of the antenna beam. It may be noted that there is a symmetrical relationship between the scanning and the motion effects. This symmetry may be made more evident by the following change of variables. In the expressions for ϵ_r let:

$$\theta = \omega_a t \tag{18}$$

$$\beta = \frac{f}{\omega_a} \tag{19}$$

In the expressions for ϵ_x, let:

$$\beta = \frac{2v_x t}{\lambda} = \frac{2vt \sin(\theta - \theta_0)}{\lambda} \tag{20}$$

$$\theta = \frac{\lambda f}{2v_x} = \frac{\lambda f}{2v \sin(\theta - \theta_0)} \tag{21}$$

The variable t may be interpreted physically as time, and the variable f may be interpreted physically as frequency. Then as functions of time, the expressions are:

$$\epsilon_r = \frac{\frac{1}{2} \int_{-\infty}^{+\infty} \left| A\left(\omega_a \left(t + \frac{T}{2}\right)\right) - A\left(\omega_a \left(t - \frac{T}{2}\right)\right) \right|^2 dt}{\int_{-\infty}^{+\infty} |A(\omega_a t)|^2\, dt} \tag{22}$$

$$\epsilon_x = \frac{\frac{1}{2}\int_{-\infty}^{+\infty} \left| F\left(\frac{2v_x}{\lambda}\left(t-\frac{T}{2}\right)\right) - F\left(\frac{2v_x}{\lambda}\left(t+\frac{T}{2}\right)\right)\right|^2 dt}{\int_{-\infty}^{+\infty} \left| F\left(\frac{2v_x t}{\lambda}\right)\right|^2 dt} \quad (23)$$

As functions of frequency, the expressions are:

$$\epsilon_r = \frac{2\int_{-\infty}^{+\infty} \left| F\left(\frac{f}{\omega_a}\right)\right|^2 \sin^2(\pi Tf)\, df}{\int_{-\infty}^{+\infty} \left| F\left(\frac{f}{\omega_a}\right)\right|^2 df} \quad (24)$$

$$\epsilon_x = \frac{2\int_{-\infty}^{+\infty} \left| A\left(\frac{\lambda f}{2\, v_x}\right)\right|^2 \sin^2(\pi Tf)\, df}{\int_{-\infty}^{+\infty} \left| A\left(\frac{\lambda f}{2\, v}\right)\right|^2 df} \quad (25)$$

The functions which are integrated in a typical case are illustrated in Figures 4 and 5. Using time as a variable, one has in the scanning case the difference between two displaced antenna patterns, while in the motion case one has the difference between two displaced aperture functions. Using frequency as the variable, one has in each case a sine function which represents the effect of delaying one signal and subtracting it from another, and another function which represents the spectrum of the signal fluctuations. In the motion case the spectrum has the shape of the antenna pattern, while in the scanning case the spectrum has the shape of the transform of the antenna pattern. It is of interest to note that for any aperture of finite width, a, the frequency spectrum for scanning fluctuations lies entirely below a given frequency,

$$f_{max} = \frac{a\,\omega_a}{\lambda} \quad (26)$$

RESULTS

Expressions for ϵ_r , ϵ_x , and ϵ_y have been worked out assuming the following azimuth antenna patterns (round trip):

Rectangular

$$F(\theta) = \begin{cases} 0 & \theta < -\tfrac{1}{2}\Delta\theta \\ 1 & -\tfrac{1}{2}\Delta\theta < \theta < +\tfrac{1}{2}\Delta\theta \\ 0 & +\tfrac{1}{2}\Delta\theta < \theta \end{cases} \tag{27}$$

Gaussian

$$F(\theta) = \exp\left[-2\left(1.777 \frac{\theta}{\Delta\theta}\right)^2 \right] \tag{28}$$

Sin x/x

$$F(\theta) = \left[\frac{\sin\left(2.78 \dfrac{\theta}{\Delta\theta}\right)}{\left(2.78 \dfrac{\theta}{\Delta\theta}\right)} \right]^2 \tag{29}$$

where $\Delta\theta$ = azimuth beam width measured between half power points on the one way pattern.

These functions may also be used in the elevation direction. However, since the elevation weighting function depends on pulse shape, the rectangular and the Gaussian functions may be suitable, but the sin x/x function is not appropriate. The effective elevation beamwidth is given approximately by

$$\Delta\phi = \frac{\tfrac{1}{2}c\tau \tan\phi \sin\phi}{h} \tag{30}$$

where $\Delta\phi$ = effective elevation beamwidth

c = velocity of light

τ = pulse duration

h = altitude

ϕ = depression angle as previously defined.

The following symbols are used:

$$N = \frac{\Delta\theta}{\omega_a T} \tag{31}$$

$$M_x = \frac{\pi X \Delta\theta}{\lambda} = \frac{\pi v T \Delta\theta \sin\theta}{\lambda} \tag{32}$$

$$M_y = \frac{\pi Y \Delta\phi}{\lambda} = \frac{\pi v c T \tau \cos\theta \sin^2\phi \tan\phi}{\lambda h} \tag{33}$$

Equations for ϵ_r

Rectangular Pattern $\epsilon_r = \frac{1}{N}$ (34)

Gaussian Pattern (35)

$$\epsilon_r = 1 - e^{-(1.177/N)^2} = \left(\frac{1.177}{N}\right)^2 - \frac{1}{2!}\left(\frac{1.177}{N}\right)^4 + \frac{1}{3!}\left(\frac{1.177}{N}\right)^6 - \ldots$$

Sin x/x Pattern

$$\epsilon_r = 1-6\left(\frac{N}{2\cdot 2.78}\right)^2 \left(1 - \frac{\sin\left(\frac{2\cdot 2.78}{N}\right)}{\left(\frac{2\cdot 2.78}{N}\right)}\right) \tag{36}$$

$$= 6\left[\frac{1}{5!}\left(\frac{2\cdot 2.78}{N}\right)^2 - \frac{1}{7!}\left(\frac{2\cdot 2.78}{N}\right)^4 + \frac{1}{9!}\left(\frac{2\cdot 2.78}{N}\right)^6 - \ldots\right] \tag{37}$$

Equations for ϵ_x

Rectangular Pattern (38)

$$\epsilon_x = 1 - \frac{\sin 2M_x}{2M_x} = \frac{1}{3!}(2M_x)^2 - \frac{1}{5!}(2M_x)^4 + \frac{1}{7!}(2M_x)^6 - \ldots$$

Gaussian Pattern (39)

$$\epsilon_x = 1 - e^{-(Mx/1.177)^2} = \left(\frac{M_x}{1.177}\right)^2 - \frac{1}{2!}\left(\frac{M_x}{1.177}\right)^4 + \frac{1}{3!}\left(\frac{M_x}{1.177}\right)^6 - \ldots$$

Sin x/x Pattern

$$\epsilon_x = \begin{cases} 6\left(\frac{M_x}{2.78}\right)^2\left(1 - \frac{|M_x|}{2.78}\right) & 0 < \frac{|M_x|}{2.78} < \frac{1}{2} \quad (40) \\ 1 - 2\left(1 - \frac{|M_x|}{2.78}\right)^3 & \frac{1}{2} < \frac{|M_x|}{2.78} < 1 \quad (41) \\ 1 & 1 < \frac{|M_x|}{2.78} \quad (42) \end{cases}$$

$$\epsilon_x = \begin{cases} 6\left(\frac{x}{a}\right)^2 \left(1 - \frac{|x|}{a}\right) & 0 < \frac{|x|}{a} < \frac{1}{2} \quad (43) \\ 1 - 2\left(1 - \frac{|x|}{a}\right)^3 & ½ < \frac{|x|}{a} < 1 \quad (44) \\ 1 & 1 < \frac{|x|}{a} \quad (45) \end{cases}$$

where a = width of uniformly illuminated aperture.

Equations for ϵ_y are the same as those for ϵ_x except that M_x is replaced by M_y. Also the sin x/x case is not appropriate for ϵ_y

The last component, ϵ_z., is usually negligible. Its value assuming a rectangular pattern is:

$$\epsilon_z = \frac{1}{10}\left(\frac{\pi Z (\Delta\theta)^2}{\lambda}\right)^2 - \frac{1}{64 \cdot 54}\left(\frac{\pi Z (\Delta\theta)^2}{\lambda}\right)^4 + \ldots \quad (46)$$

The power series expressions are more convenient for numerical calculations than the closed form expressions. Usually the first term of the series will give the result with sufficient accuracy. It may be noted that, for the most part, the coefficients are roughly the same regardless of the assumed shape of the beam.

At any point on the scope, some one of the components is likely to be larger than the others. Figure 6 indicates the regions where each component is likely to predominate and also lists typical first terms in a form which shows the angular dependence. The X term is important in the regions to the side. The Y component becomes of importance along the ground track where the X component goes to zero. It is large however only where the depression angle is large and is therefore important at high altitudes. The scanning component is dependent on the rate of rotation of the antenna and is independent of the scope position. This component may limit the performance at long range and along the ground track where the X and Y components both become small. In this region also, the Z component may be appreciable.

It is of interest that the X component, which generally represents the largest fluctuations in the case of airborne radar, can be expressed as a function of x displacement divided by aperture width. For example, using Equation 43, one finds that to attenuate ground clutter at 90 degrees from the ground track by 20 decibels, the distance travelled by the aircraft between pulses must not be greater than 4% of the antenna width. This is true regardless of the wavelength or beamwidth.

ACKNOWLEDGEMENT

The work reported on was supported in part by the U.S. Army Signal Corps, Evans Signal Laboratory, under Contract D.A. 36-039-sc-5446, and in part by the General Electric Company. The author wishes to express special thanks to Mr. Martin Santa, of the General Electric Company, for his help in formulating the mathematics presented here.

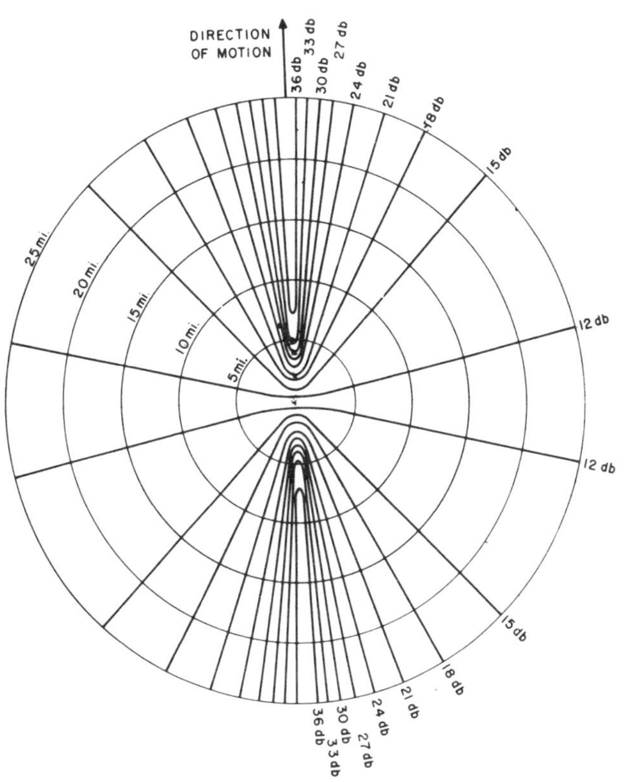

BEAM WIDTH 3.0 DEGREES PULSE LENGTH 3/4 μSEC.
GROUND SPEED 250 KNOTS PRF 2000
ALTITUDE 20,000 FEET WAVELENGTH 3.2 CM.
ANTENNA ROTATION 12 RPM

FIG. 1 TYPICAL GROUND CLUTTER ATTENUATION FOR AIRBORNE MTI

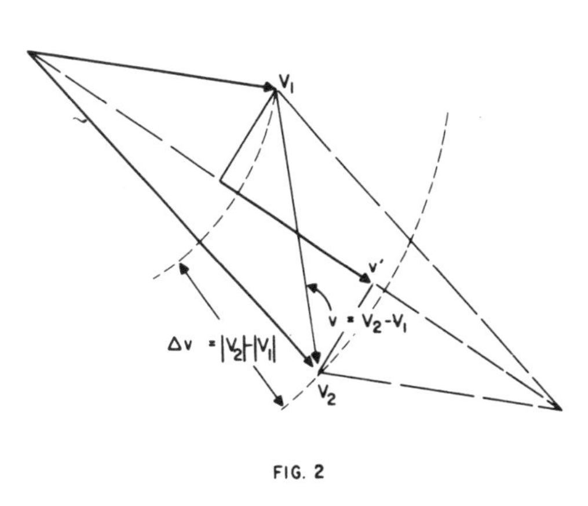

FIG. 2

VECTOR DIAGRAM SHOWING NEAR EQUALITY BETWEEN LENGTHS OF $|V_2| - |V_1|$ AND THE COMPONENT OF $V_1 - V_2$ IN PHASE WITH $V_1 + V_2$

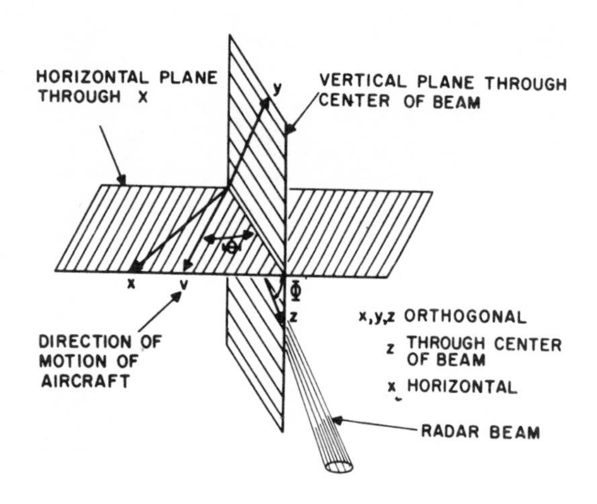

FIG. 3 - LOCATION OF COORDINATE AXES

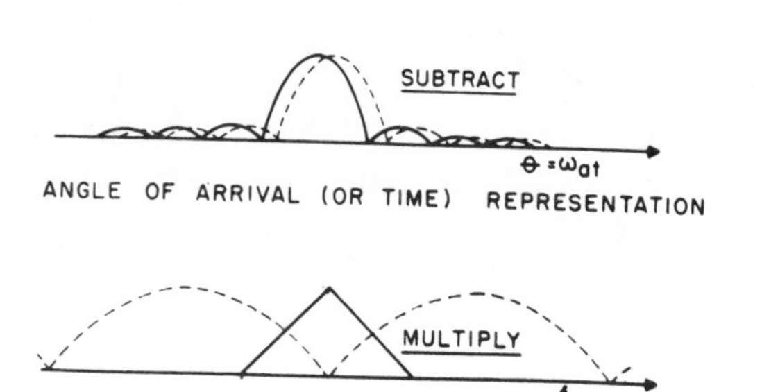

FIG. 4 SCANNING EFFECTS

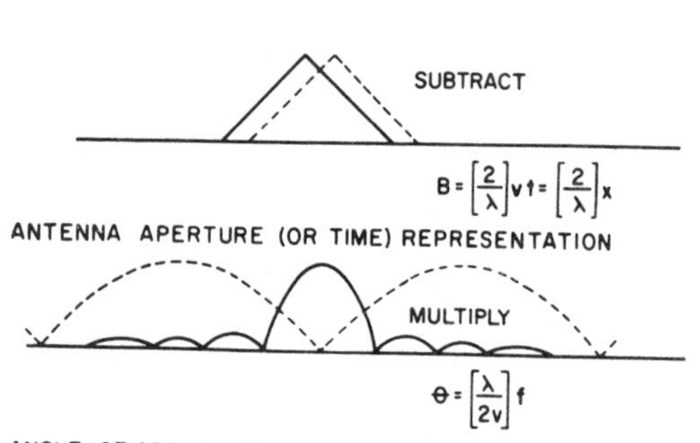

FIG. 5 MOTION EFFECTS

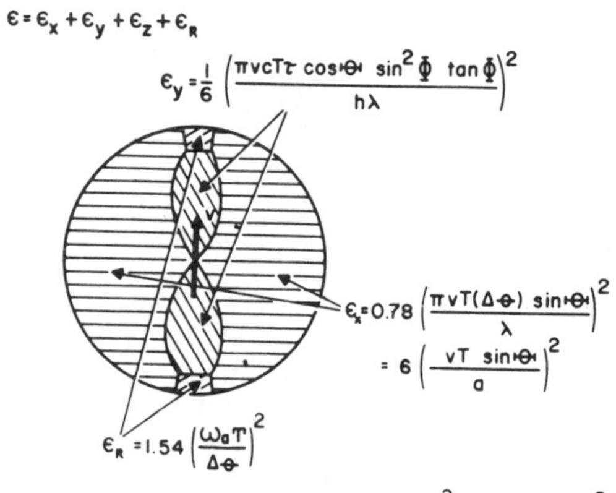

FIG. 6 - COMPONENTS OF MEAN SQUARE RESIDUAL GROUND CLUTTER AND REGIONS WHERE EACH COMPONENT IS LIKELY TO BE LARGEST

7.2

AEW RADAR ANTENNAS

T. ap Rhys and G. Andrews

AGARD Conference Preprint No. 139, Antennas for Avionics, 1974, pp. 12-1 – 12-16.

ABSTRACT

The environment in which Airborne Early Warning (AEW) radars must be able to operate, in the future, presents some severe problems to the designer in many important respects. These include the need to operate over land, the need to detect very small targets (<$1m^2$) at long range and the need to be able to operate in the presence of jamming.

It is shown that the single most important factor influencing the ability of the AEW radar system to operate in this future environment is the antenna's sidelobe performance. In general, this sidelobe performance is dominated by the airframe on which it is mounted. Careful design is needed, considering both aerodynamic as well as electrical aspects, before an overall optimal system configuration can be determined.

Once the antenna sidelobe performance has reached an acceptable level, it is then found that the next limitation is that due to errors in the compensation of platform motion. Recent advances in digital processing technique offer many advantages in connection with moving-target indication (MTI) but these can only be realized when the platform-motion compensation is sufficiently accurate. The various system considerations governing this performance are discussed and the requirements reflecting on the antenna performance described in detail.

1. Introduction

A paper[1] was presented to the Eighth Symposium of the Avionics Panel in 1964 which dealt with some of the basic considerations in the design of an airborne early-warning radar, with particular reference to that used on the E-2A carrier-based system. This used a low prf mainly because of its feature of unambiguous range. It is of interest here to examine the applicability of this type of technology to a larger aircraft, specifically a modified Lockheed Electra turbo-prop, designated as the P-3 by the U.S. Navy.

Of great importance in such a radar is the detection performance and this is largely determined by that of its AMTI sub-system. This, in turn, is critically dependent on the effectiveness of the platform motion compensation. What is in mind is a system in which the velocity components normal to the antenna aperture are corrected by some variant of TACCAR* and these components parallel to the antenna aperture by some technique similar to DPCA**.

It is of interest to determine what are the ultimate bounds on the detection performance of such a system preparatory to quantifying their effects and establishing an optimal design. It is well known that certain types of terrain give rise to large clutter returns which can effectively mask the returns from small targets and prevent their detection. Coherent integration, multiple-delay cancellation and improved system characteristics can be used to improve the detection performance, but each exacts a price in system-cost so that the problem reduces to one of determining the improvement in performance and its associated cost in hardware for each addition. It is shown below that the antenna sidelobe performance has a crucial effect on the system performance, in general, and determines the lowest operating frequency that can be used.

2. AMTI Considerations

MTI improvement factor is defined as the ratio of the output signal-to-clutter ratio to the input signal-to-clutter ratio. It has been derived in Ref. 2 for single, double, and triple-delay cancellers and elsewhere for an n-stage MTI (Ref. 3). Figure 1 shows the MTI improvement factor (I_n) as a function of the ratio of the clutter spectral width (σ_c) to the pulse repetition frequency (PRF). The system characteristics which limit the MTI improvement factor can be related to an effective clutter spectral width and to a shift in the mean of the spectrum. The total clutter spectral width is given by taking the square root of the sum of the squares of all the individual spectral widths. This assumes statistical independence between the individual contributions.

*TACCAR is an acronym for Time Average Clutter Coherent Radar, A system developed by Lincoln Laboratory of MIT. By general usage TACCAR now refers to the technique which was the special feature of that system.

**DPCA is the acronym for Displaced Phase Center Antenna, a technique which simulates electronically the motion of the antenna aperture in its own plane.

The achievable MTI improvement factor for an operating AEW system is limited by the internal motion of the clutter itself and also by system considerations. Specifically, these limitations arise from:

1. Platform motion
2. Antenna sidelobe levels
3. Antenna scanning modulation
4. System stability

To determine the relative importance of each of these contributions, they are considered separately and illustrated in Fig. 1.

A typical value for the spectral width due to internal motion of overland clutter is 0.2 meters/second (Ref. 4, Chapter 17). Figure 1 shows that the MTI improvement factor for this condition is 40 dB for a single canceller, 78 dB for a double canceller, and 114 dB for a triple canceller. If the system were limited by internal motion then all practical requirements could be met easily by the use of a double canceller.

Platform motion can be resolved into two components, one parallel to the antenna aperture and one perpendicular to the antenna aperture. These components have the effect of both spreading the spectrum of the clutter returns and shifting the mean of the spectrum. The relationship for the spreading of the spectrum is given in Ref. 4, Chapter 18. Using typical parameters (a platform velocity of 300 knots, an aperture of 25 feet and a prf of 300 Hz), Fig. 1 shows that the MTI improvement factor is about 15 dB for a single canceller, 25 dB for a double canceller and 35 dB for a triple canceller with the antenna aperture parallel to the velocity vector. For this case, many practical requirements cannot be met with either a single, double or triple canceller. Rather than go to higher order cancellers, other processing techniques should be employed.

With the antenna aperture perpendicular to the velocity vector, the dominant effect is the shifting of the mean of the clutter spectrum. The mean of the doppler spectrum is determined by the platform velocity and the transmitted frequency and generally requires compensation for successful MTI operation.

Since the antenna sidelobes essentially cover 360°, the doppler shifts received through the sidelobes range from $-v_p$ (v_p is the platform velocity) to $+v_p$. This doppler spread can cover the entire MTI ambiguous frequency band. As a result, these sidelobe returns are not cancelled and the MTI improvement factor is limited by the integrated mainlobe-to-sidelobe ratio. This is the ratio of the integrated energy entering the system through the mainlobe to that entering through the sidelobes for the two-way antenna pattern.

The effect of antenna scanning modulation on the clutter spectral width is given in Ref. 2. It is affected by the antenna scanning, the antenna aperture, the transmitter frequency, and the PRF. Most requirements can be met without compensation for this effect.

System stability is greatly simplified by digital processing which allows one clock to control the entire system including the MTI delays. However, the pulse-to-pulse phase instability of the transmitter is not eliminated by digital processing, so that the performance of the system could be limited by the transmitter stability. Excluding the transmitter, a clock stability of 10^{-8} is within the state of the art. The corresponding MTI improvement factor (referring to Fig. 1) is 22 dB for an single canceller, 42 dB for double canceller, and 62 dB for a triple canceller. Again, most requirements can be met with one or other of these cancellers.

The conclusion is that the factors which limit the improvement factor are antenna sidelobes and platform motion.

It has been seen that unless a correction is made for the shift in the mean of the clutter spectrum, then MTI is impossible. TACCAR makes this correction by phase-locking the system to the clutter returns. A phase-lock loop is gated on at a selected range interval. The time constant of this phase-lock loop must be large compared with the interpulse period in order to maintain pulse-to-pulse coherency needed for successful MTI action. This results in a single correction for all ranges. However, the average doppler changes with range since the angle of arrival varies with range for an aircraft flying at some altitude above the earth's surface.

The effect of this incomplete correction results in a smaller MTI improvement factor. This decrease in the improvement factor is derived elsewhere (Ref. 3). It is a function of the ratio of the correction error and the clutter spectral width. For a typical set of parameters, these losses due to TACCAR are illustrated by Fig. 2 when the antenna is pointed in the direction of the aircraft velocity vector.

DPCA is a technique that compensates for the component of aircraft velocity parallel to the antenna aperture. This is accomplished by either physically or electrically displacing the phase center of the antenna in the direction opposite to the velocity component. A detailed discussion of DPCA, on which the following is based, is given elsewhere (Ref. 5).

A single canceller MTI can be almost perfectly compensated by DPCA with essentially no loss. It can be shown (Ref. 5) that for double, triple or higher-order cancellers, only the first canceller is compensated. This results in an imperfect compensation so that the effects of platform motion are not completely eliminated. This is illustrated in Fig. 3 for a typical set of parameters.

The MTI improvement factor for a single canceller with DPCA coincides with the perfectly compensated curve. The improvement factors for double and triple cancellers are much less than the perfectly compensated curves.

If the system stability can be increased from 10^{-8} to 10^{-9}, then the improvement factor will be limited by scanning modulation. From Fig. 3, a UHF system is limited by scanning modulation and DPCA losses when the antenna is pointed broadside to about 30 dB for a single canceller, 50 dB for a double canceller and 62 dB for a triple canceller.

The use of coherent integration can ease the requirements on the AMTI (Ref. 6). This technique corresponds to doppler filtering as used in a pulsed-doppler radar, except that for a low-PRF radar only a very limited number of pulses can be integrated before the return from a moving target moves out of the range cell. Typically sixteen pulses can be integrated. This corresponds to a theoretical integration gain of 12 dB for a perfectly coherent signal in "white" noise. However, neither of these conditions applies. First, a perfectly coherent signal (or target return) implies that the doppler of the return is such that its frequency is at the center of one of the filters in the bank of filters formed by the coherent integration. Second, the clutter residue from the MTI is not "white". The problem then is to determine how much gain in signal-to-clutter ratio can be expected from this technique.

The integration of 16 pulses with no weighting provides 16 contiguous filters covering the frequency band from zero to the pulse repetition frequency. With no weighting, the filters have a (sin x)/x frequency response. Since the doppler of the target returns are not known, they are assumed to have a uniform probability of occurring anywhere in a particular filter between the cross-over points of adjacent filters. Averaging the signal gain over this band results in a loss in coherent integration gain of about 1.1 dB for "white" noise. If weighting functions are used to get some desired filter sidelobe level additional losses occur depending on the weighting function. Therefore, a net gain of about 10 dB could be expected for "white" noise.

Since the clutter residue from the antenna sidelobes cover all frequencies in the band from zero to the pulse repetition frequency, this 10 dB gain can be applied to this residue as well as other components of the clutter residue. Reference 6 shows that the improvement factor achieved against mainlobe clutter (which is not white) is much more than 10 dB. It is typically about 24 dB depending somewhat on the type of MTI used.

As will be shown below, the antenna sidelobes can be reduced when the transmitter frequency is increased. However, this solution affects the doppler processing. Over land, at L-band with a system stability factor of 10^{-9} the effects of internal motion, scanning modulation, and system stability are illustrated by Figs. 3 and 4. Referring to these figures it can be seen that the effect of scanning modulation dominates the other effects. For a double canceller these limitations are about 25 dB for scanning modulation and TACCAR losses (antenna at + 90°). If the TACCAR losses and the DPCA losses could be eliminated, then the limitation to the MTI improvement factor caused by scanning modulation at L-band is about 55 dB for a triple canceller (see Fig. 4). Reduction of these losses, are discussed in Refs. 3 and 5, consist of making TACCAR corrections at additional points in range and applying a DPCA correction to all cancellers.

In summary, to insure adequate performance, a typical set of system specifications would be:

1. Integrated mainlobe-to-sidelobe ratio - 60 dB (two-way)
2. System stability factor - 10^{-9} (pulse-to-pulse)
3. Transmitter stability factor - 10^{-9} (pulse to pulse)
4. Triple canceller MTI
5. 16-pulse coherent integration
6. TACCAR correction at 3 points in range
7. DPCA correction of all 3 cancellers

3. Antenna Considerations

It is helpful (Ref. 4) to separate the improvement factor associated with the antenna (I_A) into two parts -- that due to the energy entering through the sidelobes (I_{SL}) and that due to imperfect platform motion compensation over the mainlobe (I_{DPCA}).

$$1/I_A = 1/I_{SL} + 1/I_{DPCA}$$

The improvement factor determined by the energy entering through the sidelobes is equal to the integrated mainlobe-to-sidelobe ratio.

$$I_{SL} = \int_{-\pi}^{\pi} E^4(\theta)\, d\theta \;/\; \int_{SL} E^4(\theta)\, d\theta$$

Consequently it is directly related to the antenna performance when the antenna is mounted on the aircraft and in independent of other system factors. The mainlobe-to-sidelobe ratio is a quality factor for the antenna when used in this type of radar.

The improvement factor associated with the motion compensation over the arc corresponding with the beamwidth of the mainlobe is dependent on the shape of the sum and difference mainlobe patterns but also on the signal processing used. It is susceptible to improvement by the use of higher-order cancellation and motion compensation and so, in principle, its influence on system performance can be made negligible.

In general, when a large radar antenna is mounted on an aircraft, there is a deterioration in performance. This becomes evident in increased sidelobe levels resulting in decreased values of mainlobe-to-sidelobe ratio and also in distortion of the mainlobe patterns resulting in poor compensation of the velocity components parallel to the antenna aperture. These effects arise from the presence of the aircraft structure close to the antenna. Three ways in which the antenna field can be perturbed are (a) blockage of the antenna aperture by parts of the aircraft such as the wings and tails, (b) scattering from surfaces such as the tops of the wings, and (c) excitation of currents in the aircraft's surface by the near fields of the antenna. All three of these sources of pattern perturbation tend to decrease in effect with increasing frequency whereas it is advantageous otherwise to decrease the operating frequency. Consequently, for optimal design, it is necessary to determine the lowest frequency at which the aircraft perturbations will permit the required improvement factor to be achieved.

A great deal depends on the aircraft design. To minimize blockage and scattering the wing-tips, nose and tail should be so located that their depression angles from the horizontal plane through the antenna should be as large as possible. To minimize direct excitation by the near-fields of the antennas, the distance of the antenna from the nearest conducting surface, expressed in wavelengths, must be maximized.

The P-3 aircraft (see Fig. 5) is an attractive candidate for such an AEW platform, except for the effect of its large tail which bisects the antenna beam as it sweeps around the rear of the aircraft.

In Fig. 5 is pictured a P-3 with a rotodome antenna mounted on it. Such an antenna consists of an antenna array and corporate feed integrated with a radome, the whole structure being rotated mechanically as one unit. Possible frequencies of operation being considered for this system are UHF and L-band.

4. Scale-Model Experiments

A scale-factor of 1/7 was chosen for these experiments. This conveniently put the scale-model frequencies into S-band and X-band. Three scale-model antennas were procured: a ten-element array antenna and a twelve-element array antenna, both operating at S-band and therefore corresponding to full-scale UHF antennas, and a slotted-waveguide array antenna operating at X-band, and therefore corresponding to a full-scale L-band antenna.

A schematic diagram of the ten-element array antenna is shown in Fig. 6. It can be seen that symmetrically placed elements are taken to hybrid couplers, the outputs of which are summed in separate networks to give sum- and difference-pattern output ports. The element weights were chosen to give Dolph-Chebyshev sum patterns, with a peak sidelobe level of -32 dB. The elements themselves are Yagi arrays split in the vertical plane so that the antenna could be considered as a twenty-element array antenna, equally well. The elements are spaced a wavelength apart with grating lobes minimized by shaping of the element patterns.

A 1/7 scale-model of the P-3 aircraft was constructed for these tests and is shown in Fig. 7 mounted on the antenna range. This was a ground-range carefully levelled and cleared so that sidelobe levels can be measured accurately to -60 dB and usefully down to -80 dB. The measured data were recorded directly on magnetic tape to permit processing afterwards using a large-scale digital computer. In general, the antenna patterns were plotted to a convenient format and the integrated mainlobe-to-sidelobe ratio calculated and recorded at the same time. The results for an elevation angle of -2° are shown in Fig. 8. These are plotted on polar graph paper to show the effect of the aircraft geometry on the side-lobe performance. The effect of the tail can be seen to be large. Also shown is the free-space performance of the antenna. It can be seen that the performance did not deteriorate greatly when the antenna was mounted on the aircraft except over the tail. Also shown is the DPCA performance, assuming the use of a double canceller. It can be seen that this actually is inferior to the sidelobe performance near broadside. These same results are presented differently in Fig. 9. This makes clear the basic nature of the sidelobe limitation on the improvement factor.

The elevation characteristics of this model are shown in Fig. 10. The flat behavior is to be noted. The behavior over the nose (0°) is typical but that over the tail (180°) is still reasonably good.

The twelve-element array antenna was carefully designed for good sidelobe performance. Its main-lobe-to-sidelobe ratio in free space is some 10 dB better than that of the ten-element array antenna. It achieved this by reducing the spacing between elements so as to eliminate the grating lobes and by care-ful attention to the detailed design of the central elements. Its schematic diagram is shown in Fig. 11. It can be seen there that only the sum pattern is implemented in this antenna. A photographic view of this antenna is shown in Fig. 12.

The polar diagram measured for this experiment is shown in Fig. 13 to the same reference as that of Fig. 8. The free-space performance is now represented by the center point of the diagram. The general behavior is much improved except over the tail. This can be seen better in Fig. 14. There is now a larger discrepancy between free-space and on-aircraft performance indicating that direct excitation is now becoming a significant factor. If this could be suppressed, excellent performance could be achieved outside the tail region.

Figure 15 confirms the generally flat behavior available in elevation. This means that good performance can be maintained down to short-range.

The slotted-waveguide array was designed to a Dolph-Chebyshev characteristic with a peak sidelobe level of -46 dB. The free-space performance was measured to be somewhat inferior to this and with a mainlobe-to-sidelobe ratio some 20 dB superior to that of the ten-element array antenna. The schematic diagram of this array is shown in Fig. 16. It consists of eight waveguides (sticks) excited uniformly by a single manifold. A view of the completed antenna is shown in Fig. 17.

Some of the results obtained with this antenna are shown in Fig. 18. Characteristics for elevation angles of -2°, -8°, and -12° are shown. The effect of the wing can be observed in the characteristic for -12° and the effects of the nose and engine nacelles can be observed in varying degree in all cases. The effect of the tail is comparable to that observed in the UHF case except that it is restricted to a narrower sector because of the narrower antenna beam.

The elevation characteristics for this L-band experiment are shown in Fig. 19. It can be seen that the free-space performance of this antenna decreases quite rapidly with increasing depression angle. This has been traced to the presence of a two-peaked sidelobe at a wide angle, symmetrically placed around the azimuthal principal plane. The source of this sidelobe has been attributed to an error made in the construction of the manifold. The effect of this error is significant on the on-aircraft characteristics, particularly at the larger depression angles. The characteristics for the 100° look-angle is typical. This deterioration affects performance at the shorter ranges such as 20 miles or so.

Discussion

It was found useful to establish a theoretical background for these experiments by computing the antenna patterns of the various antennas in free space and also when mounted on the aircraft. This latter was done by establishing an aperture at a distance from the aircraft sufficient to clear all obstacles and then to compute the field in this aperture due to each element with suitable provision made for reflection and shadowing by the aircraft structure. These contributions were than summed and the Fourier Transform taken to determine the far-field pattern. It was found from the results of these computations that tolerances had to be met of about 5° in phase and 0.5 dB in amplitude for each element. Also, the major effects over the tail and nose are due to shadowing. Elsewhere at UHF, direction excitation by the near fields of the antenna was evident.

The most serious problem in this system is the poor performance over the tail. This occurs over a narrow sector and may not be to significant operationally. Two ways of dealing with this would be (a) to replace the metallic tail with a dielectric one designed to minimize blockage and reflection, (b) re-design the aerodynamic structure so that the tail does not intrude into the antenna beam. This would be done by adopting a multiple tail or by placing the rotodome on top of an elongated single tail. The latter approach involves too drastic a modification of the aerodynamic design, so that the multiple tail is preferred.

The best solution appears to be the adoption of a multiple-tail design with each component made from dielectric materials in such a way as to minimize reflection and shadowing. This should be sufficient to bring the performance up to an appropriate level.

Conclusions

The detection performance achievable with a low-prf airborne radar is limited by direct excitation of the aircraft at UHF. This limitation is no longer a factor at L-band but care must be taken to minimize reflection and scattering.

References

1. "Radar Techniques for Detection, Tracking and Navigation," Ed. W. T. Blackband, Gorden and Breach Science Publishers, Chapter 12.

2. Barton, D.K., "Radar System Analysis," Prentice-Hall, Englewood Cliffs, N. J., 1964

3. Andrews, G. A., "Airborne Motion Compensation Techniques, Evaluation of TACCAR," NRL Report 7407, April 12, 1972.

4. Skolnik, M. I., "Radar Handbook," McGraw-Hill, N. Y., 1970.

5. Andrews, G. A., "Airborne Motion Compensation Techniques, Evaluation of DPCA," NRL Report 7426, July 20, 1972.

6. Andrews, G. A., "Performance of Cascaded MTI and Coherent Integration Filters in a Clutter Environment," NRL Report 7533, March 27, 1973.

Acknowledgements

This work was supported under Airtask A360-5333/058B/3F12-141-601 under the direction of Naval Air Systems Command. The assistance is acknowledged of Mr. J. Tyzkiewicz (Naval Air Systems Command), Mr. D. L. Ringwalt, Mr. F. M. Staudaher, Mr. Carl Olson (NRL), Mr. D. Lloyd (RCA), Mr. B Sichelstiel, Mr. K. Ramsey and Mr. R. Vogelsang (Westinghouse Electric Corporation).

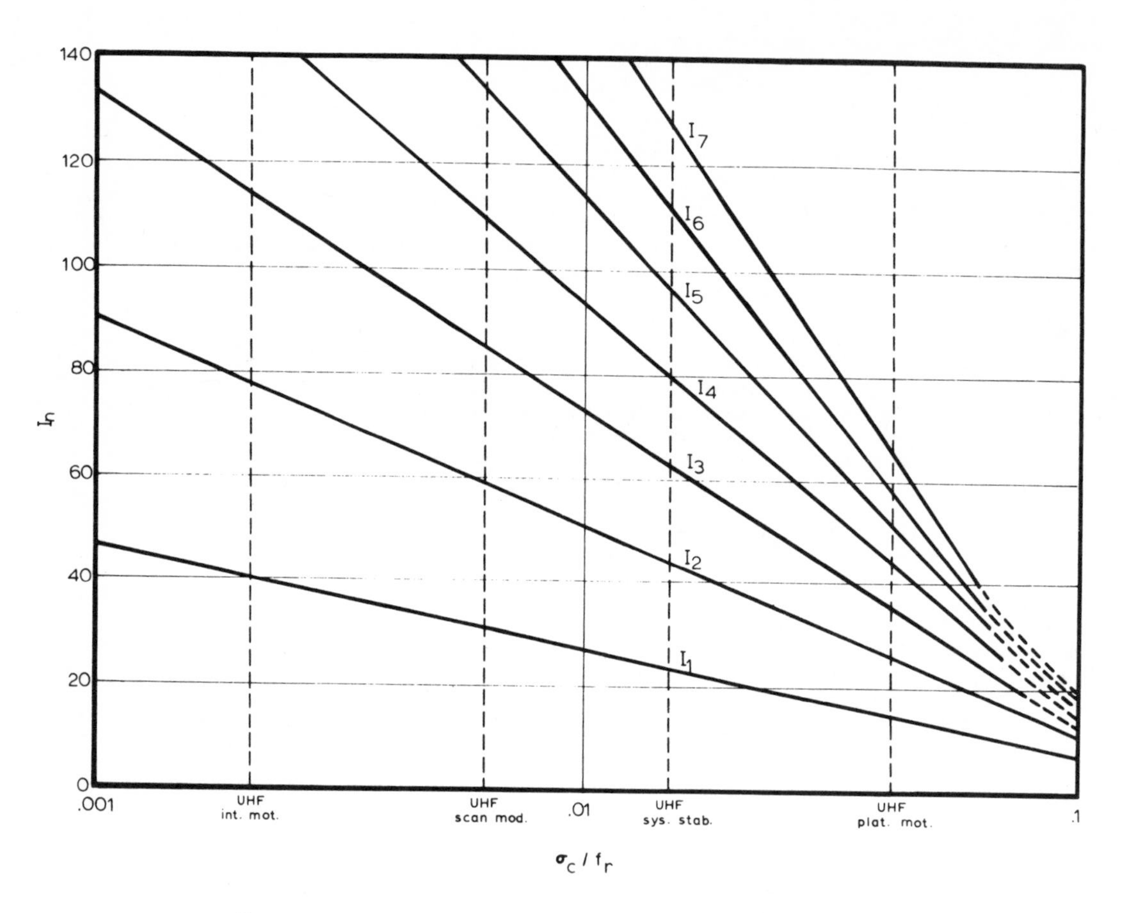

Fig.1 MTI improvement factor (I_n). n = number of delays

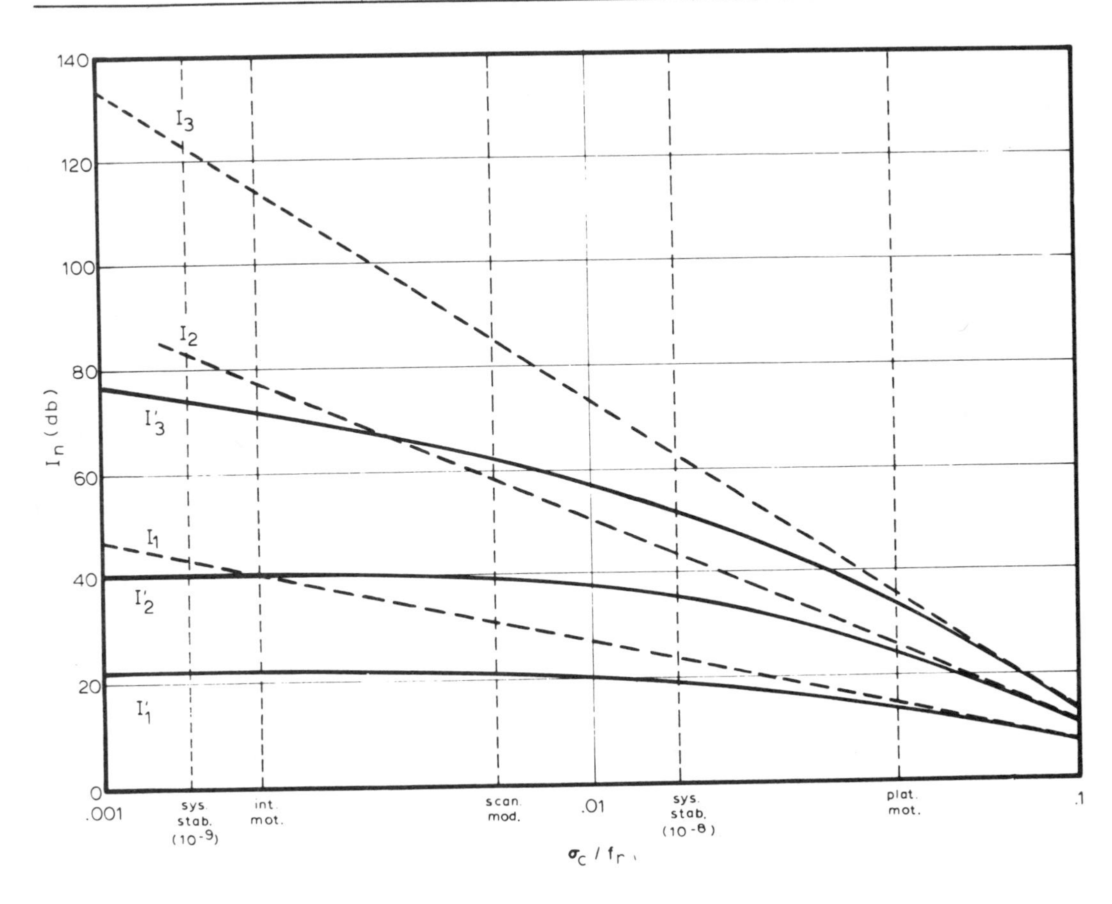

Fig.2 MTI improvement factor with no platform motion (I_n) and with platform motion and taccar losses at UHF (I'_n). n = number of delays

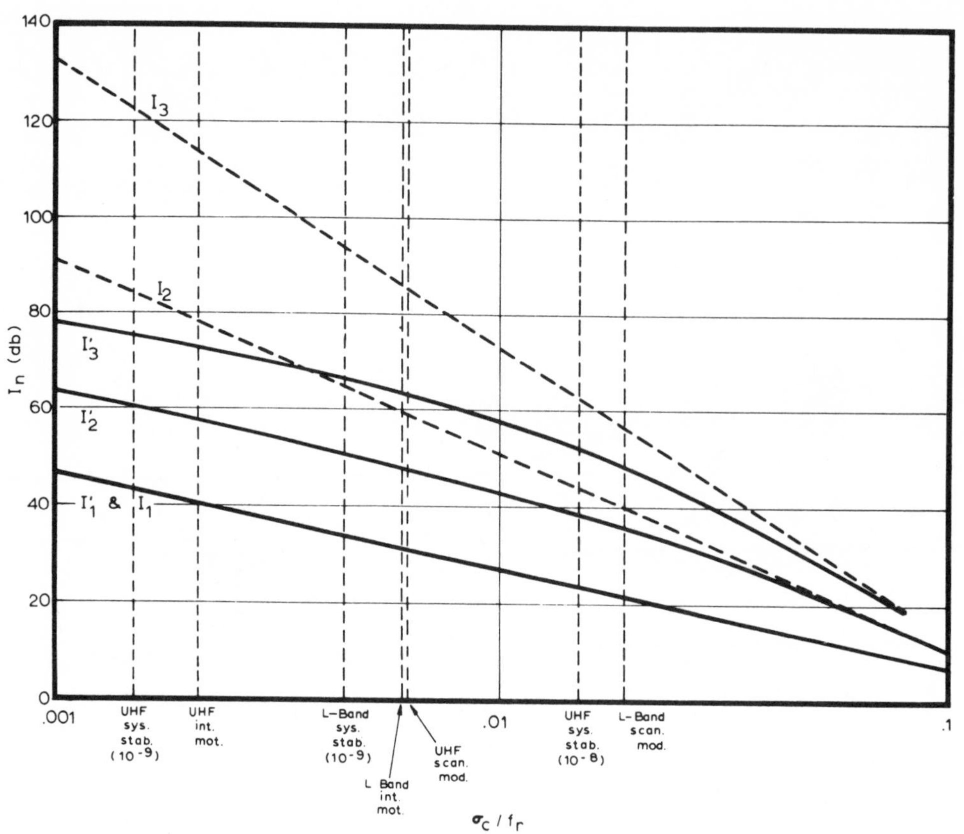

Fig.3 MTI improvement factor with no platform motion (I_n) and with platform motion and DPCA losses (I'_n). n = number of delays

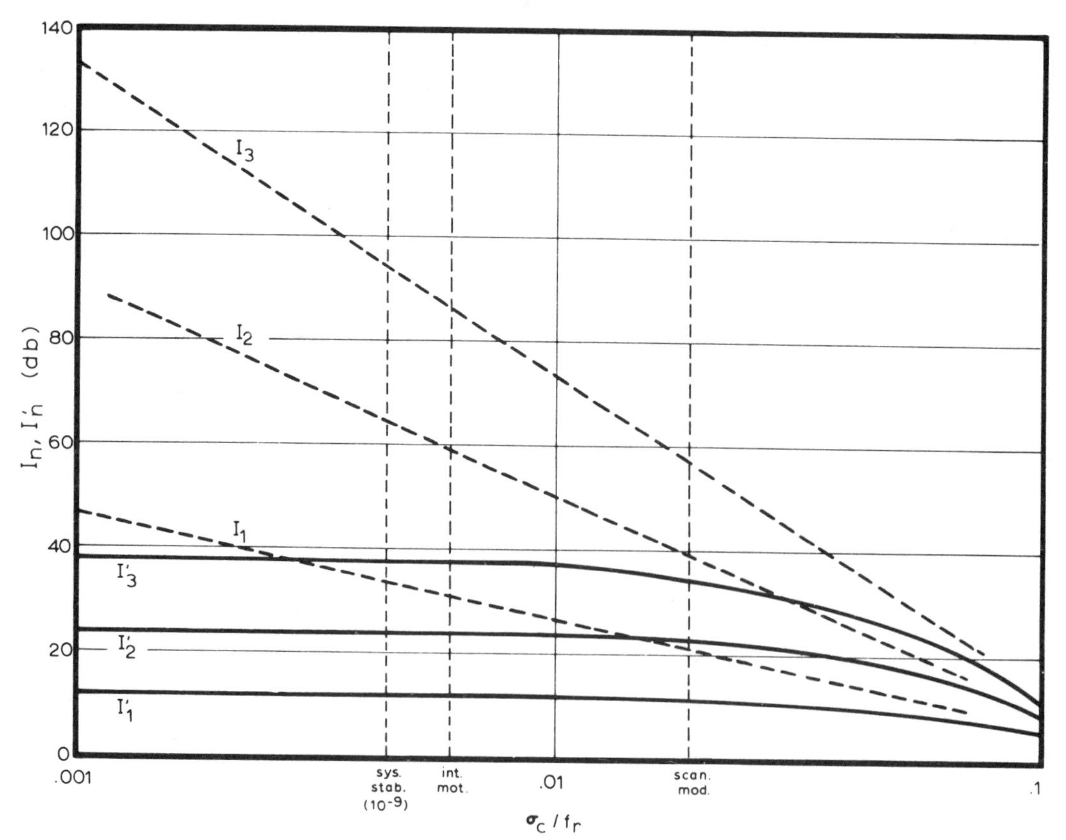

Fig.4 MTI improvement factor with no platform motion (I_n) and with platform motion and taccar losses at L-band (I'_n). n = number of delays

Fig.5 The P-3 aircraft

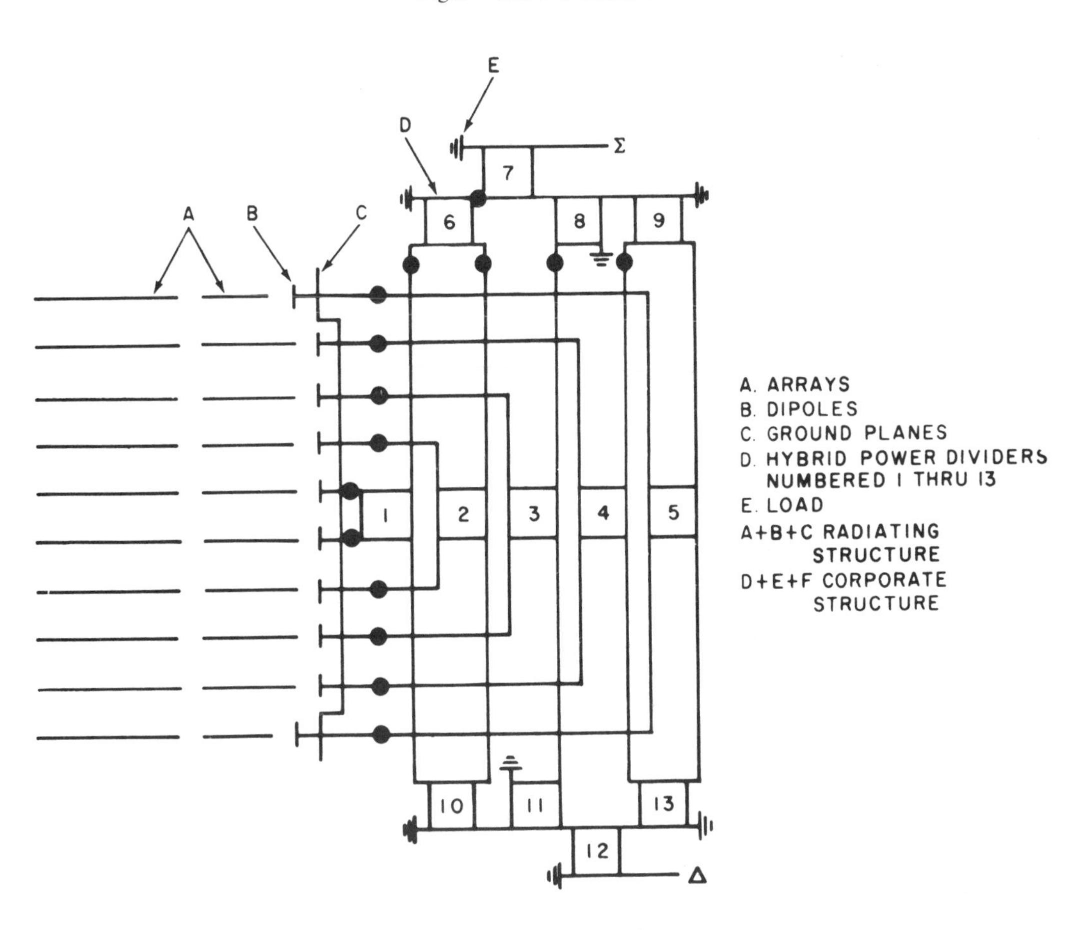

Fig.6 Schematic of the ten-element array

Fig.7 The scale-model of the P-3

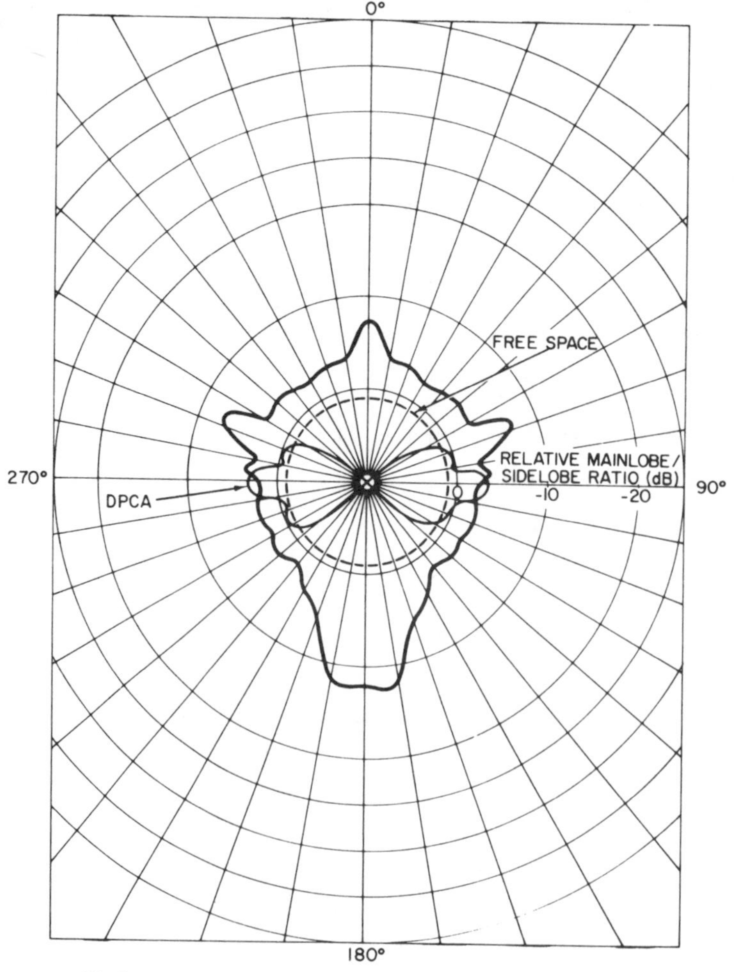

Fig.8 Antenna characteristics (−2°) ten-element array

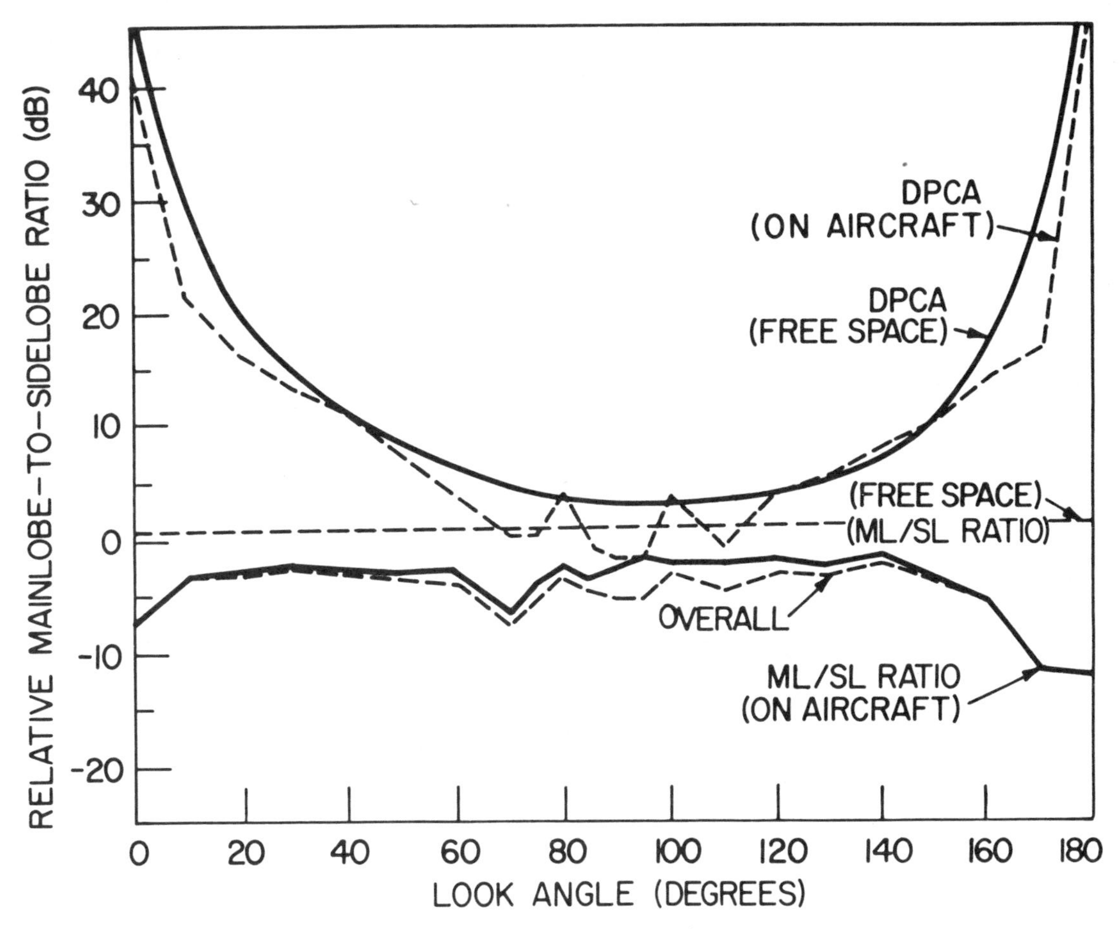

Fig.9 Antenna characteristics: ten-element array

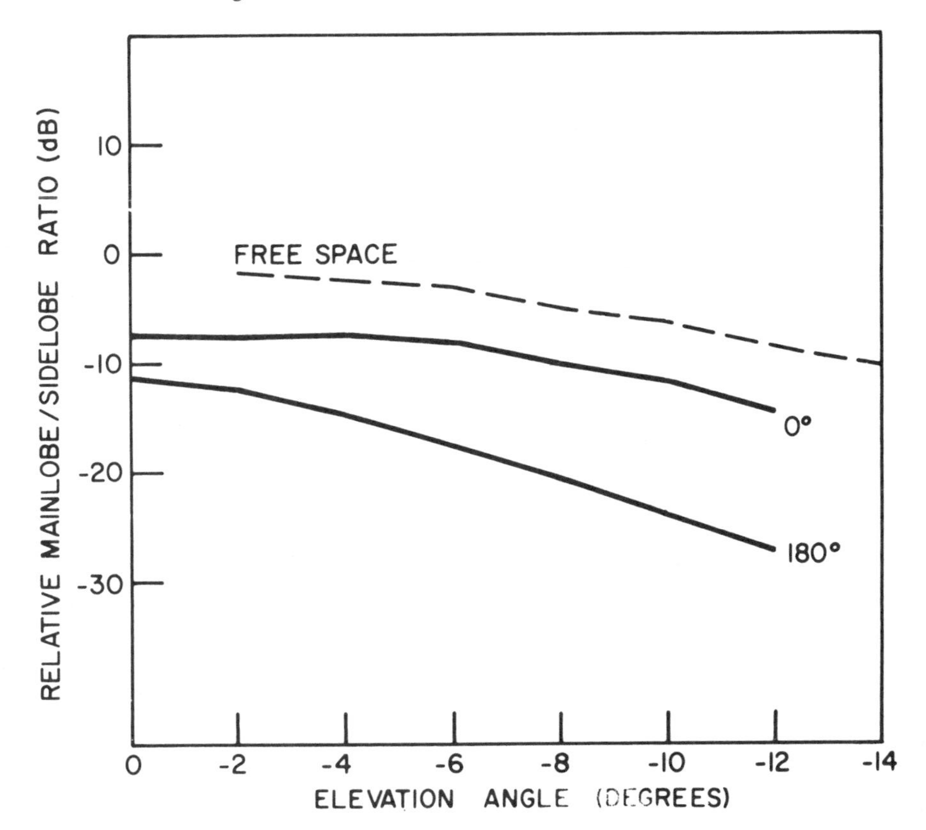

Fig.10 Elevation characteristics: ten-element antenna

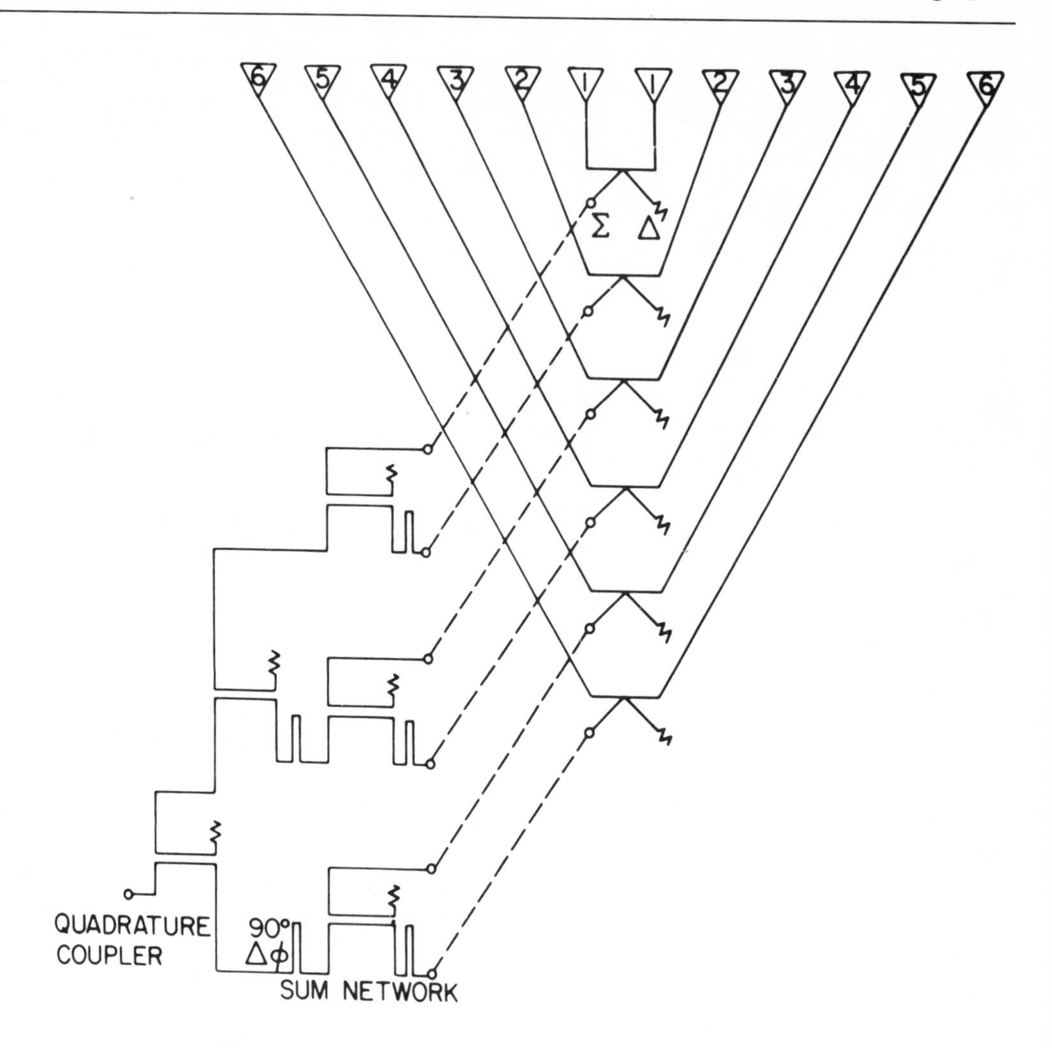

Fig.11 Schematic diagram of the twelve-element antenna

Fig.12 The twelve-element antenna

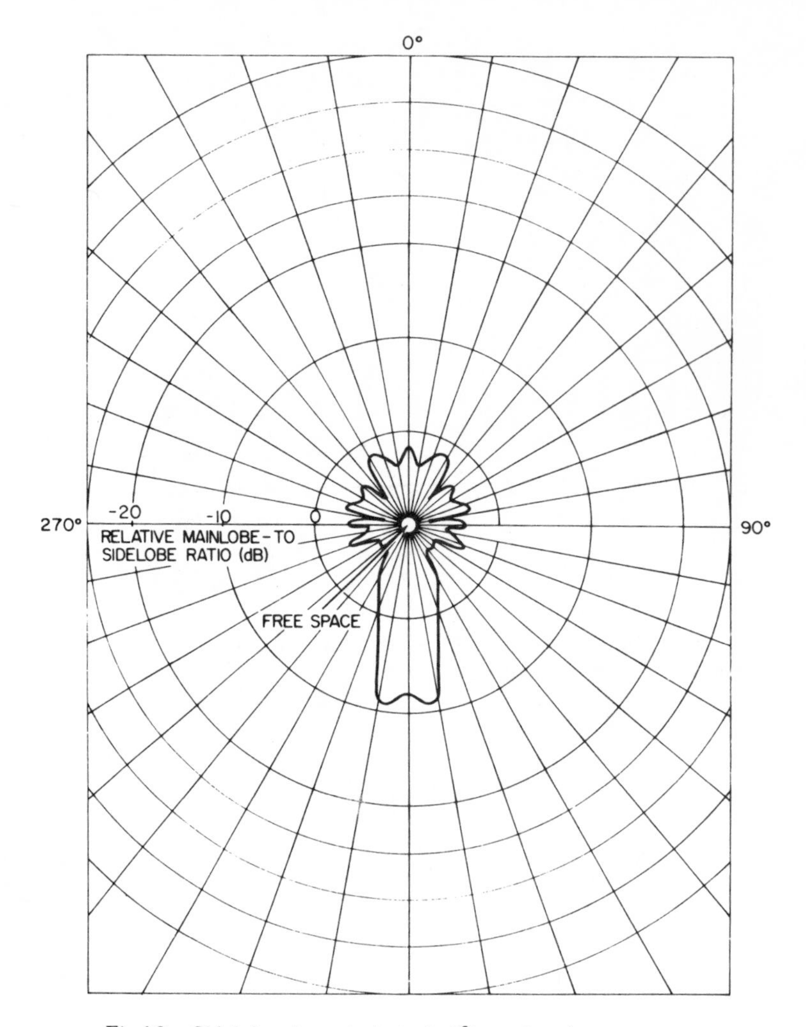

Fig.13 Sidelobe characteristic (−2°) twelve-element array

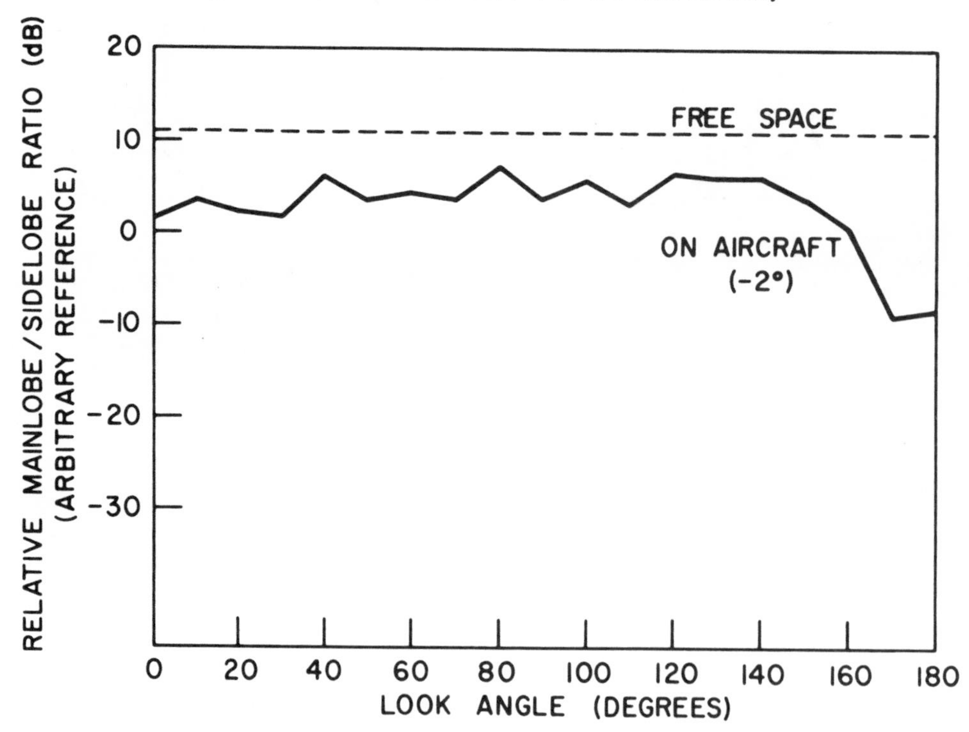

Fig.14 Antenna characteristics: twelve-element array

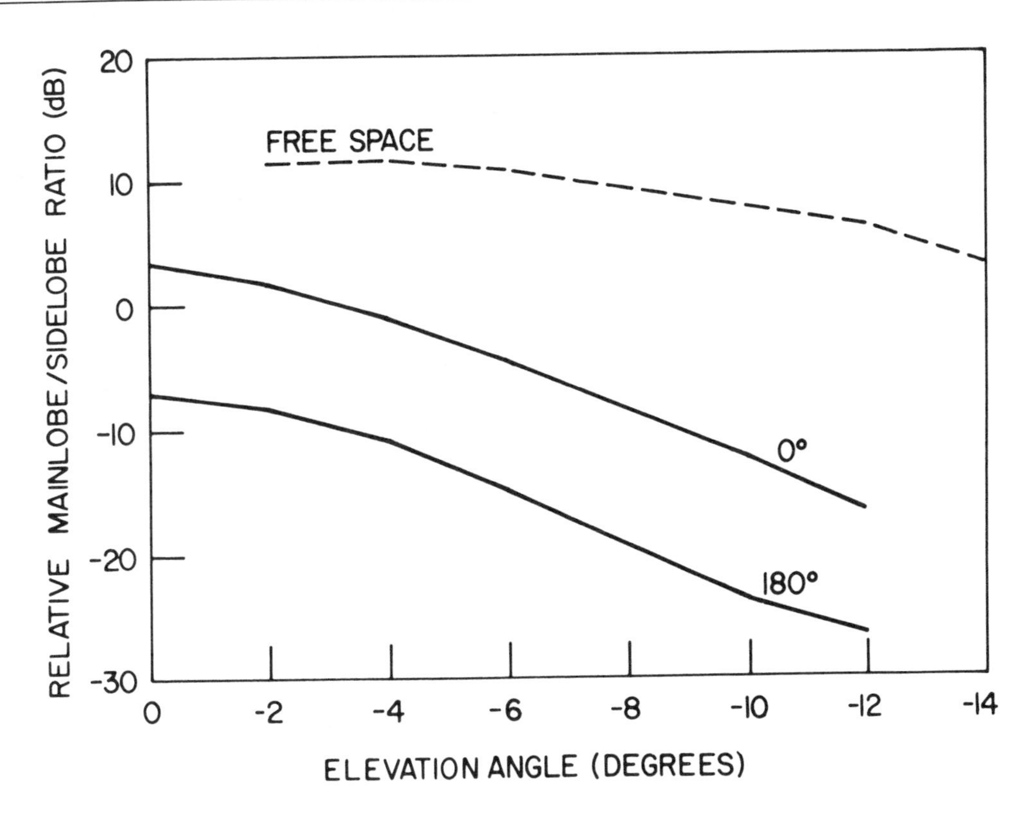

Fig.15 Elevation characteristics: twelve-element antenna

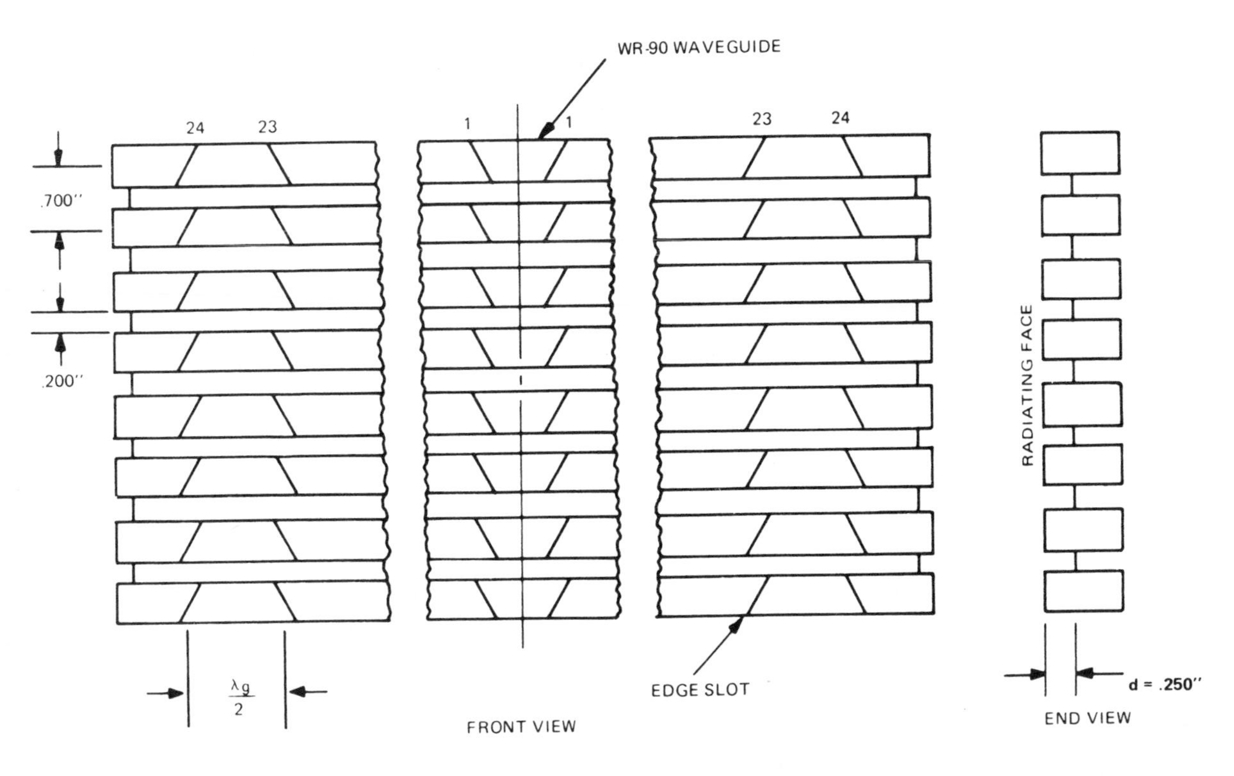

Fig.16 Schematic diagram of the slotted-waveguide antenna

Fig.17 The slotted-waveguide array antenna

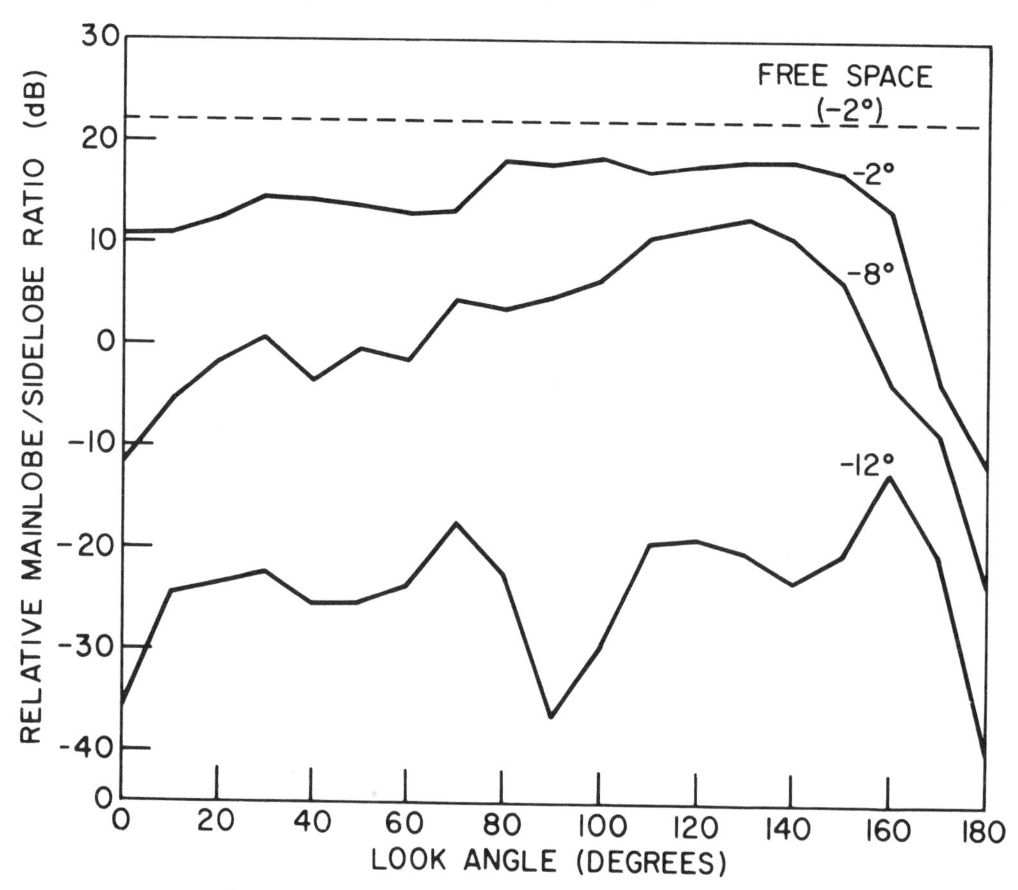

Fig.18 Sidelobe characteristics: slotted-waveguide array

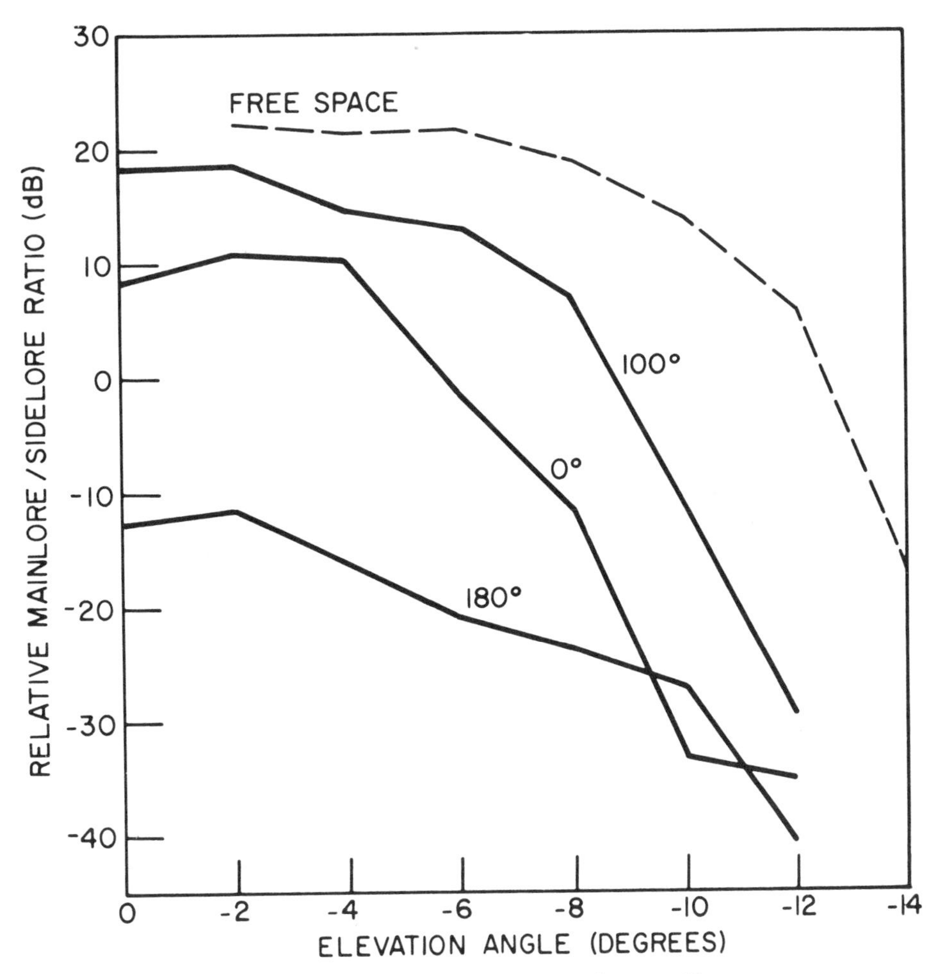

Fig.19 Elevation characteristics: slotted-waveguide array

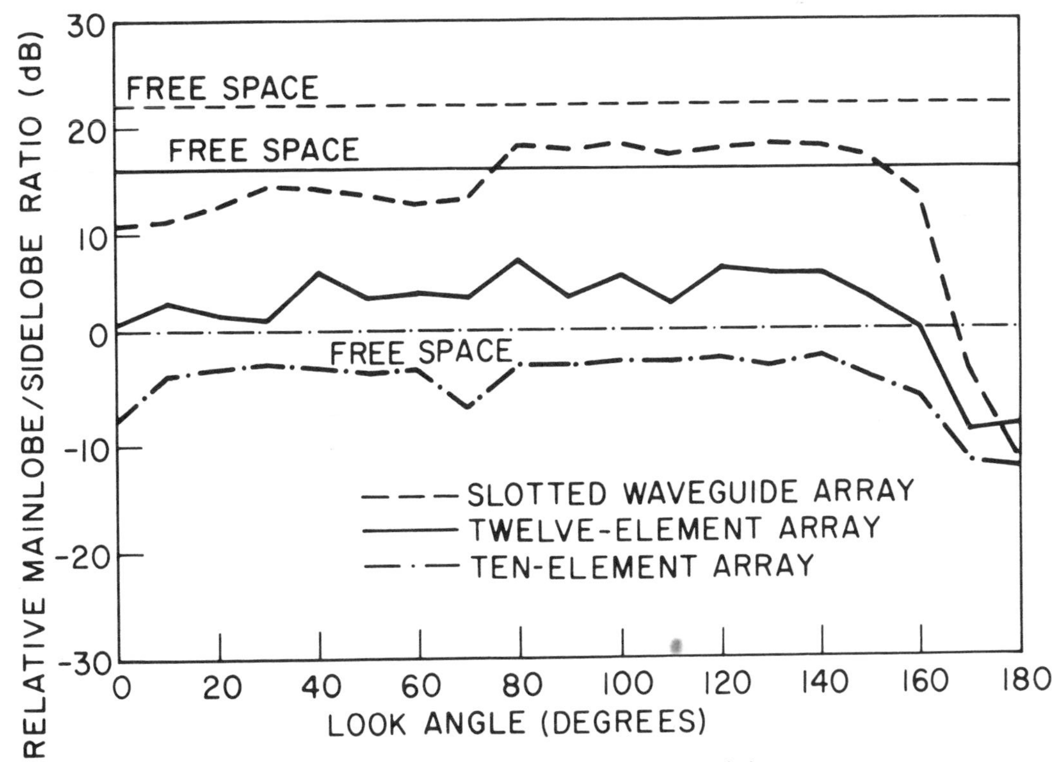

Fig.20 Comparison of sidelobe characteristics

7.3

THEORY OF ADAPTIVE RADAR

L. E. Brennan and I. Reed

IEEE Transactions on Aerospace and Electronic Systems, Vol. AES-9, No. 2, March 1973, pp. 237-252.

Theory of Adaptive Radar

L.E. BRENNAN, Member IEEE

I.S. REED
Technology Service Corporation
Santa Monica, Calif. 90401

Abstract

This paper reviews the principles of adaptive radar in which both the spatial (antenna pattern) and temporal (Doppler filter) responses of the system are controlled adaptively. An adaptive system senses the angular-Doppler distribution of the external noise field and adjusts a set of radar parameters for maximum signal-to-interference ratio and optimum detection performance. A gradient technique for control of the radar array/filter weights is described and shown to generate weights which asymptotically approach optimum values. Simulation results illustrate the convergence rate of adaptive systems and the performance improvement which can be achieved.

I. Introduction

Adaptive array Doppler processors are currently of interest for a variety of applications in the radar, communication, and sonar fields. In an adaptive processor, a separate coherent output is obtained from each element (or subarray) of a phased-array receiving antenna. The output of each element channel is sampled and multiplied by a complex weight, i.e., adjusted in both amplitude and phase, and these weighted outputs are added coherently to form Doppler-compensated receiving beams. The complex weights of this space-time filter are controlled by adaptive loops. By this means, the illumination function and the Doppler reject filters of the receiving array antenna are controlled adaptively.

Manuscript received February 28, 1972.

This work was supported by the Naval Electronic Systems Command under Contract N00039-71-C-0120, and by the Naval Air Systems Command under Contract N00019-70-C-0387.

Such adaptive systems are designed to continuously maximize the probability of detection. This maximization is shown to be equivalent to a maximization of a generalized signal-to-noise ratio. The adaptive processor senses both the receiver noise in the individual space-time channels and the external noise field in performing this optimization. External noise may include both discrete sources in the sidelobe region (e.g., interfering signals in a communication system or jammers in a radar system) and continuously distributed noise (e.g., sky noise in radio astronomy or clutter in a radar system).

The theory of adaptive arrays which maximizes the S/N ratio and an adaptive array implementation were first discussed by Applebaum [14] of Syracuse University Research Corporation. Widrow [12] and associates at Stanford University have contributed extensively to the theory of adaptive arrays, with emphasis on systems which minimize mean-square error. In this paper, a theory is developed for the adaptive processor which maximizes the probability of detection for a fixed false-alarm rate.

The theory of optimum arrays and adaptive array implementations are also of interest in other fields. The large aperture seismic array for detection of underground explosions employed a signal detection algorithm based on the noise covariance matrix which is similar to the algorithm discussed here. Application to seismic work is discussed in a series of Lincoln Laboratory reports. Similar adaptive systems and optimization theory have been used in sonar.

II. Optimal Detection of Signal in Noise

Let a radar (or sonar), possibly moving, emit an electromagnetic (sonic) signal. Suppose an echo from a target at a given range delay is received by a set of N sensors. Let each sensor yield K samples of the received signal in time. Denote the expected sensor time-sampled data from the target by the column vector

$$S = \begin{pmatrix} s_1 \\ s_2 \\ \vdots \\ s_n \end{pmatrix} \tag{1}$$

where s_k is a complex number, the sampled phase and envelope for $k = 1,2, \cdots ,n$, and $n = N \cdot K$ is the total number of space-time samples.

Vector S in (1) is called the signal vector. If the target is moving with respect to the radar (sonar), S would contain the expected relative Doppler and polarization (in the case of radar) phase factors, both with regard to the sensor's special position and its sampling time.

If vector S is "tuned" to a moving target at position R, returns from stationary targets at other positions act as clutter or noise to the detection process. Other noise components include receiver noise, interference, sky noise, etc. Let the column vector

$$N = \begin{pmatrix} n_1 \\ n_2 \\ \vdots \\ n_n \end{pmatrix} \tag{2}$$

denote the total return for noise alone, i.e., the return assuming there was no target at the position **R** of interest. The sum of vectors S and N is the signal-plus-noise vector

$$Z = S + N. \tag{3}$$

The radar (or sonar) problem is to detect the presence of signal S in noise N.

To detect S in (3), one puts the received vector through a "filter" with weights,

$$W = \begin{pmatrix} w_1 \\ w_2 \\ \vdots \\ w_n \end{pmatrix} \tag{4}$$

The output of filter W is the scalar

$$X = \sum_{k=1}^{n} w_k z_k = W_T Z \tag{5}$$

where T denotes matrix transpose. The expected value of X is zero for noise alone and

$$EX = \sum_{k=1}^{n} w_k E z_k = \sum_{k=1}^{n} w_k s_k = W_T S \tag{6}$$

for signal plus noise. Similarly, the noise power or variance of X is

$$\sigma^2 = E|X|^2 - |EX|^2 = W_T{}^* E N^* N_T W = W_T{}^* M W \tag{7}$$

where the asterisk denotes complex conjugation and M is the covariance matrix of the noise process, i.e.,

$$M = EN^* N_T = (En_j{}^* n_k). \tag{8}$$

Distributed backscatter from clutter and other interference sources is approximately Gaussian (e.g., see [1] and [2]). Assume the components of vector N are distributed jointly, either as $2n$ real Gaussian variates or, equivalently, as n complex variates with complex covariance M, as given by (8) (see [3] and [4]). Then the output X of the filter of (5) has the probability density

$$P\{X|Z = N\} = \frac{1}{2\pi\sigma^2} \exp\left(\frac{-|X|^2}{2\sigma^2}\right) \tag{9}$$

for noise alone. For signal plus noise, the probability density is

$$P\{X|Z = S + N\} = \frac{1}{2\pi\sigma^2} \exp\left(\frac{-|X - W_T S|^2}{2\sigma^2}\right) \tag{10}$$

where σ^2 is given by (7).

The optimum decision criterion for detecting the signal S in the presence of noise N is the likelihood ratio test. In terms of the densities (9) and (10), the likelihood ratio for this test is

$$\begin{aligned} L &= \frac{\{P \ X|Z = S + N\}}{P\{X|Z = N\}} \\ &= \exp\left[\frac{1}{2\sigma^2}(S_1 X^* + S_1{}^* X - S_1 S_1{}^*)\right] \\ &= \exp\left[\frac{1}{2\sigma^2}(2|S_1||X| \cos(\theta - \delta) - |S_1|^2)\right] \end{aligned} \tag{11}$$

where

$$S_1 = W_T S$$

and

$$\theta = \arg(X); \qquad \delta = \arg(S_1).$$

Since the positions of radar targets are, in general, not known with certainty, the phase θ of the returned signal-plus-noise process in (11) can be assumed to be distributed uniformly from 0 to 2π radians. If the likelihood ratio L depends on a parameter such as θ, and if the probability density $P(\theta)$ is known, then the best test for detection is the optimum Bayes test (see [5]). This test is formed by averaging L with respect to θ and by comparing the result with a threshold c. That is, if

$$\int L(X|\theta) P(\theta)\, d\theta \geqslant c > 0 \tag{12}$$

The signal is said to be detected; if this quantity is less than c, X is said to result from noise alone.

Since $P(\theta) = 1/2\pi$, by (11) and (12),

$$\begin{aligned} &\exp\left(\frac{-|S_1|^2}{2\sigma^2}\right) \int_0^{2\pi} \exp\left[|S_1||X| \cos(\theta - \delta)\right] \frac{d\theta}{2\pi} \\ &= \exp\left(\frac{-|S_1|^2}{2\sigma^2}\right) I_0(|S_1||X|) \geqslant c \end{aligned} \tag{13}$$

is the optimum Bayes test for detection, where $I_0(x)$ is the modified zero-order Bessel function of the first kind. Since

$I_0(x)$ is a monotonically increasing function of its argument x, and since c is a positive constant, test (13) is equivalent to a much simpler test for detection, namely,

$$|X| \geqslant k \tag{14}$$

where constant k is related to c in (13) by

$$\exp\left(\frac{-|S_1|^2}{2\sigma^2}\right) I_0(|S_1|k) = c.$$

The probability of a false alarm P_F and the probability of detection P_D for test (14) are, by (9) and (10),

$$P_F = \text{Prob}\left\{|X| > k|Z = N\right\}$$

$$= \int_k^\infty \int_0^{2\pi} \frac{1}{2\pi\sigma^2} \exp\left(\frac{-r^2}{2\sigma^2}\right) r\,dr\,d\theta$$

$$= \exp\left(\frac{-k^2}{2\sigma^2}\right) \tag{15}$$

and

$$P_D = \text{Prob}\left\{|X| > k|Z = S + N\right\}$$

$$= \int_k^\infty \int_0^{2\pi} \exp\left(\frac{-1}{2\sigma^2}|re^{i\theta} - ae^{i\delta}|^2\right) \frac{r\,dr\,d\theta}{2\pi\sigma^2}$$

$$= \frac{1}{\sigma^2} \int_k^\infty I_0\left(\frac{ra}{\sigma^2}\right) \exp\left(\frac{-r^2 + a^2}{2\sigma^2}\right) r\,dr \tag{16}$$

where

$$a = |S_1| = |W_T S|$$

and

$$\sigma^2 = W_T{}^* M W.$$

The above probability of detection can be expressed in terms of the Q function. The Q function was defined by Marcum and Swerling [6, p. 159] as

$$Q(\alpha,\beta) = \int_\beta^\infty v \exp\left[\frac{-(v^2 + \alpha^2)}{2}\right] I_0(\alpha v)\, dv. \tag{17}$$

If one makes the substitutions

$$v = \frac{r}{\sigma},\quad \alpha = \frac{a}{\sigma},\quad \beta = \frac{k}{\sigma}$$

in (16), by (17) the probability of detection is

$$P_D = Q(\alpha,\beta) \tag{18}$$

where β is the normalized threshold

$$\beta = \frac{k}{\sigma} = \frac{k}{\sqrt{W_T{}^* M W}} \tag{19}$$

and α^2 is the output or integrated signal-to-noise ratio

$$\alpha^2 = \left(\frac{S}{N}\right)_o = \frac{|W_T S|^2}{W_T{}^* M W} \tag{20}$$

Similarly, by (15), the false-alarm probability is

$$P_F = \exp\left(\frac{-\beta^2}{2}\right) \tag{21}$$

in terms of the normalized threshold β, as defined by (19).

In radar (or sonar) detection systems, one endeavors to restrict the number of false alarms due to noise, clutter, or interference. Mathematically, this is done by setting P_F equal to some small constant, usually in the range from 10^{-5} to 10^{-7}. If P_F is equal to such a constant, then, by (21), the normalized threshold β is determined from

$$\beta = \sqrt{2 \log \frac{1}{P_F}} \tag{22}$$

By (21) and (18), the probability of detection can be expressed as

$$P_D(\alpha) = Q\left(\alpha, \sqrt{2 \log \frac{1}{P_F}}\right)$$

$$= \int_{\sqrt{2 \log \frac{1}{P_F}}}^\infty v \exp \frac{-(v^2 + \alpha^2)}{2} I_0(\alpha v)\, dv \tag{23}$$

in terms of the signal-to-noise ratio α^2 and the false-alarm probability P_F. Equation (23) demonstrates the fact that if P_F is fixed, the probability of detection is a function only of α.

We want next to show that $P_D(\alpha)$ is a monotone increasing function of α. By (20), α is a function of the weight vector W, so that α will vary with the components of the filter W used to detect S. If $P_D(\alpha)$ can be shown to increase with α, the probability of detection will be highest for that filter W which maximizes α.

The derivative of $P_D(\alpha)$ with respect to α is, by (23) and the two identities [7, p. 79]

$$\frac{d}{dx} I_0(x) = I_1(x)$$

and

$$\frac{d}{dx} x I_1(x) = x I_0(x), \tag{24}$$

given by

$$\frac{d}{d\alpha}\left[P_D(\alpha)\right] = \int_{\sqrt{2 \log \frac{1}{P_F}}}^\infty \exp \frac{-(v^2 + \alpha^2)}{2} \left[v^2 I_1(\alpha v) - \alpha v I_0(\alpha v)\right] dv$$

$$= \int_{\sqrt{2 \log \frac{1}{P_F}}}^\infty -d\left\{\exp\left\{\frac{-(v^2 + \alpha^2)}{2}\right\} [v I_1(\alpha v)]\right\} = \beta \exp \frac{-(\beta^2 + \alpha^2)}{2} I_1(\alpha\beta) \tag{25}$$

where

$$\beta = \sqrt{2 \log \frac{1}{P_F}} > 0 .$$

By (25) and the fact that $I_1(x) > 0$ for all $x > 0$,

$$\frac{d}{d\alpha} P_D(\alpha) > 0$$

for all $\alpha > 0$. Thus, $P_D(\alpha)$ is a monotone increasing function[1] of the output signal-to-noise ratio α^2. To maximize $P_D(\alpha)$ with respect to the filter W, it is only necessary to maximize α^2.

To maximize α^2, as given by (20), use the fact that M is a Hermitian matrix.[2] Since M is Hermitian, there exists a unitary matrix P which diagonalizes M, i.e.,

$$PMP^{-1} = \Lambda \tag{26}$$

where

$$\Lambda = \begin{pmatrix} \lambda_1 & 0 & \cdot & \cdot & \cdot & 0 \\ 0 & \lambda_2 & \cdot & \cdot & \cdot & 0 \\ \cdot & \cdot & & & & \cdot \\ \cdot & \cdot & & & & \cdot \\ \cdot & \cdot & & & & \cdot \\ 0 & 0 & \cdot & \cdot & \cdot & \lambda_n \end{pmatrix}$$

and $\lambda_k > 0$ are the eigenvalues of M for $k = 1,2,\cdots,n$. In terms of Λ and P, the square root of the matrix M is given by

$$M^{1/2} = P^{-1}\Lambda^{1/2}P \tag{27}$$

where $\Lambda^{1/2}$ is defined as

$$\Lambda^{1/2} = \begin{pmatrix} \sqrt{\lambda_1} & 0 & \cdot & \cdot & \cdot & 0 \\ 0 & \sqrt{\lambda_2} & \cdot & \cdot & \cdot & 0 \\ \cdot & \cdot & \cdot & & & \cdot \\ \cdot & \cdot & & & & \cdot \\ \cdot & \cdot & & & & \cdot \\ 0 & 0 & \cdot & \cdot & \cdot & \sqrt{\lambda_n} \end{pmatrix}$$

It is easily verified that the square of $M^{1/2}$, as given by (27), equals M.

The signal-to-noise ratio α^2 can be bounded above in the following manner. Note first that

$$\alpha^2 = \frac{|W_T S|^2}{W_T{}^*MW} = \frac{|W_T M^{1/2*} M^{-1/2*} S|^2}{W_T{}^*MW}$$

$$= \frac{\left|\sum_{j=1}^{n} a_j b_j\right|^2}{W_T{}^*MW}$$

where

$$M^{-1/2*} = (M^{1/2*})^{-1}$$

$$a_j = \sum_{k=1}^{n} w_k M_{kj}{}^{1/2*}$$

$$b_j = \sum_{k=1}^{n} M_{jk}{}^{-1/2*} s_k .$$

But, by the Schwarz inequality,

$$\left|\sum_{j=1}^{n} a_j b_j\right|^2 \leqslant \sum_{j=1}^{n} |a_j|^2 \sum_{j=1}^{n} |b_j|^2 .$$

Thus,

$$\left(\frac{S}{N}\right)_o = \alpha^2 \leqslant \frac{\sum_{j=1}^{n} |a_j|^2 \sum_{j=1}^{n} |b_j|^2}{W_T{}^*MW}$$

$$= \frac{(W_T{}^*M^{1/2})(W_T{}^*M^{1/2})_T^* (M^{-1/2}S^*)_T(M^{*-1/2}S)}{W_T{}^*MW}$$

$$= \frac{(W_T{}^*MW)(S_T{}^*M^{*-1}S)}{W_T{}^*MW}$$

$$= SM^{-1}S^*. \tag{28}$$

The bound on α^2, shown by (28), can be attained if one lets

$$W = kM^{-1}S^* \tag{29}$$

where k is a complex constant, not equal to zero. This can be seen by substituting (29) into (24) as follows:

$$\alpha^2 = \frac{|W_T S|^2}{W_T{}^*MW}$$

$$= \frac{|k|^2 |S_T M^{-1} S^*|^2}{(k^* S_T M^{-1})(M)(kM^{-1}S^*)}$$

$$= S_T M^{-1} S^* .$$

Thus, the maximum of the output signal-to-noise ratio with respect to the weight vector W, that is,

$$\max_W \left(\frac{S}{N}\right)_o = \max_W \alpha^2 = S_T M^{-1} S^*, \tag{30}$$

is attained if vector W is given by (29). Hence, by the monotonicity of the Q function with respect to α^2, the probability of detection is maximized if the maximum

[1] This method of proving $P_D(\alpha)$ is a monotone function of α was suggested by G. W. Lank.

[2] M can be assumed nonsingular due to the presence of receiver noise.

value of α^2, (30), is substituted for α^2 into (23). The maximum value of P_D is given by

$$\max_w P_D = Q\left(\sqrt{S_T{}^*M^{-1}S^*},\ \sqrt{2\log\frac{1}{P_F}}\right). \tag{31}$$

This value for the maximized probability of detection is attained if the vector of weights is given by (29). The above results are summarized in the following, now obvious, theorem.

Theorem 1: Assume a radar (or sonar) transmits a waveform and receives n space-time samples. For noise alone, say hypothesis H_0, the receiver observes the vector $Z = N$ where N is the n component column vector defined by (2). For signal plus noise, say hypothesis H_1, the receiver observes the vector $Z = S + N$, where S is the n-component column vector defined by (1). Let W be the filter vector, the n-component column vector of weights, defined by (4). Then the response of the filter to the received observables is

$$X = \sum_{k=1}^{n} w_k z_k = W_T Z$$

where w_k and z_k are the kth components of W and Z, respectively, and W_T denotes the transpose of vector W. Finally, suppose that the components of N are jointly distributed Gaussian variates, and that the covariance matrix is $M = EN^*N_T$, where the asterisk denotes complex conjugation.

With the above definitions and assumptions, the filter which gives the maximum probability of detection P_D for a fixed probability of false alarm P_F is

$$W = kM^{-1}S^*$$

where k is a nonzero complex number. The probability of detection for this optimum filter is

$$P_D = Q\left(\sqrt{S_T{}^*M^{-1}S},\ \sqrt{2\log\frac{1}{P_F}}\right)$$

in terms of the false-alarm probability P_F, where $Q(\alpha, \beta)$ is the Q function defined by (17).

III. A Special Optimum Processor and Its Steering Signal S

In the last section, an optimum detection criterion was derived for the detection of a signal in nonwhite noise. It was assumed that the noise was both temporally and spacially colored. The data set was composed of data from N sensors or array elements, which were spacially displaced. The signals received on these N array elements were sampled at K points in time. As a consequence, the total received data set was a vector of $n = N \cdot K$ complex numbers, each representing the phase and envelope of a particular array element at a sampled point in time. This vector was denoted by Z.

In Theorem 1 it was shown that, in theory, the best filter W to detect the presence of a signal S in Z was proportional to $M^{-1}S^*$, where M is the covariance matrix of the noise-only vector N. This detection criterion, though known for some time (e.g., see [8] and [9]), is seldom used in practice. The covariance matrix M is rarely known, a priori. Even when it has been possible to estimate M, a direct computation of M^{-1} is often too costly and difficult to accomplish in real time. As we shall see, the practical difficulties in achieving optimum detection can, in many instances, be remedied by making such criteria adaptive.

The optimum detection criterion of Theorem 1 can often be achieved adaptively if the noise field is slowly varying and if the unwanted noise processes can be separated or filtered from the desired signals. In active radar (or sonar) detection, the latter requirement is met if the desired signals occur infrequently and do not compete significantly with the incoming noise energy.

For ease of description, adaptive systems are introduced here with a specific application in mind. Adaptive processing to achieve optimum detection is applicable to the airborne radar problem. Consequently, we initially restrict our attention to the development of an adaptive array-Doppler processor for aircraft.

Suppose for the coherent airborne radar problem that there is available from each receiving array element a separate coherent output. That is, a vector

$$V(t) = \begin{pmatrix} v_1(t) \\ v_2(t) \\ \vdots \\ v_N(t) \end{pmatrix}$$

of N complex video processes is observable from the moving-platform array of N sensors or array elements. For simplicity, further suppose the radar periodically transmits pulses at times $0, T, 2T, 3T. \cdots$. The generalization to a staggered pulse train is straightforward.

If Δ is the pulse length of the radar, there are roughly Δ seconds between independent samples of the received $V(t)$ process defined by (32). More generally, from the sampling theorem, $\Delta = 1/2B$, where B is the bandwidth of the received process $V(t)$. For this choice of Δ, the time T between successive pulses is divided up into $[T/\Delta]$ independent sampling times where $[x]$ denotes greatest integer less than x. These sampling times, if multiplied by $c/2$, are called range-sampling rings or, more simply, range rings.

Next assume that a memory or delay media is available to store K consecutive returns from a given range ring. In fact, assume that K vectors,

$$V(t), V(t-T), V(t-2T), \cdots, V(t-(K-1)T),$$

are available at the sampling times $t = k\Delta (k = 0, 1, \cdots, [T/\Delta])$. For the jth range ring, there are stored $n = K \cdot N$ coherent samples, the components of the vector Z. This vector for the jth range ring is denoted by

$$Z(j) \text{ for } j = 0, 1, 2, \cdots \tag{33}$$

where Z is defined by (3). Thus, in a radar sweep, there are $[T/\Delta]$ independent coherent samples of the n-dimensional vector $Z(j)$. In the next section it will be shown how these independent samples may be used to adaptively adjust the weights of filter W.

To design an adaptive array-Doppler processor for a moving-platform array radar, which approximates the performance of the optimum space-time filter, we need first discuss the nature of signal S, used in the optimum detection criterion (29). For simplicity, we assume T, the time between successive pulses, is sufficiently long so that clutter reflections from different range rings are not received simultaneously. That is, the amount of "second-time-around" return is minimal.

Each receiver array sensor, say the rth sensor, has a range-sampling system associated with it. That is, for radar range $d = t_0 C/2$, where $t_0 = k\Delta$ and $0 \leqslant k \leqslant [T/\Delta]$, the rth sensor produces the following sequence of samples:

$$(\cdots v_r(t_0-2T), v_r(t_0-T), v_r(t_0), v_r(t_0+T), v_r(t_0+2T), \cdots) \quad (34)$$

for $r = 1, 2, \cdots, N$.

If there is no signal, only noise, the real and imaginary parts of each complex number $v_r(t_0-kT)$ of (34) can be assumed to be independent and to be Gaussian with standard deviation σ and with a mean equal to zero. A returned signal, excluding the carrier, has for the rth sensor the form

$$s_r(k) = b_r e^{ik\gamma} \quad \text{for } r = 1, 2, \cdots, N \quad (35)$$

where the angle γ is the Doppler phase shift per sweep time T, related to the relative velocity v of the target, the angular carrier frequency $\omega = 2\pi f$, and the velocity of light c by

$$\gamma = -\frac{2v}{c}\omega T. \quad (36)$$

The quantity b_r is the complex number

$$b_r = A_r \exp(i\phi_r + i\delta) \quad \text{for } r = 1, 2, \cdots, N \quad (37)$$

where $A_r > 0$ is the amplitude of the signal at the rth sensor and δ is a constant phase factor. The phase angles ϕ_r for $r = 1, 2, \cdots, N$ are the relative signal phase factors, depending both on the direction of the target and on the physical positions of the array sensors with respect to one another.

To illustrate the quantitites ϕ_r and A_r, suppose the array is linear with sensor element spacing d. Then is the signal arrives at an angle ψ with respect to the array normal,

$$\phi_r = \left(\frac{2\pi rd}{\lambda}\right) \sin\psi \quad \text{for } r = 1, 2, \cdots, N \quad (38)$$

where $\lambda = c/f$ is the radar wavelength. Hence, for a linear array, the expected returned signal from a moving target is of the form

$$s_r(k) = A_r \exp\left[i\left(\frac{2\pi rd}{\lambda}\right)\sin\psi + ik\left(\frac{4\pi T}{\lambda}\right)v + i\delta\right] \quad (39)$$

for $r = 1, 2, \cdots, N$ and $k = 1, 2, \cdots, K$, where ψ is the direction of the target from a normal to the array, v is the relative velocity of the target with respect to the moving platform, and the A_r are proportional to the amplitude taper weights of the array.

Both wanted and unwanted returned echoes from an airborne radar have the form of (35), or in the special case of a linear array, the form of (39). Generally, the Doppler phases of potential unwanted returns from fixed ground targets (ground clutter) are computable as a function of aircraft velocity angle from direction of travel and distance from aircraft. Thus, to separate moving targets such as other aircraft from ground clutter, one matches to a signal vector S which is *distinct* from the known clutter vector S'. A signal vector S, which is sufficiently distinct from a signal return S' from an unwanted return, is called a "steering" signal.

To specifically illustrate the concept of a steering signal S, suppose the number of time samples K is equal to 2. Further, assume one wants to detect a target, moving with respect to the ground in a direction normal to the direction of platform velocity. By (39), a clutter target normal to the direction of motion has a signal vector proportional to the vector, namely,

$$\begin{aligned} S_T' &= (s_1'(1), s_2'(1), \cdots, s_N'(1), s_1'(2), s_2'(2), \cdots, s_N'(2)) \\ &= (1, 1, \cdots, 1, 1, 1, \cdots, 1) \end{aligned} \quad (40)$$

where, for simplicity, we have assumed the amplitude taper is uniform; i.e., $A_k = 1$ for $k = 1, 2, \cdots, N$. Similarly, a target moving with a relative velocity

$$v = \frac{\lambda}{4T} \pm \frac{k\lambda}{2T} \quad \text{for } K = 0, 1, 2, \cdots$$

where $\lambda/2T$ is the "blind speed" of the radar has a vector proportional to

$$S_T = (1, 1, \cdots, 1, -1, -1, \cdots, -1). \quad (41)$$

Since

$$S_T{}^*S' = 0,$$

vectors S and S' are orthogonal, and the mean-square difference or separation between S and S' is a maximum. That is,

$$\sqrt{||S' - S||} = \sqrt{||S||} + \sqrt{||S'||}$$

where

$$||S|| = S_T{}^*S. \quad (42)$$

The steering signal or vector S, given by (41), differs as far as possible from the clutter return vector S', given by (40). If the steering signal S is used in the optimum filter (29), the filter is tuned to a vector which is in the *opposite* direction, vector-wise, to clutter-like signals with vectors close in direction to S' Since clutter vectors in the antenna's main beam have directions close to S', choosing S as a steering vector has the effect of minimizing or "cancelling" the main-beam clutter. In the next sections, we will see how such a criterion can be realized adaptively.

IV. A Recursive Algorithm for Optimum Detection

To find the weight vector or filter W which yields the maximum probability of detection, it was shown in Section II that one need only maximize the signal-to-noise ratio

$$\alpha^2(W) = \frac{|W_T S|^2}{W_T{}^* M W}$$

with respect to W. If M, the noise covariance matrix, is known, the optimum filter was shown to be a weight vector proportional to $M^{-1}S^*$, where S^* is the desired steering signal.

It was pointed out in the last section that the optimum filter is rarely implemented, usually because M is not known explicitly. M is generally a function of a changing noise environment, and only short time estimates of M are obtainable. Moreover, explicit matrix inversions of estimates of M often are not practical in real time. One remedy for such a difficulty is to develop a recursive relaxation algorithm which simultaneously estimates M and recursively computes W. Our present purpose is to devise such an adaptive algorithm.

The most celebrated relaxation technique for computing a local maximum is the method of steepest ascent. To find the optimum filter W, it is only necessary to maximize α^2, the signal-to-noise ratio given by (20), with respect to W. We next show that this maximization can be accomplished by the method of steepest ascent, and then how this criterion can be made adaptive.

By (20), α^2 is a function of both vector W and its conjugate W^*. That is, functionally,

$$\begin{aligned}\alpha^2 &= F(W, W^*)\\ &= F(U + iV, U - iV)\\ &\equiv G(U, V)\end{aligned}$$

where U and V are real n–dimensional vectors, defined by

$$\begin{aligned}W &= U + iV\\ U_T &= (u_1, u_2, \cdots, u_n)\\ V_T &= (v_1, v_2, \cdots, v_n).\end{aligned}$$

Consider the effect of a small step $\delta(U, V)$ on the function $G(U, V)$ from the point (U, V) in $2n$–dimensional real space. To first order in δU and δV, the general Taylor series yields

$$\begin{aligned}\delta G &= G(U + \delta U, V + \delta V) - G(U, V)\\ &= \sum_{j=1}^{n} \frac{\partial G}{\partial u_j}\delta u_j + \frac{\partial G}{\partial v_j}\delta v_j\\ &= \sqrt{(\delta u_1)^2 + \cdots + (\delta u_n)^2 + (\delta v_1)^2 + \cdots + (\delta v_n)^2}\left(\sqrt{\left(\frac{\partial G}{\partial u_1}\right)^2 + \cdots + \left(\frac{\partial G}{\partial u_n}\right)^2 + \left(\frac{\partial G}{\partial v_1}\right)^2 + \cdots + \left(\frac{\partial G}{\partial v_n}\right)^2}\right)\cos\theta\end{aligned}$$

where θ is the angle between the $2n$–dimensional vector $\delta(U, V) = (\delta u_1, \cdots, \delta u_n, \delta v_1, \cdots, \delta v_n)$ and the gradient vector

$$\text{Grad } G = \left(\frac{\partial G}{\partial u_1}, \cdots, \frac{\partial G}{\partial u_n}, \frac{\partial G}{\partial v_1}, \cdots, \frac{\partial G}{\partial v_n}\right). \tag{43}$$

The maximum value of δG is obtained by setting $\cos\theta = 1$. Thus, the direction of steepest ascent from (U, V) is a step $\delta(U, V)$ in the direction of the gradient vector Grad G given by (43). Usually, $\delta(U, V)$ is taken to be proportional to Grad G.

The above criterion of steepest ascent is re-expressible in terms of $\alpha^2 = F$ and the vectors W and W^* of complex numbers, as follows. The direction of steepest ascent of the function $F(W)$ from the vector

$$W_T = U_T + iV_T$$

is a step

$$\delta W_T = \delta U_T + i\delta V_T = \delta(u_1 + iv_1, u_2 + iv_2, \cdots, u_n + iv_n)$$

in the direction of the "complex" gradient vector

$$(\tilde{\nabla}F)_T = \left(\frac{\partial F}{\partial u_1} + i\frac{\partial F}{\partial v_1}, \frac{\partial F}{\partial u_2} + i\frac{\partial F}{\partial v_2}, \cdots, \frac{\partial F}{\partial u_n} + i\frac{\partial F}{\partial v_n}\right) \tag{44}$$

where the subscript T denotes transpose. The real derivatives in (44) can be expressed in terms of complex derivatives by the relationships (see, for example, [10])

$$\frac{\partial F}{\partial u_j} = \frac{\partial F}{\partial w_j} + \frac{\partial F}{\partial w_j{}^*}$$

and

$$\frac{\partial F}{\partial v_j} = \left(\frac{\partial F}{\partial w_j} - \frac{\partial F}{\partial w_j{}^*}\right)i$$

where $w_j = u_j + iv_j$ for $j = 1, 2, \cdots, n$. Thus,

$$\begin{aligned}(\tilde{\nabla}F)_T &= \left(\frac{\partial F}{\partial w_1} + \frac{\partial F}{\partial w_1{}^*} - \frac{\partial F}{\partial w_1} + \frac{\partial F}{\partial w_1{}^*}, \cdots, \frac{\partial F}{\partial w_n} + \frac{\partial F}{\partial w_n{}^*} - \frac{\partial F}{\partial w_n} + \frac{\partial F}{\partial w_n{}^*}\right)\\ &= 2\left(\frac{\partial F}{\partial w_1{}^*}, \frac{\partial F}{\partial w_2{}^*}, \cdots, \frac{\partial F}{\partial w_n{}^*}\right)\end{aligned} \tag{45}$$

is the complex gradient in terms of complex derivatives.

Let

$$W(j) = U(j) + iV(j)$$

be the values of the weight vector W at the jth step in the above method of steepest ascent or "hill climbing" process. Since the direction of steepest ascent of function $\alpha^2 = F$ is in the gradient direction, the value of the weight $W(j + 1)$ is determined by the relation

$$W(j+1) = W(j) + \frac{1}{2}\mu(j)\tilde{\nabla}F[W(j)] \qquad \text{for } j = 1, 2, 3, \cdots \quad (46)$$

where $\tilde{\nabla} F[W(j)]$ is the "complex" gradient (45) evaluated at $W(j)$, the value of weight vector W at the jth step. In (46), the best choice of $\mu(j)$ is the value of a real variable μ which maximizes the function

$$F[W(j-1) + \mu\tilde{\nabla}F(W(j-1))] \qquad \text{for } j = 1, 2, 3 \cdots .$$

The initial condition of the system of equations given by (46) is assumed to be some initial estimate $W(0)$ of the filter for optimum detection.

The complex gradient F in (46) can be computed by applying formula (45) to $\alpha^2 = F$, as given by (42). By the product rule for differentiation,

$$\frac{\partial F}{\partial w_l^*} = \frac{S_l^*(W_T S)}{W_T^* M W} - \frac{(W_T^* S)(W_T S)}{(W_T^* M W)^2}\left(\sum_{j=1}^{n} w_j m_{lj}\right)$$

where $m_{jl} = En_j^* n_l$, the (j,l)th element of the covariance matrix M (8). Thus, by (45),[2]

$$\tilde{\nabla}F = 2\frac{S^*(W_T S)}{W_T^* M W} - 2\frac{(W_T^* S^*)(W_T S)}{(W_T^* M W)^2}(MW)$$

$$= 2\left(\frac{W_T S}{W_T^* M W}\right)\left[S^* - \left(\frac{W_T^* S^*}{W_T^* M W}\right) M W\right] \cdot \quad (47)$$

By Theorem 1, the signal-to-noise ratio $\alpha^2 = F$, as a function of W, is bounded above, and its maximum is attained if $W = kM^{-1}S^*$, where k is a nonzero complex number. Thus, in (46), if M is assumed known and μ is chosen to be a constant, it is straightforward to apply known theorems to show that vector $W(j)$ approaches a stationary point as a limit (e.g., see [11]). Hence, if the initial estimate $W(0)$ of W is sufficiently close in norm to the optimum filter for some value of complex scalar k, then $W(j)$ approaches $kM^{-1}S^*$ in the limit. Here the norm of a vector is the usual inner product of the vector with itself [see (42)].

The steepest ascent algorithm (46), with (47) substituted for $\tilde{\nabla}F$, is not a linear function of the vector $W(j)$. This nonlinearity is found chiefly in the coefficients

$$a(W(j)) = \frac{W_T(j)S}{W_T^*(j)MW(j)} \qquad (48)$$

of (47) for $\tilde{\nabla}F$ for $j = 1, 2, 3, \cdots$. However, suppose in algorithm (46) that $W(0)$ is chosen so that $W(j)$ approaches in the limit $kM^{-1}S^*$; that is,

$$W(j) \to kM^{-1}S^* \qquad (49)$$

as $j \to \infty$. Then,

$$a(W(j)) \to \lim_{j\to\infty} \frac{W_T(j)S}{W_T^*(j)MW(j)}$$

$$= \frac{kS_T^* M_T^{-1} S}{(kM^{-1}S^*)_T^* M(kM^{-1}S^*)} = \frac{1}{k^*}$$

$$\equiv a. \qquad (50)$$

That is, if $W(j) \to kM^{-1}S^*$, then $a(W(j)) \to 1/k^* = a$, where a is a complex scalar.

The result given in (50) suggests a method for linearizing the steepest ascent algorithm (46). This is to replace coefficients $a(W(j))$ by a limiting value. The importance of such a linearization to practical applications cannot be overemphasized. In some adaptive systems, linearization supplies the reduction in computational complexity needed to make the system practical.

If the limiting value in (50) is used for $a(W(j))$, and if $\mu(j)$ is set equal to a constant μ, then algorithm (42) becomes the linearized algorithm

$$W(j+1) = W(j) + \mu a[S^* - a^* M(j) W(j)] \qquad (51)$$

where $M(j)$ is a statistical estimate of the covariance matrix (8). To determine $M(j)$, the input data is assumed to be the sequence of independent vector $Z(j)$ for $j = 0, 1, 2, \cdots$, given by (33) in the last section.

If no prior knowledge of the covariance is available, the best estimate of the $n \times n$ covariance matrix must be

$$M(j) = Z^*(j) Z_T(j) \qquad (52)$$

for $j = 1, 2, \cdots$, where $Z(j)$ is the jth sample of vector Z of KN complex components, given by (3). Although $Z(j)$ was defined in the last section for only a finite number of range samples, $j = 0, 1, 2, \cdots, [T/\Delta]$, we assume in (52) that the range of j has been extended indefinitely. In the radar problem, for example, this extension to an unlimited number of independent input data vectors $Z(j)$ can be accomplished by extending the definition of $Z(j)$ to include successive, but independent, radar sweeps of such data sets. In the next section, we will show that the mean value of the weight vector $W(j)$ converges if $M(j)$ is defined by (52). As we shall see, the adaptive capability of this algorithm and its generalizations will justify its being called an adaptive algorithm for optimum detection.

[2] This expression for $\tilde{\nabla}F$ can also be obtained directly by replacing the w_n by $u_n + iv_n$ in (20) and forming the partial derivatives $\partial F/\partial u_n$ and $\partial F/\partial v_n$ for substitution in (44).

V. The Convergence in the Mean of Adaptive Criteria

In the last section a recursive algorithm was devised for computing the optimum weight vector. This algorithm was based on the well-known method of steepest ascent. A linearization of algorithm (46) lead to the first-order difference equation (51) for computing the best filter vector W. Using the statistical independence of estimates (52) for $j = 0, 1, 2, \cdots$, the next theorem supplies a proof that the expected or mean value of the solution to (51) converges in norm to the optimum filter. Similar theorems have been proved by others (e.g., see [12]).

Theorem 2: Let $Z(j)$ be the jth sample of vector Z of $n = KN$ complex components, the input signal-plus-noise process, (3). Further, suppose $Z(j)$ is a sequence for $j = 0, 1, 2, \cdots$ of independent range samples, as discussed following (52). Finally, let the jth estimate of the covariance matrix be $M(j) = Z^*(j)Z_T(j)$ in accordance with (52). Then, the expected value of the steady-state solution to (51), i.e., to the system

$$W(j+1) = W(j) + \mu a[S^* - a^*M(j)W(j)],$$

converges and is equal to $k\overline{M}^{-1}S^*$, where $k = 1/a^*$ and $\overline{M} = EM(j)$ for all j. Convergence of (51) is guaranteed as long as $0 < \mu < 2k^2/\max \lambda_j$, where λ_j for $j = 1, 2, \cdots, n$ are the eigenvalues of $\overline{M}$.

Proof: If $W(0)$ is the initial condition for the difference equation for the weight vector $W(j)$, the solution to (51) is easily verified to be

$$W(j) = W(0)\prod_{k=0}^{j-1}(I - \mu|a|^2M(k)) + \mu a\sum_{l=1}^{j-1}\left[\prod_{k=j-l}^{j-1}(I - \mu|a|^2M(k))\right]S^* + \mu aS^* \tag{53}$$

for $j = 1, 2, \cdots$, where I is the unit $n \times n$ matrix. The expected value of $W(j)$ is

$$EW(j) = (I - \mu|a|^2\overline{M})^jW(0) + \mu a\sum_{l=0}^{j-1}(I - \mu|a|^2\overline{M})^lS^* \tag{54}$$

for $j = 1, 2, \cdots$.

To study the convergence of (54) we use the norm defined by (42). Associated with the norm is the classical inner product for a unitary space, defined by

$$(X, Y) = X_T^*Y \tag{55}$$

where X and Y are n–dimensional column vectors. Evidently, in this notation $||X||^2 = (X, X)$. If T is an $n \times n$ matrix transformation operating on X, we define the norm of T to be

$$||T|| = \sup_{||X||<1} ||TX||. \tag{56}$$

This norm yields the uniform topology in Euclidian space [13]. If T_1 and T_2 are $n \times n$ transformation operating on vector X, the following inequalities between norms, as defined by (57), are readily demonstrated:

$$||T_1X|| \leqslant ||T_1||\,||X||$$
$$||T_1T_2|| \leqslant ||T_1||\,||T_2||$$
$$||T_1 + T_2|| \leqslant ||T_1|| + ||T_2||. \tag{57}$$

The inequalities given by (57) can be extended to any number of transformations by induction.

Let us suppose the norm [see (56)] of $I - \mu|a|^2\overline{M}$ satisfies the inequality

$$||I - \mu|a|^2\overline{M}|| < 1, \tag{58}$$

and consider the limit of $EW(j)$ as j approaches infinity. If the limit exists, it must equal the infinite series

$$\begin{aligned} EW(\infty) = \lim_{j\to\infty} EW(j) &= \mu a\sum_{l=0}^{\infty}(I - \mu|a|^2\overline{M})^lS^* \\ &= \mu a(I - (I - \mu|a|^2\overline{M}))^{-1}S^* \\ &= \frac{1}{a^*}\overline{M}^{-1}S^* = k\overline{M}^{-1}S^*, \end{aligned} \tag{59}$$

the desired optimum matched filter. To show convergence, consider the norm of difference of (54) and (59), using (57) and (58):

$$\begin{aligned} &||EW(j) - EW(\infty)|| \\ &= ||\mu a\sum_{l=j}^{\infty}(I - \mu|a|^2\overline{M})^lS^* \\ &\quad -(I - \mu|a|^2\overline{M})^jW(0)|| \\ &\leqslant \mu|a|\sum_{l=j}^{\infty}||I - \mu|a|^2\overline{M}||^l||S^*|| \\ &\quad + ||I - \mu|a|^2\overline{M}||^j||EW(0)|| \\ &= \frac{\mu|a|\;||I - \mu|a|^2\overline{M}||^j}{1 - ||I - \mu|a|^2\overline{M}||}||S^*|| \\ &\quad + ||I - \mu|a|^2\overline{M}||^j||W(0)||. \end{aligned} \tag{60}$$

Since $||I - \mu|a|^2M|| < 1$, the right side of (60) approaches zero as $j\to\infty$. Thus, $EW(j)$ converges in norm to the optimum matched filter.

To find conditions which yield inequality (58) for convergence, let us compute the norm of an arbitrary Hermitian matrix H in terms of its eigenvalues. Since H is Hermitian, it has the spectral representation

$$H = \sum_{j=1}^{n} \lambda_j E_j$$

where the λ_j are eigenvalues and the $E_1, E_2, \cdots, E_j$ are orthogonal projection matrixes. Thus, by (55),

$$||HX||^2 = (HX, HX) = \sum_{j,k=1}^{n} \lambda_j \lambda_k (E_j X, E_k X)$$

$$= \sum_{j,k=1}^{n} \lambda_j \lambda_k (E_j E_k X, X)$$

$$= \sum_{j=1}^{n} \lambda_j^2 (E_j^2 X, X)$$

$$= \sum_{j=1}^{n} \lambda_j^2 (X_j, X_j) = \sum_{j=1}^{n} \lambda_j^2 ||X_j||^2$$

where X_j is the projection of X along E_j. Since

$$||X||^2 = \sum_{j=1}^{n} ||X_j||^2 ,$$

we have

$$||HX||^2 \leqslant \max_k \lambda_k^2 \; ||X||^2$$

Thus, by (56),

$$||H|| = \max_k \lambda_k \tag{61}$$

holds for all Hermitian matrixes H.

We now apply (61) to the matrix of (58). Since $\overline{M}$ is Hermitian, the matrix $I - \mu|a|^2\overline{M}$ is likewise Hermitian. If λ_j is an eigenvalue of $\overline{M}$, $\overline{M}X_j = \lambda_j X_j$ where X_j is the corresponding eigenvector. But,

$$(I - \mu|a|^2\overline{M})X_j = (1 - \mu|a|^2\lambda_k)X_j.$$

Thus, the eigenvalues of $I - \mu|a|^2\overline{M}$ are the quantities $1 - \mu|a|^2\lambda_j$ for $j = 1, 2, \cdots, n$. Evidently, the norm for $I - \mu|a|^2\overline{M}$ is

$$||I - \mu|a|^2\overline{M}|| = \max_j |\mu|a|^2\lambda_j - 1|.$$

The theorem is now evident if one substitutes this into inequality (58).

For many applications, such as the moving-platform radar problem discussed in Section III, adaptive algorithm (51) is best regarded as an approximation to a system of differential equations. The equivalent algorithm as a vector differential equation has the form

$$\frac{d}{dt} W(t) + \frac{\mu}{\Delta} |a|^2 M(t)W(t) = \frac{\mu}{\Delta} aS^* \tag{62}$$

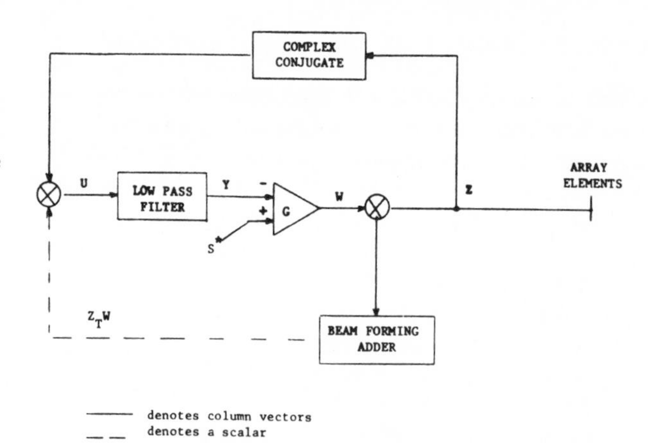

Fig. 1. Adaptive array with first-order loops.

where Δ is the time between independent samples, or, roughly, the radar pulse width for the airborne radar problem (see Section III).

Adaptive systems can be implemented in either analog or digital form. In either type of implementation, the jth component of system (62) can be realized by the block diagram shown in Fig. 1. The entire adaptive array of n such first-order loops is best described by the following three operational vector equations:

$$W = G(S^* - Y)$$

$$\tau\dot{Y} + Y = U$$

$$U = Z^*(Z_T W) = MW \tag{63}$$

where τ is the time constant of the low-pass filters of each loop and where, for generality, G is a diagonal matrix, diag $\{g_1, g_2, \cdots, g_n\}$, of possibly different amplifier gains, $g_1, g_2, \cdots, g_n$. In some cases, this degree of freedom in the amplifier gains is important to the convergence problem.

Note in (63) that the estimate of the covariance matrix M is given by

$$M = Z^* Z_T. \tag{64}$$

This corresponds to the estimate suggested in (52). Equations (63) combine into the single vector differential equation

$$\tau\dot{W} + W + GMW = GS^* \tag{65}$$

of n components where M is given by (64). Evidently, system (65) corresponds to system (62) if one associates the constants τ, G with constants μ, Δ and $|a|^2$ by the relation

$$\frac{1}{\tau} G = \left(\frac{\mu}{\Delta}\right) |a|^2 I$$

where I is the identity matrix, and if one lets the steering vector S^* in (65) be the steering vector in (62) divided by the complex constant a^*. With these identifications, (65) only differs from (62) by the replacement of the integration operation in (62) by an approximate integration, that associated with a set of low-pass filters, each with time constant τ. As we shall see next, the latter approximation leads to a slight bias error in the estimate of the optimum matched filter.

In an adaptive array, the central-loop time constant τ is chosen to be much greater than the time Δ between independent samples of the input process $V(t)$ (see Section III). Hence, it can be assumed that the weight vector W varies much more slowly than the input process. Thus, to a good approximation, the weights W are statistically independent of the instantaneous estimate of the covariance matrix given by (64). With this assumption in mind, the expected value of the weight vector, $\overline{W}$, is, from (65), given by the vector differential equation

$$\tau\dot{\overline{W}} + (G\overline{M} + I)\overline{W} = GS^* \tag{66}$$

where $\overline{M}$ is the Hermitian covariance matrix

$$\overline{M} = EZ^*(t)Z_T(t). \tag{67}$$

To solve (66), first denote the matrix $(G\overline{M} + I)$ by F and observe that $G^{-1/2}FG^{1/2}$ is Hermitian where

$$G^{1/2} = \text{diag}\,(g_1^{1/2}, g_2^{1/2}, \cdots, g_n^{1/2}).$$

Since

$$G^{1/2}(G^{-1/2}FG^{1/2})G^{-1/2} = F,$$

matrix $G^{-1/2}FG^{1/2}$ is similar to F. Thus, matrixes $G^{-1/2}FG^{1/2}$ and F have the same eigenvalues. Since $G^{-1/2}FG^{1/2}$ is Hermitian, there exists a unitary transformation P which diagonalizes the matrix. That is,

$$\mathbf{P}(G^{-1/2}FG^{1/2})\mathbf{P}^{-1} = \Gamma = \text{diag}(\gamma_1, \gamma_2, \cdots, \gamma_n) \tag{68}$$

where $\gamma_1, \gamma_2, \cdots, \gamma_n$ are the eigenvalues of F. If we multiply (66) on the left by $G^{-1/2}$, we obtain

$$\tau(G^{-1/2}\dot{\overline{W}}) + (G^{-1/2}FG^{1/2})(G^{-1/2}\overline{W}) = G^{1/2}S^*.$$

Multiplying by $\mathbf{P}$ and letting

$$X = P(G^{-1/2}\overline{W}), \tag{69}$$

system (66) reduces the following system of differential equations:

$$\tau\dot{X} + \Gamma X = PG^{1/2}S^* \equiv B \tag{70}$$

where Γ is the diagonal matrix (68) of eigenvalues of F.

The solution to the system of n linear differential equations (70) is

$$x_j(t) = \frac{b_j}{\gamma_j} + \left[x_j(0) - \frac{b_j}{\gamma_n}\right]\exp\left(-\frac{\gamma_j}{\tau}t\right) \tag{71}$$

for $j = 1, 2, \cdots, n$, where

$$X_T(0) = (x_1(0), x_2(0), \cdots, x_n(0))$$

is the initial condition. The solution of system (66), using (69), is

$$\overline{W}(t) = G^{1/2}P^{-1}X(t) \tag{72}$$

where $X_T(t) = (x_1(t), x_2(t), \cdots, x_n(t))$ with $x_j(t)$ given by (71) for $j = 1, 2, \cdots, n$.

TABLE I

Simulation of Adaptive Receiving Array

Parameters Assumed
10-element linear array
Element patterns isotropic over $-\pi/2 \leq \theta \leq \pi/2$
Half-wave spaced elements
Transmit pattern $E(\theta) = \dfrac{\sin\left[\left(\frac{\pi N}{2}\right)\sin\theta\right]}{\sin\left(\frac{\pi}{2}\sin\theta\right)}$
Uniformly illuminated transmit array
30 scatterers in sidelobe region, equally spaced
No interference for $-\theta_1 < \theta < \theta_1$
θ_1 variable
Each receiving element weight controlled adaptively
Simulate 1600 independent sets of input signals (range resolution cells)
No receiver noise

If the amplifier gains of the gain matrix G are equal, i.e.,

$$G = gI$$

where g is a positive scalar and I the identity matrix, then the eigenvalues of $(G\overline{M} + I)$ are related to the eigenvalues of $\overline{M}$ by

$$\gamma_n = g\lambda_n + 1$$

where λ_n are the eigenvalues of $\overline{M}$. In this case, if $g\lambda_{min} \gg 1$, it is easily seen that $\overline{W} \to \overline{M}^{-1}S^*$ (see [15]). If the gains of matrix G are not equal, it likewise can be shown that $\overline{W}$ approaches the optimum matched filter, except for a bias which depends on the magnitudes of the elements in the gain matrix. It can be shown that this bias term is negligible as long as the minimum gain constant is sufficiently large.

VI. Simulation of Adaptive Array Antenna

The analysis of Section V provides a direct method for computing the steady-state performance of an adaptive array antenna. We consider the case $K = 1$, so that $n = N \cdot 1 = N$ in (1). When the time constants of the array are large, the input noise field is smoothed over a long time interval and the weights approach the optimum values of (29). The resulting antenna patterns can be computed directly from the steady-state weights of (29). In many radar problems, the external noise field is varying, so that the adaptive loop convergence rate is important. In such cases, simulation is a useful tool for exploring the adaptive loop convergence rate and noise.

This section presents the results of a digital computer simulation of a 10-element adaptive array. Some of the parameters assumed in the simulation are listed in Table I. The external noise field is presumed to consist of 30 clutter scatterers. Since the number of scatterers (or interference sources) exceeds the number of array elements, it is not possible for the array to place a discrete null at each interference angle. As we shall see, the adaptive array attempts to reduce the antenna sidelobe level over a wide angular region.

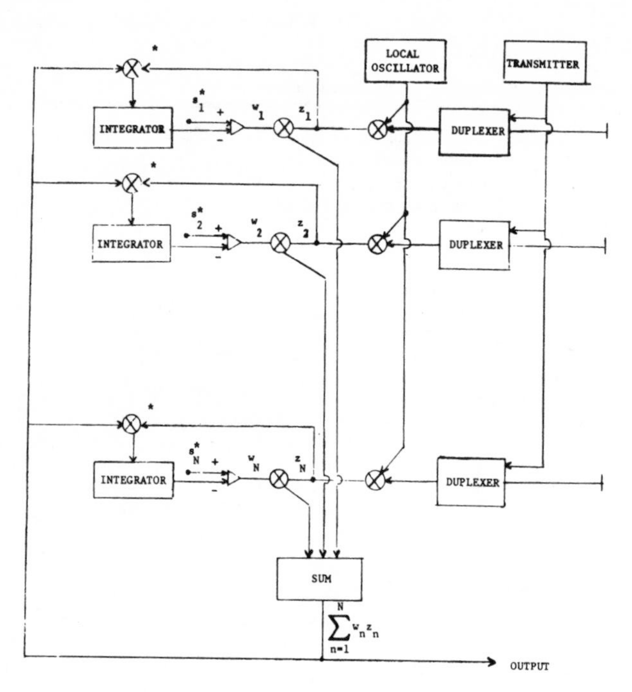

Fig. 2. Adaptive array radar.

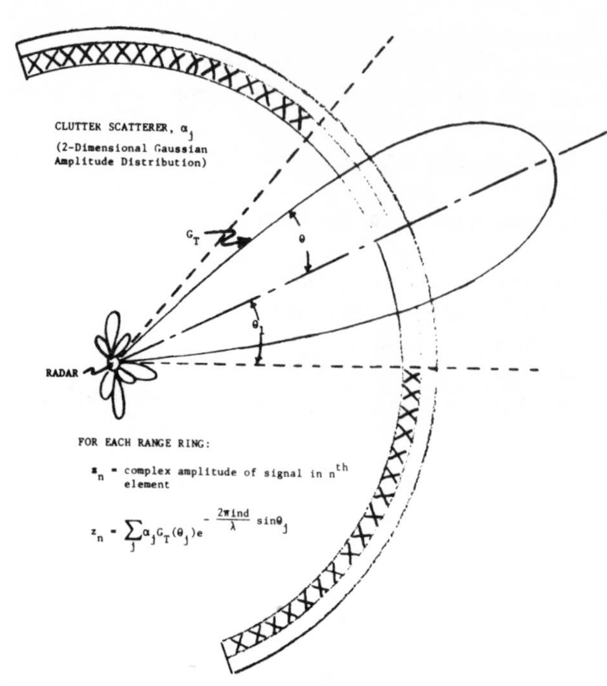

Fig. 3. Simulation of interference in a range ring.

Fig. 4. Adaptive array gain.

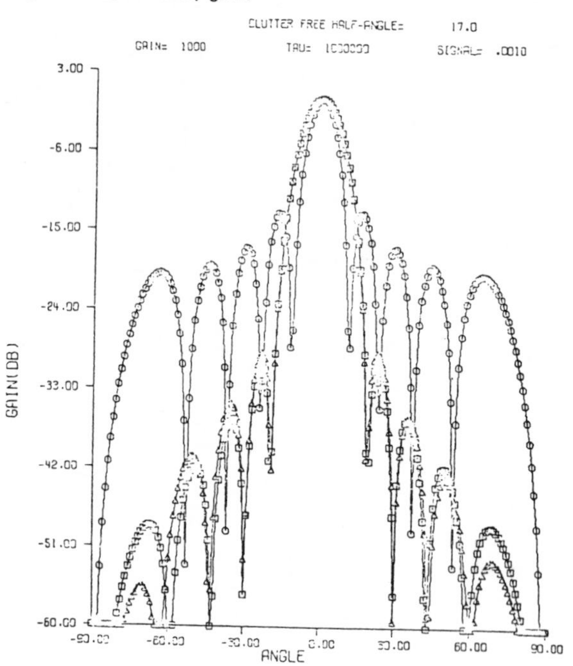

Fig. 5. Projectograph plot of adaptive array gain.

IMPROVEMENT IN SIGNAL-TO-SIDELOBE CLUTTER RATIO:

STEADY STATE	27.3 dB
1600 ITERATIONS	25.7 dB

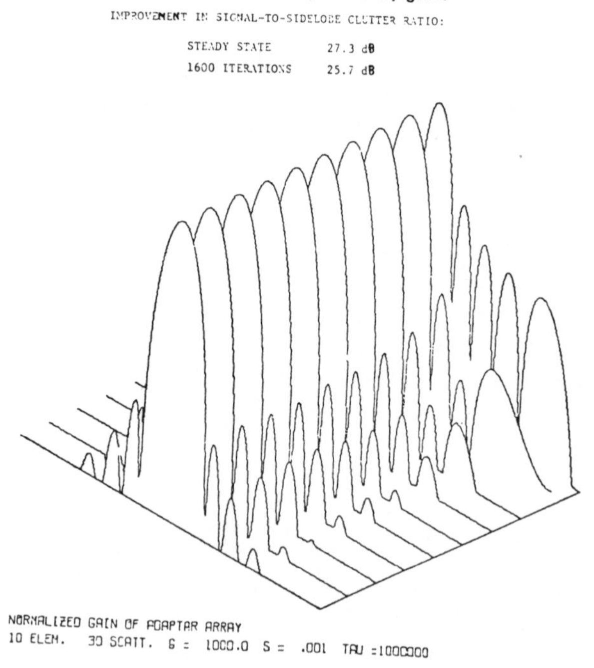

An active pulsed radar was simulated with clutter scatterers in the sidelobe angular region. It was found that scatterers at 30 equally spaced angles is roughly sufficient to represent continuously distributed clutter. A similar simulation procedure could be used to represent 30 discrete interference sources at these angles. No main-beam interference was included in the simulation; i.e., the sources are all at angles outside the main-beam region of $-\theta_1 \leqslant \theta \leqslant \theta_1$. The width of this angular region, $2\theta_1$, is a parameter in the simulation.

A possible method of implementing an active radar with an adaptive receiving array is shown in functional block diagram form in Fig. 2. A duplexer is included in each element channel so that the same array can be used for transmission and reception. A common local oscillator heterodynes each of the received signals to IF for adaptive beam forming. In the simulation of this type of radar, a uniformly illuminated transmit array was assumed with the transmit beam normal to the linear array.

At the beginning of each simulation run, the integrator outputs in the control loops were set to zero. Since the desired look angle for receive is normal to the array (the direction of the transmit beam), steering vector components s_n are equal (see Section III). The initial

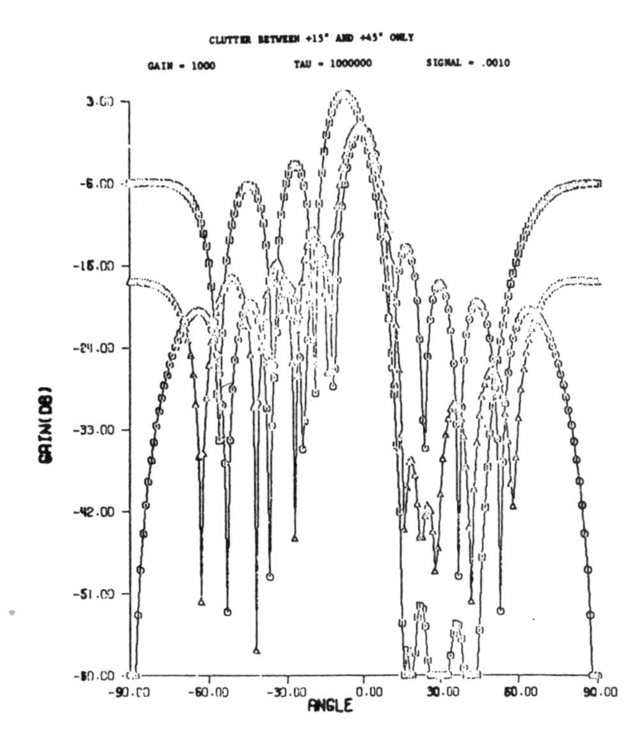

Fig. 6. Adaptive array gain, nonuniform clutter distribution.

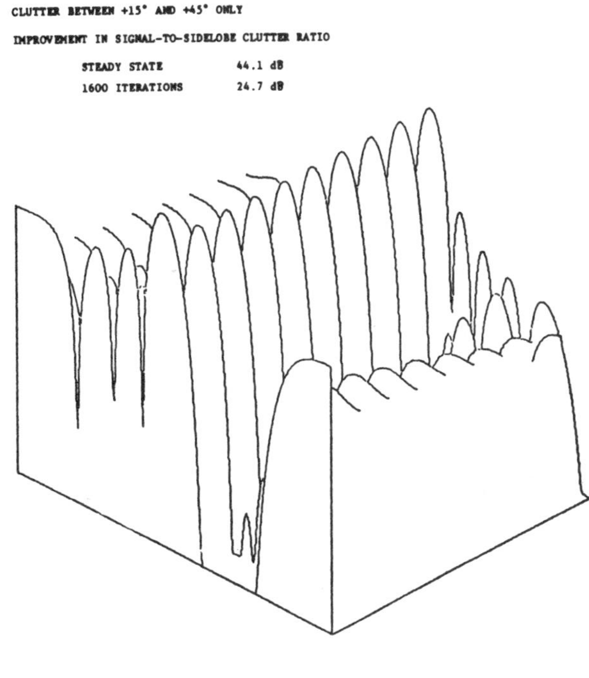

Fig. 7. Projectograph plot of adaptive array gain, nonuniform clutter distribution.

weights on the receiving array were all the same, $w_n = s_n{}^*G$. The integrator (low-pass filter) time constant τ and the amplifier gain G were identical in all loops and are inputs to the simulation program.

After the loop weights were initialized in the simulation, the interference process for the first sample interval (i.e., range ring of clutter) was generated. Each of the 30 scatterers was represented by a random complex number selected from a zero-mean, two-dimensional Gaussian process. This is a good representation of either random noise from a jammer or distributed clutter. The voltages z_n were obtained, as indicated in Fig. 3, by summing coherently the returns from the 30 sources. This summation must carefully preserve the relative phases of the 30 signals in the 10 element outputs, since this coherence is sensed by the adaptive array. Transmit antenna gain varies with angle and is included in the computation.

After the first set of 10 z_n is computed, the adaptive loops are iterated, as indicated in Fig. 2 and (63). This produces a modified set of weights, w_n. The clutter process is then generated randomly for the next range ring, the z_n computed, and the loops iterated again. In each of the simulation results to be presented, this procedure was repeated 1600 times. The results of each simulation run are presented in two formats, those of Figs. 4 and 5.

In the first figure of each pair (Fig. 4), the interference-free half-angle θ_1, loop gain G, time constant τ, and steering signal amplitude S are indicated. Three antenna patterns are plotted in Fig. 4. The upper curve (marked with circles) is the transmit pattern and initial receiving pattern. The curve marked with squares is the steady-state solution, i.e., the optimum receiving pattern obtained from weights $W = M^{-1}S^*$. The curve marked with triangles shows the pattern obtained after 1000 iterations of the adaptive loops.

Note the interleaving of nulls between the transmit pattern and adaptive receiving pattern in Fig. 4. Peaks of receiving array gain occur at the nulls of the transmit pattern. The adaptive receiving array senses the angular variations in scatterer intensity due to the transmit pattern and adjusts the receiving pattern to minimize the total sidelobe clutter output.

The simulation results are shown in more detail in the time history plot of Fig. 5. Two quantitative figures for the improvement in S/N (ratio of main-beam gain to sidelobe clutter output) are shown: the steady-state improvement obtained with the optimum weights of (29), and the improvement obtained after 1600 simulated iterations of the adaptive loops. Note that, by comparison with the uniformly illuminated receiving array, a large improvement in S/N is obtained. The base of the plot is 45 dB below the peak gain. The curve to the rear is the initial receiving pattern in decibels. Successive curves show the receiving pattern for the adaptive array at 200 range ring intervals. The eight curves in the middle thus span the interval from 200 to 1600 loop iterations. The final curve in front is the steady-state array pattern corresponding to optimum array weights. This same format will be used in presenting each of the simulation results.

Figs. 6 and 7 show the response of an adaptive array radar when the clutter is restricted to a smaller angular region between 15 and 45 degrees. Note that the sidelobes are reduced in this angular interval. The adaptive array again senses the angular distribution of interference and

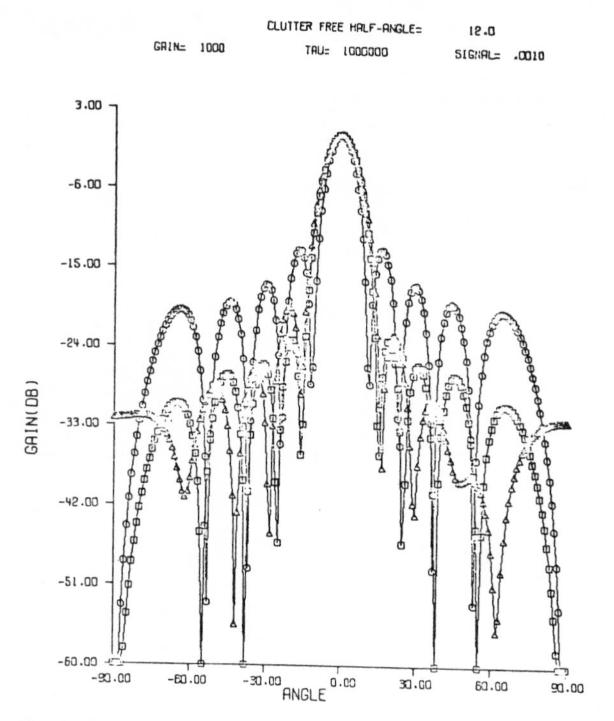

Fig. 8. Adaptive array gain, narrow clutter-free angular region.

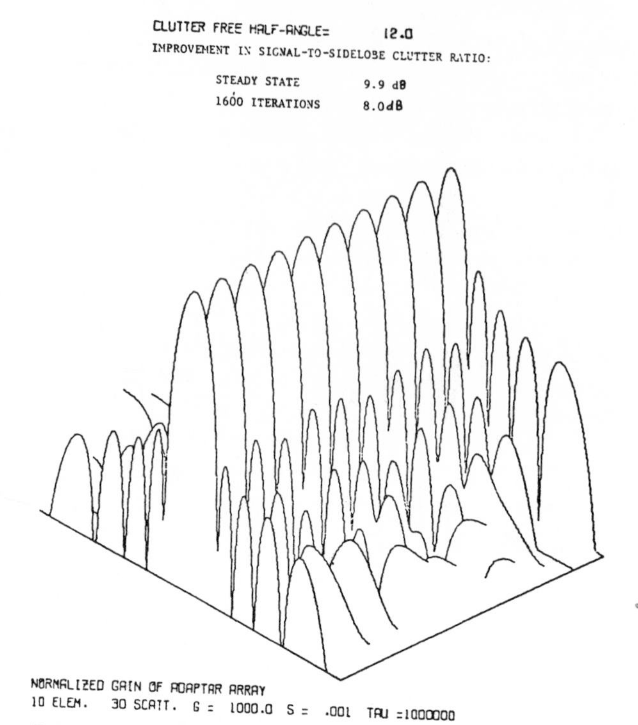

Fig. 9. Projectograph plot of adaptive array gain, narrow clutter-free angular region.

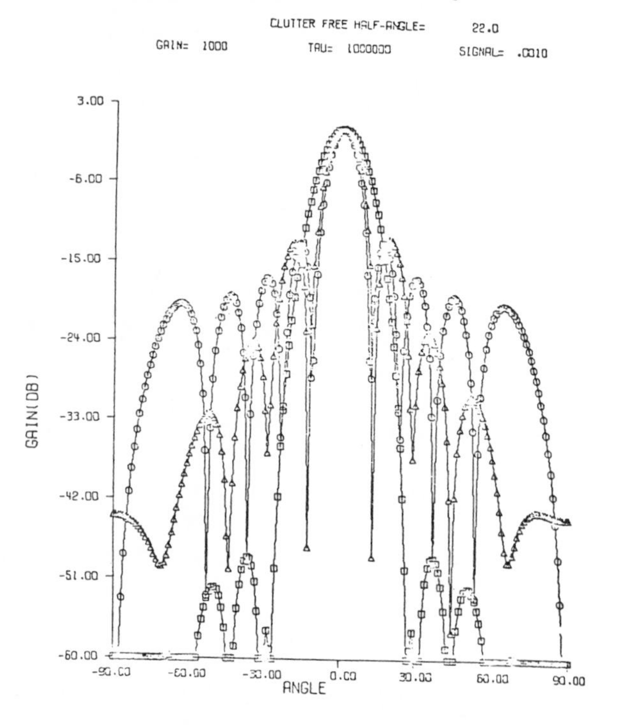

Fig. 10. Adaptive array gain, wide clutter-free region.

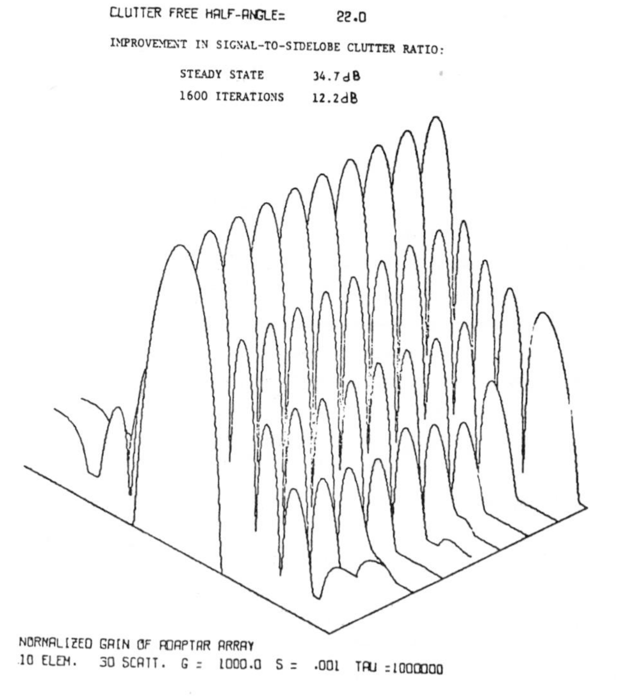

Fig. 11. Projectograph plot of adaptive array gain, wide clutter-free region.

reduces the sidelobes selectively to improve the output signal-to-noise ratio.

The convergence rate of an adaptive array to optimum weights and the noisiness of the control loops are a function of the external noise field, as well as of the loop parameters G and τ. This variation of transient response is illustrated in Figs. 8 through 11, which present simulation results for cases where the clutter-free half-angle is 12 and 22 degrees, respectively. The other parameters in these simulations are the same as in the reference case of Figs. 4 and 5 where the clutter-free half-angle is 17 degrees. Figs. 8 and 9 show that when the interference-free angle is small, the steady-state and transient performance of the adaptive array are both poor. This is not surprising; it is known from array theory (e.g., the Dolph-Chebychev theory) that it is not possible to obtain low sidelobe levels without some widening of the main beam. The interference-free region is too small in this example to permit beam widening without passing some main-beam interference.

The result in Figs. 10 and 11 for a wide clutter-free region shows that excellent performance is possible in this case. The steady-state pattern (marked with squares in Fig. 10) has very low sidelobes outside the region from −22 to +22 degrees. However, the transient response is poor in this example. Widening the interference-free region from 17 degrees (Figs. 4 and 5) to 22 degrees (Figs. 10 and 11), while improving the steady-state array performance, has reduced the convergence rate of the loops very significantly.

This variation in array performance with distribution of the interference field can be explained as follows. The differential equation describing adaptive array response, derived in (65), is

$$\frac{\tau}{G}\dot{W} + \left(M + \frac{I}{G}\right)W = S^* \tag{73}$$

where we have assumed G is a scalar and I is the identity matrix.

Since M is Hermitian, there is a unitary transformation U which diagonalizes the convariance matrix. In terms of the transformed quantities

$$\omega = UW$$
$$\mathscr{S}^* = US^*$$
$$\Lambda = UMU^{-1}, \tag{74}$$

(73) becomes

$$\frac{\tau}{G}\dot{\omega} + \left(\lambda + \frac{I}{G}\right)\omega = \mathscr{S}^* \tag{75}$$

where Λ is a diagonal matrix with elements equal to the eigenvalues of M. In these normal coordinates, (75) corresponds to a set of N independent equations

$$\frac{\tau}{G}\dot{\omega}_n + \left(\lambda_n + \frac{1}{G}\right)\omega_n = \mathscr{S}_n^*$$

with the solutions

$$\omega_n = \left(\omega_{n0} - \frac{\mathscr{S}_n^*}{\lambda_n + 1/G}\right)\exp\left(-\frac{G\lambda_n + 1}{\tau}t\right) + \frac{\mathscr{S}_n^*}{(\lambda_n + 1/G)} \tag{76}$$

where the ω_{n0} are initial values of the ω_n. The quadratic form $W_T{}^*MW$ is non-negative for any W, so the eigenvalues λ_n are all non-negative. It follows that the mean values of the weights ω converge to their respective steady-state values $[\omega_n = \mathscr{S}_n^*/(\lambda_n + 1/G)]$ for any positive value of G. The rate of convergence is determined by the effective time constants $\tau/(G\lambda_n + 1)$.

These equations can be used to compute the transient response of an adaptive array. The procedure is:

1) Compute the covariance matrix of the interference field.

2) Compute the eigenvalues of M and the unitary transformation U. Standard computer routines are available for these computations.

3) Transform the initial weights and steering signals to normal coordinates, as in (74).

4) Compute the transient response in normal coordinates using (76).

5) Transform the weights back to W space, using $W = u^{-1}\omega$.

This computation was performed for the simulation runs and gave a good analytical estimate of the simulation results in cases of small loop noise.

While the eigenvalues λ_n are a function of the incident noise field at the array, the loop parameters can be selected in designing an adaptive system. This suggests the possibility of using large loop gain G and small time constant τ to achieve rapid convergence, even in cases of small eigenvalues. The ω_n converge to their steady-state values exponentially, so each w_n converges as the sum of N exponentials. Using a large G/τ ratio is not a useful solution to the loop convergence problem, however, when the covariance matrix has widely different eigenvalues. Increasing G/τ leads to very noisy loops in these cases.

It is shown in [15] that the total noise power in the array output is

$$P_N = \overline{W}_T{}^*M\overline{W}\left[1 + \frac{G}{2\tau}\sum_{n=1}^{N}\lambda_n\right] \tag{77}$$

where $\overline{W}$ is the average value to which the array weights converge in the case of no loop noise. The quantity $\overline{W_T}{}^*M\overline{W}$ is the output noise power in the absence of loop noise. Equation (77) shows that the output power is increased by the fraction $G/2\tau) \sum \lambda_n$ due to loop noise. When the covariance matrix contains both large and small eigenvalues, it is impossible to find a G/τ ratio which yields rapid convergence of the array weights without excessive loop noise. The subject is discussed in more detail in [15].

References

[1] E. J. Kelly and E. C. Lerner, "A mathematical model for the radar echo from a random collection of moving scatterers," M.I.T. Lincoln Lab., Lexington, Rep. 123, June 15, 1956.

[2] J. L. Wong, I. S. Reed, and Z. A. Kaprielian, "A model for the radar echo from a random collection of rotating dipole scatterers," *IEEE Trans. Aerospace and Electronic Systems*, Vol. AES-3, pp. 171-178, March 1967.

[3] E. J. Kelly, I. S. Reed, and W. L. Root, "The detection of radar echoes in noise, I", *J. Soc. Indust. Appl. Math.*, vol. 8, pp. 309-341, June 1960.

[4] I. S. Reed, "On a moment theorem for complex Gaussian processes," *IRE Trans. Information Theory*, vol. IT-8, April 1962.

[5] I. Selin, *Detection Theory*. Princeton, N.J.: Princeton University Press, 1965.

[6] J. I. Marcum and P. Swerling, "Studies of target detection by pulsed radar," *IRE Trans. Information Theory*, vol. IT-6, April 1960.

[7] G. N. Watson, *Theory of Bessel Functions*, 2nd ed. London: Cambridge University Press, 1966.
[8] D. G. Childers and I. S. Reed, "On the theory of continuous array processing," *IEEE Trans. Aerospace and Electronic Systems*, vol. AES-1, June 1965.
[9] H. L. Van Trees, "Optimum signal design and processing for reverberation limited environments," *IEEE Trans. Military Electronics*, vol. MIL-9, July-October 1965.
[10] W. F. McGee, "Complex Gaussian noise moments," *IEEE Trans. Information Theory*, vol. IT-17, March 1971.
[11] M. J. D. Powell, "A survey of numerical methods for unconstrained optimization," *SIAM Rev.*, vol. 12, no. 1, January 1970.
[12] B. Widrow *et al.*, "Adaptive antenna systems," *Proc. IEEE*, vol. 55, December 1967.
[13] E. Hille, "Functional analysis and semi-groups," *Am. Math. Soc. College Publ.*, vol. 31, 1948.
[14] S. P. Applebaum, "Adaptive arrays," Syracuse University Research Corp., dept. SPL-769, June 1964.
[15] L. E. Brennan, E. L. Pugh, and I. S. Reed, "Control-loop noise in adaptive array antennas," *IEEE Trans. Aerospace and Electronic Systems*, vol. AES-7, March 1971.

Lawrence E. Brennan (S'47–A'51–M'57) was born in Oak Park, Ill., on January 29, 1927. He received the B.S. and Ph.D. degrees in electrical engineering from the University of Illinois, Urbana, in 1948 and 1951, respectively.

Most of his work since leaving the University of Illinois has been concerned with radar and infrared systems. From 1958 to 1967, he was with the Department of Electronics of The RAND Corporation, Santa Monica, Calif. From April 1962 to October 1963, on leave of absence from RAND, he was employed by the SHAPE Technical Center, The Hague, Netherlands. In June 1967 he joined Technology Service Corporation, Santa Monica, as a Senior Scientist. His current research interests include radar detection theory, adaptive systems, and phased array antennas.

S. Reed, for a photograph and biography, please see this issue, page 165.

RAPID CONVERGENCE IN ADAPTIVE ARRAYS
I. Reed, J. Mallett and L. Brennan
IEEE Transactions on Aerospace and Electronic Systems, Vol. AES-10, No. 6, November 1974, pp. 853-863.

7.4

Rapid Convergence Rate in Adaptive Arrays

I.S. REED, Fellow, IEEE
J.D. MALLETT, Member, IEEE
L.E. BRENNAN, Member IEEE
Technology Service Corporation
Santa Monica, Calif. 90401

Abstract

In many applications, the practical usefulness of adaptive arrays is limited by their convergence rate. The adaptively controlled weights in these systems must change at a rate equal to or greater than the rate of change of the external noise field (e.g., due to scanning in a radar if step scan is not used). This convergence rate problem is most severe in adaptive systems with a large number of degrees of adaptivity and in situations where the eigenvalues of the noise covariance matrix are widely different. A direct method of adaptive weight computation, based on a sample covariance matrix of the noise field, has been found to provide very rapid convergence in all cases, i.e., independent of the eigenvalue distribution. A theory has been developed, based on earlier work by Goodman, which predicts the achievable convergence rate with this technique, and has been verified by simulation.

I. Introduction

Convergence rate in adaptive systems has been found to be impractically slow with some adaptive implementations. A technique for achieving very rapid convergence in *all* cases is described here. Section II reviews the theory of adaptive radar and presents a new and concise derivation of the optimum weight equation. Section III presents the theory of convergence rate in systems employing the sample covariance matrix for weight computation. These theoretical results were verified by computer simulation, as discussed in Section IV.

Manuscript received October 29, 1973. Copyright 1974 by *IEEE Trans. Aerospace and Electronic Systems,* vol. AES-10, no. 6, November 1974.

This work was supported in part by M.I.T. Lincoln Laboratory under the direction of Dr. B. Gold and in part by the Naval Air Systems Command under Contract N00019-73-C-0093, monitored by J. Willis.

II. Optimum and Adaptive Array Processing

Consider an array of M elements. Let $x_k(t)$ be the process, received by the kth element for $K = 1, 2, \cdots, M$. Sample this set of processes at times $t_1, t_2, \cdots, t_L$. For example, for a pulsed radar it is usual to sample periodically at range $R = ct_1/2$, where c is the velocity of light and t_1 is measured from the time of transmission of a pulse. The intervals between consecutive samples, $t_n - t_{n-1}$, are equal to the pulse repetition period. Denote this sampled data set by the column vector

$$X = \begin{pmatrix} x_1 \\ x_2 \\ \cdot \\ \cdot \\ \cdot \\ x_N \end{pmatrix}$$

x

where N is the product of M and L; i.e., $N = ML$.

For detection, we choose between two hypotheses, the noise-only hypothesis H_0 and the signal-plus-noise hypothesis H_1. For an array, the signal is the set of voltages $(s_1, s_2, \cdots, s_N)$ an electromagnetic wave from the selected search direction induces on the M receiving elements. The expected value of X, given H_0 (noise only), is

$$EX = 0$$

and the expected value of X, given H_1 (signal plus noise) is

$$EX = S$$

where S is the column vector

$$S = \begin{pmatrix} s_1 \\ s_2 \\ \cdot \\ \cdot \\ \cdot \\ s_N \end{pmatrix}$$

For hypothesis H_1, the noise" is

$N = X - S.$

To illustrate the signal vector S, let the array be linear with element spacing d. If a plane wave of carrier frequency f impinges on the array at an angle ψ to the array normal, the voltages s_k at the kth element with the carrier removed by heterodyning are

$$s_k = A_k \exp [ik(2\pi d/\lambda) \sin \psi + i\delta$$

$$\text{for } k = 1, 2, \cdots, N \qquad (1)$$

where $\lambda = c/f$ is the wavelength, δ is a constant phase factor, and c is the velocity of light. A_k for $k = 1, 2, 3, \cdots, N$ is the wave amplitude times a possible amplitude taper, imposed on the elements of the array.

In order to best detect the presence of a signal S, such as (1), one designs a filter which is "tuned to" S in such a manner that the effects of noise and interference are minimized. As we shall see, one criterion which accomplishes this is the maximum signal-to-noise principle. It has been shown elsewhere [1] that if the noise and interference approximate Gaussian processes, then a maximization of the signal-to-noise ratio is equivalent to a maximization of the probability of detection. It is in this sense that the maximum signal-to-noise criterion is an optimal principle.

The filter, needed to detect S, is a weighted sum of the components of X. If the weight vector is

$$W = \begin{pmatrix} w_1 \\ w_2 \\ \cdot \\ \cdot \\ \cdot \\ w_N \end{pmatrix},$$

the output of the filter is the scalar

$$y = \sum_{i=1}^{N} \overline{w_i} x_i = W^* X \qquad (2)$$

where $\overline{w_i}$ denotes complex conjugation and W^* denotes the conjugate transpose of the column matrix W. The first- and second-order statistics of y, assuming hypothesis H_1, are the mean

$$Ey = W^* S \qquad (3)$$

and the variance

$$\text{Var}(y) = E\,|y|^2 - |Ey|^2$$

$$= EW^* NN^* W = W^* MW \qquad (4)$$

where M is the covariance matrix, given by

$$M = ENN^* = (En_i \overline{n_j}) \qquad (5)$$

for $i, j = 1, 2, \cdots, N$, where the n_j are the noise components of X_j. Clearly, $M^* = M$, so that M is Hermitian.

The noise vector N includes receiver noise of $\psi_0 = kT$ watts per unit bandwidth over bandwidth B for the kth antenna element, where $k = 1, 2, \cdots, N$. Thus, if the sampling times $t_1, t_2, \cdots, t_L$ are also separated sufficiently in time, the covariance matrix has the form

$$M = \psi_0 BI + M_0$$

where I is the $N \times N$ identity matrix and M_0 is the covariance matrix of the external noise only, e.g., sky noise, external interference, etc. Since M_0 is Hermitian and positive semidefinite, there exists a unitary matrix U which diagonalize M_0, i.e.,

$$U^* M_0 U = \Lambda = \text{diag}\,(\lambda_1, \lambda_2, \cdots, \lambda_N)$$

where $\lambda_1, \lambda_2, \cdots, \lambda_N$ are the non-negative eigenvalues of M_0, the covariance matrix of the external noise. Since $U^* = U^{-1}$, the unitary matrix U also diagonalizes M. That is,

$$U^* MU = U^* [\psi_0 BI + M_0] U = \psi_0 BI + \Lambda$$

$$= \text{diag}\,(\psi_0 B + \lambda_1, \psi_0 B + \lambda_2, \cdots, \psi_0 B + \lambda_N).$$

Hence, the eigenvalues of M, $\psi_0 B + \lambda_k$, for $k = 1, 2, \cdots, N$ are positive and M is a positive definite Hermitian matrix; i.e.,

$$X^* MX > 0$$

for all vectors $X \neq 0$.

The output signal-to-noise ratio for a complex process associated with narrowband noise is the expected signal power out of filter y divided by the noise power Var (y) given in (4). Thus, by (3) and (4),

$$(S/N)_0 = |Ey|^2/\text{Var}\,(y) = |W^* S|^2/W^* MW \qquad (6)$$

where M is the covariance matrix of the noise, (5), and S is the signal vector; e.g., for a linear array the signal would have the form of (1).

The "best" or optimum filter W_0 is obtained by maximizing the signal-to-noise ratio given in (6). Though this maximization is done elsewhere (see [1] and [2]) it is instructive to find W_0 by another method. Here we use the following easily proved fact. If X and Y are two N-component column vectors and M is a positive definite $N \times N$ Hermitian matrix, then

$$(X, Y) = X^* MY \qquad (7)$$

is an inner product of the vector space of N-component column vectors. An important property of an inner product is the Schwartz inequality

$$|(X, Y)|^2 \leqslant (X, X)(Y, Y). \tag{8}$$

If the signal-to-noise ratio (6) is re-expressed in terms of the inner product (7), we have

$$(S/N)_0 = |(W, M^{-1}S)|^2/(W, W)$$

$$\leqslant [(W, W)(M^{-1}S, M^{-1}S)]/(W, W)$$

$$= (M^{-1}S, M^{-1}S) = S^*M^{-1}S \tag{9}$$

where the bound on the right follows from the Schwartz inequality (8). The bound on the right in (9) can be attained if one lets the filter W in (6) be given by

$$W_0 = kM^{-1}S \tag{10}$$

where k is a complex number. To see this, substitute (10) in (6) as follows:

$$(S/N)_0 = |W^*S|^2/W^*MW$$

$$= [1/|k|^2]\left\{[|k|^2|(S^*M^{-1})S|^2]/\right.$$

$$\left.\cdot [(S^*M^{-1})M(M^{-1}S)]\right\}$$

$$= (S^*M^{-1}S)^2/S^*M^{-1}S = S^*M^{-1}S.$$

Thus W_0, as given by (10), is the optimum filter, the optimum set of filter weights, for the detection of S in the presence of noise M.

In practice, one generally does not know a priori the clutter and interference situation. When it is known and an interference model can be constructed, the "best" set of antenna weights is computed in accordance with W_0, as given by (10). However, if the interference field changes, for example, by the presence of moving near-field scatterers, antenna errors, interference, and jamming, one must continually update or *adapt* the weight vector W_0 to meet the varying conditions.

The weights, the components of W_0, are adapted to new situations by processing a number of independent samples of received data. The different methods of processing the data to estimate W_0 require different numbers of independent samples for the same level of performance. That is, the convergence to achieve a certain signal-to-noise ratio is dependent on the adaption criterion. The purpose of this paper is to compare the convergence rates of the already familiar Widrow and Applebaum recursive algorithms (see [1] and [2] for the development of these algorithms) with the method of direct sample matrix inversion of M, where M is the estimate of the covariance matrix M using K independent samples of data. Both computed and theoretical results will be used to make this comparison. Finally, some discussion of the trade-offs between equipment complexity and the rate of convergence will be given.

III. Rapid Adaptivity By Sample Matrix Inversion

In this section we show how the adaptive array-filter processor can be achieved by first estimating the covariance matrix M, using K samples. Next, this estimate $\hat{M}$ of M is inverted to form, finally, the filter $k\hat{M}^{-1}S$. In the next section, this method of adaptivity will be compared by example with the LMS (least-mean-square) algorithm of Widrow [2] and the MSN (maximum signal-to-noise)[1] algorithm of Applebaum [1].

Suppose the noise process impinging on the receiving elements is observed through the N-dimensional vector X for K *independent* samples. In a range sampling radar, for example, these samples might be the returns from K different and independent range intervals. Let $X^{(j)}$ for $j = 1, 2, 3, \cdots, K$ denote the jth sample of vector X. For noise only, $X^{(1)}, X^{(2)}, \cdots, X^{(K)}$ are K independent and identically distributed N-variate complex vectors. Each of these vectors is distributed with probability density

$$P(X) = (\pi)^{-N}|M|^{-1}\exp(-X^*M^{-1}X)$$

where $|M|$ is the determinant of covariance matrix M [3], [4]. Thus, the joint probability density of $X^{(1)}, X^{(2)}, \cdots, X^{(K)}$ is

$$P(X^{(1)}, X^{(2)}, \cdots, X^{(K)}) = (\pi)^{-NK}|M|^k$$

$$\cdot \exp -(\sum_{i=1}^{K} X^{(i)*}M^{-1}X^{(i)}) \tag{11}$$

where M is the covariance or complex moment matrix, (5),

$$M = EXX^*$$

where X is the sampled-data vector for noise alone.

In order to obtain the estimate $\hat{M}$ to substitute for M in (10) of the optimum filter W_0, one uses the maximum-likelihood principle. The maximum-likelihood estimate of M has been shown by Goodman [3, Theorem 4.1] to be

$$\hat{M} = 1/K \sum_{j=1}^{K} X^{(j)}X^{(j)*}$$

$$= 1/K \sum_{j=1}^{K} (x_r^{(j)}\overline{x}_s^{(j)}) \tag{12}$$

where $x_r^{(j)}\overline{x}_s^{(j)}$ denotes the elements in the rth row and sth column of matrix $X^{(j)}X^{(j)*}$. This result is obtained by maximizing the logarithm of the likelihood function

$$L = P(X^{(1)}, X^{(2)}, \cdots, X^{(K)}),$$

as given by (11). $\hat{M}$, as given by (12), is simply the arithmetic average of the K sample matrixes $X^{(j)}X^{(j)*}$ for $j = 1, 2, \cdots, K$. M is called the sample covariance matrix.

[1] In general, this algorithm maximizes the ratio of signal to noise, where the noise term includes both receiver noise and all external noise components.

If one chooses $\hat{M}$ for M in the optimum matched filter criterion (10), the filter has the form

$$\hat{W} = k\hat{M}^{-1}S \tag{13}$$

where $\hat{M}$ is given by (12). Suppose $\hat{W}$ as given above is used to filter and detect the presence of a signal S in another independent sample of vector X. By (2), substituting $\hat{W}$ for W, the output of this filter is

$$\hat{y} = \hat{W}^* X. \tag{14}$$

Let us now compute the output signal-to-noise ratio $(S/N \mid \hat{W})_0$, assuming that the estimated filter $\hat{W}$ is held fixed.

The first- and second-order statistics of $\hat{y}$, holding $\hat{W}$ fixed, i.e., the conditional first and second statistics conditioned on $\hat{W}$, are computed as follows:

$$E(\hat{y} \mid \hat{W}) = \hat{W}^* S$$

and

$$\begin{aligned} \text{Var}\,(\hat{y} \mid \hat{W}) &= E(|\hat{y}|^2 \mid \hat{W}) - |E(\hat{y} \mid \hat{W})|^2 \\ &= \hat{W}^* [E(X-S)(X-S)^*]\,\hat{W} \\ &= \hat{W}^* M \hat{W} = S^* (\hat{M})^{-1} M (\hat{M})^{-1} S. \end{aligned}$$

Thus, by (6), the output signal-to-noise ratio, conditioned on $\hat{W}$, is

$$\begin{aligned} (S/N \mid \hat{W})_0 &= [E(\hat{y} \mid \hat{W})]^2 / \text{Var}\,(\hat{y} \mid \hat{W}) \\ &= (S^* \hat{M}^{-1} S)^2 / S^* \hat{M}^{-1} M \hat{M}^{-1} S. \end{aligned} \tag{15}$$

Since $\hat{M}$, by (12), is a function of the random data vector $X^{(j)}$ for $j = 1, 2, \cdots, K$, the above-conditioned signal-to-noise ratio in (15) is a random variable. It is convenient in dealing with this random variable to normalize it with respect to its upper bound, namely, the right side of (9). Denote this normalized signal-to-noise ratio by $\rho(\hat{M})$. That is, let

$$\begin{aligned} \rho(\hat{M}) &= (S/N \mid \hat{W})_0 / S^* M^{-1} S \\ &= (S^* \hat{M}^{-1} S)^2 / [(S^* M^{-1} S)(S^* \hat{M}^{-1} M \hat{M}^{-1} S)]. \end{aligned} \tag{16}$$

Evidently, the random variable ρ lies in the interval $0 \leqslant \rho(\hat{M}) \leqslant 1$.

In [4], Capon and Goodman used the Goodman theory [3] of the complex Wishart distribution to find the probability distribution of certain estimators, similar to the random variable $\rho(\hat{M})$ defined above. Also, Goodman, in unpublished notes, found that the distribution of the inverse of $\rho(\hat{M})$ satisfied an incomplete beta function distribution. By a simple change of variable, the probability density of the normalized signal-to-noise ratio $\rho(\hat{M})$ is likewise a beta function distribution, namely,

$$\begin{aligned} P(\rho) = &\{K!/[(N-2)!(K+1-N)!]\} \\ &\cdot (1-\rho)^{N-2} \rho^{K+1-N} \end{aligned} \tag{17}$$

where $0 \leqslant \rho \leqslant 1$ and $K \geqslant N$. An outline of the derivation of (17) is included in the Appendix.

The expected value of $\rho(\hat{M})$ can be computed, using (17), to be

$$E(\rho(\hat{M})) = (K + 2 - N)/(K + 1). \tag{18}$$

This is the expected loss in power ratio if only K samples of data were used to estimate $\hat{M}$ in (12). Expressed in decibels, this expected loss in power ratio is

$$\begin{aligned} \text{loss} &= -10 \log_{10} \{E[\rho(\hat{M})]\} \\ &= -10 \log_{10} [(K + 2 - N)/(K + 1)]. \end{aligned} \tag{19}$$

If one wishes to maintain an average loss ratio of better than one-half (less than 3 dB) by (17) and (10), at least $K = 2N - 3 \cong 2N$ samples of data are needed. This useful "rule-of-thumb" requirement on the number of data samples will be shown in the next section to agree with results obtained by simulation. In general, the loss in decibels compared to the optimum, given in (19), supplies the trade-off between performance and the number of samples needed for filter adaptation in the method of sample matrix inversion (SMI).

IV. Verification of Theoretical Results

A. Comparison of Adaptive Criteria by Simulation

By (18), if the number of samples K gets large, the expected value of $\rho(\hat{M})$ approaches one. Thus, $\hat{M}$, the sample covariance matrix, converges to the covariance matrix M, given in (5). When $\hat{M}$ is generated by computer simulation, using a set of fixed clutter points with a different random phase and amplitude applied to each point of a sample, it likewise will approach M when the number of samples K is large.

The question of how many samples are required to approach the covariance matrix $\hat{M} \to M$ was addressed using a computer simulation. The simulation was of a moving platform radar with four antenna elements and two sampling pulses ($N = 8$ complex inputs). The interference clutter consisted of 30 scatterers. These scatterers were spaced equally on a semicircle of large radius centered on the array. The array lying on the semicircle diameter was imagined to move a distance of 0.2 wavelengths along its length between pulses to simulate platform motion. The scatterers were illuminated on each pulse by a broadside

TABLE I

4 Elements; 2 Pulses [$N = 2 \times 4$]; Spacing = 0.5 λ
Interpulse motion = 0.2 λ/pulses; Steady state gain = 62.2 dB

No.of Samples K	SMI Theory*	SMI Simulation (MSN)	Adaptive Loops (MSN)
5		28.4	15.3
10		55.6	15.5
15	59.7	58.9	15.8
20	61.0	58.6	17
800			37
2000			38
∞	62.2		

*Values given by 62.2 ± 10 $\log_{10}$ [$(K - N + 2)/(K$ ¼ $1)$].

transmitter beam formed by the four equally weighted antenna elements.

The ideal or steady-state covariance matrix M was first computed by making use of the known positions of clutter and elements to sum the covariance matrixes for each clutter point. This yields the steady-state or ideal matrix M, where each matrix element M_{mn} is

$$\begin{aligned} M_{mn} &= E\left\{V_m \bar{V}_n\right\} = E\left\{\Sigma V_{mj} \Sigma \bar{V}_{nj}\right\} + R_m \delta_{mn} \\ &= E\left\{\sum_{j=1}^{J} V_{mj} \bar{V}_{nj} + \left[\sum_{k=1}^{J} V_{mk} \sum_{\substack{l=1 \\ l \neq k}}^{J} \bar{V}_{nl}\right]\right\} + R_m \delta_{mn} \\ &= \sum_{j=1}^{J} V_{mj} \bar{V}_{nj} + R_m \delta_{mn} \end{aligned}$$

where

$E\ \{\ \}$ = expected value
$\bar{V}_{mj}$ = conjugate voltage at mth antenna element from jth clutter point; m and n range over all antenna elements for each pulse
$\delta_{mn} = 1, m = n$
$= 0, m \neq n$
R_m = average value of receiver noise at each element on each pulse.

From this, the optimum filter weights were obtained from (10), using matrix inversion. These weights, in turn, were then used to compute the signal-to-clutter ratio [the signal-to-noise ratio given in (6)]:

$$S/C = |W^* S|^2 / W^* M W. \tag{20}$$

The MTI gain is defined as the S/C ratio, given by (20), divided by a normalizing constant, the ratio of signal-to-clutter returned on one pulse, assuming uniform antenna weights. The steady-state maximum MTI gain for the above problem was computed to be 62.2 dB. A sample covariance matrix $\hat{M}$ was then computed using different numbers of samples k, where

$$\hat{M}_{mn} = \sum_{i=1}^{K} V_{mi} \bar{V}_{ni}. \tag{21}$$

For each sample i, every clutter point is given a random phase and amplitude and the complex voltages are added at each element for each pulse. Weights were obtained by inverting the sample covariance matrix $\hat{M}$, as in (13), and these weights were used to compute the MTI gain, using (20), where M is the steady-state expected value covariance matrix. Some results are shown in column 3 of Table I. The simulated results in column 3 show the rapid convergence to the steady-state value of 62.2 dB and the close comparison with theoretical prediction column 2 for a single run. The experiment was performed several times with different random numbers for determining clutter phase and amplitude, with similar results.

For comparison, some results are shown in column 4 of Table I, using the maximum signal-to-noise (MSN) loops of Applebaum [1]. More results for the MSN algorithm are shown in Fig. 1. The MSN algorithm and least-mean-square (LMS) algorithm of Widrow [2] both derive the next set of weights from the previous set using a recursive relation, which, in digital form, can be expressed as

$$W_i(j+1) = W_i(j)[1 - (1/\tau)] + (g/\tau)e(j)V_i^*(j) - (S_i^*/\tau)$$

where

$$e(j) = P(j) - \sum_{i=1}^{N} W_i(j) V_i(j)$$

is the difference between the received signal at the output and the transmitted pilot signal $P(j)$ (present only in the LMS case), and where g is the loop gain, τ is a filter smoothing constant, and S_i^* are the steering vectors. For LMS, $S_i^* = 0$. For MSN, $S_i^* \neq 0$ and $P(j) = 0$. The MSN algorithm usually converges more rapidly because the steering vectors are supplied. These algorithms require essentially only one complex multiply per sample for each channel, i.e., $g/\tau e(j) V_i^*(j)$, and are therefore simple compared to the direct inversion method.

Both algorithms yield optimum steady-state weights and maximum MTI gain in the limit. The number of samples required to approach the limit using these approximations

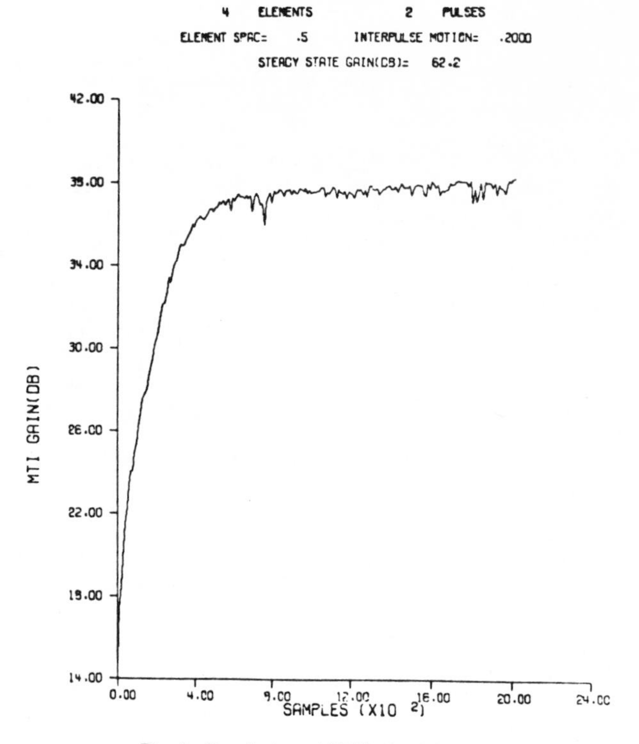

Fig. 1. Simulation of MSN algorithm.

may be very large—10^9 or more samples—and there are frequently plateaus, similar to that shown in Fig. 1 around 38 dB, that are significantly below the optimum performance of, in this case, 62 dB.

The slow convergence of the MSN and LMS loops can be understood in terms of the eigenvalues of the noise covariance matrix. A solution for the transient response of these algorithms has been found by transforming the vector equations into normal coordinates [1]. The time response is then given by a sum of exponentials in which the time constants are a function of the eigenvalues of the covariance matrix. When the eigenvalues differ by orders of magnitude, the performance converges to different levels in widely different numbers of samples.

Convergence using these algorithms is highly example-dependent. In some cases it is very fast. This usually occurs when a discrete set of noise sources is present and the number is less than the number of adaptive loops. When interference is widely distributed, as in this example, there often exists a good solution, but the number of samples required to attain it can be very high. Using the SMI method, convergence is independent of the noise environment and depends only on the number of adaptive elements.

B. SMI With Signal Present

Adaptive systems do not converge fast when weights are formed with signal present. In radar applications, it is usually possible to form the interference covariance matrix with signal absent. To determine the effect of having some signal present, experiments were run with signals of varied amplitude present in one of the samples used to form the sample covariance matrix. The weights determined from this covariance matrix were used with the expected value covariance matrix to determine MTI gain. Some results of this experiment are shown in Table II.

The small degradation in MTI gain and in the output S/C can be seen as a function of input clutter-to-signal ration. When large signals are present, the output S/C ratio is degraded, but appears to be enough for detection.

C. Estimating Relative Complexity

An advantage of the MSN algorithm is the simplicity of its implementation, especially when analog techniques are employed. The shift to digital processing, however, makes more sophisticated processing possible, and preferable in some cases. Not only can performance be achieved that was not practical before, but in many cases it can be done more cheaply. While the MSN algorithm is very simple to implement conceptually, its convergence rate can be very slow, as discussed earlier. The SMI technique, on the other hand, converges with the smallest possible number of samples, but requires more processing steps, and usually more accuracy, i.e., bits. To compare these techniques when digital processing is used, we can compare the number of complex multiplications and the amount of storage required as a way of estimating cost.

TABLE II

4 Elements; 2 Pulses; Spacing = 0.5 λ
Interpulse motion = 0.2 λ/pulse; Steady-state gain = 62.2 dB

Signal Present in One Sample

C/S (dB) Input	10 Samples MTI Gain	10 Samples S/C Out	20 Samples MTI Gain	20 Samples S/C Out	∞ Samples MTI Gain	∞ Samples S/C Out
75	55.6	-19.4	58.6	-16.4	62.2	-13
55	57.9	2.9	59.6	4.6	62.2	7
35	44.4	9.4	56.8	21.8	62.2	27

Usually when loops are used, all the incoming data is processed. This may not be practical or desirable, however, when using a sample covariance matrix. In the latter case, advantage can be taken of the small number of samples required and the fact that new weights are required only as often as there is change in position or distribution of the interference. These changes are often of the order of 0.01 to 1 seconds, rather than microseconds, which reduces the cost of SMI processing. Spatial interferers, for example, do not usually change angular direction rapidly and if several at different locations blink on and off, both will be nulled so long as the sampling interval is either long or short compared to the blink rate. Motion of the antenna itself causes a fast change, but this can sometimes be avoided by using electronic scan.

Formation of the sample covariance matrix requires $SN(N+1)/2$ complex multiplications, where S is the number of samples required for convergence and N is the total number of weights, i.e., antenna elements times pulses for a space time processor. To invert the Hermitian matrix requires $(N^3/2 + N^2)$ complex multiplications, and to form each set of weights requires another N^2 multiplications. If only one set of weights is required, i.e., no filter banks or multiple beams are needed, then the weights can be found by solving the set of equations $W = \hat{M}^{-1} S$, which requires about $N^3/6$ complex multiplications. Substituting $S = 2N$ for the number of samples required to be within 3 dB of optimum, the total number of complex multiplications required to obtain a set of weights is approximately $N^3 + N^3/6 = 7/6N^3$ for SMI, compared with NS for the MSC algorithm, where S can be very large. The rate of multiplication, a measure of processing cost, is the number of multiplications divided by the time interval between required weight changes. Often the multiplication rates can be quite low for very good performance.

In addition to the multiplication rate, the accuracy or number of bits required must be considered in estimating complexity. The MSN algorithm requires very low multiplication accuracy, while SMI requires a high accuracy in forming the sample matrix and solving the equations for the weights.

D. Conclusion

Forming a sample covariance matrix and solving for the weights provides a very fast rate of convergence. This rate is dependent only on the number of weights and is independent of the noise and interference environment. It has been found to be orders of magnitude faster than the MSN or LMS algorithms for a number of practical problems. Because of the small number of samples required and the often slow rate of required weight change, it may be economical to implement in many practical radar, communications, and sonar applications.

Appendix

Distribution of $\rho(\hat{M})$

The purpose of this appendix is to outline the method for finding the distribution, (17), of the normalized conditional output signal-to-noise ratio $\rho(\hat{M})$ given in (16). This development is based in part on a paper of Capon and Goodman [4], and on unpublished material made available to one of the authors by N.R. Goodman.

In [3], Goodman shows that the joint distribution of the elements of the estimated covariance matrix (12)

$$\hat{M} = 1/K \sum_{j=1}^{K} x^{(j)} x^{(j)*}$$

is given by the central complex Wishart distribution

$$P(A) = (|A|^{K-N}/I(M) \exp[-tr(M^{-1}A)] \qquad (22)$$

where $A = K\hat{M}$ tr(X) denotes the trace of X, i.e.,

$$\text{tr } X = \text{tr } (x_{ij}) = \sum_{i=1}^{N} x_{ii},$$

M denotes the covariance matrix (5), and $I(M)$ denotes the constant

$$I(M) = \pi^{(1/2)N(N-1)}\Gamma(K)\Gamma(K-1) \cdots \Gamma(K-N+1)|M|^K. \qquad (23)$$

The volume element, associated with density (22), is

$$dV = dA_{11} dA_{22} \cdots dA_{NN} dR(A_{12}) dI(A_{12}) \cdot dR(A_{13}) dI(A_{13}) \cdots dR(A_{N-1,N}) dI(A_{N-1,N})$$

where $R(A_{ij})$ and $I(A_{ij})$ denote the real and imaginary parts of A_{ij}, respectively. The complex Wishart density is defined only over a domain D_A of $N \times N$ matrixes A for which A is positive definite. To understand the complex Wishart distribution of Goodman, it is helpful to compare its development with that of the real Wishart distribution, given by Cramer [5].

The fundamental parameters of the central complex Wishart distribution (22) are K, the sample size, N, the number of dimensions of vector X, and M the $N \times N$ covariance matrix of X under the noise-only hypothesis H_0. Thus, following Capon and Goodman [4], one designates a central complex Wishart distribution with these parameters by $CW(K, N; M)$.

The following theorem is stated by Capon and Goodman [4]. It will be required repeatedly in the reduction process used to find the distribution of $\rho(\hat{M})$.

Theorem: If $N \times N$ matrix A is $CW(K, N; M)$, i.e., complex Wishart distribution with parameters K, N and covariance matrix M, and C is a nonsingular $N \times N$ matrix, then $B = CAC^*$ is $CW(K, N; CMC^*)$.

One wishes to find the distribution of the normalized signal-to-noise ratio in (16), i.e.,

$$\rho = (S^* \hat{M}^{-1} S)^2 / [(S^* M^{-1} S)(S^* \hat{M}^{-1} M \hat{M}^{-1} S)] \tag{24}$$

where $\hat{M}$ is the sample average estimate (12) of the covariance matrix M. To accomplish this, follow Capon and Goodman [4] by making a sequence of transformations which leave the random variable ρ in (16) fixed.

First, let $A = M^{-1/2} S$. Then, by (16),

$$\rho = (A^* \hat{M}_1^{-1} A)^2 / [(A^* A)(A^* \hat{M}_1^{-1} \hat{M}_1^{-1} A)] \tag{25}$$

where

$$\hat{M}_1 = M^{-1/2} \hat{M} M^{-1/2}.$$

The distribution of $\hat{M}$ is given by the complex Wishart distribution, defined in (22). That is, $\hat{M}$ is $CW(K, N; M)$ complex Wishart distributed with parameters K, N and covariance matrix M. By (25),

$$E\hat{M}_1 = M^{-1/2}(E\hat{M})M^{-1/2} = M^{-1/2} M M^{-1/2} = I$$

where I denotes the identity matrix. If, in the above theorem, one lets A be $\hat{M}$ and C be $M^{-1/2}$, then B must be $\hat{M}_1$ and, by the theorem, $\hat{M}_1$ is $CW(K, N; I)$, where I is the $N \times N$ identity matrix.

Next, apply a transformation to (25) which normalizes vector A. That is, let

$$B = (A^* A)^{-1/2} A.$$

Then, ρ becomes

$$\rho = (B^* \hat{M}_1^{-1} B)^2 / B^* \hat{M}_1^{-1} \hat{M}_1^{-1} B \tag{26}$$

where $B^* B = 1$.

The above two transformations on ρ are the same that Capon and Goodman use in [4]. Following them further, use the fact that $B^* B = 1$ to construct an $N \times N$ unitary matrix U (a "rotation" in N-dimensional unitary space) which sends B into the column vector

$$P = (1, 0, 0, \cdots, 0)^*. \tag{27}$$

In other words, since B is a unit vector, there exists a unitary transformation U such that

$$B = UP \tag{28}$$

where P is the unit vector in (27). The application of transformation (28) to (26) yields

$$\begin{aligned} \rho &= [(UP)^* \hat{M}_1^{-1} (UP)]^2 / [(UP)^* \hat{M}_1^{-1} UU^* \hat{M}_1^{-1} (UP)] \\ &= [P^* (U^* \hat{M}_1 U)^{-1} P]^2 / [P^* (U^* \hat{M}_1 U)^{-1} (U^* \hat{M}_1 U)^{-1} P] \\ &= [P^* C^{-1} P]^2 / P^* C^{-2} P \end{aligned} \tag{29}$$

where

$$C = U^* \hat{M}_1 U \tag{30}$$

for the normalized signal-to-noise ratio ρ. By (22),

$$EC = U^* E\hat{M}_1 U = U^* U = I.$$

Thus, by the above theorem, C is $CW(K, N; I)$ distributed.

As suggested in [4], partition matrixes C, C^{-1} as

$$C = \begin{pmatrix} C_{11} & C_{12} \\ C_{21} & C_{22} \end{pmatrix}, \quad C^{-1} = \begin{pmatrix} C^{11} & C^{12} \\ C^{21} & C^{22} \end{pmatrix} \tag{31}$$

where C_{11} and C^{11} are scalars (1×1 matrixes).

Using the fact that $CC^{-1} = I$, this expression for ρ can be expressed as

$$\rho = 1/[1 + (C_{12} C_{22}^{-1})(C_{12} C_{22}^{-1})^*] \tag{32}$$

in terms of only the matrixes C_{22} and C_{12} of the partition (31) of matrix C.

To find the distribution of ρ, as given in (32), start with the distribution of C. Since C is $CW(K, N; I)$ distributed, by (22), the joint probability density of the element of C is

$$P(C) = (|C|^{K-N} / I(I_N)) \exp[-\operatorname{tr}(C)] \tag{33}$$

where I_N is the N-dimensional identity matrix and $I(I_N)$ is the normalization constant given by (23). By a theorem of Frobenius, the partitioned matrix C can be factored as

$$C = \begin{pmatrix} 1 & C_{12} \\ 0 & C_{22} \end{pmatrix} \begin{pmatrix} C_{11} - C_{12} & C_{22}{}^{-1} C_{21}, & 0 \\ C_{22}{}^{-1} C_{21} & , & I \end{pmatrix}.$$

Hence, by Laplace's rules for computing determinants,

$$|C| = |C_{22}|(C_{11} - C_{12}C_{22}{}^{-1}C_{12}{}^{*}). \quad (34)$$

Using (34) in (33) and the fact that

$$\text{tr } C = C_{11} + \text{tr } C_{22},$$

one gets

$$P(C) = (|C_{22}|^{K-N}/I(I_N))(C_{11} - C_{12}C_{22}{}^{-1}C_{12}{}^{*})^{K-N}$$
$$\cdot \exp[-(C_{11} - C_{12}C_{22}{}^{-1}C_{12}{}^{*})]$$
$$\cdot \exp[-(C_{12}C_{22}{}^{-1}C_{12}{}^{*} + \text{tr } C_{22})]. \quad (35)$$

Equation (35) corresponds to (17) in the paper by Capon and Goodman [4]. Following them, next make the change of variables

$$D_{11} = C_{11} - C_{12}C_{22}{}^{-1}C_{12}{}^{*}$$
$$D_{12} = C_{12}$$
$$D_{22} = C_{22} \quad (36)$$

in (35). Since the Jacobian of this transformation is one,

$$P(D_{11}, D_{12}, D_{22}) = (|D_{22}|^{K-N}/I(I_K))D_{11}{}^{K-N}$$
$$\cdot \exp[-D_{11} - D_{12}D_{22}{}^{-1}D_{12}{}^{*}$$
$$- \text{tr } D_{22}]$$
$$= P(D_{11})P(D_{12}, D_{22}) \quad (37)$$

where

$$P(D_{11})\, K_1 D_{11}{}^{K-N} \exp(-D_{11}),$$

a chi-square distribution with $2(K - N + 1)$ degrees of freedom [4, eq. (19)], and

$$P(D_{12}, D_{22}) = K_2 |D_{22}|^{K-N} \exp[-(D_{12}D_{22}{}^{-1})D_{12}{}^{*}$$
$$+ \text{tr } D_{22}] \quad (38)$$

where K_1 and K_2 are normalization constants. Substituting (36) in (32),

$$\rho = 1/[1 + (D_{12}D_{22}{}^{-1})(D_{12}D_{22}{}^{-1})^{*}] \quad (39)$$

is the normalized signal-to-noise ratio in terms of the $N-1$-dimensional vector $D_{12}D_{22}{}^{-1}$. Next, in (38), let

$$E_{12} = D_{12}D_{22}{}^{-1}$$
$$E_{22} = D_{22}. \quad (40)$$

By the special form of this transformation, the Jacobian reduces to

$$\partial(E_{12}, E_{22})/\partial(D_{12}, D_{22}) = \begin{vmatrix} \partial E_{12}/\partial D_{12} & \partial E_{12}/\partial D_{22} \\ 0 & I \end{vmatrix}$$
$$= |\partial E_{12}/\partial D_{12}| = \partial(E_{12})/\partial(D_{12})$$

where $\partial E_{12}/\partial D_{12}$, etc., denotes, symbolically, the array or matrix partial derivatives associated with the transformation from D_{12} to E_{12}, etc. E_{12}, D_{12}, and $D_{22}{}^{-1}$ are expressible in real and imaginary parts as

$$E_{12} = U + iV$$
$$D_{12} = X + iY$$
$$D_{22}^{-1} = F + iG.$$

Then,

$$E_{12} = D_{12}D_{22}{}^{-1} = XF - YG + i(XG + YF).$$

Hence,

$$U = XF - YG$$
$$V = XG + YF,$$

so that

$$\partial(E_{12})/\partial(D_{12}) = \partial(U, V)/\partial(X, Y) = \begin{vmatrix} F & -G \\ G & F \end{vmatrix}$$
$$= |F + iG|^2 = |E_{22}|^{-2}.$$

Thus, if one applies transformation (40) to (38), then

$$P(E_{12}, E_{22}) = K_2 J^{-1} [(E_{12}, E_{22})/(D_{12}, D_{22})]\, |E_{22}|^{K-N}$$
$$\cdot \exp\{-[E_{12}(E_{22}E_{12})^{*} + \text{tr } E_{22}]\}$$
$$= K_2 |E_{22}|^{K-N+2}$$
$$\cdot \exp[-\text{tr}(I + E_{12}{}^{*}E_{12})E_{22}] \quad (41)$$

is the joint probability density of the components of E_{12} and E_{22}. The last step in (41) follows from the fact that $E_{12}(E_{22}E_{12})^{*}$ is scalar, and, as a consequence,

$$E_{12}(E_{22}E_{12})^{*} = \text{tr}(E_{12}{}^{*}E_{12}E_{22}).$$

Observe that the right side of (41) is, if matrix $I + E_{12}{}^*E_{12}$ is held constant, of the same form as the complex Wishart distribution developed by Goodman [3]. In fact, it has the form the distribution would have if it were $CW(K+1, N-1; [I+E_{12}{}^*E_{12}]^{-1})$ distributed. By (22), such a complex Wishart distribution of E_{12} would be

$$P(E_{22}) = K_3 \mid E_{22} \mid^{K-N+2} \mid I + E_{12}{}^*E_{12} \mid^{K+1} \cdot \exp\left[-\operatorname{tr}\left\{I + E_{12}{}^*E_{12}\right\} E_{22}\right]$$

where K_3 is the normalizing constant. Hence, (41) has the form

$$P(E_{12}, E_{22}) = K_3 \left[1/\mid I + E_{12}{}^*E_{12} \mid^{K+1}\right] P(E_{22}).$$

Since the integral of $P(E_{22})$ over domain $D_{E_{22}}$ is unity (see [4, Theorem 5.1]),

$$P(E_{12}) = K_4 \left[1/\mid I + E_{12}{}^*E_{12} \mid^{K+1}\right] \tag{42}$$

where K_3 is the constant of normalization.

The joint probability density of

$$E_{12} = D_{12}D_{22}{}^{-1} = C_{12}C_{22}{}^{-1},$$

as given by (42), can be further reduced in complexity by observing that $E_{12}{}^*E_{12}$ is a Hermitian matrix of rank one. Since the determinant of a Hermitian matrix equals the product of its eigenvalues,

$$\mid I + E_{12}{}^*E_{12} \mid = 1 + E_{12}E_{12}{}^*$$

and, by (42), the joint probability density of the elements of E_{12} is now given by

$$P(E_{12}) = K_4 \left[1/(1 + E_{12}E_{12}{}^*)^{K+1}\right]. \tag{43}$$

Since E_{12} in (43) is a row vector of $N-1$ complex numbers, the "inner" product $E_{12}E_{12}{}^*$ can be written as a sum of squares. Changing to polar coordinates in $2N-2$ dimensions, (43) with its associated volume element is

$$P(E_{12})dV = K_3 \cdot \left[1/(1+r^2)^{K+1}\right] r^{2N-3}\, dr d\Omega \tag{44}$$

where

$$r^2 = E_{12}E_{12}{}^*$$

and $d\Omega$ is a differential of solid angle in $2N-2$ dimensions. Integrating over all solid angles yields the density

$$P(r) = K_4 \left[r^{2N-3}/(1+r^2)^{K+1}\right] \tag{45}$$

of the "radial" coordinate, where constant K_4 is K_3 times the surface "area" of the unit "sphere" in $2N-2$ dimensions.

In terms of r, the normalized signal-to-noise ratio ρ in (32) is given by

$$\rho = 1/(1+r^2). \tag{46}$$

Changing variables in (45) from r to ρ yields

$$P(\rho) = K_5 (1-\rho)^{N-2} \rho^{K+1-N} \tag{47}$$

where $0 \leqslant \rho \leqslant 1$ is the probability density of ρ. The normalization constant K_5 in (47) is obtained from the well-known integral

$$\int_0^1 t^m (1-t)^n \, dt = n!m!/(m+n+1)!$$

where n and m are integers greater than -1.

Using this in (47) yields, finally, the desired probability distribution of ρ, due to Goodman, namely,

$$\operatorname{Prob}\left\{\rho \leqslant X\right\} = \left\{K!/[(N-2)!(K+1-N)!]\right\} \cdot \int_0^X (1-\rho)^{N-2} \rho^{K+1-N} \, d\rho \tag{48}$$

where $0 \leqslant X \leqslant 1$. The probability distribution of the normalized signal-to-noise ratio ρ is an incomplete beta function distribution. From this fact, the first and second moments of this distribution are

$$E\rho = (K+2-N)/(K+1)$$

and

$$E\rho^2 = [(K+2-N)(K+3-N)]/[(K+2)(K+1)]$$

This result for the first moment is used to define the expected loss in power ratio, given in (19).

Acknowledgment

The authors appreciate the useful suggestions of M. Labitt of M.I.T. Lincoln Laboratory. They also wish to thank Dr. R. Goodman for the use of some unpublished notes on the complex Wishart distributions.

References

[1] L.E. Brennan and I.S. Reed, "Theory of adaptive radar," *IEEE Trans. Aerospace and Electronic Systems,* vol. AES-9, March 1973.

[2] B. Widrow *et al.*, "Adaptive antenna systems," *Proc. IEEE,* vol. 55, December 1967.

[3] N.R. Goodman, "Statistical analysis based on a certain multivariate complex Gaussian distribution," *Ann. Math. Stat.,* vol. 34, pp. 152-177, March 1963.

[4] J. Capon and N.R. Goodman, "Probability distributions for estimators of the frequency-wavenumber spectrum," *Proc. IEEE,* vol. 58, pp. 1785-1786, October 1970.

[5] H. Cramer *Mathematical Methods of Statistics*; Princeton University Press, Princeton (1946).

Irving S. Reed was born in Seattle, Wash., on November 12, 1923. He received the B.A. and Ph.D. degrees in mathematics from the California Institute of Technology, Pasadena, in 1944 and 1949, respectively.

He was associated with the M.I.T. Lincoln Laboratory, Lexington, from 1951 to 1960. From 1960 to 1963 he was a Senior Staff Member of The RAND Corporation, Santa Monica, Calif. He has been a Professor of Electrical Engineering at the University of Southern California, Los Angeles, since 1963. He is also a Consultant to RAND and is associated with Technology Service Corporation, Santa Monica. His interests include mathematics, computer design, coding theory, stochastic processes, and information theory.

John D. Mallett (S'41–A'43–M'55) was born in New York, N.Y., on August 31, 1917. He received the A.B. degree in physics from Princeton University, Princeton, N.J., in 1941, and the M.S. degree from the Massachusetts Institute of Technology, Cambridge, in 1946.

At the Sperry Gyroscope Company, Long Island, N.Y., from 1941 to 1949, he worked on fire control systems and radar. From 1949 to 1972 he was a staff member in the Electronics Department of the RAND Corporation, specializing in radar and system studies. Since 1972 he has been with Technology Service Corporation, Santa Monica, Calif., as a Senior Scientist.

Lawrence E. Brennan (S'47–A'51–M'57) was born in Oak Park, Ill., on January 29, 1927. He received the B.S. and Ph.D. degrees in electrical engineering from the University of Illinois, Urbana, in 1948 and 1951, respectively.

Most of his work since leaving the University of Illinois has been concerned with radar and infrared systems. From 1958 to 1967, he was with the Department of Electronics of The RAND Corporation, Santa Monica, Calif. From April 1962 to October 1963, on leave of absence from RAND, he was employed by the SHAPE Technical Center, The Hague, Netherlands. In June 1967 he joined Technology Service Corporation, Santa Monica, as a Senior Scientist. His current research interests include radar detection theory, adaptive systems, and phased array antennas.